PREMIER OIL LIBRARY

NOT BE STOLEN PROPERTY OF RICHARD BACKHOUSE

Development, Evolution and Petroleum Geology
of the Wessex Basin

Geological Society Special Publications

Series Editors

A. J. FLEET

A. C. MORTON

A. M. ROBERTS

It is recommended that reference to all or part of this book should be made in one of the following ways.

UNDERHILL, J. R. (ed.) 1998. *The Development, Evolution and Petroleum Geology of the Wessex Basin*. Geological Society, London, Special Publications, **133**.

HAWKES, P. W., FRASER, A. J. & EINCHCOMB, C. C. G. 1998. The tectono-stratigraphic development and exploration history of the Weald and Wessex Basins, southern England. *In:* UNDERHILL, J. R. (ed.) *The Development, Evolution and Petroleum Geology of the Wessex Basin*. Geological Society, London, Special Publications, **133**, 39–66.

GEOLOGICAL SOCIETY SPECIAL PUBLICATION NO. 133

The Development, Evolution and Petroleum Geology of the Wessex Basin

EDITED BY

J. R. UNDERHILL
The University of Edinburgh, UK

1998

Published by

The Geological Society

London

THE GEOLOGICAL SOCIETY

The Society was founded in 1807 as The Geological Society of London and is the oldest geological society in the world. It received its Royal Charter in 1825 for the purpose of 'investigating the mineral structure of the Earth'. The Society is Britain's national society for geology with a membership of around 8000. It has countrywide coverage and approximately 1000 members reside overseas. The Society is responsible for all aspects of the geological sciences including professional matters. The Society has its own publishing house, which produces the Society's international journals, books and maps, and which acts as the European distributor for publications of the American Association of Petroleum Geologists, SEPM and the Geological Society of America.

Fellowship is open to those holding a recognized honours degree in geology or cognate subject and who have at least two years' relevant postgraduate experience, or who have not less than six years' relevant experience in geology or a cognate subject. A Fellow who has not less than five years' relevant postgraduate experience in the practice of geology may apply for validation and, subject to approval, may be able to use the designatory letters C Geol (Chartered Geologist).

Further information about the Society is available from the Membership Manager, The Geological Society, Burlington House, Piccadilly, London W1V 0JU, UK. The Society is a Registered Charity, No. 210161.

Published by The Geological Society from:
The Geological Society Publishing House
Unit 7, Brassmill Enterprise Centre
Brassmill Lane
Bath BA1 3JN
UK
(*Orders*: Tel. 01225 445046
Fax 01225 442836)

First published 1998

The publishers make no representation, express or implied, with regard to the accuracy of the information contained in this book and cannot accept any legal responsibility for any errors or omissions that may be made.

© The Geological Society 1996. All rights reserved. No reproduction, copy or transmission of this publication may be made without written permission. No paragraph of this publication may be reproduced, copied or transmitted save with the provisions of the Copyright Licensing Agency, 90 Tottenham Court Road, London W1P 9HE. Users registered with the Copyright Clearance Center, 27 Congress Street, Salem, MA 01970, USA: the item-fee code for this publication is 0305-8719/98/$10.00.

British Library Cataloguing in Publication Data
A catalogue record for this book is available from the British Library.

ISBN 1-897799-99-3

Distributors
USA
AAPG Bookstore
PO Box 979
Tulsa
OK 74101-0979
USA
(*Orders*: Tel. (918) 584-2555
Fax (918) 560-2652)

Australia
Australian Mineral Foundation
63 Conyngham Street
Glenside
South Australia 5065
Australia
(*Orders*: Tel. (08) 379-0444
Fax (08) 379-4634)

India
Affiliated East-West Press PVT Ltd
G-1/16 Ansari Road
New Delhi 110 002
India
(*Orders*: Tel. (11) 327-9113
Fax (11) 326-0538)

Japan
Kanda Book Trading Co.
Tanikawa Building
3-2 Kanda Surugadai
Chiyoda-Ku
Tokyo 101
Japan
(*Orders*: Tel. (03) 3255-3497
Fax (03) 3255-3495)

Typeset by Aarontype Ltd, Unit 47, Easton Business Centre, Felix Road, Bristol BS5 0HE, UK

Printed by The Alden Press,
Osney Mead, Oxford, UK

Contents

Preface	vii
Acknowledgements	viii

UNDERHILL, J. R. & STONELEY, R. Introduction to the development, evolution and petroleum geology of the Wessex Basin — 1

Hydrocarbon habitat

BUCHANAN, J. G. The exploration history and controls on hydrocarbon prospectivity in the Wessex Basins, Southern England, UK — 19

HAWKES, P. W., FRASER, A. J. & EINCHCOMB, C. C. G. The tectono-stratigraphic development and exploration history of the Weald and Wessex Basins, Southern England — 39

BUTLER, M. The geological history of the Wessex Basin: a review of new information from oil exploration — 67

Stratigraphic syntheses

AINSWORTH, N. R., BRAHAM, W., GREGORY, F. J., JOHNSON, B. & KING, C. The lithostratigraphy and biostratigraphy of the latest triassic to earliest cretaceous of the English Channel and its adjacent areas — 87

AINSWORTH, N. R., BRAHAM, W., GREGORY, F. J., JOHNSON, B. & KING, C. A proposed latest Triassic to earliest Cretaceous microfossil biozonation for the English Channel and its adjacent areas — 103

COLE, D. C. & HARDING, I. C. Sequence palynology in the Wessex basin: the Lower Jurassic strata (Sinemurian to Pliensbachian stages) of the Dorset Coast, England — 165

Regional studies

LAW, A. Regional uplift in the English Channel: quantification using sonic velocity data — 189

BRAY, R. J., DUDDY, I. R. & GREEN, P. F. Multiple heating episodes in the Wessex Basin: implications for geological evolution and hydrocarbon generation — 201

MCMAHON, N. A. & TURNER, J. D. Erosion, subsidence and sedimentation response to the Early Cretaceous uplift of the Wessex Basin and adjacent offshore areas — 217

Structural studies

HARVEY, M. J. & STEWART, S. A. Influence of salt on the structural evolution of the Channel Basin — 243

SMITH, C. & HATTON, I. R. Inversion tectonics in the Lyme Bay–West Dorset area of the Wessex Basin — 267

BEELEY, H. S. & NORTON, M. G. The structural development of the Central Channel High: constraints from section restoration — 283

HUNSDALE, R. & SANDERSON, D. J. Fault distribution analysis: an example from Kimmeridge Bay, Dorset, UK — 299

MILIORIZOS, M. & RUFFELL, A. Kinematics and geometry of the Watchet–Cothelstone–Hatch Fault System: implications for the structural history of the Wessex Basin and adjacent areas — 311

Sedimentological advances

RUFFELL, A. Tectonic accentuation of sequence boundaries: evidence from the Lower Cretaceous of southern England — 331

HESSELBO, S. P. The Basal Wealden of Mupe Bay: a new model — 349

GOLDRING, R., ASTIN, T. R., MARSHALL, J. E. A., GABBOTT, S. & JENKINS, C. D. Towards and integrated study of the depositional environment of the Bencliff Grit (Upper Jurassic) of Dorset — 355

Petroleum geochemistry

BIGGE, M. A. & FARRIMOND, P. Biodegradation of oils in the Wessex Basin: a complication of correlation — 373

PARFITT, M. & FARRIMOND, P. The Mupe Bay Oil Seep: a detailed organic geochemical study of a controversial outcrop — 387

Oil field case histories

MCKIE, T., AGGETT, J. & HOGG, A. J. Reservoir architecture of the upper Sherwood Sandstone, Wytch Farm Field, southern England — 399

EVANS, J., JENKINS, D. & GLUYAS, J. G. The Kimmeridge Oil Field: an enigma demystified — 407

Index — 415

Preface

Despite being one of the best known and most visited field areas in the United Kingdom, a full, integrated and unified assessment of the Wessex Basin has long been overdue. The purpose of this Special Publication is to use not only surface outcrops but also information provided by subsurface data, largely obtained in the successful quest for commercial hydrocarbons, to provide insights into the development, evolution and petroleum geology of the Wessex Basin. Whilst several of the papers contained in the book are the products of a highly successful two-day conference entitled 'The Development and Evolution of the Wessex Basin and Adjacent Areas', held at the Geological Society at Burlington House on 27 and 28 June 1995, others are included so as to provide a balanced and more comprehensive overview of the main aspects of the basin's geological history.

Given its status as volumetrically the most important UK onshore petroleum province, a particular emphasis has been placed upon understanding the hydrocarbon habitat in the basin. What becomes apparent from reading contributions contained within this Special Publication is that the 'Wessex Basin hydrocarbon province' is so limited in spatial extent that the term might be considered a misnomer (since over 95% of known reserves actually resides in one field). However, what is equally evident is that the apparently unique nature of the Wytch Farm oilfield may only be explained via a full understanding of the whole basin's development and evolution. Conversely, it is only with the unique insights afforded by the use of well-calibrated seismic data that provide important clues that have allowed significant advances to be made into understanding the controls on basin development, the stratigraphic response and sedimentary fill history in the basin.

Given the breadth and depth of the contributions in all of the areas covered in the volume, I am sure that this Special Publication will become an important reference for petroleum geologists, academic researchers, students and anyone else who has an active interest in the basins of southern Britain.

John R. Underhill
The University of Edinburgh

Acknowledgements

I would like to thank all the participants and contributors to the volume for their assistance in putting this Special Publication together. The original Geological Society Petroleum Group meeting programme from which this Special Publication draws much of its material consisted of 31 talks, three poster contributions and a one-day fieldtrip to visit classic outcrops of the Otter Sandstone Formation, the exposed analogue to the Sherwood Sandstone Group reservoir in Wytch Farm. As well as Arco's financial and logistic support of the fieldtrip, financial sponsorship for the meeting was gratefully received from five other oil companies: Amoco, Brabant Petroleum, British Gas, British Petroleum, Kerr McGee and one service company: Jebco. Their money was not only put towards the costs associated with the running of the original meeting but has also been used to provide financial support for colour plates and fold-outs contained in this volume. The meeting also benefited greatly from the support of Lynx Information Systems Ltd who curate the UK Onshore Geophysical Library. Earth Images Ltd are acknowledged for permission to produce the satellite image on the front cover of the volume. Finally, I would like to thank Angharad Hills and Alan Roberts for their help in steering the book through the final stages to eventual publication.

John R. Underhill
The University of Edinburgh

Introduction to the development, evolution and petroleum geology of the Wessex Basin

JOHN R. UNDERHILL[1] & ROBERT STONELEY[2]

[1] *Department of Geology & Geophysics, The University of Edinburgh, Grant Institute, Kings Buildings, West Mains Road, Edinburgh EH9 3JW, UK*
(e-mail: jru@glg.ed.ac.uk)
[2] *1A Pelham Court, 145 Fulham Road, London SW3 6SH, UK*

Abstract: Despite containing the largest known onshore oilfield in western Europe, the Wessex Basin hydrocarbon province appears to be extremely limited spatially and it currently only consists of three producing oilfields: Wytch Farm, Wareham and Kimmeridge. The main factor which controls hydrocarbon prospectivity in the area appears to be preservation of oil accumulations originally sited in Mesozoic tilted fault-blocks. The extensional palaeostructures of Wytch Farm and Wareham are interpreted to have been charged by upwards migration of oil from mature Liassic source rocks situated across the Purbeck–Isle of Wight fault system in the Channel (Portland–Wight) sub-basin prior to, and unaffected by, either significant effects of intra-Cretaceous (Albian–Aptian) easterly tilting or by Tertiary tectonic inversion. To date, only the small Kimmeridge oilfield, which is situated in the core of a periclinal fold created in response to structural inversion, suggests that any hydrocarbon remigration into younger structural inversion structures has taken place.

Basin definition

The Wessex Basin as defined here consists of a system of post-Variscan sedimentary depocentres and intra-basinal highs that developed across central southern England and adjacent offshore areas (Figs 1 & 2). Given its limits in age and area, the Wessex Basin may be considered to represent a series of extensional sub-basins that form a component part of a more extensive network of Mesozoic intracratonic basins that covered much of NW Europe (Ziegler 1990). Like many of the other basins around the British Isles (e.g. the Weald Basin, Southern North Sea, Cleveland Basin etc.), the Wessex Basin also records the effects of Cenozoic intraplate contraction and structural inversion of basin-bounding and intra-basinal faults.

Onshore the geological boundaries to the Wessex Basin are such that it covers a similar area to the ancient kingdom of the West Saxons and includes the present counties of Hampshire and Dorset, together with parts of East Devon, Somerset and Wiltshire. The basin is bound to the southwest and west by the Armorican and Cornubian Massifs, to the north by the London Platform (otherwise known as the London–Brabant massif) and to the south by the Central Channel High (Fig. 2). Its northeastern and northwestern boundaries are less precisely defined. Although the distinction between the Wessex Basin and the Weald Basin of Sussex, Surrey and Kent is usually taken to be marked by a fundamental change in subsurface geology running NW along Southampton Water, the WNW–ESE trending Chalk (South Downs) outcrop is included in the descriptions herein

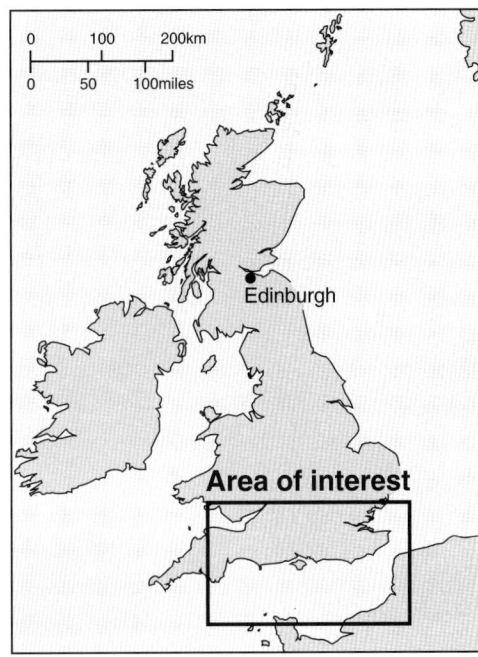

Fig. 1. Location map indicating the area covered by Fig. 2.

UNDERHILL, J. R. & STONELEY, R. 1998. Introduction to the development, evolution and petroleum geology of the Wessex Basin. *In*: UNDERHILL, J. R. (ed.) *Development, Evolution and Petroleum Geology of the Wessex Basin*, Geological Society, London, Special Publications, **133**, 1–18.

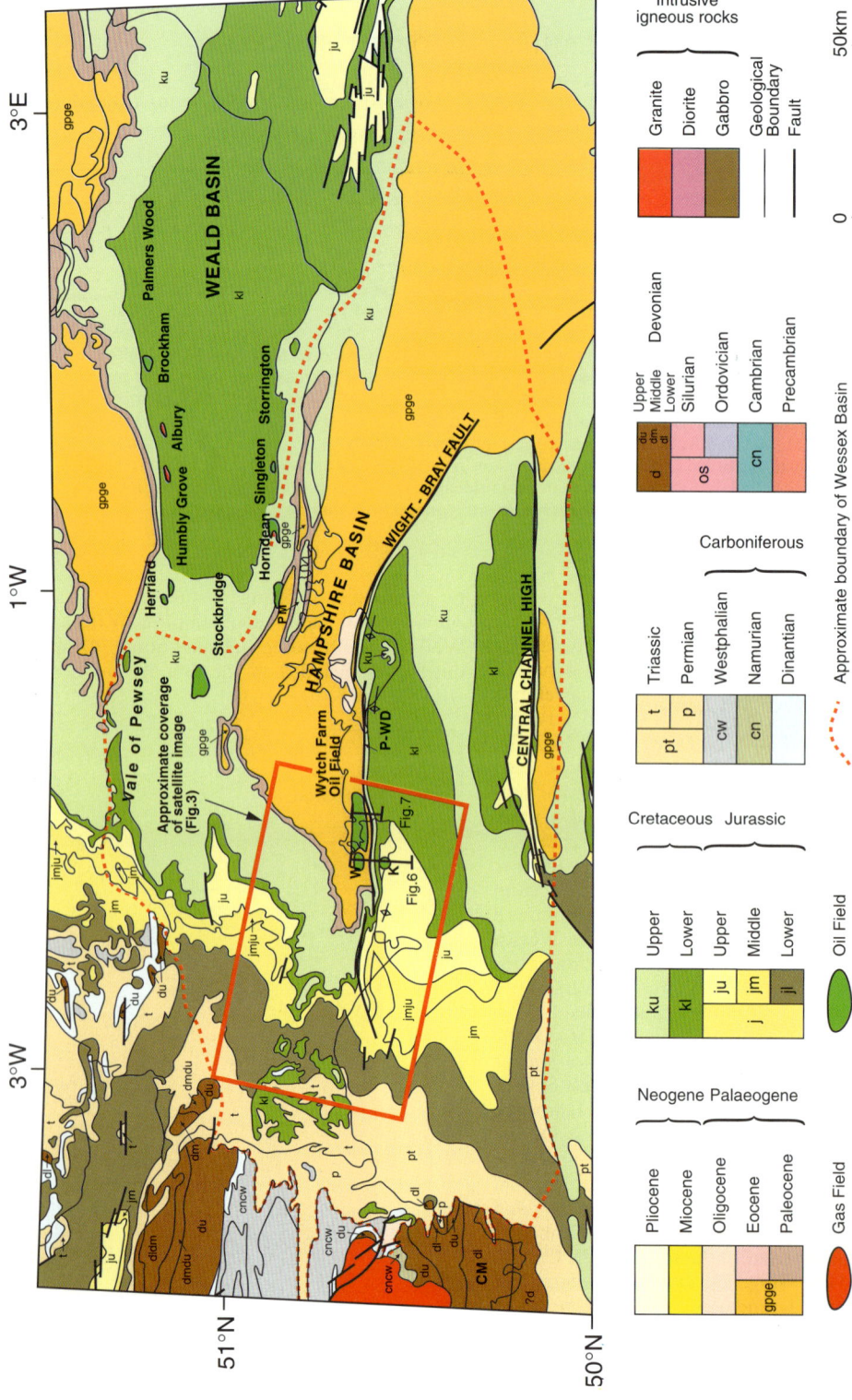

Fig. 2. Geological outcrop and subcrop patterns in southern England and the English Channel. The limits of the Wessex Basin as defined in this paper are indicated. W, Wareham oilfield; K, Kimmeridge oilfield. The position of the seismic and/or geological sections used in Figs 6 & 7 and the approximate limits of the satellite image used in Fig. 3 are shown.

so as to include the Portsdown Anticline in the basin (Fig. 2). Although areas to the northwest of the basin have affinities with the Wessex Basin (e.g. Bristol Channel Basin), its NW limit is taken to be marked by a poorly defined boundary running from the Quantock Hills across the Central Somerset Trough south of the Mendips to the western extension of the London Platform (Fig. 2).

Structural components

The Wessex Basin itself can be subdivided into a number of component parts, bounded primarily by several important exposed or buried tectonic elements, the most significant of which are given below.

The *Pewsey fault system* and *Central Channel High* which are taken to define the northern and southern margins of the Wessex Basin respectively.

The *Purbeck–Isle of Wight Disturbance* (Figs 2, 3 & 4) together with the underlying Mesozoic Purbeck–Isle of Wight fault system, effectively separates the *Channel* (or *Portland–Wight*) *Basin* to the south from the *South Dorset Shelf* and *Hampshire–Dieppe* (or Cranborne–Fordingbridge) intrabasinal highs.

Other structural elements are also wholly intrabasinal. Two E–W-trending extensional faults define a narrow *South Dorset Basin* (otherwise known as the Winterborne Kingston Trough or Cerne Basin) within the *South Dorset Shelf*. The *Wardour* and *Portsdown fault systems* represent important sets of intra-basinal extensional growth faults prior to their reverse reactivation in the Tertiary. Finally, the largely subsurface NNW–SSE-trending *Watchet–Cothelstone–Hatch* fault system transects the basin (Miliorizos & Ruffell this volume).

During Tertiary times, the main site of deposition differed from those of the important Mesozoic basins. Following the latest Cretaceous–early Tertiary inversion, sedimentation was mainly restricted to the *Hampshire Basin* which lay above the site of the former Hampshire–Dieppe High (Plint 1982, 1983,

Fig. 3. Satellite image of the South Dorest area showing the significant topographic effect created by the steep limbs of northward facing monoclinal folds formed in response to Tertiary structural inversion. The approximate location of the satellite image is shown on Fig. 2. Produced with permission of Earth Images Ltd.

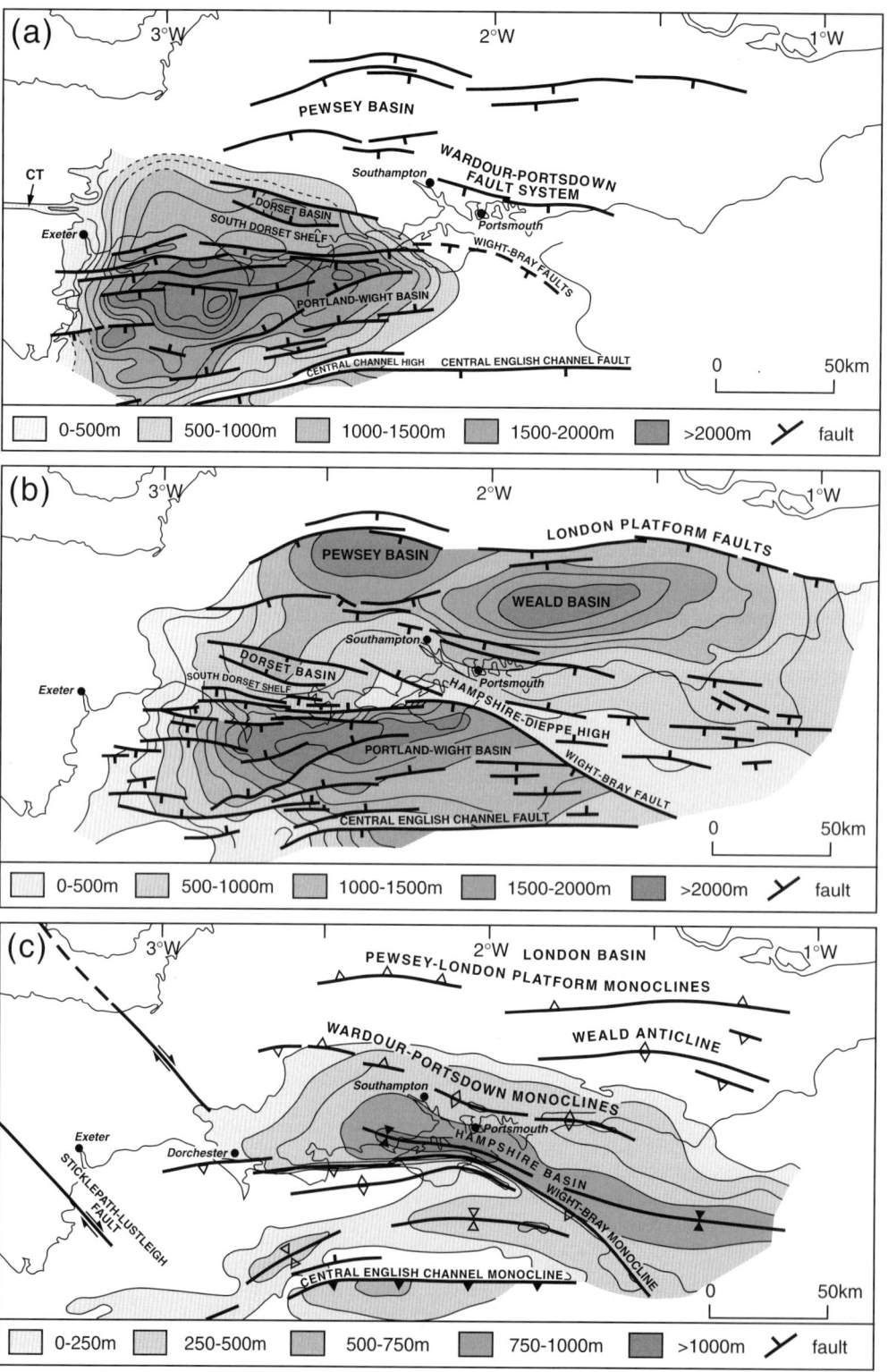

Fig. 4. Sedimentary depocentres and the main structural elements of southern England. (a) Permo-Triassic (CT, Crediton Trough); (b) Jurassic–Cretaceous; (c) Cenozoic. (Modified after Hamblin *et al.* 1992.)

1988). Although temporally and spatially distinct from the main sites of Mesozoic basin development, the Hampshire Basin is still considered an integral part of the Wessex Basin since it records the syn- and post-contractional deformation and sedimentation history of the region.

Stratigraphic framework

The basic stratigraphy and structure of the Wessex Basin are well displayed in the coastal cliffs and inland districts of South Dorset, east Devon and the Isle of Wight. These outcrops, and deductions made from them, enable predictions to be made about the possible occurrence of oil and gas in the subsurface, which can then be tested by reference to information now available from exploratory wells and from seismic data (e.g. Stoneley, 1982; Chadwick 1986; Penn et al. 1987; Selley & Stoneley 1987).

Temporally, the sedimentary history of a distinctive Wessex Basin post-dates the development and closure of the Devono-Carboniferous Proto-Tethys or Rheic Ocean (Glennie & Underhill 1998). The occurrence of major thrust faults, intense folding, regional metamorphism and intrusion of major granitic batholiths (such as the Dartmoor Granite) all attest to the severity of Variscan collisional processes. The deformed Devono-Carboniferous sediments lie beneath a marked unconformity which represents the effective lower limit to reservoir potential in the Wessex Basin (Smith 1993).

Extensional basin development and its component sedimentary fill history began in the Permian within the Variscan fold-and-thrust belt hinterland and continued until the Late Cretaceous in the Wessex Basin (*sensu stricto*; Fig. 4a & b). The sedimentary fill of the successor Hampshire Basin is entirely Tertiary with the youngest sediments being of Oligocene age (Fig. 4c). Except for localized volcanics close to the base (Exeter Volcanic Series), the basin is wholly devoid of igneous rocks. In general terms, the Permian–Oligocene succession may be separated into three internally conformable but unconformity bound mega-sequences (e.g. Hawkes et al. this volume): the Permian to Lower Cretaceous, Upper Cretaceous and Tertiary megasequences.

Permian–Lower Cretaceous megasequence

The Permian to Lower Creataceous interval has been the subject of intensive stratigraphic and sedimentological studies. For example, Ainsworth et al. (this volume a & b) provide a detailed biostratigraphic calibration and lithostratigraphic subdivision of the megasequence and its component parts. Although some sequence stratigraphic studies of the Jurassic section have focused upon the recognition and correlation of sequence boundaries (e.g. Coe 1995, 1996; Hesselbo & Jenkyns 1995), more recent efforts have attempted to define maximum flooding surface-bound, genetic stratigraphic sequences or depositional episodes (*sensu* Galloway 1989; Underhill & Partington 1993, 1994). In particular, Cole & Harding (this volume) use palynofacies to define genetic stratigraphic sequences in the Lower Jurassic with a view to enabling comparison with those defined in the North Sea and adjacent areas (e.g. Partington et al. 1993).

The acquisition of significant subsurface data together with the recent advances in sequence stratigraphic methods support earlier interpretations that subdivided the Permian–Lower Cretaceous megasequence into three component parts (Fig. 5).

The lowest division consists of Permian and Triassic continental (red bed) sediments, in which all desert environments are represented. Deposition was intially restricted to intramontane basins such as the Crediton Trough that developed due to extensional collapse of the former Variscan mountain belt. Although there is much alternation, the sequence is characterised by two large-scale fining upward trends: conglomerates are confined to the Permian and the supposed lowermost Triassic, (both of which were deposited in more or less restricted intermontane depressions) and pass up into mudstones ascribed to the Aylesbeare and Mercia Mudstone Groups respectively (Fig. 5).

The Permian, Exmouth and Dawlish Sands of the east Devon coast pass up into mudstones of the Aylesbeare Group which are in turn sharply overlain by Early Triassic Budleigh Salterton Pebble Beds (Fig. 5). The latter alternate with, and pass up into, sandstones ascribed to the Otter Sandstone Formation. Together the Budleigh Salterton Pebble Beds and the Otter Sandstone comprise the widespread and important Triassic Sherwood Sandstone Group (Fig. 5). The upper part of the sequence is formed by the extensive argillaceous Mercia Mudstone Group, known in the subsurface to include localised evaporites. The Penarth Group at the top of the Triassic succession heralds the effects of widespread Liassic marine transgression that led to the re-establishment of marine waters in the area for the first time since the Carboniferous.

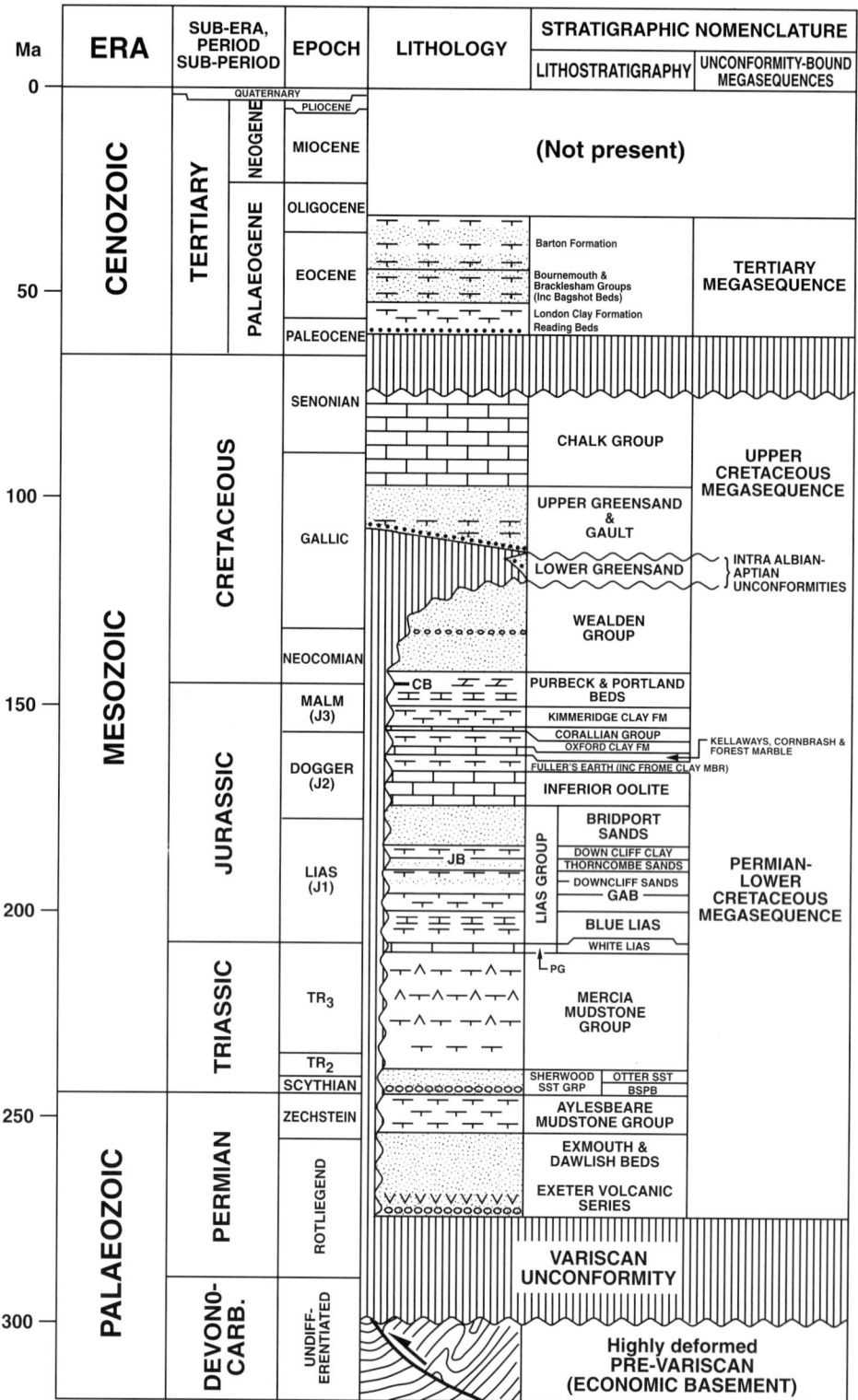

Fig. 5. Generalized stratigraphical column for the Wessex Basin illustrating the main megasequences and stratigraphic nomenclature currently used in the basin (JB, Junction Bed; CB, Cinder Bed; GAB, Green Ammonite Beds; PG, Penarth Group; BSPB, Budleigh Salterton Pebble Beds).

The middle division contains of the dominantly marine sediments of Jurassic age. It consists largely of a broadly cyclic repetition of shallow marine mudrocks, sandstones and limestones. Many formations have been defined and mapped but, with the exception of local facies variations, particularly in some of the carbonates, all appear to be remarkably widespread in the basin. The Jurassic contains all of the potential source rocks in the region, and one major reservoir (the Bridport Sands at the top of the Lower and base of the Middle Jurassic) and a number of minor potential reservoir formations including the Corallian Bencliff Grit (Allen & Underhill 1989, 1990; Goldring et al. this volume), Frome Clay and Cornbrash. The top of the succession records a major marine regression and the highest of the limestones, the Portland Limestone, passes up into sabkha and brackish water sediments (the Purbeck), and thence into the entirely non-marine Lower Cretaceous Wealden Group.

The sediments of the Wealden Group have a distribution in Dorset apparently localised along the strike south of major syn-sedimentary faults. They are essentially of fluvial origin, although lacustrine environments are well represented in the considerably thicker succession in the Isle of Wight. Evidence exists for continued extensional movement on several of the E–W-trending faults during the Early Cretaceous. For example, Ruffell & Garden (1997) document contemporaneous tectonic control by the Isle of Wight Fault on sediment dispersal in the Lower Greensand Group.

Upper Cretaceous megasequence

The Upper Cretaceous megasequence is separated from the underlying Permian–Lower Cretaceous megasequence by an important Albian–Aptian unconformity, which is marked by the progressive westerly truncation of Mesozoic and Permian strata. The lowest part of the Upper Cretaceous megasequence consists of westerly-onlapping, diachronous, marine, and commonly glauconitic sandstones which pass up into the familiar Chalk which shows evidence of thinning in the far west.

Lateral variation in thickness and Chalk lithofacies, including the development of slumps, slip scars and local lacunae have all been recorded in the Wessex Basin (Gale 1980; Mortimore & Pomerol 1997). Their close spatial association with areas that subsequently became axes of inversion suggests that the E–W-trending buried faults that either remained active in extension or, more likely, began to show the effects of contractional reactivation during the late Cretaceous.

Tertiary megasequence

The Tertiary succession is separated from the underlying megasequence by an important, but subtle, regional disconformity. The stratigraphic break covers the Maastrichtian and most of the Paleocene. The overlying sediments that comprise the Tertiary megasequence consist of nearshore marine and non-marine sediments and are largely confined to the area east of Dorchester (Fig. 4c). The megasequence reaches a maximum thickness of over 600 m in the north of the Isle of Wight as proven by the Sandhills-1 borehole. Depositional facies analysis of Upper Eocene (Priabonian) sections demonstrates that the basin was dominated by lacustrine and brackish lagoonal environments which were recharged by marine waters through a restricted inlet in the east Solent area (Hamblin et al. 1992).

Development and evolution of structural styles

Cross section geometries

Surface outcrops highlight the importance of several important zones of disturbance affecting the Chalk (Arkell 1947; e.g. landsat image of Fig. 3). Integration of seismic data with borehole data and field observations enable the construction of representative structural cross-sections (e.g. Figs 6 & 7; Stoneley 1982). The sections not only demonstrate the occurrence of east-west trending extensional growth faults beneath the present zones of disturbance but also in other areas of the basin. The sections also serve to illustrate the role that many of these extensional faults had on depositional thicknesses and structural geometries during the Triassic, Jurassic and Early Cretaceous. Extension not only controlled structural geometries at a basin-scale but also appears to have been important at more local outcrop scales too (e.g. Hunsdale & Sanderson this volume).

Of the major E–W zones of Mesozoic extension, by far the most significant are the system of structures that comprise the intra-basinal Purbeck–Isle of Wight fault system. Variation in structural styles seen along the length of the individual, en-echelon fault segments that collectively form the Purbeck–Isle of Wight fault system may in part be due to the presence of

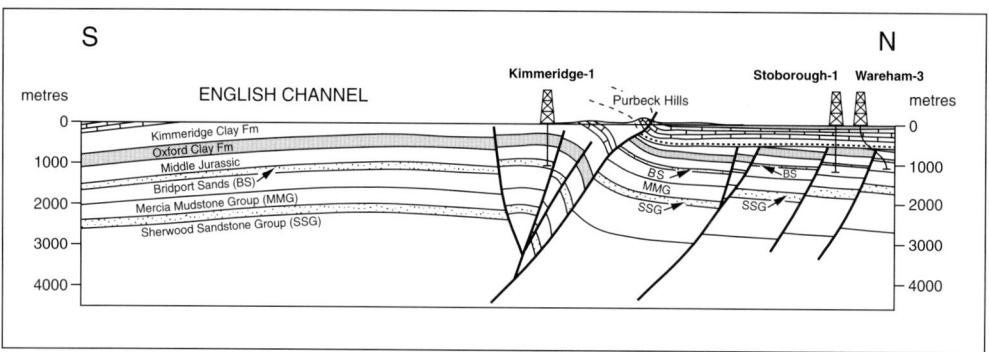

Fig. 6. North–south cross-section through Kimmeridge, Stoborough and Wareham illustrating the nature and extent of the structural inversion related folding in the hangingwall and Mesozoic extensional tilted fault blocks in the footwall of the Purbeck Fault.

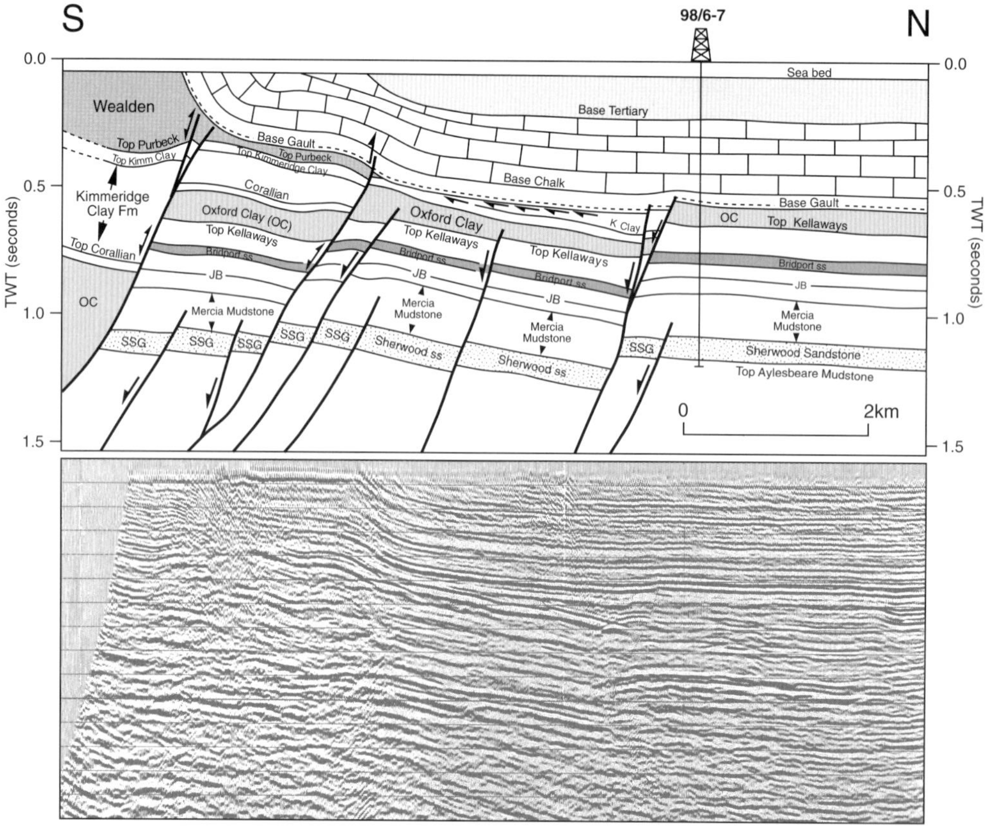

Fig. 7. North–south seismic line and interpreted cross-section through the offshore extension of the Wytch Farm oilfield. The presence of a second monoclinal fold at the base of the Late Cretaceous demonstrates that at least one tilted fault block was affected by the structural inversion process in addition to the more obvious presence of the major structural inversion-related hangingwall anticline. (OC, Oxford Clay; JB, Junction Bed; SSG, Sherwood Sandstone Group.)

easy slip (decollement) horizons in the Triassic in western parts of the basin (Stewart et al. 1996; Harvey & Stewart this volume). Several workers believe that there is evidence that syn-sedimentary roll-over anticlines developed in the hanging walls to many of the extensional faults, particularly those that have a more listric geometry due to the presence of a salt decollement at depth (e.g. Selley & Stoneley 1987)

Extension on component fault systems of the Wessex Basin appears to have largely ceased during Early Cretaceous times, and except for those characterised by unusual Chalk lithofacies, most were essentially inactive during Late Cretaceous deposition which is generally considered to represent the period of post-rift sedimentation. Total displacement on parts of the Purbeck–Isle of Wight fault system are now known to have had a cumulative pre-Tertiary displacement in excess of 2 km.

Towards the end of the Cretaceous and early in the Tertiary, initiation or continuation of south to north compression led to pronounced reversal of movement and structural inversion on many of the formerly extensional Wessex Basin fault segments including those bounding the Central Channel High (Beeley & Norton this volume). Structural inversion along the Purbeck–Isle of Wight fault system led to relative uplift of the Channel (Portland–Wight) Basin. The amount of uplift has been estimated to be approximately 1.5 km from outcrop geology which is consistent with estimates derived from sonic velocity data (Law this volume) and apatite fission track data (Bray et al. this volume). The inversion had the effect of modifying former extensional structures along the whole length of the basin (e.g. Butler this volume; Smith & Hatton this volume), with the creation of major northwards-verging monoclinal folds above the reactivated faults (e.g. the Purbeck–Isle of Wight Disturbance), modification of any pre-existing roll-over anticlines and the formation of periclinal folds in hangingwall locations and the initiation of local thrusting in the post-rift sediments (Underhill & Paterson 1998). Outwith the areas affected by Tertiary fault reactivation, the dips in the Upper Cretaceous are for the most part very gentle.

Strike-section geometries

In marked contrast to the N–S-trending dip sections, east–west (strike-parallel) cross-sections demonstrate a much simpler structural picture. There is little evidence of fault-controlled

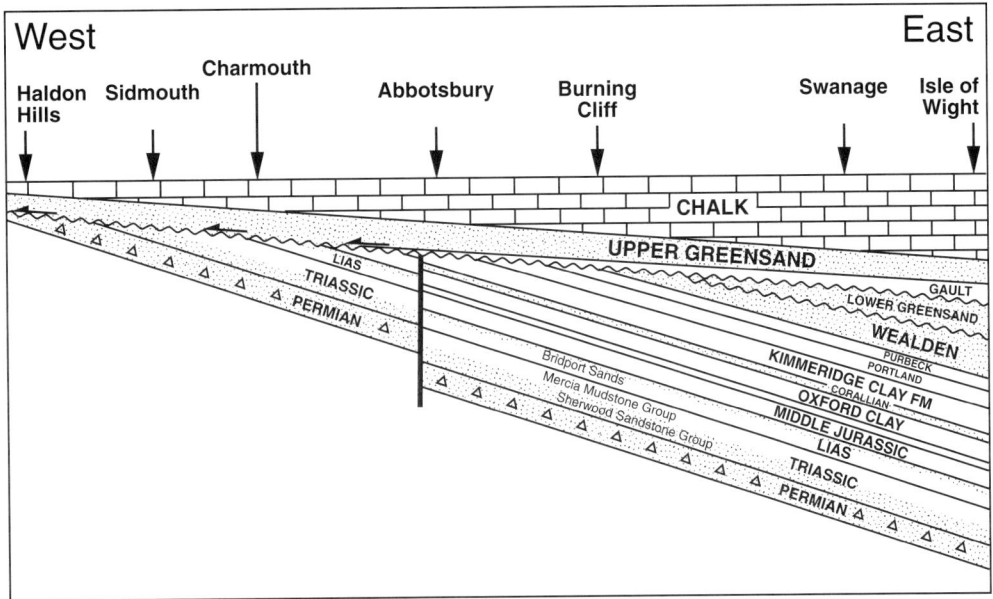

Fig. 8. Diagrammatic representation of the east–west effects of, and progressive westerly subcrop of, Jurassic–Permian strata beneath the intra-Cretaceous unconformity across the Wessex Basin. The occurrence of this particular unconformity is interpreted to have been detrimental to the area's hydrocarbon prospectivity since it probably inhibited source rock maturation in western areas as well as emparting a pronounced easterly tilt to previously formed extensional tilted fault blocks.

depositional thickness changes along strike during deposition of the Jurassic–Early Cretaceous. However, marked stratigraphic onlap and pinch-out of Permian and Triassic sedimentary units (including the Sherwood Sandstone Group; Fig. 3a) is recorded towards the basin margin (e.g. Butler this volume).

E–W-trending structural cross-sections do, however, demonstrate that an important easterly tilt was emparted to the basin during the Early Cretaceous (Fig. 8). The underlying Permian–Jurassic sequence was subjected to erosion and shows progressive truncation towards the west (e.g. McMahon & Underhill 1996; McMahon & Turner this volume), the derived sediment probably mostly passing eastwards into the Weald Basin. Subsequent westerly transgression during the Aptian and the Cenomanian eventually led to cover of the truncated Jurassic, Triassic and Permian above the megasequence bounding unconformity.

Hydrocarbon habitat

For any area to become a successful hydrocarbon province, a number of factors must be satisfied. There must be one or more suitable *reservoirs* capped by coherent *seals*. There must be a candidate *source rock* with sufficiently high total organic carbon content. Well-defined *trap* configurations must exist, be they obvious structural closures or more subtle stratigraphic features. Even given all the of the above, an area would remain unprospective without source rock *maturation* due to burial, the existence of suitable *migration routes* from source to trap and, most importantly, appropriate *timing* of reservoir-bearing trap formation relative to hydrocarbon migration. Finally, *preservation* of any accumulation must be maintained.

That the Wessex Basin contains producing oilfields attests to the fact that all the factors controlling hydrocarbon prospectivity have been met, at least locally. Integration of onshore outcrops with subsurface (seismic and borehole) data enables consideration of each of the essential requirements which have combined to make selected parts of the Wessex Basin highly prospective. Each of these essential requirements will now be reviewed in turn.

Reservoirs

Potential reservoirs are present in many parts of the succession, some of major and some of apparently more minor significance. Review of onshore exposures suggests that the basin's main reservoir potential is provided by siliciclastic units. However, some minor carbonate reservoirs also exist.

In the lower part of the outcropping sequence, the best potential would be provided by the Permian aeolian sands of East Devon. They have surface porosity up to 40% and are possibly equivalent to the Rotliegend sandstones of the Southern North Sea. The Permian sandstones, however, are not generally considered to have a high reservoir potential because it is believed that they either do not extend sufficiently far to the east in the subsurface or lie below structural closure to be of significance in the basin.

Sandstones are frequent in the Lower Triassic succession. They reach their optimum development in the Triassic Sherwood Sandstone Group. The formation is extensively exposed between Budleigh Salterton, Otterton and Sidmouth, where its thickness approaches 120 m (Lorsong & Atkinson 1995). The unit comprises an almost continuous, high net:gross arkosic sandstone body with limited mudstone lenses. Facies analysis suggests that the Sherwood Sandstone Group consists of braided alluvial deposits and perennial sheetflood sandstones, with the local development of distributary channels and a couple of aeolian sandstones interbedded with subsidiary playa lake and floodplain mudstone lenses (e.g. McKie *et al.* this volume and references therein). The latter occasionally contain palaeosols (e.g. Purvis & Wright 1991; Wright *et al.* 1991). At Wytch Farm, the Sherwood Sandstone Group is interpreted to have been deposited in a mixed fluvial and lacustrine proximal braided alluvial plain setting (Dranfield *et al.* 1987; Bowman *et al.* 1993; McKie *et al.* this volume). Significantly, the Sherwood Sandstone Group forms the main reservoir unit (c. 150 m thick) in the Wytch Farm Field (McKie *et al.* this volume), where it is generally similar to the outcrops, even though the facies associations appear not to be continuous in the subsurface in between. Porosities up to nearly 30% have been measured with permeabilities in the darcy range.

The stratigraphically lowest potential reservoir in the Lower Jurassic interval is the Thorncombe Sands which lie towards the top of the Lias Group and are exposed at Watton Cliff near Bridport. Since they are relatively thin (23 m) and very fine-grained sandstones with porosity only in the range 7–10% and permeability of approximately 25–30 mD, the Thorncombe Sands have not excited as much interest as either the underlying Sherwood Sandstone Group or the overlying, Bridport Sandstone. However, they are believed to contain oil in the vicinity of Wytch Farm.

Of more significance are the diachronous 41 m thick Bridport Sands which lie at the top of the Lias Group and extend up into the Aalenian. The sands are fairly uniformly very fine to fine-grained, clean with outcrop porosities up to 15% and permeability up to 250 mD (and up to 32% and 800 mD at Wytch Farm). They are interrupted by numerous 0.33m thick cemented beds which could be barriers to cross-formational fluid flow. The sands are believed to be shallow marine, whilst the origin of the cemented layers is still debatable (Bryant *et al.* 1988). They are capped by a thin (3 m) limestone representative of the Inferior Oolite which, where fractured, has yielded oil in the vicinity of Wareham. The Bridport Sands (*c.* 70 m thick) form the upper reservoir in the Wytch Farm Field (Colter & Harvard 1981).

Thin limestones in the Middle Jurassic, the Forest Marble and the Cornbrash, are similar to the Inferior Oolite in that they too have little natural porosity. However, the Frome Clay Member does contain oil at Wytch Farm and the Cornbrash has been proven to act as a minor reservoir in the Kimmeridge oilfield where it has been extensively fractured (Evans *et al.* this volume).

In the Upper Jurassic, exposures of the mixed siliciclastic–carbonate sequence of the Corallian Group suggest that it could offer some reservoir potential in the subsurface, particularly the approximately 3 m thick fine-grained sandstones known as the Bencliff Grit. Interestingly, at outcrop at Osmington Mills, the sandstones show signs of having been extensively impregnated with oil and still provide a live seepage (Allen & Underhill 1989). Despite this, however, there are no known occurrences of Corallian oil-bearing reservoirs in the subsurface to date.

The Portland Limestone at the top of the Jurassic is porous in two facies: there is still some primary porosity in oolitic grainstones and secondary dissolution occurs in bioclastic packstones. It could conceivably provide a target beneath the Upper Cretaceous unconformity to the north of the Purbeck–Isle of Wight Disturbance if it could be located seismically, although the carbonate develops a more chalky facies inland from the coast.

Restricted fluvial sands in the continental Wealden Group might conceivably provide minor reservoirs, but it is unlikely that they are sealed in the subsurface in southern Dorset. The Upper Cretaceous and Tertiary do not have any significant reservoir potential in Dorset since carbonates of the Chalk Group usually lack porosity and effective permeability and the Tertiary clastics occur at or near surface.

Seals

All of the major potential reservoir sandstones are each covered by a substantial thickness of mud rocks which have good sealing potential. The Permian sandstones are overlain by mudstones of the Aylesbeare Group. The Sherwood Sandstone Group is overlain by the red silty mudstones and local evaporites of the Upper Triassic Mercia Mudstone Group, which is over 350 m thick near Sidmouth. The Bridport Sands and Inferior Oolite are overlain by 190 m of the bentonitic clays of the Bathonian Fuller's Earth.

Source rocks

Potential source rocks are confined to the Jurassic and occur at three main levels. In the vicinity of Lyme Regis and Charmouth, and also in the subsurface to the east, black shales of the Lower Lias reach some 100 m in thickness (House 1993). In the lower part of the section they alternate with pale, very fine-grained limestones approximately 30–40 cm thick. Although marginally immature at outcrop, a total organic carbon content (TOC) of up to some 8% has been measured (Ebukanson & Kinghorn 1985): the organic matter is predominantly Type II algal material.

Potential source rock horizons also occur in the Middle and Upper Jurassic. Some source rock potential has been proven to occur in the lower part (Upper Callovian) of the Oxford Clay, on the shore of the Fleet Lagoon and in abandoned brickpits in the neighbourhood of Chickerell some 7 km WNW of Weymouth, as well as in the subsurface to the east.

The Kimmeridge Clay in the Upper Jurassic reaches a combined total of some 520 m in outcrops at Burning Cliff and at Kimmeridge Bay itself. The succession consists of interbedded black anoxic shales and fine dolomitic limestones. TOCs up to about 20% have been measured (Farrimond *et al.* 1984), although some 70% has been recorded from the 1 m thick Kimmeridge Oil Shale just above the middle of the section. Like the Liassic source rock intervals, Type II organic matter is marginally immature at outcrop (vitrinite reflectance equivalent 0.48%; Ebukanson & Kinghorn 1985).

Traps

Two main possible trap types predominate in the Wessex Basin: periclinal closures mainly related to structural inversion and buried, tilted extensional fault-blocks. To date, no evidence exists for the presence of any stratigraphic plays. The periclinal closures are seen running parallel to, on the south side of, and up to 2 km from the Purbeck–Isle of Wight line of inversion (e.g. Kimmeridge). These are limited in length and largely correspond to the centres of individual fault segments. Whilst the southerly and along plunge dips are very gentle, the north flanks of the periclines commonly steepen into the main fault. Although these periclines were clearly uplifted and no doubt modified, possibly even breached, by the Tertiary structural inversion and uplift, it has been argued that some may have developed as hanging wall rollovers during extension at least in the late Jurassic–early Cretaceous (Stoneley 1982). However, most smaller anticlines close to the disturbance to the west are now believed to have formed largely during Tertiary inversion (including the Poxwell Anticline; Mottram & House 1956).

Although early exploration was directed towards the periclinal traps (Buchanan this volume), the advent of seismic data led to the initial recognition and ultimate successful test for hydrocarbons in buried tilted extensional fault blocks lying to the north of the Purbeck–Isle of Wight fault system in the Wareham area. It is the occurrence of these tilted fault blocks adjacent to the active kitchen area to the south that provides the main structural plays to the basin including the Wytch Farm and Wareham oilfields (Fig. 8; Colter & Harvard 1981). A series of such tilted fault-blocks and terraces have been defined in the basin. In almost all cases, faults defining the blocks north of the Purbeck–Isle of Wight Disturbance remain in net extension and appear to have been little affected by the structural inversion process other than possibly by converting former channels of migration into seals.

Maturity

At outcrop, all of the potential source rocks are immature for oil generation. However, there is ample evidence for source rock maturation having occurred prior to tectonic inversion in the Tertiary. Basin modelling supported by well data suggest that the Lower Lias has reached peak oil generation throughout the area to the south of the Abbotsbury–Purbeck–Isle of Wight disturbance and that, in the east, the Oxford Clay just entered the oil window (e.g. Penn *et al.* 1987, fig. 4). To the east of Swanage, the Lower Lias has been buried deeply enough to raise the possibility of significant gas generation.

The expectation that maturation has occurred is supported not only by the presence of producing oilfields but also the occurrence and distribution of seepages. There are a number of biodegraded oil seeps and impregnations in Jurassic and Wealden beds along the coast between Osmington Mills and Worbarrow Bay east of Lulworth Cove (Bigge & Farrimond this volume), and gas seepages have been reported from the sea-floor south of Swanage (Stoneley 1992).

The oil seep in the Wealden at Mupe Bay has proved to be a particular focus of research, and its genesis and evolution have been subject to considerable debate (e.g. Selley & Stoneley 1987; Hesselbo & Allen 1991; Miles *et al.* 1993, 1994; Kinghorn *et al.* 1994; Wimbledon *et al.* 1996; Bigge & Farrimond this volume; Parfitt & Farrimond this volume; Hesselbo this volume). A channel sandstone contains boulders of mineralogically similar sandstones bound together by black residual oil. These have been taken by some to be evidence for the existence of eroded and reworked sediment impregnated by a nearby Early Cretaceous palaeo-oil seepage (Selley & Stoneley 1987; Kinghorn *et al.* 1994; Wimbledon *et al.* 1996). As the channel sandstone also contains a live, light oil, it has been argued that the outcrop marks the site not only Early Cretaceous but also of recent seepage. Irrespective of its exact evolution, however, the Mupe Bay exposures together with the other known seeps have been successfully typed to a Lower Lias source (Cornford *et al.* 1988).

Migration

Given the likelihood that the proven Lower Lias source rock only reached maturity in the kitchen area south of the Purbeck–Isle of Wight disturbance, charge into the South Dorset Shelf was reliant upon leakage across the Purbeck–Isle of Wight fault system and upwards migration to backfill tilted reservoirs in the footwall to the extensional fault segments (Fig. 9). Such a mechanism is envisaged for the filling history both of the Sherwood Sandstone Group and Bridport Sands together with other minor reservoirs, within the tilted fault blocks located to the north of the kitchen area. Support for such a mechanism comes from the occurrence of the numerous oil seeps in locations where they could be fed by migration up the plane of faults. BP's Kimmeridge oilfield may be an important exception to this rule, but it remains the only producing oilfield in the hangingwall to the reactivated Purbeck fault (Fig. 6; Evans *et al.* this volume).

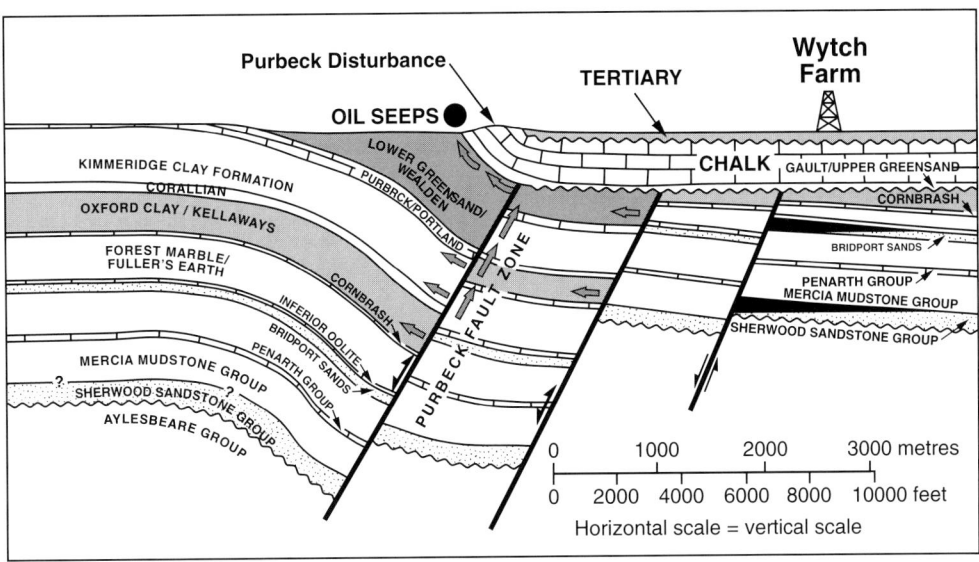

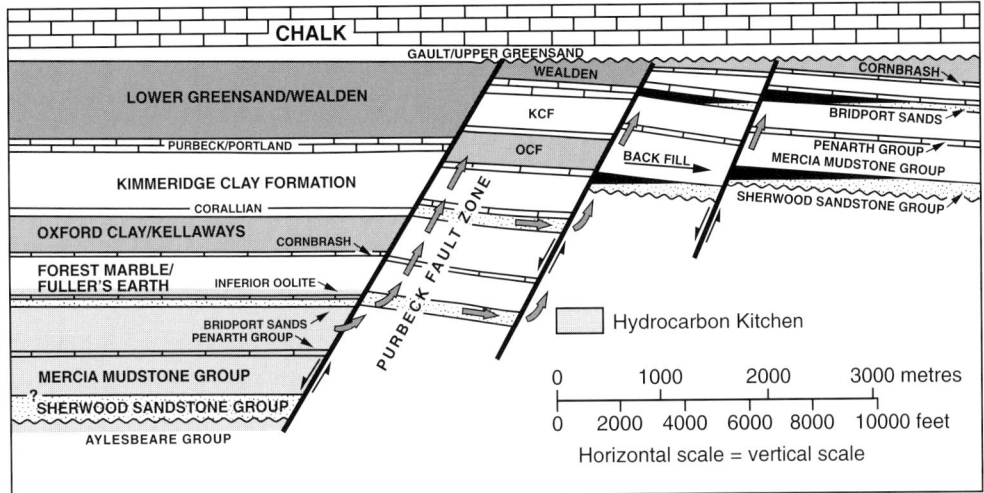

Fig. 9. North–south present-day and Late Cretaceous (restored) cross-sections though Wytch Farm depicting the main controls on hydrocarbon maturation and migration in the area. Modified after Colter & Harvard (1981). (KCF, Kimmeridge Clay Fm; OCF, Oxford Clay Fm.)

Timing of generation, migration and entrapment

Generation of hydrocarbons from the Channel (Portland–Wight) kitchen area occurred prior to Tertiary inversion. Analysis of burial histories for wells located in the Channel (Portland–Wight) depocentre, and the supposed fossil seepage at Mupe Bay, suggest that the Lower Lias entered the oil window by Early Cretaceous times with peak generation in the Middle to Late Cretaceous (Fig. 9). Whilst source rock maturation is likely to have been inhibited or stopped in western areas affected by Albian–Aptian tilting (i.e. west of Lulworth), the main Channel (Portland–Wight) kitchen probably continued to generate hydrocarbons. That kitchen area was probably only switched off when uplift south of the Purbeck–Isle of Wight fault system commenced some time between the Campanian and the Late Palaeocene (Fig. 9).

The disposition of known accumulations requires that the faults that acted as channels of migration were sealed over by the Upper Cretaceous at the time of maximum migration to prevent escape to surface, and that they subsequently became sealing probably due to the onset of compression (e.g. Selley & Stoneley 1987). As in many sedimentary basins affected by tectonic inversion, structures like the periclines that formed during the tectonic inversion process (e.g. Poxwell) would have been reliant upon late-stage remigration for them to be successful plays. As the relative disappointment of exploratory well results sited on such periclines suggests that such re-migration was of relatively minor importance in the Wessex Basin.

Finally, understanding of the structural inversion history, superimposed on a very gentle overall eastwards plunge towards the centre of the sub-basin, is critical to the interpretation of the petroleum geology because it implies that long-established unbreached palaeostructures are likely to be restricted to the eastern parts of the basin lying north of the principal axes of inversion.

Preservation of accumulation

Once an accumulation is in place, it must be preserved intact and not permitted to escape. Although many remain undeformed, some palaeostructures that existed at the end of the Cretaceous appear to have been affected by the structural inversion process and either partially or totally reconfigured by the effects of fault reactivation and basin inversion during the Tertiary. Indeed, whilst Tertiary deformation appears to have been particularly severe in the immediate vicinity of the Purbeck–Wight Disturbance, large areas to the north seem to have remained largely unaffected (e.g. the fault block containing the Wytch Farm oilfield) suggesting that the footwall to the Purbeck–Wight fault system acted as a rigid buttress with most of the compressional strain taken up on it or within the Channel Basin (Underhill & Paterson 1998). Intense fracturing in the Chalk, exposed along the Purbeck–Isle of Wight Disturbance (e.g. at Lulworth Cove) might have created a seal risk for the fault block immediately to the north, which has indeed been found to be water-bearing although with residual oil staining. More than a few hundred yards away to the north, however, where the Chalk remains undeformed, hydrocarbon accumulations have remained unbreached as is evident at Wareham and Wytch Farm.

Limits to the hydrocarbon play fairway

Assessment of the above factors leads to the conclusion that there may be effective limits to the main hydrocarbon play fairway. The interpreted limits to the main Sherwood Sandstone Group and Bridport Sands plays are governed by the respective pinch-out of the lower, and shale-out of the upper reservoir interval towards the east, the line to the west of which either the source rock never reached maturity or was affected by intra-Cretaceous uplift, and to the south by the faults and the tilted fault blocks that they define that show the effects of contractional reactivation. The resultant play fairway thus appears to be limited to the area covered by the Isle of Purbeck (north of the Purbeck Disturbance), Poole Harbour and Bournemouth Bay, but appears not to extend to Solent or to northern areas of the Isle of Wight. The northerly limit to the play fairway, however, is not well defined but is probably controlled by the distance that oil could migrate from the kitchen area and hence, is dependant upon the availability of migration routes. Although the presence of oil at Stockbridge (Fig. 2) might, however, be taken as evidence that some oil was able to migrate long distances to the north of the Channel (Portland–Wight) depocentre, or from the more local Pewsey basin, it is more likely that the field is sourced from the Weald Basin (Butler & Pullan 1990).

Oil production and the Wytch Farm oilfield

Despite the profound bias resultant from oil production rates from one field, Wytch Farm, volumetrically oil production from the Wessex basin has now overtaken that from the East Midlands and Weald Basin, thus making it Britain's most prolific hydrocarbon area.

To date three producing fields have been discovered in the Wessex Basin: Wytch Farm, Wareham and Kimmeridge. The first two of these occur within the Sherwood Sandstone play fairway defined above. It is worth noting that although often shown as a separate accumulation, the Arne discovery is now considered to form an integral part of the Wytch Farm oilfield. Furthermore, the offshore 98/7-2 discovery may also represent an easterly extension of the field rather than a separate accumulation since it has the same oil–water contact.

Of the discoveries to date, however, by far the most significant is the Wytch Farm oilfield which is the largest onshore field in western Europe. Although the field was discovered when the oil-bearing Bridport Sands were penetrated in 1974, its full potential was not realised until 1977 when an appraisal well was deepened to test for possible hydrocarbons in the Sherwood Sandstone Group (Fig. 7; Colter & Harvard 1981; Buchanan this volume).

Volumetrically the Wytch Farm oilfield largely comprises two main reservoir units: an upper, Bridport Sands reservoir and a lower Sherwood Sandstone Group reservoir. A minor contribution also comes from the Middle Jurassic Frome Clay Member. The field is now thought to have had a stock tank oil initially in place (STOIIP) of 924 MMbbls and contain oil reserves slightly in excess of 428 MMbbls of which just under one-half remain to be produced (239 MMbbls have been produced up until 31/12/97). The main Sherwood Sandstone Group reservoir had a STOIIP estimate of 754 MMbbls of which 397 MMbbls were thought to be recoverable (52% recovery efficiency). The Bridport Sandstone is believed to have contained a STOIIP of 120 MMbbls of which 27 MMbbls (23%) is thought to be recoverable, and the Frome Clay Member had an estimated STOIIP of 50 MMbbls and approximate reserves of 4 MMbbls (8% recovery efficiency).

STOIIP and reserve estimates for Wytch Farm dwarf those of the other two producing fields and other discoveries in the area. Current estimates suggest that the Wareham field had a STOIIP of 21 MMbbls of which 5 MMbbls are thought to be recoverable. Kimmeridge has a STOIIP of 10 MMbbls, reserves of 3.2 MMbbls of which 3 MMbbls have been produced to date. The 98/7-2 discovery is believed to contain an additional 20 MMbbls in reserves which are currently not included with the Wytch Farm statistics.

Development of Wytch Farm, which initially took place in the 1980's, has more recently been extended to include that part of the stucture that runs offshore beneath Poole Harbour and Bournemouth Bay (see McClure *et al.* 1995, fig. 2). Innovative drilling technology including the use of highly deviated extended reach wells has been used to drain the field (McClure *et al.* 1995; Hogg *et al.* 1996; McKie *et al.* this volume) whilst at the same time avoiding the need to put rigs or other infrastructure in such an environmentally sensitive area.

Conclusions

(1) The Wessex Basin and the successor Hampshire Basin contain a Permian–Oligocene sedimentary fill which may be subdivided into three unconformity-bound megasequences, each of which record important phases in the development and evolution of the basin.

(2) The lower, Permian–Lower Cretaceous megasequence records several phases of extensional faulting which led to the creation of numerous intra-basinal depocentres, tilted fault-blocks and terraces. It also contains at least two main sealed reservoir intervals in the Sherwood Sandstone Group and the Bridport Sands and at least three potential source rock intervals in the Lias Group, the Oxford Clay and Kimmeridge Clay Formation.

(3) Burial of the Liassic source intervals in areas south of the Purbeck–Wight intra-basinal segmented, extensional fault system led to maturation and migration of hydrocarbons from Early Cretaceous times. Whilst hydrocarbon generation was probably arrested in western areas that experienced intra-Cretaceous (Albian–Aptian) uplift, maturation and migration continued in eastern areas until Late Cretaceous times. Neither the Kimmeridge Clay nor Oxford Clay formations appear to have been matured for hydrocarbon generation anywhere in the basin at any time.

(4) Late Cretaceous and Tertiary compression led to contractional reactivation and structural inversion along the line of many of the former extensional structures (e.g. Purbeck–Isle of Wight fault system), uplift of the Channel (Portland–Wight) sub-basin and formation of north-facing monoclines and numerous periclinal folds.

(5) An understanding of the tectono-stratigraphic development and evolution of the Wessex Basin helps determine why the basin's

hydrocarbon prospectivity appears to be concentrated in the east Purbeck, Poole Harbour and Bournemouth Bay areas. It is only there that the extensional palaeostructures containing Sherwood Sandstone Group and Bridport Sands reservoirs have been unaffected by the *pronounced* effects either Albian–Aptian easterly tilting or by Tertiary tectonic inversion.

(6) As in other structurally inverted sedimentary basins in which the kitchen area has been switched off, hydrocarbon charge into subsequent, inversion-related periclinal folds relies upon remigration from breached palaeostructures. Despite numerous exploratory wells, it appears that re-migration did not play a major role in the basin. Nevertheless, some periclines modified during the structural inversion episode may, as at Kimmeridge, have retained from Cretaceous times or received oil by limited re-migration.

Both authors are greatly indebted to their many academic and oil company colleagues for input into their current understanding of the petroleum geology of the Wessex Basin. R.S. would particularly like to thank R. Selley for introducing him to the area and for his companionship on numerous fieldtrips. S. J. Davies, J. Evans and A. Ruffell are thanked for providing advice on the text and G. White is acknowledged for drafting the diagrams. Earth Images Ltd are acknowledged for permission to produce the satellite image in the paper.

References

AINSWORTH, N. R., BRAHAM, W., GREGORY, F. J., JOHNSON, B. & KING, C. 1998a. The lithostratigraphy and biostratigraphy of the latest Triassic to earliest Cretaceous of the English Channel and its adjacent areas. *This volume*.

——, ——, ——, & ——1998b. A proposed latest Triassic to earliest Cretaceous lithostratigraphic classification for the English Channel and its adjacent areas. *This volume*.

ALLEN, P. A. & UNDERHILL, J. R. 1989. Swaley cross-stratification produced by unidirectional flows, Bencliff Grit (Upper Jurassic), Dorset, UK. *Journal of the Geological Society, London*, **146**, 241–252.

—— & ——1990. Discussion on swaley cross-stratification produced by unidirectional flows, Bencliff Grit (Upper Jurassic), Dorset, UK [reply]. *Journal of the Geological Society, London*, **147**, 398–400.

ARKELL, W. J. 1947. *The Geology of the Country around Weymouth, Swanage, Corfe and Lulworth*. Memoir of the Geological Survey.

BEELEY, H. S. & NORTON, M. G. 1998. The structural development of the Central Channel High: constraints from section restoration. *This volume*.

BIGGE, M. A. & FARRIMOND, P. 1998. Biodegradation of seep oils in the Wessex Basin: a compilation for correlation. *This volume*.

BOWMAN, M. J. B., MCCLIRE, N. M. & WILKINSON, D. W. 1993. Wytch Farm oilfield: deterministic reservoir description of the Triassic Sherwood Sandstone. *In*: PARKER, J. R. (ed.) *Petroleum Geology of Northwest Europe: Proceedings of the 4th Conference*, 1513–1518. The Geological Society, London.

BRAY, R. J., GREEN, P. F. & DUDDY, I. R. 1998. Multiple heating episodes in the Wessex basin: Implications for geological evolution and hydrocarbon generation. *This volume*.

BRYANT, I. D., KANTOROWICZ, J. D. & LOWE, C. F. 1988. The Origin and Recognition of Laterally Continuous Carbonate-Cemented Horizons in the Upper Lias Sands of Southern England. *Marine and Petroleum Geology*, **5**, 108–133.

BUCHANAN, J. G. 1998. The exploration history and controls on hydrocarbon prospectivity in the Wessex Basins, Southern England, UK. *This volume*.

BUTLER, M. 1998. The Geological History of the southern Wessex Basin: a review of new information from oil exploration. *This volume*.

BUTLER, M. & PULLAN, C. P. 1990. Tertiary structures and hydrocarbon entrapment in the Weald basin of southern England. *In*: HARDMAN, R. F. P. & BROOKS, J. (eds) *Tectonic Events Responsible for Britain's Oil and Gas Reserves*. Geological Society, London, Special Publications, **55**, 371–391.

CHADWICK, R. A. 1986. Extensional tectonics in the Wessex Basin, southern England. *Journal of the Geological Society, London*, **143**, 465–488.

COE, A. L. 1995. A comparison of the Oxfordian successions of Dorset, Oxfordshire and Yorkshire. *In*: TAYLOR, P. D. (ed.) *Field Geology of the British Jurassic*. Geological Society, London, 151–172.

——1996. Unconformities within the Portlandian Stage of the Wessex basin and their sequence-stratigraphical significance. *In*: HESSELBO, S. P. & PARKINSON, D. N. (eds) *Sequence Stratigraphy in British Geology*. Geological Society, London, Special Publications, **103**, 109–143.

COLE, D. C. & HARDING, I. C. 1998. Use of palynofacies analysis to define Lower Jurassic (Sinemurian to Pliensbachian stages) genetic stratigraphic sequences in the Wessex Basin, England. *This volume*.

COLTER, V. S. & HAVARD, D. J. 1981. The Wytch Farm Oilfield, Dorset. *In*: ILLING, L. V. & HOBSON, G. D. (eds) *Petroleum Geology of the Continental Shelf of North-West Europe*. Hayden & Son, London, 494–503.

CORNFORD, C., CHRISTIE, O., ENDRESSEN, U., JENSEN, P. & MYHR, M. 1988. Source Rock and Seep Oil Maturity in Dorset, Southern England. *Organic Geochemistry*, **13**, 399–409.

DRANFIELD, P., BEGG, S. H. & CARTER, R. R. 1987. Wytch Farm Oilfield: reservoir characterisation of the Triassic Sherwood Sandstone for input to reservoir stimulation studies. *In*: BROOKS, J. & GLENNIE, K. W. (eds) *Petroleum Geology of northwest Europe*. Graham & Trotman, London, 494–503.

EBUKANSON, E. J. & KINGHORN, R. R. F. 1985. Kerogen Facies in the Major Mudrock Formations of Southern England and the Implication on the Depositional Environments of their Precursors. *Journal of Petroleum Geology*, **8**, 435–462.

EVANS, J., JENKINS, D. & GLUYAS, J. 1998. The Kimmeridge Bay Oilfield: an enigma demystified. *This volume*.

FARRIMOND, P., COMET, P., EGLINTON, G., EVERSHED, R. P., HALL, M. A., PARK, D. W. & WARDROPER, A. M. K. 1984. Organic Geochemical study of the Upper Kimmeridge Clay of the Dorset type area. *Marine and Petroleum Geology*, **1**, 340–354.

GALE, A. S. 1980. Penecontemperoneous folding, sedimentation and erosion in Campanian Chalk near Portsmouth, England. *Sedimentology*, **27**, 137–151.

GALLOWAY, W. E. 1989. Genetic stratigraphic sequences in basin analysis I: Architecture and genesis of flooding surface bounded depositional units. *American Association of Petroleum Geologists Bulletin*, **73**, 125–142.

GLENNIE, K. W. & UNDERHILL, J. R. 1998. Origin, development and evolution of structural styles. *In*: GLENNIE, K. W. (ed.) *Petroleum Geology of the North Sea*. Blackwell Scientific Publications, Oxford, in press.

GOLDRING, R., ASTIN, T. R., MARSHALL, J. E. A., GABBOTT, S. & JENKINS, C. D. 1998. Towards an integrated study of the depositional environment of the Bencliff Grit (Upper Jurassic) of Dorset. *This volume*.

HAMBLIN, R. J. O., CROSBY, A., BALSON, P. S., JONES, S. M., CHADWICK, R. A., PENN, I. E. & ARTHUR, M. J. 1992. *United Kingdom offshore regional report: the geology of the English Channel*. HMSO, London.

HARVEY, M. & STEWART, S. A. 1998. Influence of salt on the structural evolution of the Channel Basin. *This volume*.

HAWKES, P. W., FRASER, A. J. & EINCHCOMB, C. C. G. 1998. The tectono-stratigraphic development and exploration history of the Weald and Wessex basins, Southern England. *This volume*.

HOUSE, M. R. 1993. *The Geology of the Dorset Coast*. (2nd edn). Geologists' Association Guide, **22**.

HESSELBO, S. P. 1998. The basal Wealden of Mupe Bay: a new model. *This volume*.

—— & ALLEN, P. A. 1991. Major erosion surfaces in the basal Wealden Beds, Lower Cretaceous, south Dorset. *Journal of the Geological Society, London*, **148**, 105–113.

—— & JENKYNS, H. C. 1995. A comparison of the Hettangian to Bajocian successions of Dorset and Yorkshire. *In*: TAYLOR, P. D. (ed.) *Field Geology of the British Jurassic*. Geological Society, London, 105–150.

HOGG, A. J. C., MITCHELL, A. W. & YOUNG, S. 1996. Predicting well productivity from grain size analysis and Logging While Drilling. *Petroleum Geoscience*, **2**, 1–15.

HUNSDALE, R. & SANDERSON, D. J. 1998. Fault distribution analysis: an example from Kimmeridge Bay, Dorset. *This volume*.

JENKYNS, H. C. & SENIOR, J. R. 1991. Geological Evidence for Intra-Jurassic Faulting in the Wessex Basin and Its Margins. *Journal of the Geological Society, London*, **148**, 245–260.

KINGHORN, R. R. F., SELLEY, R. C. & STONELEY, R. 1994. Discussion on the Mupe Bay oil seep demythologized? *Marine and Petroleum Geology*, **11**, 124.

LAW, A. 1998. Regional uplift in the English Channel: quantification using sonic velocity data. *This volume*.

LORSONG, J. A. & ATKINSON, C. D. 1995. *Sedimentology and Stratigraphy of Lower Triassic Alluvial Deposits, East Devon Coast*. Excursion Guide of the Petroleum Group of the Geological Society.

McCLURE, N. M., WILKINSON, D. W., FROST, D. P. & GEEHAN, G. W. 1995. Planning extended reach wells in Wytch Farm Field, UK. *Petroleum Geoscience*, **1**, 115–127.

McKIE, T., AGGETT, J. & HOGG, A. J. C. 1998. Reservoir architecture of the upper Sherwood Sandstone, Wytch Farm Field, southern England. *This volume*.

McMAHON, N. A. & TURNER, J. D. 1998. Erosion, subsidence and sedimentation response to the early Cretaceous uplift of the Wessex basin and adjacent offshore areas. *This volume*.

—— & UNDERHILL, J. R. 1995. The regional stratigraphy of the southwest United Kingdom and adjacent offshore areas with particular reference to the major intra-Cretaceous unconformity. *In*: CROKER, P. F. & SHANNON, P. M. (eds) *The Petroleum Geology of Ireland's Offshore Basins*. Geological Society, London, Special Publications, **93**, 323–325.

MILES, J. A., DOWNES, C. J. & COOK, S. E. 1993. The fossil oil seep in Mupe Bay, Dorset: a myth investigated. *Marine and Petroleum Geology*, **10**, 58–70.

——, —— & —— 1994. Discussion on the Mupe Bay oil seep demythologized? [reply]. *Marine and Petroleum Geology*, **11**, 125–126.

MILIORIZOS, M. & RUFFELL, A. 1998. Kinematics and geometry of the Watchet-Cothelstone-Hatch Fault system: implications for the structural history of the Wessex Basin and adjacent areas. *This volume*.

MORTIMORE, R. & POMEROL, B. 1997. Upper Cretaceous tectonic phases and end Cretaceous inversion in the Chalk of the Anglo-Paris Basin. *Proceedings of the Geologists' Association*, **108**, 231–255.

MOTTRAM, B. H. & HOUSE, M. R. 1956. The Structure of the Northern Margin of the Poxwell Pericline. *Proceedings of the Dorset Natural History and Archaeological Society*, **76**, 129–135.

PARFITT, M. & FARRIMOND, P. 1998. The Mupe Bay Oil Seep: a detailed organic geochemical study of a controversial outcrop. *This volume*.

PARTINGTON, M. A., COPESTAKE, P., MITCHENER, B. C. & UNDERHILL, J. R. 1993. Biostratigraphic calibration of genetic stratigraphic sequences in the Jurassic-lowermost Cretaceous (Hettangian to Ryazanian) of the North Sea and adjacent areas.

In: PARKER, J. R. (ed.) *Petroleum Geology of Northwest Europe: Proceedings of the 4th Conference*, The Geological Society, London, 371–386.

PENN, I. E., CHADWICK, R. A., HOLLOWAY, S., ROBERTS, G., PHARAOH, T. C., ALLSOP, J. M., HULBERT, A. G. & BURNS, I. M. 1987. Principal features of the hydrocarbon prospectivity of the Wessex-Channel Basin. *In*: BROOKS, J. & GLENNIE, K. W. (eds) *Petroleum Geology of northwest Europe*. Graham & Trotman, London, 109–118.

PLINT, A. G. 1982. Eocene sedimentation and tectonics in the Hampshire Basin. *Journal of the Geological Society*, **139**, 249–254.

—— 1983. Facies, Environments and Sedimentary Cycles in the Middle Eocene Bracklesham Formation of the Hampshire Basin: Evidence for Global Sea-Level Changes. *Sedimentology*, **30**, 625–653.

—— 1988. Global eustacy and the Eocene sequence in the Hampshire Basin, England. *Basin Research*, **1**, 11–22.

PURVIS, K. & WRIGHT, V. P. 1991. Calcretes related to phreatophytic vegetation from the Middle Triassic Otter Sandstone of South West England. *Sedimentology*, **38**, 539–551.

RUFFELL, A. 1998. Tectonic accentuation of sequence boundaries: evidence from the Lower Cretaceous of southern England. *This volume*.

—— & GARDEN, R. 1997. Tectonic controls on the variation in thickness and mineralogy of pebblebeds in the Lower Greensand Group (Aptian–Albian) of the Iske of Wight, southern England. *Proceedings of the Geologists' Association*, **108**, 215–229.

SELLEY, R. C. 1992. Petroleum seepages and impregnations in Great Britain. *Marine and Petroleum Geology*, **9**, 226–244.

—— & STONELEY, R. 1987. Petroleum Habitat in South Dorset. *In*: BROOKS, J. & GLENNIE, K. W. (eds) *Petroleum Geology of North West Europe*. Graham & Trotman, London, 139–148.

SMITH, C. & HATTON, I. R. 1998. Inversion tectonics in the Lyme Bay-West Dorset area of the Wessex Basin. *This volume*.

SMITH, N. J. P. 1993. The case for exploration of deep plays in the Variscan fold belt and its foreland. *In*: PARKER, J. R. (ed.) *Petroleum Geology of Northwest Europe: Proceedings of the 4th Conference*, 667–675. The Geological Society, London.

STEWART, S. A., HARVEY, M. J., OTTO, S. C. & WESTON, J. 1996. Influence of salt on fault geometry: examples from the UK salt basins. *In*: ALSOP, G. I., BLUNDELL, D. J. & DAVISON, I. (eds) *Salt Tectonics*. Geological Society, London, Special Publication, **100**, 175–202.

STONELEY, R. 1982. The Structural Development of the Wessex Basin: Implications for Exploration. *Proceedings of the Ussher Society*, **8**, 1–6.

UNDERHILL, J. R. & PATERSON, S. 1998. Genesis of tectonic inversion structures: seismic evidence for the development of key structures along the Purbeck–Isle of Wight Disturbance. *Journal of the Geological Society, London*, in press.

—— & PARTINGTON, M. A. 1993. Jurassic thermal doming and deflation: the sequence stratigraphic evidence. *In*: PARKER, J. R. (ed.) *Petroleum Geology of Northwest Europe: Proceedings of the 4th Conference*, 337–345. The Geological Society, London.

—— & —— 1994. Use of maximum flooding surfaces in determining a regional tectonic control on the Intra-Aalenian ("Mid Cimmerian") Sequence Boundary: Implications for North Sea basin development and Exxon's Sea-Level Chart. *In*: POSAMENTIER, H. W. & WIEMER, P. (eds) *Siliciclastic Sequence Stratigraphy*. American Association of Petroleum Geologists Memoirs, **58**, 449–484.

WIMBLEDON, W. A., ALLEN, P. & FLEET, A. J. 1996. Penecontemporaneous oil-seep in the Wealden (early Cretaceous) at Mupe Bay, Dorset, UK. *Sedimentary Geology*, **102**, 213–220.

WRIGHT, V. P., MARRIOTT, S. & VANSTONE, S. D. 1991. A "Reg" palaeosol from the Lower Triassic of South Devon: stratigraphic and palaeoclimatic implications. *Geological Magazine*, **128**, 517–523.

ZIEGLER, P. A. 1990. *Geological Atlas of Western and Central Europe*. Shell Internationale Petroleum Mij. Elsevier, Amsterdam.

The exploration history and controls on hydrocarbon prospectivity in the Wessex basins, southern England, UK

JAMES G. BUCHANAN

British Gas Exploration and Production Limited, 100 Thames Valley Park Drive, Reading, Berkshire, RG6 1PT, UK
Present address: Conoco Inc., 600 North Dairy Ashford, Houston, TX 77079, USA

Abstract: The Wessex basins were formed during Mesozoic extension and were subsequently modified by Cretaceous uplift and Alpine inversion events. The basin geometry and evolution of the area is strongly controlled by the long-lived fault systems which cross the area.

Two primary plays are recognized within the Wessex basins; the Triassic Sherwood Sandstone and the Jurassic Bridport Sandstone. These plays contribute the vast majority of the recoverable reserves in the area. The Triassic play consists of a Sherwood Sandstone Group reservoir, a Liassic mudstone source rock and a Mercia Mudstone Group regional seal. Hydrocarbons were generated from the Late Jurassic to Tertiary in the main kitchen area and migrated into a range of fault-related traps. The main risks on the play are reservoir quality together with the timing and the route of hydrocarbon migration into valid traps.

The Jurassic play consists of the Bridport Sandstone Formation which is also sourced from the Liassic mudstones and the top seal is provided by the Fuller's Earth Formation. The dominant risk on this play is the extent and quality of reservoir facies. In the onshore domain trap definition, using two-dimensional seismic data, is an additional risk on both plays.

The basins of southern England have been explored for hydrocarbons for over 50 years. The initial phase of exploration focused on the onshore and tested anticlinal structures which had been mapped at surface. Onshore activity reached a peak in the successful drilling by British Gas of the Jurassic and Triassic reservoirs in the Wytch Farm Field in 1973 and 1977, respectively.

The first offshore well in the United Kingdom Continental Shelf (UKCS) was drilled off Lulworth Cove by BP in 1963. More recently, offshore exploration concentrated on fault-related traps within or on the edges of the Portland–Wight Basin. Although offshore activity has been reduced in the last 5 years, the 14th UKCS Licensing Round rekindled interest in the area as companies were awarded previously unlicensed acreage for exploration.

Successful future exploration in the Wessex basins will require a more complete understanding of reservoir development and three-dimensional basin evolution. Future drilling will focus on untested fault bounded prospects and stratigraphic traps with suitable reservoir quality.

The Wessex basins are a classic area of geology which has been studied by generations of students from British universities (Arkell 1933, 1947; Wilson et al. 1958; Evans 1990) (Fig. 1). The area provides excellent teaching examples of carbonate and siliciclastic sedimentology, stratigraphy (both bio and litho), structural geology, petroleum geology and basin analysis (Stoneley & Selley 1986; Evans 1990; House 1993). The accessibility and challenging complexity of the geology has resulted in the region being an area of continued research into many of the above topics (Evans 1990; Milner et al. 1990; Hamblin et al. 1992; Hesselbo & Palmer 1992; McMahon & Underhill 1995; Ruffell 1991, 1992; Underhill & Paterson 1998: Underhill & Stoneley this volume).

This paper is focused on a discussion of the hydrocarbon system of the offshore and onshore parts of the Portland–Wight Basin (PWB) and associated areas (Fig. 2). The prospective sedimentary units (Permian–Tertiary) in the basin rest unconformably upon Variscan metamorphic basement (Colter & Havard 1981; Scott & Colter 1987; Penn et al. 1987). The Permo-Triassic red-bed siliciclastics are overlain by Jurassic and Cretaceous marine siliciclastics and carbonates (Fig. 3). The Tertiary strata consist of locally thick sequences of both marine and non-marine sediments.

Hydrocarbon exploration and production in the area has been dominated by the discovery of the Wytch Farm oil field in eastern Dorset which has reserves of approximately 300 million barrels (Colter & Havard 1981; Bowman et al. 1993). There are two primary plays in the area consisting of the Triassic Sherwood Sandstone Group and Jurassic Bridport Sandstone Formation reservoirs, respectively. Both units are overlain by regional widely developed mudstone units which act as effective top seals. Hydrocarbons are sourced from organic-rich Liassic

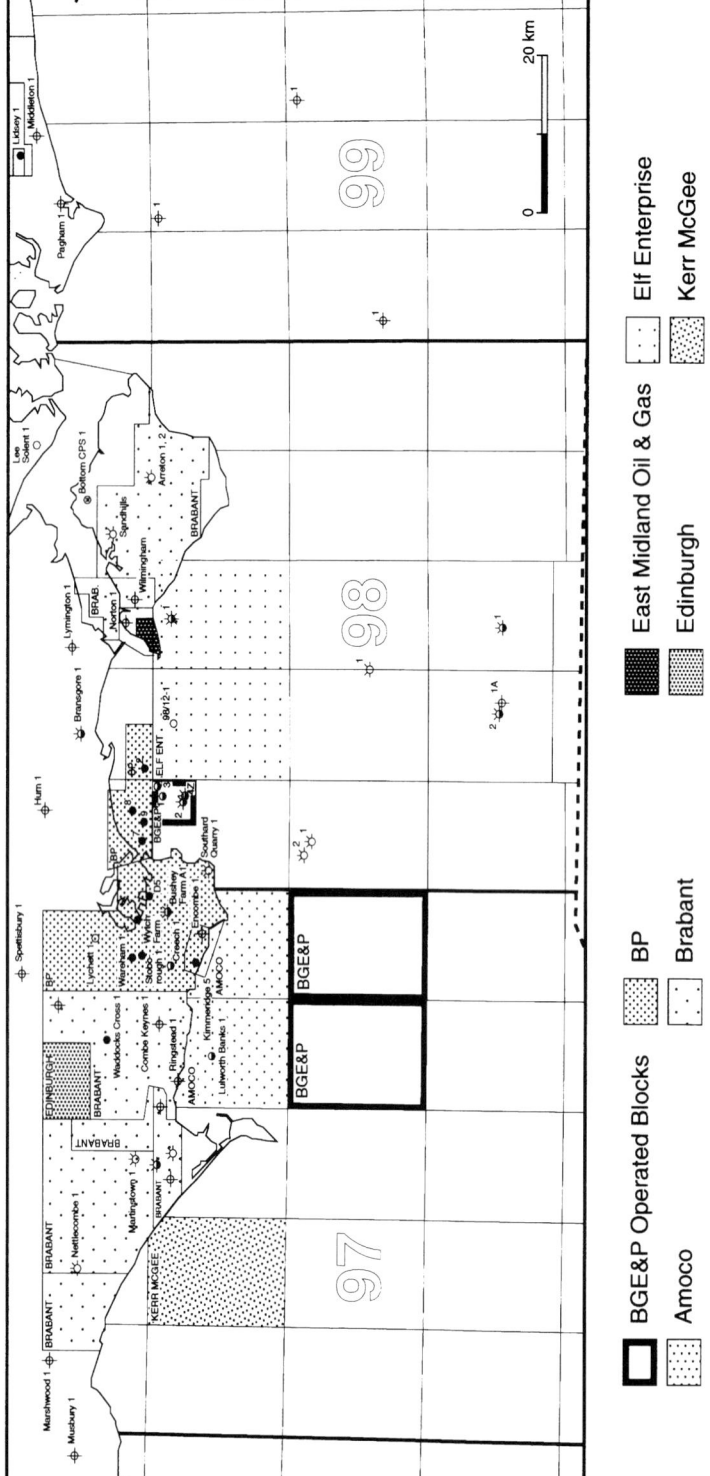

Fig. 1. The southern England area (onshore/offshore) showing the 1995 petroleum license position.

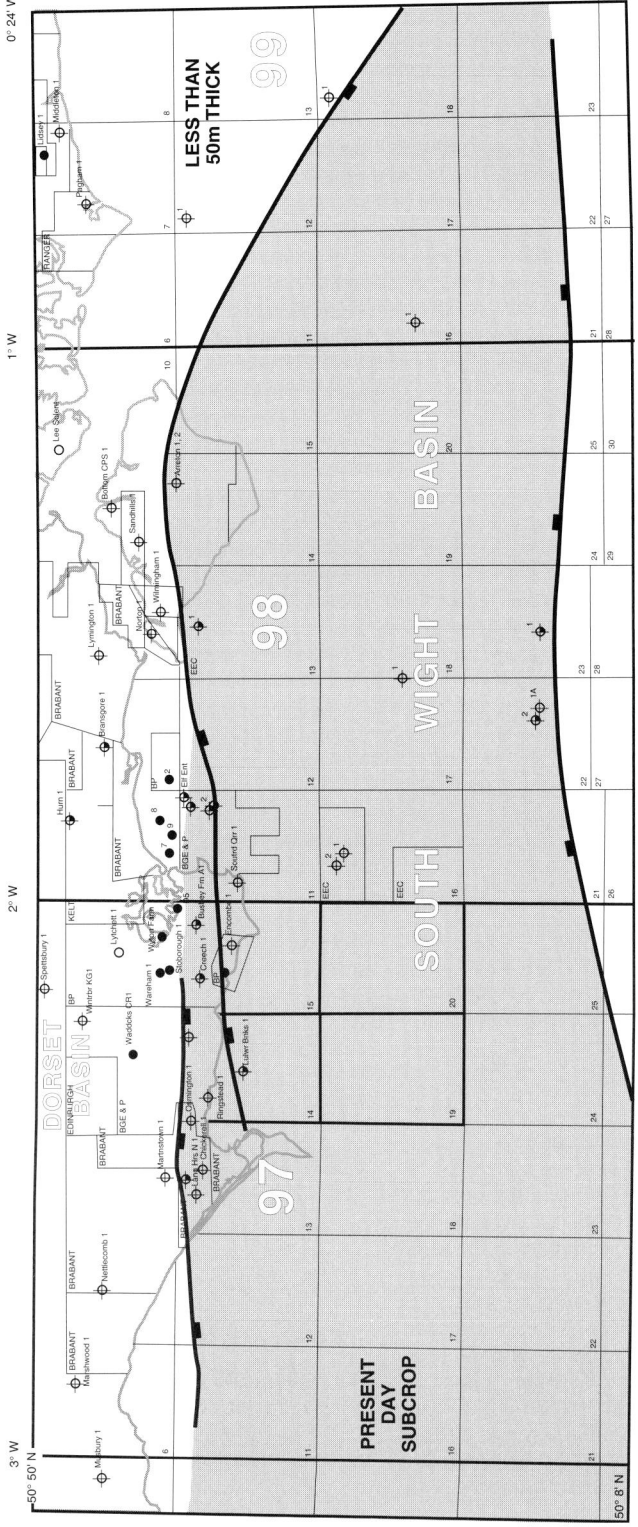

Fig. 2. The Portland–Wight Basin with the major tectonic elements highlighted. (The position of cross-sections A–C presented in Fig. 10 are shown.)

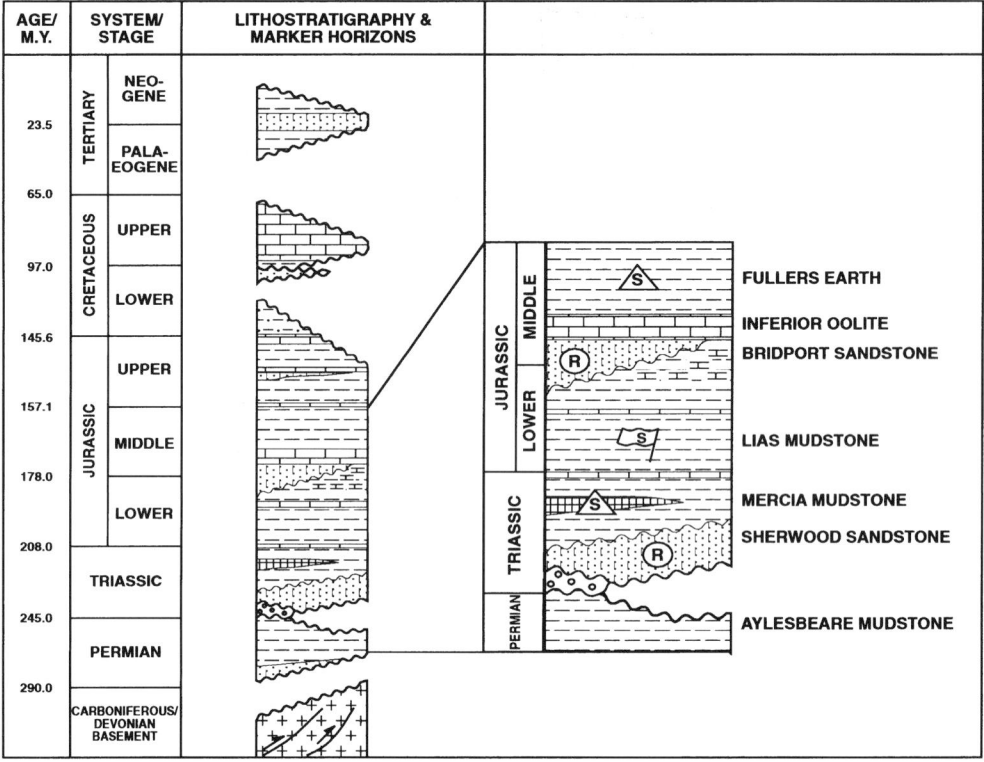

Fig. 3. Summary stratigraphic column for the Portland–Wight Basin with the major elements of the hydrocarbon plays highlighted. The Triassic Sherwood Sandstone reservoir has a Mercia Mudstone seal and a Liassic mudstone source. The Bridport Sandstone reservoir is overlain by the Fuller's Earth Formation seal and is also sourced from the Lias.

mudstones generated from a kitchen area within the PWB. Traps are formed as extensional tilted fault blocks or horsts and rarely show evidence of extensive inversion tectonics (Fig. 4).

Basin configuration

The details of the structural evolution and basin-fill of the Portland–Wight and associated basins are still the focus of much research and only the points relevant to the hydrocarbon system are discussed in this paper (Chadwick 1985a,b, 1986; Evans 1990; Hamblin et al. 1992).

The present-day configuration of the area was produced by the interaction of the Mesozoic–Tertiary basin-forming process and Alpine compressional tectonics on Variscan basement geometries (Stoneley 1982; Chadwick 1985a; Karner et al. 1987; Underhill & Stoneley this volume) (Fig. 2).

The dominant east–west structural trends in the PWB were formed during the Variscan orogeny and extensionally re-activated during the Permo-Triassic and Jurassic (Chadwick et al. 1983). The major fault systems which bound the PWB are linked by northwest-trending structures, for example, in the east by the Pays de Bray system. These faults have a major control on the evolution of the PWB both the initial periods of extension, uplift or compression (Cheadle et al. 1987; Lake & Karner 1987; Chadwick 1993). During the basin history the sedimentation and/or erosion has been controlled by movement on these major fault systems (Fig. 4).

Petroleum exploration

The exploration history of the area extends over more than 50 years with the first exploration well, Poxwell-1, being drilled in 1937 (Fig. 1). In the 1950s and 1960s onshore wells were drilled with variable success at surface anticlines which were often associated with major faults.

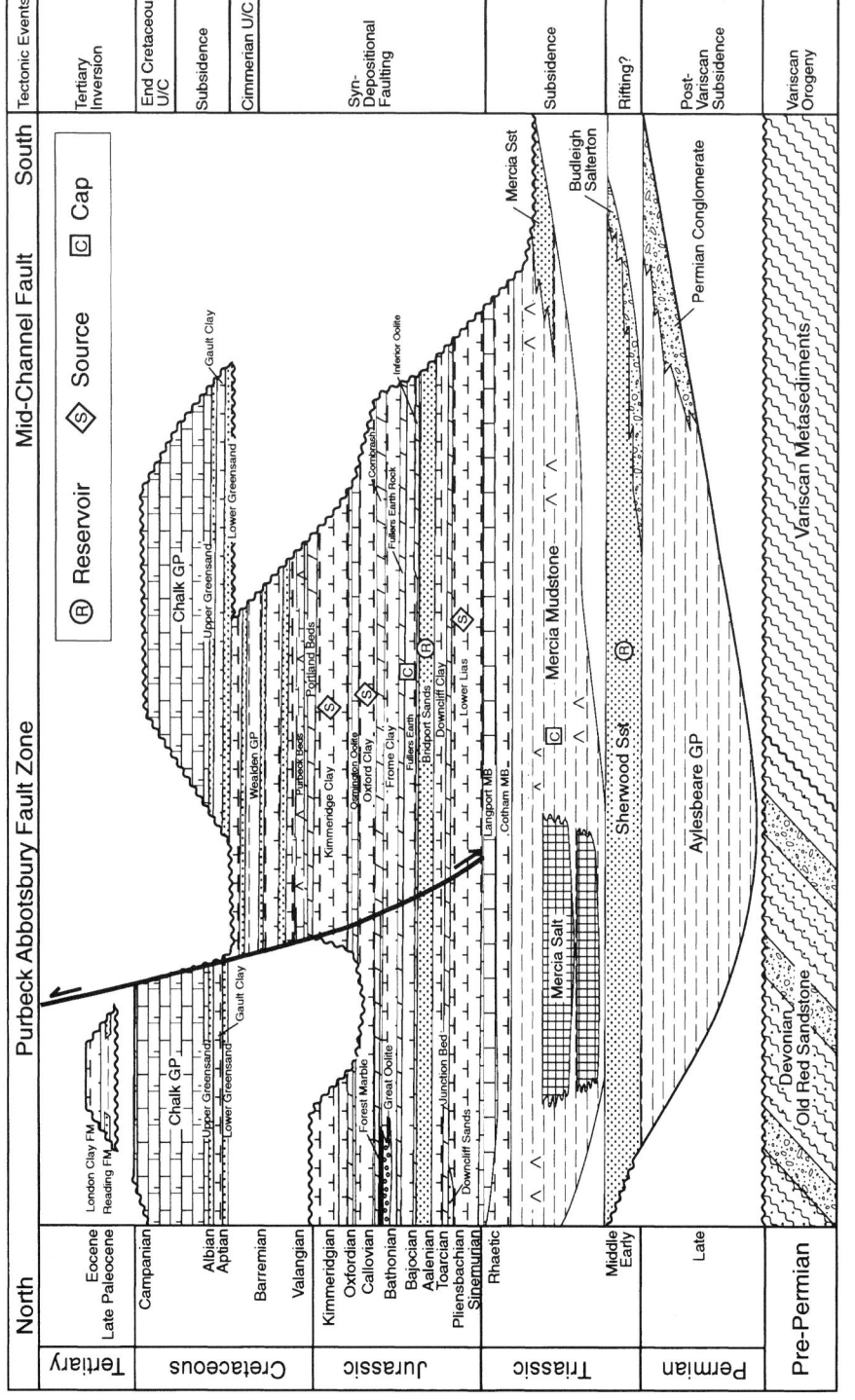

Fig. 4. A detailed stratigraphic chart for the Portland–Wight Basin showing the sediment distribution above the Variscan basement. The diagram illustrates the variability in strata distribution due to the complex basin evolution.

The main exploration success during this period was the discovery of the Kimmeridge oilfield on the Dorset coast which is still in production at present.

Offshore exploration commenced with drilling from a ship of the Lulworth Banks-1 well by BP in 1963 (Fig. 1). It was a milestone in the British exploration industry as it was the first offshore well drilled in the United Kingdom Continental Shelf (UKCS). The well was a relatively shallow test of Jurassic targets in a broad open structure which was identified by onshore structural studies and confirmed by sub-sea mapping. The offshore exploration continued in the late 1970s and 1980s with wells mainly targeted at structures associated with the major fault systems which bound the basins. Several wells (e.g. block 98/11) were drilled in or on the footwall of the Portland–Wight Fault System (PWF) with variable amounts of success. On the southern margin of the basin in Blocks 98/22 and 98/23 on the Mid-Channel Fault System (MCF) similar plays types were drilled. During this period wells were also drilled on the fault-related traps in the central part of the Portland–Wight Basin.

By far the most important discovery in the southern England basins is the Wytch Farm field (Colter & Havard 1981; Dranfield et al. 1987; Bowman et al. 1993) (Fig. 5). The field was discovered in 1973 when the Jurassic Bridport Sandstone reservoir was drilled. Further appraisal of the Bridport reservoir drilling continued until 1977 when the deeper Triassic Sherwood Sandstone reservoir was discovered. The majority of the reserves lie in the Triassic reservoir and the field as a whole is believed to have recoverable reserves of 300 million barrels (Bowman et al. 1993) (Fig. 6). Both reservoirs are now on production and the field development programme continues. In the interests of reduced environmental impact and cost savings, BP have decided that the offshore portion on the Triassic reservoir will be produced by the drilling of several long reach wells from existing production locations (McClure et al. 1995).

The discovery and appraisal of large reserves from the Wytch Farm field obviously encouraged exploration throughout the basin. Subsequent wells were drilled to test the along-strike extension of Wytch Farm plays both onshore to the west and offshore to the east. The onshore exploration areas are now dominated by focused operators such as Brabant Petroleum, who have developed a large acreage position in the basin.

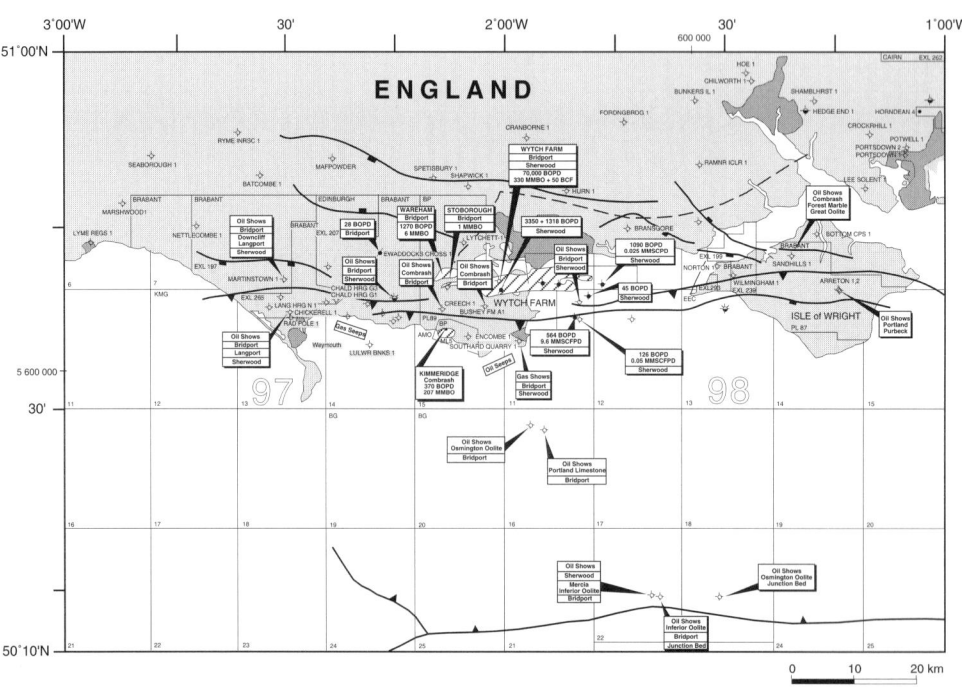

Fig. 5. The hydrocarbon occurrences in the Portland–Wight Basin (only wells released before 1996 are shown).

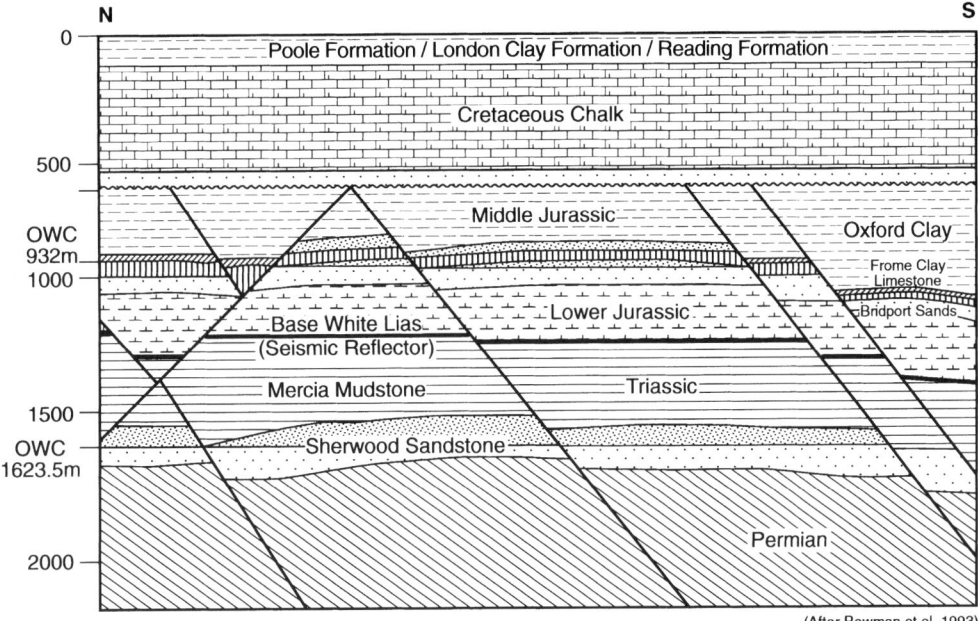

Fig. 6. North–South cross-section through the Wytch Farm Field illustrating the structural styles and the oil–water contact in the Bridport and Sherwood reservoirs (after Bowman *et al.* 1993).

Offshore exploration programmes have continued with, for example, the Elf group acquiring the first three-dimensional seismic over the northern part of 98/12.

The 14th licensing round in 1993 opened previously unlicensed acreage in the Isle of Portland area (Fig. 1). The blocks have been licensed to British Gas (97/19 and 97/20), Amoco (97/14) and Kerr–McGee (97/12) operated groups. The results of the drilling programmes in these waters are awaited with great interest.

Play concepts

As stated in the introduction, there are two primary plays within the PWB area. There is the deeper Triassic Sherwood Sandstone play and an overlying Jurassic Bridport Sandstone play. In addition to these targets, secondary reservoirs are present within the carbonates of the Jurassic (e.g. Frome Clay and Cornbrash) and the siliciclastic units within the Triassic Mercia Mudstone Group (Bowman *et al.* 1993; Underhill & Stoneley this volume).

The widespread distribution of the hydrocarbon shows and accumulations illustrates the validity of the plays and constrains the known extent their limits (Fig. 5). Fields on production are limited to northern part of the PWB in an area close to the PW fault zone. To the west of Wytch Farm, several small oil fields such as Wareham, Kimmeridge and Stoborough have reserves in the Bridport or Cornbrash. Both oil and gas discoveries were made in the offshore block 98/11. There have been numerous shows recorded across the basin which, although some are enigmatic, the overall impression has been to encourage the exploration effort.

Triassic Sherwood play: reservoir/top seal

The Sherwood Sandstone Group comprises an approx. 100–300-m thick red-bed succession which was deposited during the early Triassic in semi-arid conditions in a complex alluvial–fluvial environment (Holloway *et al.* 1989) (Fig. 4). Dominantly arkosic sediment was transported from the south and southwest from the Variscan highlands or the east from the Brabant Massif by a series of complex, braided alluvial systems (Fig. 7). Three facies associations commonly have been recorded from the Sherwood Sandstone which are named by their dominant facies; channel sandstones, sheetflood and playa/floodplain. The best reservoir properties are recorded in the channel sandstones. Reservoir properties can be excellent (especially in the channel sandstone facies), as illustrated by

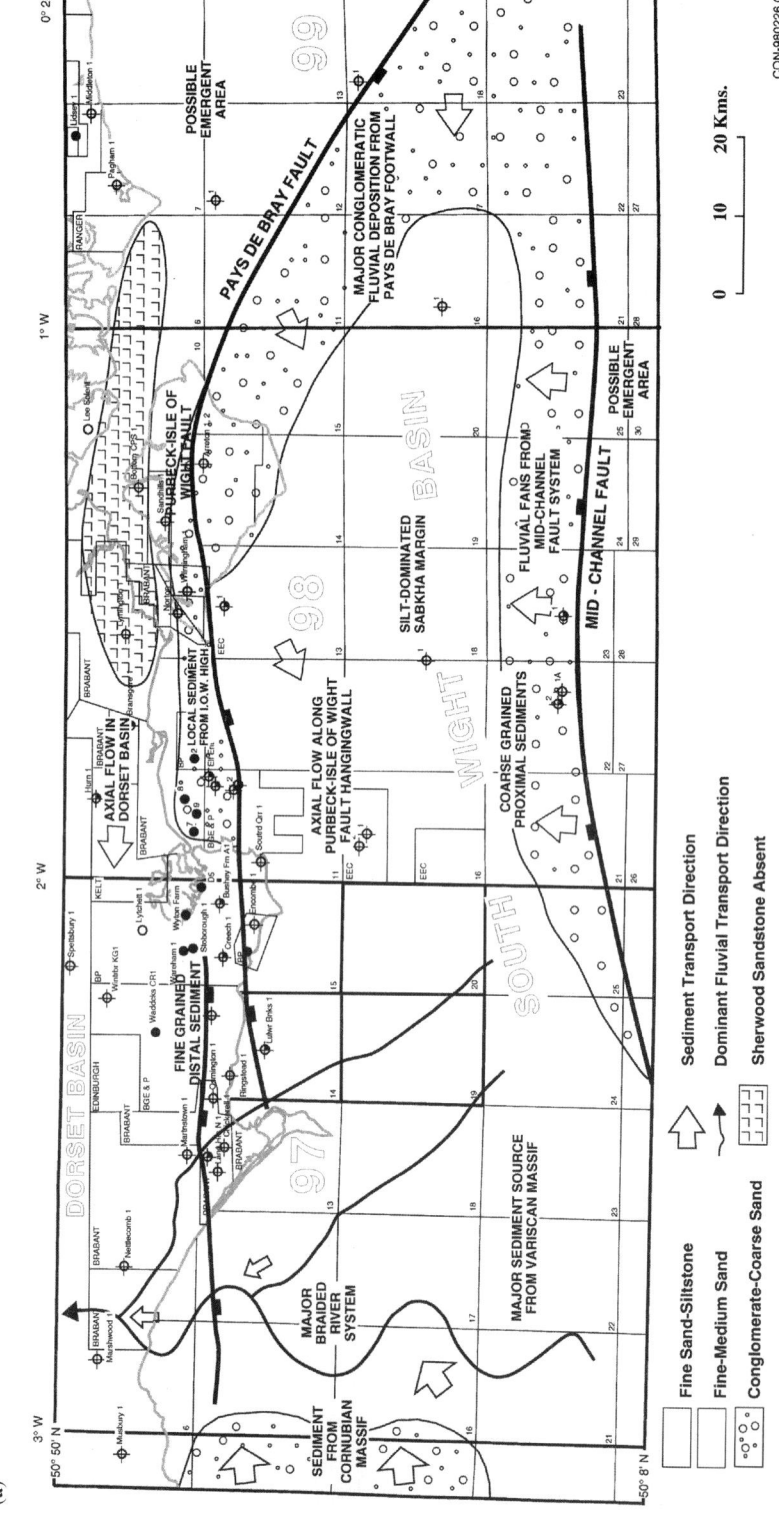

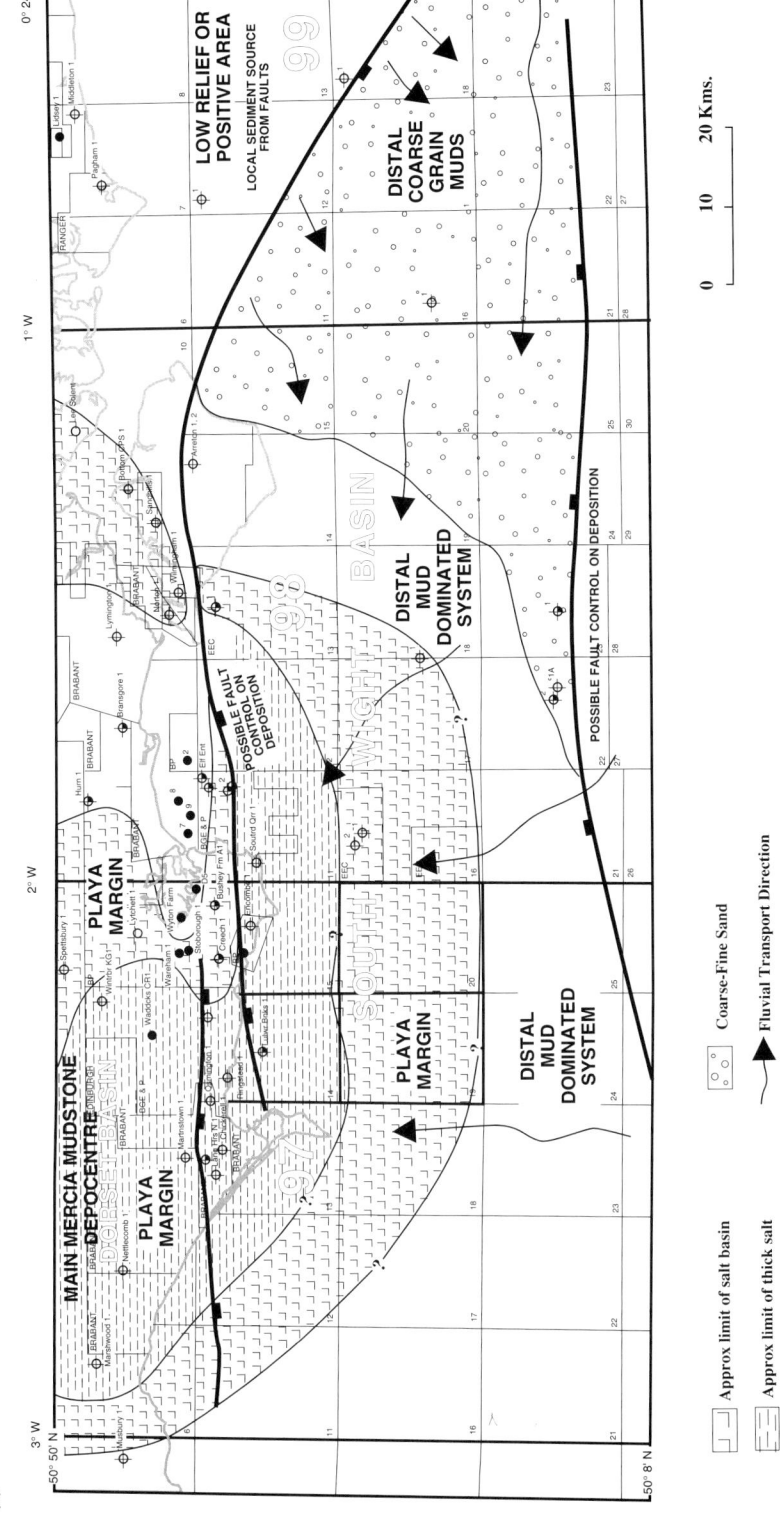

Fig. 7. (a) Summary palaeogeography map for the Sherwood Sandstone. Sand is mainly sourced from the Variscan highlands to the south and transported into the Portland–Wight Basin by major fluvial systems. (b) Summary palaeogeography map for the Mercia Mudstone Group. The mud-dominated unit has also a primary sediment source from the south. An evaporitic basin in centred to the northwest of the basin and shows strong evidence of fault control in sediment distribution.

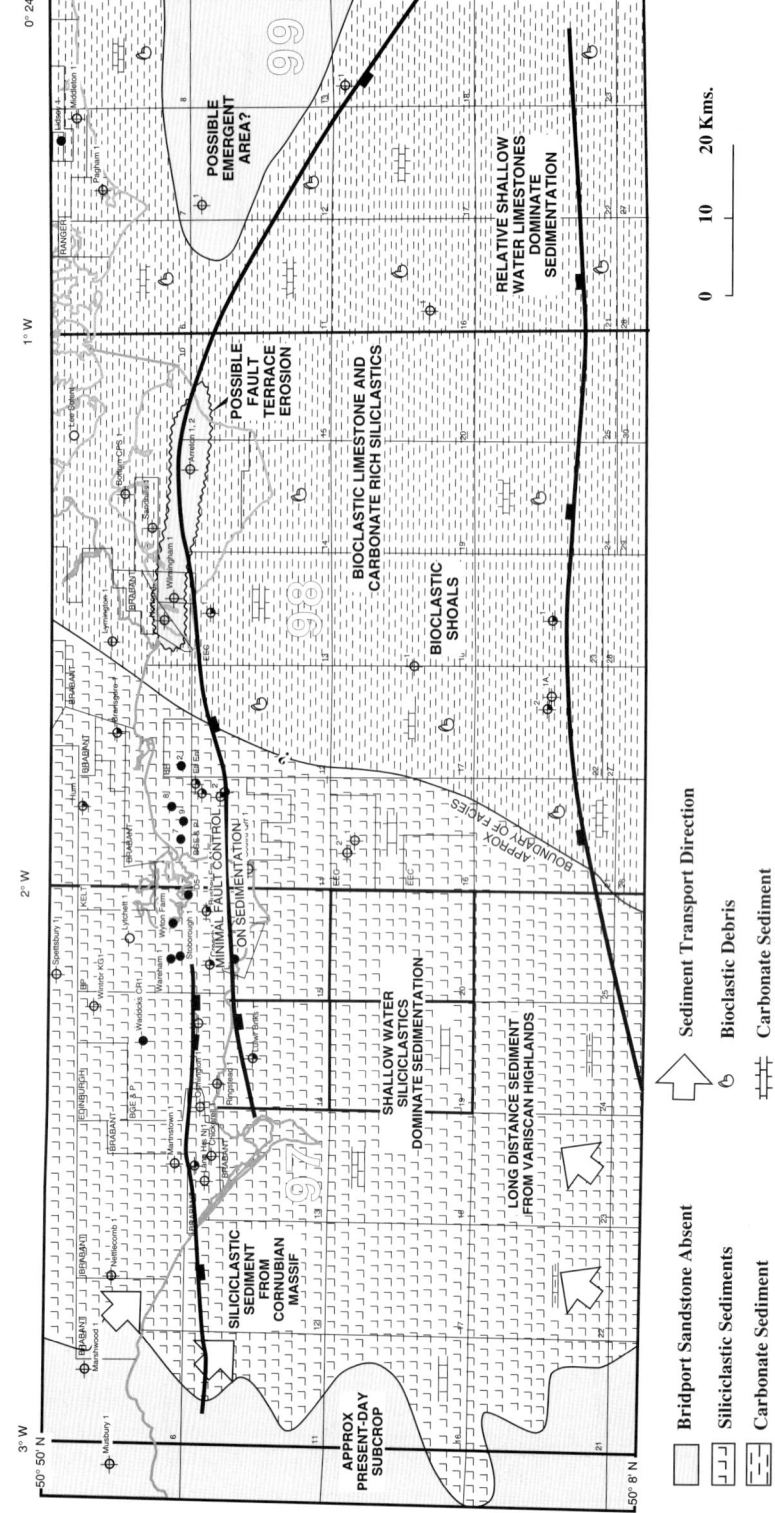

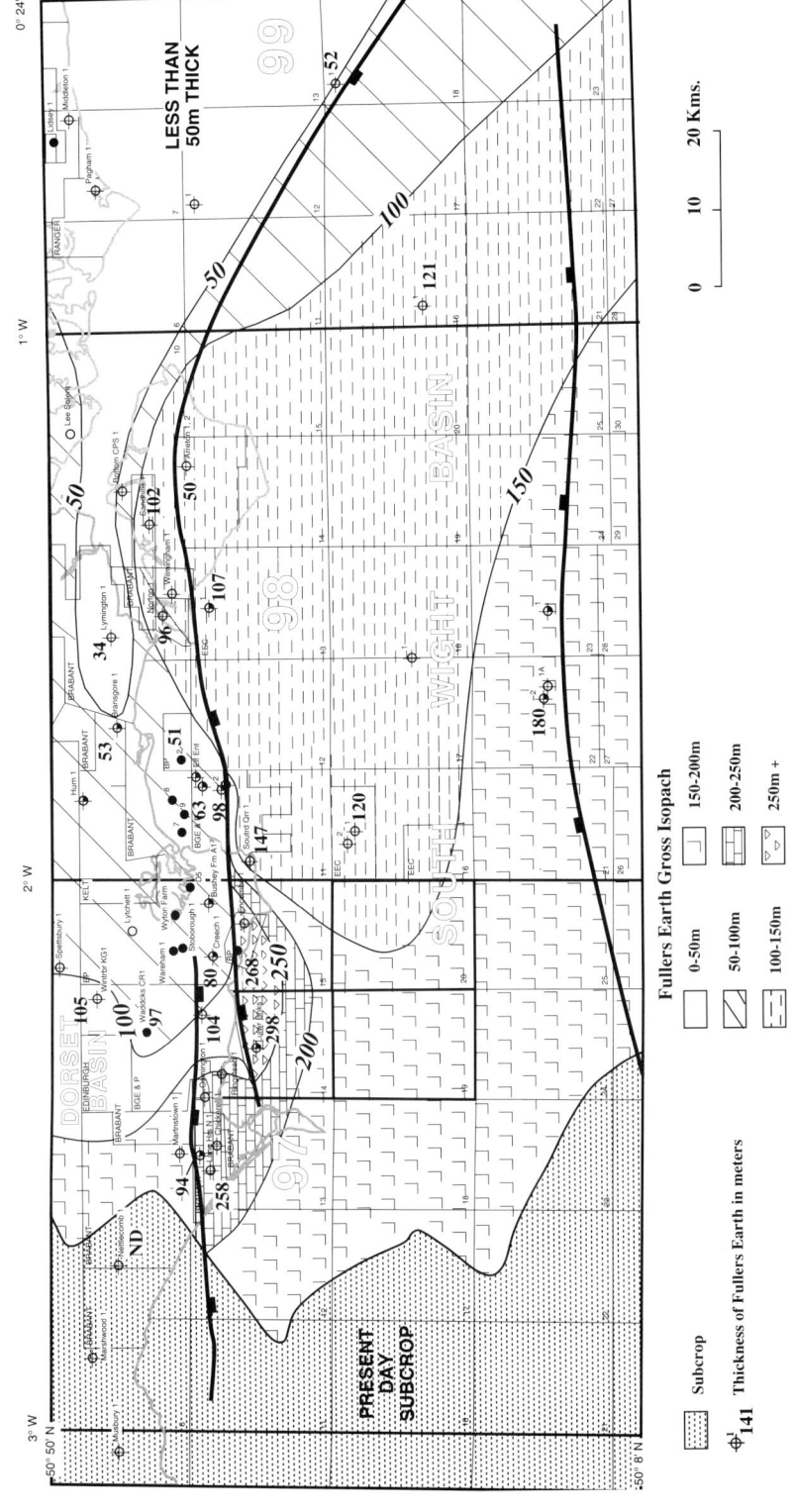

Fig. 8. (a) Summary palaeogeography map for the Bridport Sandstone. The eastern part of the basin is dominated by tight carbonate units; however, the western areas consist of reservoir quality, shallow water, fine-grained siliciclastic sediments. (b) Gross isopach of the Fuller's Earth Formation. In general, the unit thickens westwards across the basin to its present-day outcrop position. The Portland–Wight Fault Zone has a major control on the sediment distribution with the thickest sediments occurring in the hanging wall of the fault.

the Wytch Farm test rates, but calcrete horizons can often reduce the vertical permeability (Bowman et al. 1993). However, in other parts of the basin, gross sandstone development is poorer and the presence of the mudstone units, and poorly sorted conglomerates, in addition to the calcretes can reduce overall reservoir quality substantially.

Top seal and often lateral seal for hydrocarbons in the Sherwood Sandstone is provided by the mudstones and evaporites of overlying Mercia Mudstone Group (Lott et al. 1982) (Fig. 7). This unit is thickly developed across the entire area (approx. 100–600+ m) and is evaporite-rich in the western part of the PWB. The Mercia Mudstone has proven hydrocarbon sealing properties as evidenced by the accumulations in the Wytch Farm field and the 98/11 discoveries.

Jurassic Bridport reservoir/top seal

The Bridport Sandstone (approx. 25–100+ m) consists of very fine to medium-grained sandstones deposited under shallow marine conditions (Fig. 8). It is a markedly diachronous unit with local evidence of fault-controlled depositional patterns (Torrens 1969; Jenkyns & Senior 1991; Callomon & Cope 1993). The Bridport Sandstone has good reservoir properties in the western part of the basin, where the unit is dominated by siliciclastic sediments. In the eastern part of the PWB the non-reservoir calcareous mudstones are present. The siliciclastic reservoir is also down-graded due to the presence of well-cemented layers or doggers which form permeability barriers to hydrocarbons (Kantorowicz et al. 1987; Bryant et al. 1988; Bjorkum & Walderhaug 1993).

The thickly developed (max. 300+ m) mudstones of Fuller's Earth Formation acts as the top seal for the hydrocarbons reservoired in the Bridport Sandstone. The production from the Bridport Sandstone in Wytch Farm and other smaller fields testify to the effectiveness of the Fuller's Earth Formation as a top and lateral seal.

Reservoirs also lie within the carbonates of the Jurassic (e.g. the Cornbrash) but production to date from these units is limited and they remain a secondary exploration target.

Source units

The Jurassic organic potential has been studied by Ebukanson & Kinghorn (1986). Geochemical analysis (Fig. 9) indicates that three Jurassic three units have sufficient total organic content (TOC) values to be source rock for the hydrocarbons in the basin namely the Kimmeridge Clay, Oxford Clay and Liassic mudstones.

However, maturity studies and burial history modelling indicate that only the Liassic mudstones are sufficiently mature to have generated substantial amounts of oil and gas (Ebukanson & Kinghorn 1986) (Fig. 9). The Lias has TOC values of greater than 2% and the dominant kerogen is Type II with minor amounts of Type III. This has been confirmed by oil/gas geochemical studies on hydrocarbons produced from fields in the PWB and surrounding area.

Trap/structures

The PWB area has a long and complex structural evolution, with several periods of extension and uplift or inversion (Butler & Pullen 1990). These tectonics have produced a range of structural trap styles (Fig. 10). Many of these traps have been tested by the exploration programmes in the area (Butler & Pullen 1990).

Three structural trap styles are commonly present within the PWB area.

(1) The main trap type are east–west-trending tilted fault blocks and horsts which are still in the net extension at reservoir level.
(2) There are also inversion-related fold/fault structures which are possible valid traps.
(3) Salt tectonics within the evaporites of the Mercia Mudstone Group in the western part of the basin have produced major décollements, detached structures and swells or domes in the post-Triassic units. The domes or swells could be valid traps for Jurassic reservoirs.

The complexity of the basin evolution has led to traps with contrasting style and origin being superimposed on one another. The structural evolution each prospect has to be studied in detail to fully understand the trapping mechanism.

Controls on Sherwood Sandstone play

Reservoir/seal doublet

The Sherwood Sandstone Group has the potential to be an effective reservoir in many parts of the PWB. The main controls on reservoir quality are the primary depositional lithofacies and diagenesis. The coarser-grained, channelized units and sheetflood deposits have good moderate initial porosity and permeability. The vertical

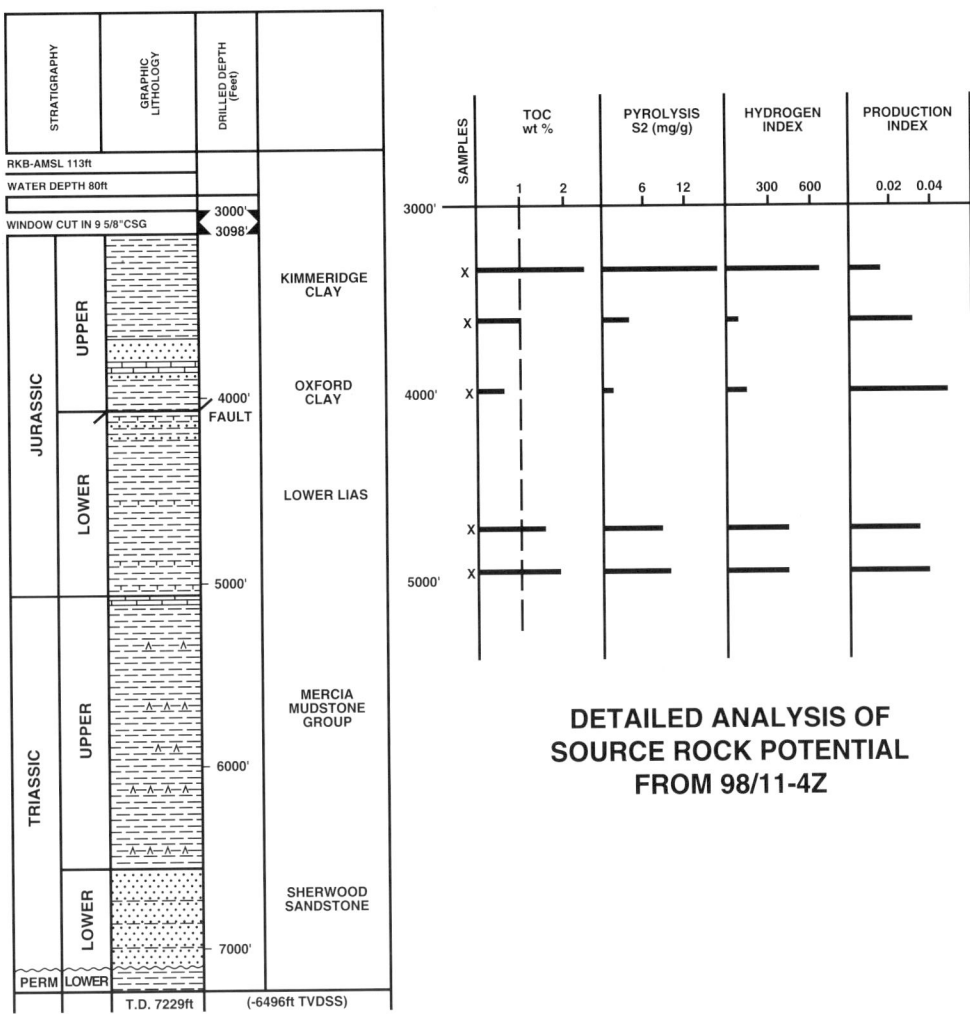

Fig. 9. (a) An analysis of the source rock potential from well 98/11-4z. The Kimmeridge Clay, Oxford Clay and the Lias have sufficient organic matter to be potential source rocks. (b) Summary map of the interpreted Lias maturity within the Portland–Wight Basin. The kitchen area lies in the hanging wall of the Portland–Wight Fault Zone where the late oil/early gas maturities are reached.

permeabilities are reduced due to the presence of mudstone interbeds and burial diagenesis (Purvis & Wright 1991; Bowman *et al.* 1993). The distribution of the lithofacies offshore and in the central part of the basin are unclear due to the lack of well penetrations and available core material.

The Mercia Mudstone Group has excellent sealing properties and is known to be effective in the Wessex and other basins. The Mercia Mudstone is thickly developed across the area and therefore the top seal risks on the Sherwood Sandstone targets are low.

Source rock

The PWB has three possible source rock units. The producing fields and numerous shows confirm the presence of a working hydrocarbon system in the area. Regional geochemical studies and burial history modelling has shown that

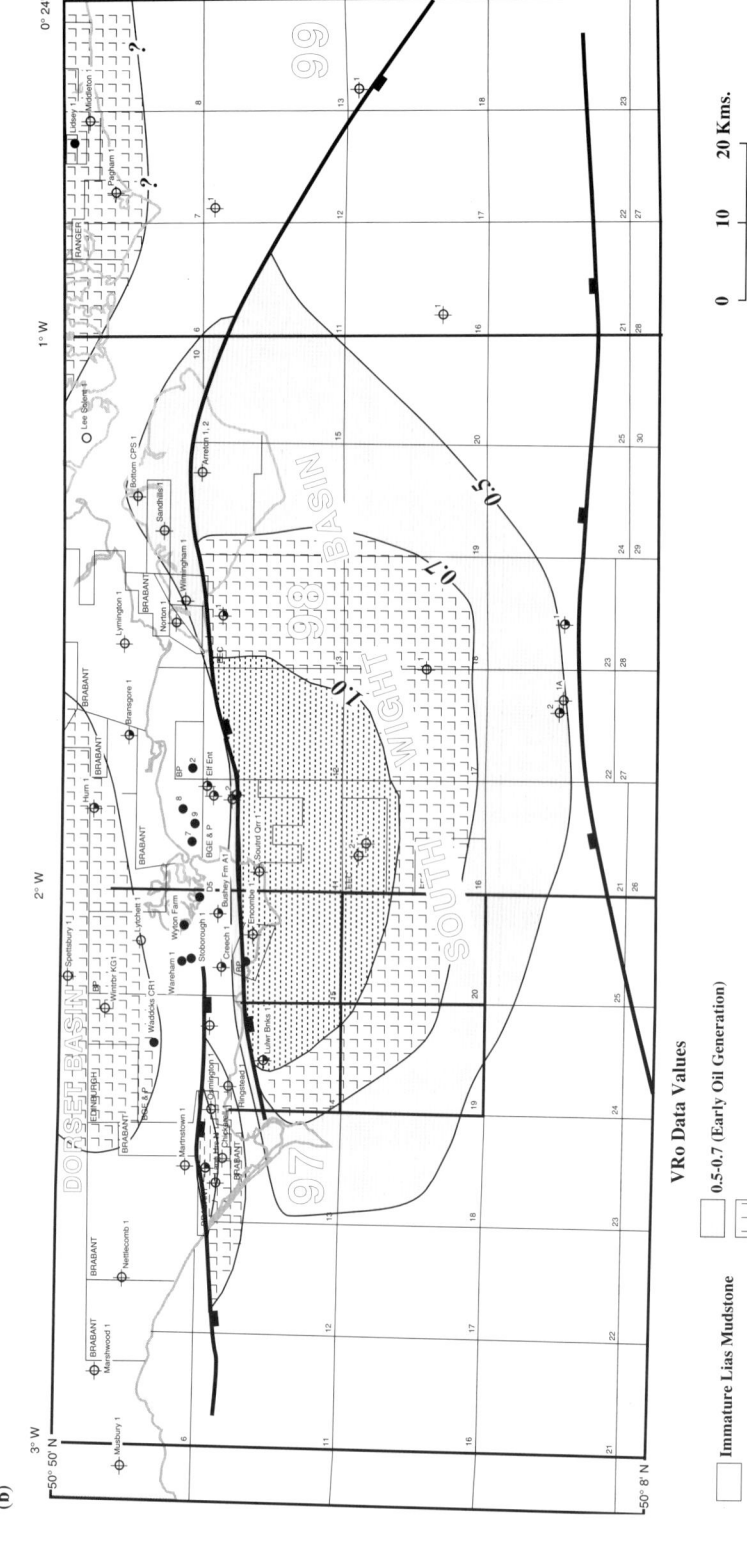

Fig. 9. (continued)

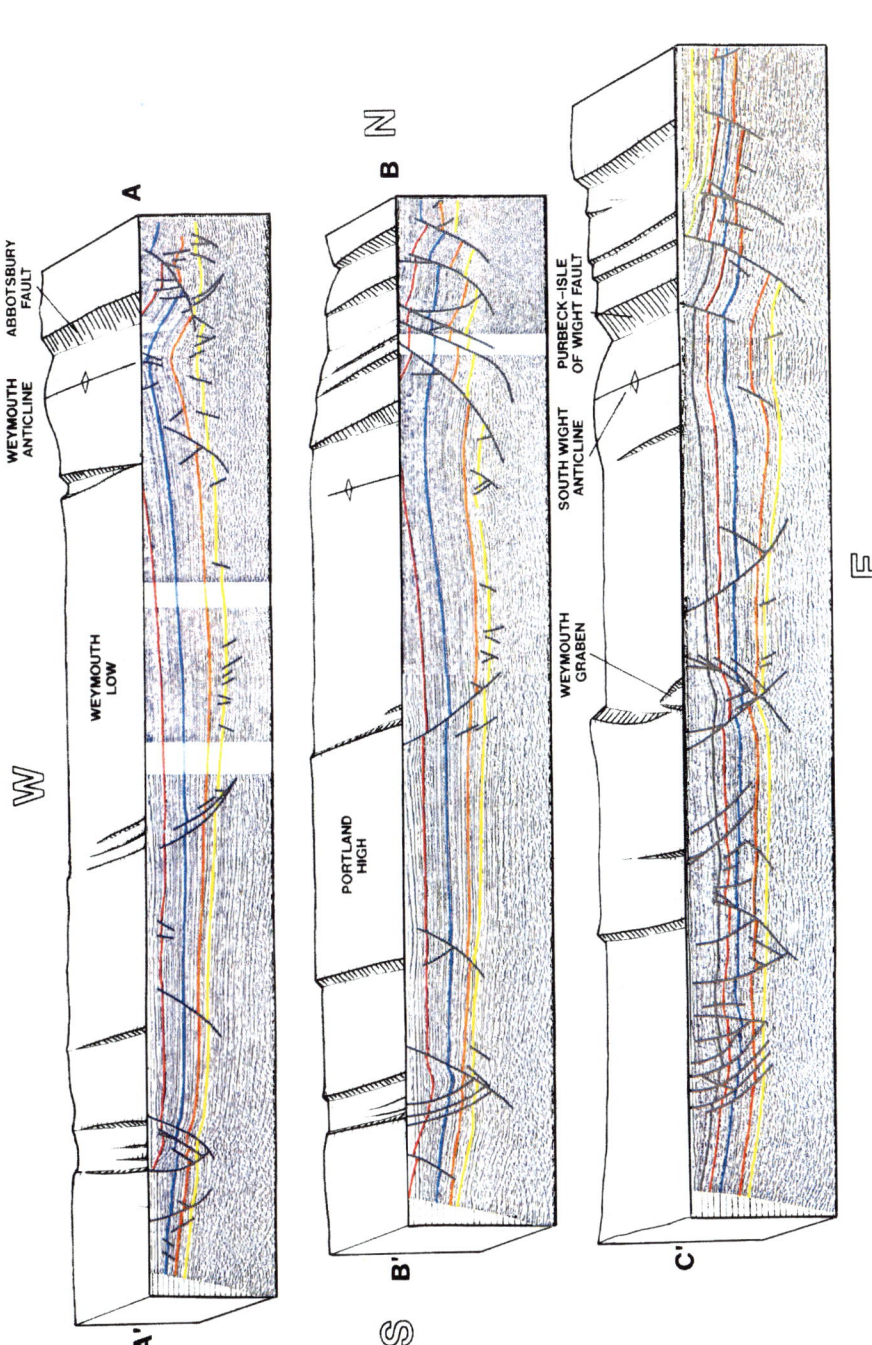

Fig. 10. Three composite seismic sections across the Portland–Wight Basin (locations/objections is shown in Fig. 2). The sections illustrate the structural style variations across the basin.

the Lower Lias mudstones are the source for the majority of hydrocarbons in the area. The source kitchen is believed to lie in the hanging wall of the PW fault system.

Migration

One of the major risks with the play lies in the migration of the hydrocarbons from the Liassic source rocks into the underlying Triassic reservoir. The migration route requires fault movement to produce juxtaposition of reservoir and carrier beds or place the Liassic lower than the Sherwood Sandstone to allow vertical and/or lateral hydrocarbon migration. A migration route therefore requires faults with substantial displacement (hundreds of metres) to be valid.

Trap definition and formation/deformation

The tectonics in the PWB has produced several trap types as discussed above. These structures were produced by different tectonic processes and at separate times during the evolution of the area. The relationship between the timing of migration and trap formation is an important control on the definition of a valid trap in the PWB.

In terms of reserves, the most successful trap style is the tilted fault block or horst in the footwall of the bounding faults of the basin. The best example of this trap type is the Wytch Farm field, which lies in an east–west-trending horst to the north of the PW fault zone. The substantial extensional throw on the major fault zone has allowed the juxtaposition of source rock and the carrier or reservoir beds enabling lateral/vertical migration in to the trap. At reservoir levels, the Wytch Farm horst structure is still in extension i.e. it has not been inverted past the null point (Cooper & Williams 1989) (Fig. 6).

The inversion structures were produced during the Tertiary compressional tectonics related in part to the Alpine orogeny (Lake & Karner 1987; Chadwick 1993; Glennie & Underhill 1998). The main risk with these structures is the timing of hydrocarbon migration relative to trap formation. The traps may have formed post-generation and/or migration, and for the Sherwood Play the migration route is very problematic. The inversion tectonics may also breach previously formed traps allowing hydrocarbons migrate elsewhere or to escape to surface. The success of this trap type is poorly documented.

There are problems with trap definition at pre-Rhaetic levels on seismic data in general and onshore data in particular. The top Sherwood Sandstone reflector can be difficult to define with confidence in seismic profile due to lack of acoustic impedance contrast.

However, with better acquisition and processing techniques (especially with statics correction) these problems largely have been reduced on more modern seismic surveys. The confidence in the interpretation of Triassic reflectors is further reduced in the western part of the basin due to the effect of salt tectonics (Butler 1997 this volume).

Controls on Bridport Sandstone play

Reservoir/seal doublet

Reservoir quality is the major risk on the Bridport Sandstone play in the area as a whole. The main controls on the reservoir area primary deposition and burial diagenesis (Bryant *et al.* 1988; Hesselbo & Palmer 1992; Bjorkum & Walderhaug 1993) (Fig. 8).

The onshore portion of the area in the footwall of the PWF and Abbotsbury fault system have good reservoir properties, as recorded in the Wytch Farm field. The eastern part of the PWB has very poor reservoir as the sandstone lithofacies has not been deposited, and the unit is dominated by the fine-grained carbonates which are non-reservoir. In the western offshore area of the basin the same onshore lithofacies are believed to extend southwards.

The presence of heavily cemented layers and doggers, as seen in the coastal outcrops within the Bridport Sandstone, act as major vertical permeability barriers. The effective permeability is also reduced by increased burial diagenesis which occludes the porosity.

The Fuller's Earth Formation is widely distributed across the area and is known as a valid and effective top seal. The three producing Bridport fields highlight the quality of the Fuller's Earth Formation as a proven top seal. The Inferior Oolite which lies directly above the Bridport may act as a top seal or non-reservoir zone.

Source rock

The source rock for the hydrocarbons in the Bridport Sandstone play is also the Liassic mudstones (Colter & Harvard 1981).

Migration

Hydrocarbon migration does not require fault movement allowing juxtaposition of the source rock and reservoir. As the Bridport Sandstone directly overlies the Liassic, simple vertical migration or relatively short lateral migration is required is required to fill the available traps.

Trap definition and formation/deformation

The first phases of drilling were targeted at structures which defined on-surface geological mapping. The introduction of seismic data allowed the identification of sub-surface traps. Initially, trap definition was limited due to the complexity of structure and problems with seismic acquisition and processing techniques. In recent years the improved seismic data have allowed interpretation of Jurassic reflectors with confidence.

The main trapping styles are tilted fault and horst blocks. The majority of faults on valid traps are in net extension with only minor amounts of inversion on the bounding faults. Inversion tectonics can produce well-defined structures but these are not hydrocarbon bearing. The timing of inversion post-dates the migration of hydrocarbons as the uplift switches off the hydrocarbon kitchen.

In the western part of the PWB, the evaporite horizons within the Mercia Mudstone Group can be a major controlling part in deformation style (Butler 1997 this volume). The salt acts as a detachment between the Sherwood Sandstone and the Jurassic units. In many cases, this can mean the structure mapped from seismic data at Top Rhaetic or shallower reflectors is not representative of the structure at top Sherwood Sandstone level.

Future

Future successful exploration in the PWB will require continued refinement of the models for both hydrocarbon migration system and the reservoir distribution and effectiveness. There is potential for stratigraphic trapping owing to the large number of reservoir units within the basin and the presence of a proven hydrocarbon system.

There is potential for the development of new plays by the revisiting of previously explored areas and using better data quality to identify hitherto unrecognized traps. The prospectivity of higher risk areas to the west and south of the basin also requires further analysis.

Environmental considerations are very important in any exploration, production and abandonment programme in these basins. The oil/gas industry will have to carry out all its efforts with a minimal impact of the environment. The costs of these measures may preclude the further development of parts of the basins.

Conclusions

There is a long history of oil/gas exploration in the southern England basins which has rewarded with abundance of encouragement from shows and small hydrocarbon columns. Unfortunately, to date, there has been only one major field discovered in the basin.

The PWB is a dynamic basin in four dimensions. From the Variscan orogeny to the present day the basin has had a complex sedimentology and tectonic evolution. The major controls on hydrocarbon prospectivity in the basin are reservoir distribution and effectiveness and play dynamics. Exploration wells have failed due to lack of reservoir quality in the targeted Triassic or Jurassic intervals. Where a dry well has found good reservoir within a valid structure failure is likely to be due to the lack of an effective hydrocarbon migration route or trap formation post-migration.

Thanks to BGE&P management for permission to publish this paper. This paper reports and builds on the work of generations of BGE&P geologists, geophysicists and engineers who have worked on the Portland–Wight Basin. In particular, personal thanks go to my colleagues N. Vivian, R. Blow, C. Singleton, A. Law, N. McMahon, N. Anderson and P. Coleman who helped me recognize the complexity of the geological and hydrocarbon systems of the Wessex basins. Reviews by R. Wrigley and N. Anderson have helped improve the paper considerably. R. Buchanan is thanked for her help in the editing of earlier versions of this paper.

References

ARKELL, W. J. 1933. *The Jurassic System in Great Britain*. Oxford University Press, London.
—— 1947. *Geology of the Country Around Weymouth, Swanage, Corfe and Lulworth*. Memoir of the Geological Survey of Great Britain, HMSO, London.
BJORKUM, P. A. & WALDERHAUG, O. 1993. Isotopic composition of a calcite-cemented layer in the lower Jurassic Bridport Sands, Southern England: Implications for formation of laterally extensive calcite-cemented layers. *Journal of Sedimentary Petrology*. **63**, 678–682.
BOWMAN, M. B. J., MCCLURE, N. M. & WILKINSON, D. W. 1993. Wytch Farm oilfield: deterministic reservoir description of the Triassic Sherwood Sandstone. *In*: PARKER, J. R. (ed.) *Petroleum Geology of Northwest Europe: Proceedings of the 4th Conference*. Geological Society, London, 1513–1517.

BRYANT, I. D., KANTOROWICZ, J. D. & LOVE, C. F. 1988. The origin and recognition of laterally continuous carbonate-cemented horizons in the Upper Lias Sands of Southern England. *Marine and Petroleum Geology*, **5**, 108–133.

BUTLER, M. 1997. The geological history of the Wessex Basin – a review of new information from oil exploration. *This volume.*

—— & PULLEN, C. P. 1990. Tertiary structures and hydrocarbon entrapment in the Weald Basin of Southern England. *In*: HARDMAN, R. F. P. & BROOKS, J. (eds) *Tectonic Events Responsible for Britain's Oil and Gas Reserves*. Geological Society, London, Special Publications, **55**, 371–391.

CALLOMON, J. H. & COPE, J. C. W. (eds) 1993. *Arkell International Symposium. The Jurassic of Dorset Excursion Guide*. University College, London.

CHADWICK, R. A. 1985a. Permian, Mesozoic and Cenozoic structural evolution of England and Wales in relation to the principles of extension and inversion tectonics. *In*: WHITTAKER, A. (ed.) *Atlas of Onshore Sedimentary Basins in England and Wales: Post-Carboniferous Tectonics and Stratigraphy*. Blackie, Glasgow, 9–25.

——1985b. Cenozoic sedimentation, subsidence and basin inversion. *In*: WHITTAKER, A. (ed.) *Atlas of Onshore Sedimentary Basins in England and Wales: Post-Carboniferous Tectonics and Stratigraphy*. Blackie, Glasgow, 61–63.

——1986. Extension tectonics in the Wessex Basin, Southern England. *Journal of the Geological Society, London*, **143**, 465–488.

——1993. Aspects of basin inversion in Southern Britain. *Journal of the Geological Society, London*, **150**, 311–322.

——, KENOLTY, N. & WHITTAKER, A. 1983. Crustal structure beneath Southern England from deep seismic reflection profiles. *Journal of the Geological Society, London*, **140**, 893–911.

CHEADLE, M. J., MCGEARY, S., WARNER, M. R. & MATTHEWS, D. H. 1987. Extensional structures on the western UK continental shelf. *In*: COWARD, M. P., DEWEY, J. F. & HANCOCK, P. L. (eds) *Continental Extensional Tectonics*. Geological Society, London, Special Publications, **28**, 445–465.

COLTER, V. S. & HAVARD, D. J. 1981. The Wytch Farm Oil Field, Dorset. *In*: ILLING, L. V. & HOBSON, G. D. (eds) *Petroleum Geology of the Continental Shelf of North-west Europe*. Heyden, London, 494–503.

COOPER, M. A. & WILLIAMS, G. W. 1989. *Inversion Tectonics*. Geological Society, London, Special Publications, **44**.

DRANFIELD, P., BEGG, S. H. & CARTER, R. R. 1987. Wytch Farm Oilfield: reservoir characterisation of the Triassic Sherwood Sandstone for input into reservoir simulation studies. *In*: BROOKS, J. & GLENNIE, K. W. (eds) *Petroleum Geology of North West Europe*. Graham & Trotman, London, 149–160.

EBUKANSON, E. J. & KINGHORN, R. R. F. 1986. Maturity of organic matter in the Jurassic of Southern England and its relation to the burial history of the sediments. *Journal of Petroleum Geology*, **9**, 259–280.

EVANS, C. D. R. 1990. *United Kingdom Offshore Regional Report: The Geology of the Western English Channel and its Western Approaches*. HMSO, London.

GLENNIE, K. W. & UNDERHILL, J. R. 1998. Origin, development and evolution of structural styles. *In*: GLENNIE, K. W. (ed.) *Petroleum Geology of the North Sea*. Blackwell Scientific Publications, Oxford, in press.

HAMBLIN, R. J. O., CROSBY, A., BALSON, P. S., JONES, S. M., CHADWICK, R. A., PENN, I. E. & ARTHUR, M. J. 1992. *United Kingdom Offshore Regional Report: The Geology of the English Channel*. HMSO for the British Geological Survey, London.

HESSELBO, S. P. & PALMER, T. J. 1992. Reworked early diagenetic concretions and the bioerosional origin of a regional discontinuity within the British Jurassic marine mudrocks. *Sedimentology*, **39**, 145–165.

HOLLOWAY, S., MILODOWSKI, A. E., STRONG, G. E. & WARRINGTON, G. 1989. The Sherwood Sandstone Group (Triassic) of the Wessex Basin, southern England. *Proceedings of the Geological Association*, **100**, 383–394.

HOUSE, M. R. 1993. The geology of the Dorset coast. *In*: LISTER, C. J. & GREENSMITH, J. T. (eds) *Geologists' Association Guide No. 22.*

JENKYNS, H. C. & SENIOR, J. R. 1991. Geological evidence for intra-Jurassic faulting in the Wessex Basin and its margins. *Journal of the Geological Society, London*, **148**, 245–260.

KANTOROWICZ, J. D., BRYANT, I. D. & DAWANS, J. M. 1987. Controls on the geometry and distribution of carbonate cements in Jurassic sandstones: Bridport Sands, Southern England and Viking Group, Troll Field, Norway. *In*: MARSHALL, J. D. (ed.) *Diagenesis of Sedimentary Sequences*. Geological Society, London, Special Publications, **36**, 103–118.

KARNER, G. D., LAKE, S. D. & DEWEY, J. F. 1987. The thermo-mechanical development of the Wessex Basin, southern England. *In*: HANCOCK, P. L., DEWEY, J. F. & COWARD, M. P. (eds) *Continental Extensional Tectonics*. Geological Society, London, Special Publications, **28**, 517–536.

KNOX, R. W. O'B., MORTON, A. C. & LOTT, G. K. 1982. Petrology of the Bridport Sands in the Winterborne Kingston Borehole, Dorset. *In*: RHYS, G. H., LOTT, G. K. & CALVER, M. A. (eds) *The Winterborne Kingston Borehole, Dorset, England*. Report of the Institute of Geological Science No. 81/3, 107–121.

LAKE, S. D. & KARNER, G. D. 1987. The structure and evolution of the Wessex Basin, southern England: an example of inversion tectonics. *Tectonophysics*, **137**, 347–378.

LOTT, G. K., SOBEY, R. A., WARRINGTON, G. & WHITTAKER A. 1982. The Mercia Mudstone Group (Triassic) in the western Wessex Basin. *Proceedings of the Ussher Society*, **5**, 340–346.

MCCLURE, N. M., WILKINSON, D. W., FROST, D. P. & GEEHAN, G. W. 1995. Planning extended reach wells in Wytch Farm Field, UK. *Petroleum Geoscience*, **1**, 115–127.

MCMAHON, N. A. & UNDERHILL, J. R. 1995. The regional stratigraphy of the southwest United Kingdom and adjacent offshore areas with particular reference to the major intra-Cretaceous unconformity. *In*: CROCKER, P. F. & SHANNON, P. M. (eds) *The Petroleum Geology of Ireland's Offshore Basins*. Geological Society, London. Special Publications, **93**, 323–325.

MILNER, A. R., GARDINER, B. G., FRASER, N. C. & TAYLOR, M. A. 1990. Vertebrates from the Otter Sandstone Formation of Devon. *Palaeontology*, **33**, 873–892.

PENN, I. E., CHADWICK, R. A., HOLLOWAY S., ROBERTS, G., PHARAOH, T. C., ALLSOP, J. M., HULBERT, A. E. & BURNS, I. M. 1987. Principal features of hydrocarbon prospectivity of the Wessex-Channel Basin, UK. *In*: BROOKS, J. & GLENNIE, K. (eds) *Petroleum Geology of North West Europe*. Graham & Trotman, London, 108–118.

PURVIS, K. & WRIGHT, V. P. 1991. Calcretes related to phreatophytic vegetation from the Middle Triassic Otter Sandstone of South West England. *Sedimentology*, **38**, 539–551.

RUFFELL, A. H. 1991. Geophysical correlation of the Aptian and Albian formations in the Wessex Basin of Southern England. *Geological Magazine*, **128**, 67–75.

—— 1992. Early to mid-Cretaceous tectonic and unconformities of the Wessex Basin (Southern England). *Journal of the Geological Society, London*, **149**, 443–454.

SCOTT, J. & COLTER, V. S. 1987. Geological aspects of current onshore Great Britain exploration plays. *In*: BROOKS, J. & GLENNIE, K. (eds) *Petroleum Geology of North West Europe*. Graham & Trotman, London, 95–107.

STONELEY, R. 1982. The structural development of the Wessex Basin. *Journal of the Geological Society, London*, **139**, 545–552.

—— & SELLEY, R. C. 1986. *A Field Guide to the Petroleum Geology of the Wessex Basin*. Department of Geology, Imperial College, London.

TORRENS, H. S. (ed.) 1969. *International Field Symposium on the British Jurassic. Excursion No. 1, Guide for Dorset and South Somerset*. University of Keele, A1–A71.

UNDERHILL, J. R. & PATERSON, S. 1998. Genesis of tectonic inversion structures: seismic evidence for the development of key structures along the Purbeck–Isle of Wight Disturbance. *Journal of the Geological Society, London*, in press.

—— & STONELEY, R. 1998. Introduction to the development, evolution and petroleum geology of the Wessex basin. *This volume*.

WILSON, V., WELCH, F. B. A., ROBBIE, J. A. & GREEN, G. W. 1958. *Geology of the Country Around Bridport and Yeovil*. Memoirs of the Geological Survey of Great Britain, 22–67.

The tectono-stratigraphic development and exploration history of the Weald and Wessex basins, Southern England, UK

P. W. HAWKES,[1] A. J. FRASER[2] & C. C. G. EINCHCOMB[3]

[1] *Enterprise Oil, Grand Buildings, Trafalgar Square, London WC2N 5EJ, UK*
(e-mail: Paul.Hawkes@london.entoil.com)
[2] *BP Exploration Inc., BP Plaza, 200 Westlake Park Boulevard, Houston, TX 77079, USA*
[3] *BP Exploration, Chertsey Road, Sunbury on Thames, Middlesex TW16 7LN, UK*

Abstract: Exploration drilling for hydrocarbons in southern England commenced over 50 years ago prompted by numerous seepages along the Dorset coast and gas shows in water boreholes. The two main depocentres, the Weald and Wessex basins, exhibit many similarities in both tectonic and stratigraphic evolution through Early Triassic–Tertiary times, reflecting the regional influences of Atlantic Margin rift–subsidence processes and subsequent Tertiary (Alpine) inversion tectonics. Given these similarities, the development of the hydrocarbon play system in both these areas can be compared using a common tectono-stratigraphic framework which has been identified and calibrated using well and seismic data. In detail, facies variability recognized within individual depositional units highlights differences in both source and reservoir distribution between the Weald and Wessex areas, which have in turn exerted fundamental controls on the hydrocarbon prospectivity of the two basins. These differences are influenced by the position of major basin-bounding fault systems and by the changing importance of the Cornubian–Armorican and the London–Brabant Massifs as clastic provenance areas through time. Sequence isopachs illustrate the same discrete episodes of rift-related subsidence in both areas. These rifting episodes led to the development of extensional trapping geometries particularly during Late Jurassic–Early Cretaceous times. Subsequent Late Cretaceous thermal subsidence led to the formation of mature source kitchen areas focused within the hanging walls of the earlier extensional faults. Hydrocarbon expulsion from the source kitchen areas ceased during Tertiary uplift and subsequent cooling. Inversion movements on the old extensional faults led to the development of hanging wall anticlines which rely upon remigration mechanisms, including exsolution of gas from uplifted Liassic source rocks, to receive hydrocarbon charge. To date over 300 million barrels of oil and 100 bcf of gas have been discovered.

This paper describes the results of a major regional study carried out at BP Exploration's UK land offices in Eakring during the late 1980s. At that time BP had amassed a substantial database of modern well and seismic data in addition to the results of exploration activity dating back to the 1930s. The facies and isopach maps presented in this paper represent summarized versions of the detailed maps produced for a large regional report on southern England which BP have used to focus exploration in the basin over the past 10 years.

The authors have adopted a seismostratigraphic approach, constrained by both well and outcrop data tied to modern seismic, to describe the geological development of the province from Permo-Triassic to Recent times. Four plate sequences are recognized: (i) early Atlantic (Late Permian–Triassic); (ii) Atlantic (Early–Late Jurassic); (iii) Biscay (Late Jurassic–Late Cretaceous); and (iv) Alpine (Late Cretaceous–Tertiary). All are related to major plate margin processes in the Atlantic and Tethyan provinces

and, importantly, exert a fundamental control on the development of the hydrocarbon system.

The megasequences are divided into a series of unconformity-bound depositional sequences. The latter form the basis for both palaeofacies and isopach mapping which are used to assess the hydrocarbon habitat in southern England. Our key objective has been to constrain exploration risk at a basin scale by building: (i) a predictive model for reservoir, seal and source rock distribution; (ii) evaluating trap timing; and (iii) modelling charge/fill history of existing discoveries and remaining exploration prospects.

Exploration history

One of the earliest reports of hydrocarbons from a well in southern England is from Heathfield in Sussex. A borehole was drilled in 1896 near the station to provide water for use on the railway. When the well reached a depth of 100 m, in Kimmeridgian strata, a strong gas odour was noted which on ignition produced a 5 m flare at the top of the borehole. The well has the distinction of being completed as England's first natural gas well with production of 1000 scf/day used to provide gaslight for the station.

BP's involvement in exploration in southern England dates back to 1934. The exposures of Lower Cretaceous and Jurassic rocks along the Dorset coast from Weymouth to Swanage were examined by geologists for oil seeps and stain. Rich oil residues were found in Wealden Sands and Purbeck Limestones cropping out along the Dorset coast at Durdle Door, Dungy Head, Lulworth Cove and Hope and Worbarrow Bays.

Structures in the Hampshire Basin were rated highly and exploration prospects were developed on the basis of the seeps noted above and the presence of large surface structures such as the Portsdown, Winchester and Kimmeridge anticlines. The Portsdown Anticline, a large gentle structure about 30 km long × 5 km wide, was the prospect chosen by BP as the first deep well in Britain following the newly issued 1934 Petroleum Act Exploration Licences. The well reached a total depth of 2000 m in the Triassic. There were some oil shows and several drill stem tests were conducted but no free oil was recovered. A second well on the Portsdown Anticline and subsequent tests on the Henfield, Poxwell and Kingsclere anticlines were similarly unsuccessful.

The first test of the Kimmeridge Anticline was spudded on 4 December 1936. After drilling problems the well was respudded on 7 April 1937. The resident geologist, the then BP Chief Geologist Sir Peter Kent, noted the presence of live oil in fractures in a core cut in the Corallian limestone at 2522 m. The well, Broad Bench No. 1 was plugged and abandoned and it would be another 22 years before the full significance of this 'discovery' would be appreciated.

The post-war search for hydrocarbons in southern England began with the drilling of a deep well on the Isle of Wight. The well, Arreton No. 1 spudded on 12 October 1952 on an anticline identified by gravity surveying. The well reached a TD of 1573 m in the Inferior Oolite on 19 April 1953 without encountering any productive horizons.

The next hydrocarbon prospect to be tested was the Ashdown Anticline, a large structure over 30 km long × 7 km wide, located in the centre of the Weald Basin in north Sussex. It was hoped that significant gas reserves might be encountered in Jurassic reservoirs in view of the earlier natural gas discoveries at Heathfield. The well, spudded on 3 June 1954, encountered gas shows in the Purbeck and Portland beds and gas was produced on test from the Corallian at a rate of 20 000 scf/day.

The Kellaways Sandstone yielded small quantities of oil and gas with formation water. The

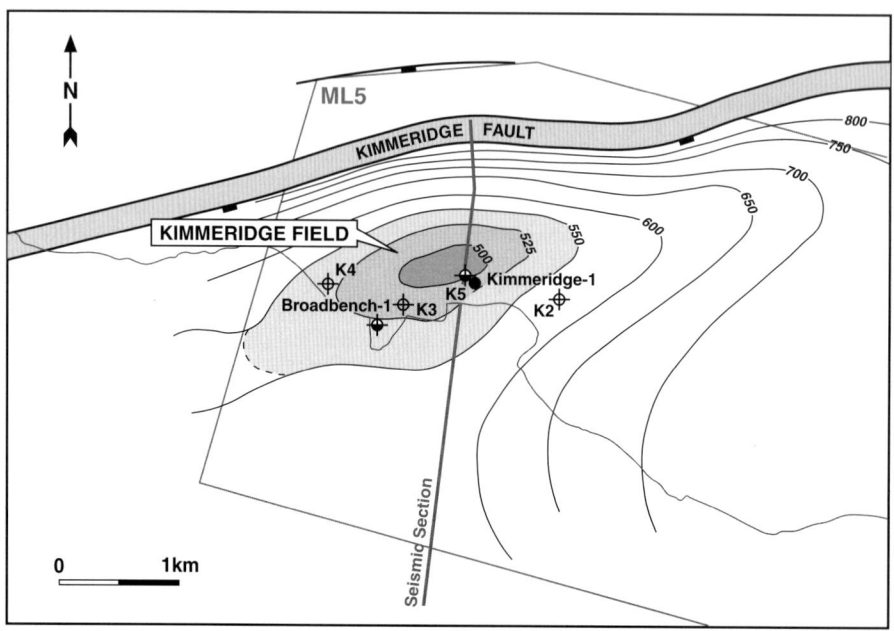

Fig. 1. Depth to the Top Cornbrash reservoir (metres), Kimmeridge Anticline, Kimmeridge Bay. Location of seismic section used in Fig. 2 is indicated (cf. Evans *et al.* this volume).

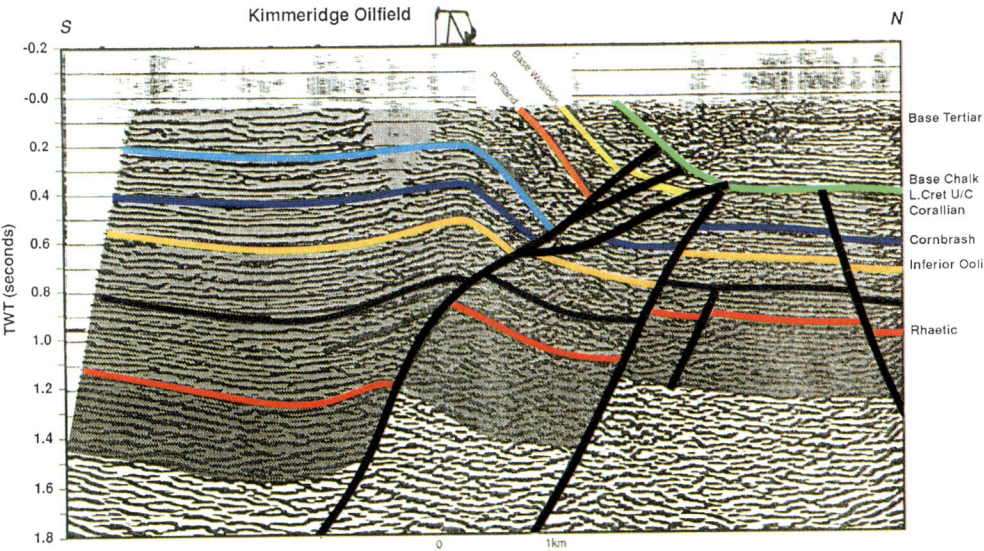

Fig. 2. North–south-trending seismic section through the Kimmeridge Anticline, showing the approximate position of the Kimmeridge oilfield that forms the subject of the paper by Evans *et al.* this volume).

Kellaways oil was significant in that it represented the first time that crude oil had been recovered from the Jurassic of southern England. A second well on the structure recovered small amounts of oil from the deeper Lias section as well as the Kellaways Sandstone.

The lack of success in finding commercial hydrocarbon accumulations in the Jurassic of southern England continued until the drilling of the Kimmeridge oil discovery well early in 1959. Following a re-evaluation of the structure in 1957, a second well was located on a cliff

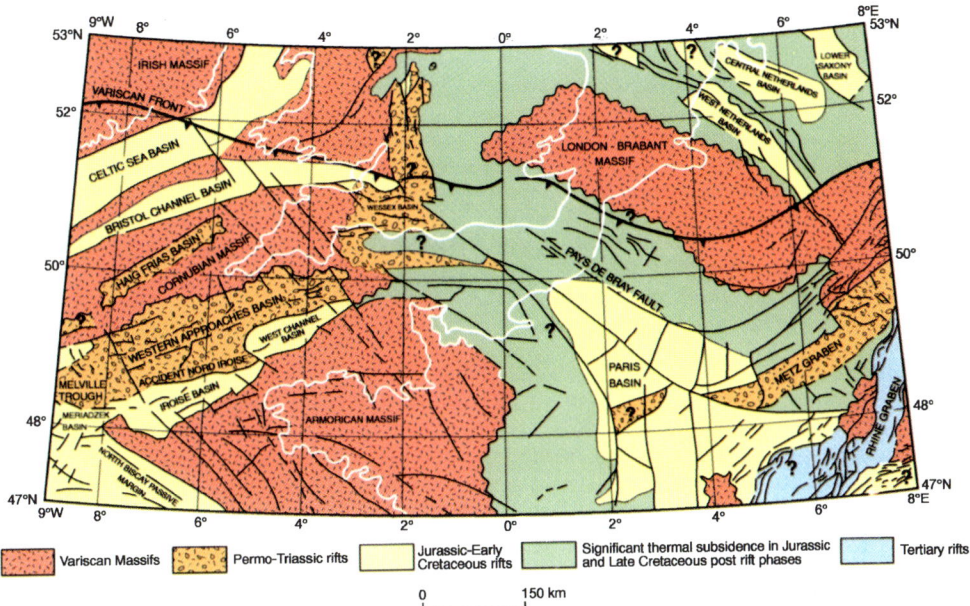

Fig. 3. Regional setting for the Wessex Basin depicting the main post-Carboniferous tectonic elements of the Western Approaches and the Anglo-Paris Basin.

over-looking Kimmeridge Bay less than 500 m from the original Broad Bench site (Fig. 1). The Kim-meridge-1 well spudded on 12 February 1959 with the Corallian as the main target. There was no oil in the Corallian and the well was deepened through the Oxford Clay to the Cornbrash Limestone. On test, the Cornbrash produced at over 500 barrels of oil/day from a highly fractured under pressured reservoir. More details can be found in a companion paper by Evans et al. (this volume).

The Kimmeridge No. 1 well was put on production on 16 January 1961, since when it has produced some 2.5 million barrels of oil. This volume is significantly in excess of the structurally mapped closure (Fig. 1). Seismic studies conducted by BP Eakring have since suggested that the accumulation is being actively recharged via a footwall Bridport reservoir in juxtaposition with the Cornbrash across the Kimmeridge Fault (Fig. 2). Since 1964 oil production has been assisted by pumping.

A marine seismic survey undertaken in Weymouth Bay during 1956 indicated the presence of an anticlinal culmination in the Lulworth Banks area. The discovery at Kimmeridge provided the impetus to test the Lulworth Banks Anticline and a second marine seismic survey was undertaken in 1960 to further delineate the structure. The well was located in 20 m of water 4 km southwest of Lulworth Cove and about 12 km from the Kimmeridge oilfield.

The Lulworth Banks well was spudded on 23 September 1963 – the UK's first offshore exploration well, drilled more than 1 year before exploration in the North Sea commenced. The well reached a total depth of 762 m in the Bridport sands without encountering significant hydrocarbons. Small amounts of gas (1750 scf/day) were produced from an open hole test of the Kellaways, Cornbrash and Forest Marble, and minor amounts of oil were recovered along with formation water from the Bridport Sands.

The early drilling in southern England had been focused on anticlinal features identified from surface geological mapping and some primitive geophysical techniques. With the advent of modern reflection seismic, more complex Jurassic fault block structures lying below relatively undeformed Cretaceous chalk and Tertiary clastics were identified. The first of these to be tested was at Wareham. The Wareham No. 1 well spudded on 18 October 1964 and reached a total depth of 1746 m in Triassic Sherwood Sandstones. Upper Jurassic intervals were non-productive but oil was recovered at 853 m from a fine-grained, calcareous sandstone at the top of the Bridport Sands.

The discovery at Wareham was a fault block structure bounded to the south by the down-to-the-south throwing Litton–Cheney Fault. Some 10 years later the Wytch Farm discovery was made on the adjoining fault block to the south, again in Liassic Bridport Sands. Reserves in Wytch Farm were estimated at 50 million barrels recoverable, by far the largest UK onshore discovery. However, the full significance of the Wytch Farm structure was not realized until 1977 when the Wytch Farm D5 well was deepened to test the Triassic Sherwood Sandstone Group reservoir, the presence of which had been demonstrated by the Wareham No. 1 well in 1964. Reserves at Wytch Farm have recently been revised upwards from previous estimates of 300 million barrels (Bowman et al. 1993) to slightly in excess of 428 million barrels (Underhill & Stoneley this volume) making it the largest onshore oil discovery in Western Europe.

The discovery at Wytch Farm led to a resurgence of interest in exploration in southern England. With the benefit of modern seismic data, explorers realized that the surface geology of the Weald and Wessex basins differed significantly from the structure at depth. This led to the discovery of significant oil fields in the Middle Jurassic Great Oolite in the Weald Basin at Humbly Grove, Horndean, Stockbridge, Storrington and Singleton. Additional discoveries at Palmers Wood, Godley Bridge and Brockham in Upper Jurassic reservoirs in the north of the Weald Basin have indicated that fault block structures on the margins of the Weald and Wessex basins still represent important exploration targets in southern England.

Regional geological setting of the Weald and Wessex basins

Variscan inheritance

The structural development of southern England during the Mesozoic has its origins in the events of the late Carboniferous–early Permian culmination of the Variscan orogeny (Chadwick et al. 1983). At this time southern England lay in the foreland of a northwards-verging collision zone with large-scale thrust and nappe development. Evidence from northern France and Belgium some 300 km along strike (Fig. 3) suggests the presence of a considerable chain of mountains running east–west across the region at this time. The development of anthracites in the Kent coalfield points to some 5000 m of burial during this phase. The Varsican orogeny imparted a strong east–west (thrusts) and northwest–southeast (lateral transfers) structural grain across the province. In the area of the present-day Western Approaches the east–west trends swing round into a more northeast–southwest direction.

During subsequent Permian and Mesozoic rifting and inversion, Variscan trends were periodically reactivated in an extensional or compressional sense (Chadwick et al. 1983). A similar reactivation of Caledonian collisional grain during Carboniferous rifting and inversion was noted by Fraser et al. (1990) in northern England. This is, therefore, a fairly common occurrence.

Regional setting

When we look at the Permian to present-day geological history of southern England, it is important, first, to put it in its regional plate tectonic context and, second, to compare it with the tectono-stratigraphic evolution of adjacent provinces such as the Western Approaches, Celtic Sea and the Paris Basin. A summarized chronostratigraphic diagram for the region highlights the similarity of both facies and stratigraphies observed in each area during the Permian and Mesozoic (Fig. 4). During this time the province was influenced by major plate margin processes active in the Tethyan, Biscay and Atlantic regions. Although not directly involved in any of these plate margin processes, southern England and adjacent provinces were all to some degree affected by these processes. Hence, we would argue that all were subject to similar tectonic events and controlled at the megasequence level by the evolution of rifted margins in Tethys, Biscay and the North Atlantic.

Plate cycles

The major plate margin processes we observe are categorized as plate sequences and are described in Figs 5–7.

The early Atlantic plate sequence from Late Permian to Late Triassic records two phases of unsuccessful attempts to open the North Atlantic. Two large-scale fining-upwards cycles characterized by coarse continental clastics at the base, culminating in basin-scale evaporite events were developed. The sandstones at the base of these cycles (Permian Rotleigendes, Triassic Sherwood/Bunter) constitute important exploration targets throughout the province. The overlying evaporites provide effective seals where developed. The sequence is significantly source poor and this is an important consideration when exploring these targets.

The base of the Jurassic across the province is marked by a rapid and extensive marine transgression. This is tentatively related to widespread rifting in western Tethys culminating in break-up and the formation of oceanic crust in the Toarcian. This 'syn-rift' package forms the most important source interval for liquid hydrocarbons in the Western Approaches, southern England and the Paris basin.

The Atlantic plate sequence (Rhaetic–Tithonian) records the successful rifting of the Central Atlantic south of the Azores fracture zone. Pulsed rifting has generated a series of regionally correlable shallowing upwards sequences (J1–J6) across the province. In areas starved of clastic sediment supply, these sequences culminated in shallow-water carbonate deposition. When a clastic source was active, sandstones dominated the interval. These units form important reservoir targets throughout the region. The earlier mudstone-dominated parts of the cycles contain locally important source facies.

The Cretaceous, Biscay plate cycle records the successful rifting and spreading of the Bay of Biscay and Rockall areas during this time. The effects are most noticeable in the Western Approaches and southern England, proximal to the centre of rifting.

The widespread Cimmerian unconformity observed across the province in Apto-Albian times records the break-up unconformity in Biscay and Rockall, when rifting gave way to the generation of oceanic crust. The marked erosional unconformity observed across the province at this time has often been associated with inversion. Clearly in an extensional regime, we have to consider a rift-related explanation. The most reasonable solution envisages a major heat pulse at this time leading to massive thermal uplift of both the basins and margins. Basaltic volcanics of this age have been recorded in wells in the French sector of the Western Approaches and at outcrop in southern England. No regionally significant hydrocarbon reserves have been associated with this plate sequence, its significance lying in the differential burial it exerts on Jurassic source facies and the formation of tilted fault block traps during the Early Cretaceous rift phase.

The Alpine plate sequence records the north–south closure of western Tethys and the subsequent collisions of Spain with southern

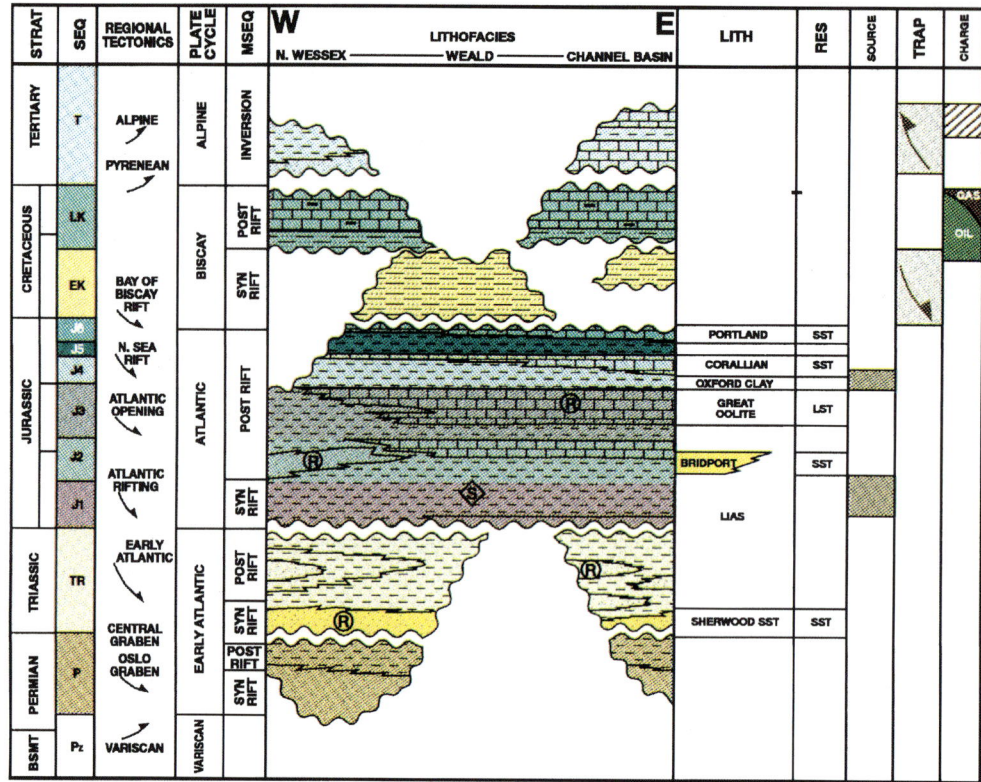

Fig. 4. Chronostratigraphic diagram depicting the main plate cycle seismic megasequences, their component sequences, relevant Jurassic lithostratigraphic elements and their respective reservoir, source, trap and charge potential. The J numbers ascribed to the Jurassic megasequences are used in subsequent figures.

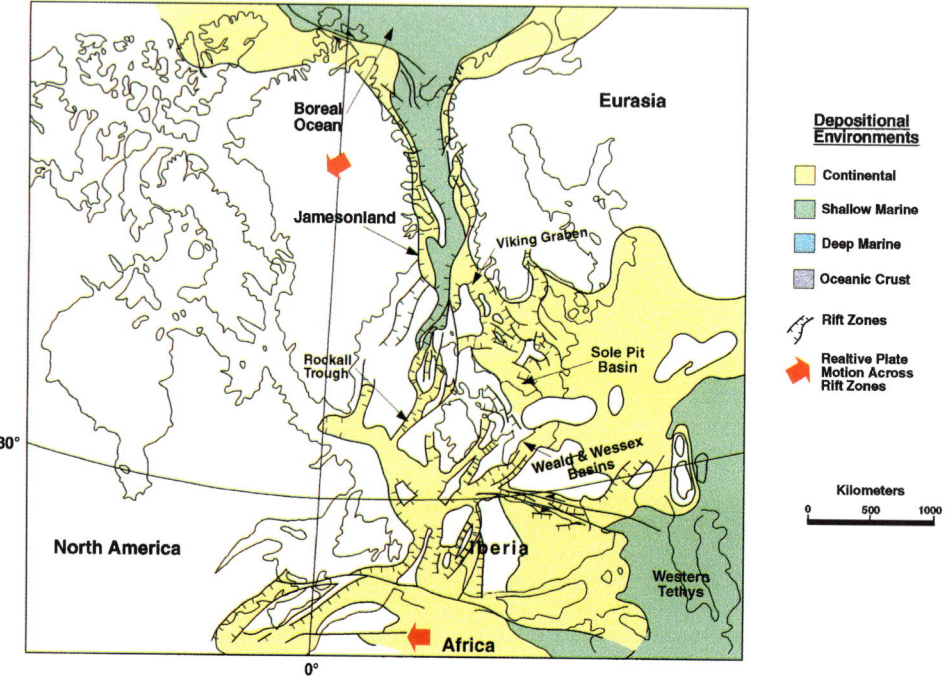

Fig. 5. Late Norian (210 Ma), early Atlantic megasequence plate tectonic reconstruction for northwest Europe.

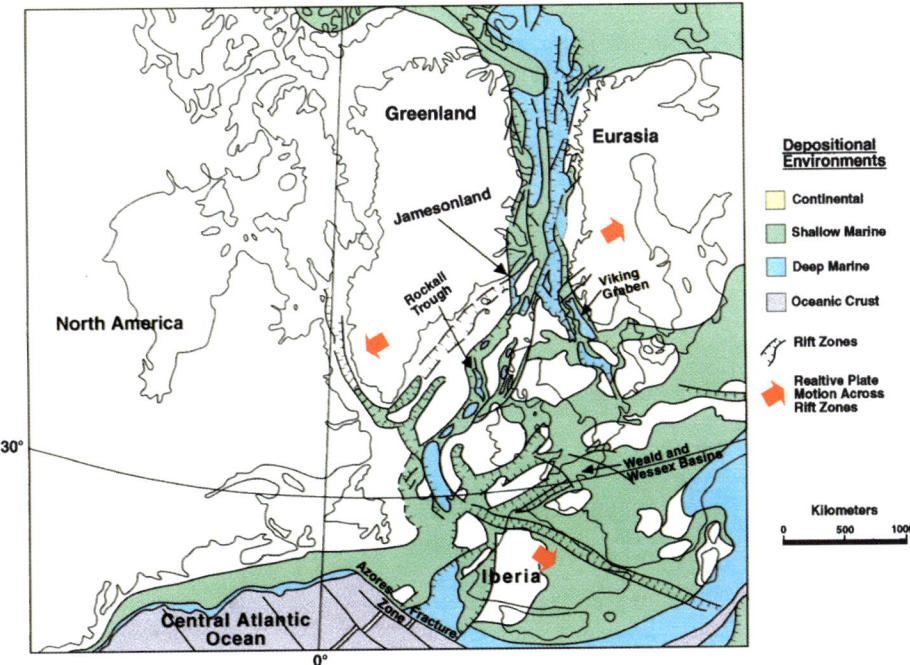

Fig. 6. Tithonian (150 Ma), Atlantic megasequence plate tectonic reconstruction for northwest Europe.

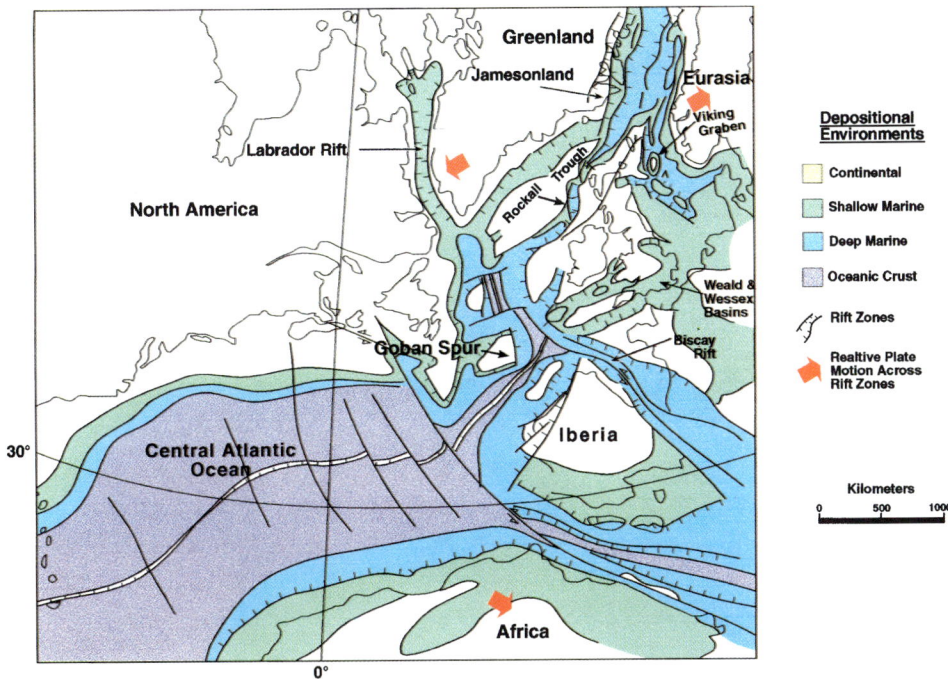

Fig. 7. Early Aptian (125 Ma), Biscay megasequence plate tectonic reconstruction for northwest Europe.

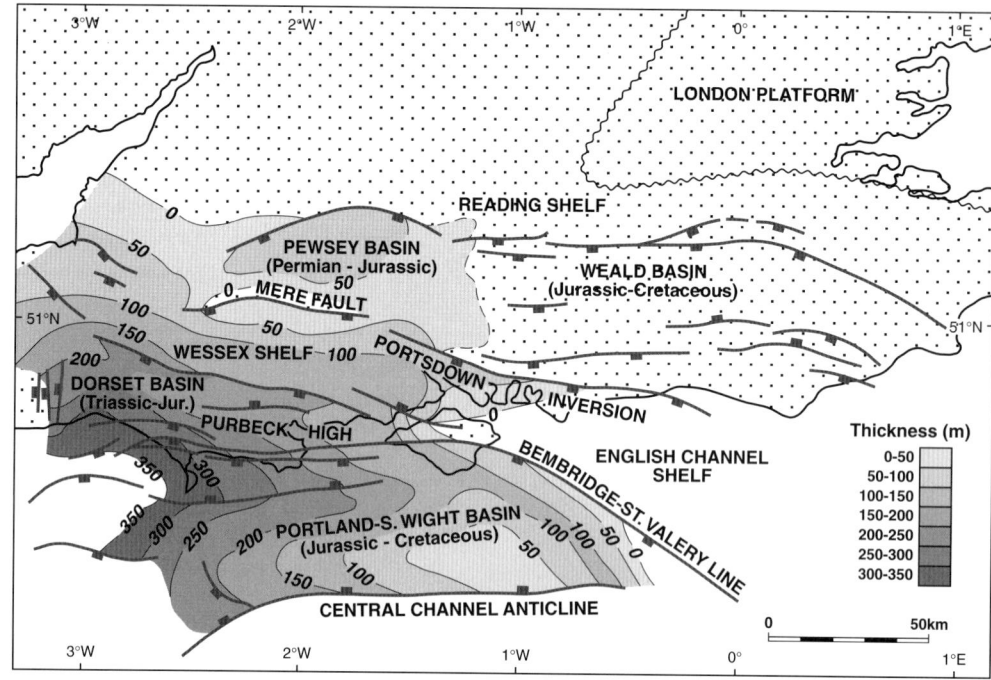

Fig. 8. Gross isopach map for the Triassic Sherwood Sandstone Group.

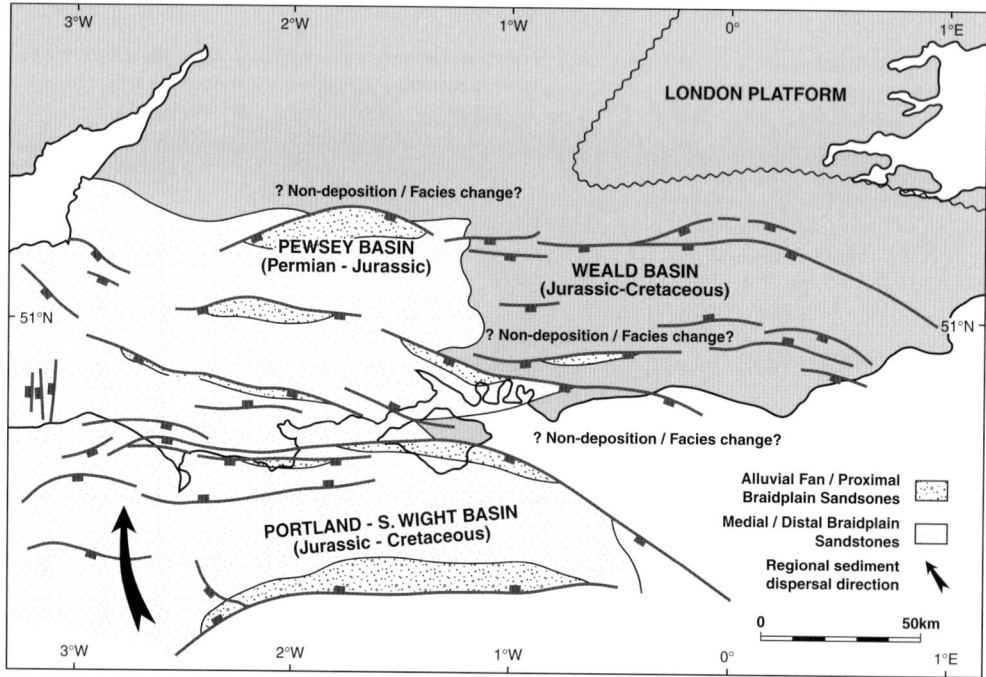

Fig. 9. Lithofacies map for the Triassic Sherwood Sandstone Group showing the role and importance of extensional faults in controlling deposition of alluvial fan and proximal braidplain sandstones (cf. Butler 1997 this volume).

France and Africa with Europe. In simple terms the onset of the plate sequence can be related to a marked change in the movement of the African continent in the Late Cretaceous from west–east to south–north (Dewey et al. 1989). These collisions exerted strong compression in the foreland and the consequent inversion of east–west trending Jurassic–Cretaceous basins in the Western Approaches (Ruffell 1995), Southern England (Stoneley 1982: Penn et al. 1987; Underhill & Stoneley this volume) and the Paris Basin (BRGM 1980). The effects are also noted in the Sole Pit Trough (Van Hoorn 1987a), Broad Fourteens Basin (Van Wijhe 1987) and the Celtic Sea Basin (Tucker & Arter 1987; Van Hoorn 1987b; Petrie et al. 1989). The Alpine plate sequence represents an important structure-forming event across the province with the development of broad-scale inversion anticlines in the hanging walls of earlier extensional faults (Underhill & Paterson 1998). The Alpine plate sequence records a period of uplift and erosion centred on the inversion axes.

Tectono-stratigraphic evolution

We have adopted a seismo-stratigraphic approach constrained by both well and outcrop data to describe the structural and stratigraphic evolution of southern England from Permo-Trassic to Recent times. Based on our analysis of plate margin processes in the Atlantic and Tethyan provinces described in the previous section, four plate sequences have been identified.

Early Atlantic plate sequence

Major rifting in the late Permian which may have continued into the Triassic was located on a series of northwest–southeast and east–west faults (Fig. 8). Sedimentation was continental and was dominated by aeolian and fluvial sediments with evaporites deposited during the Upper Triassic in the central part of the basin (Figs 9 and 10). Sand provenance is believed to have been from the Armorican Massif to the south with localized input from subsidiary fault systems such as the Portsdown and Ryme Intrinseca faults. Isopachs on the Triassic Sherwood sandstone reservoir show it thickening markedly to the southeast from the onshore Dorset area (Fig. 8). There is little evidence for Permo-Triassic deposition in the area of the Weald Basin. The log correlation diagram for the Triassic sequence shows a marked thinning onto the Sandhills High in the east of the Wessex Basin (Fig. 10). The subdivision of the Triassic sequence follows the lithostratigraphic schemes developed by Lott et al. (1982) and Ivimey-Cook et al. (1980). The paucity of biostratigraphic control and the poor resolution and understanding of facies relationships both vertically and laterally preclude the detailed sequence stratigraphic breakdown of the interval carried out for the overlying Jurassic.

Atlantic plate sequence

The onset of rifting in the Early Jurassic (Lias) along a series of east–west-trending faults was accompanied by a widespread marine transgression. One is tempted to relate this rifting event to the westerly propagation of oceanic spreading in the western Tethys.

The Atlantic plate sequence has been divided into six shallowing-upwards depositional sequences (Fig. 4). We argue that each cycle is initiated by pulsed propagation of Atlantic rifting. The top of the sequences are typically represented by either sandstones or carbonates depending on location with respect to active clastic source areas.

Sequence J1. Rifting and regional subsidence of the Weald and Wessex basins during the early Jurassic led to a widespread marine transgression and the establishment of a mud-dominated shelf over the whole of southern England (Fig. 11). The Lias J1 sequence forms an important oil-prone source rock interval in the Wessex Basin. In the Weald Basin the interval is generally characterized by a higher terrestrial input due to its proximity to the emergent London–Brabant Massif. As a consequence of this, Lias source rock potential is poorer and more gas prone. Oil-prone source potential improves to the southwest away from the London–Brabant Massif, with oil-prone potential only locally developed within the Weald Basin in the hanging wall of faults active during sequence J1 deposition, e.g. the Ashdown Fault.

The J1 sequence shows little reservoir potential with poor quality sands (Thorncombe Sands equivalent) restricted to the eastern part of the Weald Basin, coincident with the area in which source rock quality is poorest (Fig. 12).

Sequence J2. The boundary between the J1 and J2 sequences is marked by a major angular

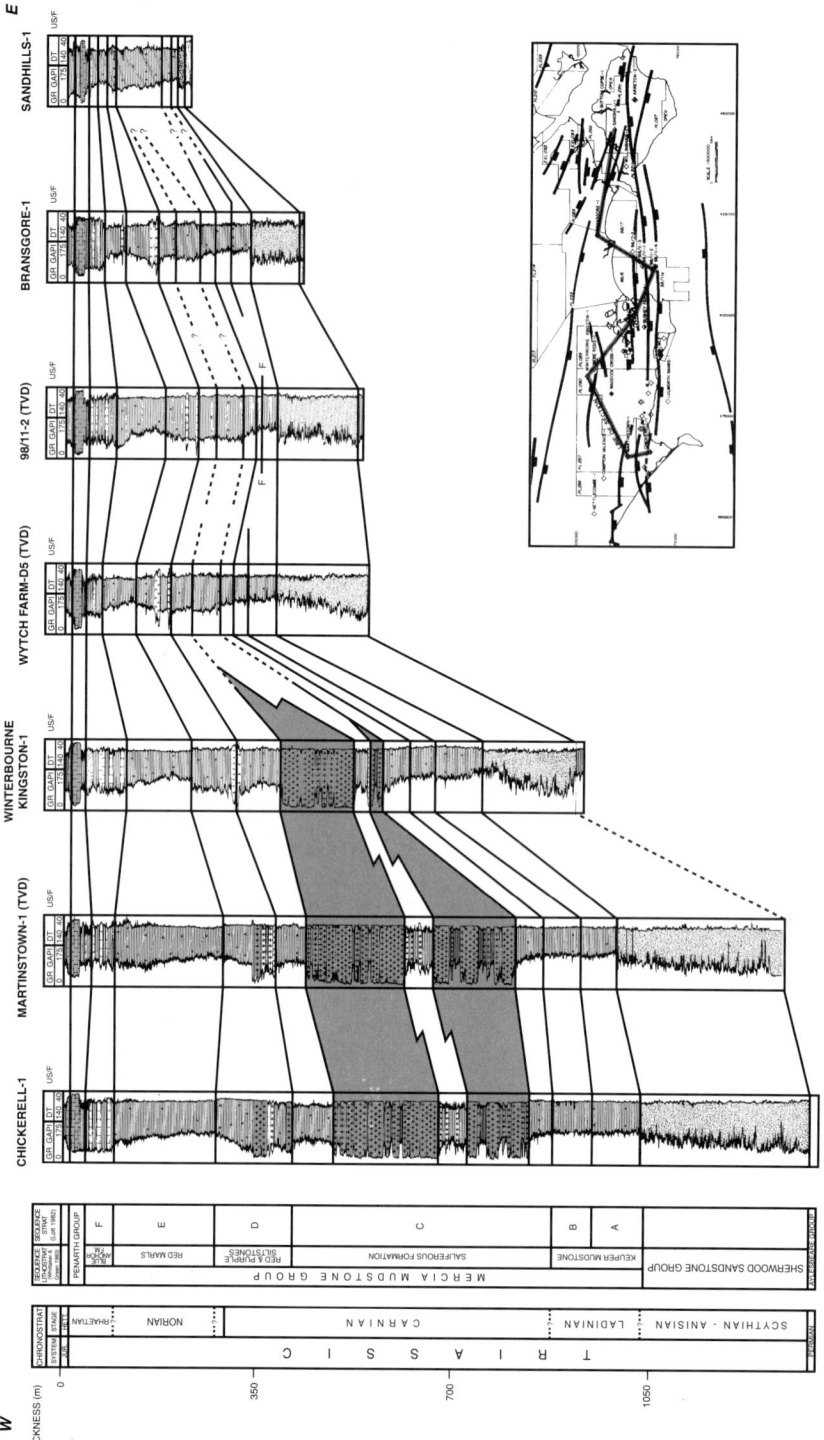

Fig. 10. East–west-trending electrical well log correlation showing the progressive easterly thinning of the Triassic succession. Note that the evaporites which characterize the Mercia Mudstone Group in western parts of the basin do not extend to the Wytch Farm oilfield area.

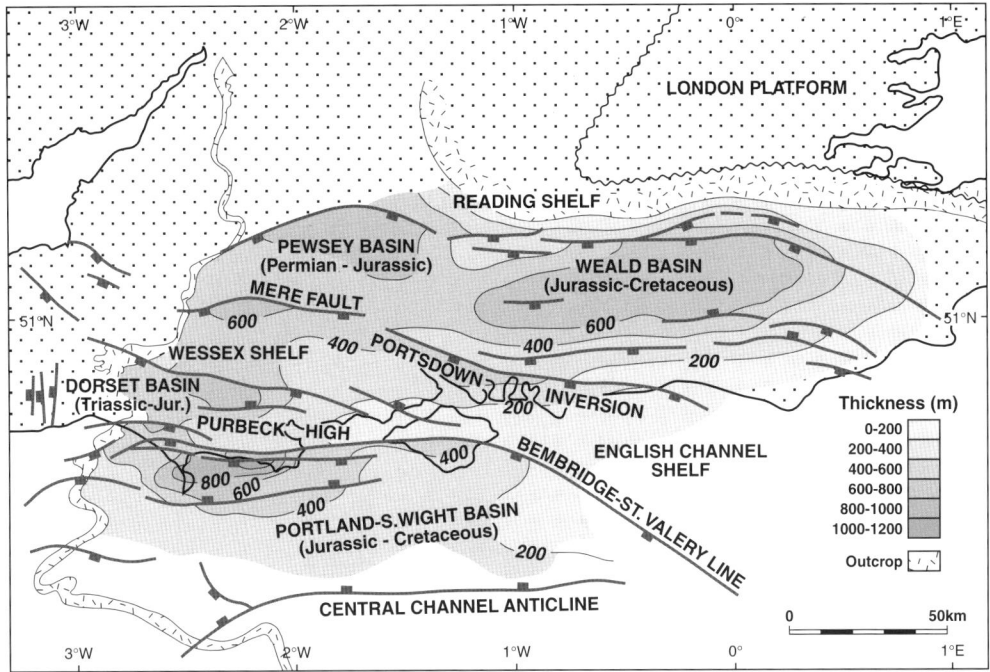

Fig. 11. Gross isopach map for the Early Jurassic Lias Group showing the importance that some extensional faults had in controlling thickness variation.

unconformity at the top of the Junction Bed of Toarcian age (Fig. 4). The top of the J2 sequence is marked by the Inferior Oolite in both the Wessex and Weald basins. It is fairly well developed in the sediment-starved Weald, but is thin and sometimes difficult to identify above the Bridport Sands in the Wessex Basin. The Bridport Sands (Figs 13–15) form an important reservoir horizon in the Wessex Basin. The sands prograded in a southeasterly direction from the Cornubian Massif to the west.

Sequence J3. The subdivision of the Middle Jurassic (J3) sequence is based upon the lithostratigraphic scheme developed from coastal exposures (Arkell 1947). The relationship of Middle Jurassic lithostratigraphy and wireline-log signature was established by Penn (1982) with the drilling of the BGS Winterborne Kingston Borehole and illustrated in Fig. 16 by the dashed correlation lines. It is argued, however, that the Middle Jurassic lends itself to a meaningful subsequence division suggesting the presence of four transgressive–regressive cycles (A–D) – their development possibly related to local or regional tectonic controls. These cycles can be traced eastwards from the basinal Wessex Basin section into the carbonate dominated platform interval characteristic of the Weald Basin, e.g. Marchwood-1. The Great Oolite carbonate platform which covers most of the Weald Basin during the Middle Jurassic is a major reservoir in the basin (Figs 17 and 18). The basinal equivalent in the Wessex Basin (Forest Marble) forms a secondary reservoir target in the Wytch Farm area. The presence of a well-developed carbonate platform during J3 times is related to the switching off of a clastic source in the area as drainage patterns switched into the North Sea at this time.

Sequence J4. A major transgression during Callovian times (Sequence J4) led to the widespread deposition of the Oxford Clay over both the Weald and Wessex basins. The Oxford Clay forms a major regional seal on Middle Jurassic Great Oolite reservoirs in the Weald Basin. It is also an important oil-prone source rock, although it is probably immature over most of the Wessex Basin area. The sequence is capped by

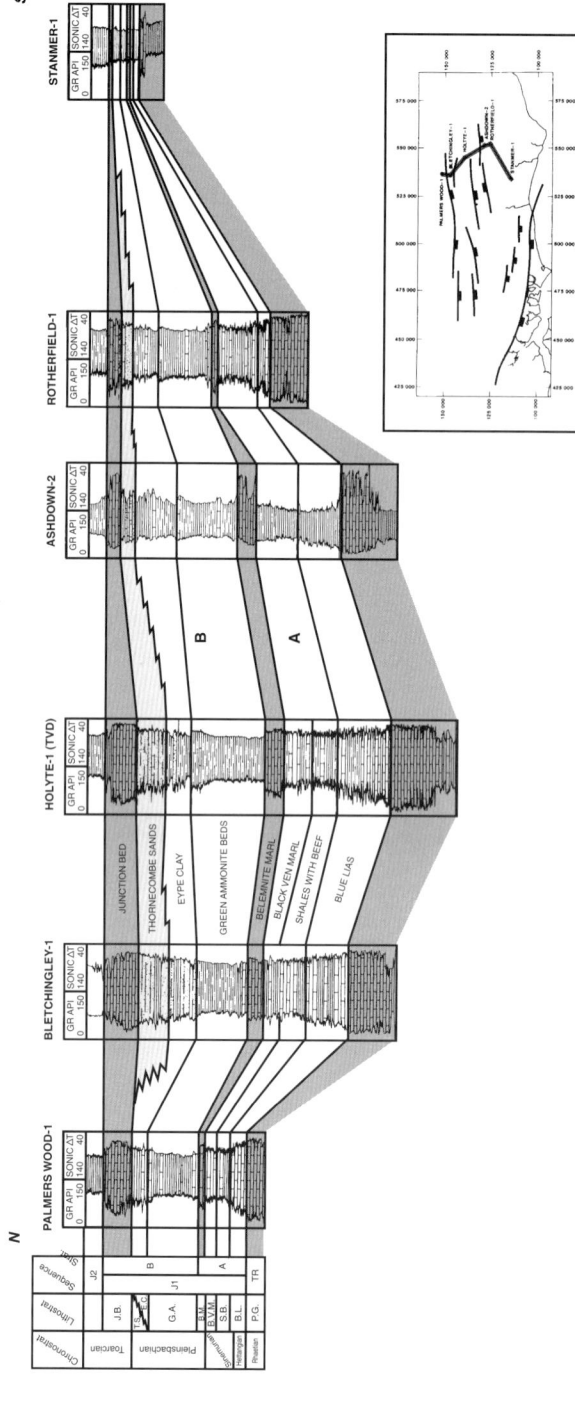

Fig. 12. North–south-trending electrical well log correlation of the Lower Jurassic (Rhaetian–Toarcian) J1 sequence, Weald Basin.

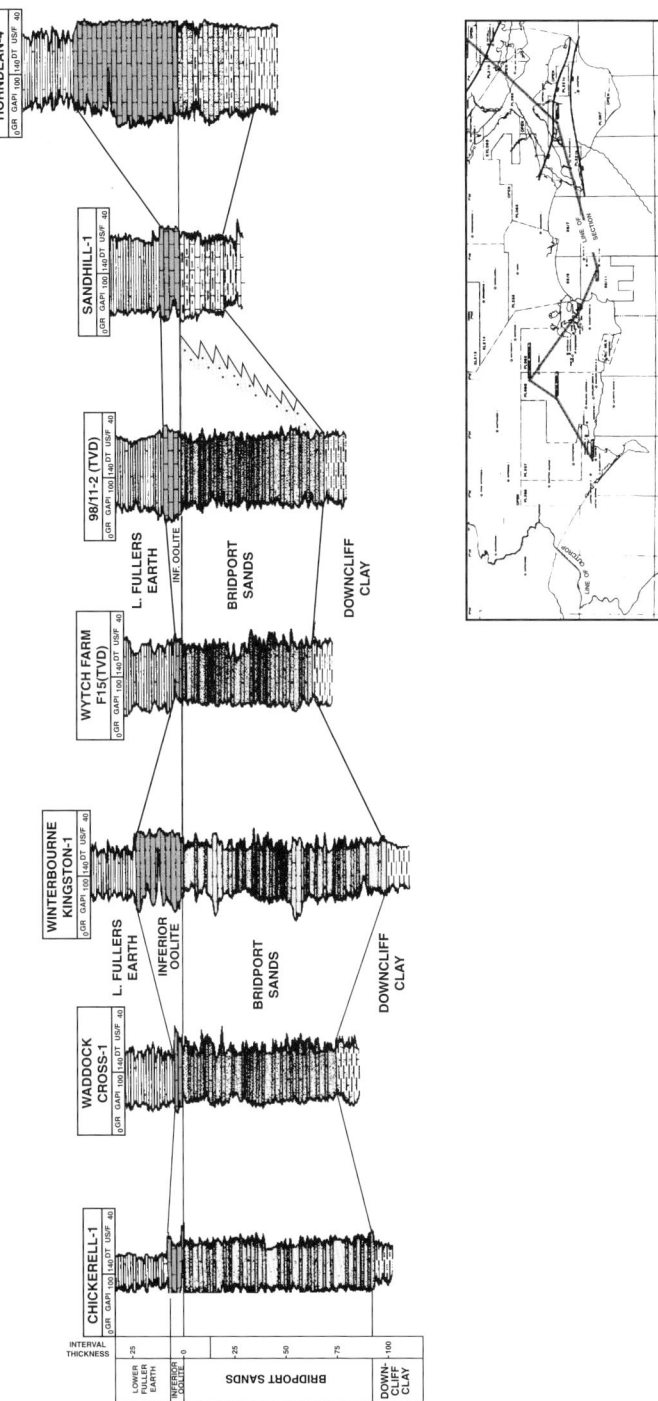

Fig. 13. East–west-trending electrical wireline log correlation of the Lower–Middle Jurassic, Bridport Sandstone and Inferior Oolite succession. The major sandstone to carbonate facies change that occurs between offshore block 98/11 and the Isle of Wight (Sandhills-1 well) is highlighted (cf. Fig. 15).

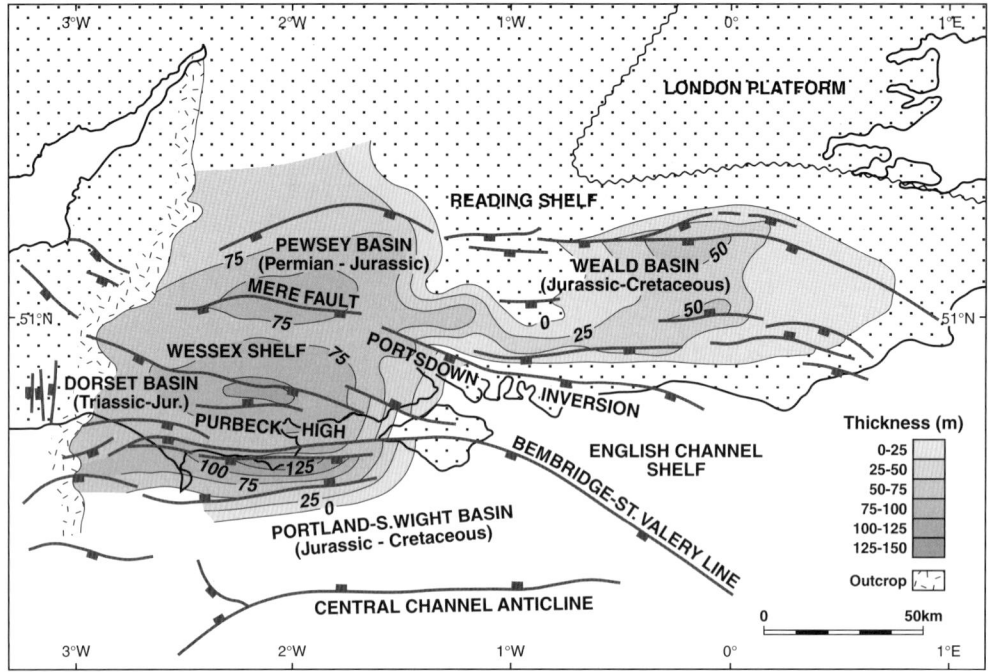

Fig. 14. Gross isopach map for the Bridport Sandstone showing the persistent role that extensional faults had in controlling deposition.

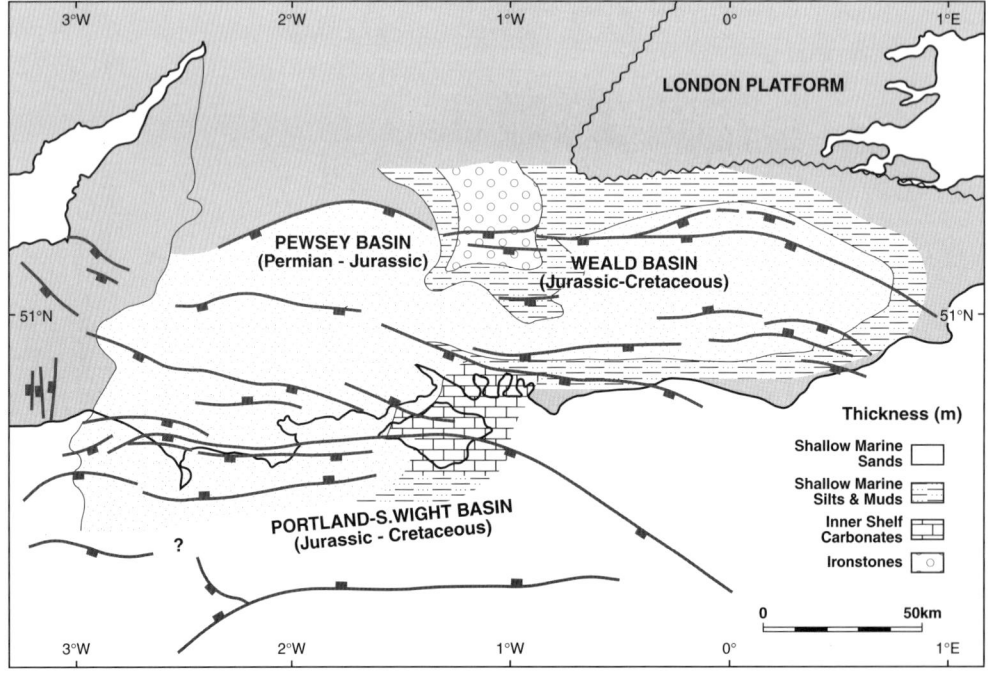

Fig. 15. Lithofacies map for the Bridport Sandstone showing the limits of coarse-grained clastic deposition.

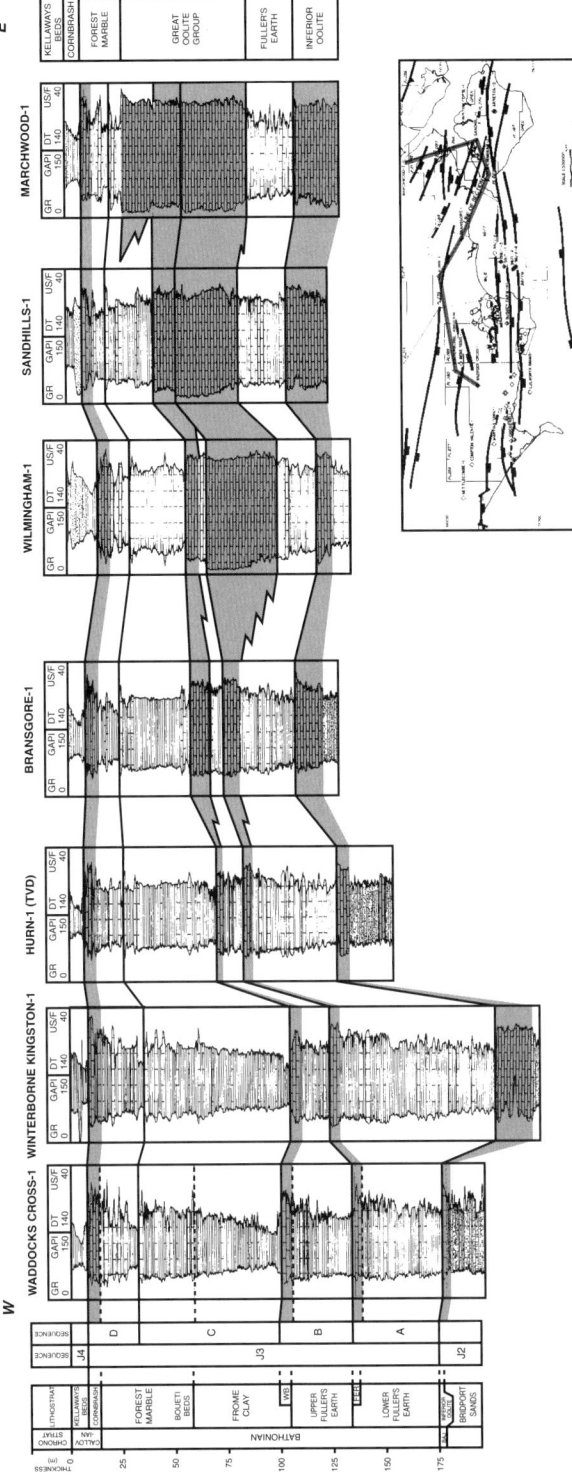

Fig. 16. East–west-trending electrical well log correlation panel for the Middle Jurassic (J3) Fuller's Earth, Frome Clay and Forest Marble sedimentary sequence showing the effect of local facies changes across the basin.

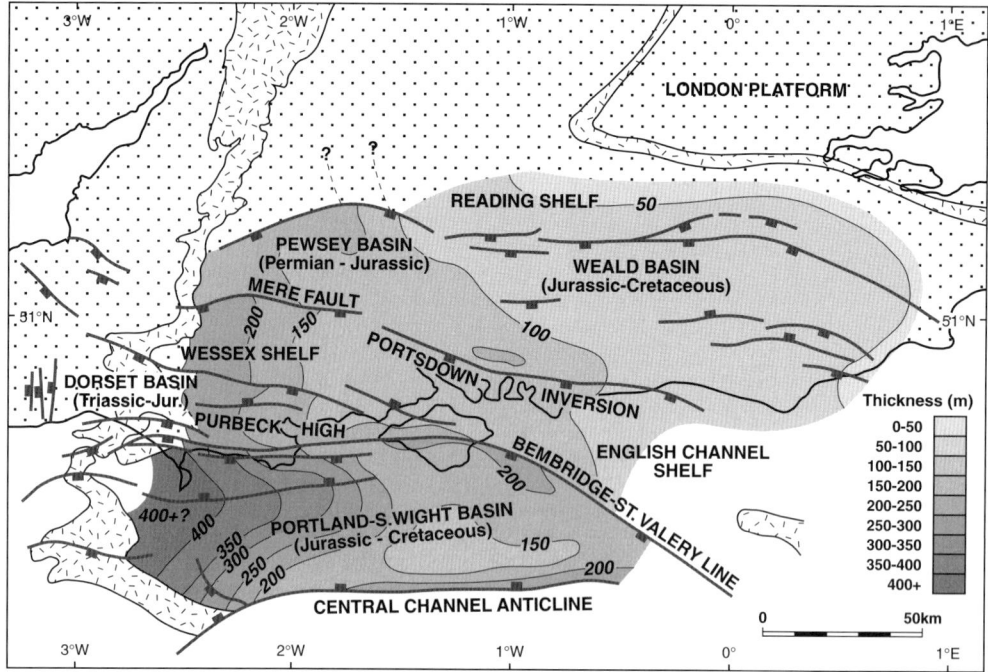

Fig. 17. Gross isopach map for the Middle Jurassic (J3) sequence.

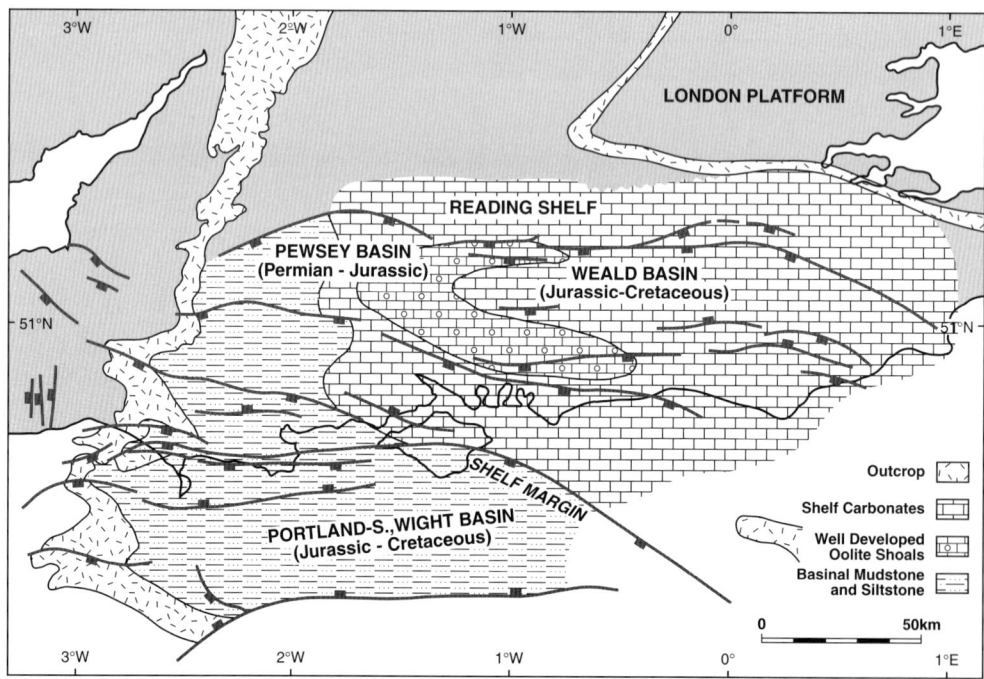

Fig. 18. Lithofacies map for the Middle Jurassic (J3) sequence showing the depositional limits of the Great Oolite play fairway.

the Corallian Sandstones in the north of Weald Basin which were sourced from the emergent London Platform to the north. The Corallian Sands form an important reservoir in the Palmers Wood oil field (Butler & Pullan 1990). Elsewhere in the basin the Corallian is represented by carbonate facies (Figs 19 and 20).

Sequence J5. Sequence J5 represents the lower part of the Kimmeridge Clay Formation and is separated from the overlying J6 sequence by a distinctive middle Kimmeridge carbonate and a marked regional unconformity which is particularly well developed in the north of the Weald Basin adjacent to the London Platform. The Kimmeridge Clay is a dark oil-prone organic-rich mudstone, but is immature over the whole of southern England.

Sequence J6. The deposition of the upper part of the Kimmeridge Clay (sequence J6) occurred in response to renewed rifting and regional subsidence during late Kimmeridge times. Facies development within the sequence exhibits an overall shallowing upwards profile (Fig. 21), from the marine mudstones and shallow marine sands of the Kimmeridge Clay and Portland Sandstone, respectively, through to the marginal marine/brackish depositional system of the Purbeck Beds. The Kimmeridge Clay again shows good oil-prone source rock potential but is thermally immature over both the Weald and Wessex basins. The Portland Sandstone forms an excellent reservoir horizon and is the producing interval at Brockham and Godley Bridge. It was sourced from the London Platform to the north (Figs 22 and 23).

The Purbeck Anhydrite forms an excellent regional top seal. The Purbeck Beds are predominantly argillaceous but occasional thin sands may be developed which exhibit reservoir potential. Their distribution is difficult to predict: a sandy limestone interval within the Purbeck Beds forms the gas reservoir at Albury. Evaluation of the J6 sequence within the Wessex Basin is hampered by the fact that in most wells the Apto-Albian break-up unconformity has eroded most of the Lower Cretaceous and Upper Jurassic. Full preservation of this sequence in the Wessex Basin is limited to the hanging wall

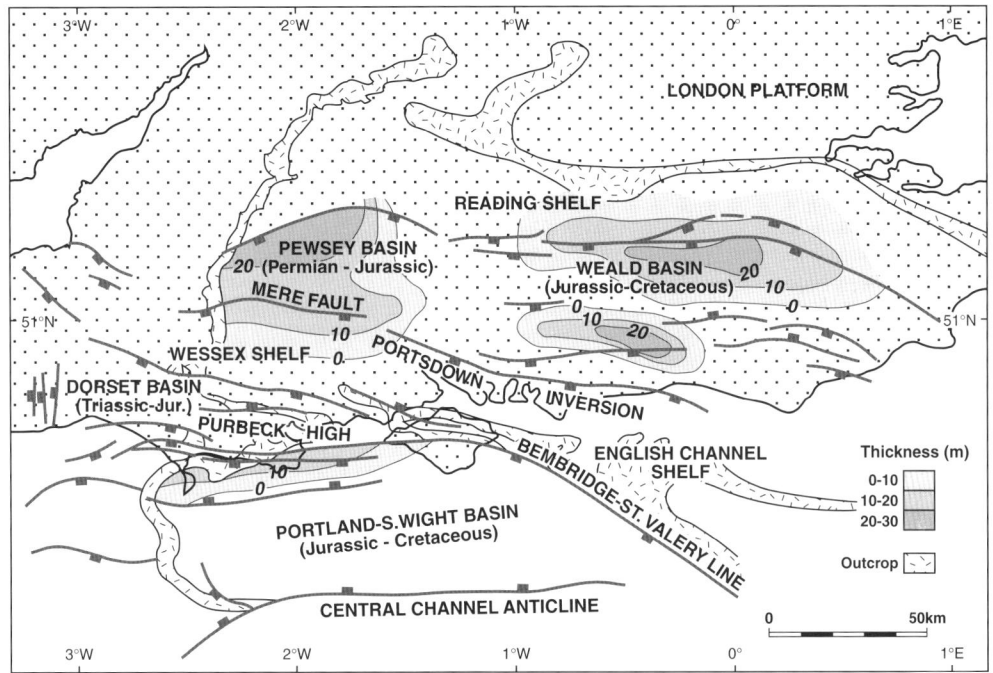

Fig. 19. Gross isopach map for the Corallian showing the continued role that extensional faults had in controlling deposition.

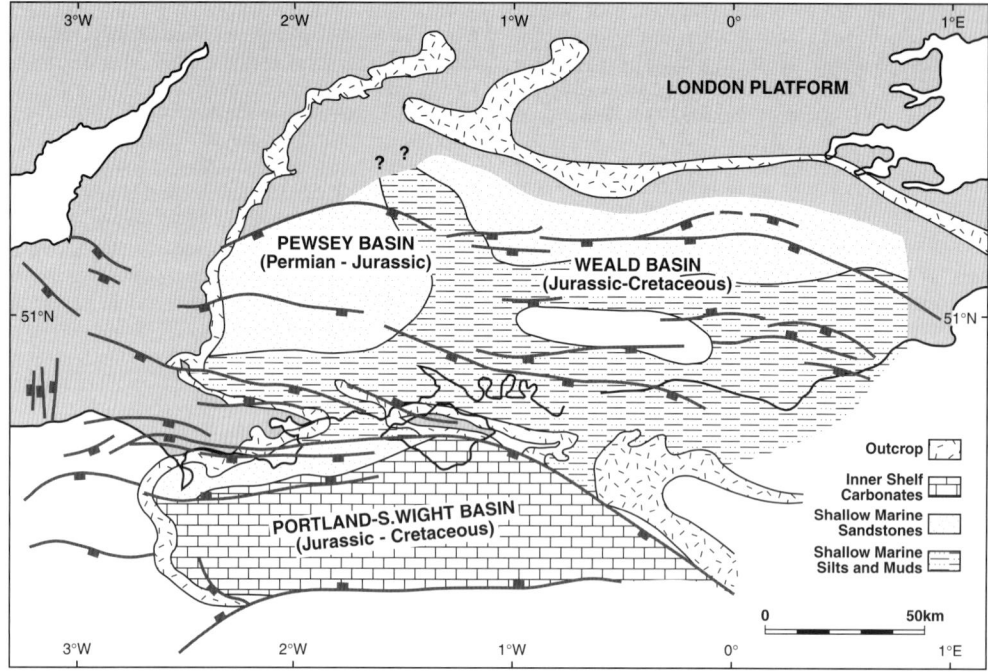

Fig. 20. Lithofacies map for the Corallian showing the limits of carbonate and clastic deposition in the Weald and Wessex basins.

of faults active during the Lower Cretaceous rifting event associated with opening of the Bay of Biscay.

Biscay plate sequence

During the Early Cretaceous the Weald and Wessex (Portland–South Wight) basins were subject to a phase of major renewed extension, probably associated with active rifting along the Biscay continental margin (Fig. 7). Faults initiated during Lower Jurassic extension were reactivated, although the site of the major depocentres migrated eastwards (compare Figs 11 and 24). A series of lacustrine–braidplain–alluvial fan deposits prograded into the basins, predominantly shed from active fault scarps to the north (Fig. 25).

As a result of Tertiary inversion and erosion, deposits of this plate sequence occur at outcrop within the central part of the Weald Basin and as erosional remnants within the hanging wall of the Portland–Kimmeridge Fault System along coastal exposures in Dorset.

Occasional gas and oil shows and seeps have been observed within Lower Cretaceous sands, but due to their shallow depth of burial and the fact that they lie above the regional seal to the basin (Purbeck Anhydrite), they are not regarded as being important hydrocarbon exploration targets.

In terms of the evolution of the basin as a hydrocarbon province, rifting, which occurred during the early Cretaceous, led to the subsidence and the early maturation of Jurassic source rock intervals within the hanging walls of active faults (e.g. the central Weald and Portland–South Wight basins). Rifting also led to the formation of tilted fault block and horst structures which form the main hydrocarbon traps in the basin.

The onset of oceanic spreading in the Bay of Biscay in the Aptian coincides with the switch from active rifting to thermal subsidence in the Weald and Wessex area which continued into the Late Cretaceous (Fig. 26). The basin was transgressed, and widespread shallow-marine conditions were established above a marked angular break-up unconformity with the deposition of the Lower Greensand during the Aptian (Fig. 27). At the beginning of the Albian, further regional subsidence led to the deposition of the Gault Clay. The Gault Clay and the Upper

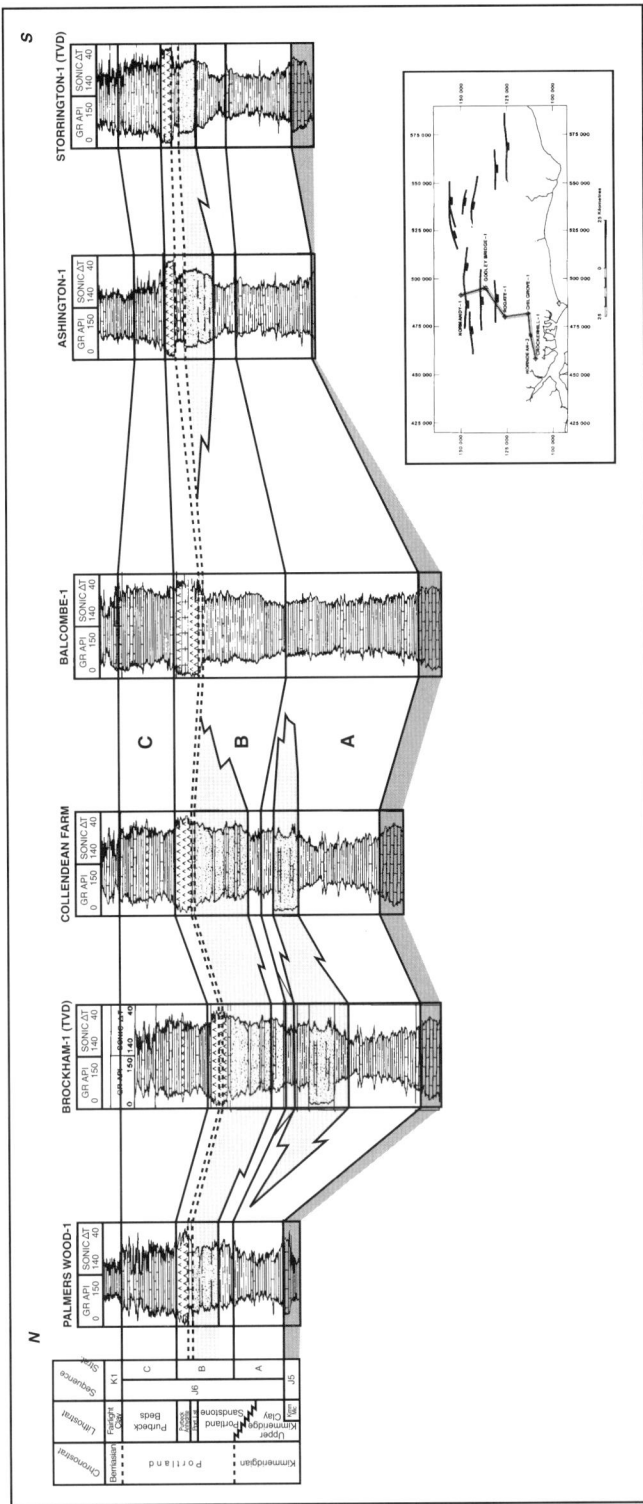

Fig. 21. North–south-trending electrical well log correlation of the Upper Jurassic, Kimmeridgian and Portlandian (J6) sequence in the Weald Basin.

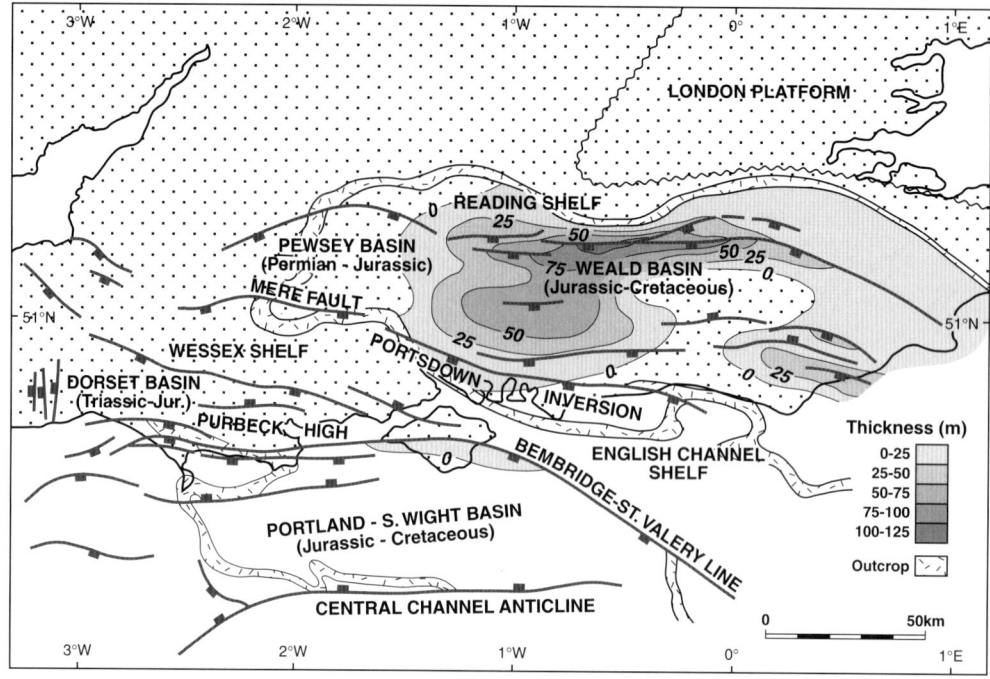

Fig. 22. Gross isopach map for the Portland Sandstone in the Weald and Wessex basins.

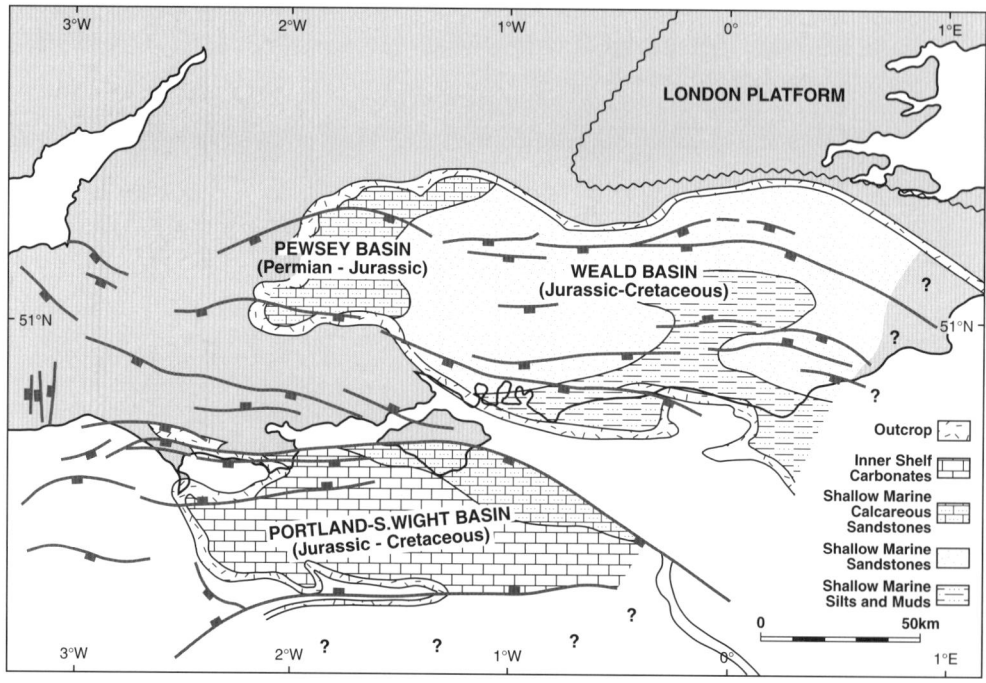

Fig. 23. Lithofacies map for the Portland Group showing the limits of carbonate and clastic deposition in the Weald and Wessex basins.

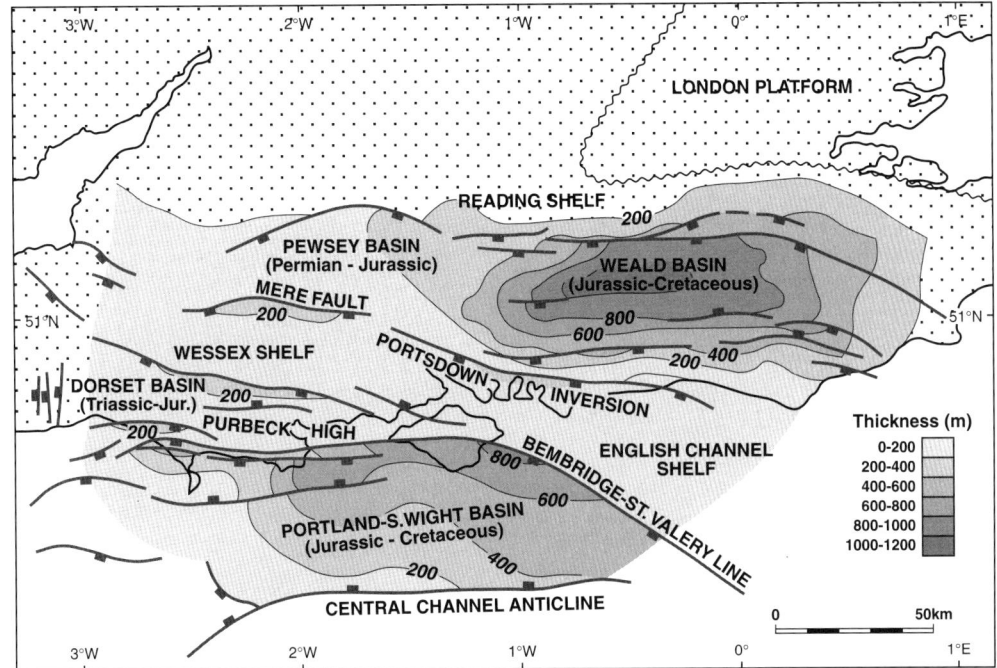

Fig. 24. Gross isopach map for the Wealden (Biscay cycle) sequence in the Weald and Wessex basins.

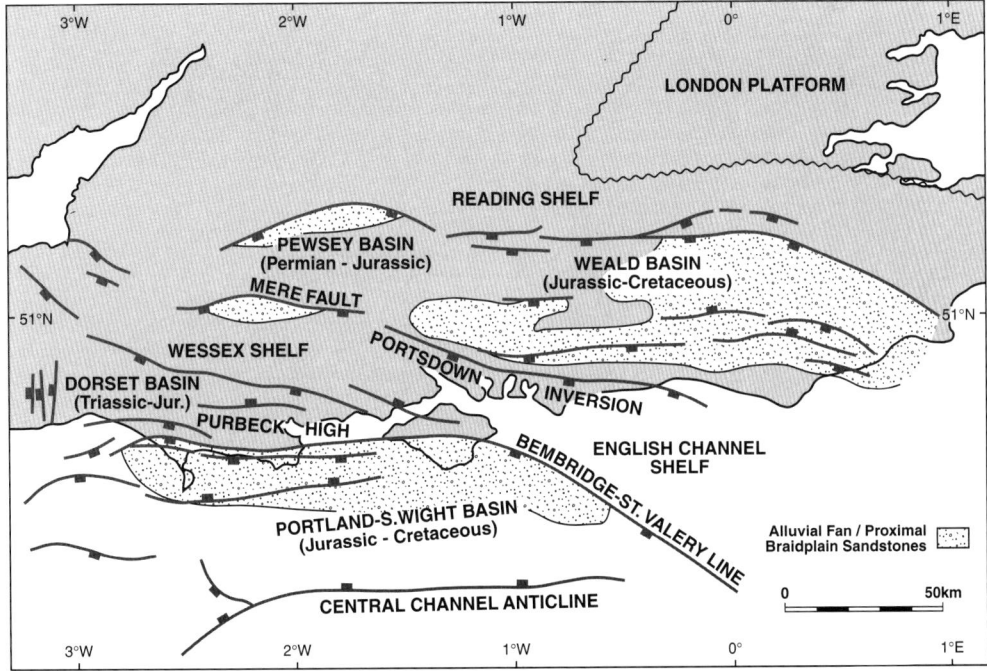

Fig. 25. Lithofacies map depicting the aerial distribution of the Ashdown sands in the Weald and Wessex basins.

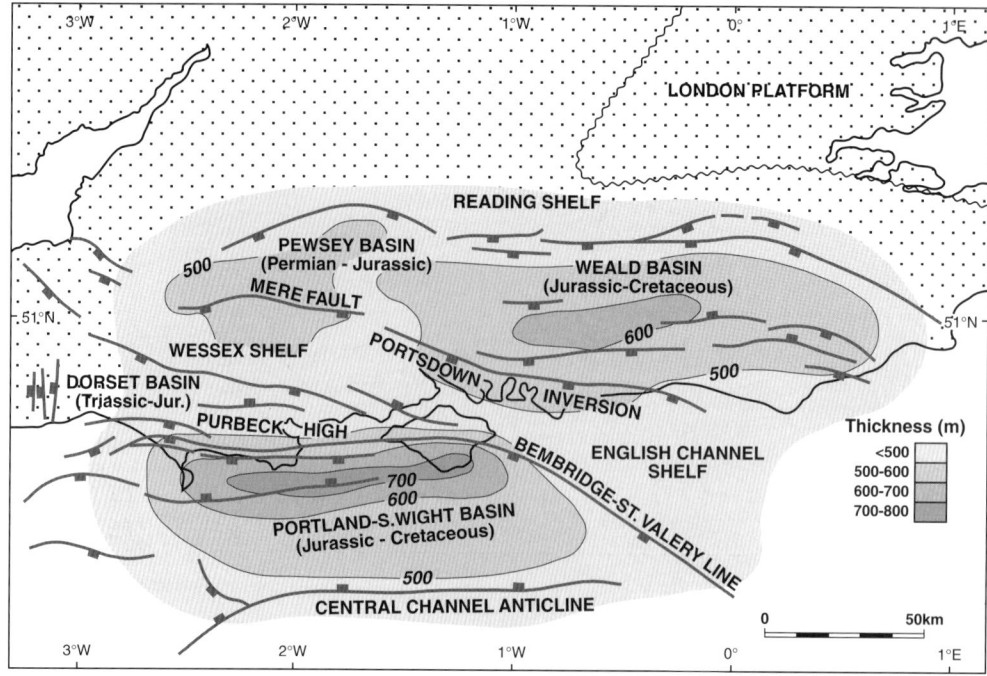

Fig. 26. Gross isopach map of the Lower Greensand–Chalk sequence showing the main sedimentary depocentres present during the Late Cretaceous.

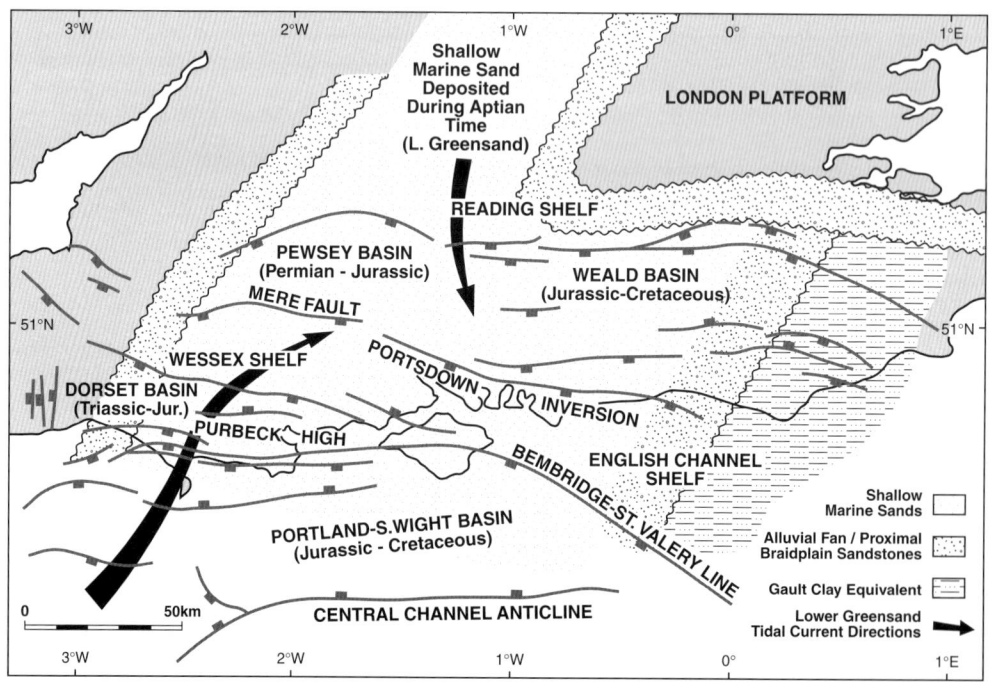

Fig. 27. Lower Greensand and Gault lithofacies map showing the distribution of shallow marine and proximal (alluvial fan) sandstones in the Weald and Wessex basins.

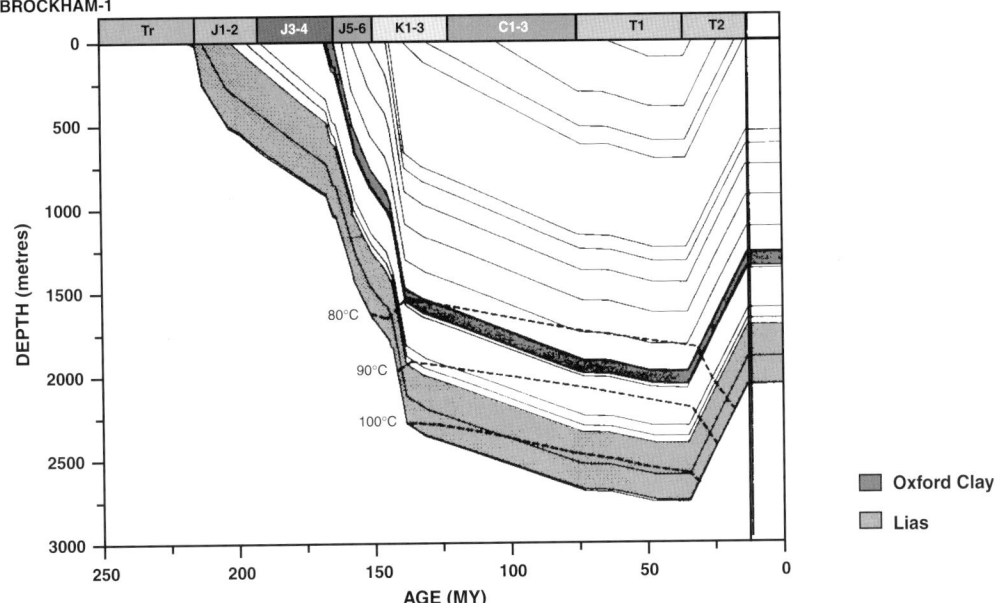

Fig. 28. Burial history–subsidence curve for the Brockham-1 well demonstrating that whilst the Oxford Clay and Lias Group both entered the oil-generating window, the Kimmeridge Clay Formation remained immature throughout Jurassic and Cretaceous subsidence.

Greensand represent a single transgressive–regressive cycle. Regional evidence shows that the Upper Greensand is diachronous as a lithostratigraphic unit and forms a nearshore equivalent to the Gault Clay to the west of the Wessex Basin.

The late Cretaceous marks a period of further thermal subsidence and marine transgression. Clastic input to the basin was shut off following deposition of the Upper Greensand by drowning of the surrounding hinterlands. The Upper Cretaceous is therefore characterized by the deposition of pelagic carbonates (the Chalk). The Lower Chalk represents cycles of chalky limestones and interbedded marls. The alternations occur at 1–5 m intervals and are not easily resolvable on wireline correlation. The Upper Chalk is characterized by decreasing detrital clay content but increasing development of flint horizons.

Although a major aquifer in the region the chalk has no economic significance as a hydrocarbon reservoir due to its shallow depth of burial and lack of top seal (it is at outcrop level over most of the study area), its importance was in contributing to the burial of the Lias and Oxford Clay source rocks during maximum burial of the basin in Late Cretaceous–Early Tertiary times (Fig. 28).

Alpine plate sequence

The rapid northward drift of Africa in the Late Cretaceous culminated in the onset of the Alpine orogeny from latest Maastrichtian times. The resulting uplift in the region of the French and Swiss Alps is well documented. The effects in southern England were less severe but still had an important impact on the hydrocarbon potential of both the Weald and Wessex basins. The main impact of the Alpine event was to invert earlier east–west-trending Lias and Early Cretaceous extensional faults. Evidence fom vitrinite reflectance, shale velocity and basin backstripping studies suggests over 1500 m of uplift and subsequent erosion in the east-central part of the Weald Basin and the generation of numerous inversion anticlines in the hanging walls of the major basin bounding faults across southern England (e.g. the Kimmeridge Anticline, Fig. 2).

The isopach and lithofacies maps for the Alpine sequence (Figs 29 and 30) show how the inversion has changed the polarity of the basin with the Tertiary sediments now deposited on the footwalls of the original Mesozoic Weald and Wessex rifts forming the Hampshire and London Basins as mini 'foreland' basins. Evidence for erosion of the evolving chalk inversion anticlines is provided by flints which were deposited in the

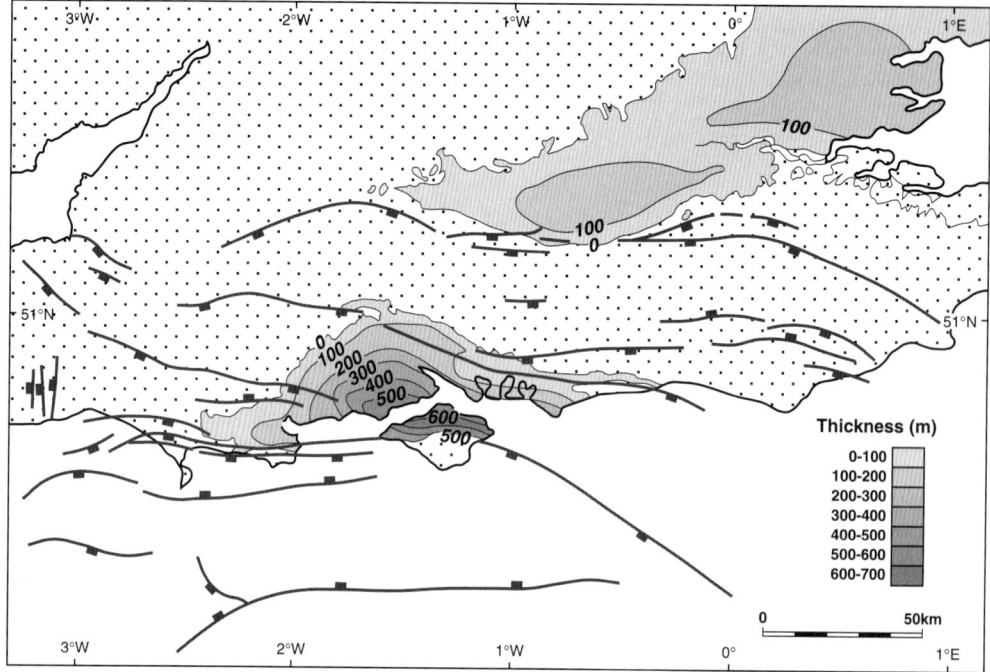

Fig. 29. Gross isopach map for the Tertiary (Alpine) sequence.

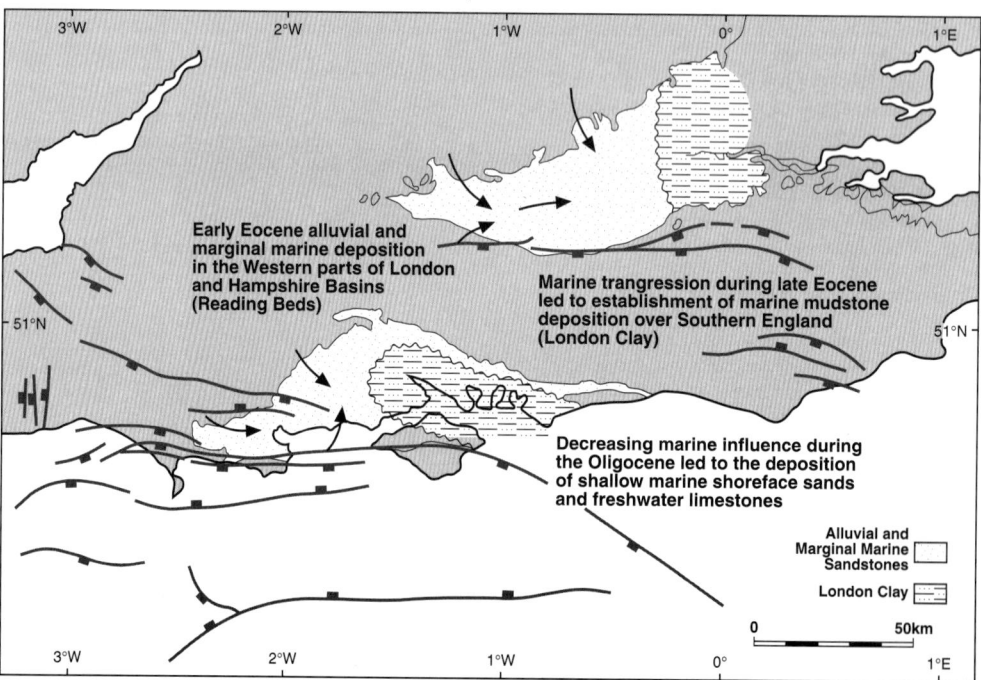

Fig. 30. Lithofacies map for the Tertiary (Alpine) sequence showing the aerial extent of the Hampshire and London basins.

Palaeocene of the Hampshire Basin adjacent to the Purbeck disturbance (Plint 1983).

Latest Palaeocene alluvial and marginal marine deposition took place in the western parts of the London and Hampshire basins (Reading Beds). A marine transgression during the early Eocene led to the establishment of marine mudstone deposition over southern England (London Clay). Decreasing marine influence during the Oligocene led to the deposition of shallow marine shoreface sands and freshwater limestones.

Play types

Prior to 1960, most wells in southern England tested Tertiary anticlinal structures whereas the majority of the Jurassic–Early Cretaceous tilted fault block and horst plays were drilled after 1960. There was early exploration interest in the anticlinal structures which can be defined on the basis of surface geological mapping and were the earliest defined traps in the area. The tilted fault block/horst plays lie below the Apto-Albian break-up unconformity, and cannot be located directly by surface mapping.

Exploration interest shifted from Tertiary anticlines to earlier formed fault block structures during the 1960s. The main reasons for the change in exploration emphasis were:

(i) continued failure of wells on Tertiary inversion structures;
(ii) new discoveries in earlier formed structures;
(iii) improved definition of structures at depth with the use of modern multifold seismic data; and
(iv) improved understanding of the geology and, in particular, the plumbing system of the Weald and Wessex basins.

Progressively, stratigraphically deeper targets were identified. Early interest was concentrated on Upper Jurassic Corallian and Portlandian reservoirs. However, deeper Great Oolite, Bridport and Sherwood targets became attractive as a result of indications that reservoir quality persisted at depth, improved seismic imaging and earlier identification of shows in these intervals.

Trap types

Two distinct trap types are recognized (Fig. 31).

(i) Jurassic–Early Cretaceous tilted fault blocks/horsts: these structures pre-date hydrocarbon generation in the basin.

(ii) Tertiary inversion anticlines: these structures post-date hydrocarbon generation in the basin.

The fault block/horst play is by far the most successful in the Weald and Wessex basins. In the Wessex Basin many wells were drilled a significant distance from mature source (greater than 12.5 km). When these wells are taken out of the figures, the historical risk for traps with Jurassic targets is assessed as 1 in 3 and 1 in 5 at the Triassic level.

The Tertiary inversion anticline play has only been successful at Kimmeridge (Fig. 2) in the Wessex Basin, even though most wells on inversion structures have been drilled in areas of mature source rocks. Perhaps the secondary charge mechanism noted earlier for Kimmeridge is significant here. Elsewhere only residual oil shows and non-commercial gas accumulations have been encountered, particularly in the Weald Basin, where the majority of inversion trap tests occur.

Reasons for prospect failure

The key reasons for past exploration well failures in southern England can be analysed within the tectono-stratigraphic framework described above.

(1) Influence of Tertiary inversion. Not only was hydrocarbon generation switched off but pre-existing accumulations were breached and/or remigrated to newly formed traps.
(2) Many unsuccessful wells were drilled on structures which were never on a migration pathway. This observation underlines the importance of understanding migration pathways in a basin where only a limited volume of hydrocarbons have been generated compared with the volume of traps within the migration limits of the source kitchen.
(3) Wells encountered poorer reservoir quality than expected. In the Wessex Basin, the Corallian (J4) interval reservoir quality is very poor and has never produced on test. The Cornbrash limestone (J3) produces at very poor rates (less than 20 barrels of oil/day) except where local fracturing occurs such as at Kimmeridge (up to 500 barrels of oil/day from an underpressured reservoir). The Bridport Sands (J2) reservoir silts-out eastwards towards the Isle of Wight and southwards

across the Portland–South Wight Basin. The Great Oolite passes into the non-reservoir Fuller's Earth facies to the west of the Weald Basin (although production from thinner carbonate equivalents of the Frome Clay Limestone occurs at Wytch Farm) and the Sherwood Sandstone Group reservoir quality appears to deteriorate eastwards and southwards from the Wytch Farm horst block, mainly associated with diagenesis due to increasing pre-inversion palaeo-burial depths.

(4) Trap definition. It is clear in the light of modern seismic data that many wells have not been drilled on valid traps. In the past onshore seismic data in southern England have been of very poor quality. Within the Purbeck–Isle of Wight Fault Zone persistent poor data quality has failed to resolve the complex structure of this zone. The pre-Jurassic is characterized by low-amplitude, discontinuous reflectors and lies within the realm of interference from long-period multiples. Data quality and depth conversion within fault shadows can lead to unreliable structural mapping in such zones. It can therefore be difficult to accurately define whether or not a structure is closed at depth or for that matter the exact location of a closed structure.

Conclusions

Exploration drilling for hydrocarbons in southern England commenced over 50 years ago, prompted by numerous seepages along the Dorset coast and small gas discoveries in water boreholes. To date, over 300 million barrels of oil and 100 bcf of gas reserves have been discovered.

The Weald and Wessex basins exhibit many similarities in both tectonic and stratigraphic evolution through Early Triassic–Tertiary times, reflecting the regional influences of Atlantic Margin rift–subsidence processes and subsequent Tertiary (Alpine) inversion tectonics. These processes which have impacted on both basins can be described in terms of four plate sequences. These sequences highlight both similarities and differences in the hydrocarbon habitat of the two basins (Fig. 31).

The early Atlantic plate sequence (late Permian–Late Triassic) records two unsuccessful attempts to open the North Atlantic. It is significant in the Wessex Basin for the deposition of the Sherwood Sandstone reservoir.

The Atlantic plate cycle (Rhaetic–Tithonian) records the successful rifting of the Central Atlantic and is characterized by the widespread deposition of the Lias and Oxford Clay source intervals across southern England. The important reservoir intervals of the Bridport Sands

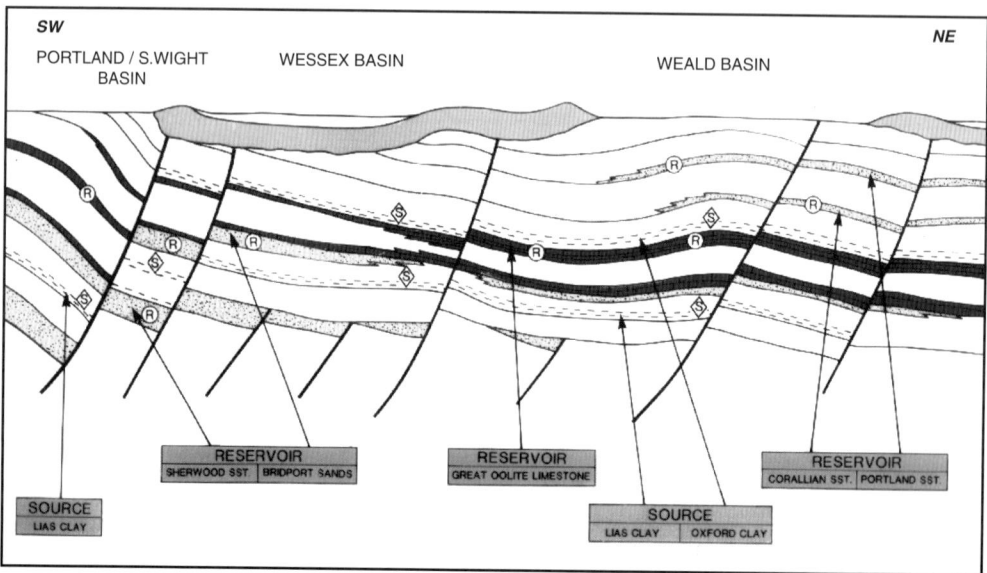

Fig. 31. Schematic northeast–southwest-trending cross-section to demonstrate the main play types present in the Weald and Wessex basins (see text for details).

(Wessex), Great Oolite limestone (Weald), Corallian and Portland sandstones (Weald) were also deposited at this time.

The Biscay plate sequence (Creataceous) records the successful rifting and spreading of the Bay of Biscay, and represents the primary trap forming event in southern England. Maximum maturation of source intervals occurred during the post-rift thermal subsidence phase in Late Cretaceous (Chalk deposition) times.

Finally, the Alpine plate sequence (Tertiary) is characterized by regional uplift and formation of inversion anticlines, breaching of palaeostructures and remigration of hydrocarbons in both the Weald and Wessex basins.

This paper is based on a major BP in-house study of the petroleum geology of southern England carried out at our former UK Land offices in Eakring. We would like to thank the British Petroleum Co. for permission to publish this paper. Thanks also to our colleagues I. Simpson and S. Thompson for their earlier contributions to the study.

References

ARKELL, W. J. 1947. *The Geology of the Country around Weymouth, Swanage, Corfe and Lulworth*. Memoir of the Geological Survey, UK.

BOWMAN, M. B. J., MCCLURE, N. M. & WILKINSON, D. W. 1993. Wytch Farm Oilfield: deterministic reservoir description of the Triassic Sherwood Sandstone. *In*: PARKER, J. R. (ed.) *Petroleum Geology of Northwest Europe: Proceedings of the 4th Conference*. Geological Society, London, 1513–1517.

BRGM. 1980. *Synthese Géologique du Bassin de Paris*, Vol. II. Atlas Memoire Bureau de Recherches Géologies et Minieres, 102.

BUTLER, M. 1998. The geological history of the Wessex Basin – a review of new information from oil exploration. *This volume*.

—— & PULLAN, C. P. 1990. Tertiary structures and hydrocarbon entrapment in the Weald Basin of southern England. *In*: HARDMAN,, R. F. P. & BROOKS, J. (eds) *Tectonic Events Responsible for Britain's Oil and Gas Reserves*. Geological Society, London, Special Publications, 55, 371–391.

CHADWICK, R. A., KENOULTY, N. & WHITTAKER, A. 1983. Crustal structure beneath southern England from deep seismic reflection profiles. *Journal of the Geological Society, London*, 140, 893–912.

DEWEY, J. F., HELMAN, M. L., TURCO, E., HUTTON, D. H. W. & KNOTT, S. D. 1989. Kinematics of the western Mediterranean. *In*: COWARD, M. P., DIETRICH, D. & PARK, R. G. (eds) *Alpine Tectonics*. Geological Society, London, Special Publications, 45, 265–283.

EVANS, J., JENKINS, D. & GLUYAS, J. G. 1997. The Kimmeridge Bay oilfield: an enigma demystified. *This volume*.

FRASER, A. J., NASH, D. F., STEELE, R. P. & EBDON, C. C. 1990. A regional assessment of the intra-Carboniferous play of northern England. *In*: BROOKS, J. (ed.) *Classic Petroleum Provinces*. Geological Society, London, Special Publications, 50, 417–440.

IVIMEY-COOK, H. C., HORTON, A., GAUNT, G. D., WARRINGTON, G., GREEN, G. W., WHITTAKER, A., DONOVAN, D. T. & POOLE, E. G. 1980. The Triassic-Jurassic boundary in Great Britain. Discussion and Reply. *Geological Magazine*, 117, 617–620.

KARNER, G. D., LAKE, S. D. & DEWEY, J. F. 1987. The thermal and mechanical development of the Wessex Basin, southern England. *In*: COWARD, M. P., DEWEY, J. F. & HANCOCK, P. L. (eds) *Continental Extensional Tectonics*. Geological Society, London, Special Publications, 28, 517–536.

LOTT, G. K., SOBEY, R. A., WARRINGTON, G. & WHITTAKER, A. 1982. The Mercia Mudstone Group (Triassic) in the western Wessex Basin. *Proceedings of the Ussher Society*, 5, 340–346.

PENN, I. E. 1982. Middle Jurassic stratigraphy and correlation of the Winterborne Kingston Borehole, Dorset. *In*: RHYS, G. H., LOTT, G. K. & CALVER, M. A. (eds) *The Winterborne Kingston Borehole, Dorset, England*. Report of the Natural Environmental Research Council, Institute of Geological Sciences No. 81–3, 53–76.

——, CHADWICK, R. A., HOLLOWAY, S., ROBERTS, G., PHAROAH, T. C., ALLSOP, J. M., HULBERT, A. M. & BURNS, I. M. 1987. Principal features of the hydrocarbon prospectivity of the Wessex–Channel Basin, UK. *In*: BROOKS, J. & GLENNIE, K. (eds) *Petroleum Geology of North West Europe*. Graham & Trotman, London, 109–118.

PETRIE, S. H., BROWN, J. R., GRAINGER, P. J. & LOVELL, J. P. B. 1989. Mesozoic history of the Celtic Sea Basins. *In*: TANKARD, A. J. & BALKWILL, H. R. (eds) *Extensional Tectonics and Stratigraphy of the North Atlantic Margin*. American Association of Petroleum Geologists Memoir, 46, 433–444.

PLINT, A. G. 1982. Eocene sedimentation and tectonics in the Hampshire Basin. *Journal of the Geological Society, London*, 139, 249–254.

RUFFELL, A. 1995. Evolution and hydrocarbon prospectivity of the Brittany Basin (Western Approaches Trough), offshore north-west France. *Marine and Petroleum Geology*, 12, 387–407.

STONELEY, R. 1982. The structural development of the Wessex Basin. *Journal of the Geological Society, London*, 139, 545–552.

TUCKER, R. M. & ARTER, G. 1987. The tectonic evolution of the North Celtic Sea and Cardigan Bay basins with special reference to basin inversion. *Tectonophysics*, 137, 291–307.

UNDERHILL, J. R. & PATERSON, S. 1998. Genesis of tectonic inversion structures: seismic evidence for the development of key structures along the Purbeck–Isle of Wight Disturbance. *Journal of the Geological Society, London*, in press.

—— & STONELEY, R. 1982. Introduction to the development, evolution and petroleum geology of the Wessex Basin. *This volume*.

VAN HOORN, B. 1987a. Structural evolution, timing and tectonic style of the Sole Pit inversion. *Tectonophysics*, **137**, 239–284.

VAN HOORN, B. 1987b. The South Celtic Sea/Bristol Channel Basin: origin, deformation and inversion history. *Tectonophysics*, **137**, 309–334.

VAN WIJHE, D. H. 1987. Structural evolution of inverted basins in the Dutch offshore. *Tectonophysics*, **137**, 171–219.

The geological history of the southern Wessex Basin – a review of new information from oil exploration

MALCOLM BUTLER

Burnside House, Church Road, Paddock Wood, Kent TN12 6HD, UK

Abstract: Mapping of intervals between the base of the Aylesbeare Mudstone (Permian or early Triassic) and the top of the Penarth Group (early Jurassic) in the Dorset–Isle of Wight area indicates that the deposition of these formations was influenced strongly by northwest–southeast-trending faults, probably related to underlying Variscan elements. There are significant differences between the distribution of the Aylesbeare Mudstone and that of later Triassic formations. An unconformity is recognized within the Sherwood Sandstone Group, separating the syn-rift sequences of Aylesbeare Mudstone, and overlying Budleigh Salterton Pebble Beds and anhydrites from the post-rift Otter Sandstone and Mercia Mudstone (middle–late Triassic). It is postulated that the top of the Sherwood Sandstone Group becomes progressively younger from the Mid Dorset Platform towards the Isle of Wight, steadily replacing the lower Mercia Mudstone. Halite beds are present within the middle Mercia Mudstone of Dorset, although they are absent over much of the Mid Dorset Platform and areas to the east of Wytch Farm. East–west-trending faults became dominant in the Jurassic, and there is strong seismic evidence for syn-depositional movement on these faults during the Lias across the entire Dorset–Isle of Wight area. In Dorset, many of these east–west faults are listric, hading out into the Mercia Mudstone halite and creating very different structural styles above and below the salt. Major syn-depositional movement is thought to have also taken place along these faults during Kimmeridgian and Portlandian times, creating a series of rollover anticlines in the hanging walls. Seismic evidence indicates that the early Cretaceous sequences of the Portland–South Wight Trough thinned rapidly onto the footwalls of these east–west faults and that very little deposition took place north of the Purbeck Disturbance, either in Dorset or on the Isle of Wight. Several of these east–west faults reversed their throw during compression associated with the mid-Tertiary inversion. Reverse displacement estimates range from 50 m on the Litton Cheney Fault to 1320 m on the Purbeck Disturbance in the Isle of Wight. Offsets of the inversion trends may be related to underlying northwest–southeast-trending Variscan faults which influenced the margins of the Jurassic and early Cretaceous basins, but strike-slip movements may have accentuated some of these offsets during the Tertiary. These offsets appear to be dextral to the west of Wytch Farm oilfield and sinistral in the area to the east. Structural styles suggest that there may also have been strike-slip movements along east–west-trending faults.

The Dorset–Isle of Wight region is a classical area of British geology and has been studied by generations of geological students. Over the years, numerous geologists have published explanations of the complex structures exposed along the coast (for example, Arkell 1947), but it is only with the advent of modern seismic techniques that oil company geologists have been able to gain some insight into the broader (and deeper) picture. Even then, seismic events beneath the strong reflector formed by the top of the Penarth Group (White Lias, early Jurassic) have been very difficult to resolve in the south Dorset area until recent years. Seismic data recorded by the Brabant group since 1990 have provided the first clear indications of pre-Jurassic structure in the area between Bridport and Wareham, often with clear imaging of Sherwood and Aylesbeare events, and have given new insights into the deep structure of the Isle of Wight. Exploration wells operated by Brabant, BP and British Gas in this region have helped to tie together the seismic data, but it should be stressed that very little age dating is available for beds below the Lias: correlations are based on lithostratigraphy in the knowledge that some boundaries are almost certainly diachronous.

This paper presents some of the new stratigraphic and structural information obtained as a result of the recent oil exploration work and discusses the regional significance of this information. Other papers in this volume describe the Wessex Basin stratigraphic sequence and its hydrocarbon occurrences in some detail and this information is not repeated here.

Structural framework

The present-day outcrop pattern is dominated by the effects of Tertiary inversion movements

BUTLER, M. 1998. The geological history of the southern Wessex Basin – a review of new information from oil exploration. *In*: UNDERHILL, J. R. (ed.) *Development, Evolution and Petroleum Geology of the Wessex Basin*, Geological Society, London, Special Publications, **133**, 67–86.

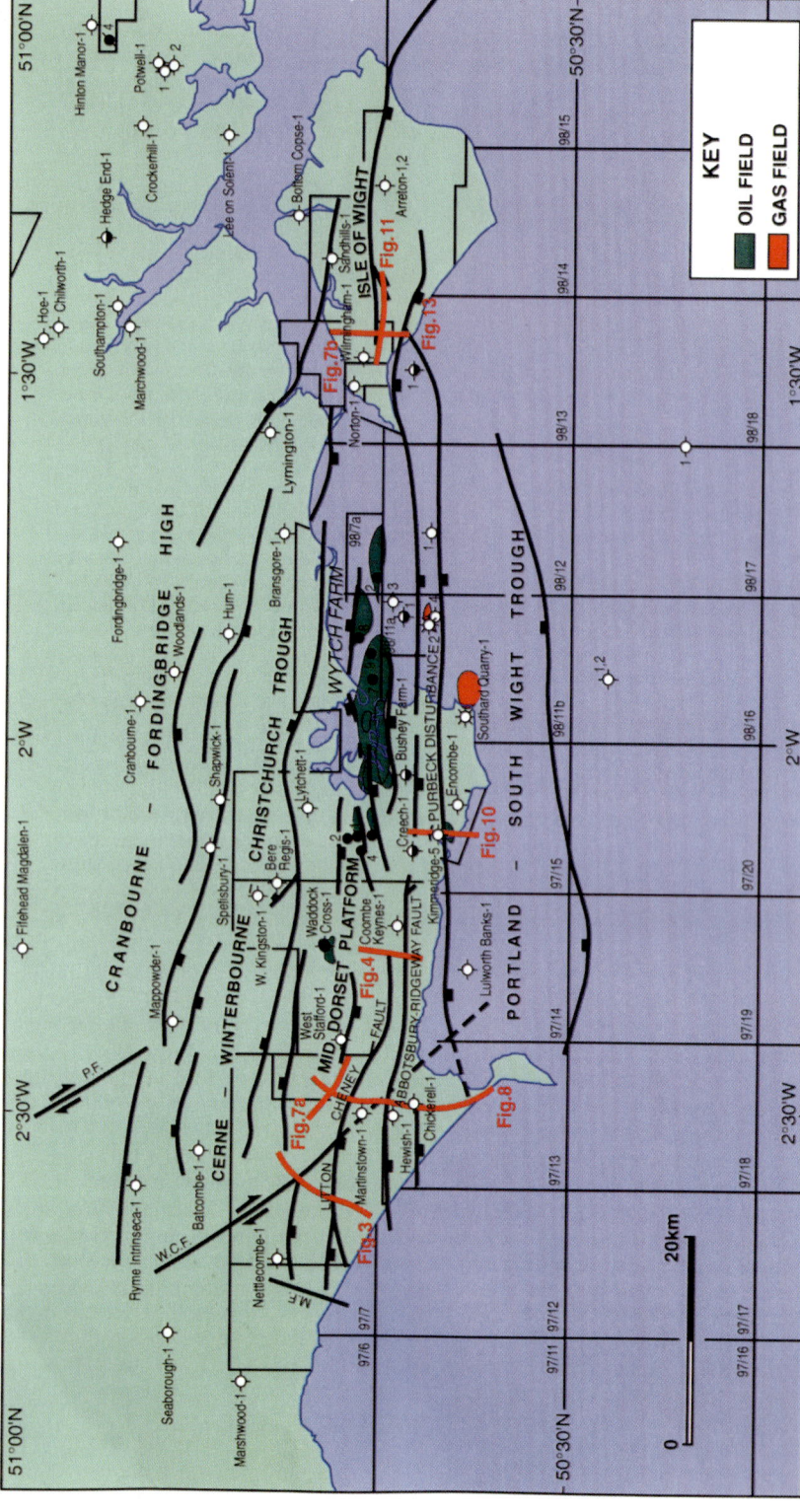

Fig. 1. Location map, showing main structural elements. Note that fault positions are those mapped at base Jurassic level. M.F., Mangerton fault; W.C.F., Watchet–Cothelstone Fault extension; P.F., Poyntington Fault. Note that the sense of movement shown on the two strike-slip faults is that of the Variscan and it is not intended to infer active fault movement at any other time except during the Tertiary.

and by the major east–west-trending faults along which growth took place in the Jurassic and early Cretaceous (Underhill & Paterson 1998) (Fig. 1). Crossing these east–west lineations are a number of major northwest–southeast trends, including the extensions of the Watchet–Cothelstone and Poyntington faults of Somerset and probably related to major late Variscan faults (Chadwick 1993). These Variscan trends seem to have had a control over sedimentation during the Triassic and can be related to offsets and transfer zones in the Tertiary inversions. They can rarely, if ever, be mapped as continuous faults on seismic data onshore, although Smith & Hatton (this volume) and Harvey & Stewart (this volume) were able to map sets of northwest–southeast- and northeast–southwest-trending faults in the offshore Lyme Bay area. To the west of the studied area, a number of surface faults have orientations ranging from north–south to northeast–southwest (for example, the Mangerton Fault, Fig. 1). Westhead & Woods (1994) carried out detailed mapping of the Chalk at outcrop around Dorchester (close to West Stafford-1, Fig. 1) and noted the occurrence of minor faults with these orientations and with displacements of the order of 30–60 m. These have not been mapped on modern seismic data in this area and their significance is not yet understood.

Aylesbeare Mudstone

The earliest post-Variscan sediments known in the south Dorset region are those of the Aylesbeare Mudstone (and associated basal conglomerates). These have traditionally been referred to by oil company geologists as 'Permian', although they may be entirely of earliest Triassic Scythian age (Hamblin *et al.* 1992). The Permian breccia and volcanic series which outcrop in the Exeter region have not been noted in Dorset, although they may be present beneath some of the thicker Aylesbeare sequences. Breccias such as those seen in Wytch Farm wells are considered to be locally derived and of the same age as the Aylesbeare Mudstone. New seismic data, acquired over the past 5 years, and new well information have enabled the isopachs produced by Whittaker (1985) to be refined in this area. The new, somewhat conjectural, isopachs of the Aylesbeare Group are shown in Fig. 2 and are based largely on seismic data in the western part of the area and largely on well data to the east. It is important to note that seismic imaging of the base of the Aylesbeare Group is generally poor to the east of Waddock Cross and south of the main inversion faults: here, interpretation remains difficult and is often subjective.

Borehole information suggests that the distribution of the Aylesbeare Mudstone is controlled to the north by the faults forming the northern boundary of the Cerne–Winterbourne–Christchurch Trough. To the east, the boundary appears to be formed by a northwest–southeast-oriented lineation, cutting across the tip of the Isle of Wight and forming the northeast boundary to a major depocentre offshore Wytch Farm. Although there are few basement penetrations, there appears to be a coincidence between the edge of the Aylesbeare Mudstone depocentre in south Dorset and the boundary between probably Devonian phyllitic basement to the south and Old Red Sandstone to the north and east. It is therefore likely that the northeastern margin of the Aylesbeare Mudstone in this region is related to extension and relaxation along the line of a major Variscan thrust system, associated with a phase of late Permian or early Triassic rifting and differential subsidence.

In the western part of the region, north of the Abbotsbury–Ridgeway Fault and west of Waddock Cross, the base of the Budleigh Salterton Pebble Beds (resting unconformably on the Aylesbeare Mudstone) and the base of the Aylesbeare Mudstone (resting unconformably on Palaeozoic basement) both form relatively strong reflectors (note the sonic contrast in Fig. 6). It is therefore possible to form a more detailed view of the controls on Aylesbeare deposition in this area. Figure 2 demonstrates that the Aylesbeare Mudstone thins rapidly westwards from the depocentre in Bournemouth Bay, in excess of 1000 m of sediment beneath Wytch Farm thinning to less than 100 m beneath the western part of the Mid Dorset Platform. At this time, the Mid Dorset Platform appears to have been bounded to the west by a major northwest–southeast-trending fault system, giving rise to a half-graben on the southwest side, in which in excess of 800 m of Aylesbeare accumulated (Fig. 3). This northwest–southeast-trending fault lies along the projection of the Watchet–Cothelstone Fault, a major Variscan lineation recognized by Whittaker (1972), Chadwick (1986, 1993) and other authors, and there is indication that strike-slip movement took place along this trend during the Tertiary. It is tempting to suggest that this fault formed by reactivation along the line of a deep Variscan fault which defines the separation between the Upper Carboniferous 'Culm' noted in the wells at Seaborough and Nettlecombe and

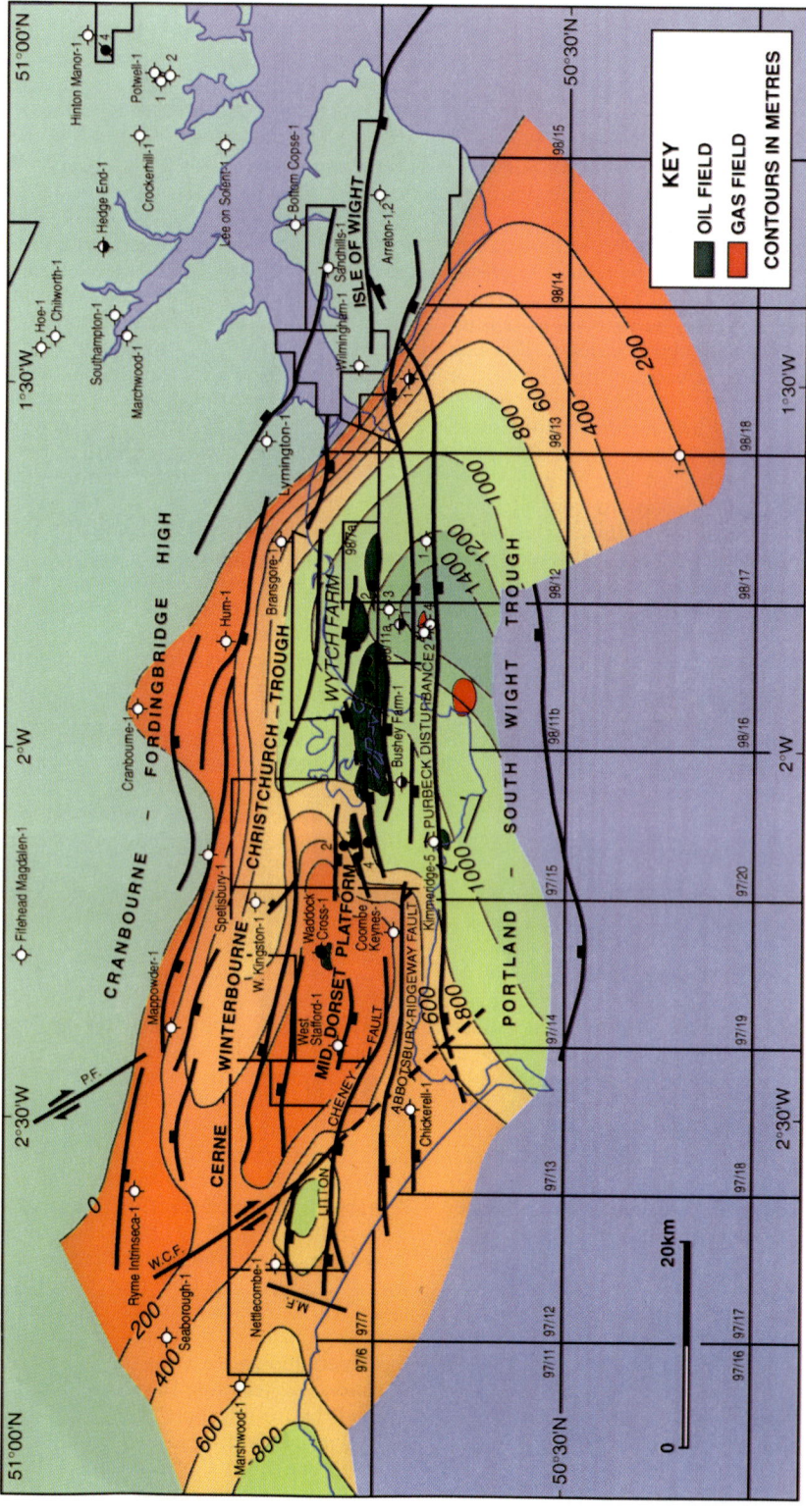

Fig. 2. Conjectural isopach map of the Aylesbeare Mudstone, based largely on seismic picks to the west and well information to the east. Exploration wells which provide control at this level are shown on the map. Faults are those shown in Fig. 1 at base Jurassic level.

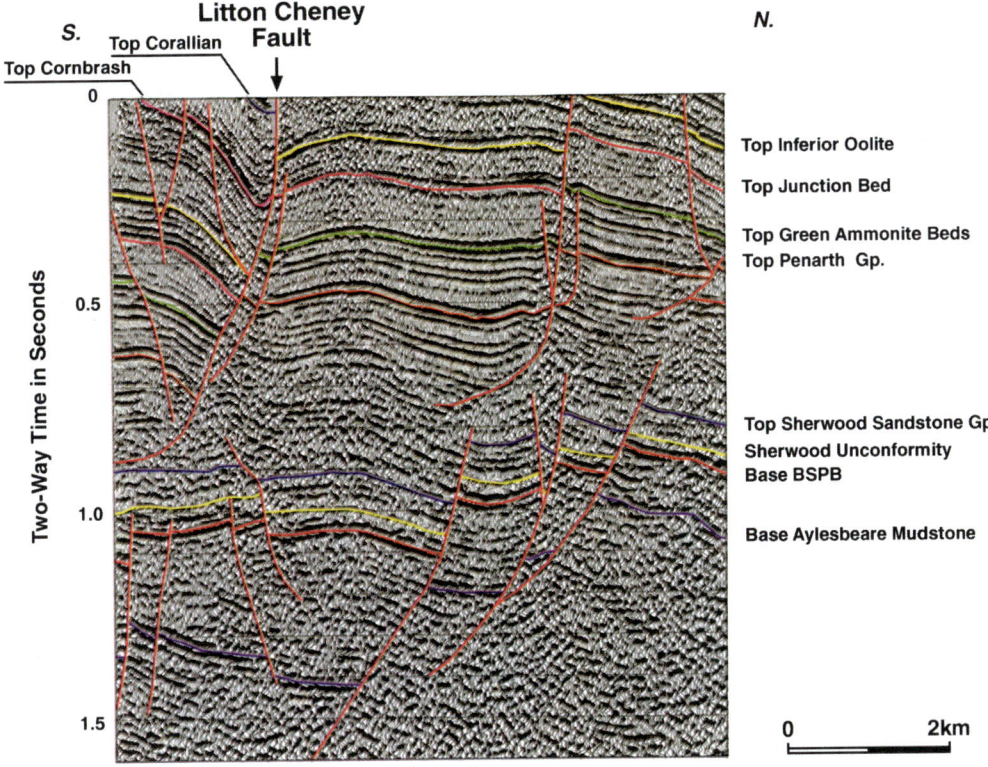

Fig. 3. Seismic line Abbotsbury–Compton Valence. BSPB, Budleigh Salterton Pebble Beds. For location, see Fig. 1.

the probable Devonian phyllites found in the wells at Ryme Intrinseca, Mappowder and Wytch Farm: unfortunately, there is presently no basement control point between Nettlecombe and Wytch Farm.

Sherwood Sandstone Group

The overall thickness of the Sherwood Sandstone Group in the Dorset–Isle of Wight area does not appear to have been affected greatly by contemporaneous faulting (Fig. 5; cf. Hawkes et al. this volume). It is interesting to note that, although the Sherwood Sandstone Group thickens into the site of the Aylesbeare half-graben to the west, it is not affected by the Aylesbeare isopach thick in the Wytch Farm area. Instead, the Sherwood thins consistently from Wytch Farm to the east.

The basal beds of the Sherwood Sandstone Group at outcrop in east Devon are the Budleigh Salterton Pebble Beds (BSPB), which unconformably overlie the Aylesbeare Mudstone. Smith & Edwards (1991) carried out detailed studies on the BSPB in outcrop and considered the relationship of similar conglomerates seen in wells around Southampton and the Isle of Wight. Seismic and well evidence supports the views of Warrington et al. (1980) that the BSPB are not present in the Wytch Farm area as a result of erosion, although they may not have been deposited over much of the eastern part of the Dorset–Isle of Wight area. Several seismic lines in the area around Dorchester appear to show an unconformity within the Sherwood Sandstone Group and, passing eastwards onto the Mid Dorset Platform, the high-amplitude event associated with the contact between the BSPB and the Aylesbeare Mudstone appears to subcrop this unconformity (Fig. 7a). This is borne out by correlation of well logs, which shows the BSPB to disappear between the Martinstown and Waddock Cross wells (Fig. 6). Seismic evidence appears to show that the lower part of the Sherwood Sandstone Group, including the BSPB, was affected by faulting which occurred before the intra-Sherwood unconformity (Figs 3, 4 and 7). This may have been contemporaneous faulting, although

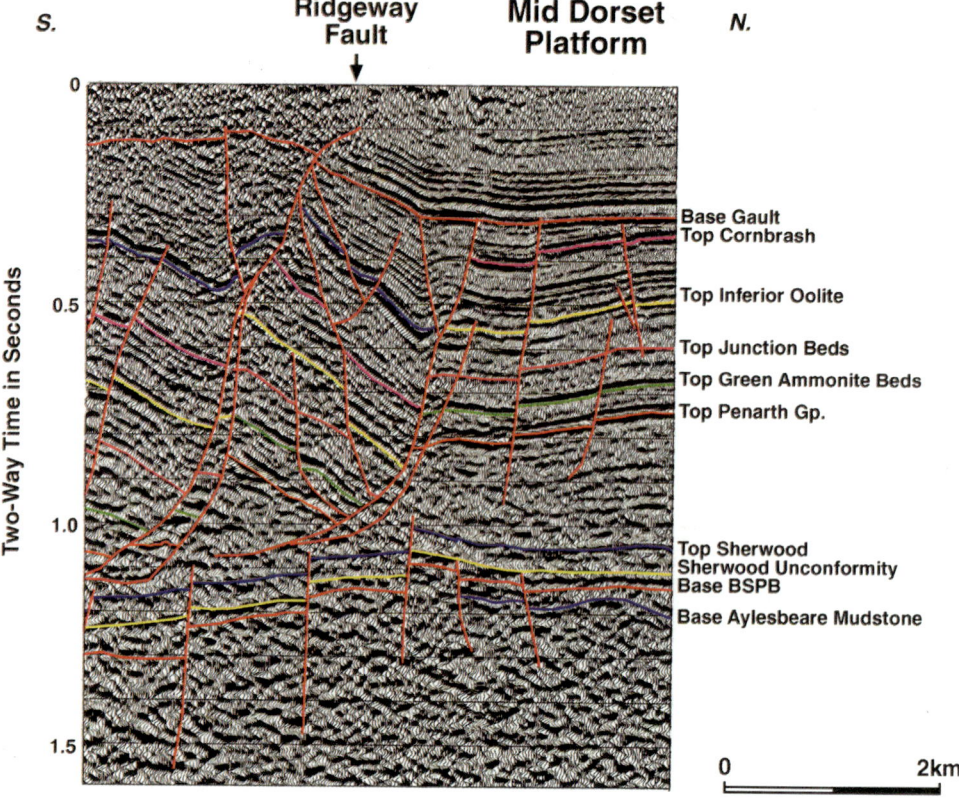

Fig. 4. Seismic line Chaldon–Ridgeway area. For location, see Fig. 1.

any related thickness changes are of much lower magnitude than those which affected the Aylesbeare Mudstone. Smith & Edwards (1991) discuss the exotic nature of the clasts distributed by the BSPB streams and the wide regional distribution of these beds, with a possible lateral extent of 250 km. The seismic evidence tends to support their view that the BSPB are not the result of rapid syn-sedimentary subsidence of the Wessex Basin.

The intra-Sherwood unconformity marks an apparent major change in tectonic style and it is possible that it may be of similar age to the Hardegsen Unconformity of the southern North Sea. The correlation in Fig. 6 demonstrates that the upper part of the Sherwood Sandstone Group (equivalent to the Otter Sandstone) thickens westward off the Mid Dorset High and there is some evidence for this in Fig. 7a. Martinstown-1 and other wells to the west record an additional sequence between the Otter Sandstone and the BSPB. In Martinstown-1 it consists of coarse sandstones and conglomerates, and in Nettlecombe-1 and Seaborough-1 it consists of anhydrites and siltstones. These sequences are considered to have been deposited in a local basin, probably coincident with the Aylesbeare half-graben, and are interpreted to lie beneath the intra-Sherwood unconformity.

The Otter Sandstone equivalent can be seen to lie unconformably on the BSPB and the Aylesbeare, and appears little affected by syn-sedimentary faulting. It post-dates the early Triassic rifting phase of the Wessex Basin and its distribution and that of most of the overlying Mercia Mudstone Group is probably related to post-rifting thermal relaxation subsidence. This interpretation is at variance with the views of Smith & Edwards (1991) that the Otter Sandstone heralds a phase of Triassic rifting in the Wessex Basin and those of Hawkes *et al.* (this volume) that the Sherwood Sandstone represents a syn-rift sequence in the Channel Basin.

Well correlation carried out by Brabant in the Waddock Cross–Wareham–Wytch Farm area strongly suggests that the top of the Sherwood Sandstone Group becomes younger to the east,

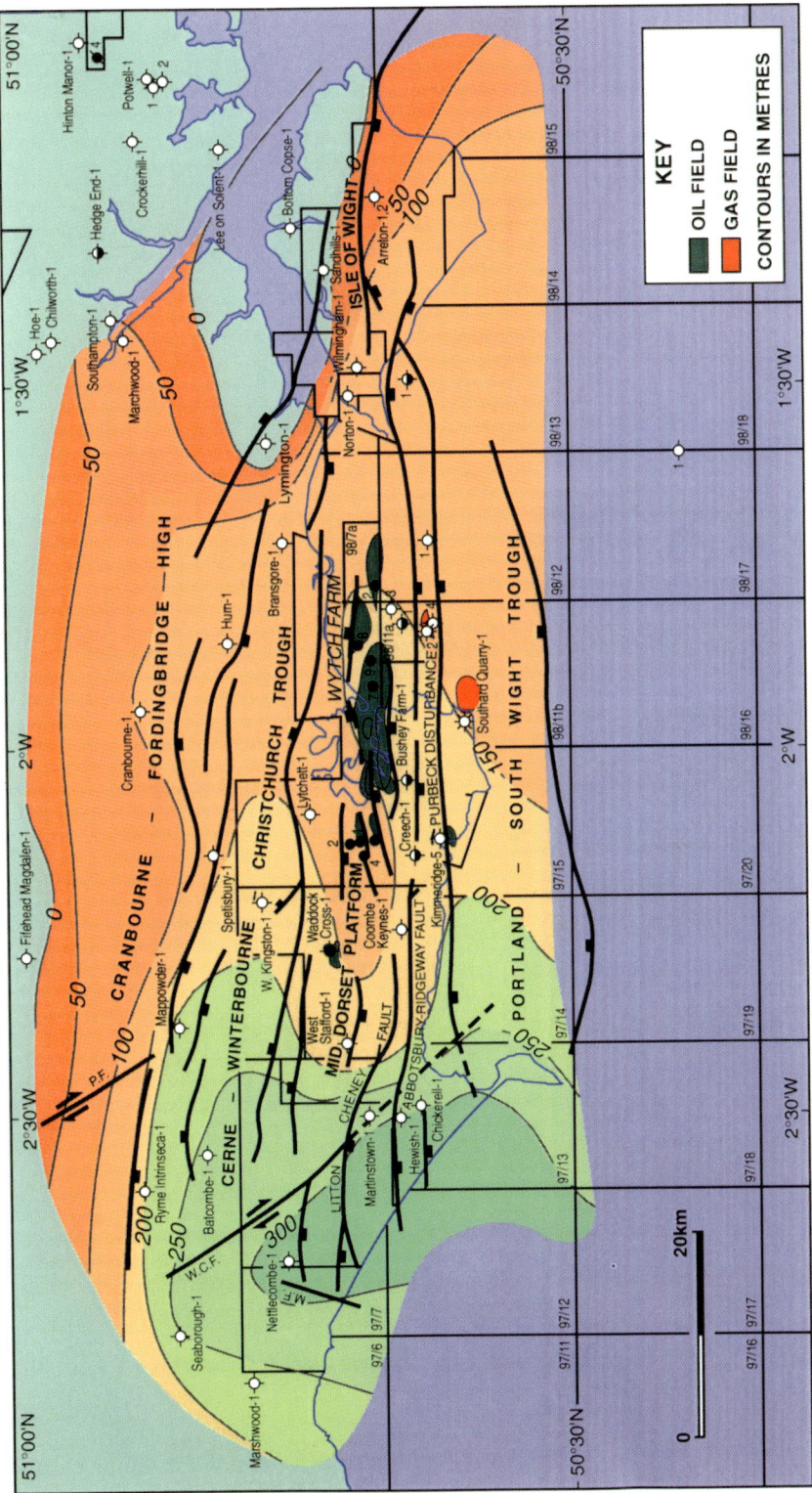

Fig. 5. Isopach map of the Sherwood Sandstone Group, based on seismic and well information. Exploration wells which provide control at this level are shown on the map. Faults are those shown on Fig. 1 at base Jurassic level.

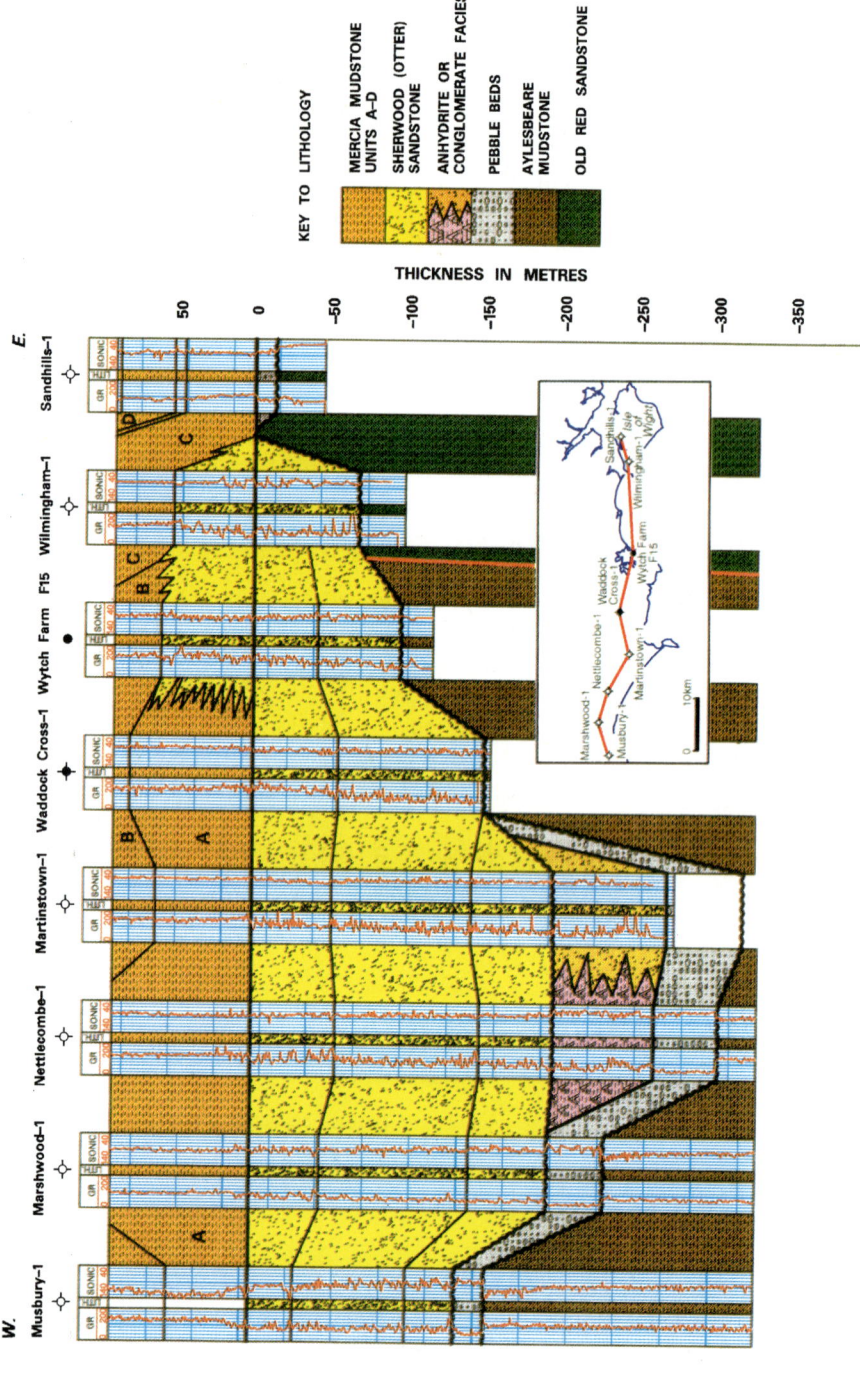

Fig. 6. Well correlation of the Sherwood Sandstone Group; Marshwood-1 to Sandhills-1. No horizontal scale.

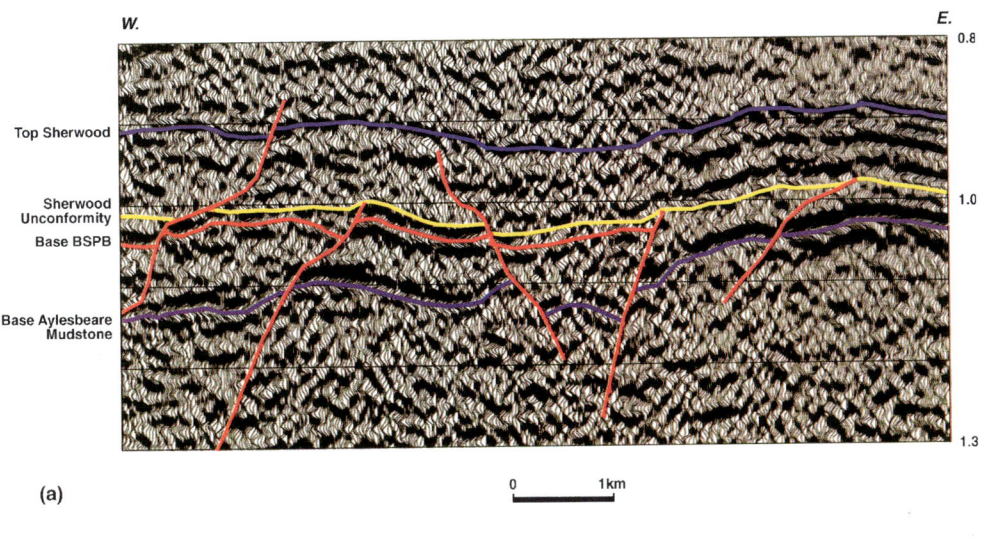

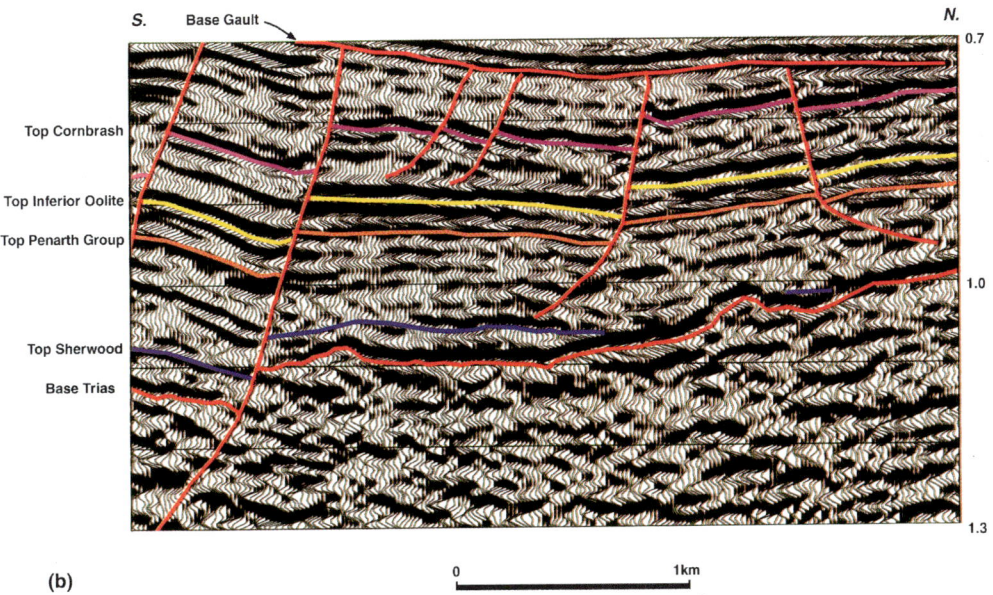

Fig. 7. (a) Seismic line Dorchester area; (b) seismic line Isle of Wight. For location, see Fig. 1.

being replaced by lower Mercia Mudstone to the west (Fig. 6). Further to the east, well correlation suggests that the lower Mercia Mudstone may onlap onto the top of the Sherwood Sandstone Group on the Isle of Wight High. The seismic resolution is not sufficient to prove or disprove onlap at this level, but it is considered more likely that progressive easterly replacement of the lower Mercia Mudstone by Otter Sandstone facies has occurred, as shown in Fig. 6. In this case, the base of the Sherwood Sandstone Group also becomes younger to the east and the basal pebble beds seen in Sandhills-1 may not be related to the BSPB. The progressive onlap of the Mercia Mudstone Group around the margins of the Wessex Basin can be seen in outcrop in the Mendip Hills and has been described in the Fifehead Magdalen area by Evans & Chadwick (1994). The seismic line in Fig. 7b demonstrates the onlap of the Sherwood Sandstone Group and the Mercia Mudstone onto basement topography in the Isle of Wight. The underlying rocks here consist of Old Red Sandstone facies and, although it is difficult to fully resolve the detail of the rather low-frequency seismic event, it seems likely that onlap has occurred onto an undulating erosion surface, with high blocks representing palaeo-hills, rather than the results of contemporaneous faulting.

Mercia Mudstone

The Mercia Mudstone of much of south Dorset includes deposits of halite and andydrite and acts as a décollement zone, into which later listric faults hade. For this reason, it is often difficult to recontruct its original depositional thickness to determine the extent to which it was affected by contemporaneous faulting. Figure 8, in particular, shows clearly the separation in tectonic styles between the pre- and post-Mercia Mudstone sections. The post-Mercia Mudstone section is dominated by listric faults which appear to have begun their growth during the early Lias and to have continued growth sporadically through the early Jurassic and early Cretaceous. These faults sole out into the Mercia Mudstone evaporite group and their continued growth has been facilitated by mobility of the halite beds. Although these do not form piercement structures, migration of halite into pillows has given rise to significant variation in Mercia Mudstone thickness associated with the hanging walls of listric faults.

The most obvious syn-sedimentary thickening in this interval appears to occur into the Cerne–Winterbourne–Christchurch Trough, across bounding faults to the north and south, and westwards of the Mid Dorset Platform into the Nettlecombe area (Fig. 9). Thus, Mercia Mudstone deposition appears to have been affected to some extent in Dorset by the major east–west fault trends which controlled the overlying Jurassic and Early Cretaceous sedimentation, but the influence of the northwest–southeast lineation in west Dorset was still marked. As noted above, the area to the south of the Mid Dorset Platform is much affected by halokinesis and it is difficult to determine original depositional thicknesses and, thus, the extent of contemporaneous growth across the southerly bounding faults. Minor halokinesis has also occurred on the Mid Dorset Platform: the northern end of Fig. 4 shows part of a small salt pillow. Figures 3, 4 and 8 show that the Sherwood and older rocks were dislocated by faulting which appears to die out within the mid Mercia Mudstone evaporite group. However, there is no obvious growth within the lower Mercia Mudstone and it is probable that the majority of these faults are of much later age, being related to phases of extension in the early Jurassic and in the late Jurassic to early Cretaceous.

Figure 9 demonstrates that the entire Mercia Mudstone interval thins from west to east along the Mid Dorset Platform and continues to thin across Bournemouth Bay onto the Isle of Wight High. This is interpreted as being due, in part, to the progressive replacement of the lower beds of the Mercia Mudstone by Otter Sandstone facies towards the east. The major Jurassic faults of the Isle of Wight appear to have had little effect on the Mercia Mudstone. Although the seismic resolution is not good at base Mercia level in the Portland–South Wight Trough, a comparison between Sandhills-1, on the crest of the Isle of Wight High, and Arreton-2, south of the major bounding fault, indicates very similar thicknesses of Mercia Mudstone. In both these wells, the lower part of the Mercia Mudstone is interpreted as absent, as a result of transition to Otter Sandstone facies and overlap on the margin of the basin (see Fig. 7b). Halite is absent in Wytch Farm wells and in other parts of the Mid Dorset High, but is present in the Cerne–Winterbourne–Christchurch Trough to the north and in the Portland–South Wight Trough to the south, at least as far east as offshore block 98/11.

Lower Jurassic

In their study of the evidence from outcrop for intra-Jurassic faulting in the Wessex Basin, Jenkyns & Senior (1991) concluded that movements were concentrated in two phases: Hettangian to Bajocian, corresponding to the early rifting phase of the Central Atlantic, and latest Oxfordian to Portlandian, corresponding to the early rifting phase of the North Atlantic. The seismic evidence certainly confirms the presence of intra-Jurassic faulting, which has caused rapid thickness changes within the sequence from the top of the Penarth Group to the top of the Inferior Oolite across east–west-trending faults. Figure 8 demonstrates obvious growth within the Lias below the Green Ammonite Beds across the Litton Cheney Fault. This interval is estimated to be 180 m thick in the footwall of the fault and to attain 260 m in the hanging wall. There is apparently little growth across the Abbotsbury Fault, but the section from the base of the Lias to the Inferior Oolite appears to

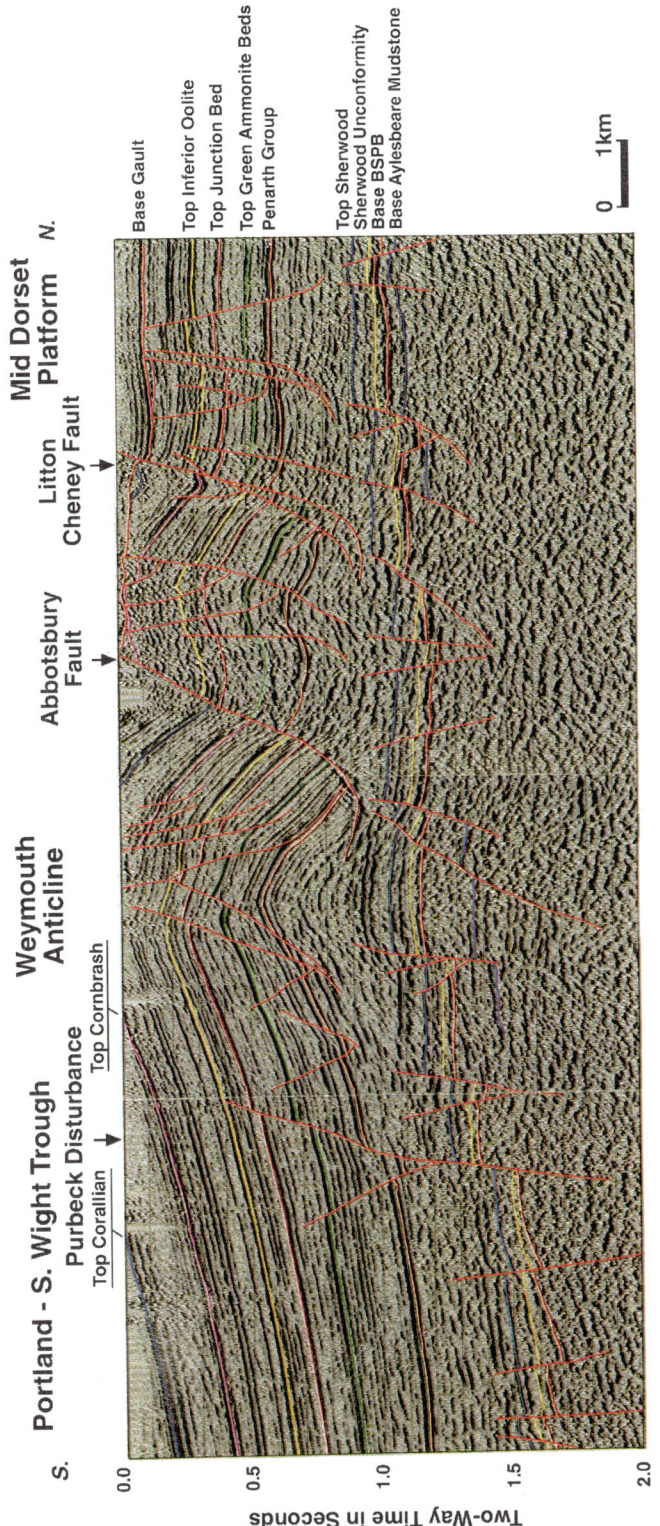

Fig. 8. Seismic line Portland–Dorchester. For location, see Fig. 1.

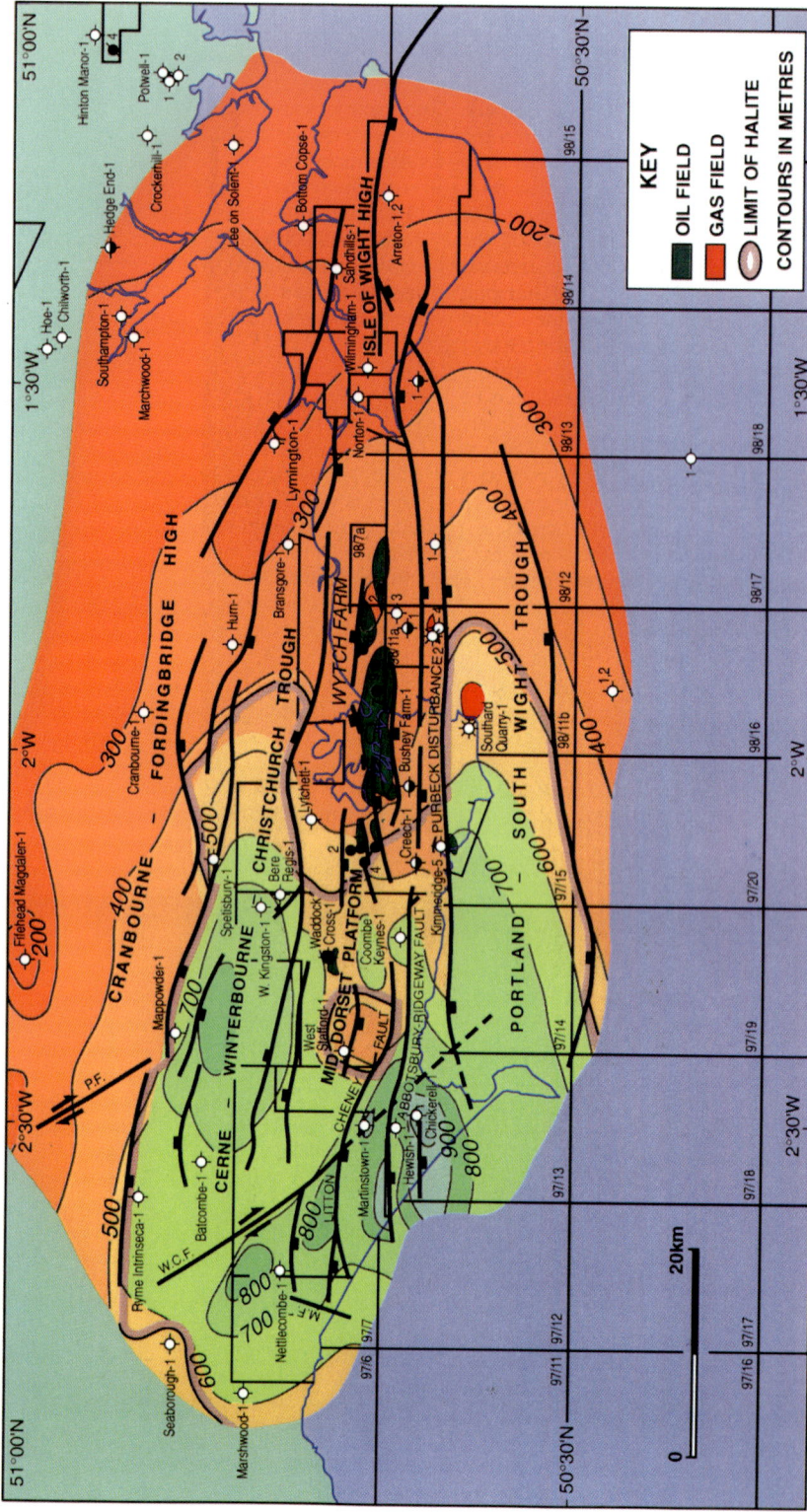

Fig. 9. Isopach map of the combined Mercia Mudstone and Penarth Groups, based on seismic and well information. Exploration wells which provide control at this level are shown on the map. Faults are those shown in Fig. 1 at base Jurassic level.

thicken rapidly to the south within the segment between the Abbotsbury Fault and the extension of the Purbeck Disturbance and there is some indication of early movement of the underlying Mercia halite. Further growth occurs across the extension of the Purbeck Disturbance within the entire Lias section (Fig. 8) and this is the only place on this line where contemporaneous fault growth can be seen in the section between the top of the Junction Bed and the top of the Inferior Oolite. On this line, the Lias interval reaches a maximum thickness in the hanging wall of the Purbeck Disturbance of some 1220 m (compared with some 450 m on the Mid Dorset Platform at the northern end of the line), before thinning gently towards the south. It should be noted that apparent thinning of Jurassic intervals into the downthrown side of the Abbotsbury Fault is an artifact generated by increases in velocity towards the fault zone: depth conversion indicates minor growth across this fault within the lower Lias. The maximum Liassic thickness noted in this area occurs in Weymouth Bay to the south of Kimmeridge. The section through Kimmeridge-5 (Fig. 10) indicates a thickness of some 1515 m to the south of this well. Thickening within the Lias section also occurs across the faults bounding the Cerne–Winterbourne–Christchurch Trough, both to the north and to the south, and westwards along the Mid Dorset Platform. The Inferior Oolite–Penarth Group section is some 300 ms (approximately 410 m) thick in the footwall of the Litton Cheney Fault in Fig. 8, but has thickened to some 350 ms (approximately 480 m) in the same position in Fig. 3. Indeed, it is probable that the Mid Dorset Platform plunged to the west throughout the Jurassic. In the eastern part of the study area, the Lias is very much reduced in thickness, but it again forms the focus of fault growth. Figures 11 and 13 demonstrate Liassic growth across the main Purbeck Disturbance on the Isle of Wight, and Chadwick (1993, Fig. 8) demonstrates more dramatic growth across the Disturbance offshore, between Purbeck and the Isle of Wight.

Upper Jurassic and Lower Cretaceous

Major growth across the Litton Cheney and Abbotsbury Faults and the Purbeck Disturbance is postulated to have begun again in the Kimmeridgian and continued through to Lower Greensand times. Unfortunately, early Cretaceous erosion of the footwall sides of these faults makes it generally impossible to estimate the amount of contemporaneous growth. There is clear evidence of fault control on sedimentary thicknesses in the Weald Basin at this time (Butler & Pullan 1990) and the outcrop evidence summarized by Jenkyns & Senior (1991) indicates that a similar situation prevailed in Dorset. Figure 11 demonstrates fault-related thickness changes in the lower Kimmeridgian section of the Isle of Wight, where there is no underlying

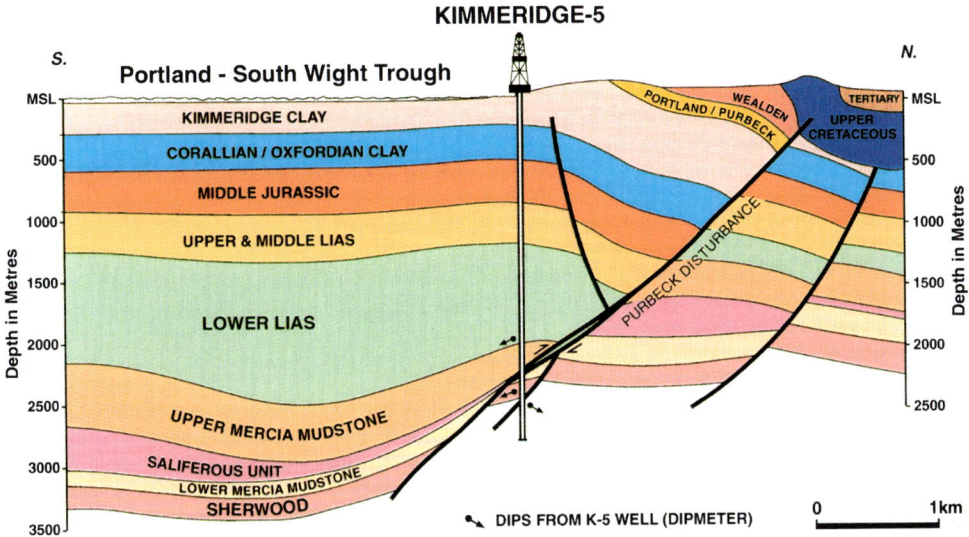

Fig. 10. Diagrammatic north–south cross-section through well Kimmeridge-5.

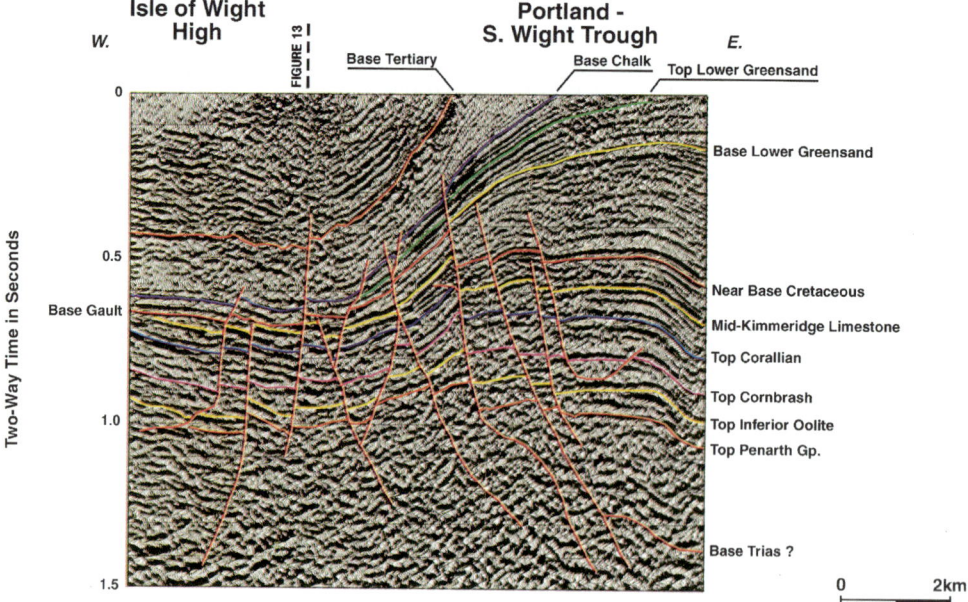

Fig. 11. Seismic line Isle of Wight. For location, see Fig. 1.

Mercia halite. In contrast to the Weald Basin, where there is little Mercia Mudstone and no halite, renewed extension in the lower Kimmeridgian of Dorset gave rise to further movement on major fault systems which haded out into the Mercia Mudstone and caused the halite within this interval to become mobile. Growth on the Abbotsbury–Ridgeway Fault was compensated for by swelling of halite immediately to the south, causing the main growth of the Weymouth Anticline (Fig. 8). A similar, smaller scale, anticline developed between the Abbotsbury Fault and the Litton Cheney Fault and can be clearly dated as pre-Upper Cretaceous in age by its relationship to the Base Gault (Albian) unconformity in Fig. 8. It is also interesting to note that minor faults haded out both into the non-halite bearing Mercia Mudstone and into the Fuller's Earth in the Isle of Wight (Fig. 7b). Notwithstanding contemporaneous movement on the bounding faults, seismic evidence indicates that a thick Kimmeridge Clay sequence was deposited over the Mid Dorset High and the Isle of Wight High. Remnants of these beds are commonly seen preserved in half-graben beneath the Albian unconformity in both areas.

The original distribution of the Lower Cretaceous in Dorset is even more problematical than that of the Upper Jurassic. Ruffell & Batten (1994) note that there is evidence for uplift and erosion at three levels: very early Cretaceous, mid-Aptian; and early Albian. There is a very obvious unconformity across areas north of the Purbeck Disturbance, above which basal Gault (Albian) overlies beds ranging in age from Cornbrash to Kimmeridge Clay on the Isle of Wight High and Mid Dorset Platform. Further to the west, strong regional tilting which took place before Gault deposition has caused the unconformity to cut down into the Triassic (and eventually to the Palaeozoic in Devon) and progressive westerly onlap is demonstrated by the fact that the Gault is eventually overlapped by the Upper Greensand in the western part of the area studied. The original depositional thickness of both Wealden and Lower Greensand beds is an important factor in reconstructing the burial history and source rock maturation of the Dorset–Isle of Wight region, particularly in the western part of the area. Unfortunately, pre-Gault erosion has removed much of the evidence and the original thicknesses must be inferred from isolated remnants in the hanging walls of faults and by analogy with areas further east.

In the Isle of Wight area, Fig. 11 provides valuable evidence which suggests that both Wealden Beds and Lower Greensand thin rapidly onto the Isle of Wight High and it seems unlikely that any great thicknesses were deposited over the high. This line does not show any significant angularity at the boundaries between the Wealden, Lower Greensand and Gault, although outcrop studies suggest discontinuities between these sequences. Similar relationships between these beds appear to be shown in the offshore area to the southeast of Wytch Farm by Chadwick (1993 Fig. 8) and no Wealden Beds have been reported in Wytch Farm wells. Studies of the Weald Basin (Butler & Pullan 1990, Ruffell 1992) indicate that the Lower Cretaceous sequences thin rapidly onto the margins of the London Platform and the Portsdown High and that little, if any, Wealden Beds were deposited across the top of the London Platform.

On the basis of these analogies and the regional westerly thinning which can be mapped in the Weald Basin, it seems likely that the deposition of a thick sequence of Wealden Beds and the Lower Greensand was restricted in Dorset to the area south of the Purbeck Disturbance. In outcrop in south Dorset, the Wealden appears to thin rapidly towards the west, from 716 m at Swanage, south of Wytch Farm, to 166 m at Lulworth and 107 m at Bincombe, near Weymouth. However, the Lulworth section lies within the zone of the Purbeck Disturbance and may represent the thickness on an intermediate fault block, rather than the true basinal thickness (cf. Fig. 11). In addition, much of this westerly thinning may be the result of Aptian or Albian erosion.

It is probable that thinner Wealden Beds were also deposited outside this main depocentre in a series of half-graben formed by the east–west-oriented listric faulting. Thus, Wealden Beds can be postulated to have been deposited in a number of half-graben which are partially preserved beneath the Albian unconformity on the Mid Dorset Platform and in the area to the west, where halokinesis commonly assisted listric fault growth, although much of the evidence has since been removed by the regional tilting. There is some evidence for listric fault growth along the northern margin of the Mid Dorset Platform, in the Bere Regis area, and it is possible that Wealden Beds were deposited here. It seems less likely that these beds were deposited throughout the remainder of the western part of the Cerne–Winternourne–Christchurch Trough, particularly as they are not preserved in wells drilled within the eastern part of the Trough. In reconstructions which have been carried out in Fig. 8, it has been assumed that the 107 m of Wealden preserved at Bincombe were also present in the hanging wall of the Abbotsbury Fault on this line before Tertiary uplift. This may be too conservative an estimate for the pre-inversion thickness and almost certainly underestimates the thickness before pre-Gault uplift. It is possible that the high velocities noted in the hanging wall of the Abbotsbury Fault indicate much deeper burial before this regional tilting.

In the Isle of Wight area, the Carstone, at the base of the Gault Clay, can be seen from outcrop and well information to lie with an erosional base, but no apparent angularity, on the underlying Lower Greensand south of the Isle of Wight Disturbance, but to overstep across the Isle of Wight High. Here it forms the base of the post-unconformity sequence, apparently without any noticeable thickness change across the main Disturbance fault (Ruffell 1992). Further to the west, the evidence is complicated by strong regional tilting as well as the erosion which took place before deposition of the Gault Clay and Upper Greensand. Offshore to the southeast of Wytch Farm, in well 98/11-1, the Gault lies unconformably on Wealden Beds. This well appears to lie on a subsidiary downthrown block immediately north of the main Purbeck Disturbance (Underhill & Paterson 1998). However, at Swanage, onshore immediately to the southwest of this well and on the other side of the Purbeck Disturbance, in excess of 80 m of Lower Greensand are present, compared with 246 m at the western end of the Isle of Wight. Ruffell & Batten (1994) document the thinning of the Lower Greensand from Swanage to the west, until it disappears entirely at Lulworth, close to the southern end of Fig. 4. These authors also describe facies changes which indicate that this thinning marks the margin of the original depositional basin.

Tertiary inversion

Regional uplift and inversion of the main Jurassic and early Cretaceous basins of southern England took place within the Tertiary (Underhill & Stoneley this volume). There is evidence of contemporaneous fault movements affecting Upper Cretaceous Chalk in Dorset (Drummond 1970, Westhead & Woods 1994) and it is has been postulated that these mark the beginning of inversion movements. Certainly, there was significant erosion of the top of the Chalk in early Tertiary times (Plint 1982). However, seismic evidence over the Isle of Wight High indicates that the deepest erosion of the Chalk took place beneath what is now the deepest part of the Tertiary Basin and the Chalk thickens southwards, towards the main inversion structure (see Fig. 13). Butler & Pullan (1990) and Chadwick

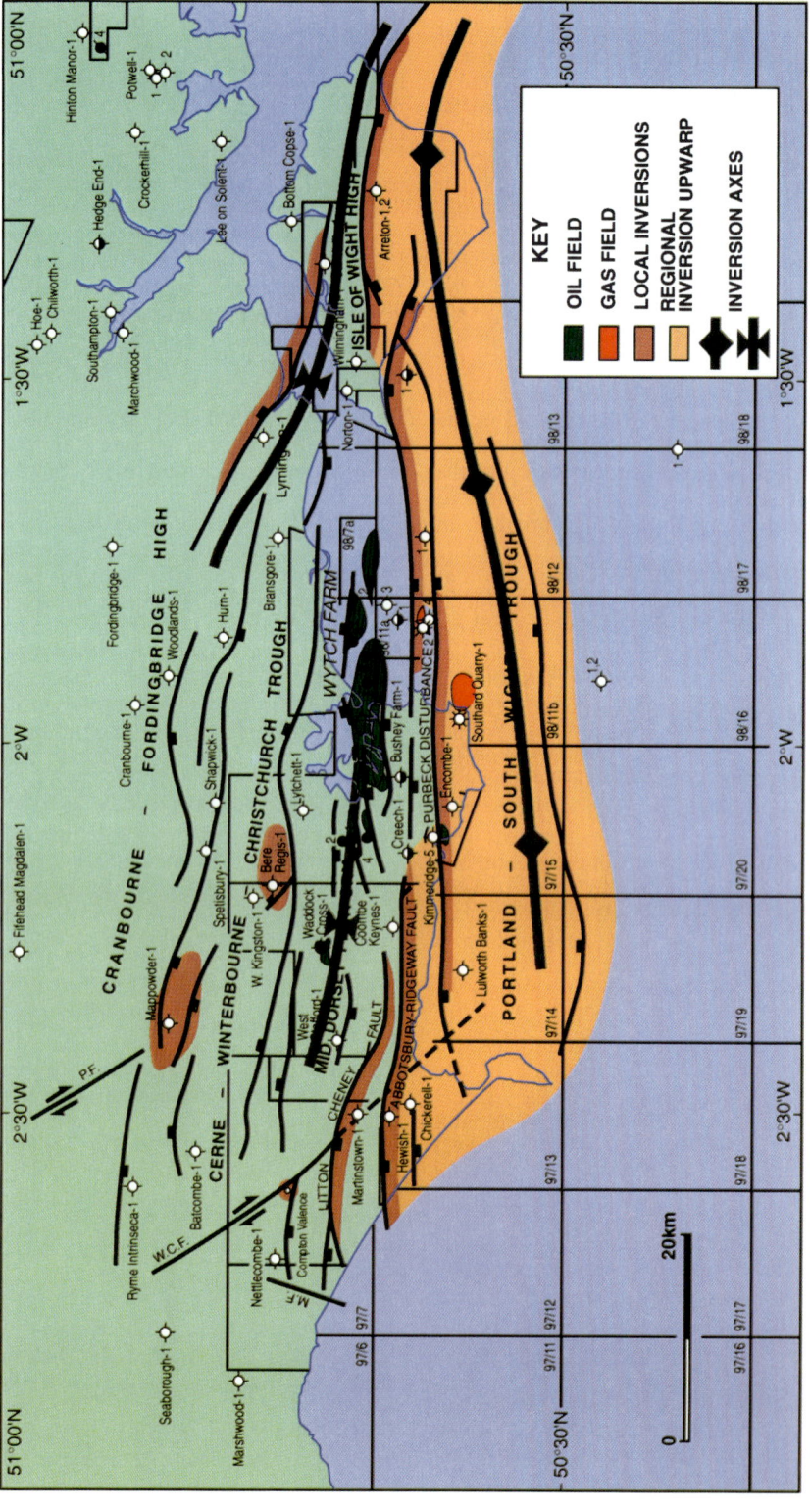

Fig. 12. Map showing Tertiary inversion elements. Faults are those shown in Fig. 1 at base Jurassic level.

(1993) discussed the evidence pointing to the age of the inversion and concluded that the major, intense inversion movements took place during the Miocene. Figure 12 shows the main areas of inversion in the Dorset–Isle of Wight region. The broad regional upwarping of the Portland–South Wight Trough, with associated intense folding and reversal of movements along the northern bounding faults, has been well documented in the past (for example, Stoneley 1982, Chadwick 1993). Less well documented are the gentle inversions noted in two areas on the margins of the Cerne–Winterbourne–Christchurch Trough and the generation of the Porchfield Anticline by inversion of the down-to-the-north fault north of Sandhills-1 and Lymington-1 (illustrated by Chadwick 1993, fig. 12). The availability of modern seismic data enables the identification and dating of structures which lie beneath the Chalk and it is therefore possible to determine more accurately which structures are the result of Tertiary inversion and which pre-date the Upper Cretaceous. However, seismic ray-paths become very convolute in the zones of intense inversion and it is often still difficult to image the beds in these areas (see Fig. 13).

Figure 8 shows, near its northern end, the late Jurassic–early Cretaceous rollover structure to the south of the Litton Cheney Fault and also demonstrates the extent to which the hanging wall has been inverted. This appears to have taken the form of reverse movement up the fault zone, with a reverse displacement of between 50 and 75 m. It is interesting to note that, at depth, the reverse throw appears to have occurred along a different path from the original major normal throw. Unusually, there also appear to have been reverse movements along faults close to the crest of the hanging wall rollover to the south. This may possibly have been the result of halokinesis associated with Tertiary compression. However, this structure lies close to the projected line of the Watchet–Cothelstone Fault (Fig. 12). The main inversion is offset northwards from the Purbeck Disturbance along this lineation to the Abbotsbury Fault, and the reverse faulting here may be the result of transpression. The presence of a strike-slip fault is inferred south of the Litton Cheney Fault because of structural offsets, but it can be mapped on seismic profiles to the north of this fault. It passes through the Compton Valence structure, an almost circular uplift of Chalk and older beds which has been postulated to have been caused by halokinesis (Falcon and Kent 1950, Stoneley 1982) or by movement of mobile clays in the Jurassic, and it is probable that this is a transpressional feature. Chadwick (1993) has discussed the importance of strike-slip faults, offsetting the inversion trends in the Wessex Basin and their probable relationship to underlying Variscan faults. The Watchet–Cothelstone Fault extension, through Compton Valence, is the only major northwest–southeast-trending fault which it has been possible to map on Brabant seismic data, but the sub-parallel Poyntington Fault has been mapped at the surface to the northeast and other zones in which the main inversion transfers between the east–west fault trends may also be related to underlying Variscan faults with this orientation. However, the zones of intense inversion are generally related to zones of greatest Jurassic thickness and offsets which may have been intensified by Tertiary strike-slip movement were already present along the Jurassic and early Cretaceous basin margins. In particular, the offset of the main inversion on the Isle of Wight appears to have taken place along a northwest–southeast line, parallel to the major lineation which is inferred to define the depositional boundary of the Aylesbeare Mudstone, further to the west (Figs 2 and 12). Seismic evidence indicates that this offset controlled deposition during early Cretaceous time and it therefore pre-dates the Tertiary movements. These offsets appear have a dextral sense of movement in Dorset, but Fig. 12 demonstrates that a major offset to the south of the offshore Wytch Farm extension and the offset on the Isle of Wight have a sinistral sense of movement.

A modelling study undertaken in Fig. 8 by Harvey (unpublished report for Brabant Petroleum Limited) concluded that minimum Tertiary shortening across the northern margin of the Portland–South Wight Trough was 900 m. Reverse heave of the Abbotsbury Fault itself was computed at 250 m, with 250 m of reverse throw and 350 m of displacement up the fault plane. As noted above, this assumes that very little Wealden remained in the hanging wall before the Gault was deposited. Figure 4 shows with reasonable clarity the nature of the inversion on the Ridgeway Fault. The secondary listric fault to the north of the Ridgeway Fault is interpreted as an extension of the Litton Cheney Fault: it shows strong growth before the Upper Cretaceous unconformity, but only very minor inversion. The Ridgeway Fault, in contrast, appears to thrust through the base of the Chalk and an antithetic reverse fault has developed to the south, causing a 'pop-up' at Corallian level. The morphology of this feature suggests that there may be an east–west strike-slip component to some of the inversion movements.

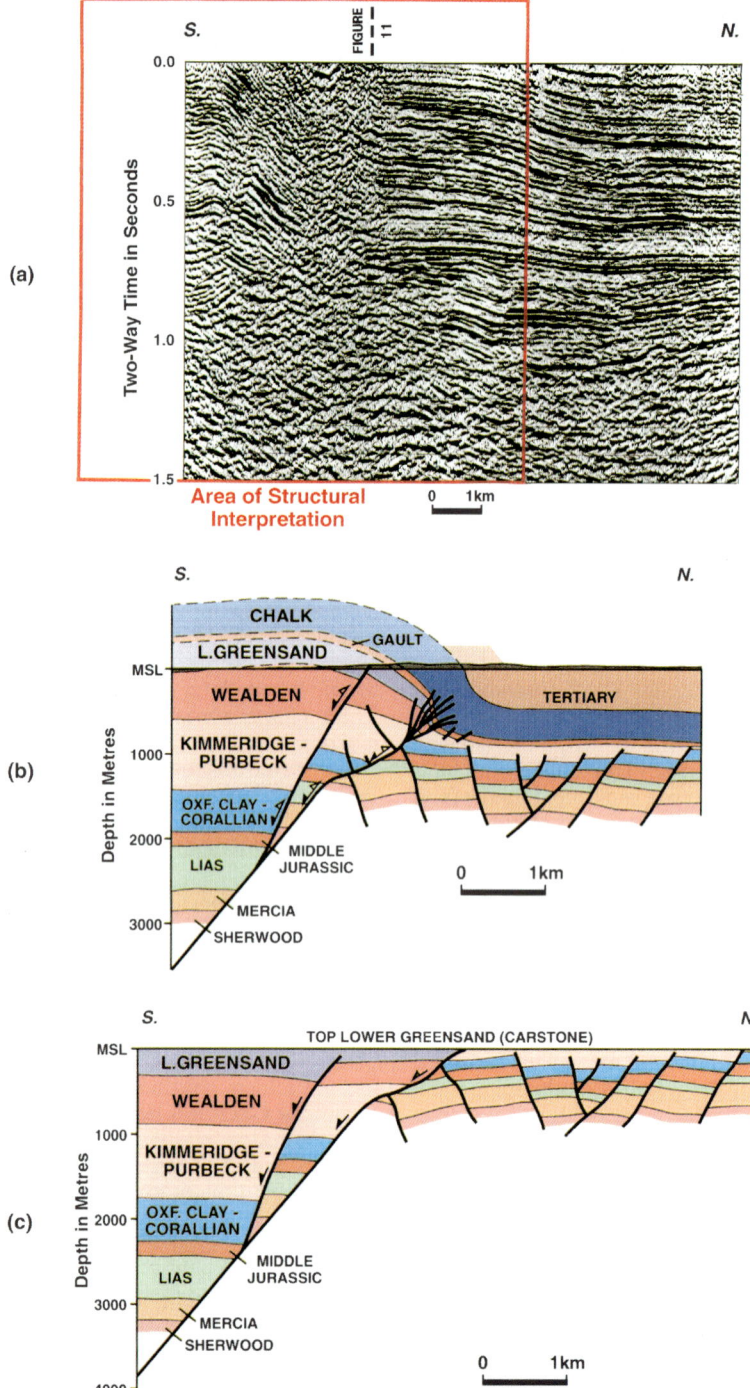

Fig. 13. (**a**) Seismic line Isle of Wight; (**b**) present-day depth interpretation of southern part of line; (**c**) reconstruction to Albian time (not decompacted). For location, see Fig. 1.

An estimate of the minimum reverse movement along the Purbeck Disturbance can be obtained from the results of the Kimmeridge-5 well (Fig. 10). This well encountered a section of grey shale immediately above the Sherwood Sandstone Group. Analysis of this shale indicates that it is unlike any Mercia Mudstone seen in this area and may be of Liassic age. If this is the case, it probably represents a sliver of rock downthrown during normal movement of the major fault and left behind when movement was reversed along a slightly different path. Assuming this to be the lowermost Lias would indicate minimum heave of some 380 m across the Purbeck Disturbance, with a minimum 380 m of reverse throw and 480 m of reverse displacement along the fault. A rather simplistic, non-decompacted reconstruction of Fig. 13, on the Isle of Wight, demonstrates estimated uplift of 1050 m on the south side of the Purbeck–Isle of Wight Disturbance, with reverse heave of some 800 m across the fault zone itself and combined reverse displacement along the two fault planes of some 1320 m.

The work of Sibson (1995) has indicated that the reversal of movement along listric faults with the range of steep dips shown by the seismic lines illustrated here would encounter significant frictional resistance. Without pulsing large volumes of fluids up the fault plane during compression, footwall cutoff thrusts would be expected to develop. It is interesting that these have rarely been interpreted on any of the recent seismic data in Dorset or the Isle of Wight. On the other hand, the absence of reported seeps on the Isle of Wight is at variance with any contention that large volumes of fluid moved up the Purbeck–Isle of Wight Disturbance during mid-Tertiary inversion, particularly as the Lias of Arreton-2 appears to have good source rock potential and is clearly mature. It may be that some of the dip-slip frictional resistance has been overcome by a strike-slip component to the inversion.

The structure of the Isle of Wight Monocline is particularly difficult to image (Fig. 13). It is very tempting to draw a major reverse fault through the seismic data on dip lines, but the outcrop information suggests that there is no such fault. It appears that the reverse movement along the pre-Upper Cretaceous fault is taken up by folding of the Gault, Chalk and lower Tertiary, but it is likely that some of the movement is accommodated along a splay of sub-horizontal faults through the base of the fold. Such faults have been described by Phillips (1964) in coastal outcrops of Chalk in south Dorset.

Discussion

Much of the structural and stratigraphic information described above has resulted from the interpretation of seismic data recorded in the area since 1990. Unfortunately, recent seismic coverage remains sparse over much of the area and more data are needed, particularly to confirm the distribution of the Triassic sequences and the details of minor fault trends. It is possible that minor breaks at base Gault level on the Mid Dorset Platform have been assumed to be static problems when processing the data and have therefore been edited out. Further work is necessary to determine the regional significance of faults with northeast–southwest trends in this area.

Age dating remains a major concern when considering the nature of the intra-Sherwood unconformity and the apparent transition from Otter Sandstone to Mercia Mudstone facies towards the east. Fossil evidence is unlikely to be sufficient to assist in this, but detailed work on chemostratigraphy may offer a solution.

The pre-Gault tilting and erosion in the west (Underhill & Stoneley this volume) poses problems in determining the maturation histories of potential oil source rocks and it is important to be able to reconstruct the original depositional thicknesses of Upper Jurassic and Lower Cretaceous intervals. Stoneley (1982) refers to an observation of P. E. Kent that abundant Kimmeridgian-age ammonites occur within the Aptian. It would be instructive to carry out a detailed study of exotic clasts within the entire Wealden and Lower Greensand sequence of this area to attempt to determine the chronology of erosion.

The extent to which Wessex Basin traps may have been modified, or destroyed, by inversion movements is of major concern to oil explorationists. In particular, the magnitude of the reverse displacement on the major bounding faults and the mechanisms which enabled this to take place are of great interest. Detailed studies of fault planes, their slickenslides and their mineralization would assist in determining the volume of fluid which may have pulsed up the planes during inversion and whether there was an oblique-slip element. In addition, detailed study of micro-fractures and mineralization within Jurassic shales may give some evidence for pore fluid build-up at the time of inversion.

The availability of modern seismic and well data has provided important new information on the geological history of the Wessex Basin.

However, new information raises new questions and much work remains to be done before the structure and stratigraphy of the area is fully understood.

The purpose of this paper is to share some of the information and ideas which the Brabant group has derived from its exploration programme over the past 6 years. This would not be possible without the approval of Brabant Petroleum Limited and its partners in this venture and thanks are due to BP Exploration Operating Company, British Gas Exploration and Production Limited, Canyon Resources Company and Yates Company (UK). The author is grateful to the reviewers, R. Stoneley and M. Harvey, for their helpful comments and suggestions. The assistance of those members of staff of Brabant who have assisted in preparing this presentation is gratefully acknowledged, particularly that of J. Floodpage and J. White. However, the ideas presented here are those of the author and do not necessarily represent those of the companies involved.

References

ARKELL, W. J. 1947. *The Geology of the Country around Weymouth, Swanage, Corfe and Lulworth*. Memoir of the Geological Survey, UK.

BUTLER, M. & PULLAN, C. P. 1990. Tertiary structures and hydrocarbon entrapment in the Weald Basin of southern England. *In*: HARDMAN, R. F. P. & BROOKS, J. (eds) *Tectonic Events Responsible for Britain's Oil and Gas Reserves*. Geological Society, London, Special Publications, **55**, 371–391.

CHADWICK, R. A. 1986. Extension tectonics in the Wessex Basin, southern England. *Journal of the Geological Society, London*, **143**, 465–488.

——1993. Aspects of basin inversion in southern Britain. *Journal of the Geological Society, London*, **150**, 311–322.

DRUMMOND, P. V. O. 1970. The mid-Dorset swell: Evidence of Albian–Cenomanian movements in Wessex. *Proceedings of the Geologists' Association*, **81**, 679–714.

EVANS, D. J. & CHADWICK, R. A. 1994. Basement–cover relationships in the Shaftesbury area of the Wessex Basin, southern England. *Geological Magazine*, **131**, 387–394.

FALCON, N. L. & KENT, P. E. 1950. Chalk rock of Dorset: More evidence of salt? *Geological Magazine*, **87**, 302–303.

HAMBLIN, R. J. O., CROSBY, A., BALSON, P. S., JONES, S. M., CHADWICK, R. A., PENN, I. E. & ARTHUR, M. J. 1992. *United Kingdom Offshore Regional Report: The Geology of the English Channel*. HMSO for the British Geological Survey, London

HARVEY, M. & STEWART, S. A. 1998. Influence of salt on the structural evolution of the Channel Basin. *This volume*.

HAWKES, P. W., FRASER, A. J. & EINCHCOMB, C. C. G. 1998. The tectono-stratigraphic development and exploration history of the Weald and Wessex basins, southern England. *This volume*.

JENKYNS, H. C. & SENIOR, J. R. 1991. Geological evidence for intra-Jurassic faulting in the Wessex Basin and its margins. *Journal of the Geological Society, London*, **148**, 245–260.

PHILLIPS, W. J. 1964. The structures in the Jurassic and Cretaceous rocks on the Dorset coast between White Nothe and Mupe Bay. *Proceedings of the Geologists' Association*, **75**, 373–406.

PLINT, A. G. 1982. Eocene sedimentation and tectonics in the Hampshire Basin. *Journal of the Geological Society, London*, **139**, 249–254.

RUFFELL, A. H. 1992. Early to mid-Cretaceous tectonics and unconformities of the Wessex Basin (southern England). *Journal of the Geological Society, London*, **149**, 443–454.

—— & BATTEN, D. J. 1994. Uppermost Wealden facies and Lower Greensand Group (Lower Cretaceous) in Dorset, southern England: correlation and palaeoenvironment. *Proceedings of the Geologists' Association*, **105**, 53–69.

SIBSON, R. H. 1995. Selective fault reactivation during basin inversion: potential for fluid redistribution through fault-valve action. *In*: BUCHANAN, J. G. & BUCHANAN, P. G. (eds) *Basin Inversion*. Geological Society, London, Special Publications, **88**, 3–19.

SMITH, C. & HATTON, I. R. 1997. Inversion tectonics in the Lyme Bay–West Dorset area of the Wessex Basin, UK. *This volume*.

SMITH, S. A. & EDWARDS, R. A. 1991. Regional sedimentological variations in Lower Triassic fluvial conglomerates (Budleigh Salterton Pebble Beds), southwest England: some implications for palaeogeography and basin evolution. *Geological Journal*, **26**, 65–83.

STONELEY, R. 1982. The structural development of the Wessex Basin. *Journal of the Geological Society, London*, **139**, 545–552.

UNDERHILL, J. R. & PATERSON, S. 1998. Genesis of tectonic inversion structures: seismic evidence for the development of key structures along the Purbeck–Isle of Wight Monocline. *Journal of the Geological Society, London*, in press.

—— & STONELEY, R. 1998. Introduction to the development, evolution and petroleum geology of the Wessex Basin. *This volume*.

WARRINGTON, G., AUDLEY-CHARLES, M. G., ELLIOT, R. E., EVANS, W. B., IVIMEY-COOK, H. C., KENT, P. E., ROBINSON, P. L, SHOTTON, F. W. & TAYLOR, F. M. 1980. *A Correlation of Triassic Rocks in the British Isles*. Special Report of the Geological Society, London No. 13.

WESTHEAD, R. K. & WOODS, M. A. 1994. Anomalous Turonian–Campanian Chalk deposition in south Dorset; the influence of inherited pre-Albian structures. *Proceedings of the Geologists' Association*, **105**, 81–89.

WHITTAKER, A. 1972. The Watchet Fault – a post-Liassic transcurrent reverse fault. *Bulletin of the Geological Survey of Great Britain*, **41**, 75–80.

——. (ed.) 1985. *Atlas of Onshore Sedimentary Basins in England and Wales: Post-Carboniferous Tectonics and Stratigraphy*. Blackie, Glasgow.

A proposed latest Triassic to earliest Cretaceous microfossil biozonation for the English Channel and its adjacent areas

NIGEL R. AINSWORTH,[1] WILLIAM BRAHAM,[2] F. JOHN GREGORY,[3] BEN JOHNSON[4] & CHRISTOPHER KING[5]

[1] *39 De Tany Court, St. Albans, Hertfordshire AL1 1TU, UK*
(e-mail: Nigel_R_Ainsworth@compuserve.com)
[2] *11 Corner Hall, Hemel Hempstead, Hertfordshire HP3 9HN, UK*
[3] *32 Inkerman Road, St. Albans, Hertfordshire AL1 3BB, UK*
[4] *StratLab Ltd, 299 North Deeside Road, Cults, Aberdeen AB15 9PA, UK*
[5] *41 Montem Road, Newmalden, Surrey KT3 3QU, UK*

Abstract: Three detailed biostratigraphic zonations, based on ostracods, foraminiferids and dinocysts, are proposed for the latest Triassic through to the earliest Cretaceous of the Portland–Wight Basin and its adjacent areas. A total of 112 zones–zonules (34 ostracod zones/subzones, 25 foraminiferid zones, 53 dinoflagellate zones, subzones and zonules) are recognized. All three zonation schemes are based on samples obtained from both outcrop (the Dorset coast) and boreholes, the latter from ditch cuttings, with some conventional and sidewall core samples, in association with published data, and have where possible been calibrated to the standard ammonite zones recognized throughout northwest Europe.

The three biostratigraphic zonation schemes (ostracods, foraminiferids, dinocysts/miospores) illustrated in Fig. 1 and discussed below are based on samples obtained both at outcrop (the Dorset coast) and the subcrop, in association with numerous microfossil publications. All three zonations have where possible been calibrated to the standard ammonite zones recognized throughout northwest Europe. The authors have used taxa which are, as far as possible, common and geographically widespread. The three schemes described below are intended to be applied to borehole sections, which are generally represented by ditch cuttings samples, in association with rarer sidewall and conventional cores, and therefore emphasize the highest downhole occurrences (extinctions) of particular taxa and major assemblage changes. In sections where there may be good conventional core and/or sidewall core coverage, inceptions (lowest downhole occurrences) and/or acmes can be determined.

Ostracod zonal scheme

The zonation of the marine Hettangian to Portlandian (OJ zones) has in part been extracted from a number of sources, including Whatley (1965), Field (1968), Kilenyi (1969, 1978), Christensen & Kilenyi (1970), Morris (1983), Bate (1978), Bate & Sheppard (1982), Sheppard (1982), Wilkinson (1982), Morris (1983), Fuller (1984), Lord & Bown (1987), Partington *et al.* (1993), and unpublished outcrop and subcrop data. The ostracod zonation of the non-marine Purbeck (PW zones) is calibrated to that described by Anderson (1971, 1985) and Kilenyi & Neale (1978) from the Weald and its adjacent areas.

Zone OJ1

Top: defined on common/abundant specimens of *Ogmoconchella aspinata*.

Assemblage: characterized by a generally low-diversity ostracod fauna, which is dominated by the above taxon.

Notable occurrences include *Pseudomacrocypris subtriangularis* and *Cardobairdia* sp. B *sensu* Ainsworth.

Age: latest Rhaetian–Early Sinemurian (pre *planorbis–bucklandi* ammonite zones). Occurring within the Blue Lias.

Zone OJ2

Top: defined on the highest occurrence of *Ogmoconchella aspinata*.

Assemblage: a sparse ostracod fauna is present, dominated by the above index taxon. Notable occurrences include *Bairdia* sp. 1 *sensu* Ainsworth, *Cytherelloidea circumscripta*, *C. pulchella*, *Ogmoconcha hagenowi*, *Ogmoconchella nasuta* and *O. telata*.

Age: Early Sinemurian (*semicostatum–turneri* ammonite zones). Occurring within the topmost Blue Lias and Shales with Beef.

Zone OJ3

Top: defined on the highest occurrence of *Klinglerella* gr. *triebeli*

Assemblage: a generally sparse assemblage dominated by well-ornamented (ribbed) ostracods. Notable highest occurrences include *Isobythocypris elongata, Klinglerella intermedia, K. variabilis, Ogmoconchella danica, O.* cf. *mouhersensis, Paratrachycythere pseudotubulosa sensu* Park Ms., *Patellacythere gruendeli* and *Progonoidea reticulata*

Age: latest Early–Late Sinemurian (uppermost *turneri–raricostatum* ammonite zones). Occurring within the Black Ven Marls.

Zone OJ4

Top: defined on the highest occurrence of *Pleurifera vermiculata*.

Assemblage: dominated by *Ogmoconchella* spp.

Age: Early Pliensbachian (*jamesoni*–lower *ibex* ammonite zones). Occurring within the Belemnite Marls.

Zone OJ5

Top: defined on the highest occurrence of *Gammacythere klingleri*.

Assemblage: dominated by *Ogmoconcha* and *Ogmoconchella* spp.

Age: Early Pliensbachian (upper *ibex* ammonite zone). Occurring within the Belemnite Marls.

Zone OJ6

Top: defined on the highest occurrence of *Gammacythere ubiquita*.

Assemblage: dominated by *Ogmoconchella* spp. Notable highest appearances include *Gramannicythere bachi, Ogmoconcha amalthei* form A, *Paradoxostoma pusillum, Pleurifera harpa* and *Polycope* spp. Similar to Zone OJ5, the more abundant assemblages occur in the onshore sections.

Age: Early Pliensbachian (*davoei* ammonite zone). Occurring within the Green Ammonite Beds.

Zone OJ7

Top: defined on the highest occurrence of *Wicherella semiora semiora*.

Assemblage: dominated by *Ogmoconchella* spp. Notable highest occurrences include *Liasina vestibulifera* and *Ogmoconchella transversa*. The more abundant assemblages occur in the onshore sections.

Age: earliest Late Pliensbachian (lower *margaritatus* ammonite zone). Occurring within the Eype Clay and lateral equivalents.

Zone OJ8

Age: defined on the highest occurrence of *Ogmoconchella* spp.

Assemblage: dominated by metacopid ostracods. Ostracods are often rare and/or poorly preserved due to the often silty or calcareous nature of the lithologies. Notable highest occurrences include *Cytherelloidea anningi, Ektyphocythere quadrata, Grammannella apostolescui, Nanacythere simplex, Ledahia septenaria, Ogmoconcha amalthei, Ogmoconchella aequalis, O. bispinosa, O. gruendeli, O. pseudospinosa* and *Pseudohealdia truncata*.

Age: Late Pliensbachian (upper *margaritatus*–*spinatum* ammonite zones). Occurring within the Middle Lias Clays, the argillaceous sequences within the Thorncombe Sands and the Marlstone Rock Bed.

Zone OJ9

Top: defined on the highest occurrence of *Cytherella toarcensis* and/or *Cytherelloidea cadomensis*.

Assemblage: ostracods are rare to moderately abundant within the Down Cliff Clay. In the overlying Bridport Sands (equivalent) a marked decrease in abundance is observed, with few forms recorded. Notable new occurrence include *Acrocythere herrigi sensu* Park Ms., *Bairdia ohmerti, Cytheropteron alafastigatum, Ektyphocythere bizoni, E. intrepida, Kinkelinella costata, Lophodentina cultrata, Otocythere callosa, Praeschuleridea pseudokinkelinella* and *P. ventriosa*.

Age: Early–Late Toarcian (*falciferum–levesquei* ammonite zones). Occurring within the Junction Bed, the Down Cliff Clay and the Bridport Sands.

Comment: at present, no ostracod faunas have been recorded from the distinctive 'Paper Shales' unit.

Zone OJ10

Top: defined on the highest occurrence of *Kinkelinella sermoisensis* and/or *Praeschuleridea* gr. *ventriosa*.

89

ngusta and *Praes-*

Bathonian (upper
zones). Occurring
er Fuller's Earth
he Upper Fuller's
Beds of the Frome

est occurrence of
nsis.
by a diverse and
ia. Notable highest
latourella bullata,
ncentrica, Nophre-
lea quadrata, Rec-
lea (*Eoschuleridea*)
utiplicata. Notable
this zone include
and *Glyptocythere*

s ammonite zone).
e Clay, the Great
ble (Boueti Bed).

est occurrence of
a.
by a moderately
od fauna. Notable
e *Fastigatocythere*
blakeana, Glabella-
e oscillum, G. penni,
pinqua, Micropneu-
ocytheris fullonica,
ogonocythere stilla,
trigonalis and *Ter-*

us ammonite zone).
st Marble and the
otsbury Cornbrash

occurrence of *Fasti-*
A *sensu* Wilkinson

d by sparse and

per *macrocephalus–*
e zones). Occurring
ition.

Zone OJ18

Top: defined on the highest occurrence of *Terquemula flexicosta lutzei.*

Assemblage: a moderately common and diverse ostracod fauna occurs throughout the interval. Notable new occurrences include *Lophocythere caesa caesa, L. caesa* subsp. A *sensu* Wilkinson, *L. interupta interupta, Praeschuleridea batei, P. caudata* and *Procytheropteron* sp. A *sensu* Wilkinson.

Age: late Middle–Late Callovian (*coronatum–lamberti* ammonite zones). Occurring within the Peterborough and Stewartby Members of the Oxford Clay Formation.

Zone OJ19

Top: defined on the highest occurrence of *Lophocythere scabra bucki* and/or the lowest occurrence of common *Schuleridea triebeli.*

Assemblage: a moderately poor ostracod fauna is present within this interval. Notable new appearances include *Fuhrbergiella horrida horrida, Glabellacythere dolbra, Nophrecythere cruciata intermedia, Schuleridea parva* Fuller Ms. and *Terquemula redcliffensis* Fuller Ms.

Age: Early Oxfordian (*mariae–cordatum* ammonite zones). Occurring within the Weymouth Member of the Oxford Clay Formation and the Nothe Grit.

Zone OJ20

Top: defined on the highest occurrence of *Macrodentina (M.) whatleyi.*

Assemblage: a very diverse and abundant fauna occurs throughout the interval, dominated by *Galliaecytheridea* spp. and *Schuleridea triebeli.* Newly occurring taxa include *Cytheropteron* sp. A *sensu* Wilkinson, *Hekistocythere osmingtonensis* Fuller Ms., *Hutsonia* sp. C Glashoff, *Macrodentina (M.) tenuistriata, Procytherura tenuicostata, Progonocythere multipunctata* and *Pseudoperissocytheridea parahieroglyphica.*

Age: Middle Oxfordian (*densiplicatum–tenuiserratum* ammonite zones). Occurring within the Redcliff and Osmington Oolite Formations.

Zone OJ21

Top: defined on the highest occurrence of *Vernoniella sequana.*

Assemblage: dominated by *Galliaecytheridea* spp., many of which extend up into the overlying Early Kimmeridgian. Other notable highest occurrences include *Galliaecytheridea posterospinosa* Fuller Ms., *G. ventrocauda* Fuller Ms., *Lophocythere cruciata oxfordiana, Pleurocythere oxfordiana, Procytherura tenuicostata, Pseudohutsonia tuberosa, Terquemula multicostata* and *Vernoniella caletorum.*

Age: Late Oxfordian (*glosense–rosenkrantzi* ammonite zones). Occurring within the Trigonia clavellata, Sandsfoot and Ringstead Formations, and the Ampthill Clay Equivalent.

Zone OJ22

Top: defined on the highest occurrence of *Macrodentina (M.) cicatricosa.*

Assemblage: a diverse and common ostracod fauna occurs throughout this interval. Other notable highest occurrences at this level include *Amphicythere confundens, A. pennyi, A. sphaericulata, Eocytheropteron decoratum, Galliaecytheridea fragilis, G. postrotunda, Mandelstamia (M.) angulata* and common/abundant *Schuleridea triebeli.*

Age: Early Kimmeridgian (*baylei–cymodoce* ammonite zones). Occurring within the Lower Kimmeridge Clay.

Zone OJ23

Top: defined on the highest occurrence of *Galliaecytheridea dissimilis.*

Assemblage: comprising a common, diverse ostracod fauna. Many of the taxa from the overlying interval extend into this zone.

Age: Early Kimmeridgian (lower *mutabilis* ammonite zone). Occurring within the Lower Kimmeridge Clay.

Zone OJ24

Top: defined on the downhole reappearance of *Galliaecytheridea* spp., including the highest occurrence of *G. elongata.*

Assemblage: comprising a moderately common and diverse ostracod fauna. Other notable highest occurrences include *Dicrorygma (Orthorygma) reticulata, Galliaecytheridea dorsetensis, Macrodentina (Polydentina) proclivis, Mandelstamia (M.) rectilinea,* rare *Nodophthalmocythere tripartita* and *Schuleridea triebeli*

Age: Early Kimmeridgian (upper *mutabilis*–middle *eudoxus* ammonite zones). Occurring within the Lower Kimmeridge Clay.

Zone OJ25

Top: defined on the highest occurrence of *Mandelstamia maculata*.

Assemblage: characterized only by this species. Below the last occurrence of this taxon, a barren interval is recognized, corresponding to the lower *wheatleyensis* to upper *eudoxus* ammonite zones. This barren interval is recognizable throughout the Dorset region (Christensen & Kilenyi 1970; Kilenyi 1978).

Age: Late Kimmeridgian (middle *wheatleyensis* ammonite zone). Occurring within the Upper Kimmeridge Clay.

Zone OJ26

Top: defined on the highest occurrence of *Schuleridea moderata*.

Assemblage: characterized by a sparse ostracod fauna. *Galliaecytheridea spinosa* has its lowest occurrence in the upper part of this zone. Another notable event includes the highest occurrence of *Mandelstamia tumida*.

Age: Late Kimmeridgian (middle *huddlestoni–pectinatus* ammonite zones). Occurring within the Upper Kimmeridge Clay.

Zone OJ27

Top: defined on the highest occurrence of *Galliaecytheridea spinosa*.

Assemblage: this taxon dominates the poorly diverse ostracod fauna. It occurs in association with the highest occurrences of *Aaleniella (Danocythere) inornata* and *Dicrorygma (Orthorygma) brotzeni*.

Age: latest Kimmeridgian (uppermost *pectinatus–rotunda* ammonite zones). Occurring within the Upper Kimmeridge Clay.

Zone OJ28

Top: defined on the highest occurrence of *Galliaecytheridea compressa*.

Assemblage: the index taxon dominates this poorly diverse ostracod zone. Other characteristic taxa include the closely related species *Galliaecytheridea polita*.

Age: latest Kimmeridgian–earliest Portlandian (*fittoni–albani* ammonite zones). Occurring within the uppermost Kimmeridge Clay and the lowermost part of the Portland Sand Formation.

Zone OJ29

Top: defined on the highest occurrence of *Macrodentina reticulata*.

Assemblage: characterized by the appearance of a sparse marine ostracod fauna. Other characteristic forms at this level include *Macrodentina transiens* and *Protocythere serpentina*.

Age: Early Portlandian (*glaucolithus–anguiformis* ammonite zones). Occurring within the Portland Sand and Portland Limestone Formations.

PW zones

Zone PW1

Top: defined on the highest occurrence of *Fabanella boloniensis*.

Assemblage: this taxon is generally a long-ranging (Portlandian–Late Barremian) form. However, within the Portland–Wight Basin it is a common taxon within the Lower Purbeck of Late Portlandian age. Although rare, other new highest occurrences include *Cypridea dunkeri carinata* and *Mantelliana purbeckensis*. Both of these species are more frequently recorded in the onshore boreholes.

Age: Late Portlandian. Occurring with the Lower Purbeck Beds.

Zone PW2

Top: defined on the highest occurrence of *Cypridea granulosa granulosa*.

Assemblage: ostracods faunas are similar to the overlying interval. Other notable highest occurrences, however, include *Cypridea dunkeri dunkeri, C. posticalis, C. tumenscens acrobeles, C. tumenscens peltoides, Klieana alata*, and common *Bisulocypris dilatata, Damonella ellipsoidea, D. pygmaea* and *Theriosynoeum forbesii*.

Age: latest Portlandian. Occurring within the lower Middle Purbeck Beds below the Cinder Beds.

Zone PW3

Top: defined on the highest occurrence of *Cypridea vidrana*. Two subzones can be recognized.

Subzone PW3a. Top: defined on the highest occurrence of *Cypridea granulosa fasciculata*.

Assemblage: characterized by an abundant and diverse ostracod fauna. Other notable highest appearances include *Cypridea coelnothi, C. primaeva, C. sagena, C. tumenscens tumescens* and *Scabriculocypris trapezoides*.

Age: Berriasian. Occurring within the Middle Purbeck Beds, above the Cinder Bed.

Subzone PW3b. Top: defined on the highest occurrence of *Cypridea vidrana*.

Assemblage: characterized by an abundant and moderately diverse ostracod fauna. Other notable new highest occurrences include *Bisulocypris forbesii*, *B. verrucosa*, *Cypridea altissima*, *C. amisia* and *C. tuberculata langtonensis*.

Age: Berriasian. Occurring within the topmost Middle Purbeck Beds above the Cinder Bed.

Zone PW4

Top: defined on the highest occurrence of *Cypridea setina*.

Assemblage: characterized by an abundant and moderately diverse ostracod fauna. Other notable new highest occurrences include *Cypridea alata*, *C. bimammata*, *C. brevirostrata*, *C. dolabrata*, *C. setina rectidorsata*, *C. setina setina*, *C. wicheri* and *C. wolburgi*.

Age: Berriasian–Early Valanginian. Occurring only in the Upper Purbeck Beds.

Foraminiferid zonal scheme

The foraminiferid zonation (FJ zones) is taken from a number of sources, including Barnard (1950, 1953), Cifelli (1959), Coleman (1979, 1980, 1982), Barnard *et al.* (1981), Medd (1982), Morris (1983), Copestake & Johnson (1984, 1989), Lord & Bown (1987), Copestake (1989), Morris & Coleman (1989), Shipp (1989), Partington *et al.* (1993) and unpublished outcrop and subcrop data.

Zone FJ1

Top: defined on a poorly diverse *Eoguttulina liassica* dominated assemblage.

Age: latest Rhaetian. Occurring within the Langport Member.

Zone FJ2

Top: defined on the highest occurrence of *Dentalina langi* and/or *Frondicularia terquemi* subsp. A *sensu* Barnard.

Assemblage: dominated by nodosariids and *Eoguttulina liassica*. *Frondicularia brizaeformis* is common throughout this interval. *Lingulina tenera collenoti* and *L. tenera subprismatica* also have highest occurrences in this zone.

Age: Hettangian (*planorbis–angulata* ammonite zones). Occurring in the Blue Lias.

Zone FJ3

Top: defined on the downhole reappearance of common/consistent *Involutina liassica*.

Assemblage: dominated by nodosariids.

Age: Early Sinemurian (*bucklandi* ammonite zone). Occurring within the Blue Lias.

Zone FJ4

Top: defined on the highest occurrence of *Reinholdella margarita*

Assemblage: dominated by nodosariids and *Reinholdella margarita*. Other notable occurrences within this interval include *Astacolus semireticulata*, *Frondicularia brizaeformis*, and influxes of both *Lingulina tenera tenera* and *Neobulimina* sp. 2 *sensu* Bang.

Age: Early Sinemurian (*semicostatum*–lower *turneri* ammonite zones). Occurring within the topmost Blue Lias and Shales with Beef.

Zone FJ5

Top: defined on the highest occurrence of *Planularia inaequistriata*.

Assemblage: dominated by nodosariids. This zone is often not sharply defined due to the rare occurrence of this taxon.

Age: latest Early Sinemurian (upper *turneri* ammonite zone). Occurring within the Shales with Beef.

Zone FJ6

Top: defined on the highest occurrence of *Marginulinopsis quadricostata*.

Assemblage: dominated by nodosariids. Other notable new occurrences include *Annulina metensis* and *Nodosaria issleri*. The *Lenticulina muensteri* plexus is often very abundant within this interval

Age: latest Early–Late Sinemurian (uppermost *turneri–raricostatum* ammonite zones). Occurring within the Black Ven Marls.

Zone FJ7

Top: defined on an influx of *Involutina liassica*.

Assemblage: characterized by large numbers of long-ranging nodosariids (*Dentalina* spp.,

Lenticulina spp., *Lingulina* spp.). *Brizalina liassica* has its highest occurrence within this zone.

Age: Early Pliensbachian (*jamesoni–davoei* ammonite zones). Occurring within the Belemnite Marls and Green Ammonite Beds.

Zone FJ8

Top: defined on the highest occurrence of *Dentalina matutina*.

Assemblage: throughout the study area, this interval generally yields a sparse foraminiferid fauna due to the often sandy–silty lithologies of the Middle Lias. Other notable highest occurrences include *Dentalina terquemi*, *Frondicularia terquemi bicostata*, *Frondicularia terquemi sulcata* forms C, E, F, G, *Frondicularia terquemi terquemi*, *Lingulina tenera pupa*, *Lingulina tenera tenuistriata*, *Marginulina prima spinata*, *Marginulinopsis speciosa* and *Saracenaria sublaevis*. A notable marker is the occurrence of common specimens of *Lenticulina acutiangulata* within the *subnodosus* subzone of the *margaritatus* ammonite zone.

Age: early Late Pliensbachian (*margaritatus* ammonite zone). Occurring within the Middle Lias.

Zone FJ9

Top: defined on the lowest occurrences of *Nodosaria regularis robusta* and *Lenticulina dorbignyi*.

Assemblage: this zone is an interval zone between the lowest occurrences of *Nodosaria regularis robusta* and *Lenticulina dorbignyi* and the highest occurrence of *Dentalina matutina*. Benthic foraminiferids are generally very rare, dominated by long-ranging *Lenticulina* spp. and *Lingulina tenera tenera*, due to the interval being represented by the condensed limestone facies of the Junction Bed.

Age: Late? Pliensbachian–earliest Late Toarcian (*spinatum?–variabilis* ammonite zones). Occurring within the Marlstone Rock Bed and the Junction Bed.

Zone FJ10

Top: defined on the highest occurrence of *Nodosaria regularis robusta*.

Assemblage: characterized by moderately common *Lenticulina* spp. Faunas occurring within the Bridport Sands and the Lower Inferior Oolite are often sparse and very poorly preserved. *Lenticulina dorbignyi* and *Saracenella mochrasensis* are generally rare, and therefore are not used as zonal markers. *Vaginulina listi* has its highest occurrence within lower part of this zone. Its highest appearance in this region is envisaged to be Late Toarcian, however, it has not been used as an index taxon due to its limited geographical occurrence in the present study.

Age: Late Toarcian–earliest Bajocian (*thouarsense–discites* ammonite zones). Occurring within the upper part of the Junction Bed, the Down Cliff Clay, the Bridport Sands and the Lower Inferior Oolite.

Zone FJ11

Top: defined on the lowest occurrence of *Planularia eugenii*.

Assemblage: this zone is an interval zone between the lowest occurrence of *Planularia eugenii* and the highest occurrence of *Nodosaria regularis robusta*. The faunas are dominated by long-ranging benthic species.

Age: Early–Late Bajocian (*laeviuscula*–lower *parkinsoni* ammonite zones). Occurring within the topmost Lower to Upper Inferior Oolite.

Zone FJ12

Top: defined on the highest occurrence of *Planularia eugenii*.

Assemblage: characterized by very rich and diverse foraminiferid faunas, comprising epistominids and nodosariids. Comparable to Coleman's Faunule A.

Age: latest Bajocian–earliest Bathonian (upper *parkinsoni–zigzag* ammonite zones). Occurring within the topmost Upper Inferior Oolite and the lowermost Lower Fuller's Earth Clay.

Zone FJ13

Top: defined on the occurrence of dominantly agglutinating foraminiferid assemblages.

Assemblage: dominated by low-diversity foraminiferid faunas, which occasionally include rich horizons of agglutinating foraminiferids (*Trochammina globigeriniformis*, *Verneuilinoides* sp.). Comparable to Coleman's Faunule B1.

Age: late Early–early Middle Bathonian (*tenuiplicatus–progracilis* ammonite zones). Occurring within the Lower Fuller's Earth Clay.

Zone FJ14

Top: defined on the highest occurrence of *Lenticulina tricarinella* and/or *Lenticulina dictyodes*.

Assemblage: characterized by a diverse foraminiferid fauna, dominated by the nodosariids

(*Lenticulina* spp. and *Nodosaria* spp.), in association with often common *Ammobaculites coprolithiformis*. Comparable to Coleman's Faunules B2 and C3. Those wells drilled in the Portland–Wight Basin generally yield poorly diverse faunas which are similar to the overlying zone (Zone FJ15).

Age: Middle–earliest Late Bathonian (*subcontractus–hodsoni* ammonite zones). Occurring within the Fuller's Earth Rock, the Upper Fuller's Earth Clay, the Wattonensis Beds of the Frome Clay and Great Oolite.

Zone FJ15

Top: defined on the highest occurrence of *Epistomina stelligera* and *E. regularis*.

Assemblage: dominated by epistominids, within a moderately diverse foraminiferid assemblage. However, those wells drilled in the Portland–Wight Basin generally yield sparser faunas. Notable occurrences also include *Reinholdella crebra* and *Trocholina nodulosa*. Comparable to Coleman's Faunule C4.

Age: Late Bathonian (*orbis* ammonite zone). Occurring within the Frome Clay, the Great Oolite and the basal Forest Marble.

Zone FJ16

Top: defined on the highest occurrence of *Massilina dorsetensis* and/or *Tetrataxis* sp. *sensu* Coleman.

Assemblage: dominated by nodosariids. Highest occurrences include *Cornuspira liasina*, *Nodosaria pectinata* and *Vaginulina legumen*. Comparable to Coleman's Faunule D.

Age: latest Bathonian (*discus* ammonite zone). Occurring within the Forest Marble and the Berry Member of the Abbotsbury Cornbrash Formation.

Zone FJ17

Top: defined on the occurrence of a predominantly agglutinating foraminiferid assemblage.

Assemblage: dominated by agglutinating foraminiferids, including *Ammobaculites agglutinans*, *Trochammina globigeriniformis* and *Verneuilinoides tryphera*. Calcareous foraminiferids are rare, but include *Eoguttulina liassica*, *Lenticulina muensteri* and *L. varians*.

Age: Early Callovian (uppermost *macrocephalus*–middle *calloviense* ammonite zones). Occurring within the Kellaways Formation.

Zone FJ18

Top: defined on the highest occurrence of *Reinholdella lutzei*.

Assemblage: dominated by nodosariids, in association with common epistominids and *Ammobaculites coprolithiformis*. Other notable taxa include the highest occurrence of *Epistomina stellicostata*.

Age: latest Early–earliest Late Callovian (upper *calloviense*–middle *athleta* ammonite zones). Occurring within the Peterborough and lowermost Stewartby Members of the Oxford Clay Formation.

Zone FJ19

Top: defined on the highest occurrence of *Lenticulina quenstedti* and/or *Epistomina mosquensis*.

Assemblage: dominated by *Ammobaculites coprolithiformis* and *Lenticulina* spp. Epistominids are also often common. Notable highest occurrences include *Miliospirella lithuanica*, *Ophthalmidium compressa*, *O. strumosum*, *Tristix triangularis*.

Age: Early Oxfordian (*mariae–cordatum* ammonite zones). Occurring in the Weymouth Member of the Oxford Clay Formation and Nothe Grit Formation.

Zone FJ20

Top: defined on the highest occurrence of *Ammobaculites coprolithiformis*.

Assemblage: often dominated by agglutinating taxa, notably *Ammobaculites coprolithiformis*, in association with the calcareous benthic forms *Lenticulina muensteri* and *L. subalata*. Notable new occurrences include *Nubeculinella bigoti*, *Recurvoides obskiensis*, *Quinqueloculina horelli*, *Spirillina tenuissima*, *Vaginulina anomala sensu* Gordon and *V. pasquetae*.

Age: Middle–Late Oxfordian (*densiplicatum–rosenkrantzi* ammonite zones). Occurring within the Redcliff, Osmington Oolite, Trigonia clavellata, Sandsfoot, Ringstead Formations and the Ampthill Clay Equivalent.

Zone FJ21

Top: defined on the highest occurrence of *Haplophragmoides canui*.

Assemblage: dominated by long-ranging nodosariids and epistominids.

Age: earliest Kimmeridgian (*baylei* ammonite zone). Occurring within the basal Lower Kimmeridge Clay.

Zone FJ22

Top: defined on the highest occurrence of *Ammobaculites subaequalis*.

Assemblage: diverse and often abundant foraminiferid assemblages occur throughout this interval. Notable highest occurrences include *Ammobaculites subaequalis*, *Epistomina ornata*, *E. parastelligera* and *Valvulina meentzeni*.

Age: Early Kimmeridgian (*cymodoce–eudoxus* ammonite zones). Occurring within the Lower Kimmeridge Clay.

Zone FJ23

Top: defined on the highest occurrence of abundant *Lenticulina uralica*.

Assemblage: characterized by common calcareous benthic foraminiferids dominated mainly by *Lenticulina* spp. A second influx of *Lenticulina uralica* may also be present, approximating to the *scitulus* ammonite zone. Rare specimens of *Reinholdella pseudorjasanensis* occur in the Portland–Wight Basin and its adjacent areas. Where recovered, it is typical of the *autissiodorensis* to basal *scitulus* ammonite zones.

Age: latest Early–Late Kimmeridgian (*autissiodorensis*–lower *hudlestoni* ammonite zones). Occurring within the topmost Lower and Upper Kimmeridge Clay.

Zone FJ24

Top: defined on the highest occurrence of abundant, often crushed *Haplophragmoides* spp.

Subzone FJ24a. Top: defined on the lowest occurrence of common/abundant *Haplophragmoides* spp.

Age: Late Kimmeridgian (upper *hudlestoni* ammonite zone). Occurring within the Upper Kimmeridge Clay.

Subzone FJ24b. Top: defined on the highest occurrence of abundant *Haplophragmoides* spp.

Assemblage: characterized by an influx of agglutinating foraminiferids. Calcareous benthic taxa are absent.

Age: Late Kimmeridgian (lower *pectinatus* ammonite zone). Occurring within the Upper Kimmeridge Clay.

Zone FJ25

Top: defined on the occurrence of *Lenticulina muensteri* and/or *L. uralica*.

Assemblage: characterized by a moderately sparse foraminiferid assemblage. The calcareous benthic taxa dominate the assemblages. Notable highest occurrences including the agglutinating foraminiferids *Ammobaculites alaskensis* and *Haplophragmoides alaskensis*, and the calcareous benthic taxa *Citharina serratocostata*, *Marginulina costata*, *Pseudonodosaria vulgata* and *Saracenaria oxfordiana*.

Age: Late Kimmeridgian–earliest Portlandian (*pallasioides–albani* ammonite zones). Occurring within the Upper Kimmeridge Clay and the lower part of the Portland Sand Formation.

Dinocyst zonal scheme

The Rhaetian–Portlandian sediments investigated in this study have been systematically organized with reference to the following zonal scheme. In part, this scheme is based on Rawson & Riley (1982), Riley & Fenton (1982), Woolham & Riding (1983), Riley *et al.* (1989), Riding & Thomas (1992), Partington *et al.* (1993), and unpublished data (particularly that of L. A. Riley whilst at Gearhart and Paleo Services). For a review of dinocyst studies on British Mesozoic sections see Powell (1992) for the Triassic, and Riding & Thomas (1992) for the Jurassic.

Zone DJ1

Top: defined by the highest occurrence of *Rhaetogonyaulax rhaetica*.

Age: Rhaetian. Occurring within the Westbury and Lilstock Formations.

Zone DJ2

Top: defined by the lowest occurrence of *Nannoceratopsis gracilis* and *Luehndea spinosa*.

Subzone DJ2a. Top: defined by the highest occurrence of *Dapcodinium priscum*.

Assemblage: A two-fold subdivision is recognized based on variations in abundance of *Dapcodinium priscum*. Two 'zonules' are recognized.

'Zonule' DJ2a1. Top: based on the occurrence of increased numbers of *Dapcodinium priscum*.

Age: Hettangian (*planorbis–angulata* ammonite zones). Occurring within the Blue Lias.

'Zonule' DJ2a2. Top: based on the highest occurrence of *Dapcodinium priscum*.

Age: earliest Sinemurian (*bucklandi* ammonite zone). Occurring within the uppermost Blue Lias.

Subzone DJ2b. Top: defined on the lowest occurrence of *Liasidium variabile*.

Assemblage: this subzone corresponds to the interval immediately below the lowest occurrence of *Liasidium variabile* and immediately above the highest occurrence of *Dapcodinium priscum*.

Age: Early Sinemurian (*semicostatum–turneri* ammonite zones). Occurring within the uppermost Blue Lias, the Shales with Beef and the lowermost Black Ven Marls.

Subzone DJ2c. Top: defined by the highest occurrence of *Liasidium variabile*.

Assemblage: based on fluctuations in numbers of *Liasidium variabile*. Three 'zonules' are recognized.

'Zonule' *DJ2c1.* Top: defined immediately below the lowest occurrence of abundant *Liasidium variabile*.

Assemblage: this 'zonule' is defined on the interval immediately below the lowest occurrence of abundant *Liasidium variabile* down to the lowest occurrence of *L. variabile*.

Age: early Late Sinemurian (*obtusum* ammonite zone). Occurring within the lowermost Black Ven Marls.

'Zonule' *DJ2c2.* Top: defined by the influx of *Liasidium variabile*.

Assemblage: this 'zonule' corresponds to the stratigraphic acme of *Liasidium variabile*.

Age: Late Sinemurian (*oxynotum* ammonite zone). Occurring within the Black Ven Marls.

'Zonule' *DJ2c3.* Top: defined by the highest occurrence of *Liasidium variabile*.

Age: latest Sinemurian (*raricostatum* ammonite zone). Occurring within the uppermost Black Ven Marls.

Subzone DJ2d. Top: defined by the lowest occurrence of *Nannoceratopsis gracilis* and *Luehndea spinosa*.

Assemblage: this subzone represents an interval immediately below the lowest occurrence of *Luehndea spinosa* and *Nannoceratopsis gracilis* and above the highest occurrence of *Liasidium variable*.

Age: Early Pliensbachian (*jamesoni–davoei* ammonite zones). Occurring within the Belemnite Marls and the Green Ammonite Beds.

Zone DJ3

Top: defined by the highest occurrence of *Eyachia prisca*.

Remarks: recognition of the subzones within the Junction Bed and the Lower Inferior Oolite is often difficult due to the condensed nature of these units.

Subzone DJ3a. Top: defined on the highest occurrence of consistent/common *Luehndea spinosa*.

Age: Late Pliensbachian (*margaritatus–spinatum* ammonite zones). Occurring within the Middle Lias Clays, the argillaceous sequences within the Thorncombe Sands and the Marlstone Rock Bed.

Subzone DJ3b. Top: defined by an influx of sphaeromorphs/*Spheripollenites* spp.

Assemblage: in addition to abundant sphaeromorph/*Spheripollenites* spp. clusters, this assemblage is also associated with a major increase in amorphous organic matter (AOM). Forms such as *Eyachia prisca*, *Parvocysta* spp., *Phallocysta eumekes* and *Susadinium scrofoides* do not occur below the uppermost part of this subzone. The lowermost part of this subzone yields AOM free assemblages with sporadic records of *Luehndea spinosa*.

Age: Early Toarcian (*tenuicostatum*–lower *bifrons* ammonite zones). Occurring within the upper Marlstone Rock Bed and the lower part of the Junction Bed.

Subzone DJ3c. Top: defined by the lowest occurrence of *Nannoceratopsis dictyambonis* and *Wallodinium cylindricum*.

Assemblage: this subzone is an interval subzone between the lowest occurrence of *Nannoceratopsis dictyambonis* and *Wallodinium cylindricum* and the highest occurrence of abundant sphaeromorphs/*Spheripollenites* spp. Significant numbers of *N. gracilis/senex*, *Parvocysta* spp., *Phallocysta eumekes*, *Moesidinium railneaui*, *Eyachia prisca*, *Susadinium scrofoides* and *Reutlingia cardobarbata* may occur within this subzone.

Age: latest Early–Late Toarcian (upper *bifrons–thouarsense* ammonite zones). Occurring within the Junction Bed.

Subzone DJ3d. Top: defined by the highest occurrence of *Eyachia prisca*.

Assemblage: *Nannoceratopsis senex/gracilis* figures prominently in the assemblages from this subzone. Associated species include *Dapsilidinium deflandrei*, *Eyachia prisca*, *Facetodinium* spp., *Mancodinium* spp., *Moesidinium railneaui*, *Parvocysta* spp. and *Scriniocassis weberi*. Five 'zonules' are recognized.

'Zonule' *DJ3d1.* Top: defined by an influx of *Nannoceratopsis gracilis/senex*.

Age: latest Toarcian (*levesquei* ammonite zone). Occurring within the uppermost Junction Bed, the Down Cliff Clay and the lower Bridport Sands.

'Zonule' DJ3d2. Top: defined by the highest occurrence of the *Parvocysta* suite.

Age: earliest Aalenian (lower *opalinum* ammonite zone). Occurring within the upper Bridport Sands.

'Zonule' DJ3d3. Top: defined by the highest occurrence of *Scriniocassis weberi*.

Age: Aalenian (upper *opalinum–concavum* ammonite zones). Occurring within the Lower Inferior Oolite.

'Zonule' DJ3d4. Top: defined by the highest occurrence of *Moesidinium railneaui*.

Age: earliest Bajocian (*discites* ammonite zone). Occurring within the Lower Inferior Oolite.

'Zonule' DJ3d5. Top: defined by the highest occurrence of *Eyachia prisca*.

Assemblage: the highest occurrence of *Nannoceratopsis triceras* is noted within this 'zonule'.

Age: Early Bajocian (*laeviuscula–sauzei* ammonite zones). Occurring within the topmost Lower and Middle Inferior Oolite.

Zone DJ4

Top: defined by the highest occurrence of *Acanthaulax crispa* and *Nannoceratopsis gracilis*.

Remarks: subdivision of this zone is often difficult due to the occurrence of the carbonate facies of the Inferior Oolite Group.

Subzone DJ4a. Top: defined by the highest occurrence of common *Nannoceratopsis gracilis*.

Assemblage: similar to the overlying subzone, dominated by species of *Dichadogonyaulax/ Ctenidodinium* and *Energlynia*. *Diacanthum filapicatum*, *Durotrigia daveyi* and *Nannoceratopsis gracilis* are also prominent components of the assemblages.

Age: latest Early–earliest Late Bajocian (*humphriesanum–subfurcatum* ammonite zones). Occurring within the Middle Inferior Oolite.

Subzone DJ4b. Top: defined by the highest occurrence of *Acanthaulax crispa* and *Nannoceratopsis gracilis*.

Assemblage: dominated by *Ctenidodinium* spp./*Dichadogonyaulax* spp. and *Energlynia* spp., together with *Diacanthum filapicatum* and *Sentusidinium* spp.

Age: Late Bajocian–earliest Bathonian (*garantiana–zigzag* ammonite zones). Occurring within the Upper Inferior Oolite and lowermost Lower Fuller's Earth Clay.

Zone DJ5

Top: defined by an influx of *Lithodinia* spp., in association with *Nannoceratopsis spiculata*.

Remarks: several subdivisions of this zone can be recognized. However, in areas where limestones predominate in the succession then recognition of the subzones often proves to be difficult.

Subzone DJ5a. Top: defined by the highest occurrence of *Lithodinia valensii*.

Assemblage: the overall composition of the assemblages from this subzone is comparable to the overlying DJ5b subzone. An alternative pick for the top of this subzone in the absence of *L. valensii* is immediately below the lowest occurrence of *Meiourogonyaulax reticulata* and *Pareodinia prolongata*.

Age: earliest–latest Middle Bathonian (uppermost *zigzag*–lower *morrisi* ammonite zones). Occurring within the Lower Fuller's Earth Clay and Fuller's Earth Rock.

Subzone DJ5b. Top: defined by the highest occurrence of *Hystrichogonyaulax regale*.

Assemblage: dominated by species of *Ctenidodinium* and *Dichadogonyaulax*, with localized influxes of *Chytroeisphaeridia chytroeides* and fluctuating numbers of *Ellipsoidictyum* spp., *Dingodinium minutum*, *Pareodinia* spp. (*P. ceratophora*, *P. tripartita*, *P. prolongata*) and *Valensiella* spp. (*V. ovula*, *V. ampulla*, *V. vermiculata*). *Energlynia acollaris* is prominent in the assemblage towards the base of the subzone.

Age: latest Middle–Late Bathonian (upper *morrisi–discus* ammonite zones). Occurring within the Fuller's Earth Rock, Upper Fuller's Earth Clay, the Frome Clay, the Great Oolite, the Forest Marble and the Berry Member of the Abbotsbury Cornbrash Formation.

Subzone DJ5c. Top: defined by an influx of *Lithodinia* spp., in association with *Nannoceratopsis spiculata*.

Assemblage: dominated by *Lithodinia* spp., in association with abundant *Dichadogonyaulax gochtii*. In addition, *Aldorfia aldorfensis* and *Ctenidodinium combazii* have their highest occurrences in this subzone.

Age: Early Callovian (*macrocephalus–*lower *calloviense* ammonite zones). Occurring within the Fleet Member of the Abbotsbury Cornbrash and the Kellaways Formations.

Zone DJ6

Top: defined by the lowest occurrence of abundant *Mendicodinium groenlandicum*.

Subzone DJ6a. Top: defined by the highest occurrence of common *Nannoceratopsis pellucida* and common *Kalyptea stegasta.* Two 'zonules' are recognized.

'Zonule' *DJ6a1.* Top: defined by the highest occurrence of abundant *Chytroeisphaeridia hyalina.*

Age: late Early Callovian (upper *calloviense* ammonite zone). Occurring within the Peterborough Member of the Oxford Clay Formation.

'Zonule' *DJ6a2.* Top: defined by the highest occurrence of common *Nannoceratopsis pellucida* and *Kalyptea stegasta.*

Assemblage: dominated by *Chytroeisphaeridia cerastes, C. hyalina, Kalyptea stegasta, Nannoceratopsis pellucida, Polystephanephorus paracalathus* and locally *Ctenidodinium continuum, C. ornatum* and *Palaeostomocystis tornatilis.* The highest occurrence of *Dichadogonyaulax gochtii* is within this 'zonule'.

Age: early Middle Callovian (*jason* ammonite zone). Occurring within the Peterborough Member of the Oxford Clay Formation.

Subzone DJ6b. Top: defined by the lowest occurrence of *Mendicodinium groenlandicum.*

Assemblage: this unit is an interval subzone. Its top lies immediately below the lowest occurrence of the *Mendicodinium groenlandicum* acme and its base immediately above the highest occurrence of common *Nannoceratopsis pellucida* and common *Kalyptea stegasta.* Abundant AOM occurs within this and the underlying subzone in Oxford Clay facies.

Age: late Middle Callovian (*coronatum* ammonite zone). Occurring within the Peterborough Member of the Oxford Clay Formation.

Zone DJ7

Top: defined by an influx of *Rigaudella aemula.*

Subzone DJ7a. Top: defined by an influx of *Mendicodinium groenlandicum.* Two 'zonules' are recognized.

'Zonule' *DJ7a1.* Top: defined by the highest occurrence of *Energlynia acollaris.*

Assemblage: the assemblage composition of this 'zonule' is similar to the overlying DJ7a2 'zonule'. In addition to the above index taxon, *Eodinia pachytheca* and *Dichadogonyaulax sellwoodii* have their highest occurrence at the top of this zonule. Abundant *M. groenlandicum* does not occur below this 'zonule'.

Age: early Late Callovian (*athleta* ammonite zone). Occurring within the uppermost Peterborough Member and the Stewartby Member of the Oxford Clay Formation.

'Zonule' *DJ7a2.* Top: defined by an influx of *Mendicodinium groenlandicum.*

Assemblage: also dominated by abundant *Rigaudella aemula, Gonyaulacysta jurassica, Hystrichogonyaulax cladophora, Ctenidodinium ornatum* and more locally *Palaeostomocystica tornatilis, Ctenidodinium continuum* and *Surculosphaeridium* spp. *Pareodinia prolongata* has its highest occurrence at the top of this zonule.

Age: latest Callovian (*lamberti* ammonite zone). Occurring within the Stewartby Member of the Oxford Clay Formation.

Subzone DJ7b. Top: defined by the highest occurrence of *Wanaea fimbriata.*

Assemblages: The assemblages occurring in this subzone are similar to the overlying DJ7c subzone. *Ctenidodinium continuum, Polystephanephorus paracalathus* and *Chytroeisphaeridia hyalina* have highest occurrences with this subzone.

Age: earliest Oxfordian (*mariae* ammonite zone). Occurring within the Weymouth Member of the Oxford Clay Formation.

Subzone DJ7c. Top: defined by an influx of *Rigaudella aemula.*

Assemblage: dominated by *Acanthaulax senta, Gonyaulacysta jurassica, Hystrichogonyaulax cladophora, Scriniodinium galeritum,* and more locally *Compositosphaeridium costatum/polonicum* and species of *Surculosphaeridium.*

Age: late Early Oxfordian (*cordatum* ammonite zone). Occurring within the uppermost Oxford Clay (Weymouth Member) Formation and the Nothe Grit Formation.

Zone DJ8

Top: defined by the occurrence of increased numbers of *Scriniodinium* spp.

Zone DJ8a. Top: defined by the highest occurrence of *Acanthaulax senta.*

Assemblage: the assemblages characteristic of this zone are comparable to those in the overlying zone being dominated by *Gonyaulacysta jurassica, Hystrichogonyaulax cladophora, Scriniodinium crystallinum, S. galeritum* and *Systematophora* spp.

Age: early Middle Oxfordian (*densiplicatum* ammonite zone). Occurring within the Redcliff Formation.

Subzone DJ8b. Top: defined by the highest occurrence of *Rigaudella aemula.*

Age: late Middle Oxfordian (*tenuiserratum* ammonite zone). Occurring within the Osmington Oolite Formation.

Subzone DJ8c. Top: defined by the highest occurrence of *Compositosphaeridium costatum/polonicum*.

Age: earliest Late Oxfordian (*glosense* ammonite zone). Occurring within the Trigonia clavellata Formation and the Ampthill Clay Equivalent.

Subzone DJ8d. Top: defined by the occurrence of increased numbers of *Scriniodinium* spp.

Assemblage: this subzone is characterized both by an increase in numbers and in diversity of *Scriniodinium* spp. including *S. crystallinum*, *S. luridum* and *S. galeritum*. Additional elements of the assemblage include abundant *Gonyaulacysta jurassica*, common *Hystrichogonyaulax cladophora*, *Leptodinium* spp. (*L. subtile*, *L. mirabile*, *Leptodinium* sp. FG) and *Sentusidinium* spp. Significant numbers of *Ctenidodinium chondrum* may also be present.

Age: late Oxfordian (*serratum–regulare* ammonite zones). Occurring within the Ampthill Clay Equivalent and Sandsfoot Formation.

Zone DJ9

Top: defined by the highest occurrence of *Scriniodinium crystallinum*.

Subzone DJ9a. Top: defined by an increase in numbers of *Hystrichogonyaulax cladophora*.

Assemblage: characterized by significant numbers of *Systematophora* spp. and abundant *Gonyaulacysta jurassica*.

Age: latest Oxfordian (*rosenkrantzi* ammonite zone). Occurring within the Sandsfoot and Ringstead Formations, and the Ampthill Clay Equivalent.

Subzone DJ9b. Top: defined by the highest occurrence of *Scriniodinium crystallinum*.

Age: earliest Kimmeridgian (*baylei* ammonite zone). Occurring within the lowermost Lower Kimmeridge Clay.

Zone DJ10

Top: defined by the highest persistent occurrence of *Gonyaulacysta jurassica*.

Subzone DJ10a. Top: defined by an increase in numbers of *Gonyaulacysta jurassica*.

Age: Early Kimmeridgian (lower *cymodoce* ammonite zone). Occurring within the Lower Kimmeridge Clay.

Subzone DJ10b. Top: defined by the highest persistent occurrence of *Scriniocassis dictyotum*.

Age: Early Kimmeridgian (upper *cymodoce*–lower *mutabilis* ammonite zones). Occurring within the Lower Kimmeridge Clay.

Subzone DJ10c. Top: defined by the highest persistent occurrence of *Gonyaulacysta jurassica*.

Age: Early Kimmeridgian (upper *mutabilis* ammonite zone). Occurring within the Lower Kimmeridge Clay.

Zone DJ11

Top: defined by the highest occurrence of increased numbers of *Oligosphaeridium patulum*.

Subzone DJ11a. Top: defined by the highest occurrence of *Scriniodinium luridum*. Three 'zonules' are recognized.

'Zonule' *DJ11a1*. Top: defined by an influx of *Perisseiasphaeridium pannosum*.

Assemblage: This 'zonule' represents the stratigraphic acme of *P. pannosum*. Abundant *G. paeminosum* also occur.

Age: late Early Kimmeridgian (lower *eudoxus* ammonite zone). Occurring within the Lower Kimmeridge Clay.

'Zonule' *DJ11a2*. Top: defined by an influx of *Geiselodinium paeminosum*.

Assemblage: this 'zonule' represents the upper part of the stratigraphic acme of *G. paeminosum*.

Age: late Early Kimmeridgian (upper *eudoxus*–lower *autissiodorensis* zones). Occurring within the uppermost Lower Kimmeridge Clay.

'Zonule' *DJ11a3*. Top: defined by the highest occurrence of *Scriniodinium luridum*.

Age: latest Early Kimmeridgian (upper *autissiodorensis* ammonite zone). Occurring within the uppermost Lower Kimmeridge Clay.

Subzone DJ11b. Top: defined by the highest occurrence of increased numbers of *Oligosphaeridium patulum*. Five 'zonules' are recognized.

'Zonule' *DJ11b1*. Top: defined by the highest occurrence of increased numbers of *Perisseiasphaeridium pannosum*.

Age: Late Kimmeridgian (lower *elegans* ammonite zone). Occurring within the lowermost Upper Kimmeridge Clay.

'Zonule' *DJ11b2*. Top: defined by the highest occurrence of *Geiselodinium paeminosum*.

Age: Late Kimmeridgian (upper *elegans* ammonite zone). Occurring within the lowermost Upper Kimmeridge Clay.

'Zonule' *DJ11b3*. Top: defined the occurrence of increased numbers of *Gonyaulacysta longicornis*.

Age: Late Kimmeridgian (*scitulus* ammonite zone). Occurring within the Upper Kimmeridge Clay.

'Zonule' *DJ11b4*. Top: defined by the highest occurrence of *Gonyaulacysta longicornis*.

Age: Late Kimmeridgian (*wheatleyensis*–lower *hudlestoni* ammonite zones). Occurring within the Upper Kimmeridge Clay.

'Zonule' *DJ11b5*. Top: defined by the occurrence of increased numbers of *Oligosphaeridium patulum*.

Age: Late Kimmeridgian (upper *hudlestoni* ammonite zone). Occurring within the Upper Kimmeridge Clay.

Zone DJ12

Top: defined by the highest occurrence of *Oligosphaeridium patulum*.

Subzone DJ12a. Top: defined by the highest occurrence of *Perisseiasphaeridium pannosum*.

Age: Late Kimmeridgian (*pectinatus* ammonite zone). Occurring within the Upper Kimmeridge Clay.

Subzone DJ12b. Top: defined by the highest occurrence of *Oligosphaeridium patulum* and the lowest occurrence of *Muderongia duxburyii*.

Assemblage: the highest occurrence of *Egmontodinium ovatum* also occurs within this subzone.

Age: Late Kimmeridgian (*pallasoides*–lower *rotunda* ammonite zones). Occurring within the Upper Kimmeridge Clay.

Zone DJ13

Top: defined by the highest occurrence of *Glossodinium dimorphum*.

Subzone DJ13a. Top: defined immediately below the lowest occurrence of common *Muderongia duxburyii*

Age: Late Kimmeridgian (upper *rotunda* ammonite zone). Occurring within the Upper Kimmeridge Clay.

Subzone DJ13b. Top: defined by the highest occurrence of common *Muderongia duxburyii*.

Age: latest Kimmeridgian–earliest Portlandian (*fittoni–albani* ammonite zones). Occurring within the uppermost Upper Kimmeridge Clay and lowermost Portland Sand Formation.

Subzone DJ13c. Top: defined by the highest occurrence of *Muderongia duxburyii*.

Age: Early Portlandian (*glaucolithus–okusensis* ammonite zones). Occurring within the Portland Sand Formation.

Subzone DJ13d. Top: defined by the highest occurrence of *Senoniasphaera jurassica*.

Age: Early Portlandian (*kerberus* ammonite zone). Occurring within the upper part of the Portland Limestone Formation.

Subzone DJ13e. Top: defined by the highest occurrence of *Glossodinium dimorphum* and *Dichadogonyaulax pannea*.

Age: late Early Portlandian (*anguiformis* ammonite zone). Occurring within the uppermost Portland Limestone Formation.

Zone DJ14

Top: defined by the highest occurrence of *Cribroperidinium* sp. A *sensu* Davey, 1982.

Subzone DJ14a. Top: defined by the highest occurrence of *Egmontodinium polyplacophorum*.

Age: latest Early Portlandian (*oppressus* ammonite Zone). Occurring within the lowermost Lower Purbeck Beds.

Purbeck–Wealden unzoned

Top: defined by the occurrence of miospore-dominated assemblages.

Assemblage: the microfloras from the uppermost Purbeck Group are dominated by long-ranging environmentally influenced miospores of limited stratigraphic value.

Conclusions

An integrated microfossil biostratigraphy, based on ostracods, foraminiferids and dinocysts, for the latest Triassic through to the earliest Cretaceous has been applied to the Portland–Wight Basin and its adjacent areas. A total of 112 zones–zonules (34 ostracod zones/subzones, 25 foraminiferid zones, 53 dinoflagellate zones, subzones and zonules) are recognized. All three zonation schemes have, wherever possible, been calibrated to the standard ammonite zones recognized throughout northwest Europe, based on outcrop material. The microfossil biostratigraphy has allowed detailed correlation with wireline log and lithological data to elucidate a regional stratigraphic framework for study area (see Ainsworth this volume).

The authors would like to thank E. Powell for the draughting of Fig. 1. Some of the data used in this paper were collected by the authors and included in a commercial study for PaleoServices in 1993. The authors are grateful for the assistance of RPS Paleo in allowing this paper to be published, and wish to emphasize that the views represented herein are those of the authors.

References

AINSWORTH, N. R., BRAHAM, W., GREGORY, F. J., JOHNSON, B. & KING, C. 1998. The lithostratigraphy and biostratigraphy of the latest Triassic to earliest Cretaceous of the English Channel and its adjacent areas. *This volume.*

ANDERSON, F. W. 1971. The ostracods. *In*: ANDERSON, F. W. & BAZLEY, R. A. B. (eds) *The Purbeck Beds of the Weald (England). Bulletin of the Geological Survey of Great Britain*, **34**, 1–173.

—— 1985. Ostracod faunas in the Purbeck and Wealden of England. *Journal of Micropalaeontology*, **4**, 1–67.

BARNARD, T. 1950. Foraminifera from the Lower Lias of the Dorset coast. *Quarterly Journal of the Geological Society of London*, **105**, 347–391.

—— 1953. Foraminifera from the Upper Oxford Clay (Jurassic) of Redcliff Point, near Weymouth, England. *Proceedings of the Geologists' Association*, **64**, 183–197.

——, CORDEY, W. G. & SHIPP, D. J. 1981. Foraminifera from the Oxford Clay (Callovian–Oxfordian) of England. *Revista Espanola de Micropaleontologia*, **13**, 383–462.

BATE, R. 1978. The Jurassic Part II – Aalenian to Bathonian. *In*: BATE, R. H. & ROBINSON, E. (eds) *A Stratigraphical Index of British Ostracoda. Geological Journal*, Special Issue 213–258.

—— & SHEPPARD, L. M. 1982. Bathonian Ostracoda in the Winterborne Kingston borehole, Dorset. *In*: RHYS, G. H., LOTT, G. K. & CALVER, M. A. (eds) *The Winterborne Kingston Borehole, Dorset, England. Report of the Institute of Geological Sciences No.* **81**, 77–81.

CALLOMON, J. H. & CHANDLER, R. B. 1990. A review of the ammonite horizons of the Aalenian–Lower Bajocian stages in the Middle Jurassic of Southern England. *Memoire descrittive dela Carta geologica d'Italia*, **40**, 85–112.

CHRISTENSEN, O. B. & KILENYI, T. 1970. Ostracod biostratigraphy of the Kimmeridgian in Northern and Western Europe. *Danmarks Geologiske Undersogelse, II*, **95**, 1–65.

CIFELLI, R. 1959. Bathonian Foraminifera of England. *Bulletin of the Museum of Comparative Zoology, Harvard*, **121**, 265–368.

COLEMAN, B. E. 1979. Bathonian Foraminifera. *In*: PENN, I. E. & WYATT, R. J. (eds) *The stratigraphy and correlation of the Bathonian strata of the Bath–Frome Area. Report of the Institute of Geological Sciences No.* 78, 1–88.

—— 1980. *In*: PENN, I. E., DINGWALL, R. G. KNOX, R. W. O'B. *The Inferior Oolite (Bajocian) Sequence from a borehole in Lyme Bay, Dorset. Report of the Institute of Geological Sciences No.* 79, 1–27.

—— 1982. Lower and Middle Jurassic Foraminifera in the Winterborne Kingston borehole, Dorset. *In*: RHYS, G. H., LOTT, G. K. & CALVER, M. A. (eds) *The Winterborne Kingston borehole, Dorset, England. Report of the Institute of Geological Sciences No.* **81**, 82–88.

COPE, J. C. W., GETTY, T. A., HOWARTH, M. K., MORTON, N. & TORRENS, H. S. 1980a. *A Correlation of Jurassic rocks in the British Isles. Part One: Introduction and Lower Jurassic.* Special Report of the Geological Society, London No. 14.

——, DUFF, K. L., PARSONS, C. F., TORRENS, M. S., WIMBLEDON, W. A. & WRIGHT, J. 1980b. *A Correlation of Jurassic rocks in the British Isles. Part Two. Middle and Upper Jurassic.* Special Report of the Geological Society, London No. 15.

COPESTAKE, P. 1989. Triassic. *In*: JENKINS, D. G. & MURRAY, J. W. (eds) *Stratigraphical Atlas of Fossil Foraminifera*, 2nd edition. Ellis Horwood, Chichester, for the British Micropalaeontology Society, 97–124.

—— & JOHNSON, B. 1984. Lower Jurassic (Hettangian–Toarcian) Foraminifera from the Mochras borehole, North Wales (U.K.) and their application to a worldwide biozonation. *Benthos '83; Second International Symposium on Benthic Foraminifera Pau, April 1983*, 183–184.

—— & JOHNSON, B. 1989. The Hettangian to Toarcian. *In*: JENKINS, D. G. & MURRAY, J. W. (eds) *Stratigraphical Atlas of Fossil Foraminifera*, 2nd edition. Ellis Horwood, Chichester, for the British Micropalaeontology Society, 129–189.

FIELD, R. A. 1968. *Lower Jurassic Ostracoda from England and Normandy*. PhD thesis, University of London.

FULLER, N. G. 1984. *Upper Jurassic (Callovian and Oxfordian) Ostracoda From England and Northern France*. PhD thesis, University College of London.

KILENYI, T. 1969. The Ostracoda of the Dorset Kimmeridge Clay. *Palaeontology*, **12**, 112–160.

—— 1978. The Jurassic Part III. Callovian–Portlandian. *In*: BATE, R. H. & ROBINSON, E. (eds) *A Stratigraphical Index of British Ostracoda. Geological Journal*, Special Issue 8, 259–298.

—— & NEALE, J. W. 1978. The Purbeck/Wealden. *In*: BATE, R. H. & ROBINSON, E. (eds). *A Stratigraphical Index of British Ostracoda. Geological Journal*, Special Issue 8, 299–324.

LORD, A. R. & BOWN, P. R. (eds) 1987. *Mesozoic and Cenozoic Stratigraphical Micropalaeontology of the Dorset Coast and Isle of Wight, Southern England.* 20th European Micropalaeontological Colloquium Field Guide. British Micropalaentological Society.

MEDD, A. W. 1982. Upper Jurassic Foraminifera from the Winterborne Kingston borehole. *In*: RHYS, G. H., LOTT, G. K. & CALVER, M. A. (eds) *The Winterborne Kingston borehole, Dorset, England.* Report of the Institute of Geological Sciences No. 81, 52–53.

MORRIS, P. H. 1982. Distribution and palaeoecology of Middle Jurassic Foraminifera from the Lower Inferior Oolite of the Cotswolds. *Palaeogeography, Palaeoclimatology, Palaeoecology*, **37**, 319–347.

—— 1983. Palaeoecology and stratigraphic distribution of Middle Jurassic ostracods from the Lower Inferior Oolite of the Cotswolds, England. *Palaeogeography, Palaeoclimatology, Palaeoecology*, **41**, 287–303.

—— & COLEMAN, B. E. 1989. The Aalenian to Callovian (Middle Jurassic). *In:* JENKINS, D. G. & MURRAY, J. W. (eds) *Stratigraphical Atlas of Fossil Foraminifera*, 2nd edition. Ellis Horwood, Chichester, for the British Micropalaeontological Society, 189–236.

PAGE, K. N. 1989. A stratigraphic revision for the English Lower Callovian. *Proceedings of the Geologists' Association*, **100**, 363–382.

PARTINGTON, M. A., COPESTAKE, P., MITCHENER, B. C. & UNDERHILL, J. R. 1993. Biostratigraphic calibration of genetic stratigraphic sequences in the Jurassic–lowermost Cretaceous (Hettangian to Ryazanian) of the North Sea and adjacent areas. *In:* PARKER, J. R. (ed.) Petroleum Geology of Northwest Europe: Proceedings of the 4th Conference. Geological Society, London, **1**, 371–386.

POWELL, A. J. 1992. Dinoflagellate cysts of the Triassic System. *In:* POWELL, A. J. (ed.) A Stratigraphic Index of Dinoflagellate Cysts. Chapman & Hall, London, 1–6.

RAWSON, P. F. & RILEY, L. A. 1982. Latest Jurassic–Early Cretaceous events and 'Late Cimmerian Unconformity' in the North Sea area. *Bulletin of the American Association of Petroleum Geologists*, **66**, 2628–2648.

RIDING, J. B. & THOMAS, J. E. 1992. Dinoflagellate cysts of the Jurassic System. *In:* POWELL, A. J. (ed.) A Stratigraphic Index of Dinoflagellate Cysts. Chapman & Hall, London, 7–98.

RILEY, L. A. & FENTON, J. P. G. 1982. A dinocyst zonation of the Callovian to Middle Oxfordian succession (Jurassic) of Northwest Europe. *Palynology*, **6**, 193–202.

SHEPPARD, L. M. 1982. Bathonian ostracod correlation north and south of the English Channel with the description of two new ostracod species. *In:* NEALE, J. W. & BRASIER, M. D. (eds) *Microfossils from Recent and Fossil Shelf Seas*. Ellis Horwood, Chichester, for the British Micropalaeontological Society, 73–89.

SHIPP, D. J. 1989. The Oxfordian to Portlandian. *In:* JENKINS, D. G. & MURRAY, J. W. (eds) *Stratigraphical Atlas of Fossil Foraminifera*, 2nd edition. Ellis Horwood, Chichester, for the British Micropalaeontological Society, 237–272.

WHATLEY, R. C. 1965. *Callovian and Oxfordian Ostracoda from England and Scotland*. PhD thesis, University of Hull.

WILKINSON, I. P. 1982. Lower to Upper Jurassic Ostracoda in the Winterborne Kingston borehole, Dorset. *In:* RHYS, G. H., LOTT, G. K. & CALVER, M. A. (eds) *The Winterborne Kingston Borehole, Dorset, England*. Report of the Institute of Geological Sciences No. **81**, 39–44.

WOOLHAM, R. & RIDING, J. B. 1983. Dinoflagel-late cyst zonation of the English Jurassic. Report of the Institute of Geological Sciences No. 83, 1–42.

The lithostratigraphy of the latest Triassic to earliest Cretaceous of the English Channel and its adjacent areas

NIGEL R. AINSWORTH,[1] WILLIAM BRAHAM,[2] F. JOHN GREGORY,[3] BEN JOHNSON[4] & CHRISTOPHER KING[5]

[1] *39 De Tany Court, St. Albans, Hertfordshire AL1 1TU, UK*
(e-mail: Nigel_R_Ainsworth@compuserve.com)
[2] *11 Corner Hall, Hemel Hempstead, Hertfordshire HP3 9HN, UK*
[3] *32 Inkerman Road, St. Albans, Hertfordshire AL1 3BB, UK*
[4] *StratLab Ltd, 299 Deeside Road, Cults, Aberdeen AB15 9PA, UK*
[5] *41 Montem Road, Newmalden, Surrey KT3 3QU, UK*

Abstract: The stratigraphy of the latest Triassic through to the earliest Cretaceous of the Portland–Wight Basin and its adjacent areas may be subdivided using petrophysical (gamma-ray and interval transit time) criteria, in association with gross lithology to allow a total of 50 lithological units to be recognized. Three units occur within the latest Triassic, 46 in the Jurassic and one in the earliest Cretaceous. The lithostratigraphy can be integrated into a biostratigraphic template using published data and subsequent observations based on micropalaeontology (ostracods, foraminiferids) and palynology (dinocysts, miospores). Throughout the study area recognition of major lithostratigraphic and biostratigraphic events enables the chronostratigraphic framework of the basin to be determined, which has aided identification of regional stratigraphic breaks throughout much of the Portland–Wight Basin, including haitii in proximity to the Pliensbachian–Toarcian and the Aalenian–Bajocian boundaries, and also within the Late Oxfordian. Although the majority of the lithostratigraphic events have been deduced to be isochronous, a number of lithological units exhibit significant diachroneity. These specifically include the Frome Clay and the Great Oolite, the Kellaways Sand Member and Clay Member, and the Sandsfoot Grit Formation and the Ampthill Clay Equivalent.

Exploration of the English Channel Basin and its adjacent areas began in the early 1940s. Although many commercial wells have been drilled onshore since the 1950s, searching for oil and/or gas, few have proved to be economic due to variable porosities or complicated geological structure. The first oil was discovered onshore in 1959, in the Kimmeridge-1 well. Onshore, the two most important oilfields comprise the Wytch Farm Oilfield and the Humbly Grove Oilfield discovered in 1973 and 1980, respectively. The first exploration well to be drilled offshore was the BP Lulworth Banks-1 well in 1963 under an onshore licence. Further offshore drilling did not commence until after the 5th Offshore Exploration Licensing Round in 1977, with the spuding of the 98/22–1 well in 1979. To date (October 1996), 24 offshore exploration wells have been drilled in the Portland–Wight Basin, with some commercial success. The main prospective reservoirs comprise the Sherwood Sandstone Group and the Bridport Sands, with the hydrocarbon sources limited to the Lower and Upper Lias mudstones, the Oxford Clay and the Kimmeridge Clay (Hamblin *et al.* 1992).

The purpose of this paper is to describe the stratigraphy of the latest Triassic through to the earliest Cretaceous of the Portland–Wight Basin and its adjacent onshore areas using all released onshore and offshore exploration wells, in association with published open-file data and information gleaned from the study of the onshore exposures. A sequence stratigraphic analysis has not been undertaken in this paper because the authors did not have access to seismic, detailed sedimentological or dipmeter datasets over this structurally complex region.

In total, 18 offshore wells and 23 onshore wells have been incorporated into this study (Fig. 1). Subdivision of the lithological units has been based on petrophysical (gamma ray, sonic velocity) criteria, gross lithological analyses of ditch cuttings, sidewall cores and conventional core samples. The lithostratigraphy has been integrated into a biostratigraphic template using published data and subsequent observations based on both micropalaeontology (ostracods, foraminiferids) and palynology (dinocysts, miospores) (see Ainsworth *et al.* this volume). A total of 50 lithological units are recognized by

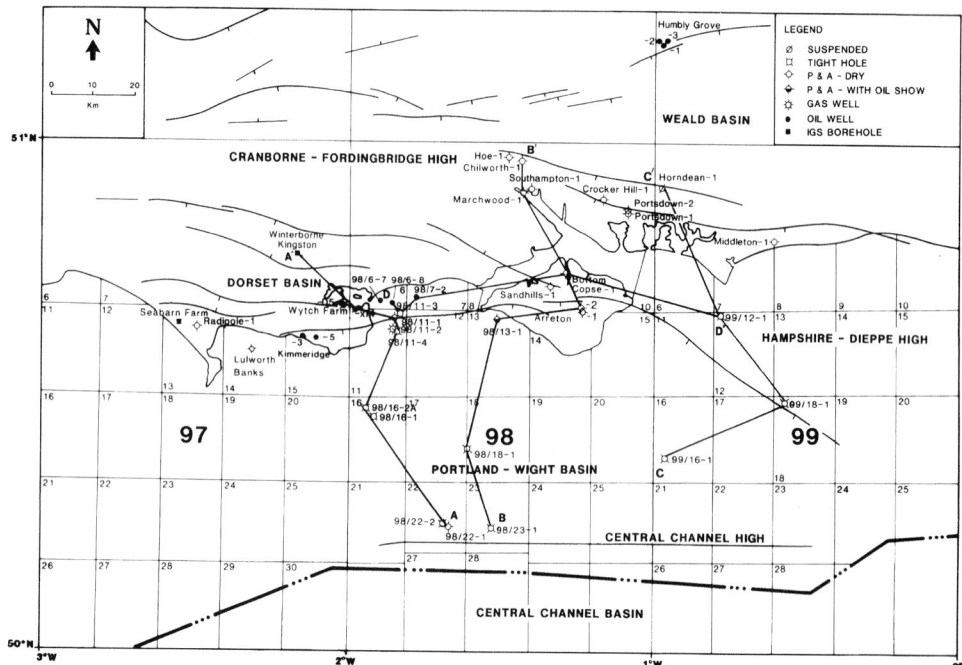

Fig. 1. Geographical location map, indicating lines of correlation, exploration wells and BGS boreholes used in the study area.

the present authors; three in the latest Triassic, 46 in the Jurassic and one in the earliest Cretaceous. A series of facies and isopachyte maps illustrating the geological history of the area are presented, in association with three north–south- and one east–west-trending lines of correlation.

There are numerous publications on the tectonic history, sedimentary, litho- and biostratigraphy of the English Channel and its surrounding areas. For a full and comprehensive overview of the region, the reader is referred to Arkell (1947), Stoneley (1982), Stoneley & Selley (1983), Whittaker et al. (1985), Bristow et al. (1991), Hamblin et al. (1992) and Underhill & Stoneley (this volume).

Geographical location and structural setting

The 41 exploration wells utilized in this study are situated within one of three basins (the Portland–Wight Basin, the Dorset Basin or the Weald Basin). All three basins are located south of the 'Variscan Front', which marks the northern boundary of the Varsican orogenic belt extending from the Bristol Channel through to Kent. Consequently, south of this boundary, the pre-Permian sediments are more intensely folded and intersected by a series of major east–west thrusts and northwest–southeast-trending steep faults. The deformation increases to the south where it becomes more intensely folded and metamorphosed towards the axis of the Variscan orogenic fold belt (Whittaker et al. 1985). These pre-Permian rocks outcrop or occur at shallow depths in the Mendips, the London Platform and in Cornubia. In between these positive areas (highs) are lows which contain Mesozoic and Tertiary sediments. Kent (1949) believed that the area south of the Variscan Front comprised a single depositional basin during the Mesozoic which he termed the 'Wessex Basin'. Recent subsurface information (boreholes and seismic data) has demonstrated that several distinct depocentres are present within the Portland–Wight Basin and its adjacent areas, separated by zones in which the Mesozoic and Tertiary sedimentary cover is thin or absent. Seismic and well data show that post-Variscan basin evolution is largely determined by the reactivation of Variscan structural elements, during alternating compressional and extensional regional tectonic regimes. For discussion of the regional stress fields and modelling of basin subsidence and inversion see Whittaker et al. (1985), Chadwick (1986), Chadwick et al. (1983), Karner et al.

(1987), Lake & Karner (1987), Butler & Pullan (1990), Ziegler (1990), Hamblin et al. (1992), Underhill & Paterson (1998) and Underhill & Stoneley (this volume).

The three basins discussed in this paper comprise fault-bounded asymmetric grabens or half-grabens, which are generally aligned in an east–west or southeast–northwest direction (Fig. 2). Both the Dorset Basin and the Portland–Wight Basin were initiated in the Early Permian, whilst the Weald Basin was created during the Early Jurassic (Whittaker et al. 1985). Intermittent movements along normal fault zones has led to differential subsidence patterns, with progressive compartmentalisation of the basins and highs. The occurrence of synsedimentary normal faults movements are associated with abrupt changes in the formation/member thickness across the growth faults, with gradual thinning due to condensing onlap up the dip-slope of the half-grabens. The major east–west extensional fault zones occurring in southern England are envisaged to reflect Variscan thrust lines. They do not comprise single faults, rather en echelon faults segments or zones, with several subparallel faults, forming transfer fault systems (Chadwick 1986). Periods of pronounced differential rates of subsidence in these basins can be seen to alternate with periods when differential subsidence is not clearly indicated, due to a temporary cessation of fault activity. Surrounding these basinal areas are noticeable highs, e.g. the Cranborne–Fordingbridge High, the London Platform, the Hampshire–Dieppe High and the South Dorset High. These positive relief structures either possess attenuated Triassic–Early Cretaceous sequences (e.g. Hampshire–Dieppe High) or remained a positive feature throughout this time (e.g. London Platform).

The earliest Triassic to earliest Cretaceous stratigraphy discussed herein is intimately involved with tectonic activity, as well as being affected by both eustatic and relative sea-level changes.

Stratigraphy

The stratigraphy of the study area is summarized in terms of lithostratigraphic units, with each unit described under the headings of age, lithology, thickness, distribution and depositional environment. The age of each unit is based wherever possible on the standard ammonite zonation described from outcrop material. Three further biozonation schemes are proposed (see Ainsworth et al. this volume) using ostracods, foraminiferids and dinocysts from

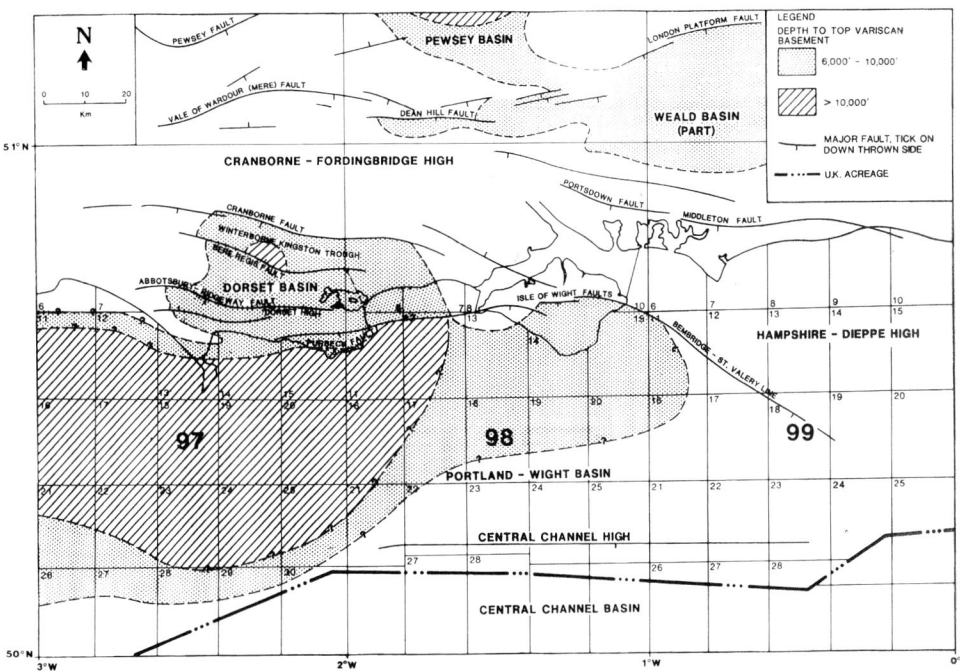

Fig. 2. Principal Mesozoic structural elements of the study area.

AGE			AMMONITE ZONES	LITHOSTRATIGRAPHY			OSTRACOD ZONES	FORAMINIFERDS ZONES	DINOCYST ZONES
				GROUP	FORMATION	MEMBER			
EARLY CRET.	EARLY VALANGINIAN 140.5			PURBECK	UPPER PURBECK BEDS		PW4		
	BERRIASIAN 145.6				MIDDLE PURBECK BEDS	CINDER BED	PW3 b / a		
							PW2		
LATE JURASSIC	PORTLANDIAN	Late	?Titanites (Paracraspedites) oppressus		LOWER PURBECK BEDS	GYPSIFEROUS BEDS	PW1		DJ14a
			Titanites anguiformis						DJ13e
		Early	Galbanites (Kerberites) kerberus	PORTLAND	PORTLAND LIMESTONE		OJ29		DJ13d
			Galbanites okusensis						
			Glaucolithites glaucolithus		PORTLAND SANDSTONE				DJ13c
			Progalbanites albani						
	KIMMERIDGIAN 152.1	Late	Virgatopavlovia fittoni			UPPER CALCAREOUS MEMBER	OJ28	FJ25	DJ13b
			Pavlovia rotunda				OJ27		DJ12b
			Pavlovia pallasioides			WHITE BAND BLACKSTONE			
			Pectinatites (Pectinatites) pectinatus				OJ26	FJ24	DJ11b5
			Pectinatites (Arkellites) hudlestoni		KIMMERIDGE CLAY	LOWER CALCAREOUS MEMBER			DJ11b4
			Pectinatites (Virgatosphinctoides) wheatleyensis					FJ23	DJ11b3
			Pectinatites (Virgatosphinctoides) scitulus						
			Pectinatites (Virgatosphinctoides) elegans						DJ11a2
			Aulacostephanus autissiodorensis				OJ24	FJ22	DJ10b
		Early	Aulacostephanus eudoxus						
			Aulacostephanoides mutabilis						
			Rasenia cymodoce				OJ22	FJ21	DJ9b
			Pictonia baylei 154.7						
	OXFORDIAN	Late	Amoeboceras rosenkrantzi	"UPPER CORALLIAN"	AMPTHILL CLAY	RINGSTEAD SANDSFOOT GRIT	OJ21		DJ9a
			Amoeboceras regulare					FJ20	
			Amoeboceras serratum		SANDSFOOT EQUIVALENT	SANDSFOOT CLAY			DJ8d
			Amoeboceras glosense			TRIG. CLAV.			DJ8c
		Middle	Cardioceras tenuiserratum	"LOWER CORALLIAN"	OSMINGTON OOLITE	NOTHE CLAY	OJ20		DJ8b
			Cardioceras densiplicatum		REDCLIFF	BENCLIFF			DJ8a
		Early	Cardioceras cordatum		NOTHE GRIT	PRESTON GRIT WEYMOUTH	OJ19	FJ19	DJ7c / DJ7b
			Quenstedtoceras mariae 157.1						
MIDDLE JURASSIC	CALLOVIAN	Late	Quenstedtoceras (Lamberticeras) lamberti		OXFORD CLAY	STEWARTBY	OJ18		DJ7a2 / DJ7a1
			Peltoceras athleta						DJ6b
		Middle	Erymnoceras coronatum			PETERBOROUGH		FJ18	DJ6a2
			Kosmoceras (Gulielmites) jason						
		Early	Sigaloceras calloviense		KELLAWAYS	KELLAWAYS SAND	OJ17	FJ17	
			Macrocephalites (M.) macrocephalus			KELLAWAYS CLAY FLEET			DJ5c
	BATHONIAN	Late	Clydoniceras (C.) discus	GREAT OOLITE	ABBOTSBURY CORNBRASH		OJ16	FJ16	
			Oppelia (Oxycerites) orbis		FOREST MARBLE	BERRY WATTONENSIS BEDS	OJ15	FJ15	DJ5b
			Procerites hodsoni		GREAT OOLITE FROME CLAY				
		Middle	Morrisiceras (M.) morrisi			UPPER FULLER'S EARTH CLAY	OJ14	FJ14	
			Tulites (T.) subcontractus		FULLER'S EARTH	FULLER'S EARTH ROCK			DJ5a
			Procerites progracilis			LOWER FULLER'S EARTH CLAY	OJ13	FJ13	
		Early	Asphinctites tenuiplicatus						
			Zigzagiceras (Z.) zigzag 166.1						
	BAJOCIAN	Late	Pakinsonia parkinsoni	INFERIOR OOLITE	UPPER INFERIOR OOLITE	HESTERS COPSE		FJ12	DJ4b
			Strenoceras (Garantiana) garantiana						
			Strenoceras subfurcatum		MIDDLE INFERIOR OOLITE			FJ11	DJ4a
		Early	Stephanoceras humphriesianum						
			Emileia (Otoites) sauzei				OJ11		DJ3d5
			Witchellia laeviuscula 173.5		LOWER INFERIOR OOLITE				DJ3d4
	AALENIAN		Hyperlioceras discites						DJ3d3
			Graphoceras concavum				OJ10	FJ10	
			Ludwigia murchisonae						DJ3d2
			Leioceras opalinum 178.0		BRIDPORT SANDS				DJ3d1
EARLY JURASSIC	TOARCIAN	Late	Dumortieria levesquei	UPPER LIAS	DOWN CLIFF CLAY		OJ9		DJ3c
			Grammoceras thouarsense		JUNCTION BED				
			Haugia variabilis						
			Hildoceras bifrons		"PAPER SHALES"			FJ9	DJ3b
		Early	Harpoceras falciferum						
			Dactylioceras tenuicostatum 187.0	MIDDLE LIAS	MARLSTONE ROCK BED		OJ8		DJ3a
	PLIENSBACHIAN	Late	Pleuroceras spinatum		MIDDLE LIAS SANDS	MIDDLE LIAS CLAY	OJ7	FJ8	
			Amaltheus margaritatus		GREEN AMMONITE BEDS		OJ6		
		Early	Prodactylioceras davoei				OJ5	FJ7	DJ2d
			Tragophylloceras ibex		BELEMNITE MARLS		OJ4		
			Uptonia jamesoni 194.5						DJ2c3
	SINEMURIAN	Late	Echioceras raricostatum	LOWER LIAS	BLACK VEN MARLS		OJ3	FJ6	DJ2c2
			Oxynoticeras oxynotum						DJ2c1
			Asteroceras obtusum						
		Early	Caenisites turneri		SHALES WITH BEEF		OJ2	FJ5 / FJ4	DJ2b
			Arnioceras semicostatum						
			Arietites bucklandi 203.5					FJ3	DJ2a2
	HETTANGIAN		Schlotheimia angulata		BLUE LIAS		OJ1	FJ2	DJ2a1
			Alsatites liasicus						
			Psiloceras planorbis 208.0			LANGPORT			
LST. TRI.	RHAETIAN				LILSTOCK	COTHAM			DJ1
					WESTBURY				

Fig. 3. Stratigraphic subdivision of the Rhaetian to Early Valanginian sediments of the Portland–Wight Basin and its adjacent areas. The three microfossil zonations are discussed in Ainsworth *et al.* (this volume). Note: the zones shown in this figure are the 'standard' zones of Cope *et al.* (1980*a, b*). The *opalinum* zone has since divided into the *opalinum* and *scissum* zones, the *murchisonae* zone into the *murchisonae* and the *bradfordensis* zones (Callomon & Chandler 1990), the *macrocephalus* zone has since been renamed the *herveyi* zone while the *calloviense* zone has divided into the *koenigi* and *calloviense* zones (Page 1989).

both outcrop and subcrop samples. These zonules, subzones and zones are mainly based on the highest downhole occurrences (extinctions) of generally common and geographically widespread microfossils, and have been integrated into the lithostratigraphy in this paper (Fig. 3)

A number of interpretative maps illustrating generalized thickness trends and facies distributions of selected stratigraphic units, in association with one east–west- and three north–south-trending lines of correlation, have also been included to help illustrate the stratigraphy of the Portland–Wight Basin and its adjacent areas (Figs 3–20).

Penarth Group

The Penarth Group of Rhaetian age can be subdivided into two formations; a lower Westbury Formation and an upper Lilstock Formation (Warrington *et al.* 1980).

Westbury Formation. Age: Rhaetian. Occurring in palynological zone DJ1 (*pars*).

Lithology: this formation is dominated by dark grey to black, organic-rich, pyritic, subfissile shales. Thin beds of light grey, bioclastic limestones also occur within the interval. Beds of white, very fine- to fine-grained sandstone occur in the lower part of the formation in a number of wells (e.g. Winterborne Kingston-1 and Kimmeridge-5).

Wireline log characteristics: the lower boundary is sharp; marked by a prominent increase in gamma-ray response, associated with a decrease in sonic velocity values, denoting a lithological change to organic-rich claystones. The upper boundary is sharp; defined by a decrease in gamma-ray response, in association with an increase in sonic velocity values, coinciding with a lithological change to calcareous claystones or limestones. The high gamma-ray and moderate sonic velocity responses for the claystones, with low gamma-ray and high sonic velocity values for the sandstone and limestone interbeds, form a spiky/serrate log motif.

Distribution and thickness: the Westbury Formation occurs throughout most of the study area, except over the Hampshire–Dieppe High (99/18-1, 99/12-1 and Middleton-1). It is fairly consistent in thickness ranging from 5 feet in 99/16-1 to a maximum of 43 feet in Winterborne Kingston-1. Thicknesses generally range between *c.* 14 and *c.* 30 feet. Thin beds of sandstone, up to *c.* 10 feet thick, are noted in a number of wells (Winterborne Kingston-1, Chilworth-1, Hoe-1, Horndean-1 and Portsdown-1).

Depositional environment: the initial Rhaetian marine transgression occurring throughout the UK spread from the southwest (Anderton *et al.* 1979) across a subdued topography, forming very wide areas of very shallow water, with little tidal influence. The base of the Westbury Formation is envisaged to represent the initial transgression across southern Britain. The high organic contents of the claystones, in association with distinctive low-diversity foraminiferid and bivalve faunas, is indicative of low-energy, dysaerobic conditions (Copestake 1989). The occurrence of thin sandstone and limestone interbeds, the latter often with disarticulated bivalves shells are envisaged to reflect higher energy, episodic marine conditions (Whittaker & Green 1983). The interbedded sandstones may represent reworking of pre-existing Triassic sediments during the initial northeast transgressive phase (Holloway 1985*a*). The occurrences in a number of wells of thick arenaceous units at the base of the Westbury Formation (e.g. Winterborne Kingston-1) are thought to represent shoreline sediments flanking the London Platform (Knox 1982). With the continuing transgression these sandstones are replaced upsequence by mudstones and shales.

Lilstock Formation. The Lilstock Formation can be subdivided into two members; a lower Cotham Member and an upper Langport Member (Warrington *et al.* 1980).

Cotham Member. Age: Rhaetian. Occurring in palynological zone DJ1 (*pars*).

Lithology: this member comprises an interbedded claystone–limestone sequence. The former consists of grey-green to brown, locally pyritic, calcareous claystones and marls, while the latter comprise light grey, often algal or argillaceous limestones.

Wireline log characteristics: the lower boundary is sharp; defined by a marked decrease in the gamma-ray response and an associated increase in sonic velocity values, coinciding with a lithological change to calcareous claystones and limestones. The upper boundary is sharp; defined by an abrupt decrease in gamma-ray response and a significant increase in sonic velocity values, corresponding to a lithological change to well-indurated limestones. The claystones of the Cotham Member possess lower gamma-ray responses than the underlying Westbury Formation, reflecting their increased carbonate content.

Distribution and thickness: the Cotham Member is a very thin unit, ranging in thickness from 5 feet in 98/23-1 through to 25 feet in 98/22-1. It occurs throughout much of the study area, except over the Hampshire–Dieppe High (99/12-1, 99/18-1 and Middleton-1).

Depositional environment: the Cotham Member was deposited during a minor, widespread regressive episode. A number of low-energy, freshwater through to marine environments are envisaged due to the occurrence of both marine and non-marine benthic faunas; these environments include both lagoonal and supratidal conditions (Kelling & Moshrif 1977; Poole 1978; Hamblin et al. 1992; Warrington & Ivimey-Cook 1992).

Langport Member. Age: Late Rhaetian. Occurring within foraminiferid zone FJ1, palynological zone DJ1 (*pars*).

Lithology: this unit is dominated by a white to light grey, microcrystalline, locally pyritic, well-indurated (recrystallized mudstone–pelletal packestone) limestones. Thin, light to dark grey, fissile shales and calcareous claystones are rare throughout the unit.

Wireline log characteristics: the lower boundary is very sharp; defined by an abrupt decrease in gamma-ray response and an associated increase in sonic velocity values, coinciding with a lithological change to well-indurated limestones. The upper boundary is sharp; defined by an abrupt increase in gamma-ray response and a corresponding decrease in sonic velocity values, coincident with a lithological change to claystones. Very low gamma-ray values and high sonic velocities are characteristic of the Langport Member, reflecting the highly indurated limestone lithologies. The Langport Member possesses a very distinctive cylindrical wireline log motif.

Distribution and thickness: the Langport Member is present throughout much of the study area, except over the Hampshire–Dieppe High (99/12-1 and 99/18-1). It is fairly uniform in thickness varying from 12 feet in Middleton-1 to 58 feet in 98/16-2A. Its thickest developments occur in the west part of the study area, thinning towards the Hampshire–Dieppe High. In Humbly Grove-1, the occurrence of sandy oolitic limestones and calcareous sandstones are envisaged to represent a marginal equivalent of the Langport Member.

Depositional environment: the calcareous sequences of the Langport Member were deposited in a shallow, low- to high-energy marine, warm, carbonate-shelf environments (Whittaker & Green 1983), starved of clastic input. The occurrence of low-diversity macro- and microfaunas is suggestive of greater than normal marine salinities (Anderson 1964; Hallam & El Shaarawy 1982; Copestake 1989). The presence of slumping in the Langport Member is indicative of coeval fault activity during this time (Holloway 1985a). However, the relatively uniform thickness of this member is attributed to the subdued topography and the fact that the major fault zones were relatively quiescent. The Langport Member equivalent occurring in Humbly Grove-1 is envisaged to have been deposited above wave-base, in shallow-marine, high-energy coastal environments.

Lias Group

The Lias Group of latest Rhaetian–Late Toarcian age is divided into three broad units; the Lower Lias, the Middle Lias and the Upper Lias. It occurs both at outcrop in the Dorset Basin, and in subcrop throughout the Channel Basin and its adjacent areas. Biostratigraphic and lithostratigraphic resolution of its constituent units is generally good throughout the sequence.

Lower Lias. Following Cope et al. (1980a) and Whittaker et al. (1985) the Lower Lias can be subdivided into five distinct lithological/wireline log units; in ascending order, they comprise the Blue Lias Formation (LL1), the Shales with Beef (LL2), the Black Ven Marls (LL3), the Belemnite Marls (LL4) and the Green Ammonite Beds (LL5).

Blue Lias Formation (LL1). Age: Latest Rhaetian–Early Sinemurian (pre-*planorbis*–*semicostatum* ammonite zones, *lyra* ammonite subzone). Occurring within ostracod zones OJ1–OJ2 (*pars*), foraminiferid zones FJ2–FJ4 (*pars*), palynological subzones DJ2a–DJ2b (*pars*).

The base of the Blue Lias, which comprises the 'pre-Planorbis Beds' of onshore outcrops, is conventionally dated as latest Rhaetian as it lies below the earliest occurrence of the Jurassic ammonite *Psiloceras planorbis* (Cope et al. 1980a).

Lithology: the Blue Lias is characterized by alternations of light to medium grey marls and calcareous claystones, in association with light grey, well-indurated, bioclastic limestones (mudstones to wackestones) and medium grey to greyish black, locally pyritic, organic-rich, fissile shales ('paper-shales'). The latter are the dominant mudrocks in the middle and upper intervals of the Blue Lias within parts of the Central English Channel (UK Quadrant 98).

Each claystone–limestone cycle comprises a laminated organic claystone ('paper-shale') with a sharp base, followed by a calcareous mudstone, then an argillaceous limestone and finally a claystone (Hallam 1975; Whittaker & Green 1983). Successive cycles vary by omission or expansion of each of these components (Weedon 1986). It is possible to subdivide the Blue Lias regionally by identification of several limestone-dominated or claystone-dominated intervals. For example, the claystone-dominated interval in the lower part of the Blue Lias can be identified in the subsurface and at outcrop within the Middle Hettangian (*liasicus* zone). Towards the base of the Blue Lias, one or more thick limestone beds are often present.

Wireline log characteristics: the lower boundary is sharp; defined by an abrupt increase in gamma-ray response and an associated decrease in sonic velocity values, coinciding with a lithological change to claystones. The upper boundary is sharp; defined by a decrease in gamma-ray response, associated with an increase in sonic velocity values, reflecting a lithological change to more calcareous claystones. The gamma-ray and sonic log motifs through the Blue Lias are characteristically highly serrated and 'spiky', reflecting the limestone–claystone interbeds. A similar gamma-ray profile has been described by Parkinson (1996) for the Blue Lias outcropping along the Dorset coast.

In general, a broad two-fold subdivision of the Blue Lias can be recognized; a lower less serrate, more mudrock-dominated section and an upper highly serrate, interbedded limestone–mudstone sequence. The gamma-ray response of the Blue Lias, however, varies throughout much of the study area. In the central part of the Portland–Wight Basin, gamma-ray values are often higher than onshore reflecting the more organic-rich nature of the claystone lithologies. In many wells in UK Quad 98 and in those onshore wells immediately north of UK Quad 98, the higher gamma-ray values occur in the middle and upper intervals, whereas in those wells situated on and close to the Isle of Wight the highest gamma-ray values occur in the lower half of the Blue Lias. Those wells situated in the eastern and northeastern regions of the study area possess more uniform gamma-ray motifs, in association with the least organic-rich higher gamma-ray shales. The pre-Planorbis Beds can often be recognized at the base of the Blue Lias by a marked decrease in gamma-ray and an associated increase in sonic velocity values, reflecting an increase in the carbonate contact of the claystones.

Distribution and thickness: there is little evidence for any significant lateral facies changes within the Blue Lias in the present study, except for a tendency to an increase in the proportion of limestones in the more attenuated sequences. Outside of the study area, within the main part of the Weald Basin, the Blue Lias passes into a nearshore facies comprising bioclastic calcarenites (Young & Lake 1988; Hamblin *et al.* 1992), while in the eastern part of the Hampshire–Dieppe High (Grove Hill-1) basal Early Sinemurian red-beds are suggestive of some emergence during the early part of the Lower Jurassic (Lake & Karner 1987; Hamblin *et al.* 1992). Similar sediments may fringe the southern and eastern margins of the Channel Basin, however, these have not yet been proved by drilling.

The Blue Lias occurs throughout most of the study area, with thicknesses varying between 47 feet in Humbly Grove-1 and 471 feet in Arreton-2. In general the Blue Lias is less than 350 feet thick. A major depocentre is indicated near the northern edge of the Portland–Wight Basin, suggested by large thicknesses of the Blue Lias (e.g. 471 feet at Arreton-2 and 372 feet at 98/13-1). South of this depocentre within the central and southern Portland–Wight Basin, there is a decrease in thickness varying between 149 (98/16-2A) and 357 feet (98/23-1). Similarly over the Purbeck–Isle of Wight Fault Zone, thicknesses decrease to between 107 (98/11-3) and 85 feet (98/6-8), while thickening into the Dorset Basin (420 feet at Winterborne Kingston-1). Moderately thick sections are recorded on the northwestern part of the Hampshire–Dieppe High (e.g. 176 feet at Hoe-1, 332 feet at Chilworth-1 and 308 feet at Horndean-1). Southeastwards along the main part of the Hampshire–Dieppe High the Blue Lias is absent or thin; in well 99/18-1 the Lower Lias is attenuated and could not be stratigraphically subdivided. Identification of a complete Lower and Middle Lias sequence in these areas by Bristow *et al.* (1991) is considered to be erroneous, based on the lithostratigraphic and biostratigraphic data evaluated during compilation of this present paper. Their conclusions were based solely on wireline log data, whereas the present study indicates that in places Lower Lias lithologies and microfaunas occur directly beneath the Junction Bed.

Depositional environment: deposition of the Blue Lias occurred during the continuing marine transgression initiated in the latest Rhaetian–earliest Hettangian. The Blue Lias was deposited over a subdued topography, with no clastic input, under similar conditions to the underlying Penarth Group. The sediments of the Blue Lias were laid down below wave-base in a low-energy, muddy bottomed, shelfal environment (Hallam

AGE			AMMONITE ZONES	GENERALISED LITHOLOGY OF THE STUDY AREA	RELATIVE SEA LEVEL
EARLY CRET	EARLY VALANGINIAN				
	140..			UPPER PURBECK BEDS	
	BERRIASIAN				
	145.6			MIDDLE PURBECK BEDS	
LATE JURASSIC	PORTLANDIAN	Late		LOWER PURBECK BEDS	
		Early	?Titanites (Paracraspedites) oppressus		
			Titanites anguiformis		
			Galbanites (Kerberites) kerberus	PORTLAND GROUP	
			Galbanites okusensis		
			Glaucolithites glaucolithus	PORTLAND SANDSTONE FM. PORTLAND LIMESTONE FM.	
			Progalbanites albani		
	KIMMERIDGIAN 152.1	Late	Virgatopavlovia fittoni		
			Pavlovia rotunda		
			Pavlovia pallasioides	"UPPER CALCAREOUS MEMBER"	
			Pectinatites (Pectinatites) pectinatus		
			Pectinatites (Arkellites) hudlestoni		
			Pectinatites (Virgatosphinctoides) wheatleyensis		
			Pectinatites (Virgatosphinctoides) scitulus	"LOWER CALCAREOUS MEMBER"	
			Pectinatites (Virgatosphinctoides) elegans	KIMMERIDGE CLAY FORMATION	
		Early	Aulacostephanus autissiodorensis		
			Aulacostephanus eudoxus		
			Aulacostephanoides mutabilis		
			Rasenia cymodoce		
	154.7		Pictonia baylei		
	OXFORDIAN	Late	Amoeboceras rosenkrantzi	RINGSTEAD FM. AMPTHILL CLAY EQUIVALENT	
			Amoeboceras regulare	SANDSFOOT FORMATION	
			Amoeboceras serratum		
			Amoeboceras glosense	TRIGONIA CLAVELLATA FORMATION	
		Middle	Cardioceras tenuiserratum	OSMINGTON OOLITE FORMATION	
			Cardioceras densiplicatum	REDCLIFF FORMATION NOTHE GRIT FORMATION	
		Early	Cardioceras cordatum	WEYMOUTH MEMBER	
	157.1		Quenstedtoceras mariae	OXFORD CLAY FORMATION STEWARTBY MEMBER	
MIDDLE JURASSIC	CALLOVIAN	Late	Quenstedtoceras (Lamberticeras) lamberti		
			Peltoceras athleta		
		Middle	Erymnoceras coronatum	PETERBOROUGH MEMBER	
			Kosmoceras (Gulielmites) jason		
		Early	Sigaloceras calloviense	KELLAWAYS FORMATION	
	161.3		Macrocephalites (M.) macrocephalus	ABB. CORNBRASH FM.	
	BATHONIAN	Late	Clydoniceras (C.) discus	FOREST MARBLE GREAT OOLITE	
			Oppelia (Oxyceritas) orbis	FROME CLAY	
			Procerites hodsoni	UPPER FULLER'S EARTH CLAY	
		Middle	Morrisiceras (M.) morrisi	FULLER'S EARTH ROCK	
			Tulites (T.) subcontractus		
			Procerites progracilis		
		Early	Asphinctites tenuiplicatus	LOWER FULLER'S EARTH CLAY	
	166.1		Zigzagiceras (Z.) zigzag		
	BAJOCIAN	Late	Pakinsonia parkinsoni		
			Strenoceras (Garantiana) garantiana	UPPER INFERIOR OOLITE	
			Strenoceras subfurcatum		
		Early	Stephanoceras humphriesanum	MIDDLE INFERIOR OOLITE	
			Emileia (Otoites) sauzei		
			Witchellia laeviuscula		
	173.5		Hyperlioceras discites		
	AALENIAN		Graphoceras concavum	LOWER INFERIOR OOLITE	
			Ludwigia murchisonae	BRIDPORT SANDS	
	178.0		Leioceras opalinum		
EARLY JURASSIC	TOARCIAN	Late	Dumortieria levesquei	DOWN CLIFF CLAY	
			Grammoceras thouarsense		
			Haugia variabilis		
			Hildoceras bifrons	JUNCTION BED	
		Early	Harpoceras falciferum		
	187.0		Dactylioceras tenuicostatum		
	PLIENSBACHIAN	Late	Pleuroceras spinatum	MARLSTONE ROCK BED	
			Amaltheus margaritatus	MIDDLE LIAS	
		Early	Prodactylioceras davoei	GREEN AMMONITE BEDS	
			Tragophylloceras ibex	BELEMNITE MARLS	
	194.5		Uptonia jamesoni		
	SINEMURIAN	Late	Echioceras raricostatum	BLACK VEN MARLS	
			Oxynoticeras oxynotum		
			Asteroceras obtusum		
		Early	Caenisites turneri	SHALES WITH BEEF	
			Arnioceras semicostatum		
	203.5		Arietites bucklandi		
	HETTANGIAN		Schlotheimia angulata	BLUE LIAS	
			Alsatites liasicus		
	208.0		Psiloceras planorbis		
LST. TRI.	RHAETIAN			PRE-PLANORBIS BEDS LILSTOCK FORMATION WESTBURY FORMATION	

1975; Whittaker & Green 1983). Dysaerobic episodes on the sea floor are, however, indicated by the restricted benthic faunas. Periods of anoxia are also suggested by the absence of benthos and the occurrence of organic-rich shales. Offshore, in the central part of the Portland–Wight Basin, anoxic/dysaerobic bottom water conditions are more common, especially within the upper half of the interval as suggested by the lack of benthic microfossils and the occurrence of dark grey shales.

The Blue Lias exhibits small-scale rhythms/alternations of limestones and mudstones (Hallam 1975). The lowest part of the cycle (a basal shale containing no benthos) is envisaged to reflect an initial rapid water deepening in association with sediment starvation. The overlying calcareous mudstone yielded macro- and microfaunas indicative of aerated bottom waters, associated with pulses of sedimentation and a shallowing water depth. A change to shallower water conditions is indicated by the development of limestones, while the topmost cycle (mudstones) are indicative of a return to somewhat deeper conditions (Whittaker & Green 1983). These rhythms are envisaged to reflect astronomically induced climatic cyclicity (House 1985, 1987).

Shales with Beef (LL2). Age: Early Sinemurian (*semicostatum* ammonite zone, *scipionianum* ammonite subzone–*turneri* ammonite zone, lower *birchi* ammonite subzone). Occurring within ostracod zone OJ2 (*pars*), foraminiferid zones FJ4 (*pars*)–FJ5, palynological subzone DJ2b (*pars*).

Lithology: the Shales with Beef comprise an interbedded claystone and shale–limestone sequence. The former comprises medium grey to dark grey, non-calcareous to calcareous, locally fissile mudrocks, while the latter consists of light grey to medium grey, microcrystalline limestones and argillaceous limestones. The Shales with Beef can be subdivided into two gross lithological packages, a lower more marly dominated sequence and an upper more argillaceous and less calcareous sequence.

Wireline log characteristics: the lower boundary is sharp; defined by a decrease in gamma-ray response and an associated increase in sonic velocity values, coinciding with a lithological change to more calcareous claystones. The upper boundary is sharp; marked by a significant increase in gamma-ray response and a corresponding decrease in sonic velocity values, reflecting the poorly calcareous nature of the claystones. The Shales with Beef display a highly serrated gamma-ray response, indicating the interbedding of calcareous claystones and limestones.

Parkinson (1996), using the hand-held spectral gamma-ray tool at outcrop along the Dorset coast, found similar low gamma-ray values to the present authors for the Shales with Beef. Both the Blue Lias and the Shales with Beef possess similar total gamma-ray counts, making differentiation between the two units extremely difficult (Parkinson 1996). However, the boundary between the units can be easily differentiated using uranium gamma-ray counts, denoted by a marked decrease in uranium counts and also low thorium/uranium ratios from the Blue Lias up into the Shales with Beef.

Distribution and thickness: the Shales with Beef occur throughout much of the present study area. Recorded thickness variations range from 37 feet in Middleton-1 to 240 feet in 98/16-1. Within the Portland–Wight Basin, the thicker sections are envisaged to occur along the northern and central regions (e.g. 240 feet at 98/16-1 and 192 feet at 98/11-2), with a thinning towards both the Central Channel (e.g. 101 feet at 98/22-2) and the Hampshire–Dieppe Highs (98 feet at 99/16-1). Over the Purbeck–Isle of Wight Fault Zone, a rapid thinning is observed (e.g. 67 feet at 98/11-1, 56 feet at 98/6-8 and 72 feet at Wytch Farm-X14). On the northwestern margin of the Hampshire–Dieppe High thicknesses range from 37 (Middleton-1) to 141 feet (Chilworth-1). Over the main part of the Hampshire–Dieppe High the Shales with Beef is absent, including well 99/12-1. The Lower Lias is attenuated in 99/18-1 and could not be stratigraphically subdivided.

Depositional environment: sediments comprising the Shales with Beef were deposited in a low-energy, marine shelfal environment, below wave-base. A number of authors (including Hallam 1981; House 1993) infer a deeper water environment than the underlying Blue Lias. Low oxygen levels are envisaged, denoted by the occurrence of pyritized, laminated organic-rich shales in association with restricted benthic micro- and macrofaunas. These conditions are often associated with sluggish water circulation,

---

Fig. 8. Generalized stratigraphy and relative sea-level curve (after House 1993). Note: the zones shown in this figure are the 'standard' zones of Cope *et al.* (1980*a*, *b*). The *opalinum* zone has since divided into the *opalinum* and *scissum* zones, the *murchisonae* zone into the *murchisonae* and the *bradfordensis* zones (Callomon & Chandler 1990), the *macrocephalus* zone has since been renamed the *herveyi* zone while the *calloviense* zone has divided into the *koenigi* and *calloviense* zones (Page 1989).

but only appear to become occasionally anoxic in the onshore sections. However, offshore, in the central part of the Portland–Wight Basin, anoxic bottom water conditions are more common suggested by the near absence of benthic microfossils and the occurrence of highly organic-rich shales.

Black Ven Marls (LL3). Age: latest Early–Late Sinemurian (*turneri* ammonite zone, upper *birchi* ammonite subzone–*raricostatum* ammonite zone). Occurring within ostracod zone OJ3, foraminiferid zone FJ6, palynological subzones DJ2b (*pars*)–DJ2c.

Lithology: this formation is dominated by dark grey to greyish black, pyritic, generally organic-rich, poorly calcareous, shales and claystones. Impersistent thin beds of light grey, well-indurated limestone and argillaceous limestone are also noted throughout the section.

Wireline log characteristics: the lower boundary is sharp; defined by a marked increase in gamma-ray response, in association with a decrease in sonic velocity values, reflecting a lithological change to poorly calcareous claystones. The upper boundary is sharp; defined by a decrease in gamma-ray response and an associated increase in sonic velocity values, reflecting a lithological change to limestones and calcareous claystones. The Black Ven Marls are characterized by serrate, high gamma-ray responses, often with the highest gamma-ray levels developed mid-way through the sequence reflecting organic-rich claystone lithologies. The highest gamma-ray values in the Lower Lias occur within this formation. Sonic log spikes reflect the presence of limestone stringers.

Similar high gamma-ray counts are described by Parkinson (1996) using portable gamma-ray spectrometry on the Black Ven Marls sequence along the onshore Dorset coast. He recorded very high uranium, thorium and potassium levels and also high thorium/potassium ratios. A major uranium peak is indicative of organic rich shales within the *obtusum* ammonite zone.

Distribution and thickness: the Black Ven Marls occur throughout much of the study area. Its thickness varies between 13 feet in 98/11-1 and 145 feet at 98/16-1. Thicknesses of the Black Ven Marls are generally within the range 50–120 feet, with the more consistently thicker sections occurring in the central and southern parts of the Portland–Wight Basin (125 feet at 98/16-2A, 113 feet at 98/18-1 and 90 feet at 98/22-2). Eastwards, the Black Ven Marls decrease in thickness (e.g. 45 feet at 98/13-1 and 40 feet at 99/16-1) and they are absent over the main part of the Hampshire-Dieppe High, including 99/12-1. The Black Ven Marls are envisaged to be partly truncated by pre-Toarcian erosion in Wytch Farm (29 feet) and the adjacent offshore areas (e.g. 13 feet at 98/11-1 and 18 feet at 98/7-2). On the northwestern edge of the Hampshire–Dieppe High moderate thicknesses of between 50 feet (Marchwood-1) and 114 feet (Chilworth-1) are recorded.

Depositional environment: the Black Ven Marls were laid down below wave-base, in a broad, shallow to moderately deep, low-energy, epicontinental sea. Periodic episodes of restricted circulation occurred, leading to low oxygen levels at the sea floor, denoted by the restricted or absence of benthic micro- and macrofaunas. Shallowing events caused by either eustatic sea level falls or tectonic change are indicated at a number of horizons in the Black Ven Marls (Hallam 1978, 1981), suggested by marked faunal turnovers and omission of strata. Both the Black Ven Marls and the underlying Shales with Beef are considered by Wignall & Hallam (1991) to represent the most prolonged phase of oxygen restriction or anoxic deposition in the onshore British Jurassic sequences.

Belemnite Marls (LL4). Age: Early Pliensbachian (*jamesoni–ibex* ammonite zones). Occurring within ostracod zones OJ4–OJ5, foraminiferid zone FJ7 (*pars*), palynological subzone DJ2d (*pars*).

Lithology: the Belemnite Marls comprise alternating beds of light grey to medium grey, calcareous, silty claystones–marls and medium dark grey to dark grey mudstones. Impersistent, light grey, microcrystalline limestone beds also occur throughout the interval.

Wireline log characteristics: the lower boundary is sharp; defined by a decrease in gamma-ray response and a corresponding increase in sonic velocity values, reflecting a lithological change to calcareous claystones and limestones. The upper boundary is sharp; denoted by an increase in gamma-ray response, in association with a decrease in sonic velocity values, coincident with a lithological change to poorly calcareous claystones. Where the Belemnite Marls are attenuated, this boundary becomes more sharply defined, suggesting the possibility of a stratigraphic break at the Belemnite Marl–Green Ammonite Beds contact.

The Belemnite Marls are characterized by serrate gamma-ray and sonic log motifs, reflecting the interbedding of mudstones and calcareous claystones. Low gamma-ray responses and high sonic velocities are characteristic of the unit, often with several prominent sonic spikes indicating thin limestone beds. The low gamma-ray responses are in agreement with those data described by Parkinson (1996), using the hand-held spectral gamma-ray along the Dorset coast. He recorded low thorium and potassium levels, and also reduced thorium/potassium ratio levels. In those wells where the Belemnite Marls are thin, two units can be recognized; a lower claystone and an upper limestone–calcareous claystone unit.

Distribution and thickness: the Belemnite Marls are recognizable throughout half of the wells in the present study area. There is considerable thickness variations throughout the study area, from a minimum of 6 feet in Hoe-1 to a maximum of 150 feet at 98/16-2A. These relative thickness changes within the Belemnite Marls are far greater than in other units of the Lower Lias due to removal by pre-Toarcian uplift and erosion. Moderately thick sequences of the Belemnite Marls occur in both the central and southern parts of the Portland–Wight Basin (e.g. 150 feet at 98/16-2A and 61 feet at 98/22-1). On the Hampshire–Dieppe High the Belemnite Marls often reduce in thickness to less than 20 feet (e.g. 9 feet at Chilworth-1 and 13 feet at Horndean-1) or are absent (e.g. 99/12-1). It is uncertain, however, if this thinning reflects sedimentary condensing or a reduction of section due to stratigraphic breaks within this interval. On the Dorset coast two minor stratigraphic breaks are recognized, one within the Belemnite Marls and one at the Belemnite Marls–Green Ammonite Beds boundary (Cope *et al.* 1980*a, b*).

Depositional environment: the Belemnite Marls are envisaged to have been deposited below wave-base in a low-energy, shallow-marine shelfal environment. Bottom water conditions were more aerobic than the underlying Black Ven Marls, as suggested by the more calcareous nature of the sediments, in association with the moderately common benthic micro- and macrofaunas. The occurrence of scours in the top section of the Belemnite Marls on the Dorset coast is suggestive of storm origins (Hesselbo & Jenkyns 1995). The decimetre alternations of light grey calcareous claystones–marls and dark grey claystones couplets may reflect climatic controls, with the regular couplets recording orbital precession cycles with a duration of 20 ka (Weedon & Jenkyns 1990; Hesselbo & Jenkyns 1995).

Green Ammonite Beds (LL5). Age: late Early Pliensbachian (*davoei* ammonite zone). Occurring within ostracod zone OJ6, foraminiferid zone FJ7 (*pars*), palynological subzone DJ2d (*pars*).

Lithology: the Green Ammonite Beds are dominated by medium–light grey to medium grey, micaceous, non- to slightly calcareous claystones and silty claystones. Rare thin, light grey limestones also occur throughout the interval.

Wireline log characteristics: the lower boundary is sharp; defined by an increase in gamma-ray response and a corresponding decrease in sonic velocity values, reflecting a lithological change to claystones. The upper boundary is sharp; defined by a decrease in gamma-ray response, in association with an increase in sonic velocity values, denoting a lithological change to silty, calcareous claystones. The Green Ammonite Beds are characterized by moderately high gamma-ray responses and moderately low sonic velocities. Gamma-ray values are generally highest in the middle of the formation, decreasing slightly towards the top and base indicating both an increase in silt and calcareous content. Occasional sonic log 'spikes' indicate thin limestone beds.

The spectral gamma-ray data recorded from the Dorset onshore (Parkinson 1996) indicate high thorium and potassium levels, and also a high thorium/potassium ratio.

Distribution and thickness: the Green Ammonite Beds are envisaged to occur in approximately half of the wells studied. Their thicknesses vary from 15 feet in 98/11-2 to 203+ feet in Radipole-1. Similar to the Middle Lias Clays, this formation occurs in four main areas, the Dorset Basin, the Portland–Wight Basin, the northwestern margin of the Hampshire–Dieppe High and the Weald Basin. The latter two areas possess the generally most expanded sequences with thicknesses varying between 92 (Middleton-1) and 204 feet (Chilworth-1). The Green Ammonite Beds are notably absent from the main part of the Hampshire–Dieppe High, and also from some wells situated along the northern edge and central parts of the Portland–Wight Basin (e.g. the Isle of Wight wells, 98/6-7, 98/11-3, 98/13-1 and 98/18-1) where they have been removed by pre-Toarcian uplift and erosion. Thin sequences of the Green Ammonite Beds are noted close to the Central Channel High (e.g. 24 feet at 98/23-1).

Depositional environment: a low-energy, shallow-marine, muddy sea floor shelfal environment is envisaged for the deposition of the Green Ammonite Beds. Dysaerobic/anaerobic conditions are indicated by the restricted/absence of benthic micro- and macrofaunas in the sediments. Hallam (1975) has suggested a slight shallowing began at the base of these sediments continuing to the base of the Middle Lias.

Middle Lias. Following Cope *et al.* (1980*a*) and Whittaker *et al.* (1985) the Middle Lias has been divided into three distinct units in the present study; in ascending order they comprise the Middle Lias Sands (ML1), the Middle Lias Clays (ML2) and the Marlstone Rock Bed.

Middle Lias Sands (ML1). This unit encompasses both the Down Cliff Sands and the Thorncombe Sands of the Dorset coast. It is extremely difficult to correlate the offshore subsurface stratigraphy with that of the onshore coastal exposures, therefore the more general name of Middle Lias Sands has been used in the present study.

Age: early Late Pliensbachian (*margaritatus* ammonite zone). Occurring within ostracod zones OJ7 (*pars*)–OJ8 (*pars*), foraminiferid zone FJ8 (*pars*), palynological subzone DJ3a (*pars*).

Lithology: the sequence is dominated by light grey to pale yellow, micaceous, sandy mudstones and claystones. Very fine to fine-grained, micaceous, quartzose or calcareous, glauconitic, silty sandstones and siltstones are also developed throughout the sequence. Clean sandstones are, however, rarely observed in the study area. There is considerable lateral and vertical variation in lithology throughout the region, with local developments of calcareous doggers, ferruginous micaceous sandy clays and sandy limestones.

Wireline log characteristics: the lower boundary is sharp to gradational; defined by a decrease in gamma-ray response, associated with an increase in sonic velocity, reflecting a lithological change to sandy mudrocks and silty sandstones. In a number of wells this boundary is transitional. The upper boundary is sharp; denoted by a decrease in gamma-ray response and an associated increase in sonic velocity values, denoting a lithological change to limestones. The Middle Lias Sands are characterized by moderately low to low gamma-ray values and relatively low sonic velocities, reflecting the high silt content of the sands. The logs may be noticeably serrated, reflecting the presence of thin calcareous beds. In the offshore Lyme Bay area the banded nature of the Middle Lias Sands gives rise to good seismic reflectors on shallow seismic profiles (Darton *et al.* 1979).

Distribution and thickness: the Middle Lias Sands are absent in many of the wells in the present study. This may reflect some contemporaneous facies patterns or more likely a phase of pre-Toarcian uplift and erosion. To date they have only been identified in the onshore wells and at outcrop, occurring in two discrete areas.

First, within the Dorset Basin, the north-western part of the Cranborne–Fordingbridge High, Dorset and Somerset, with a northern limit approximating to the Mere Fault and an eastern limit near the Hampshire–Dorset border. Owing to their interfingering with the Middle Lias Claystones, and their often transitional lower boundary, their thickness is very difficult to determine, however, average thicknesses of between 100 and 250 feet appear to be typical. On the Dorset coast, the two major sand units, the Down Cliff Sands (up to 98 feet) and the Thorncombe Sands (up to 85 feet), are separated by a thin argillaceous unit – the Margaritatus Clay Bed (Melville & Freshney 1982; Stoneley & Selley 1983). These sand packages have been proved along strike in Radipole-1, however, a large part of the Thorncombe Sands is very argillaceous. In Winterborne Kingston-1, the Middle Lias Sands (227 feet thick) comprise only a single discrete sand package. In the former it has been suggested that the reduction in thickness compared to Radipole-1 is due to eastward thinning of the lower argillaceous part of the Thorncombe Sands (Hamblin *et al.* 1992). Second, on the northwestern edge of the Hampshire–Dieppe High (e.g. 31 feet at Chilworth-1) and within the Weald Basin (e.g. 42 feet at Humbly Grove-1).

Depositional environment: the sands of the Middle Lias (Down Cliff Sands and Thorncombe Sands) are typically fine grained, cross-bedded and highly bioturbated, often containing thick-shelled bivalve species, deposited in lower shoreface to shallow offshore marine environments (Sellwood 1978; Sellwood & Wilson 1990; Callomon & Cope 1995). Their distribution pattern suggests progradation from the west ('Cornubia') and the London Platform. Small-scale sedimentary cycles are noted within the arenaceous Middle Lias, culminating in thin condensed sequences, for example the Margaritatus Stone and the Thorncombiensis Bed (Sellwood & Jenkyns 1975).

Middle Lias Clays (ML2). Age: early Late Pliensbachian (*margaritatus* ammonite zone). Occurring within ostracod zones OJ7 (*pars*)–OJ8 (*pars*), foraminiferid zone FJ8 (*pars*), palynological subzone DJ3a (*pars*).

Lithology: The principle lithologies comprise medium grey to bluish grey, silty, micaceous, calcareous claystones and mudstones. Occasional beds of light to medium grey limestone and argillaceous limestone are noted within the section. Light grey to brownish grey siltstones and fine-grained calcareous sandstones are locally developed onshore.

Wireline log characteristics: the lower boundary is sharp; denoted by a decrease in gamma-ray response and a corresponding increase in sonic velocity values indicating a lithological change to silty, calcareous claystones. The upper boundary is sharp to moderately well delimited; defined by a decrease in gamma-ray response and a corresponding increase in sonic velocity values, reflecting a lithological change to siltstones and silty sandstones. This upper boundary can be easily recognized where clean arenaceous lithologies of the Middle Lias Sands overlie silty claystones of the Middle Lias Clays. However, in many of the wells in the present study, the boundary is transitional and often very difficult to identify. The gamma-ray and sonic velocity profiles are constant throughout the unit. The occasional sonic 'spikes' indicate thin limestone beds.

Distribution and thickness: the Middle Lias Clays possess a similar distribution to that of the underlying Green Ammonite Beds; occurring in the Dorset Basin, the Portland–Wight Basin, the northwestern edge of the Hampshire–Dieppe High and the Weald Basin. Its thickness varies between 9 feet in Bottom Copse-1 to 210 feet in 98/16-1. The thickest sections of the Middle Lias Clays occur in the central and northern margins of the Portland–Wight Basin (e.g. 210 feet in 98/16-1, 197 feet at both 98/11-4 and Radipole-1, and 108 feet at 98/16-2A). Towards the Central Channel High, a rapid thinning can be observed (e.g. 23 feet at 98/23-1). Over both the northwestern margin of the Hampshire–Dieppe High and the Weald Basin the formation is thinner than that occurring in the Dorset Basin (e.g. 150 feet at Hoe-1 and 140 feet at Horndean-1), however, thicknesses increase into the centre of the Weald Syncline (Hamblin *et al.* 1992). The Middle Lias Clays are absent from much the Hampshire–Dieppe High, and also from some of the wells situated along the northern edge and the central parts of the Portland–Wight Basin (e.g. Arreton-2, Sandhills-1, 98/7-2, 98/13-1 and 99/16-1) where they have been removed by pre-Toarcian uplift and erosion

Depositional environment: the Middle Lias Clays, including the Eype Clay, possess both a wider geographical and stratigraphical range than the Middle Lias Sands. They are envisaged to have been deposited in warm, slightly deeper marine environments than the arenaceous sequences of the Middle Lias. Bottom water environments were generally aerobic due to the presence of often profuse benthic organisms. Within the main Domerian shallowing-upwards event, small-scale shallowing cycles have been described; for example the Starfish Bed and the Eype Nodule Bed. These often yield large numbers of both benthic and nektonic fossils.

Marlstone Rock Bed. The Marlstone Rock Bed is frequently included (and often confused) with the overlying Junction Bed. It is lithologically distinct and biostratigraphically older than the Junction Bed.

Age: latest Pliensbachian–earliest Toarcian (*spinatum–tenuicostatum* ammonite zones). Occurring within ostracod zone OJ8 (*pars*), foraminiferid zone FJ9 (*pars*), palynological subzone DJ3a (*pars*).

Lithology: the Marlstone Rock Bed comprises medium grey to light brown, locally oolitic, bioclastic, ferruginous sparry grainstones, wackestones and mud-supported conglomerates.

Wireline log characteristics: the lower boundary is sharp; denoted by an abrupt decrease in gamma-ray response and an associated increase in sonic velocity values, reflecting a lithological change to ferruginous limestones. The upper boundary is often difficult to identify, as it is usually overlain by limestones of the Junction Bed, which have similar log responses. Where overlain by the Down Cliff Clay, the boundary is sharp, defined by an abrupt increase in gamma-ray response, associated with a decrease in sonic velocity values, reflecting a lithological change to silty claystones. The Marlstone Rock Bed possesses a cylindrical log profile, with low gamma-ray and high sonic velocity responses.

Distribution and thickness: the condensed Marlstone Rock Bed is generally present wherever a complete Middle Lias section has been preserved. In the present study it has only been recorded in two discrete regions; the northern edge of the Portland–Wight and Dorset Basins (Radipole-1 and Winterborne Kingston-1), and the northwestern margin of the Hampshire–Dieppe High–Weald Basin (Bottom Copse-1, Horndean-1, Crokerhill-1 and Humbly Grove-1). Its thickness is always less than 12 feet.

Depositional environment: the Marlstone Rock Bed comprises a condensed unit laid down in a low-energy, sediment-starved, inner sublittoral, marine environment, which is locally near emergent (Sellwood & Wilson 1990). The presence of small poorly coated chamositic ooids in outcrop sections has been described by Hallam (1967) as indicating agitated environments.

Upper Lias. Following Cope *et al.* (1980*a*) and Whittaker *et al.* (1985) the Upper Lias can be subdivided in four units; in ascending order, they comprise, the Paper-Shales, the Junction Bed, the Down Cliff Clay (UL1) and Bridport Sands (UL2).

'Paper-Shales' and Junction Bed. Age: Late–Early Toarcian (*falciferum-levesquei* ammonite

zones, *dispansum* ammonite subzone). Occurring within ostracod zone OJ9 (*pars*), foraminiferid zones FJ9 (*pars*)-FJ10 (*pars*), palynological subzones DJ3b–DJ3d1 (*pars*).

A lithologically distinctive organic-rich, subfissile claystone unit, 8–10 feet thick, underlies the Junction Bed on the southern margin of the Channel Basin. Sediments of equivalent age and similar lithology also occur in southern England and the Midlands, and have been informally named 'Paper-Shales' (from their fissility). The Junction Bed, as typically developed, is a highly condensed unit spanning most of the Toarcian Stage.

Lithology: (1) The 'Paper-Shales' comprise dark grey to greyish black, organic-rich, non-calcareous, silty claystones and fissile shales.

(2) The Junction Bed is dominated by white, light grey to light brown, microcrystalline, bioclastic, hard (mudstone–packestone) limestones. Where the Junction Bed is thickly developed, interbeds of medium grey to dark grey, calcareous claystone and marls are also noted.

In ditch cuttings samples, the light grey bioclastic limestone of the Junction Bed is lithologically distinct from the sparry ferruginous limestone of the Marlstone Rock Bed, and this has enabled positive identification in a number of wells where wireline log characteristics are indecisive.

Wireline log characteristics: (1) 'Paper Shales': the lower boundary is very sharp; defined by an abrupt increase in gamma-ray response and a corresponding decrease in sonic velocity values, reflecting a lithological change to organic-rich claystones. The upper boundary is sharp; defined by a marked decrease in gamma-ray response and a corresponding increase in sonic velocity values, reflecting a lithological change to well-cemented limestones.

This thin unit locally underlies the Junction Bed on the southern margin of the Portland–Wight Basin. It has a very distinctive wireline log signature comprising high gamma-ray responses and very low sonic velocity values, reflecting the high organic content of the lithologies.

(2) Junction Bed: the lower boundary is sharp; defined by an abrupt decrease in gamma-ray response and a corresponding increase in sonic velocity values, reflecting a lithological change to well-indurated limestones. The upper boundary is generally sharp; defined by a marked increase in gamma-ray response and a corresponding decrease in sonic velocity values, reflecting a lithological change to claystones.

The Junction Bed is characterized by very low gamma-ray and high sonic velocity values typically forming a cylindrical log motif. In areas where the Junction Bed is more thickly developed and more generally argillaceous, separation from the Down Cliff Clay is not as well defined (e.g. in 98/13-1). These wireline log motifs are much more subdued, characterized by moderately low, coarsely serrated gamma-ray responses, indicating interbedded calcareous claystones–marls and argillaceous limestones. These thicker developments have previously not been differentiated from the overlying Down Cliff Clay (e.g. in Winterborne Kingston-1), however, their lithological characteristics and the biostratigraphic data clearly indicate that they represent a more expanded equivalent of the Junction Bed.

Distribution and thickness: the 'Paper-Shales' unit is developed at the southern edge of the Portland–Wight Basin in wells 98/22-1, 98/22-2 and 98/23-1. A very thin claystone unit, which is present between the Junction Bed and the underlying Marlstone Rock Bed in Horndean-1, Humbly Grove-1 and Bottom Copse-1, is envisaged to be equivalent to the 'Paper-Shales'. It reaches a maximum thickness of no more than 10 feet.

The Junction Bed and its expanded equivalent occurs throughout much of the present study area, except over the Hampshire–Dieppe High and in the main part of the Weald Basin. Its thickness varies from a minimum of 4 feet in Chilworth-1 to a maximum of 172 feet in 98/16-2A (Fig. 9). In many wells it rests unconformably upon the Lower Lias. Its thickest development occurs in the Portland–Wight Basin, where it reaches a maximum thickness of 172 feet in 98/16-2A, thinning in a southerly direction towards the Central Channel High (e.g. 41 feet in 98/22-2). The Junction Bed generally thins northwards over the Purbeck–Isle of Wight Fault Zone (e.g. 16 feet in 98/11-3 and 20 feet in 98/6-8), remaining thin over much of the Dorset Basin and the Hampshire–Dieppe High, with average thicknesses less than 10 feet (e.g. 5 feet at Hoe-1 and 7 feet at Horndean-1). An exception occurs in the Winterborne Kingston-1, where an expanded equivalent of the Junction Bed (41 feet) is envisaged to be present (data from Rhys *et al.* 1982, although not therein recognized as the Junction Bed). The absence of the Junction Bed in the main part of the Weald Basin may reflect lateral facies changes into the lower part of the Down Cliff Clay.

Depositional environment: The 'Paper Shales' unit represents deposition in a low-energy restricted shallow-marine environment, denoted by its monospecific microfaunas, in association with the organic-rich nature of the sediments.

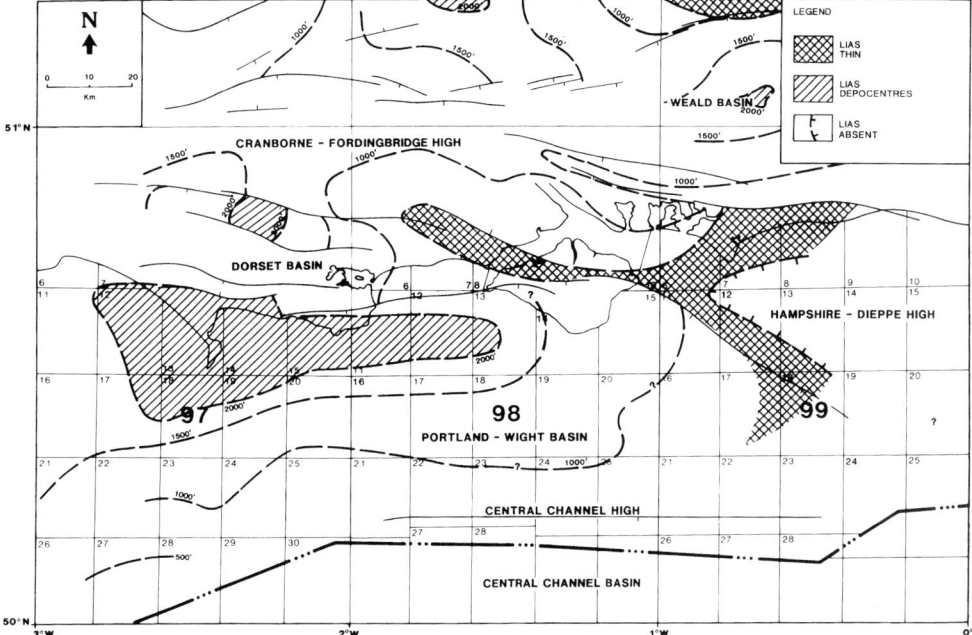

Fig. 9. Thickness and distribution of the Lower Jurassic Lias Group.

The Junction Bed comprises a highly condensed unit, containing algal stromatolites, several non-sequences and planed surfaces, deposited in a low-energy, very shallow-marine environment, similar to the underlying Marlstone Rock Bed. The presence of iron staining is indicative of negligible deposition over a long period, with no terrigenous influx. Possible extreme condensing is suggested by the occurrence of local pebble layers (Hallam 1975). The occurrence of large numbers of well-preserved, vertically orientated ammonites is suggestive of episodic rapid sedimentation (Hallam 1967).

The Junction Bed marks the last regressive phase of the Domerian sedimentary cycle (Sellwood & Wilson 1990). Throughout the Middle Lias of Britain a 'shallowing upwards' (coarsening upwards) cyclicity is noted, similar to that described for the topmost Upper Lias (Hallam 1961; Sellwood & Jenkyns 1975). In southern England this depositional cycle, in ascending order comprises the Eype Clay (mudstone), the Down Cliff and Thorncombe Sands (sandstones) and the Marlstone Rock Bed–Junction Bed (condensed, iron rich limestones).

Down Cliff Clay (UL1). Age: latest Toarcian (*levesquei* ammonite zone, *levesquei* ammonite subzone). Occurring within ostracod zone OJ9 (*pars*), foraminiferid zone FJ10 (*pars*), palynological subzone DJ3d1 (*pars*).

Lithology: the formation is dominated by bluish grey to dark grey, micromicaceous, silty, calcareous claystones. In the upper part of the interval, medium grey to medium dark grey, highly micromicaceous, locally sandy, calcareous siltstones are often present. Rare, light grey limestone stringers, some of which are argillaceous, occur throughout the interval.

Wireline log characteristics: the lower boundary is generally abrupt; defined by a sharp increase in gamma-ray response, associated with a marked decrease in sonic velocity, denoting a lithological change to silty claystones. The upper boundary varies from being sharp to gradational; defined by a decrease in gamma-ray response, in association with a slight to sharp increase in sonic velocity values. The gamma-ray responses between the Down Cliff Clay and the Bridport Sands are generally not well marked due to the high feldspar and clay content of the overlying Bridport Sands. The change in sonic velocity responses from a very serrated character in the Bridport Sands, reflecting the alternation of friable and cemented bands, to a more even response in the Down Cliff Clay, is the most useful indicator for the boundary. In Dorset and the adjacent offshore areas, this boundary is often transitional and therefore can be difficult to identify consistently, possibly partly due to interfingering between these two units. The

Down Cliff Clay is characterized by moderately high to high, finely serrated gamma-ray and moderately low sonic velocity responses, reflecting the siltstone lithologies. Within the Down Cliff Clay, several medium-scale stacked coarsening-upwards trends can be observed in some wells (e.g. Arreton-2). A thin basal unit with a relatively high gamma-ray response has been observed in the southern Channel Basin (98/22-1, 98/23-1 and 99/18-1), which probably represents a more argillaceous unit or a condensed section.

Distribution and thickness: the Down Cliff Clay is present throughout most of the study area, excluding the main part of the Hampshire–Dieppe High where it is absent (Fig. 10). On the southwesterly edge of the Hampshire–Dieppe High (99/18-1), however, an attenuated development (17 feet) does occur. The more silty developments of the Down Cliff Clay generally occur in the Portland–Wight and the Dorset basins, whereas the more argillaceous sequences are noted along the northwesterly margin of the Hampshire–Dieppe High and also the Weald Basin, reflected by a sharp contact with the Bridport Sands (e.g. in Southampton-1). In well 98/18-1, thin limestone and claystone beds are present in the Down Cliff Clay. It is most likely that the boundary between the Down Cliff Clay and Bridport Sands is regionally diachronous. However, biostratigraphical resolution is not sufficient to determine the exact nature/degree of diachroneity.

The thickest developments of Down Cliff Clay (i.e. 530 feet at 98/11-2, 594 feet at 98/16-1 and 586 feet at 98/16-2A) occur within the northern and central part of the Portland–Wight Basin. There is a progressive decrease in thickness of the Down Cliff Clay in both a southerly direction towards the Central Channel High (e.g. 65 feet at 98/22-1 and 46 feet at 99/16-1) and a westerly direction (e.g. 138 feet at Radipole-1). Movement on the Purbeck–Isle of Wight Fault Zone is indicated by a very marked northwards decrease in thickness (210 feet in 98/7-2, 273 feet in Arreton-2 and 21 feet in Sandhills-1). The development of 251 feet of Down Cliff Clay in the Winterborne Kingston-1 decreases to between 70 and 200 feet over the northwesterly margin of the Hampshire–Dieppe High (e.g. 93 feet at Chilworth-1 and 191 feet at Horndean-1). Similar thicknesses are recorded to the east in the Weald Basin (e.g. 45 feet at Humbly Grove-1).

Depositional environment: the Down Cliff Clay represents the base of the second major shallowing-upwards cycle in the Jurassic, following a rapid transgression, after the deposition of

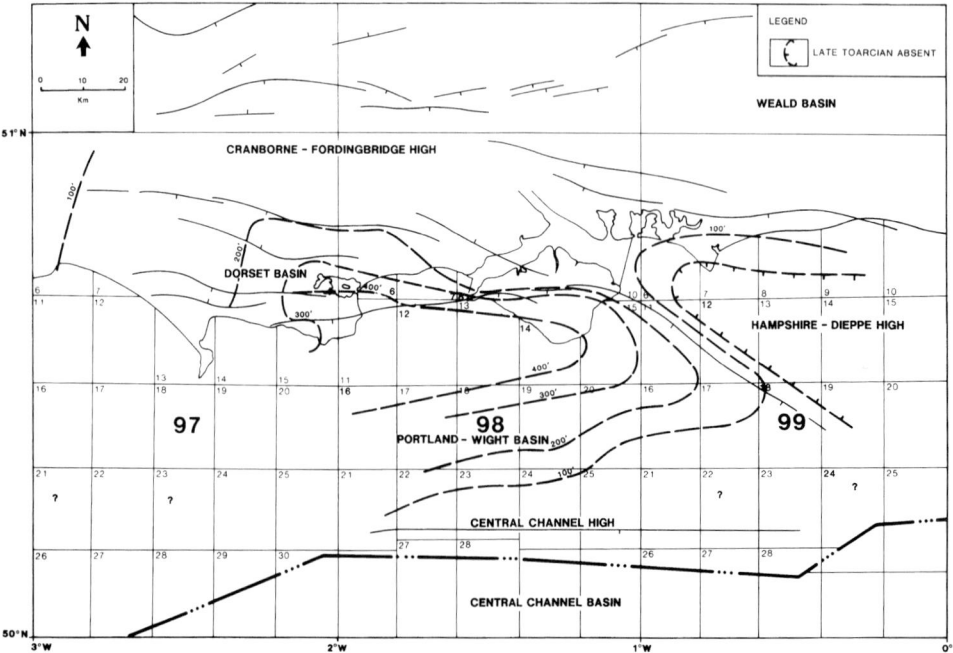

Fig. 10. Thickness and distribution of the uppermost Toarcian Downcliff Clay.

the underlying Marlstone Rock Bed–Junction Bed (Sellwood & Wilson 1990). The Down Cliff Clay deposition commenced with an influx of clay in response to climatic change and/or deepening in a basin immediately offshore of a low archipelago (Cornubia and Amorica) (Sellwood & Wilson 1990). Within the Down Cliff Clay sequence a subtle increase in grain size is noted upsequence from claystones through to silty claystones to micaceous siltstones–argillaceous siltstones suggesting a decreasing water depth. High sedimentation rates are envisaged due to the thick sediment pile accumulated during the short time span, with deposition having taken place in well-oxygenated shallow-marine environments.

Bridport Sands (UL2). Proven hydrocarbons have been described from the Bridport Sands in several localities in Dorset. They form the upper oil reservoir at Wytch Farm (Colter & Havard 1981).

Age: latest Toarcian–earliest Aalenian (*levesquei* ammonite zone, *moorei* ammonite subzone–*opalinum* ammonite zone, *opalinum* ammonite subzone). Occurring within ostracod zones OJ9 (*pars*)–OJ10 (*pars*), foraminiferid zone FJ10 (*pars*), palynological subzones DJ3d1 (*pars*)–DJ3d2 (*pars*).

Lithology: three distinct lithofacies have been recognized in this present study.

(1) The Bridport Sands facies comprises light grey to yellowish grey, very fine- to fine-grained, variably silty and argillaceous, locally glauconitic, sandstones or calcareous sandstones. At outcrop and in conventional core samples the Bridport Sands are heavily bioturbated, with a variety of burrow types picked out by clay-rich linings. At some levels they retain traces of original sedimentary structures, including cross-lamination and flaser-bedding. They comprise metre-scale alternations of friable sands and calcareous nodular sandstones. The calcareous sandstone beds are cleaner, slightly coarser-grained sediments and are rich in bioclasts. Occasional thin beds of sandy, ferruginous (chamositic), bioclastic limestone are also noted.

(2) The 'ferruginous' facies is dominated by medium grey to yellowish brown limestones. They are locally argillaceous, ferruginous, with goethite–berthierine ooliths and grains, and contain abraded bioclasts (bivalve and echinoderm debris) in a matrix of goethite, with ferroan calcite and dolomite cement. Conventional core samples from the Winterborne Kingston-1 and Hoe-1 wells also show planar cross-stratification at some levels.

(3) The 'Bridport Sands equivalent' (calcareous) facies comprises alternating beds of light grey, sandy, silty, bioclastic, microcrystalline limestones and light grey, very fine-grained, silty, argillaceous, micaceous sandstones, in association with medium grey, silty, calcareous claystones and siltstones.

Wireline log characteristics: the lower boundary is gradational to sharp; denoted by a slight decrease in gamma-ray response, associated with a slight increase in sonic velocity values, denoting a lithological change to silty–argillaceous sandstones. The upper boundary is usually well defined; denoted by a decrease in gamma-ray response and an associated decrease in sonic velocity values, reflecting a lithological change to well-indurated limestones. In some areas, however, the base of the Lower Inferior Oolite is represented by sandy limestones, and in such instances the contact with the Bridport Sands can be more difficult to differentiate.

The Bridport Sands are characterized by moderately low gamma-ray and high sonic velocity responses. The gamma-ray values are relatively high for a sandstone unit, due to the significant clay and feldspar content. The sonic log is characteristically serrate, reflecting the alternation of friable sandstones with indurated calcareous sandstone beds. Larger-scale vertical fluctuations in log characteristics indicate changes in clay content, which may be cyclic. The 'ferruginous' facies displays lower gamma-ray and higher sonic velocity responses than the typical Bridport Sands. In the 'Bridport Sands equivalent' facies, the alternation of argillaceous limestones and argillaceous sandstones produces rather irregular log motifs, with low gamma-ray and high sonic velocity responses. The limestone beds possess lower gamma-ray and higher sonic velocity responses than the sandy lithologies.

Distribution and thickness: three major facies belts are recognized in the Bridport Sands (Fig. 11).

(1) The 'Bridport Sands' facies is present throughout the northern Portland–Wight Basin, south Dorset, the Cranborne–Fordingbridge High, and as far east as the Southampton area. At the eastern margin of this area (Southampton-1 and Marchwood-1) it includes beds of ferruginous limestone; these have also been described from Winterborne Kingston-1, suggesting that they occur over a wide area. At outcrop and further south in the subsurface at Wytch Farm (Dorset Basin) and in the Portland–Wight Basin the available data suggest that only sandstones are present. The maximum thickness of the 'Bridport Sands' facies occurs in Winterborne Kingston-1 (324 feet) and Kimmeridge-3 (443+ feet). In the other parts of the present study they range in thickness from

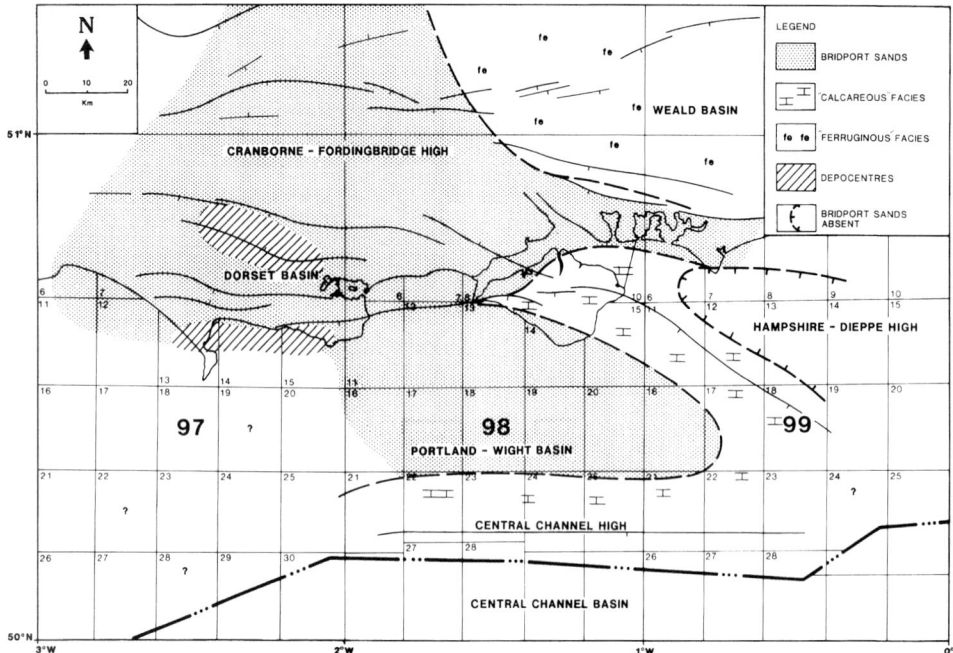

Fig. 11. Thickness and facies distribution of the uppermost Toarcian Bridport Sands.

c. 190 to c. 300 feet, thinning both in a southerly direction (e.g. 226 feet at 98/16-1 and 115 feet at 98/18-1) and westwards at outcrop on the Dorset coast. It is probable that some of these thickness variations are due to lateral facies changes with the Down Cliff Clay. The 'Bridport Sands' facies pass northwards (at outcrop in north Dorset and Somerset) into the Yeovil Sands, which are similar in lithology but cover a slightly longer stratigraphical range.

(2) In the eastern part of the study area, the Bridport Sands are represented by alternations of 'Bridport Sands' facies and 'ferruginous' facies, with the latter often dominant. This facies belt is recognized in Hoe-1, Chilworth-1, Horndean-1 and, possibly, Portsdown-1 (north-western edge of the Hampshire–Dieppe High), and throughout the western Weald Basin. The thin succession at Middleton-1 (47 feet) is, however, apparently within the Bridport Sands facies belt. Their thicknesses vary from 135 feet in Humbly Grove-1 to 270 feet in Chilworth-1. Thicknesses of between 100 and 250 feet are typical in this area.

(3) The calcareous 'Bridport Sands equivalent' facies is developed on the eastern and southern margins of the Portland–Wight Basin, including Arreton-2, Bottom Copse-1, 99/18-1 and 98/22-1. Those wells situated close to the Central Channel High (i.e. 98/22-1, 98/22-2 and 98/23-1) possess thin developments of 'Bridport Sands equivalent' (e.g. 80 feet at 98/22-2 and 64 feet at 98/23-1), whereas those wells occurring along the eastern margin have thicker developments (e.g. 162 feet at Arreton-2). The calcareous 'Bridport Sands equivalent' is absent over the main part of the Hampshire–Dieppe High.

Depositional environment: the Bridport Sands were deposited in association with a shallow-marine migrating barrier bar (Davis 1969, Davis et al. 1971) which extended east–west, with a forebar facies to the south and a back bar facies to the north. The presence of small-scale cross-bedding in the otherwise heavily bioturbated sand indicates transport and deposition by wave-generated currents. The numerous calcareous sandstone nodular beds are interpreted as storm-generated, with early diagenetic cementation, due to the high calcareous bioclastic component (Bryant et al. 1988). Deposition was relatively rapid, as evidenced by the high sedimentation rates. However, these high sedimentation rates may have been intermittent as indicated by very common localized bioturbation.

This thick sand sheet apparently resulted from rapid progradation from the west, where sand deposition commenced rather earlier than in the study area (Cope et al. 1980a), indicated by the occurrence of progressively younger faunas southwards. The Bridport Sands comprise a southwards-extending basinal infill which formed the broad sand flats onto which the overlying carbonates of the Inferior Oolite were deposited. Progradation may have been intermittent, as there is some evidence of lateral passage and interfingering between the Bridport Sands and the Down Cliff Clay. Pauses in progradation may reflect minor episodes of relative sea-level rise. This interpretation is reinforced by the occurrence of thin ferruginous limestone beds (interbeds of the 'ferruginous' facies) in Winterborne Kingston-1, which are sharp-based and form the initial phases of cleaning-upwards (interpreted as shallowing-upwards) sequences. Hoe-1 also possesses similar sequences to Winterborne Kingston-1. These appear to represent condensed units deposited during the initial phases of minor episodes of relative sea-level rise.

The 'ferruginous' facies is characterized by the presence of ferruginous ooliths and cross-stratification at some levels, indicating deposition above wave-base in an environment starved of clastic sediments. Where developed, fringing the London Platform, this may not have been a significant source of sediment during the latest Toarcian. The distribution of calcareous 'Bridport Sands equivalent' is at present poorly documented due to inadequate well data. This sequence of sediments is, however, envisaged to represent a clastic-starved facies, deposited in shallow, sublittoral environments.

Knox *et al.* (1982) suggest that both the 'ferruginous' facies and the calcareous 'Bridport Sands equivalent' were deposited during periods of relative uplift, with their sediments derived from carbonate facies located on nearby shallows.

Inferior Oolite Group

The Inferior Oolite Group comprises carbonate-dominated units subdivided at outcrop into the Lower Inferior Oolite, Middle Inferior Oolite and Upper Inferior Oolite. Each of these units is separated by erosion surfaces marking stratigraphic breaks. In the subsurface the boundary between Lower and Middle Inferior Oolite is generally difficult to recognize both lithologically and on petrophysical log criteria, however, the Upper Inferior Oolite can usually be differentiated.

Age: Aalenian–earliest Bathonian (*opalinum* ammonite zone, *scissum* ammonite subzone–*zigzag* ammonite zone, lower *yeovilensis* ammonite subzone). Occurring within ostracod zones OJ10–OJ12, foraminiferid zones FJ10 (*pars*)-FJ12 (*pars*), palynological subzones DJ3d3–DJ4b.

The Lower Inferior Oolite is of Aalenian–earliest Bajocian age, the Middle Inferior Oolite is of Early–early Late Bajocian age, while the Upper Inferior Oolite is of Late Bajocian–earliest Bathonian age.

Lithology: a three-fold subdivision is recognized.

(1) The Lower Inferior Oolite is dominated by light grey to yellow, locally ferruginous, bioclastic, sparry wackestones and packstones. A few limestone beds yield abundant limonitic ooliths and pisoliths. Towards the base of the unit there may be local thin developments of light grey to light yellow, very fine-grained, silty sandstones and calcareous sandstones.

(2) The Middle Inferior Oolite comprises light grey to buff, oolitic, bioclastic sparry packstones and grainstones, many of which are ferruginous. Thin, interbedded horizons of buff to medium grey marl occur in a number of wells. Dark grey to black chert is present at some levels.

(3) The Upper Inferior Oolite is dominated by light grey to white or buff, bioclastic, sparry packstones and grainstones. Oolites and glauconite may occur throughout the section. Beds of buff to medium grey marl and calcareous claystone may be developed throughout the unit. A basal bed with reworked limestone clasts and phosphatic pebbles is noted in a number of localities.

Wireline log characteristics: the lower boundary is generally sharp; defined by a decrease in gamma-ray response and a corresponding increase in sonic velocity values, reflecting a lithological change to oolitic or sandy limestones. The upper boundary is sharp; defined by a marked increase in gamma-ray response and an associated decrease in sonic velocity values, denoting a lithological change to claystones. In some areas, where the Fuller's Earth Clay is very calcareous and the Upper Inferior Oolite argillaceous, the contrast in the gamma-ray response may be less marked, however, the sonic log velocity still exhibits a clear break.

Low gamma-ray values, in association with moderately high sonic velocity values, characterize the Inferior Oolite Group, with thin, argillaceous beds being marked by higher sonic velocity responses. The Lower and Middle Inferior Oolite are characterized by similar low gamma-ray and high sonic velocity values to the Upper Inferior Oolite. The boundary between these units, however, cannot be consistently identified from wireline log characteristics in the present study. The base of the Upper Inferior Oolite is often marked by a very prominent gamma-ray spike, reflecting a concentration of phosphate.

Distribution and thickness: the Inferior Oolite Group occurs throughout the study area, except over the main part of the Hampshire–Dieppe High. Its thickness ranges from a minimum of 4 feet in Wytch Farm-X14 to a maximum of 361 feet in Crokerhill-1. Wireline log criteria indicates that both the Lower–Middle and Upper Inferior Oolite are present in many of the wells. In many cases the Lower–Middle Inferior Oolite are generally thicker than the Upper Inferior Oolite, although the thickness of all the units seems to be very variable regionally.

Great thicknesses of the Inferior Oolite Group occur in the northern part of the Portland–Wight Basin (e.g. 231 feet at Arreton-2) and the northwestern edge of the Hampshire–Dieppe High (e.g. 361 feet at Crokerhill-1 and 387 feet at Portsdown-1 (Taitt & Kent 1958)). The limited data available indicate the presence of a depocentre in this region (Fig. 12). There is a very rapid eastward decrease in thickness to less than 30 feet over the Purbeck–Isle of Wight Fault Zone (e.g. 16 feet at 98/7-2, 4 feet at Wytch Farm-X14 and 6 feet at outcrop on the Dorset coast); southeast of Sandhills-1 the Inferior Oolite is absent on the upthrown side of the Wight–Bray Fault (Hamblin et al. 1992). Southwards into the central part of the Portland–Wight Basin, thicknesses in the range of 20–37 feet are encountered, these rapidly increase westwards (e.g. 148 feet at 99/16-1) and towards the Central Channel High (e.g. 148 feet at 98/23-1). The Inferior Oolite Group thickens northwards into the Dorset Basin (69 feet at Winterborne Kingston-1) and continues to thicken eastwards across both the Cranborne–Fordingbridge and Hampshire–Dieppe Highs (e.g. 169 feet at Chilworth-1) (Thomas & Holliday 1982; Young & Lake 1988). This trend of increasing thickness continues into the Weald Basin, where a maximum of over 400 feet of Inferior Oolite Group has been recorded (Gallois & Worssam 1993).

Depositional environment: the Inferior Oolite Group represents the culmination of a shallowing upwards cycle which commenced with the Late Toarcian Down Cliff Clay. These sediments were envisaged to have been deposited in an intra-basinal structural high, distal to a carbonate-ramp facies developed along the London–Brabant landmass (Sellwood & Jenkyns 1975). The occurrence of oolitic–shell detrital grainstones, conglomerates ('snuff boxes') and profuse benthic faunas, including algal stromatolites and oncolites, indicates deposition in shallow, generally high-energy, marine environments

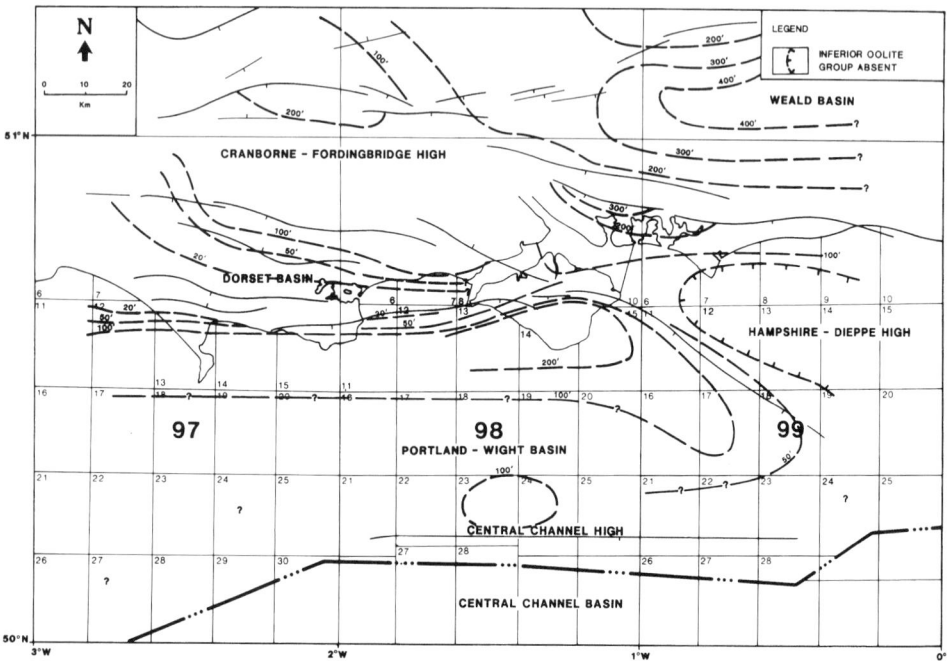

Fig. 12. Thickness and distribution of the Aalenian–lowermost Bathonian Inferior Oolite.

(Penn et al. 1980; Hollaway 1985c). Gatrall et al. (1972) suggested that these sediments were laid down in waters only tens of feet deep, upon a submarine swell or on an undulating shelf. A number of non-sequences are noted within the formation (accentuated by episodic tectonic activity) associated with the development of hardgrounds and erosion surfaces (Arkell 1933). These are followed by marine transgressions which can be identified across much of southern Britain. Examples include the Aalenian–Bajocian and the intra-Early Bajocian hardgrounds. Condensed intervals, resulting from marine transgression or sediment starvation over the highs are often enriched in iron and local phosphate. The development of calcareous sandstones in some areas indicates local or regional shallowing at times, however, it is uncertain whether this was due to structural or eustatic controls.

Great Oolite Group

Following Cope et al. (1980b) the Great Oolite Group can be subdivided into five formations; in ascending order, they comprise, the Fuller's Earth, the Frome Clay, the Great Oolite, the Forest Marble and the Abbotsbury Cornbrash Formation. Conventionally the base of the Fleet Member, the upper member of the Abbotsbury Cornbrash Formation, lies above the Great Oolite Group. However, for the purposes of this paper, it has been placed within the Group.

Fuller's Earth. The Fuller's Earth comprises two thick claystone-dominated units, the Lower and Upper Fuller's Earth Clay, separated by a thinner unit dominated by argillaceous limestones, the Fuller's Earth Rock (Cope et al., 1980b).

Age: Early–earliest Late Bathonian (*zigzag* ammonite zone, upper *yeovilensis* ammonite subzone–*hodsoni* ammonite zone, lower *quercinus* ammonite subzone). Occurring within ostracod zones OJ13–OJ14 (*pars*), foraminiferid zones FJ12 (*pars*)–FJ14 (*pars*), palynological subzones DJ5a–DJ5b (*pars*).

The Lower Fuller's Earth is of Early–earliest Middle Bathonian age, the Fuller's Earth Rock is of Middle–earliest Late Bathonian age, while Upper Fuller's Earth is of earliest Late Bathonian age.

Lithology: a three-fold subdivision is recognized.

(1) The Lower Fuller's Earth Clay comprises medium grey to dark grey, locally silty mudstones, many of which are quite calcareous. Rare, medium grey argillaceous limestone stringers also occur throughout the section.

(2) The Fuller's Earth Rock is dominated by light to medium grey, bioclastic, argillaceous limestones, interbedded with thinner medium to dark grey, silty mudstones and marls.

(3) The Upper Fuller's Earth comprises light to dark grey, locally silty, calcareous mudstones. Stringers of light grey limestone, some of which are argillaceous, occur in many well sections, more especially towards the top of the unit.

Wireline log characteristics: a three-fold subdivision is recognized:

(1) The Lower Fuller's Earth Clay: the lower boundary of the Lower Fuller's Earth Clay is well defined; denoted by an abrupt increase in the gamma-ray response associated with a sharp decrease in sonic velocity values, coinciding with a lithological change to claystones. The upper boundary is well to poorly defined; denoted by a decrease in the gamma-ray response, associated with an increase in sonic velocity values, coinciding with a lithological change to marls and limestones. The Lower Fuller's Earth Clay possesses similar log responses to the Upper Fuller's Earth Clay, comprising moderately low, slightly serrate gamma-ray and moderately high sonic velocity responses, reflecting the uniform lithologies. Towards the base of this unit the gamma-ray response increases whilst sonic velocity decreases, reflecting the less calcareous nature of the claystones.

(2) The Fuller's Earth Rock: the lower boundary is sharp to gradational; denoted by a decrease in the gamma-ray response, associated with an increase in sonic velocity values, coinciding with a lithological change to marls and limestones. The upper boundary is well to poorly defined; denoted by an increase in the gamma-ray response associated with a decrease in sonic velocity values, coinciding with a lithological change to calcareous claystones. The Fuller's Earth Rock is represented by a serrated log profile with low gamma-ray and high sonic velocity, reflecting the interbedded argillaceous limestones and calcareous claystones. The argillaceous nature of the limestones, however, only gives a very limited contrast with the claystones above and beneath. The Fuller's Earth Rock is more clearly differentiated in north Dorset and Somerset. In the Portland–Wight Basin it is only poorly developed and often cannot be recognized with any degree of certainty.

(3) The Upper Fuller's Earth: the lower boundary is well to poorly defined; denoted by an increase in the gamma-ray response, associated with a decrease in sonic velocity values,

coinciding with a lithological change to calcareous claystones. The upper boundary is generally sharp; defined by a decrease in gamma-ray response and an associated increase in sonic velocity, reflecting a lithological change to argillaceous limestones–calcareous claystones. In areas where the Wattonensis Beds are poorly developed, the Upper Fuller's Earth is differentiated from the Frome Clay by its more spiky gamma-ray response. The Upper Fuller's Earth Clay generally has serrate, moderately high gamma-ray responses, and moderately low sonic velocities, reflecting the interbedded argillaceous limestone–claystone lithologies.

Distribution and thickness: the Fuller's Earth occurs throughout the study area and in many of the wells examined. Subdivision of the Fuller's Earth into the three constituent units (Lower Fuller's Earth Clay, Fuller's Earth Rock and Upper Fuller's Earth Clay) can often be extremely tenuous due to poor wireline log differentiation and the recovery of only long-ranging microfossil taxa. This is most evident in the offshore areas, in UK Quadrant 99 and parts of UK Quadrant 98. In a number of wells (e.g. well 98/18-1) the Fuller's Earth could not be subdivided due either to the absence of the Upper Fuller's Earth Clay or, more likely, to the Fuller's Earth Rock shaling out.

A maximum thickness of almost 700 feet of Fuller's Earth occurs in Lulworth Banks-1, situated within the northern Portland–Wight Basin (Martin 1967), in an area where the seismic and sparse well data indicate a depocentre. The adjacent wells (Kimmeridge-5 and Seabarn Farm-1) also possess thick sections, 573 and 661 feet, respectively. The Fuller's Earth thins very slightly towards the Central Channel High (e.g. 419 feet at 98/16-2A and 471 feet at 98/22-2), but thins abruptly to the north across the Purbeck–Isle of Wight Fault Zone (e.g. 167 feet at 98/11-3 and 117 feet at Sandhills-1) and over the south Dorset High (Fig. 13). In the Wytch Farm area less than 200 feet of Fuller's Earth is present. There is a general eastward thinning from the Cranborne–Fordingbridge High across the northwesterly part of the Hampshire–Dieppe High (e.g. 140 feet at Bottom Copse-1, 97 feet at Chilworth-1 and 114 feet at Southampton-1) and into the Weald Basin (e.g. 100 feet at Humbly Grove-1). The Fuller's Earth is the oldest Jurassic formation present over the main part of the Hampshire–Dieppe High (e.g. 197 feet at 99/12-1), which is envisaged to have been onlapped by Jurassic sediments during the Bathonian.

Depositional environment: the Fuller's Earth was deposited in a low-energy, open marine,

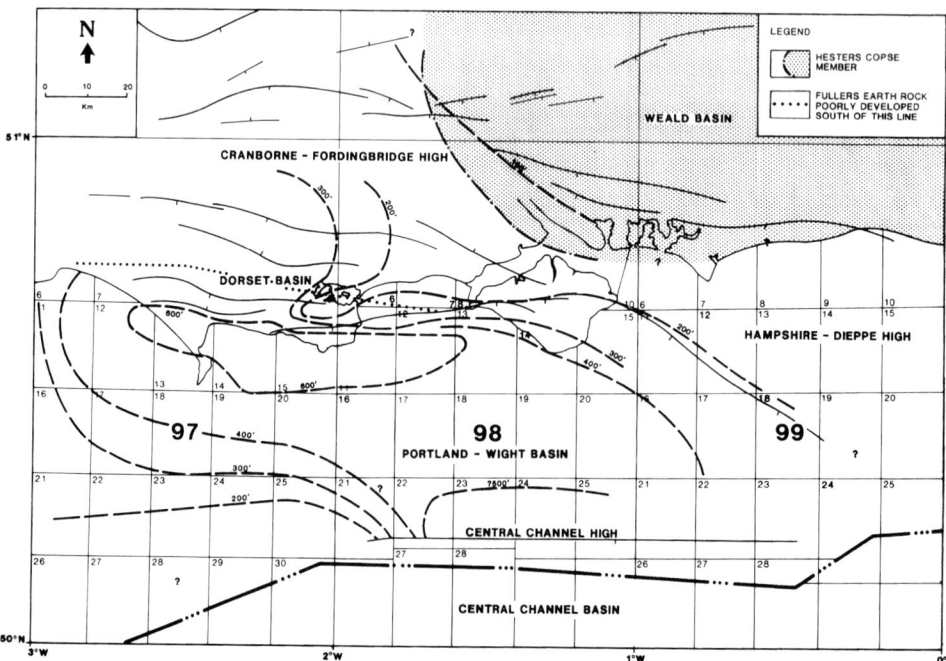

Fig. 13. Thickness and distribution of the Early to lowermost Late Bathonian Fuller's Earth Clay.

shelf environment, largely below wave-base. Benthic micro- and macrofaunas occur throughout the formation, indicating well-oxygenated bottom waters. However, episodes of restricted circulation, locally with bottom water anoxia, are indicated by the absence of microfaunas. Similar restricted environments have been described by Penn (1982) from the Fuller's Earth of the Winterborne Kingston-1.

The Fuller's Earth represents an initial deepening event before the commencement of a major regressive sedimentary cycle (House 1993), which culminated at the top of the Forest Marble. Superimposed upon this cycle are two further shallowing-upwards cycles; an underlying Lower Fuller's Earth Clay–Fuller's Earth Rock cycle and an overlying Upper Fuller's Earth Clay–Wattonensis Beds cycle (Sellwood et al. 1985). Penn (1982) also recognized a number of very minor regressive cycles, comprising a basal, silty limestone yielding a varied shelly fauna, passing upwards into mudrocks containing a less diverse benthos and finally into a minor regressive facies at the top.

Hesters Copse Member. The Hesters Copse Member (Sellwood et al. 1985) comprises a carbonate-dominated interval, which includes both sandstones and dolomitic limestones. It occurs between an underlying Fuller's Earth and an overlying Great Oolite in the Humbly Grove Oilfield. It is regarded as a member of the Fuller's Earth Clay due to its lateral passage into the upper part of the Fuller's Earth Clay.

Similar lithologies to the Hesters Copse Member are recognized over a wide area, apparently passing laterally into the Fuller's Earth Rock(?), Upper Fuller's Earth and Wattonensis Beds. Sellwood et al. (1985) recognized a representative of the Fuller's Earth Rock underlying the Hesters Copse 'Formation' in the type area.

Age: earliest Late Bathonian. Occurring within ostracod zone OJ14 (*pars*), foraminiferid zone FJ14 (*pars*), palynological subzone DJ5b (*pars*).

Lithology: The Hesters Copse Member comprises a complex interbedded sequence of sediments. Notable lithologies include white, very fine- to fine-grained, calcareous sandstones and light grey to medium grey, dolomicrites through to bioclastic dolomitic packstones. Some of the dolomites are locally oolitic or oncolitic, while others are sandy. The claystones are generally medium grey, oncolitic (characteristically with black oncolites and pisoliths) and are calcareous.

In the Humbly Grove Oilfield, the Hesters Copse Member comprises three subdivisions (Sellwood et al., 1985), in ascending order they comprise the lower Sandstone Unit, the Wackestone Unit and the Dolomitic Unit. Similar lithologies are widespread in the subsurface, however, where conventional core samples are not available the existence of a similar succession is difficult to verify. The Hesters Copse Member differs from similar aged units mainly by the presence of sandstones and dolomitic limestones.

Wireline log characteristics: the lower boundary is sharp; defined by a decrease in gamma-ray response and an increase in sonic velocity values reflecting a lithological change to sandstones. The upper boundary is sharp; defined by an increase in gamma-ray response, with a coinciding decrease in sonic velocity values reflecting a lithological change to claystones. The Hesters Copse Member is characterized by moderately low, erratic irregular gamma-ray and moderately high sonic velocity responses, reflecting the interbedded lithologies.

A prominent, low sonic spike is recognized within this member in the Humbly Grove Oilfield. It is not obvious in Humbly Grove-1, however, it is clearly defined in other wells in the area (Sellwood et al. 1985). This spike is situated in the lower part of the Wackestone Unit, close to its base. An identical log feature is present in other wells in the study area where the Hesters Copse Member is developed (e.g. Horndean-1).

Distribution and thickness: the Hesters Copse Member occurs mainly within the Weald Basin. Its western limit lies to the west of Southampton, extending southeastwards along the northern margin of the Hampshire–Dieppe High. Its thickness ranges from 23 feet in Southampton-1 to 79 feet in Horndean-1. An equivalent to the Hesters Copse Member can be seen as far south as the Isle of Wight (e.g. Sandhills-1).

Depositional environment: the Hesters Copse Member is envisaged to represent a nearshore equivalent of the Upper Fuller's Earth. The lower Sandstone Unit is envisaged to have been deposited as an offshore bar–lower shoreface sand, accumulating during a phase of increased sediment supply (Sellwood et al. 1985). Following sand deposition there was a relative rise in sea level leading to the deposition of argillaceous sediments (the Wackestone Unit and the Dolomitic Unit) formed in a quiet, wave-influenced, shelf-lagoonal environment, with storms introducing oolitic debris into the area. This member fringes the London Platform, passing downslope into the upper part of the Fuller's Earth, which was deposited in a deeper marine setting.

Frome Clay. The Frome Clay comprises two units, the Frome Clay (*sensu stricto*) and a basal calcareous unit, the Wattonensis Beds (Cope et al. 1980b).

Age: Late Bathonian (*hodsoni* ammonite zone, upper *quercinus* ammonite subzone–lower *orbis* ammonite zone). Occurring within ostracod zones OJ14 (*pars*)–OJ15 (*pars*), foraminiferid zones FJ14 (*pars*)–FJ15 (*pars*), palynological subzone DJ5b (*pars*).

Lithology: a two-fold subdivision is recognized:

(1) The Wattonensis Beds comprise an interbedded argillaceous limestone–calcareous claystone sequence. The former are light grey, generally bioclastic, with common echinoderm debris; while the latter comprise medium grey, silty, bioclastic calcareous claystones and marls.

(2) The Frome Clay is dominated by medium–light grey to dark grey, locally silty, mudstones. The calcareous content of these mudrocks is high both at the base (the Wattonensis Beds) and also in its upper third. Stringers of light grey, bioclastic argillaceous limestone occur throughout the section. In UK blocks 98/6, 98/7 and 98/11 the Frome Clay is dominated by medium–light grey marls and argillaceous limestones.

Wireline log characteristics: the lower boundary is sharp; defined at a decrease in gamma-ray response, associated with an increase in sonic velocity denoting a lithological change to calcareous claystone–argillaceous limestone. The upper boundary is sharp; defined by a decrease in gamma-ray response, associated with an increase in sonic velocity values, denoting a lithological change to limestones.

The Frome Clay (*sensu stricto*) is characterized by a progressive upsection increase in gamma-ray response, mirrored by decreasing sonic velocity, reflecting a progressive upward decrease in carbonate content. An interval with very low sonic velocities can be identified at the base in some onshore wells (e.g. Winterborne Kingston-1; Whittaker et al. 1985), corresponding to a unit of black organic-rich claystones. This unit can also be recognized in some other wells in the area (e.g. Wytch Farm-X14). The gamma-ray values of the Frome Clay are characteristically higher than those of the claystones of the underlying Fuller's Earth Clay. In UK blocks 98/6, 98/7 and 98/11 the Frome Clay possesses a flattened, low gamma-ray motif reflecting the high calcareous content of the lithologies (marls and argillaceous limestones).

The top of the Wattonensis Beds is marked by an increase in the gamma-ray log response and a corresponding decrease in the sonic velocity. The sonic log is more serrate in the Wattonensis Beds than in the overlying Frome Clay. These log motifs reflect a downward lithological change from claystones to calcareous claystones with interbedded argillaceous limestones. In the onshore sections the Frome Clay is relatively thick compared with the Wattonensis Beds, however, in the Portland–Wight Basin, the Wattonensis Beds are thicker and become more argillaceous. In some wells, particularly in the Portland–Wight Basin (e.g. 98/23-1), a thick sequence of interbedded argillaceous limestones and claystones with a markedly serrated wireline log profile lies between the Lower Fuller's Earth and the Frome Clay. This unit represents a lateral equivalent of the Wattonensis Beds which, together with the Fuller's Earth Rock and the Upper Fuller's Earth, cannot be readily differentiated.

Distribution and thickness: the Frome Clay occurs in the western and southern parts of the present study area, including Dorset and the western and southern parts of the Portland–Wight Basin (Fig. 14). Northeastwards it passes laterally into the carbonates of the Great Oolite (Martin 1967; Green & Donovan 1985; Hollaway 1985b; Whittaker 1985). In UK blocks 98/6, 98/7 and 98/11 the Frome Clay is highly calcareous, comprising marls and argillaceous limestones throughout its section. This is envisaged to reflect proximity to the boundary between the Frome Clay and the Great Oolite facies belts.

The thickness of the Frome Clay varies from 16 feet in well 99/16-1 to 246 feet in both Kimmeridge-5 and Seabarn Farm-1. The maximum thicknesses of this unit are situated towards the northern margin of the Portland–Wight Basin indicating a depocentre within this area (Martin 1967). Within the main part of the Portland–Wight Basin, the Frome Clay ranges in thickness between 164 feet at 98/23-1 and 239 feet at 98/18-1. There is a significant reduction in thickness across the Purbeck–Isle of Wight Fault Zone (e.g. 166 feet at Wytch Farm-X14, 168 feet at 98/11-1, 110 feet at 98/6-8 and 79 feet at 98/7-2) and the Dorset High. The sequence thickens up northwards in the Winterborne Kingston Trough (190 feet at Winterborne Kingston-1), while slightly reduced thicknesses are noted over the Cranborne–Fordingbridge High.

Depositional environment: the Frome Clay was deposited in low-energy, shallow shelf marine environments (with parts experiencing slightly restricted circulation), associated with moderately high sedimentation rates. Although

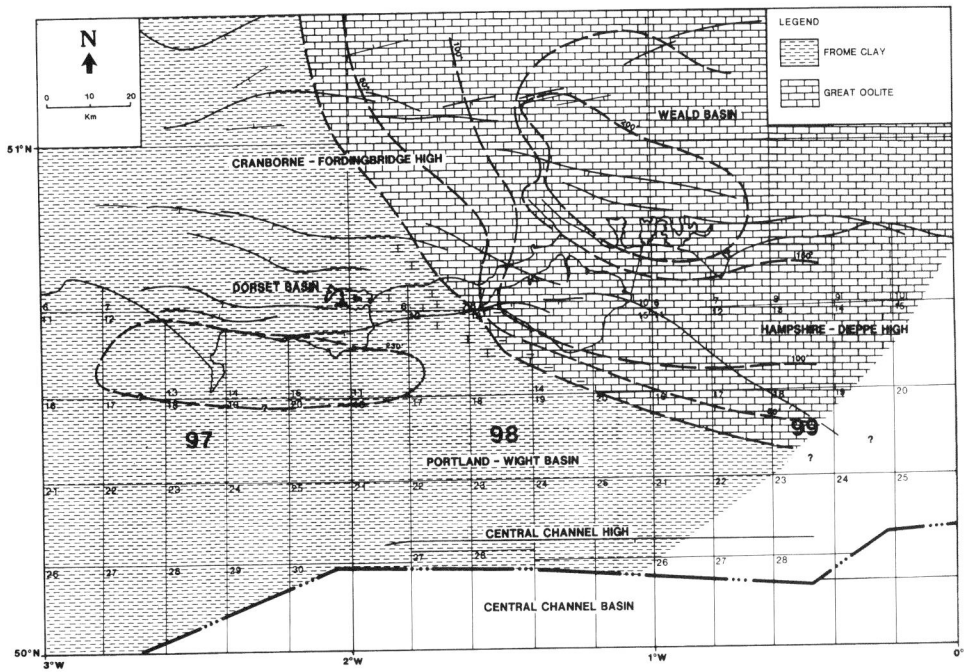

Fig. 14. Thickness and facies distribution of the Late Bathonian Great Oolite and Frome Clay.

relatively shallow-marine deposition, this formation was laid down in a deeper water setting compared to the Great Oolite Formation. The basal, silty Wattonensis Beds, containing an abundant epifaunal benthic macrofauna, are indicative of a period of limited clastic sediment supply and slow sedimentation rates (Penn 1982). This basal sequence is thought to be the end phase of a minor regressive sedimentary cycle which commenced at the base of the Upper Fuller's Earth Clay.

Great Oolite. The limestones of the Great Oolite form the main reservoir in the Humbly Grove Oilfield as well as several other localities in the Weald Basin (Sellwood *et al.* 1985, Hancock & Mithen 1987).

Age: late Bathonian (*hodsoni* ammonite zone, upper *quercinus* ammonite subzone–lower *orbis* ammonite zone). Occurring within ostracod zones OJ14 (*pars*)–OJ15 (*pars*), foraminiferid zones FJ14 (*pars*)–FJ15 (*pars*), palynological subzone DJ5b (*pars*).

Lithology: the Great Oolite is dominated by white to light grey, buff, bioclastic, oolitic grainstones and packstones. These limestones are locally argillaceous or more rarely dolomitized. Thin, light grey to medium grey calcareous claystone–marl interbeds occur throughout the section.

Wireline log characteristics: the lower boundary is sharp; denoted by an abrupt decrease in gamma-ray response, coinciding with an increase in sonic velocity values, reflecting a lithological change to oolitic limestones. The upper boundary is sharp; defined by a marked increase in gamma-ray response and an associated decrease in sonic velocity values, corresponding with a lithological change to calcareous claystones. The Great Oolite is characterized by a cylindrical, slightly serrated log motif, with very low gamma-ray values and moderately high sonic velocities. Thin intervals with slightly higher gamma-ray values indicate beds of calcareous claystone. Large-scale vertical variations in gamma-ray response indicate the positions of prograding oolite shoals, with concomittant 'cleaning-upwards' trends. The Great Oolite grades into the Frome Clay (e.g. UK block 98/11); at the distal margins of the Great Oolite, argillaceous limestone units are developed, denoted by higher gamma-ray responses (e.g. in Marchwood-1).

Distribution and thickness: the Great Oolite forms a carbonate ramp surrounding the southern and western fringes of the London Platform (Sellwood et al. 1985). In the present study, its maximum thickness (275 feet at Crokerhill-1) occurs on the northwestern margin of the Hampshire–Dieppe High (Fig. 14). It thins slightly northeastwards, extending north and east of the study area towards the London Platform (Hamblin et al. 1992). The Great Oolite rapidly thins both southwards (120 feet at 99/12-1) and southwestwards (e.g. 83 feet at Bottom Copse-1 and 97 feet at 98/13-1), passing transitionally into (98/11-1 and 98/11-2) and inter-fingering with the Frome Clay. The northern margin of the Great Oolite apparently extends from Wiltshire through western Hampshire, south of the Isle of Wight and then eastwards towards the centre of the English Channel. Thin oolitic limestone beds within the Frome Clay in a Wytch Farm well (Bristow et al. 1991), are suggestive of a wide area of interdigitation at the boundary between these facies belts.

Depositional environment: the Great Oolite was deposited on a shallow carbonate shelf, above wave-base, which gradually deepens southwards and westwards (Sellwood et al. 1985; Sellwood & Evans 1987). It comprises cross-bedded oolite dunes, with intervening bioclastic limestones, which reflect an intermittently mobile complex of oolite sand dunes with shell debris accumulated in interdune channels. In cored sections, stacked coarsening-upwards cycles with hardgrounds at the base are identified. These are interpreted as successive episodes of oolite shoal progradation under tidal influence, followed by abandonment (Sellwood et al. 1985). There is evidence for emergence at the tops of some of these cycles. On a regional scale, the Great Oolite can be seen to form a wedge prograding outwards from the London Platform during a period of relative structural stability.

Forest Marble. Age: Late Bathonian (upper *orbis–discus* ammonite zones, *hollandi* ammonite subzone). Occurring within ostracod zones OJ15 (*pars*)–OJ16 (*pars*), foraminiferid zones FJ15 (*pars*)–FJ16 (*pars*), palynological subzone DJ5b (*pars*).

Lithology: this formation encompasses a wide variety of sediment throughout the study area. It is dominated by alternations of limestones, marls and claystones. The limestones comprise light grey to brownish grey, bioclastic micrites through to grainstones, a number of which are oolitic; while the marls and claystones are light grey to medium grey, grey–green, locally silty or oolitic and often bioclastic. Thin stringers of medium grey to brown, very fine-grained, sandy limestones–calcareous sandstones are also noted in a number of wells. Plant debris is common locally.

Wireline log characteristics: the lower boundary is sharp; where it overlies the Great Oolite or the Frome Clay. In the former, it is denoted by an abrupt increase in gamma-ray response and an associated decrease in sonic velocity values, reflecting a lithological change to calcareous claystones; in the latter, it is denoted by a decrease in gamma-ray response and an associated increase in sonic velocity values reflecting a lithological change to limestones. The upper boundary can be difficult to identify in the subsurface as both this unit and the overlying Abbotsbury Cornbrash Formation comprise interbedded limestones and claystones. The upper boundary of the Forest Marble is placed at the top of the highest thick claystone bed, beneath the limestone-dominated Cornbrash. This boundary is indicated by a marked decrease in gamma-ray response and an associated increase in sonic velocity values.

The Forest Marble is characterized by a serrated/irregular log motif, indicating interbedded limestones, claystones and sandstones. The claystones exhibit moderately high gamma-ray responses due to their low calcareous content, whilst the limestones and sandstones are characterized by high sonic velocities with a spiky/serrated profile and low gamma-ray responses. A prominent unit with moderate to high gamma-ray and very low sonic velocity responses can be identified within the upper part of the Forest Marble, with an average thickness of 20–30 feet. This unit has been cored in the Winterborne Kingston-1 borehole (Rhys et al. 1982), and comprises grey calcareous claystones. This unit can also be recognized in south Dorset and within the Portland–Wight Basin.

Thin, bioclastic limestone beds within the lower part of the Forest Marble form prominent low sonic velocity spikes. Two such discrete units are recognized in south Dorset; the Boueti Bed (which defines the base of the Forest Marble) and the higher Digona Bed (Melville & Freshney 1982); these beds have been tentatively identified in some wells in the Portland–Wight Basin.

Distribution and thickness: the Forest Marble occurs throughout the study area, with thicknesses ranging from less than 5 feet at Horndean-1 to a maximum of 245 feet in Seabarn Farm-1 (Fig. 15). Its thickest developments occur along the northern edge of the Portland–Wight Basin, where thicknesses of over 200 feet in both the Seabarn Farm-1 and Lulworth

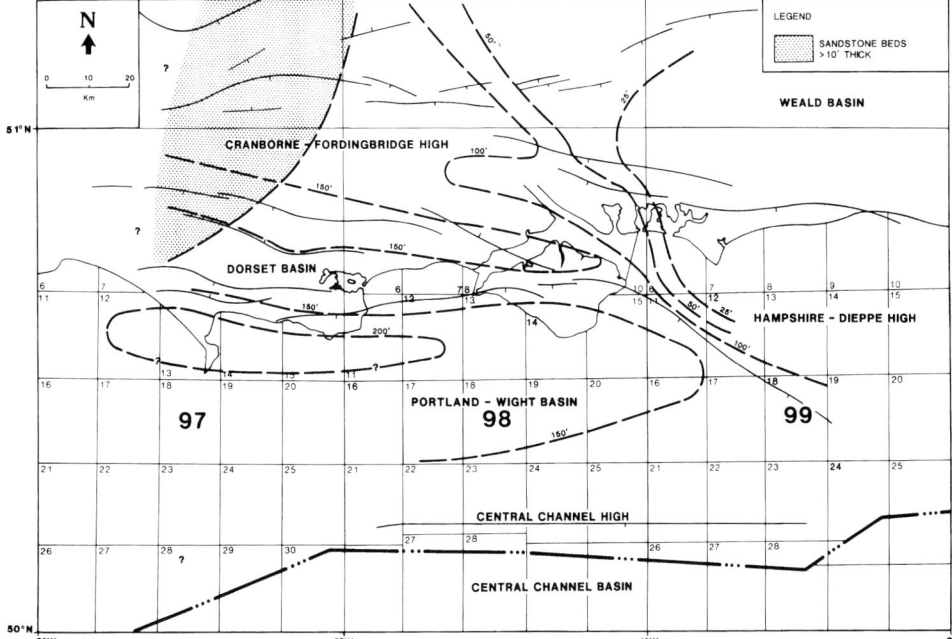

Fig. 15. Thickness and distribution of the Late Bathonian Forest Marble.

Banks-1 suggest an east–west elongated depocentre (Hamblin et al. 1992). Further south within the main part of the Portland–Wight Basin the Forest Marble ranges between 95 and 161 feet (98/22-2 and 98/16-1, respectively). The influence of both the Purbeck–Isle of Wight Fault Zone and the Dorset High is subdued, reflected by a slight thinning of the Forest Marble (e.g. 150 feet at 98/11-2 and 143 feet at 98/6-8). Thicknesses of between 100 and 180 feet are maintained over a wide area to the north over the Cranborne–Fordingbridge High. Due east of the Cranborne–Fordingbridge High, however, there is rapid eastward thinning over both the Hampshire–Dieppe High (e.g. from 165 feet at Hoe-1 and Bottom Copse-1 to 10 feet at 99/12-1 and 5 feet at Horndean-1) and the Weald Basin (e.g. 32 feet at Humbly Grove-1). This trend continues into the main part of the Weald Basin (Hamblin et al. 1992).

Laterally the thin developments of the Forest Marble occurring within parts of the study area coincide with the thick developments of the Great Oolite. These thin sequences of Forest Marble may reflect the persistence of a relatively shallow platform–ramp fringing the London Platform leading to limited differential compaction.

A detailed stratigraphical breakdown of the Forest Marble is extremely difficult due to the impersistence of many of the individual beds, however, the overall facies development is similar in most parts of the study area. Moderately thick sandstone units (the 'Hinton Sands' facies up to 50 feet thick) are developed to the northwest of the study area. In other parts of the region sandstones generally occur as very thin beds, or laminae, and lenses rather than discrete units.

Depositional environment: the Forest Marble is envisaged to reflect the topmost unit of a major regressive cycle which commenced at the base of the Fuller's Earth (Penn 1982; House 1993). Superimposed on this major regressive cycle are three shell beds (in ascending order comprising the Boueti Bed, Digona Bed and the hebridica lumachelle unit) which represent basal phases of three minor sedimentary cycles (Penn 1982). In all three instances, a decrease in both sediment supply and grain size is noted.

The lower part of the Forest Marble is envisaged to have been deposited in low-energy marine environments below wave-base. The occurrence of further thin, bioclastic limestone beds probably represents minor breaks in sedimentation or storm-generated lag deposits (Sellwood et al. 1985). The middle and upper parts of the formation show progressive shallowing, interrupted by minor transgressive episodes, with clays and silts deposited in low-energy, restricted marginal marine (bay/ lagoon) environments. In less restricted shallow-water environments, tidal influences are indicated by lenticular and flaser bedded sediments, while bioclastic limestone lenses reflect tidal channel deposits with sands deposited as tidal sandwaves. The increasing thickness of sands towards the northwest of the present study suggests clastic sourcing in this area.

Abbotsbury Cornbrash Formation. The Abbotsbury Cornbrash Formation comprises two distinct lithological units, often separated by a stratigraphic discontinuity, the Lower Cornbrash (latest Bathonian) and the Upper Cornbrash (earliest Callovian). Page (1989) has renamed the Upper Cornbrash the Fleet Member, and the Lower Cornbrash the Berry Member. These two members cannot, however, be readily be discerned in the subcrop without conventional core samples. Therefore, in the present study the Abbotsbury Cornbrash Formation has been treated as a single lithological unit.

The Kimmeridge Oilfield, located on the Dorset coast, produces hydrocarbons from a fissure system within the Abbotsbury Cornbrash Formation (Brunstrom 1963; Colter & Havard 1981).

Age: latest Bathonian–earliest Callovian (*discus* ammonite zone, *discus* ammonite subzone–*macrocephalus* ammonite zone, lower *kamptus* ammonite subzone). Occurring within ostracod zone OJ16 (*pars*), foraminiferid zone FJ16 (*pars*), palynological subzones DJ5b (*pars*)–DJ5c (*pars*).

Lithology: this formation is dominated by light grey to buff, bioclastic limestones (micrites through to packstones), some of which are argillaceous or locally sandy. Thin interbeds of light grey to dark grey, calcareous claystone and calcareous siltstone also occur throughout the section.

Wireline log characteristics: the lower boundary is sharp; denoted by a marked decrease in gamma-ray response, associated with an increase in sonic velocity values, coinciding with a lithological change to limestones. The upper boundary is well defined; marked by an increase in gamma-ray and an associated decrease in sonic velocity values, reflecting a lithological change to claystones.

Low gamma-ray values and high sonic velocities are characteristic of the Abbotsbury Cornbrash Formation. Where subdivisions can be recognized, the Fleet Member often has a serrated wireline log profile reflecting interbedding of limestones and claystones, while the Berry Member possesses a cylindrical log profile due to the less frequently interbedded lithologies.

Distribution and thickness: the Abbotsbury Cornbrash Formation is present throughout the study area with thicknesses ranging from 7 feet at 98/11-3 to 62+ feet at Seabarn Farm-1. Average thicknesses range between 20 and 40 feet. Similar to the underlying Forest Marble, the thickest developments of the Abbotsbury Cornbrash Formation occur in the northern half of the Portland–Wight Basin (e.g. 62+ feet at Seabarn Farm-1 and 43 feet at Kimmeridge-5) suggesting a depocentre in this region. Southwards towards the Central Channel High the Abbotsbury Cornbrash Formation thins abruptly (e.g. 14 feet at 98/22-2 and 10 feet at 98/22-1). A slight thinning of the formation also occurs across the Purbeck–Isle of Wight Fault Zone (e.g. 27 feet at 98/11-2 and 18 feet at 98/7-2), while thin attenuated sequences are noted across much of the Hampshire–Dieppe High (e.g. 12 feet at 99/18-1 and 18 feet at Southampton-1).

Depositional environment: the Abbotsbury Cornbrash Formation was deposited in a marine, inner sublittoral, carbonate-shelf environment, mainly below wave-base, during a period of relative quiescence (Page 1989). The occurrence of large numbers of infaunal and epifaunal benthic macrofaunal suspension feeders is suggestive of warm, shallow waters, with little or no terrestrial clastic input (House 1993). It is a is a relatively condensed unit formed during two successive marine transgressive phases, both of which were probably eustatically controlled (Arkell 1933).

Kellaways Formation

The Kellaways Beds were formally designated a formation by Page (1989). A two-fold subdivision is recognized throughout the area, a lower Kellaways Clay Member and an upper Kellaways Sand Member.

Kellaways Clay Member. Age: Early Callovian (*macrocephalus* ammonite zone, upper *kamptus* ammonite subzone–*calloviense* ammonite zone, lower *calloviense* ammonite subzone). Occurring within ostracod zone OJ17 (*pars*), foraminiferid zone FJ17 (*pars*), palynological subzone DJ5c (*pars*).

Lithology: the Kellaways Clay Member comprises medium grey to light brown, locally silty claystones, some of which are calcareous, more especially towards its base.

Wireline log characteristics: the lower boundary is very sharp; defined by an increase in gamma-ray response, coincident with a marked decrease in sonic velocity values, denoting a lithological change to claystones. The upper boundary is sharp; defined by a slight decrease in gamma-ray response, associated with a sharp increase in sonic velocity values, denoting a lithological change to siltstones and sandstones. In areas where the overlying Kellaways Sand Member is argillaceous, the gamma-ray response of the Kellaways Clay Member is not definitive, however, the sonic log maintains its character. Gamma-ray responses increase upwards from the base of the Kellaways Clay Member, while sonic velocities decrease, reflecting a decrease in carbonate content.

Distribution and thicknesses: the Kellaways Clay Member is present throughout the study area. Its thickness ranges from a minimum of 9 feet in Chilworth-1 to a maximum of 39 feet in 98/22-2. Average thicknesses are between 20 and 30 feet, with the more condensed sequences occurring in the northwestern edge of the Hampshire–Dieppe High (e.g. 17 feet at both Middleton-1 and Horndean-1) and the Weald Basin (10 feet at Humbly Grove-1). This member is absent in Sandhills-1, where it has been removed by pre-Albian uplift and erosion.

Depositional environment: the generally silty Kellaways Clay Member, which contains a rich bivalve fauna, was deposited below wave-base in a low-energy, well-oxygenated marine environment (Page 1989). Its formation was due to the initiation of a marine transgression subsequent to the deposition of the underlying shallower water Abbotsbury Cornbrash Formation.

Kellaways Sand Member. Age: late Early Callovian (*calloviense* ammonite zone, upper *calloviense*–lower *enodatum* ammonite subzones). Occurring within ostracod zone OJ17 (*pars*), foraminiferid zone FJ17 (*pars*), palynological subzone DJ5c (*pars*).

Lithology: this member generally comprises white to medium grey, very fine- to fine-grained, quartzose to locally calcareous siltstones and sandstones. In many of the wells examined both the siltstones and sandstones possess high argillaceous contents. Beds of medium to dark grey, silty claystone also occur throughout the Kellaways Sand Member. Calcareous doggers are noted in a number of wells.

Wireline log characteristics: the lower boundary is sharp; defined by a slight decrease in gamma-ray response, coinciding with a sharp increase in sonic velocity values, coincident with a lithological change to siltstones and sandstones. The upper boundary is sharp; defined by a marked increase in gamma-ray response and an associated decrease in sonic velocity, reflecting a lithological change to claystones. The Kellaways Sand Member typically exhibits a low gamma-ray response decreasing away from the base, and a moderately high sonic velocity response which increases away from the base forming a funnel log profile. Sonic log 'spikes' reflect thin, calcareous sandstone beds.

In the Portland–Wight Basin the Kellaways Sand Member comprises a more silty and argillaceous interval, with a gamma-ray response only slightly lower than in the overlying Oxford Clay Formation (e.g. in well 98/23-1). However, the sonic log velocity response is still distinctive enabling identification of this unit.

Distribution and thickness: the Kellaways Sand Member occurs throughout the present study area, with thicknesses ranging from a minimum of 17 feet in Wytch Farm-X14 to a maximum of 46 feet in Arreton-2. On average, thicknesses range between 20 and 40 feet, with generally no major thickness variations across the region. It is, however, absent in Sandhills-1 where it has been removed by pre-Albian uplift and erosion.

Depositional environment: the Kellaways Sand Member forms the upper part of a coarsening-upwards (shallowing-upwards) sequence, which culminates in the progradation of shallow, offshore sheet sands, deposited above wave-base (Page 1989). The generally finer-grained lithologies in the Portland–Wight Basin are suggestive of either greater water depths or limited clastic supply being attained within this area.

Oxford Clay Formation

The Oxford Clay was formally designated a formation by Cox et al. (1992). A three-fold subdivision of the Oxford Clay Formation using the traditional subdivisions (Lower, Middle and Upper) was endorsed as lithostratigraphic units and, respectively, named the Peterborough Member, the Stewartby Member and the Weymouth Member.

Age: latest Early Callovian–latest Early Oxfordian (*calloviense* ammonite zone, upper *enodatum* ammonite subzone–*cordatum* ammonite zone, lower *cordatum* ammonite subzone). Occurring within ostracod zones OJ18–OJ19 (*pars*), foraminiferid zones FJ18–FJ19 (*pars*), palynological subzones DJ6a–DJ7b (*pars*).

The Peterborough Member is of latest Early–earliest Late Callovian age, the Stewartby

Member is of Late Callovian age, while the Weymouth Member is of Early Oxfordian age.

Lithology: a three-fold subdivision can be recognized.

(1) The Peterborough Member is dominated by bluish to dark grey, grey brown, non-calcareous to calcareous, silty, fissile mudstones. Towards the base of the interval, the lithologies are often dominated by calcareous siltstones and silty claystones. Organic-rich (carbonaceous) claystones are often common in the upper two-thirds of the section. Stringers of medium grey, bioclastic limestone occur throughout this member. A prominent limestone band (the Acutistriatum-comptoni Bed) occurs at the top of this member.

(2) The Stewartby Member comprises medium to dark grey, silty, fissile shales and calcareous claystones. The more calcareous lithologies occur mid-way through the unit. Thin, light to medium grey, argillaceous limestones may occur throughout the interval.

(3) The Weymouth Member is dominated by bluish to dark grey, grey brown, non-calcareous to calcareous, silty, in part organic-rich, fissile mudstones. In the upper half of the interval the sediments are often less calcareous in nature.

Wireline log characteristics: a three-fold subdivision is noted.

(1) The Peterborough Member: the lower boundary is sharp; defined by an increase in gamma-ray response, and an associated decrease in sonic velocity values, corresponding with a lithological change to claystones. The upper boundary is sharp to abrupt; defined by a decrease in gamma-ray response, associated with an increase in sonic velocity, reflecting a lithological change to a well indurated limestone. The Peterborough Member exhibits a highly serrated gamma-ray response. Towards the base of the interval there is a decrease in gamma-ray response, associated with increased sonic log velocity. Intervals with high gamma-ray response and very low sonic velocity denote organic-rich claystones, while several prominent sonic 'spikes' indicate the presence of thin limestone beds.

(2) The Stewartby Member: the lower boundary is sharp; defined by a decrease in gamma-ray response and a corresponding increase in sonic velocity values, denoting a lithological change to a well-indurated limestone. The upper boundary is moderately sharp; defined by an increase in gamma-ray response, associated with a decrease in sonic velocity, corresponding to a decrease in the carbonate content of the claystones. The upper three-quarters of the Stewartby Member is characterized by a bow log profile, with the lowest gamma-ray responses and highest sonic velocities midway through the unit, reflecting the more calcareous natures at this point. The occurrence of a serrated log motif denotes thin, argillaceous limestone beds and/or concretions, interbedded within the claystones. The lower quarter of this member possesses the higher gamma-ray responses denoting the more organic-rich claystones.

(3) The Weymouth Member: the lower boundary is moderately sharp; defined by an increase in gamma-ray response, coinciding with a decrease in sonic velocity values, reflecting a decrease in the calcareous content of the claystones. The upper boundary is sharp; defined by a marked decrease in gamma-ray response, associated with an increase in sonic velocity values, reflecting a lithological change to sandstones. The Weymouth Member is characterized by serrated wireline log motifs with moderately high gamma-ray and moderately low sonic velocities. 'Sonic' spikes reflect thin limestone beds or concretions. Several high-gamma 'peaks' are often present in the upper part of the Weymouth Member. These are envisaged to reflect organic-rich claystone beds as described by Wright (1986b) from outcrop at Furzy Cliff.

Distribution and thickness: throughout the study area the three units of the Oxford Clay Formation maintain a consistency in sedimentary facies. Thicknesses of the Oxford Clay Formation vary from 129 feet at 98/16-2A to 608 feet at Winterborne Kingston-1, with average thicknesses between 300 and 550 feet. There is little evidence within the present study area for any major variations in thickness on a regional scale, with structural control appearing to have been minimal. However, a regional trend of progressive thinning from west to east is noted (e.g. 608 feet at Winterborne Kingston-1 and 576 feet in 98/18-1 to 338 feet at 99/12-1 and 371 feet at Humbly Grove-1). In all those wells which possess a complete succession of the Oxford Clay Formation, it is the Stewartby Member which consistently is the thicker unit and is also the least variable in thickness throughout the region. The Peterborough Member is generally the most inconsistent in thickness due the presence/absence of its lower silty sequences.

Depositional environment: deposition of the Oxford Clay Formation was initiated during a marine transgression commencing in the latest Early Callovian. The Oxford Clay Formation was deposited below wave-base, in a low-energy marine environment, on a wide, shallow shelf (Morris 1980; Wright 1986b). Water depths are envisaged to have been no more than a few tens of metres (Hudson & Martill 1991) with low-lying land situated nearby, indicated by the remains of large herbivore and carnivore dinosaurs, and an abundant terrestrial nutrient supply. Water column stratification did occur leading to temporary dysaerobic bottom water environments throughout the Oxford Clay Formation, however, these were more prevalent in the Peterborough Member. They did not, however, give way to anoxic bottom water conditions for any length of time as denoted by the often intensely burrowed nature of the sediments (Duff 1975; Hudson & Martill 1991). The organic-rich shales of the Peterborough Member often possess low-diversity micro- and macrofaunas due to sluggish water circulation at the sediment–water interface.

During the deposition of the Stewartby Member conditions ameliorated with deposition in more open marine, well-oxygenated environments, indicated by the more profuse benthic faunas in association with the more calcareous lithologies. A moderately low sedimentation rate is envisaged for the Stewartby Member.

Similar open marine environments occurred in the Weymouth Member, denoted by their profuse benthic micro- and macrofaunas (epifaunal bivalves). An increase in water depth has been described by Hollingworth & Wignall (1992) at the Stewartby–Weymouth Member boundary, denoted by a faunal diversity decrease. This is suggestive of either a transgression or of subsidence. The presence of carbonaceous sandy clays is indicative of tropical storms redepositing rotting vegetation and quartz sand into the shelf depths (Wright 1986b).

Nothe Grit Formation

Age: latest Early Oxfordian (*cordatum* ammonite zone, *cordatum* ammonite subzone). Occurring within ostracod zone OJ19 (*pars*), foraminiferid zone FJ19 (*pars*), palynological subzone DJ7c (*pars*).

Lithology: this unit is dominated by light grey to pale yellow, very fine-grained to medium-grained, locally argillaceous or calcareous sandstones. Medium grey, calcareous siltstones and silty claystones are also moderately common, more especially in the Portland–Wight Basin. Onshore along the Dorset coast calcareous sandy concretions up to 3 feet in width are observed.

Wireline log characteristics: the lower boundary is sharp; defined by a marked decrease in gamma-ray response and an associated increase in sonic velocity values, denoting a lithological change to arenaceous lithologies. The upper boundary is sharp; defined by a decrease in gamma-ray response and an associated increase in sonic velocity values, corresponding to a change to well-indurated limestones.

The Nothe Grit Formation is characterized by a cylindrical, serrated wireline log motif, with low gamma-ray responses and high sonic velocities, indicating the dominantly arenaceous lithologies.

Distribution and thickness: the Nothe Grit Formation occurs in many wells in the study area, with thicknesses ranging from a minimum of 9 feet in Humbly Grove-1 to a maximum of 56 feet in 98/18-1. The available data indicate a progressive northward thinning from the Portland–Wight Basin (e.g. 28 feet at 98/23-1 and 45 feet at 98/11-1) into the Dorset Basin (e.g. 22 feet at Winterborne Kingston-1) (Fig. 16). North of the Purbeck–Isle of Wight Fault Zone (e.g. 98/6-8 and 98/7-2) and over the Cranborne–Fordingbridge High the Nothe Grit Formation has been removed by pre-Albian uplift and erosion. Over much of the Hampshire–Dieppe High the formation is also envisaged to be absent, the exceptions being Bottom Copse-1 (25 feet), Arreton-2 (22 feet) and its northwestern edge (e.g. 30 feet at Chilworth-1 and 23 feet at Horndean-1).

Depositional environment: the sediments of the Nothe Grit Formation were deposited as high-energy, nearshore subtidal sands (Wright, 1986a). Two coarsening-upwards cycles have been recognized by Wright (1986a), with the upper cycle having been deposited closer to the shoreline. The top of this formation is envisaged to mark the termination of Talbot's (1973) first regressive cycle.

Corallian Group

Sediments of the Middle to Late Oxfordian Corallian Group (formerly the Corallian Beds) can be subdivided into six stratigraphic units, five of which have been reviewed by Wright (1986a). In ascending order, they comprise, the Redcliff Formation, the Osmington Oolite Formation, the Trigonia clavellata Formation, the Sandsfoot Formation, the Ringstead Formation and the Ampthill Clay Equivalent. All of these units, excluding the Ampthill Clay Equivalent, are very difficult to recognize outside of the type area (outcrop on the Dorset coast). In much of the subcrop, the upper part of the Corallian Group is predominantly represented by a clay facies, which is envisaged to be equivalent to the Late Oxfordian Ampthill Clay of the English

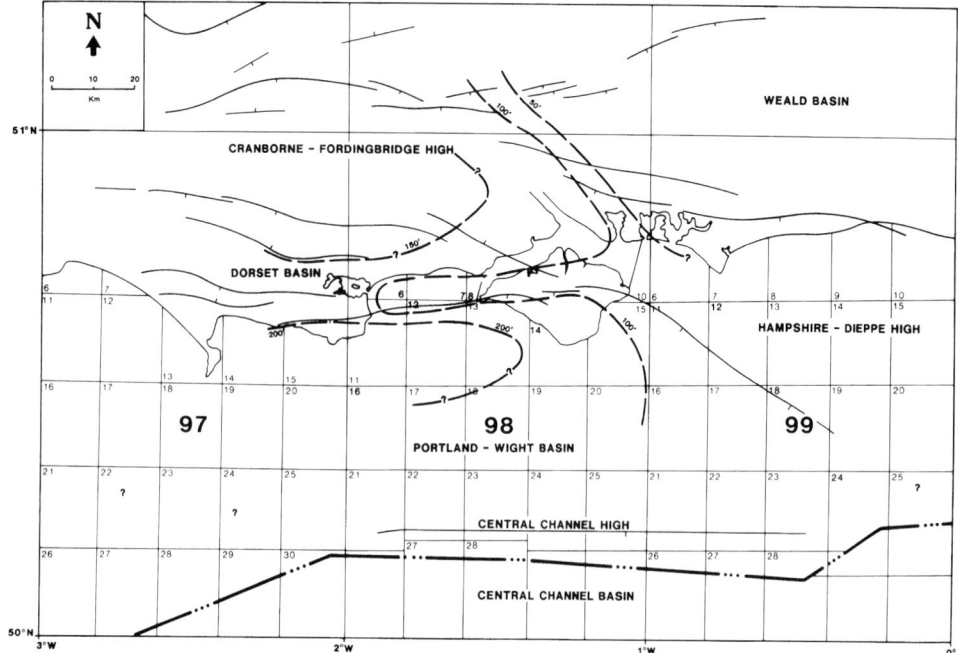

Fig. 16. Thickness and distribution of the uppermost Early to Middle Oxfordian 'lower' Corallian Group (Nothe Grit, Redcliff and Osmington Oolite Formations).

Midlands (Cope *et al.* 1980*b*). Biostratigraphic and lithostratigraphic resolution of its constituent units is generally good throughout the sequence.

The Corallian Group is composed of three shallowing up sedimentary cycles, comprising mudstones, sandstones and limestones (Arkell 1933, 1936). Talbot (1973) suggested that the boundaries between each cycle should be situated at the erosion surfaces which mark an abrupt change between moderately deep and shallow water depositional environments. A large number of sedimentological and palaeoecological studies on the Corallian Group have been undertaken since Talbot's (1973) paper, including Wright (1986*a*), Fürsich (1975, 1976, 1977), Sun (1989, 1990) and Rioult *et al.* (1990). At present no definitive depositional model has been established to explain the formation of the cycles of the whole Corallian Group on a basinwide scale.

'Lower Corallian Subgroup'

Redcliff Formation. Following Wright (1986*a*), the Redcliff Formation is subdivided into three members. In ascending stratigraphic order they comprise the Preston Grit Member, Nothe Clay Member and the Bencliff Grit Member.

Preston Grit Member. Age: earliest Middle Oxfordian (*densiplicatum* ammonite zone, lower *vertebrale* ammonite subzone). Occurring within ostracod zone OJ20 (*pars*), foraminiferid zone FJ20 (*pars*), palynological subzone DJ8a (*pars*).

Lithology: this unit comprises medium–light grey, locally oolitic, argillaceous or arenaceous, bioclastic wackestones, which often yield abundant moulds of ovoid siliceous sponge spicules (*Rhaxella perforata*).

Wireline log characteristics: the lower boundary is sharp; defined by a marked decrease in gamma-ray response and an associated increase in sonic velocity values, denoting a lithological change to well-indurated limestones. The upper boundary is sharp; defined by an abrupt increase in gamma-ray response and an associated decrease in sonic velocity values, corresponding to a lithological change to claystones.

The Preston Grit Member is characterized by a cylindrical log motif, which is most prominent in the onshore wells.

Distribution and thickness: the Preston Grit Member occurs in a number of wells in the study area, with thicknesses ranging from a minimum of 5 feet in Marchwood-1 and 98/18-1 to a maximum of 31 feet in 98/11-1 (Fig. 16). Within much of the study area, this unit maintains a consistency in thicknesses, averaging less 15 feet.

Depositional environment: the Preston Grit Member has been interpreted as a rapidly accumulating, transgressive beach deposit formed under high-energy, normal marine conditions (Wright 1986a; Coe 1995), or possibly in a lower shoreface environment (Sun 1989). The base of this member is envisaged to mark the commencement of Talbot's (1973) second regressive cycle.

Nothe Clay Member. Age: Middle Oxfordian (*densiplicatum* ammonite zone, upper *vertebrale*–lower *maltonense* ammonite subzones). Occurring within ostracod zone OJ20 (*pars*), foraminiferid zone FJ20 (*pars*), palynological subzone DJ8a (*pars*).

Lithology: this member is dominated by light to dark grey, locally silty, bioclastic claystones and calcareous claystones. Stringers of light to medium grey, bioclastic limestones (mudstones-wackestones) are also noted.

Wireline log characteristics: the lower boundary is very sharp; defined by a marked increase in gamma-ray response and an associated decrease in sonic velocity values, denoting a lithological change to claystones. The upper boundary is sharp; defined by a decrease in gamma-ray response and an increase in sonic velocity values, corresponding with a lithological change to sandstones.

The Nothe Clay Member is characterized by a slightly serrate log motif with high gamma-ray and low sonic velocity responses, reflecting the interbedding of claystones and limestones.

Distribution and thickness: the Nothe Clay Member occurs in much of the present study area. It ranges in thickness from 8 feet in Crokerhill-1 to 64 feet in Winterborne Kingston-1, with average thicknesses between 20 feet and 50 feet (Fig. 16). Thickness trends for the Nothe Clay Member show a progressive southward and eastward thinning from 56 feet at Kimmeridge-5 to 27 feet at Bottom Copse-1 and 14 feet at 98/16-1. These sediments are absent over the Cranborne–Fordingbridge High and much of the Hampshire–Dieppe High, except for the northwestern edge of the latter (e.g. 28 feet at Marchwood-1). In the Weald Basin the Nothe Clay Member is considered to be condensed (14 feet at Humbly Grove-1).

Depositional environment: the generally argillaceous sediments of the Nothe Clay Member are envisaged to have been deposited into a generally quiet, shallow-marine (nearshore subtidal to offshore shelf) sea (Wilson 1968; Talbot 1973; Fürsich 1977; Wright 1986a; Coe 1995).

Towards the top of the member Whatley (1965) envisaged a period of restriction, possibly with reduced salinities denoted by the marked decrease in ostracod microfaunas. The occurrence of thin, shelly sandstone beds is suggestive of higher-energy marine depositional episodes, transgressing from the west of the south Dorset coast (Wright 1986a).

Bencliff Grit Member. Age: early Middle Oxfordian (*densiplicatum* ammonite zone, upper *maltonense* ammonite subzone). Occurring within ostracod zone OJ20 (*pars*), foraminiferid zone FJ20 (*pars*), palynological subzone DJ8a (*pars*).

Lithology: the Bencliff Grit Member is dominated by white to pale yellow, very fine- to coarse-grained, quartzose sandstones, which are locally argillaceous and/or contain large amounts of fragmented plant debris. The sandstones decrease in their argillaceous content and/or increase in grain size upsection. Interbeds of medium to dark grey, variably calcareous, claystones and siltstones can also occur in this member.

Wireline log characteristics: the lower boundary is sharp; defined by a decrease in gamma-ray response and a corresponding increase in sonic velocity values, denoting a lithological change to sandstones. The upper boundary is moderately well defined; denoted by a slight decrease in gamma-ray response, associated with an increase in sonic velocity, corresponding with a lithological change to well indurated oolitic limestones. There is often a thin claystone unit at the top of the Bencliff Grit Member (e.g. in Marchwood-1).

This unit possesses a funnel-shaped log motif, denoting the upward grain size increase. Moderately low gamma-ray values and moderately low sonic velocities are characteristic of this member; the latter reflecting the predominantly uncemented nature of the sandstones, with sonic spikes denoting thin cemented beds or concretions.

Distribution and thickness: the Bencliff Grit Member has a sporadic distribution throughout the study area. Where recognizable, thicknesses vary between a minimum of 9 feet in 98/23-1 and Bottom Copse-1 to a maximum of 40 feet at Chilworth-1 (Fig. 16). Although isopach data are poor for the Bencliff Grit Member, the thicker sections are envisaged to occur in the Dorset Basin (29 feet at Winterborne Kingston-1) and the northwestern edge of the Hampshire–Dieppe High (e.g. 39 feet at Southampton-1 and 40 feet at Chilworth-1). In the main part of the Portland–Wight Basin the Bencliff Grit Member is generally thin (e.g. 12 feet at 98/11-3 and 15 feet at 98/23-1). Over the Cranborne–Fordingbridge High and the main part of the Hampshire–Dieppe High this member is absent. In the western Weald Basin, these sediments are difficult to differentiate due to occurrences of condensed sequences of the 'lower' Corallian Group.

Depositional environment: several interpretations have been put forward to explain the sedimentary history of the Bencliff Grit Member, ranging from deposition in a braided alluvial fan which has extended seawards (Wright 1986a), through intertidal/tidal flats (Wilson 1968; Talbot 1973; Fürsich 1975, 1977; Coe 1995), shoreface deposits (Sun 1989) to estuarine (Allen & Underhill 1989; Goldring et al. this volume). Sediment sourcing for the Bencliff Grit Member is via localized uplift to the northeast of the south Dorset coast, with sediments being reworked from earlier deposited Corallian sediments (Wright 1986a). High rates of sedimentation are envisaged due to the occurrence of large-scale cross-bedding and the presence of the extended burrow tubes.

The top of the Bencliff Grit Member is envisaged to mark the termination of Talbot's (1973) second regressive cycle.

Osmington Oolite Formation. Onshore, the Osmington Oolite Formation has been subdivided by Wright (1986a) into three members; in ascending order they comprise the Upton Member, the Shortlake Member and the Nodular Rubble Member. None of these three units can be recognized with any degree of certainty using either petrophysical logs and/or lithological criteria in the present study.

Age: late Middle Oxfordian (*tenuiserratum* ammonite zone). Occurring within ostracod zone OJ20 (*pars*), foraminiferid zone FJ20 (*pars*), palynological subzone DJ8b.

Lithology: the Osmington Oolite Formation is dominated by white to light grey, generally oolitic, bioclastic, limestones (mudstones–grainstones). Thin stringers of light grey to medium grey, locally oolitic–pisolitic calcareous claystones and mudstones, and sandy marls/sandy limestones are also noted. Ovoid siliceous sponge spicules (*Rhaxella perforata*) are locally a common consistent of the sediments.

Wireline log characteristics: the lower boundary is moderately sharp; denoted by a slight decrease in gamma-ray response and an associated increase in sonic velocity values, denoting a lithological change to well-indurated oolitic limestones. The upper boundary is sharp; denoted by an abrupt increase in gamma-ray response and an associated decrease in sonic velocity values, corresponding with a lithological change to claystones and marls. The Osmington Oolite Formation is characterized by a cylindrical wireline log motif, with very low gamma-ray responses and high sonic velocity values.

The top of the Osmington Oolite Formation forms an important regional seismic reflector (Whittaker et al. 1985).

Distribution and thickness: the Osmington Oolite Formation is present throughout the study area, with thicknesses varying between 10 feet in Humbly Grove-1 and 91 feet in well 98/23-1. Average thicknesses for the formation are within the range 40–75 feet (Fig. 16). The thickest sequences of the Osmington Oolite Formation occur within the Portland–Wight Basin (e.g. 91 feet in well 98/23-1, 73 feet in 98/18-1, 86 feet at Arreton-2, 80 feet at Kimmeridge-3). Across the Purbeck–Isle of Wight Fault Zone the formation thins or is absent (e.g. 40 feet at 98/11-3 and 0 feet at Sandhills-1), while over the Cranborne–Fordingbridge High and the main part of the Hampshire–Dieppe High these sediments are envisaged to be absent. On the northwestern edge of the latter, the formation is moderately thin (e.g. 39 feet at Southampton-1 and 24 feet at Middleton-1), while in the Weald Basin it comprises a condensed sequence (e.g. 10 feet at Humbly Grove-1).

Depositional environments: the Osmington Oolite Formation was deposited under a variety of shallow marine regimes, varying from lagoonal through subtidal, low- to high-energy shallow-marine, offshore environments (Wilson 1968; Talbot 1973; Wright 1986a; Sun 1989; Coe 1995). The presence of large numbers of oolites in the various well sections, in association with cross-bedded oolitic limestones at outcrop (Wright 1986a), is suggestive of shallow-water oolitic deltas and/or sand shoals similar to those described by Evans et al. (1973) from Abu Dhabi.

Talbot's (1973) third shallowing up cycle commences at the base the Osmington Oolite Formation, and finishes at the Osmington Oolite–Trigonia clavellata formational boundary.

'Upper Corallian, Calcareous Subgroup'

In a regional context, the 'Upper Corallian Calcareous Subgroup' (in ascending order, the Trigonia clavellata Formation, the Sandsfoot Formation and the Ringstead Formation) are envisaged to reflect a localized marginal facies development of the predominantly argillaceous Ampthill Clay Equivalent, which was deposited over topographically positive areas. Their occurrence in south Dorset and the adjacent offshore area (UK block 98/11) may be related to movement on the nearby Abbotsbury and Purbeck Faults, part of the Purbeck–Isle of Wight Fault Zone. Late Oxfordian sediments are absent over the Cranborne–Fordingbridge High, probably reflecting pre-Kimmeridgian uplift and erosion. These sediments also occur on the northern margin of the Cranborne–Fordingbridge High. In the Weald Basin they grade laterally into the Ampthill Clay facies.

Trigonia clavellata Formation. Wright (1986a) subdivided the Trigonia clavellata Formation into four members. In ascending stratigraphic order they comprise; the Sandy Block Member, the Chief Shell Beds Member, the Clay Band Member and Red Beds Member. These four members could not be recognized definitively in the subcrop in the present study.

Age: earliest Late Oxfordian (*glosense* ammonite zone). Occurring within ostracod zone OJ21 (*pars*), foraminiferid zone FJ20 (*pars*), palynological subzone DJ8c.

Lithology: the Trigonia clavellata Formation comprises a sequence of interbedded limestones and sandy calcareous claystones–marls. The former comprise light grey to white, bioclastic, oolitic mudstones to wackestones, while the latter comprise medium light to dark grey, silty to fine-grained, sandy, calcareous claystones–marls.

Wireline log characteristics: the lower boundary is sharp; defined by an abrupt increase in gamma-ray response and a sharp decrease in sonic velocity values, denoting a lithological change to calcareous claystones and marls. The upper boundary is sharp; defined by an increase in the gamma-ray response and a decrease in sonic velocity values, corresponding to a lithological change to claystones.

Distribution and thickness: the Trigonia clavellata Formation could only be identified in the southern part of the Dorset Basin and in the northern part of the Portland–Wight Basin (Kimmeridge-3 and 98/11-4), reaching a maximum thickness of 34 feet in the former well (Fig. 17).

Depositional environment: this formation was deposited under shallow-marine, low- to moderate-energy, well-oxygenated marine conditions, below wave-base. The base of the Trigonia clavellata Formation is envisaged to mark the commencement of Talbot's (1973) fourth regressive cycle.

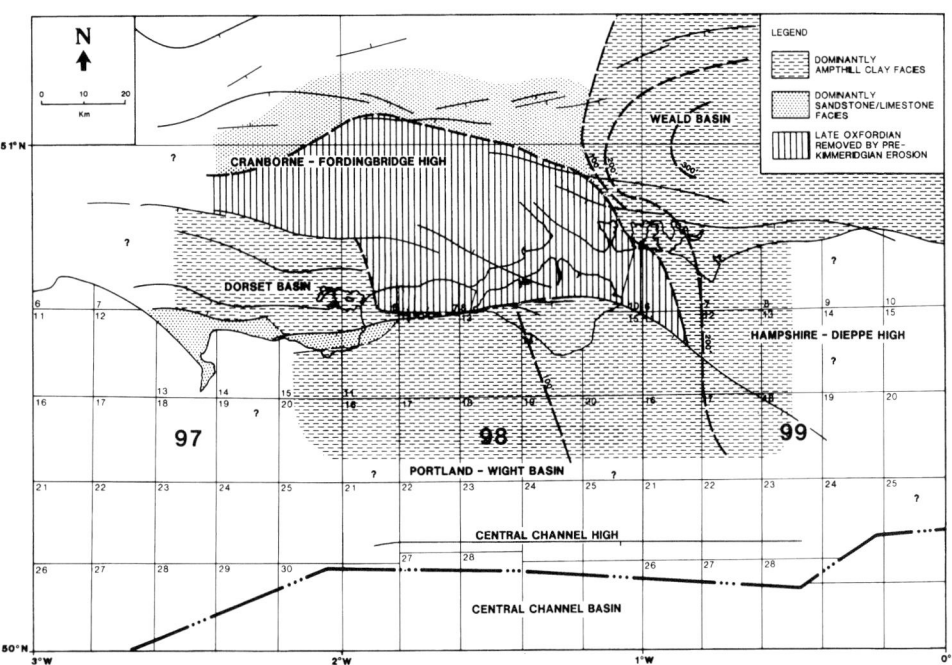

Fig. 17. Thickness and facies distribution of the Late Oxfordian 'upper' Corallian Group (Trigonia clavellata, Sandsfoot and Ringstead Formations, Ampthill Clay Equivalent).

A slight increase in water depth (offshore shelf) is envisaged for the deposition of the Chief Shell Beds Member through to the Red Beds Member, denoted by the occurrence of claystones and micrite muds, with the high-energy shelly faunas deposited onto the offshore shelf by storms (Wilson 1968; Talbot 1973; Wright 1986a; Coe 1995). Sun (1989), however, regards the Trigonia clavellata Formation as a lower shoreface deposit. Sedimentation rates for the formation is thought to be slow, more especially for the upper intervals (e.g. Red Beds Member), as indicated by the phosphatic nodules, intense bioturbation and post-mortem encrustation of shells.

Sandsfoot Formation. The Sandsfoot Formation can be subdivided into two members; a lower Sandsfoot Clay Member and an upper Sandsfoot Grit Member (Wright 1986a).

Sandsfoot Clay Member. Age: Late Oxfordian (*serratum*–lower *regulare* ammonite zones). Occurring within ostracod zone OJ21 (*pars*), foraminiferid zone FJ20 (*pars*), palynological subzone DJ8d (*pars*).

Lithology: this member is dominated by bluish to medium dark grey, bioclastic, calcareous claystones and silty mudstones. Thin, light grey, marl, limestone and sandstone horizons are also noted.

Wireline log characteristics: the lower boundary is sharp; defined by an increase in the gamma-ray response and an associated decrease in sonic velocity values, corresponding to a lithological change to claystones. The upper boundary is sharp; defined by a decrease in gamma-ray response and an increase in sonic velocity values, denoting a lithological change to sandstones or siltstones. The wireline logs exhibit vertical alternations between intervals with low-gamma–high sonic log velocities and high gamma–low sonic log velocities, reflecting the interbedded nature of the clay, sandstone and limestone lithologies.

Distribution and thickness: in the present study the Sandsfoot Clay Member is only recognized with certainty in the southern part of the Dorset Basin and in the northern part of the Portland–Wight Basin (Kimmeridge-3 and 98/11-4), reaching a maximum thickness of 44 feet (Fig. 17).

Depositional environment: a low-energy, nearshore (Wright 1986a) or offshore (Talbot 1973; Sun 1989) marine environment is envisaged for the deposition of the Sandsfoot Clay Member. The absence of profuse ammonite faunas at outcrop, compared with its lateral equivalent the Ampthill Clay, has been suggested by both Talbot (1973) and Brookfield (1978) to indicate low salinities.

Sandsfoot Grit Member. Age: Late Oxfordian (upper *regulare–rosenkrantzi* ammonite zones, lower *marstonense* ammonite subzone). Occurring within ostracod zone OJ21 (*pars*), foraminiferid zone FJ20 (*pars*), palynological subzones DJ8d (*pars*)–DJ9a (*pars*).

Lithology: the Sandsfoot Grit Member comprises pale yellow to light grey, very fine- to medium-grained, iron-stained, calcareous or quartzose sandstones and siltstones, many of which are very argillaceous. Rare chamosite ooliths have been recorded.

Wireline log characteristics: the lower boundary is sharp; defined by a decrease in the gamma-ray response and an associated increase in sonic velocity values, corresponding to a lithological change to sandstones. The upper boundary is sharp; defined by a slight increase in gamma-ray response and a decrease in sonic velocity values, denoting a lithological change to oolitic, sandy ironstones. The petrophysical logs exhibit highly serrate motifs, reflecting the variable argillaceous content of the sandstone lithologies.

Distribution and thickness: similar to the overlying Ringstead Formation, the Sandsfoot Grit Member is noted in only the southern part of the Dorset Basin and in the northern part of the Portland-Wight Basin (Kimmeridge-3, 98/11-4). No discernible thickness variations can be observed from the limited data (Fig. 17).

Depositional environment: the sediments are envisaged to have been deposited as marine, well-oxygenated subtidal bar shelly beach sands, with a relative decrease in water depth from base to top (Wright 1986a; Sun 1989; Coe 1995). The presence of chamosite in a number of the sand units onshore was possibly derived from westerly situated lagoons (Wright 1986a), with the sands reflecting either uplift in the source area (Wright, 1986a) or coastal progradation. The close similarity of the successions north and south of the Cranborne-Fordingbridge High suggests that either eustatic control may have been predominant or that similar tectonic influences effected both areas at this time.

Ringstead Formation. Onshore Dorset, the Ringstead Formation can be subdivided into two distinct units; a lower Ringstead Clay Member and an upper Osmington Mills Ironstone Member (Wright 1986a). In the present paper, these two members could not subdivided on either lithological or wireline log criteria in the subcrop.

Age: Late Oxfordian (*rosenkrantzi* ammonite zone, upper *marstonense–bauhini* ammonite subzones). Occurring within ostracod zone OJ21 (*pars*), foraminiferid zone FJ20 (*pars*), palynological subzone DJ9a (*pars*).

Lithology: the Ringstead Formation comprises medium to dark grey, silty, locally waxy, mudrocks, interbedded with beds of light to medium light grey, oolitic, bioclastic limestones (generally wackestones) and brownish yellow, sandy, oolitic ironstones.

Wireline log characteristics: the lower boundary is sharp; defined by a slight increase in gamma-ray response and a decrease in sonic velocity values, reflecting a lithological change to oolitic, sandy ironstones. The upper boundary is sharp; defined by a marked increase in gamma-ray response and a decrease in sonic velocity values, reflecting a lithological change to claystones.

Distribution and thickness: the Ringstead Formation occurs in only a small area in the present study (Fig. 17); in the southern part of the Dorset Basin and within the northern part of the Portland–Wight Basin (Kimmeridge-3 and 98/11-4). No discernible thickness variations can be observed from the limited data.

Depositional environment: the Ringstead Formation was deposited under shallow-marine, low-energy, well-oxygenated conditions as suggested by the common occurrence of benthic macrofaunas including corals. Talbot (1973), Fürsich (1977) and House (1993) suggest that this unit was laid down in a lagoonal environment, whereas a number of other authors, including Wright (1986*a*) and Sun (1989), envisage a deeper offshore shelf deposition. The presence of a diverse macrofauna, borings, encrustation, phosphatic nodules and ironrich ooids within the Osmington Mills Ironstone Member is suggestive of condensation (Coe 1995). The Ringstead–Kimmeridge Clay formational boundary is envisaged to mark the top of Talbot's (1973) fourth regressive cycle.

Ampthill Clay Equivalent. Age: Late Oxfordian (*glosense–rosenkrantzi* ammonite zones). Occurring within ostracod zone OJ21 (*pars*), foraminiferid zone FJ20 (*pars*), palynological subzones DJ8c (*pars*)–DJ9a (*pars*).

Lithology: the Ampthill Clay Equivalent is dominated by medium to dark grey, silty, locally pyritic mudstones and calcareous claystones. The claystones become progressively more calcareous up sequence. Beds of light to medium grey, pale yellow, bioclastic limestone and argillaceous limestone occur throughout the sequence, especially in the upper third. The latter may grade up to wackestone textures, and locally may be oolitic. Thin, light grey, very fine- to fine-grained calcareous sandstones are often noted in the upper sections of the Ampthill Clay Equivalent in wells situated in the Weald Basin.

Wireline log characteristics: the lower boundary is sharp; defined by an abrupt increase in gamma-ray response and an associated decrease in sonic velocity values, reflecting a lithological change to claystones. The upper boundary is gradational to sharp; defined by an increase in gamma-ray response and a corresponding decrease in sonic velocity values, reflecting a lithological change to poorly calcareous claystones. The Ampthill Clay Equivalent characteristically displays a relatively high gamma-ray response in the lower part, decreasing upsection, which is mirrored by the sonic response.

Distribution and thickness: the Ampthill Clay Equivalent occurs over a wide area in the present study. It has a minimum thickness of 24 feet in 98/11-1 and a maximum of 251 feet in Horndean-1, with average thicknesses between 50 and 200 feet in the study area (Fig. 17). The thickest sections in the study area occur on the northwestern edge of the Hampshire–Dieppe High (e.g. 233 feet at Horndean-1 and 251 feet at Middleton-1), thickening to over 300 feet in the centre of the Weald Basin. The Ampthill Clay Equivalent is absent over the Cranborne–Fordingbridge High. The thickness trends observed in the Portland–Wight Basin suggest that it originally extended over much of the Hampshire–Dieppe High, however, the Upper Jurassic has subsequently been removed by pre-Albian erosion (e.g. 99/12-1).

It is represented entirely by claystones in the Portland–Wight Basin (UK Quad 98) and Dorset Basin. However, the upper part of the Ampthill Clay Equivalent in the Weald Basin and Portland–Wight Basin (UK Quad 99) is often represented by limestones and thin sandstone beds, especially in the top third of the interval (e.g. 99/18-1 and Middleton-1). North of the Cranborne–Fordingbridge High, and also in south Dorset, the Ampthill Clay Equivalent passes laterally into the Sandsfoot and Ringstead Formations.

Depositional environment: the Ampthill Clay Equivalent is a deeper water equivalent of the Sandsfoot and Ringstead Formations having been deposited below wave-base in a moderately shallow-marine, low-energy, well-oxygenated environment. There is a general shallowing-upwards trend, which culminates in the occurrence of thin shallow-marine sandstone beds in some areas (e.g. the Weald Basin).

Kimmeridge Clay Formation.

Age: Early–Late Kimmeridgian (*baylei–fittoni* ammonite zones). Occurring within ostracod zones OJ22–OJ28 (*pars*), foraminiferid zones FJ21–FJ25 (*pars*), palynological subzones DJ9b–DJ13b (*pars*).

Lithology: six distinct lithologies occur in the Kimmeridge Clay Formation, of which the argillaceous components (claystones, bituminous shales and oil shales) are the most common. The claystones comprise medium grey to dark grey, variably calcareous and silty, bioclastic sediments. The bituminous shales are greyish black to brownish grey, kerogen-rich (10–40%), pyritic, locally bioclastic, fissile sediments, while the oil shales comprise dark brown to olive black, organic-rich (40–70%), pyritic rocks. The claystone lithologies generally increase in grain size (silty claystones–siltstones) in the uppermost part of the Kimmeridge Clay Formation where they grade into the Portland Group. Beds of white to medium light grey, bioclastic limestone and argillaceous limestone–marl are also developed. The former generally comprise coccolith limestones. Three prominent calcareous units are easily identifiable in the subsurface in the present study area; the White Stone Band, (a white coccolith-rich limestone) and the 'Lower' and 'Upper Calcareous Members' which comprise calcareous claystones–marls in the west of the study area and limestones–argillaceous limestones in the east of the area. Dolomites are also noted in the Kimmeridge Clay Formation forming distinct greyish orange, fine-grained, well-indurated sediments.

The Kimmeridge Clay Formation has been shown to comprise a cyclic sequence of small-scale sedimentary rhythms of the above lithologies (Downie 1955; Tyson *et al.* 1979; Cox & Gallois 1981; Tyson *et al.* 1979; House 1993; Tyson 1996). The patterns vary from an asymmetric (ABCABC, ABAB) cycle to an symmetric (ABCDCBA) pattern. Within this formation the coccolithic limestones reach their greatest development above the Lower Kimmeridge Clay, disappearing in the Upper Kimmeridge Clay.

Wireline log characteristics: the lower boundary is sharp to gradational; denoted by an increase in gamma-ray response, in association with a decrease in sonic velocity values, reflecting a lithological change to poorly calcareous claystones. The upper boundary of the Kimmeridge Clay Formation is sharp to gradational; defined by a decrease in gamma-ray response and an associated increase in sonic velocity values, reflecting a decrease in argillaceous content and a corresponding increase in calcareous content.

The Kimmeridge Clay Formation is characterized by its moderate to highly serrate wireline log motifs, with high gamma-ray responses and moderately high sonic velocity values. The serration reflects the fine-scale alternation of calcareous and organic claystones, with low sonic responses reflecting marls and limestone beds. The more highly organic-rich claystones (oil shales) are identified by their exceptionally high gamma-ray and low sonic velocities.

Broad-scale vertical changes in lithology can be delineated regionally and have enabled a detailed stratigraphical subdivision of the Kimmeridge Clay Formation using wireline log criteria, calibrated to cored boreholes (Whittaker *et al.* 1985). However, considerable difficulty has been encountered in the identification of these units in sections of structural complexity or where the Kimmeridge Clay Formation is unusually thin and, consequently, Whittaker *et al.*'s (1985) units have not been employed in the present study. Melynk *et al.* (1992, 1994) subdivided the Portland–Wight and Weald Kimmeridge Clay Formation into 13 distinct cycles using digitally filtered gamma-ray logs. Similar to the overlying Portland Group, many of these cycles can be consistently recorded in many of the wells in the present study.

In the present paper, four lithostratigraphic units within the Upper Kimmeridge Clay are widely recognized in the study area.

(1) 'Lower Calcareous Member': this unit comprises a moderately thin calcareous claystone in Dorset and adjacent areas. Further east it thickens and becomes an argillaceous limestone. It is characterized by low gamma-ray response and high sonic velocity log values, often with a bow profile. Age: *wheatleyensis* ammonite zone.

(2) The 'Upper Calcareous Member': this unit is of similar lithology to the 'Lower Calcareous Member', but is generally thinner. It exhibits similar low gamma-ray and high sonic velocity log responses, characteristically with the gamma-ray response increasing upwards. Age: *hudlestoni* ammonite zone. This unit forms a mappable seismic reflector in the east part of the study area (Hamblin *et al.* 1992).

(3) The Blackstone (named at outcrop in Kimmeridge Bay, Dorset; Arkell 1947): one of the thickest highly organic-rich claystone units. It possesses a very high and distinctive gamma-ray response. Age: *hudlestoni* ammonite zone.

(4) The White Stone Band (named at outcrop in Kimmeridge Bay, Dorset; Arkell 1947): this thin limestone bed is reflected by a low gamma-ray response and a high sonic velocity value. It is lithologically distinctive and can often be

identified in ditch cuttings samples. Age: earliest *pectinatus* ammonite zone.

Distribution and thicknesses: the Kimmeridge Clay Formation occurs throughout the study area, except where truncated by pre-Albian uplift and erosion (e.g. Wytch Farm-X14, 98/6-7, 98/7-2 and 99/12-1). Thicknesses vary from a minimum of 18 feet in Bottom Copse-1 to a maximum of 1902 feet at Bolney-1 (Hamblin et al. 1992). The huge thickness of the Kimmeridge Clay Formation in the latter well is incomplete due to its faulted contact with the Forest Marble. A major depocentre is envisaged to occur between the central part and the northern edge of the Portland–Wight Basin (Fig. 18) due to the presence of several very thick sections, including wells 98/11-4 (1159 feet), 98/13-1 (1864+ feet), 98/16-2A (1119 feet) and Arreton-1 (1107 feet). South of this depocentre there is a progressive southward thinning towards the Central Channel High (e.g. 826 feet at 98/18-1 and 200+ feet at 98/23-1).

A northward thinning of the Kimmeridge Clay Formation over the Purbeck–Isle of Wight Fault Zone is indicated in well 98/11-3 (332 feet) and at outcrop to the northwest from Kimmeridge Bay, to less than 700 feet within a few miles of this locality (Arkell 1947). Immediately north of this fault zone these sediments are noticeably absent (e.g. 98/6-7, 98/6-8 and 98/7-2).

Over the Cranborne–Fordingbridge High the Kimmeridge Clay Formation is often truncated. Truncation is also noted on the edges of the Hampshire–Dieppe High (e.g. 18 feet at Bottom Copse-1 and 582 feet at Chilworth-1), with complete removal by erosion of the Kimmeridge Clay Formation over the main part of the high (e.g. 99/12-1 and Sandhills-1). On the northwestern edge of the Hampshire–Dieppe High thicknesses of less than 600 feet are preserved (e.g. 588 feet at Horndean-1), increasing towards the centre of the Weald Basin to over 1900 feet (Hamblin et al. 1992). A localized depocentre appears to have existed adjacent to the Portsdown Fault Zone, with over 1100 feet of Kimmeridge Clay Formation preserved in Portsdown-1 (Taitt & Kent 1958).

Depositional environment: the Kimmeridge Clay Formation was initiated by the base Kimmeridgian transgression, with deposition onto a very extensive marine shelf (Gallois 1976; Tyson et al. 1979; Oschmann 1991). Many of its various units are correlatable across a widespread area, e.g. the White Band (Cox & Gallois 1981).

The sediments and faunas/floras of the Kimmeridge Clay Formation can be placed into number of recognizable cyclothems (Aigner 1980; Tyson et al. 1979; Cox & Gallois 1981; House 1985; Oschmann 1991; Tyson 1996).

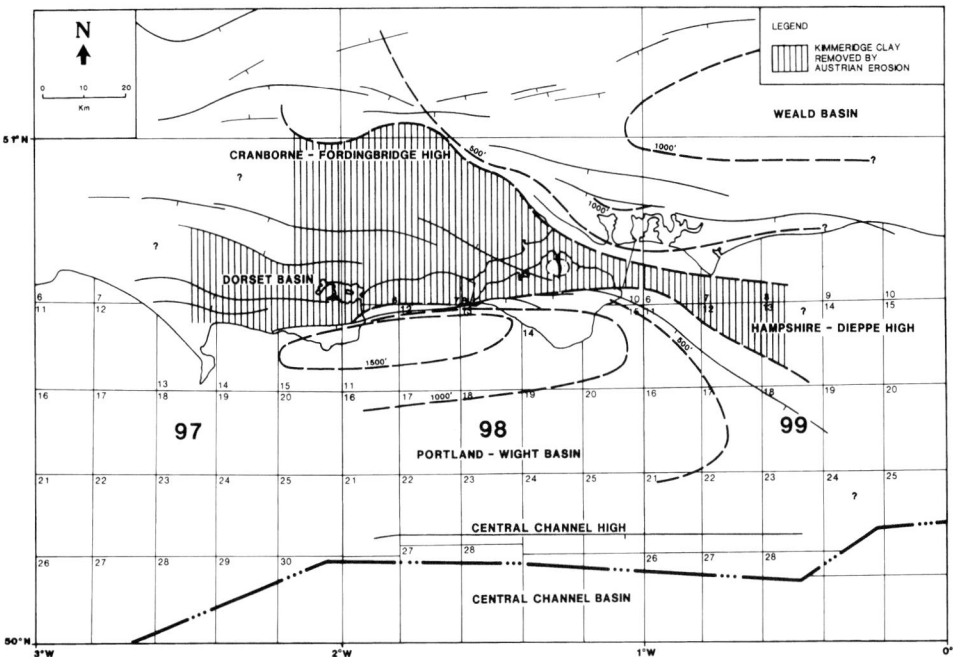

Fig. 18. Thickness and distribution of the Kimmeridgian Kimmeridge Clay Formation.

A number of models have been presented by various authors to explain their formation, for example Gallois (1978), Tyson et al. (1979) and Wignall (1989); all of which have provoked a lot of debate. Gallois (1978) considered that the monospecific character of the coccolith limestone assemblages represented formation by seasonal blooms. By direct analogy the bituminous laminated shales were considered to result from seasonal blooms of organic walled phytoplankton ('red tides'). Modern phytoplankton blooms are often associated with deep oceanic upwelling and increased nutrient supply; the former of which must have been far removed from the Kimmeridge Clay basin and suggest that this model is not entirely applicable.

In Tyson et al.'s (1979) water column model, the migration of the oxygen–hydrogen sulphide interface in a stratified water mass influences the type of sediment deposited, similar to that occurring in the present-day Black Sea. The bituminous shales were formed as the hydrogen sulphide zone moved to immediately above the water–sediment interface, whereas the oil shales were formed when the oxygen–hydrogen sulphide interface was quite high, while the coccolith limestones formed when the oxygen–hydrogen sulphide interface reached the euphotic zone. In this model the oil shales mark the deepest part of the rhythm with the coccolith limestones representing a relatively shallower environment.

Oschmann (1988) envisaged Kimmeridge Clay Formation deposition constrained by palaeogeography and palaeoclimate. He described stratification of the water mass, with production of anoxic bottom waters resulting from the influences of currents between the North Atlantic Shelf and Arctic Ocean areas. Fluctuations in bottom water oxygen levels were considered to have been influenced by four different cycles with widely varying periodicity affecting the counter current system.

Wignall (1989) has suggested deposition induced by storms, in association with a stratified water column for the Kimmeridge Clay Formation. He envisages rapid deposition from short-lived, high-velocity benthic currents. Episodes of anoxia are thought to have been terminated by storm-induced mixing of stratification and the re-establishment of water circulation. Wignall's (1989) model suggests that an inner shelf deposition existed around Boulogne and Swindon, with an inshore basin in the vicinity of Warlingham, an offshore swell at Ringstead Bay and an offshore basin at Kimmeridge, with each region possessing distinct lithological characteristics.

Various estimates of water depths during the deposition of the Kimmeridge Clay Formation have been suggested (Hallam 1992). For example, Hallam (1975) envisaged deposition under a water column in the order of tens of feet deep, whilst Tyson et al. (1979) influenced by a Black Sea analogue favoured depths in the order of hundreds of feet. Oschmann (1988) suggested a palaeodepth in the range of 150–350 feet, although Wignall's (1989) recognition of storm disturbance in organic claystones in south Dorset suggest that somewhat shallower depths may be more appropriate.

The alternations of oxic and anoxic sea floor environments may be due to long-term climatic control, perhaps reflecting Milankovitch cyclicity (Dunn 1974; House 1985; Wignall & Ruffell 1990).

Portland Group

The Portland Group (Townson 1975) corresponds approximately to the former Portland Beds (Arkell 1947). At outcrop on the Dorset coast, it was divided by Arkell (1947) and Townson (1975) into a lower Portland Sand Formation (mainly dolomitic limestones, dolomites, siltstones and minor sandstones) and an upper Portland Limestone Formation (mainly bioclastic limestones, significantly cherty in the lower part). Most of the 'sand' in the Portland Sand Formation is composed of sand-sized dolomite crystals. In the subsurface, these two formations and their constituent members cannot be identified on lithology or wireline log character due, in part, to lateral facies changes. Therefore, no formal subdivisions have been herein recognized regionally within the Portland Group.

Age: Early Portlandian (*albani–anguiformis* ammonite zones). Occurring within ostracod zones OJ28 (*pars*)–OJ29, foraminiferid zone FJ25 (*pars*), palynological subzones DJ13b (*pars*)-DJ13e.

Lithology: the Portland Group comprises a highly variable sequence of sediments, dominated by siltstones, sandstones and limestones. The siltstones are generally medium grey to medium dark grey, variably argillaceous or sandy, often possessing calcareous or dolomitic cements. These sediments are more characteristic of the basal intervals of the Portland Group. Sandstone development occurs in the lower half of the formation and comprises light grey to light brown, very fine- to fine-grained, locally glauconitic, argillaceous silty sandstones and sandstones, which possess calcareous or more often dolomitic cements. The limestones are dominated by white to light grey, bioclastic, locally oolitic, cherty or dolomitic mudstones through to packstones. The main calcareous units occur within the upper part of the Portland Group. Thinner beds of light grey to medium grey, locally bioclastic and/or dolomitic claystone and marls occur throughout the interval.

Wireline log characteristics: the lower boundary is gradational to sharp; defined by a decrease in gamma-ray response and an associated increase in sonic velocity values denoting a lithological change to siltstones or limestones–dolomites. The upper boundary is sharp; defined at the base of a gamma-ray spike, which is envisaged to represent carbonaceous claystones at the base of the Purbeck Group (see below). The upper part of the Portland Group usually exhibits lower gamma-ray responses and higher sonic velocities than the Purbeck Group reflecting a lithological change to 'cleaner' limestones.

Where fully developed in the subsurface the Portland Group is characterized by three stacked funnel-shaped signatures, each comprising an upwards-decreasing gamma-ray log motif and an upwards-increasing sonic velocity response representing an increase in carbonate or coarse clastic content. The boundaries of these cycles are often marked by prominent gamma-ray spikes which appear to represent hardgrounds. The three shallowing-upwards cycles described at outcrop by Townson (1975) cannot be consistently recognized in the subsurface. However, there is an overall upsection decrease in gamma-ray response, with an associated increase in sonic velocity, marking a decrease in the clay content of the sediments. Melynk et al. (1992, 1994) used digitally filtered gamma-ray logs in the Portland–Wight and Weald basins to subdivide the Portland Group into five distinct cycles. These cycles could be consistently recorded in many of the wells in the present study.

Distribution and thickness: the Portland Group occurs in half of the wells in the study area, with thicknesses ranging from 15 feet in well 99/18-1 to 232 feet in well 98/16-2A. On the Isle of Purbeck, the Portland Group reaches a maximum thickness of over 300 feet (Arkell 1947). Offshore, data are limited, however, the occurrences of thick sequences in wells 98/16-2A (232 feet) and 98/13-1 (216 feet) are suggestive of a major depocentre within the central and northern parts of the Portland–Wight Basin (Fig. 19). The Portland Group is envisaged to thin southwards (e.g. 138 feet at 98/18-1 and 91 feet at 99/16-1) and is envisaged to be absent close to the Central Channel High. Rapid thinning northwards over the Purbeck–Isle of Wight Fault Zone is suggested by the

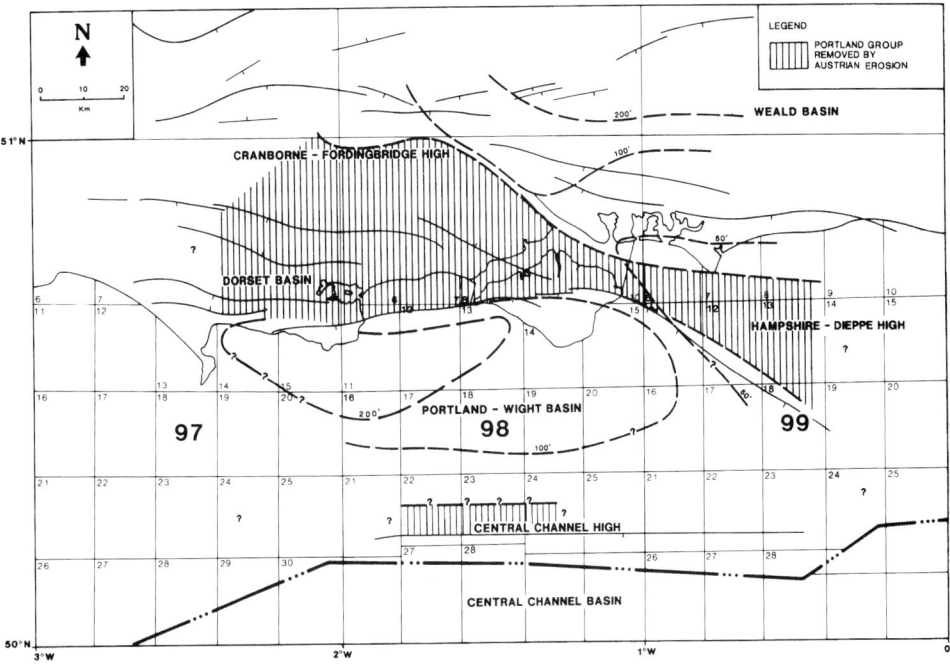

Fig. 19. Thickness and distribution of the Early Portlandian Portland Group.

northward decrease in thickness from well 98/11-4 (212 feet) to 98/11-1 (95 feet). These sediments are, however, absent immediately north of these well locations, e.g. 98/6-7, 98/6-8 and 98/7-2. There is also a thinning of the Portland Group along the outcrop in Dorset (Melville & Freshney 1982). It is absent over much of the Cranborne–Fordingbridge High due to pre-Albian uplift and erosion. On the northwestern edge of the Hampshire–Dieppe High thicknesses of between 81 (Portsdown-1) and 162 feet (Southampton-1) are recorded. Over the main part of this high the Portland Group is either highly attenuated (e.g. 15 feet at 99/18-1) or absent.

The Portland Group also occurs in the Weald Basin, with a general increase in thicknesses towards the northern part of the Weald Basin with in excess of 200 feet present (e.g. 196 feet at Humbly Grove-1). The apparent asymmetry of the Weald Basin is suggestive of more rapid depositional rates along its northern margin and may be due to a lateral facies change within the lower part of the Portland Group comprising silty claystones and mudstones (Whittaker *et al.* 1985; Hamblin *et al.* 1992).

Depositional environment: Townson (1975) recognized three major regressive and minor transgressive phases in the Portland Group, within an overall regressive sequence which commenced during the deposition of the Kimmeridge Clay. He suggested that during the deposition of the Portland Group the Dorset region was divided into two discrete areas, an eastern and western basin, separated by a northeast–southwest-trending swell. His depositional model envisaged a marine basin with a marginal carbonate ramp which sloped into deeper offshore waters, with sediment distribution controlled by depth.

Two shallowing-up cycles are recognized in the Portland Sand Formation, with deposition having commenced in relatively deep basins which were locally anaerobic on the sea bed. The upper part of two regressive cycles indicates deposition in low- to medium-energy deeper shelf and shallow-water carbonate-shelf environments. The siliciclastics, where present, indicate derivation from the west, however, the main constituent of the 'sand' is dolomite, formed in relatively deep waters by magnesium-enriched brines concentrated due to basin margin evaporation. Subsequent sinking due to heavier density concentrates material to the sea floor, dolomitizing aragonite and high-magnesium calcite muds on or within the sea bed (Townson 1975).

The Portland Limestone Formation was mainly deposited during the third shallowing-up cycle of Townson (1975) upon a low- to high-energy shallow shelf or open lagoon, as denoted by the presence of often common macrofaunas and carbonate lithologies. The presence of ooid sands indicates shallow water depths over the swell areas, while the rarity of certain benthic fossil groups may be indicative of hypersalinities (Townson 1975).

Coe (1996) has recognized three unconformity bounded sequences and also several transgressive–marine flooding episodes

Purbeck Group

The Purbeck Beds (Purbeck Group of Townson 1975) were originally divided into the Lower, Middle and Upper Purbeck Beds (Arkell 1947). Townson (1975) subdivided this group into a lower Durlston Formation and an upper Lulworth Formation, with their junction taken at the base of the Cinder Bed (Jurassic–Cretaceous boundary), a bioclastic limestone unit within the 'Middle' Purbeck Beds (Casey 1963, 1973; Melville & Freshney, 1982). Both of these formations have been further subdivided into a number of differing member nomenclatures (see Ensom 1985; Clements 1993; Westhead & Mather 1996). All of these members, including the Cinder Bed, are extremely difficult to identify both on wireline logs and in ditch cuttings samples, therefore, in the present paper, the authors have retained the original three-fold subdivision of Arkell (1947).

Lower and Middle Purbeck Beds. These two units are herein grouped together as they cannot easily be differentiated in the subsurface either on wireline log lithology or by palaeontological criteria.

Age: Late Portlandian–Berriasian (*oppressus* ammonite zone). Occurring within ostracod zones PW1–PW3, palynological subzone DJ14a.

Lithology: the lithologies are dominated by alternations of limestones–argillaceous limestones and calcareous mudstones. The former are highly variable, comprising white to medium grey mudstones grading to packstones, which are often highly bioclastic (algal or ostracodal) and locally pelloidal. The variably calcareous mudstones are light grey to medium grey, often bioclastic (ostracodal), and blocky to subfissile.

At the base of the Lower Purbeck Beds there are units of light grey to white, microcrystalline to coarsely crystalline anhydrite (Gypsiferous Beds). The Cinder Bed, a shelly, dark grey limestone containing oysters, can sometimes be identified in ditch cuttings samples.

Wireline log characteristics: the lower boundary is sharp; defined at the base of a gamma-ray spike which reflects the carbonaceous claystones at the base of the Purbeck Group. The upper boundary is sharp; defined by an increase in the gamma-ray response and a decrease in sonic velocity values corresponding with the lithological change to a calcareous claystones.

The Lower and Middle Purbeck Beds are characterized by serrate, low gamma-ray and high sonic velocity log motifs, reflecting the limestone-dominated succession, with some higher gamma–lower sonic intervals representing calcareous claystones. A prominent gamma-ray 'spike' often occurs at the base of the Lower Purbeck Beds. This is considered to be equivalent to a thin carbonaceous claystone unit, the 'Dirt Bed' (Arkell 1947), present at outcrop on the Isle of Purbeck. Above this unit there is often a thin interval of calcareous claystones with slightly higher gamma-ray responses. The Gypsiferous Beds are a distinctive unit within the lower part of the Lower Purbeck Beds characterized by a very low gamma-ray response, high sonic velocities and very high resistivity. Thin interbeds of claystone within the Gypsiferous Beds are reflected by high-gamma 'spikes'.

Distribution and thickness: the Lower and Middle Purbeck Beds occur in approximately half the wells in the present study with thicknesses ranging between 38 feet in well 99/18-1 and 531 feet in well 98/13-1. Average thicknesses vary between 100 and 300 feet with little or no facies variation. The thickest sections in the study area occur within the northern and central parts of the Portland–Wight Basin (e.g. 347 feet at 98/16-2A, 315 feet at 98/11-4, 335 feet at Arreton-2 and 531 feet at 98/13-1), the latter three of which are adjacent to the Purbeck–Isle of Wight Fault Zone (Fig. 20). These sediments progressively thin south and southeastwards into the southern part of the Portland–Wight Basin (e.g. 232 feet at 98/18-1 and 190 feet at 99/16-1) and are absent close to the Central Channel High (e.g. 98/22-1 and 98/23-1). Very rapid northward thinning over the Purbeck–Isle of Wight Fault Zone is marked by the change in thicknesses from 291 feet in well 98/11-2 and 154 feet in 98/11-1 to an absence in wells 98/6-7, 98/7-2 and 98/11-3. A similar thinning towards this fault zone is seen at outcrop in the Isle of Purbeck (Melville & Freshney 1982).

The Lower and Middle Purbeck Beds were probably also deposited to the north of the Purbeck–Isle of Wight Fault Zone, over the Cranborne–Fordingbridge High and the Hampshire–Dieppe High, although they are much thinner than in the adjacent basins. In many wells these sediments have subsequently been removed by pre-Albian uplift and erosion. They

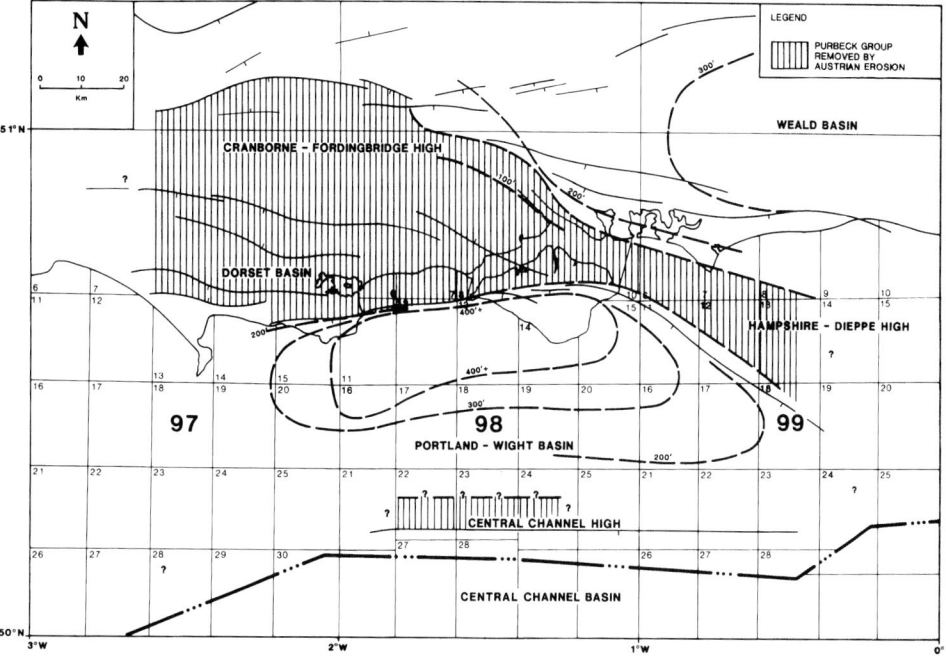

Fig. 20. Thickness and distribution of the Late Portlandian to Early Valanginian Purbeck Group.

reappear on the northwestern edge of the Hampshire–Dieppe High (e.g. 117 feet in Hoe-1 and 178 feet at Portsdown-1) thickening into the Weald Basin (e.g. 186 feet in Humbly Grove-1).

Depositional environment: a series of shallow-water environments occurred during the deposition of the Lower and Middle Purbeck Beds. The basal part of the Lower Purbeck contains a variety of sediments including gypsum and anhydrite beds, dirt beds (fossil soils), pelletoidal and algal limestones, in association with algal stromatolites, ostracods, fossil forests and various macrofaunas deposited in an intertidal–supratidal evaporitic lagoon (West 1975; Sellwood & Wilson 1990). Regressive phases are denoted by the presence of fossil soils, gypsum–anhydritic salt deposits and fossil forests, while the minor transgressive marine incursions are indicated by stromatolitic limestones, and generally profuse marine ostracod microfaunas (e.g. *Galliaecytheridea*, *Macrodentina*) and marine macrofaunas (bivalves and echinoids) recovered from dark grey shales and limestones (Anderson 1971, 1985; West 1975). A markedly seasonal semi-arid climate, with rainfall of up to 400 mm, has been suggested by West (1975) based on the occurrence of fossil forests, tree rings and the presence of evaporites which lack red beds and wind-blown sand.

Conditions generally ameliorated during the remaining part of the Lower Purbeck and into the Middle Purbeck, with environments ranging from low-energy, quasi-marine lagoonal types which were periodically hypersaline or hyposaline through to freshwater or brackish lagoonal. These conditions have been suggested by the absence/presence of the various ostracod microfaunas (Anderson 1971). Emergence has been noted by the presence of desiccation cracks and dinosaur footprints (Delair 1960; El-Shahat & West 1983). A marine transgressive episode in the lower half of the Middle Purbeck Beds (the Cinder Bed) is indicated on the occurrence of echinoids, marine bivalves and ostracods (Casey 1963; Anderson 1985).

Upper Purbeck Beds. Age: Berriasian–Early Valanginian. Occurring within ostracod zone PW4.

Lithology: the Upper Purbeck Beds are dominated by light to medium grey, bluish grey, bioclastic (ostracodal) mudstones and calcareous claystones. The uppermost argillaceous sediments of this unit are generally the least calcareous, leading to a transition between the Purbeck and Wealden rocks. Interbedded within the mudstones are white to light grey, bioclastic (often ostracodal) limestones and argillaceous limestones.

Wireline log characteristics: the lower boundary is sharp; defined by an increase in gamma-ray response, associated with a decrease in sonic velocity values, reflecting a lithological change to calcareous claystones. The upper boundary ranges from moderately sharp to sharp; defined by an increase in gamma-ray response and a more prominent decrease in sonic velocity values, reflecting a lithological change to non-calcareous claystones.

The Upper Purbeck Beds are characterized by serrated log motifs, with relatively high gamma-ray and moderately low sonic log responses reflecting variations in carbonate content. Thin limestones are represented by sonic 'spikes'. Up-section through the Upper Purbeck Beds there is a general increase in the gamma-ray response, in association with a sonic velocity decrease, reflecting the decrease in carbonate content.

Distribution and thickness: the Upper Purbeck Beds occur in approximately half of the wells in the study area, with thicknesses ranging from a minimum of 16 feet in Hoe-1 to a maximum of 268 feet in well 98/16-2A. Thickness variations across the study area are similar to those of the underlying Lower and Middle Purbeck Beds with maximum recorded thicknesses of 268 feet in well 98/16-2A, 192 feet at 98/16-1, 167 feet at 98/13-1 and 152 feet at 98/11-2, suggesting a depocentre situated in the northern and central parts of the Portland–Wight Basin (Fig. 20). To both the south and southeast of this area a progressive thinning is noted (e.g. 50 feet at 98/18-1 and 52 feet at 99/16-1), while it is absent close to the Central Channel High (e.g. 98/22-1 and 98/23-1). Across the Purbeck–Isle of Wight Fault Zone the Upper Purbeck Beds are absent (e.g. 98/6-7, 98/7-2 and 98/11-3) or highly attenuated (e.g. 40 feet at 98/11-1). Over both the Cranborne–Fordingbridge High and its continuation, the Hampshire–Dieppe High, the Upper Purbeck Beds have generally been removed by pre-Albian uplift and erosion; it is only on the northwestern edge of the latter that these sediments have been preserved (e.g. 16 feet at Hoe-1 and 51 feet at Middleton-1). On the western edge of the Weald Basin the Upper Purbeck Beds are moderately thin, however, they thicken into the centre of the basin (e.g. 97 feet in Humbly Grove-1).

Depositional environment: the Purbeck Group are a rhythmic sequence, with sedimentation controlled by cyclic variations in climate and/or relative changes in water depth (Anderson 1971, 1985), forming the end member of an extended phase of Jurassic marine sedimentation. Throughout the Upper Purbeck sequence both the macro- and microfaunas (ostracods) are commonly recovered in both the shales and limestones suggesting deposition in freshwater to brackish salinities within sheltered lagoonal/pond environments, with periodic evidence of occasional emergence denoted by desiccation mud cracks and fossil soils (Casey 1963, 1973; Anderson 1971, 1985; Sellwood & Wilson 1990). These environments occurred throughout southern England and its adjacent offshore area (Portland–Wight Basin), with palaeotemperatures similar to those occurring in the present-day Mediterranean Sea (Sellwood & Wilson 1990).

Geological history

The marine transgression in the latest Triassic led to the establishment of shallow-marine sedimentation over a wide geographic area, as indicated by the extent and uniformity of the Lower Lias. The continued lithospheric extension ('Early Cimmerian event') resulted in intermittent fault movements, continued subsidence of pre-existing basins, and the initiation and rapid subsidence of the Weald Basin (Chadwick 1986). There was progressive onlap onto the London Platform, probably due to successive phases of relative sea-level rise and subsidence (Donovan *et al.* 1979). Fault-controlled differential subsidence of the various basins had relatively little effect on lithofacies except at certain levels discussed below. The fluctuating sea-level changes led to the development of a near 'layer-cake' lithostratigraphy, dominated by relatively fine-grained sediments, and vertical lithofacies changes which are effectively synchronous over wide areas.

House (1985, 1993) considered that the Jurassic succession in Dorset could be divided into a series of sedimentary rhythms (herein termed J1–J6), which will be used as a framework to discuss the depositional history of the study area (Fig. 21). Each rhythm commences with a mudstone and finishes with a either a sandstone or limestone, and represents a large-scale shallowing-upwards sequence. Based partly on the concept of 'shallowing cycles', a schematic relative sea-level curve for the latest Triassic to earliest Cretaceous is illustrated in Fig. 8. Some variations in sediment thickness and facies may be due to low-amplitude epeirogenic movements of an epicontinental shelf. In shallow, epicontinental shelf seas, low-amplitude (less than 30 m) movements are able to produce pronounced changes in sedimentary facies. There is, however, also evidence that penecontemporaneous movements along deep-seated faults also had a marked effect on sedimentation (House 1993). Consequently, whilst eustatic/relative sea-level changes have had significant effects on deposition, some parts of the succession demonstrate the more local effects of tectonic influences (e.g. Junction Bed,

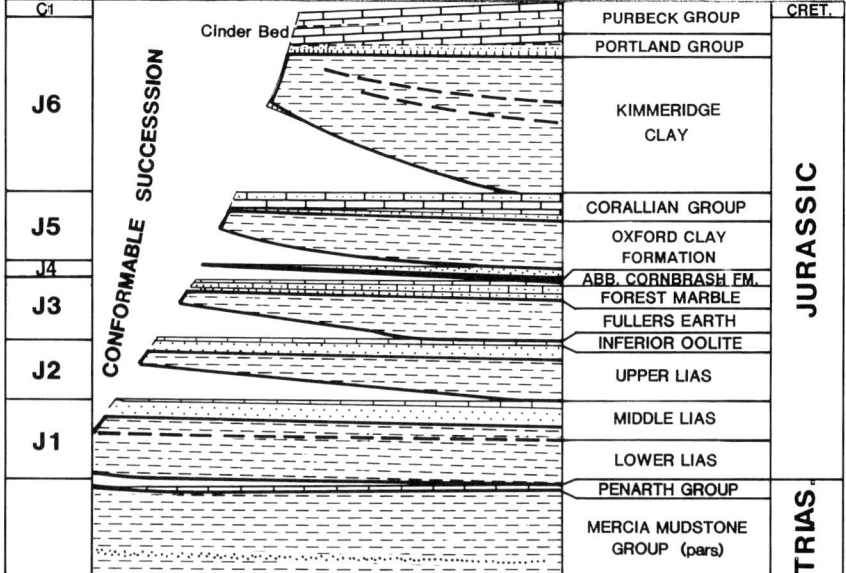

Fig. 21. Major sedimentary rhythms in the Upper Mesozoic of Dorset (after House 1985).

Inferior Oolite Group). Throughout the succession tectonically active episodes alternate with more quiescent periods.

During these 'quiet' periods progradation of higher-energy sediments from the basin margins occurred, and these episodes are the only times when significant lateral facies changes are noted. Some of the most important potential reservoir units were deposited during these quiescent periods (i.e. Bridport Sands, Great Oolite and Corallian limestones). These long-term changes in structural control probably reflect the alternation of extensional tectonics with periods characterized by thermal relaxation subsidence (Chadwick 1986).

Unit J1

Deposition of the Lias was initiated by a marine transgression during the latest Rhaetian which resulted in a broad, shallow, epicontinental shelf sea with alternations of aerobic and restricted bottom water circulations. Conditions such as this persisted throughout the Hettangian through to the Early Pliensbachian. In the study area, the Middle and Upper Lias comprise shallowing-upwards sequences with a prominent condensed component marked by the Marlstone Rock Bed–Junction Bed. The Middle Lias in the main part of the Portland–Wight Basin comprise, mudstone or calcareous claystone sequences, while in a number of wells it is cut out beneath an unconformity.

Sediments attributed to the Marlstone Rock Bed–Junction Bed show considerable evidence of having been deposited in very shallow water, and exhibit a significant number of non-sequences and examples of sedimentary condensing in the coastal area. A period of tectonic quiescence during the latest Toarcian permitted a deepening/flooding 'event' which resulted in the deposition of the Down Cliff Clay in an offshore shelf environment into which subtidal, storm-influenced Bridport Sands prograded southwards.

House (1985, 1987, 1993) recognized the Marlstone Bock Bed–Junction Bed and the Inferior Oolite Group as the culmination of two overall swallowing-upwards 'cycles' (Fig. 21). Hallam (1961) in a regional synthesis of the Lias in Britain recognized 11 sedimentary (I–XI) rhythms/cyclothems – although not all are equally well developed in the study area. Significant points of Hallam's (1961) work are illustrated in Fig. 22 and include:

(1) erosion and a slight non-sequence at the end of Cyclothem V;

(2) erosion cutting out most of the *oxynotum* ammonite zone;

(3) a non-sequence in Cyclothem VIII with uplift and shallowing in the *spinatum* ammonite zone;

(4) significant sedimentary condensing and attenuation resulting in the Junction Bed representing Cyclothems IX and X;

(5) a flooding event within the *levesquei* ammonite zone in Cyclothem XI, with inundation of the top of the Junction Bed and progradation of the Bridport Sands.

Hallam (1964) accounted for variations in sedimentation in the Blue Lias by climatic oscillation, whilst House (1985, 1987) considered the cyclicity to be due to climatic variation controlled by orbital forcing (i.e. Milankovitch cycles).

Differential subsidence continued through the Lower Lias, with dominantly argillaceous sedimentation below wave-base in shallow-marine environments. Restricted circulation at the sea floor has led to the deposition of organic-rich claystones at several levels (e.g. Blue Lias, Black Ven Marls). The lower parts of the Lower Lias, the Blue Lias and the Shales with Beef comprise interbedded thin limestones and claystones. The Lower Lias units are very widespread, with little in the way of lateral facies changes within the study area. Several minor stratigraphic breaks are recognisable at outcrop in Dorset. These include non-sequences at the following horizons.

(1) Table Ledge (*bucklandi* ammonite zone)

(2) Birchi Tabular Bed (*turneri* ammonite zone)

(3) Coinstone (*oxynotum* ammonite zone)

(4) Apoderoceras Bed (*raricostatum–jamesoni* ammonite zones)

(5) Belemnite Stone (*davoei* ammonite zone)

These 'events' cannot be identified with certainty in the subsurface due to limited biostratigraphic control. For example, the Belemnite Marls are very thin in some wells, while in others they are locally absent. The attenuation/truncation of the Belemnite Marls accordingly may represent a subregional unconformity possibly associated with the non-sequence marked by the Apoderoceras Bed on the Dorset Coast. The genesis of the above major and several minor non-sequences within the Lower Lias is considered to be related to a combination of minor tectonic influences and relative sea-level changes.

The Middle Lias initially continued the pattern of Lower Lias sedimentation, but a subsequent shallowing-upwards trend culminated in the eventual deposition of shallow-marine sands in Dorset and the northern part of the study area. These sands, which appear to

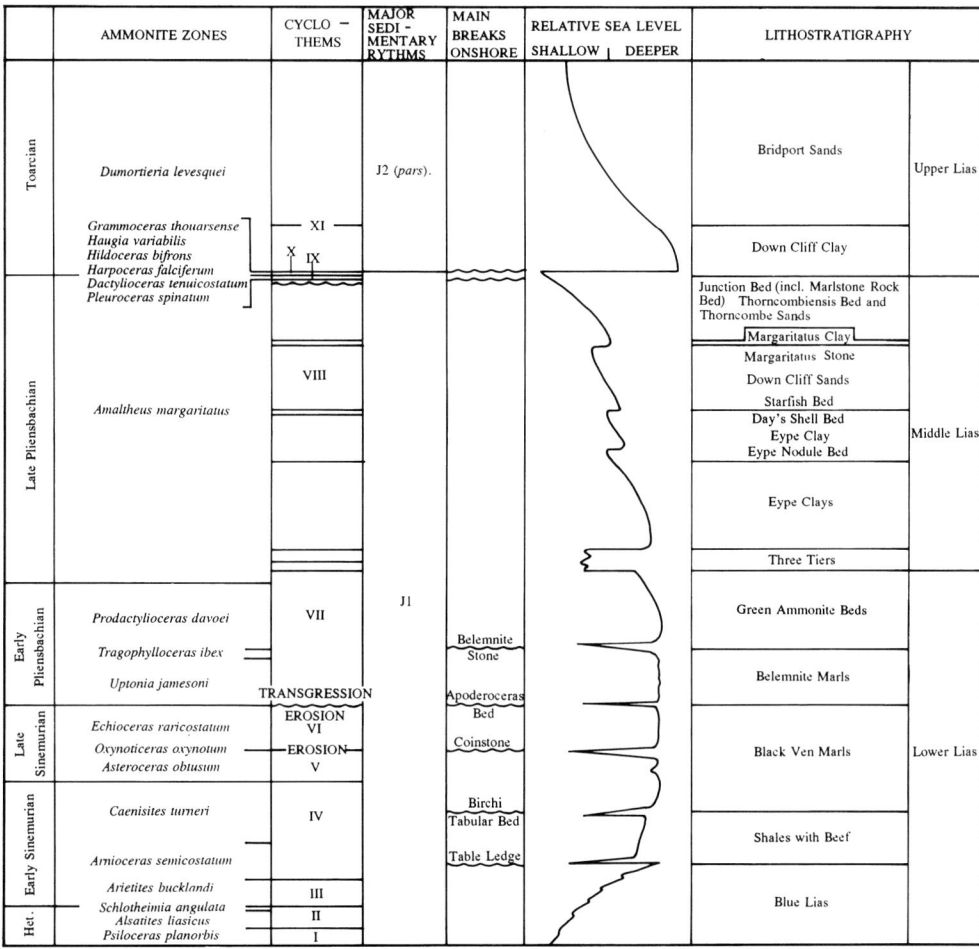

Fig. 22. Comparison of the lithostratigraphy, biostratigraphy and environmental interpretations of the Lias of Dorset (cyclothems after Hallam 1961; major sedimentary rhythms after House 1985, 1987; major non-sequences after Cope *et al.* 1980a, House, 1993).

fringe the London Platform, occur in the Dorset Basin and parts of the northern edge of the Portland–Wight Basin. They probably reflect the temporary cessation of regional subsidence, perhaps allied to a relative sea-level fall, permitting progradation from basin margins and progressive shallowing of water depths. Differential subsidence is not obvious during the Middle Lias, while major fault movement seems to have ceased or been minimal.

Over some of study area the Middle Lias shallowing-upwards trend is terminated by a very thin, condensed Late Pliensbachian ferruginous limestone unit, the Marlstone Rock Bed. This is overlain by another condensed carbonate-dominated interval of Early to Late Toarcian age, the Junction Bed. In the eastern Portland–Wight Basin, and adjacent onshore areas including the Wytch Farm area and the Isle of Wight, wireline log correlations and biostratigraphic data indicate that the Middle Lias, and in places the upper part of the Lower Lias, are absent or incomplete, with Toarcian sediments resting unconformably on older sediments. Previous interpretations (e.g. Bristow *et al.* 1991) have attempted to identify a complete Lower and Middle Lias succession in the Wytch Farm area, but this cannot be supported by the data available from the present study. This stratigraphic break, with evidence of uplift and erosion, clearly indicates a complex regional tectonic event which influenced subsequent sedimentation patterns producing condensed facies such as the Marlstone Rock and Junction Beds. Frequently the Junction Bed is closely associated with the Marlstone Rock Bed (Wilson et al. 1958; Howarth 1980). The Junction Bed contains indications of several episodes of shallow-water sedimentation and many non-sequences represented within a condensed unit of micritic and algal limestones in Dorset. Howarth (1980) has shown that the Junction Bed contains ammonites from several horizons in the Lower and lowermost Upper Toarcian, whilst Jenkyns & Senior (1991) reviewed evidence of synsedimentary faulting at outcrop and considered that it was a significant influence on deposition. Differential subsidence rates, with thicker sections in the Portland–Wight Basin and the Dorset Basin, do suggest that fault movement was a significant control on sedimentation on a regional scale.

Coeval sediments to the Junction Bed in the vicinity of the south Dorset coast may possibly be represented by claystones and argillaceous limestones (e.g. Radipole-1, BGS Borehole 74/36, Hamblin et al. 1992). In Normandy the whole of the Upper Lias is represented by a condensed sequence of sediments c. 30 feet thick, a significant part of which is correlative with the uppermost Junction Bed.

A thin unit of organic-rich claystones is locally present at the base of the Junction Bed. In the study area it is too restricted in occurrence to be of significance as a source rock, but it expands to form a major Early Toarcian source rock unit in the Paris Basin and The Netherlands. It is possible that it is well developed to the south of the Central Channel High, as it is present in wells immediately north of this structure.

Unit J2

The sediments included within this unit comprise the Down Cliff Clay, the Bridport Sands and the Inferior Oolite Group. The Down Cliff Clay represents a return to shallow-marine clastic sedimentation and was deposited during a relatively short period (one ammonite subzone). This suggests that sedimentation rates were relatively high compared to the underlying units (the Marlstone Rock Bed and Junction Bed). The widespread distribution of the Down Cliff Clay and its stratigraphic position, above the condensed limestones of the Junction Bed, point to the onset of a period of relative tectonic stability and a relative rise in sea level. The contact between the Down Cliff clay and the Bridport Sand is often gradational and reflects an upwards-coarsening (shallowing) sequence.

The Bridport Sands represent the continuance of the shallowing-upwards trend discernible in the Down Cliff Clay. The sands are shallow-marine sediments, deposited as an offshore sand shoal with some indications of storm activity within the lower part of the succession, and form part of a latest Toarcian–earliest Aalenian sand complex extending from the Portland–Wight Basin to Gloucestershire. In the present study area this sand sheet prograded rapidly from a presumed westerly source with no evidence for regional structural control, suggesting that differential fault movement was rather limited at this time.

Decreased subsidence in the Weald Basin during the Late Toarcian is reflected by the development of condensed and iron-rich sediments and clastic-starved facies. The Bridport Sands pass laterally into thinner, more argillaceous and calcareous facies over the fringes of the Central Channel and the Hampshire–Dieppe Highs. These are relatively condensed shallow-marine units by comparison with the Dorset Basin and main part of the Portland–Wight Basin. It is uncertain whether the depositional attenuation is structurally controlled, or merely reflects sediment starvation and a position at the limits of sand progradation/supply.

Renewal of tectonic activity during the Early Aalenian, possibly associated with marine transgression (cf. Rioult et al. 1991), led to the initiation of carbonate-dominated marine deposition, which persisted into the earliest Bathonian. Active structural controls on sedimentation are clearly indicated by the regional thickness variations of the Inferior Oolite Group. A highly condensed succession occurs over the Purbeck Fault Zone, with relatively thick sequences in the Weald Basin and northern parts of the Portland–Wight Basin.

The major divisions of the Inferior Oolite Group (i.e. Lower–Middle–Upper) are marked by prominent erosion planes. Arkell (1933) considered that the subdivisions of the Inferior Oolite represented phases of sedimentation, denudation and transgression. In Dorset the Inferior Oolite Group is highly condensed (Cope et al. 1980b), with several horizons comprising remainé beds with rolled ammonites and localized non-sequences which are considered to reflect the effects of penecontemporaneous tectonics.

Regionally Arkell (1933) recognized a 'Bajocian transgression' at the base of the Middle Inferior Oolite and a 'Vesulian transgression' at the base of the Upper Inferior Oolite. In general the Upper Inferior Oolite is more uniformly developed than the Lower–Middle subdivisions. The onset of Inferior Oolite Group deposition follows on conformably from the Bridport Sands in many parts of the study area, however, there is some minor diachroneity of the two facies in Dorset (Cope *et al.* 1980*b*). This, in association with evidence for continued shallow-water shelf conditions throughout the deposition of the Inferior Oolite Group, suggests that Arkell's (1933) 'transgressions' may be thought of as phases of onlap/tectonically influenced relative sea-level change rather than eustatic sea-level rise. For example, Whittaker (1985) considered that the stratigraphic breaks recognized at the Lower–Middle, Middle–Upper Inferior Oolite boundaries may be due to uplift in the North Sea area. Recently, however, Rioult *et al.* (1991) have made an outcrop sequence stratigraphic analysis of the Middle Jurassic of Normandy and Dorset and have recognized several of Haq *et al.*'s (1988) eustatic sequence boundaries.

North of the present study area, the Upper Inferior Oolite oversteps onto older Jurassic sediments. Within the study area the thicker Inferior Oolite Group sequences appear to be relatively complete, although sedimentary condensing and minor stratigraphic breaks are prominent features of the thinner successions. The Upper Inferior Oolite is generally less condensed and more uniform in facies than the Middle and Lower Inferior Oolite, possibly having been deposited in rather deeper water environments, indicating more regional subsidence with more subdued structural control.

Unit J3

The Great Oolite Group in association with the upper Fleet Member of the Abbotsbury Cornbrash Formation comprise Unit J3. Deposition of the Great Oolite Group was initiated by a major deepening which may have encompassed both an eustatic sea-level rise and the onset of a phase of subsidence and onlap. Holloway (1985*d*) comments that considerable subsidence occurred in the Dorset Basin during the Bathonian. In the western part of the study area the Great Oolite comprises an argillaceous sequence with occasional prominent carbonate units, whilst oolitic and shell detrital limestones comprise most of the succession in the eastern part of the area.

Consideration of the stratigraphic architecture of the Great Oolite Group suggests that a carbonate ramp was developed in the east and north of the study area, whilst a fault-controlled basin in the west was the locus for a series of four shallowing-upwards sequences. The documentation of significant growth faulting during the Bathonian (Chadwick 1986) suggests that the generation of these sequences owes more to tectonic rather than eustatic controls, however, in contrast, Rioult *et al.* (1991) recognized and refined some of Haq *et al.*'s (1988) sequence boundaries in Normandy, so eustatic influences cannot be entirely excluded.

The relationship between basinal argillaceous sequences and the carbonate ramp is illustrated in Fig. 23. Following the initial deepening event at the base of the Great Oolite Group the first shallowing-upwards sequence is developed through the Lower Fuller's Earth Clay culminating in the Fuller's Earth Rock. This event may correspond to the sandstone unit developed in the Hesters Copse Member of the Humbly Grove Oilfield. A second 'flooding' or deepening event is marked by the Upper Fuller's Earth Clay and by the intra-Hesters Copse oyster unit (Sellwood *et al.* 1985). The shallowing phase of this sequence is represented by the Wattonensis Beds and possibly the dolomites in the uppermost Hesters Copse Member. The third 'flooding event' is marked by the onset of clay deposition in the Frome Clay, with the culmination of this phase represented by the Boueti Bed. Coeval sediments on the carbonate ramp comprise oolitic limestones of the Humbly Grove and lower (Unit II) Herriard Members. In Dorset, an oyster bed present within the Frome Clay is envisaged to represent a minor phase of shallowing, reflecting the maximum development of the oolitic shoals of the Humbly Grove Member. A subsequent minor flooding/deepening event represented by Hoddington Member claystones is inferred to reflect the termination of this intra-Frome Clay cycle. The fourth and final 'flooding event' is marked by the base of the Forest Marble. In the Cotswolds, the Upper Rags Member of the Great Oolite facies is coeval with the lowermost part of the Forest Marble and suggests that the base Forest Marble transgression was less marked than the preceding events. This hypothesis is supported by the more restricted distribution of Unit I of the Herriard Member (e.g. Sellwood *et al.* 1985) which may represent a retrogradational setting analogous to the Upper Rags Member of the Cotswolds.

The depositional history of the Great Oolite Group may be summarized as the repetition of a series of shoaling events coupled with the construction/progradation of a carbonate ramp southwards and westwards. Subsequent deepening events may be viewed as resulting from a

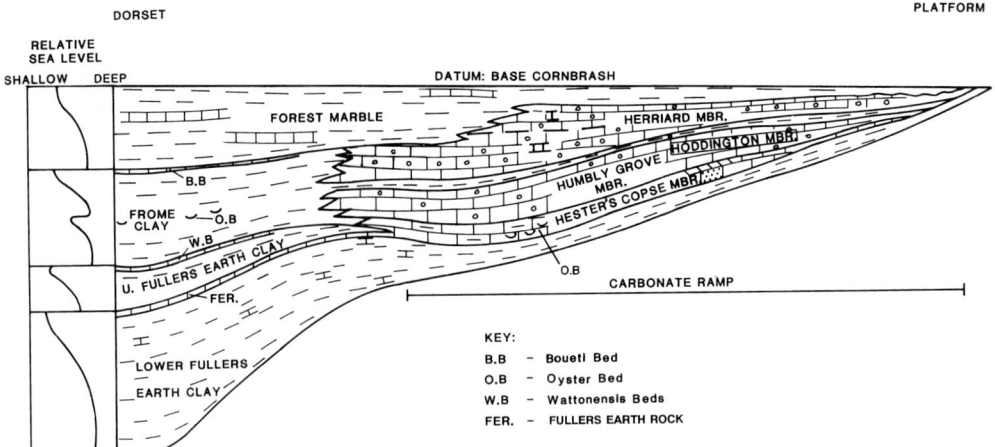

Fig. 23. Stratigraphic relations of the Great Oolite Group and relative sea level.

combination of growth faulting and rapid subsidence (tilting?) of the ramps and relative sea-level changes (Whittaker 1985; Sellwood et al. 1985; Rioult et al. 1991) rather than purely eustatic changes.

The Abbotsbury Cornbrash Formation spans the Bathonian–Callovian boundary and was considered by Arkell (1933) to be a transgressive unit. In a regional context, the Abbotsbury Cornbrash Formation represents a series of shallow-marine limestones and claystones which mark the beginning of an important marine transgressive episode (Arkell 1933; Rioult et al. 1991). In the Humbly Grove Oilfield, the lower Berry Member packstones transgress the Forest Marble and are succeeded by bioturbated clays in the upper Fleet Member (Sellwood et al. 1985). The relatively attenuated development of the Abbotsbury Cornbrash Formation by comparison with the underlying Great Oolite Group suggests that the latest Bathonian–earliest Callovian represents a period of relative stability/tectonic quiescence. Page (1989) envisages the presence of a broad shallow shelf to the north of the study area throughout deposition of both the Abbotsbury Cornbrash and Kellaways Formations. Northwest of the study area a structural high/shelf edge was present in Wiltshire, where Page (1989) records the occurrence of cross-bedded coarse sands and common hardgrounds in the Abbotsbury Cornbrash Formation. Overstep of the Abbotsbury Cornbrash Formation onto pre-Jurassic sediments on the London Platform reflects its persistence as a positive feature until the latest Bathonian–earliest Callovian. An erosional stratigraphic break is regionally present between the Berry and Fleet Members, and has been interpreted as possibly reflecting a relative sea-level fall (Bradshaw et al. 1992). Following this break in sequence, argillaceous limestones attributed to the Fleet Member spread progressively northwards (Calloman 1955). The onlap at the base of this unit may correspond to the maximum flooding surface recognized by Rioult et al. (1991) in Normandy and south Dorset.

Unit J4

Unit J4 comprises the thinnest unit recognized in this study, consisting of the Kellaways Formation. The uniform thickness of this formation indicates that by the Early Callovian much of the study area represented a shallow-marine shelf (Arkell 1933; Page 1989). The lower Kellaways Clay Member reflects a well-developed marine flooding event which represents an offshore muddy shelf onto which the Kellaways Sand Member prograded. The Kellaways Sand and Kellaways Clay Members represent one of the large-scale coarsening-upwards cycles (J4 herein) recognized by House (1985, 1993). Page (1989) notes that the Kellaways Clay Member contains a rich bivalve fauna, whilst the overlying unit exhibits opportunistic benthic assemblages and restricted ammonite faunas. This assemblage change was considered to predate the 156 Ma sequence boundary by Rioult et al. (1991). The uppermost Kellaways Sand Member exhibits small-scale cross-bedding and contains abundant trace fossils at outcrop.

In contrast, Bradshaw et al. (1992) interpreted the Kellaways Sand to be a 'basinwards' spread of sheet sand reflecting strong shelf current reworking of coastal/alluvial sands during a period of transgression.

Unit J5

This unit consists of the Oxford Clay Formation and the Corallian Group. Deposition of the lower Peterborough Member occurred in an extensive, shallow, epicontinental sea which was not undergoing agitation below wave-base (Hallam 1975). An abundance of organic matter (fed into an inshore area experiencing a warm humid climate) was rapidly oxidized favouring whole or partial stagnation whenever water circulation was minimal. Stagnation/stratification of the water column was promoted by the existence of a series of shallow sills created and maintained by mild tectonic activity which inhibited recycling of bottom waters in the intervening shallow basinal areas. The bituminous shale of the Peterborough Member represents the transgression, during the latest Early Callovian, of such a shallow epicontinental shelf sea with restricted water circulation and temporary dysaerobic conditions (Hallam 1975). Continued transgression/subsidence produced more open marine shelf sedimentation with aerated bottom waters, resulting in the deposition of the calcareous claystones of the Stewartby Member. The Peterborough–Stewartby Member contact was interpreted as the horizon of a maximum flooding surface by Rioult et al. (1991). Hollingworth & Wignall (1992) have suggested that there may be a localized but marked shallowing event in the latest Callovian associated with the development of carbonate-rich sediments (lamberti Limestone) in the uppermost Stewartby Member. In addition, Rioult et al. (1991) recognized a latest Callovian (intra-*lamberti* ammonite zone) maximum flooding surface within the uppermost part of the Stewartby Member, post-dating the *lamberti* Limestone horizon. Deposition of the upper Weymouth Member is considered to result from subsidence/transgression during the earliest Oxfordian. Certainly the Stewartby–Weymouth Member contact reflects significant changes within the basin: for example, Hollingworth & Wignall (1992) note a change from high- to low-diversity assemblages across the boundary, while Wright (1986b) notes the presence of organic-rich shales within the Weymouth Member. An intra-Early Oxfordian sequence boundary (151 Ma *mariae–cordatum* ammonite zonal contact) was recognized by Rioult et al. (1991), with a maximum flooding surface occurring within the *cordatum* ammonite zone.

A regional tectonically driven regression in the latest Early Oxfordian resulted in deposition of sandstones, claystones and limestones of the Corallian Group. This is in marked contrast to the overall trend of rising sea levels through the most of the Oxfordian postulated by Hallam (1975). In a regional context, Oxfordian sequences are dominated by the deposition of predominantly argillaceous facies (e.g. Oxford and Ampthill Clays, West Walton, Staffin Bay and Heather Formations). However, by the latest Early Oxfordian the mixed limestone, shale and sandstone facies of the Corallian Group were widespread throughout the northern edge of the Portland–Wight, Dorset and Weald basins. Bradshaw et al. (1992) considered that the maximum spread of Corallian facies occurred during the Middle Oxfordian (*tenuiserratum* ammonite zone). Since the pioneering studies of Arkell (1933, 1936), the Corallian has been extensively studied with respect to the 'cyclicity' of its constituent sediments. Several interpretations of the depositional history of the Corallian Group have been presented with varying emphasis on the tectonic and eustatic controls on deposition (e.g. Arkell 1933; Grieves 1968; Brookfield 1978; Wright 1986a; Sun 1989; Allen & Underhill 1989; Rioult et al. 1991). Sun (1989) reviewed most of the previous work and presented his own revised model which is outlined in Fig. 24. Rioult et al. (1991) compared the Corallian sequences of Normandy and south Dorset and provided a sequence stratigraphic interpretation which was significantly influenced by and modified the work of Haq et al. (1988)

Corallian sedimentation took place in a tectonically unstable period following on from basin inversion initiated during the latest Early Oxfordian (Kent 1980). Deposition of the Middle Oxfordian sequence took place in a series of extensional basins bounded by deep-seated faults with intra-basin highs providing a sediment source. Isopach data suggest that there may be a depocentre within the northern part of the Portland–Wight Basin, but otherwise the gross distribution of the Corallian is marked by low subsidence rates. There is little evidence for significant subsidence in the Weald Basin during the Middle Oxfordian except for the development of a thick carbonate sequence at its northern margins. Fault movement along the Purbeck–Isle of Wight Fault Zone may, in tandem with sea-level changes, account for what has long been considered to be the underlying influence on sedimentation of the Corallian

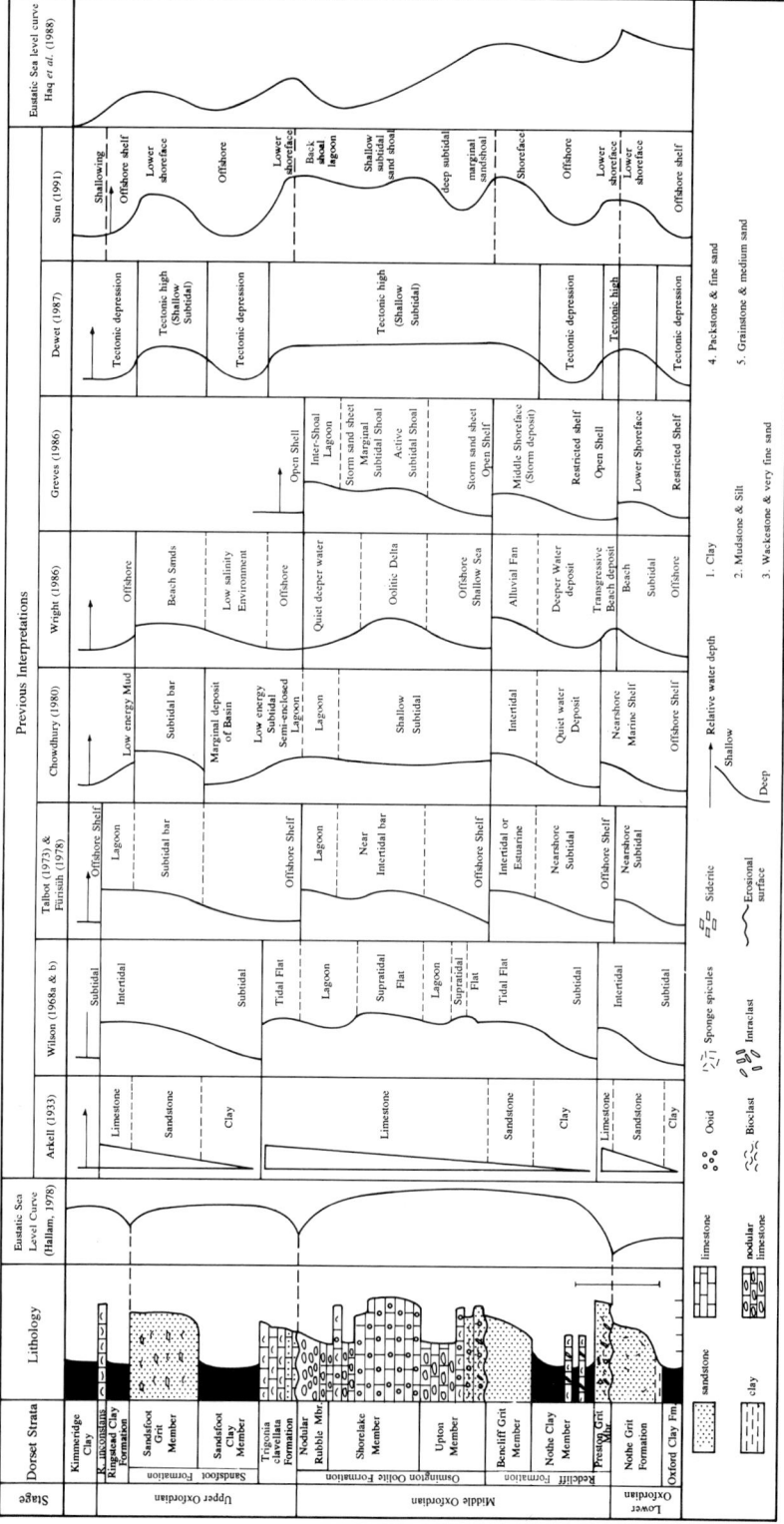

Fig. 24. Review of interpretations of the Corallian Group environmental development in Dorset (modified after Sun 1990).

in the Dorset Coast area (Brookfield 1978; Wright 1986a, b). In contrast some workers have suggested that controls on typical Corallian sedimentation were greatly influenced by eustatic processes. (e.g. Talbot 1973; Rioult et al. 1991)

The Late Oxfordian represents a period with more widespread regional subsidence and localized uplift (Wright 1986a). Rawson & Riley (1982) recognize a base Late Oxfordian transgression which correlates with the development of a transgressive systems tract within the Trigonia clavellata Formation, above the postulated position of Rioult et al.'s (1991) 147.5 Ma sequence boundary. In the Dorset Basin, and the northern part of the Portland–Wight Basin, alternating sandstones, limestones and claystones of typical shallow-marine 'Corallian' facies were deposited until late in the Oxfordian. In other parts of the study area (e.g. central and southern parts of the Portland–Wight Basin, parts of the Hampshire–Dieppe High, Weald Basin), low-energy, marine shelf clays with minor limestones and sandstones comprising an equivalent of the Ampthill Clay were deposited. In contrast, basin boundary faults in the Weald Basin were reactivated in the Late Oxfordian, and the Weald became the site of a depocentre for the first time since the Bajocian. The isopach data, however, suggest that the Portland–Wight Basin was not undergoing differential subsidence during the Late Oxfordian.

The Cranbourne–Fordingbridge and the Hampshire–Dieppe Highs underwent a phase of uplift and erosion during the Late Oxfordian. Middle Oxfordian sediments are unconformably overlain by Early Kimmeridgian claystones. It is difficult to assess the full extent or significance of this stratigraphic break due to limited well control, but comparable breaks in succession have been documented north of the study area in Oxfordshire (Cope et al. 1980b). Wright (1986a) suggested that a minor stratigraphic break may occur within the Late Oxfordian in south Dorset, whilst Cox & Gallois (1981) indicated that a widespread non-sequence occurs at the base of the Kimmeridge Clay Formation throughout wide areas of southern England. Wright (1986a) considered the progradational nature of some of the Corallian sequences as resulting from a balance between high- and low-energy sedimentation, with localized fault-generated subsidence producing regular progradation of sand interrupted by clay deposition. Phases of extensional movement were considered to be marked by clay sedimentation, whilst coarse clastic progradation marked periods of stress easement. Consequently, whilst each basin exhibits similar stratigraphic architecture and subsidence history, the detailed succession in any particular location varies with the amount of absolute subsidence and sediment availability (Wright, 1986a).

Unit J6

This unit encompasses the Kimmeridge Clay Formation, the Portland and Purbeck Groups (*pars*). The onset of the marine claystones of the Kimmeridge Clay with enhanced organic material was initiated by a base Kimmeridgian (*baylei* ammonite Zone) transgression. In Dorset, Cox & Gallois (1981) noted a minor, yet widespread, erosion surface at the base of the Kimmeridge Clay Formation. Rawson & Riley (1982), although concentrating on 'events' in the North Sea area, reviewed evidence from surrounding areas including southern England and consider that the *baylei* ammonite zone 'event' was controlled by local tectonics.

Wignall (1991) recognized a series of 10 hiatus-bounded sedimentary packets or sequences within the Kimmeridge Clay Formation, which were compared to the sequence models of Galloway and the Exxon Group. Few of the 'hiatuses' described by Wignall (1991) conform to the sequence boundaries recognized by Haq et al. (1988) on the eustatic sea-level chart. Wignall (1991) considered that the sequence boundaries resulted from falls in base level controlled by storm wave-base, with the development of transgressive–regressive cycles being symmetrical as recognized in Einsele's (1985) model of sedimentation in epicontinental seas. These changes in base level may represent either a fall in sea-level or widespread tectonic (epeirogenic) movements independent of intra-basinal tectonics (Wignall 1991; Hallam 1992). Furthermore, Wignall (1991) commented that condensed sections are significantly absent within the Kimmeridge Clay Formation, a feature which he considered to be a reflection of the small scale of Kimmeridgian basins in which progradational sediments are poorly developed.

Wignall's (1991) sedimentary hiatuses are illustrated in Fig. 25, together with those recognized by Rawson & Riley (1982). Some of the differences may be reconciled as a difference in philosophy or calibration, for example event K10 (intra-*fittoni* ammonite zone) of Wignall may equate to Rawson & Riley's (1982) *rotunda* 'event'. Similarly, event K9 (*pallasioides* ammonite zone) may equate to the *pectinatus* 'event'. Rawson & Riley (1982)

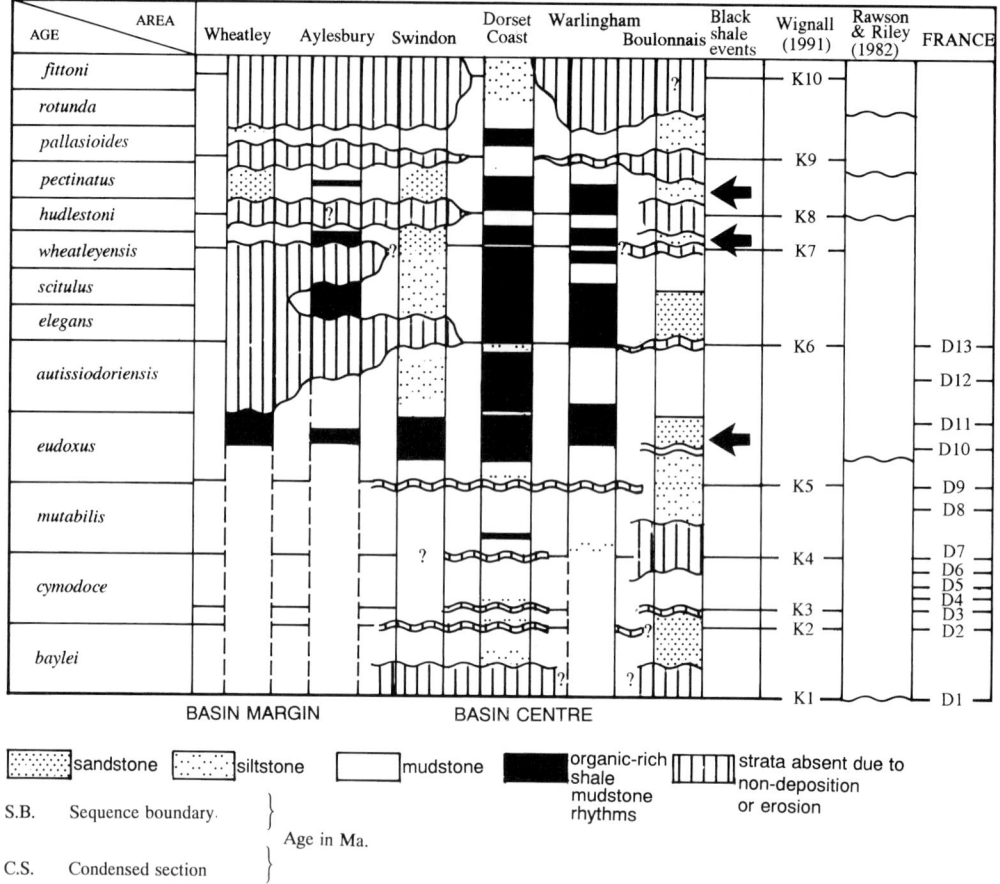

Fig. 25. Chronostratigraphic chart of some onshore Kimmeridgian sections in southern England (modified from Wignall 1991).

recognize a transgressive event within the *eudoxus* ammonite zone, associated with the onset of oil-shale deposition, whilst Wignall (1991) places his 'eudoxus event' at a non-sequence at the base of the zone. Further differences in the level of 'event' recognition may be accounted for by differences in the scale of investigation (i.e. regional v. subregional).

In the context of northwest Europe, Rawson & Riley (1982) described transgressions/flooding events in the *baylei* and *eudoxus* ammonite zones, whilst subsequent unpublished work also suggests a further flooding 'event' may also occur in the *hudlestoni* ammonite zone. The *pectinatus* and *rotunda* 'events' were considered to represent regressive events linked to tectonic effects. Townson (1975) illustrated the onset of a regressive unit within the Lingula Shales (*fittoni* ammonite zone) in the uppermost Kimmeridge Clay Formation. Rawson & Riley (1982) recog-

nized the commencement of regression in the uppermost part of the Kimmeridge Clay Formation and considered that the regression continued in tandem with phases of uplift into the Portlandian.

The rhythmic lithological alternations on a metric scale have been described from Dorset and other onshore areas, comprising sequences from laminated bituminous shale through bioturbated shales to coccolith-rich limestone. The origin of this rhythmicity has been the subject of some debate (see Gallois 1978; Tyson *et al.* 1979; Oschmann 1988). Various estimates of water depths during the deposition of the Kimmeridge Clay Formation have also been made, from tens of feet deep (Hallam 1975) through to hundreds of feet deep (Tyson *et al.* 1979). House (1985, 1987), when considering the nature of the rhythmic alternations within the Kimmeridge Clay Formation, suggested that there is some indication that orbitally forced climatic change may be a significant factor in the formation of these units. Sea level may have reached a maximum during the middle of the Kimmeridgian, with a gradual fall through the later Kimmeridgian into the Portlandian (Hallam 1978, 1988).

Thick developments of the Kimmeridge Clay Formation are present both in the Weald Basin and in the northern Portland–Wight Basin. The thickness of the Kimmeridge Clay Formation rapidly decreases towards the Cranbourne–Fordingbridge and Hampshire–Dieppe Highs where it is cut out by a non-sequence beneath the Portland Group. In the vicinity of the Marchwood-1 and Bottom Copse-1 wells, the Kimmeridge Clay Formation is truncated by the Lower Albian Carstone, the former in turn overstepping onto truncated Corallian Group sediments.

A gradual relative sea-level fall initiated during the Kimmeridgian continued during the Portlandian and was possibly accentuated by uplift of some areas such as the London–Brabant Platform. The Kimmeridge Clay Formation coarsens upwards via the dolomitic silts of the Portland Sand Formation into the marine limestones of the Portland Limestone Formation. In onshore areas, the faunal associations are suggestive of a restricted shallow-marine environment with fluctuations in salinity in strong contrast to the Kimmeridge Clay Formation. Townson (1975) revised the base of the Portland Beds and recognized three transgressive–regressive cycles, while Coe (1996) recognized three unconformities (P1–P3) bounding distinct sequences.

In Townson's (1975) model relative sea-level fall/regression coupled with the relatively arid Late Jurassic climate resulted in increased evaporation in shallow marginal areas of the basin. The denser brines produced in the marginal areas are interpreted as having settled out and flowed beneath normal salinity seawater to have become ponded in the deeper parts of the basin. Townson postulated a critical depth below which the resulting brine seawater blend became too alkaline and oxygen deficient/stagnant to support any benthos and where the lime-rich muds would have become dolomitized. Above the critical depth bottom water conditions were unsuitable for any benthos other than sponge spicules such as *Rhaxella* spp. but where no dolomitization occurred. Progressive sea-level fall resulted in shallow shoals being developed with an aerated, turbulent water column. In this setting bivalve-rich shell beds were developed from both drifted and autochthonous fauna whilst any lime muds were winnowed out by wave action. The shallowest parts of the basin developed as oolite shoals fringing isolated, periodically hypersaline lagoons with algal stromatolites and occasionally evaporites.

Townson's cycles and lithostratigraphy are illustrated in Fig. 26. The basal transgressive part of each sequence is represented by the following units (in ascending stratigraphical order).

Lingula Shales (Kimmeridge Clay Formation).
Pondfield Member (Portland Sand Formation).
Dungy Head Member (Portland Limestone Formation).

The third cycle was considered to reach its culmination in the Lower Purbeck Beds. In the Portland–Wight Basin and parts of the onshore area it is possible to recognize coarsening-upwards profiles on the wireline logs which may reflect Townson's 'regressive' sequences, although it has not been possible to consistently map out each of these in the present study area.

Rawson & Riley (1982) noted the onset of regression during the latest Kimmeridgian and considered that the combined effects of uplift and regression are marked throughout the Early Portlandian. A major phase of uplift and regression is noted during the *oppressus* ammonite zone spreading shallow-marine sands across the East Midlands Shelf, whilst the study area became the site of an enclosed basin of non-marine deposition. Synsedimentary fault movement is evident as marked by differential subsidence between the Weald Basin and the northern Portland–Wight Basin.

Coe (1996) also reviewed the Portland Group succession using sequence stratigraphic methods. Coe's interpretations differ from those of Townson (1975) in that she recognized three unconformities (P1–P3), which are considered to delimit individual sequences. Unconformity P1 is coincident with the base of the Corton Member, P2 with the basal Shell Beds and P3 with the lower part of the Portland Freestone. Within each of these unconformity bounded sequences, Coe identified transgressive/marine flooding surfaces.

The Purbeck Group lacustrine and lagoonal micritic limestones and marls, together with evaporites and soil beds, represent continuation of the 'regression' represented by the Portland Group. Townson (1975) considered the lowermost part of the Purbeck Beds to comprise the uppermost part of his Portland Group 'regressive units'. Rawson & Riley (1982) related the

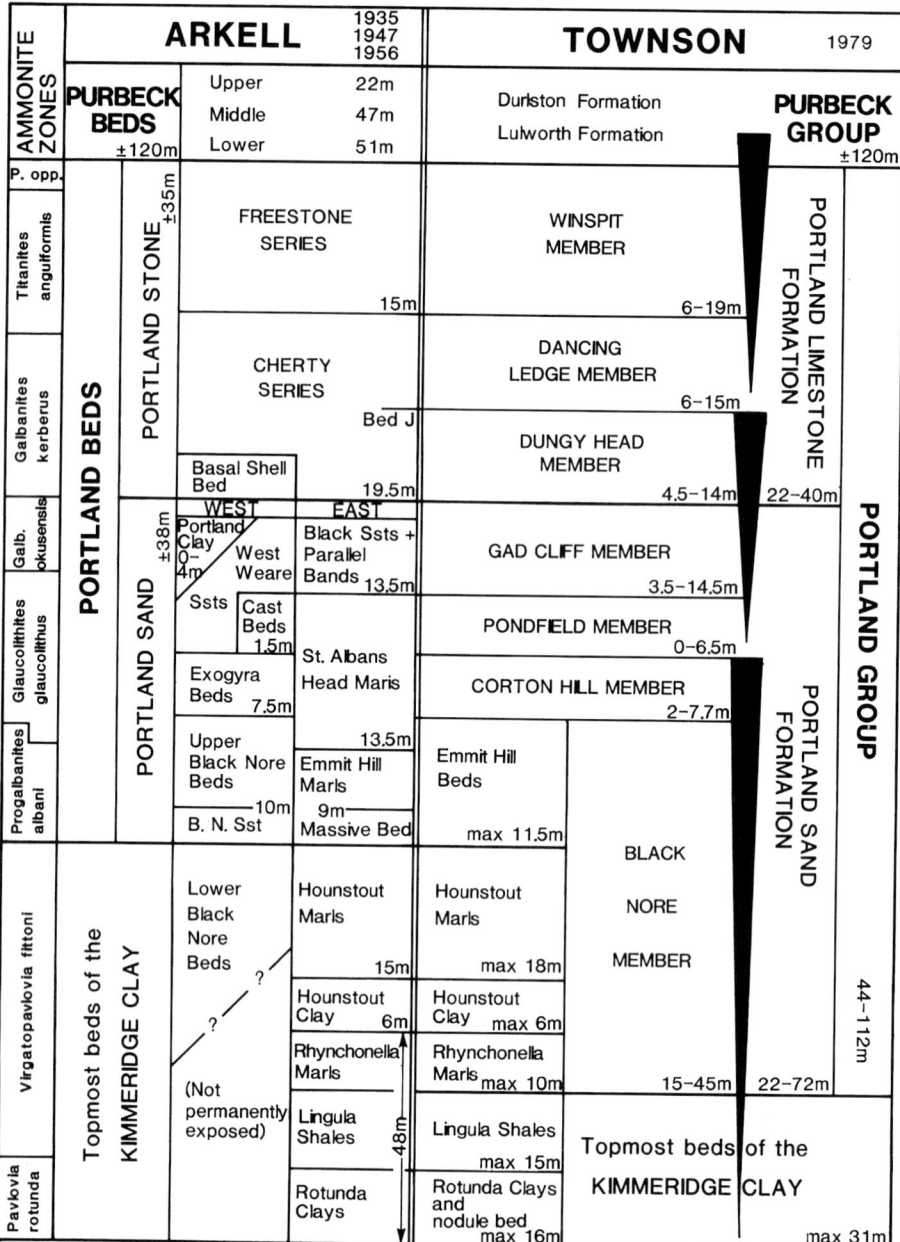

Fig. 26. Lithostratigraphy of the Portland Group of Dorset. Regressive sequences indicated by arrows (after Townson 1975).

onset of Purbeck deposition during the latest Early Portlandian to the regressive event and phase of uplift in the *oppressus* ammonite zone.

In Dorset and the Weald a prominent oyster lumachelle, the Cinder Bed, is considered to represent a marine 'incursion'. Casey (1963) equated the Cinder Bed with the base of the Cretaceous (*runctoni* ammonite zone), whilst subsequently Rawson & Riley (1982) correlated the 'Cinder Bed' with the transgressive 'event' they recognized in the Early Ryazanian (*kochi* ammonite zone). Following Casey (1963, 1973), the Cinder Bed was considered to represent the invasion of the early Ryazanian Boreal Sea into southern England. More recently, however, Wimbledon & Hunt (1983) and Morter (1984) have suggested that the Cinder Bed probably represents an inundation of a Tethyan Sea into the Anglo-Paris Basin.

Small-scale cyclicity is well represented within the Purbeck Beds, denoted by alternations of stromatolites, evaporites, soil beds, etc. The nine molluscan associations recognized by Morter (1984) have been interpreted as representing different salinities and can be matched with the ostracod faunicycles of Anderson (1971, 1985). The development of alternating freshwater–marine faunas may reflect peaks of regression–transgression, respectively, possibly representing aggradational parasequences or alternatively fluctuations in climate. The continuation of differential subsidence in the Portland–Wight, Dorset and Weald basins reflects ongoing fault movement, continuing the trend which commenced in the Portland Beds.

Conclusions

A consistent and applicable stratigraphic framework integrating lithological, petrophysical and biostratigraphic data has been achieved for the latest Triassic through to the earliest Cretaceous of the Portland–Wight Basin and its adjacent areas. The majority of the rock units present are geographically widespread, however, a number exhibit significant diachroneity. These specifically include the Frome Clay and Great Oolite, the Kellaways Clay and Sand Members, and the 'Upper Calcareous Corallian' and the Ampthill Clay Equivalent. Sand-prone units occur at various levels within the Jurassic, although most individual developments have a localized geographic distribution. Foremost amongst these include the Middle Lias Sands, the Bencliff Grit and the Sandsfoot Grit. Their distributions are envisaged to be influenced by local tectonic structuring. Tectonically controlled subsidence of the various subbasins had little effect on lithofacies except at certain horizons, for example the Junction Bed, although it has had a marked effect on sediment thickness. Regional stratigraphic breaks have been recognized, including haituses close to the Pliensbachian–Toarcian and the Aalenian–Bajocian boundaries and also within the Late Oxfordian.

The authors would like to thank the numerous friends and colleagues for their discussions, and BP Exploration for technical assistance. J. Andrews and L. E. Powell are thanked for the draughting of the figures.

Some of the data used in this paper were collected by the authors and included in a commercial study for PaleoServices in 1993. The authors are grateful for the assistance of RPS Paleo in allowing this paper to be published, and wish to emphasize that the views represented herein are those of the authors.

References

AIGNER, T. 1980. Bio-fabrics and stratinomy of the Lower Kimmeridge Clay (U. Jurassic), Dorset, England. *Jahrbuch für Geologie und Päläontologie Abhandlung*, **159**, 324–338.

AINSWORTH, N. R., BRAHAM, W., GREGORY, F. J., JOHNSON, B. & KING, C. 1998. A proposed latest Triassic to earliest Cretaceous microfossil biozonation for the English Channel and its adjacent areas. *This volume*.

ALLEN, P. A. & UNDERHILL, J. R. 1989. Swaley cross-stratification produced by unidirectional flows, Bencliff Grit (Upper Jurassic), Dorset, UK. *Journal of the Geological Society, London*, **146**, 241–252.

ANDERSON, F. W. 1964. Rhaetic Ostracoda. *Bulletin of the Geological Survey of Great Britain*, **21**, 133–174.

—— 1971. The ostracods. *In*: ANDERSON, F. W. & BAZLEY, R. A. B. (eds) *The Purbeck Beds of the Weald (England)*. Bulletin of the Geological Survey of Great Britain, **34**, 1–173.

—— 1985. Ostracod faunas in the Purbeck and Wealden of England. *Journal of Micropalaeontology*, **4**, 1–67.

ANDERTON, R., BRIDGES, P. H., LEEDER, M. R. & SELLWOOD, B. W. 1979. *A Dynamic Stratigraphy of the British Isles – A Study in Crustal Evolution*. Allen & Unwin, London.

ARKELL, W. J. 1933. *The Jurassic System in Great Britain*. Clarendon Press, Oxford.

—— 1936. The Corallian Beds of Dorset. *Proceedings of the Dorset Natural History and Archaeological Society*, **57**, 59–93.

—— 1947. *The Geology of the Country Around Weymouth, Swanage, Corfe and Lulworth*. Memoirs of the Geological Survey of Great Britain. Sheets 341, 342, 343 with small portions of sheets 327, 328 and 329.

BRADSHAW, M. J., COPE, J. C. W., CRIPPS, D. W., DONOVAN, D. T., HOWARTH, M. K., RAWSON, P. F., WEST, I. M. & WIMBLEDON, W. A. 1992. Jurassic. *In*: COPE, J. C. W., INGHAM, J. K. & RAWSON, P. F. (eds) *Atlas of Palaeogeography and Lithofacies*. Memoir of the Geological Society, London No. 13, 107–129.

BRISTOW, C. R., FRESHNEY, E. C. & PENN, I. E. 1991. *Geology of the Country around Bournemouth*. Memoir of the Geological Survey of Great Britain. Sheet 329.

BROOKFIELD, M. E. 1978. The lithostratigraphy of the Upper Oxfordian and Lower Kimmeridgian Beds of south Dorset. *Proceedings of the Geologists' Association*, **89**, 1–32.

BRUNSTROM, R. G. W. 1963. Recently discovered oil fields in Britain. *In*: *6th International Petroleum Congress, Section 1*, 11–20.

BRYANT, I. D., KANTOROWICZ, J. D. & LOVE, C. F. 1988. The origin and recognition of laterally continuous carbonate-cemented horizons in the Upper Lias sands of southern England. *Marine and Petroleum Geology*, **5**, 108–133.

BUTLER, M. & PULLAN, C. P. 1990. Tertiary structures and hydrocarbon entrapment in the Weald Basin of southern England. *In*: HARDMAN, R. F. P. & BROOKS, J. (eds) *Tectonic Events Responsible for Britain's Oil and Gas Reserves*. Geological Society, London, Special Publications, **55**, 371–392.

CALLOMON, J. H. 1955. The ammonite succession in the Oxford Clay and Kellaways Beds at Kidlington, Oxfordshire, and the zones of the Callovian. *Philosophical Transactions of the Royal Society of London*, **B239**, 215–264.

—— & COPE, J. C. W. 1995. The Jurassic geology of Dorset. *In*: TAYLOR, P. D. (ed.) *Field Geology of the British Jurassic*. Geological Society, London, 51–104.

—— & CHANDLER, R. B. 1990. A review of the ammonite horizons of the Aalenian-Lower Bajocian stages in the Middle Jurassic of Southern England. *Memoire descrittive dela Carta geologica d'Italia*, **40**, 85–112.

CASEY, R. 1963. The dawn of the Cretaceous Period in Britain. *Bulletin of the South-Eastern Union of Scientific Societies*, **117**, 1–15.

—— 1973. The ammonite succession at the Jurassic – Cretaceous Boundary in eastern England. *In*: RAWSON, P. F. & CASEY, R. (eds) *The Boreal Lower Cretaceous*. Geological Journal, Special Issue 5, 193–266.

CHADWICK, R. A. 1986. Extension tectonics in the Wessex Basin, southern England. *Journal of the Geological Society, London*, **143**, 465–488.

——, KENOLTY, N. & WHITTAKER, A. 1983. Crustal structure beneath southern England from deep seismic profiles. *Journal of the Geological Society, London*, **140**, 893–911.

CLEMENTS, R. G. 1993. Type-section of the Purbeck Limestone Group, Durlston Bay, Swanage. *Proceedings of the Dorset Natural History and Archaeological Society*, **114**, 181–206.

COE, A. L. 1995. A comparison of the Oxfordian successions of Dorset, Oxfordshire and Yorkshire. *In*: TAYLOR, P. D. (ed.) *Field Geology of the British Jurassic*. Geological Society, London, 151–172.

—— 1996. Unconformities within the Portlandian Stage of the Wessex Basin and their sequence-stratigraphical significance. *In*: HESSELBO, S. P. & PARKINSON, D. N. (eds) *Sequence Stratigraphy in British Geology*. Geological Society, London, Special Publications, **103**, 109–144.

COLTER, V. S. & HAVARD, D. J. 1981. The Wytch Farm Oilfield, Dorset. *In*: ILLING, L. V. & HOBSON, G. D. (eds) *Petroleum Geology of the Continental Shelf of North-west Europe*. Heyden, London, 893–911.

COPE, J. C. W., GETTY, T. A., HOWARTH, M. K., MORTON, N. & TORRENS, H. S. 1980a. *A Correlation of Jurassic Rocks in the British Isles. Part One: Introduction and Lower Jurassic*. Special Report of the Geological Society, London No. 14.

——, DUFF, K. L., PARSONS, C. F., TORRENS, M. S., WIMBLEDON, W. A. & WRIGHT, J. 1980b. *A Correlation of Jurassic Rocks in the British Isles. Part Two. Middle and Upper Jurassic*. Special Report of the Geological Society, London No. 15.

COPESTAKE, P. 1989. Triassic. *In*: JENKINS, D. G. & MURRAY, J. W. (eds) *Stratigraphical Atlas of Fossil Foraminifera*, 2nd edition. Ellis Horwood, Chichester, for the British Micropalaeontology Society, 97–124.

COX, B. M. & GALLOIS, R. W. 1981. *The Stratigraphy of the Kimmeridge Clay of the Dorset Type Area and its Correlation with Some Other Kimmeridgian Sequences*. Report of the Institute of Geological Sciences No. 80.

——, HUDSON, J. D. & MARTILL, D. M. 1992. Lithostratigraphic nomenclature of the Oxford Clay (Jurassic). *Proceedings of the Geologists' Association*, **103**, 343–346.

DARTON, D. M., DINGWALL, R. G. & MCCANN, D. M. 1979. *Geological and Geophysical Investigations in Lyme Bay*. Report of the Institute of Geological Sciences No. 79.

DAVIS, D. K. 1969. Shelf sedimentation: an example from the Jurassic of Britain. *Journal of Sedimentary Petrology*, **39**, 1344–1370.

——, ETHRIDGE, F. G. & BERG, R. R. 1971. Recognition of barrier environments. *Bulletin of the American Association of Petroleum Geologists*, **55**, 550–565.

DELAIR, J. B. 1960. The Mesozoic reptiles of Dorset. *Proceedings of the Dorset Natural History and Archaeological Society*, **81**, 59–85.

DONOVAN, D. T., HORTON, A. & IVIMEY-COOK, H. I. C. 1979. The transgression of the Lower Lias over the northern flank of the London Platform. *Journal of the Geological Society, London*, **136**, 165–173.

DOWNIE, C. 1955. *The Nature and Origin of the Kimmeridge Oil Shale*. PhD thesis, University of Sheffield.

DUFF, K. L. 1975. Palaeoecology of a bituminous shale: The Lower Oxford Clay of central England. *Palaeontology*, **18**, 443–482.

DUNN, C. E. 1974. Identification of sedimentary cycles through Fourier analysis of geochemical analysis data. *Chemical Geology*, **13**, 217–232.

EDWARDS, R. A. & FRESHNEY, E. C. 1985. *The Geology Around Southampton*. Memoir of the British Geological Survey Sheet 315 (England and Wales).

EINSELE, G. 1985. Responses of sediments to sea-level changes in differing subsiding storm-dominate marginal and epeiric basins. *In*: BAYER, U. & SEILACHER, A. (eds) *Sedimentary and Evolutionary Cycles*. Springer, Berlin, 68–112.

EL-SHAHAT, A. & WEST, I. M. 1983. Early and late lithification of aragonitic bivalve beds in the Purbeck Formation (Upper Jurassic–Lower Cretaceous) of southern England. *Sedimentary Geology*, **35**, 15–41.

ENSOM, P. C. 1985. An annotated section of the Purbeck Limestone Formation at Worbarrow Tout, Dorset. *Proceedings of the Dorset Natural History and Archaeological Society*, **106**, 87–91.

EVANS, G., MURRAY, J. W., BIGGS, H. E. J., BATE, R. & BUSH, P. R. 1973. The oceanography, ecology, sedimentology and geomorphology of parts of the Trucial Coast Barrier Island Complex, Persian Gulf. *In*: PURSER, B. H. (ed.) *The Persian Gulf*. Springer, Heidelberg, 233–278.

FALCON, N. L. & KENT, P. E. 1960. *Geological Results of Petroleum Exploration in Britain 1945–1957*. Memoir of the Geological Society, London No. 2.

FÜRSICH, F. T. 1975. Trace fossils as environmental indicators in the Corallian of England and Normandy. *Lethaia*, **8**, 151–172.

—— 1976. The use of macroinvertebrate associations in interpreting Corallian environments. *Palaeogeography, Palaeoclimatology and Palaeoecology*, **23**, 1–23.

—— 1977. Corallian (Upper Jurassic) marine benthic associations from England and Normandy. *Palaeontology*, **20**, 337–385.

GALLOIS, R. W. 1976. Coccolith blooms in the Kimmeridge Clay and the origin of North Sea oil. *Nature*, **259**, 473–475.

—— 1978. *A Pilot Study of Oil Shale Occurrences in the Kimmeridge Clay*. Report of the Institute of Geological Sciences No. 78.

—— & WORSSAM, R. C. 1993. *Geology of the Country Around Horsam*. Memoir of the British Geological Survey Sheet 302.

GATRALL, M., JENKYNS, H. C. & PARSONS, C. F. 1972. Limonitic concretions from the European Jurassic with particular reference to the 'snuff boxes' of southern England. *Sedimentology*, **18**, 79–103.

GREEN, G. W. & DONOVAN, D. T. 1969. The Great Oolite of the Bath area. *Bulletin of the Geological Survey of Great Britain*, **30**, 1–63.

GOLDRING, R., ASTIN, T. R., MARSHALL, J. E. A., GABBOT, S. & JENKINS, C. D. 1998. Towards an integrated study of the depositional environment of the Bencliff Grit (U. Jurassic) of Dorset. *This volume*.

GRIEVES, I. A. T. 1968. *A Depositional Model for Part of the Corallian Sequence (Oxfordian), Dorset Coast*. MSc thesis. University of Reading.

HALLAM, A. 1961. Cyclothems, transgressions and faunal change in the Lias of north-west Europe. *Transactions of the Geological Society of Edinburgh*, **18**, 124–174.

—— 1964. Origin of the limestone–shale rhythms in the Blue Lias of England: a composite theory. *Journal of Geology*, **72**, 157–169.

—— 1967. An environmental study of the Domerian and Lower Toarcian of Great Britain. *Philosophical Transactions of the Royal Society of London*, **B243**, 1–44.

—— 1975. *Jurassic Environments*. Cambridge University Press, Cambridge.

—— 1978. Eustatic cycles in the Jurassic. *Palaeogeography, Palaeoclimatology, Palaeoecology*, **23**, 1–32.

—— 1981. A revised sea-level curve for the Early Jurassic. *Journal of the Geological Society of Great Britain*, **138**, 735–743.

—— 1988. A re-evaluation of Jurassic eustasy in the light of new data and the revised Exxon Curve. *In*: WILGUS, C. K., HASTINGS, B. S., KENDALL, C. G. ST. C., POSAMENTIER, H., ROSS, C. A. & VAN WAGONER, J. (eds) *Sea-level Changes: An Integrated Approach*. Society of Economic Paleontologists and Mineralogists, Special Publications, **42**, 261–273.

—— 1992. *Phanerozoic Sea-level Changes*. Columbia University Press, New York.

—— & EL SHAARAWY, Z. 1982. Salinity reduction of the end-Triassic sea from the Alpine region into northwestern Europe. *Lethaia*, **15**, 169–178.

HAMBLIN, R. J. O., CROSBY, A., BALSON, P. S., JONES, S. M., CHADWICK, R. A., PENN, I. E. & ARTHUR, M. J. 1992. *United Kingdom Offshore Regional Report: The Geology of the English Channel*. HMSO for the British Geological Survey, London.

HANCOCK, F. R. P. & MITHEN, D. P. 1987. The geology of the Humbly Grove Oilfield, Hampshire, UK. *In*: BROOKS, J. & GLENNIE, K. (eds) *Petroleum Geology of North West Europe*. Graham & Trotman, London, 161–170.

HAQ, B. V., HARDENBOL, J. & VAIL, P. R. 1988. Mesozoic and Cenozoic chronostratigraphy and eustatic cycles. *In*: WILGUS, C. K., HASTINGS, B. S., KENDALL, C. G. ST. C., POSAMENTIER, H., ROSS, C. A. & VAN WAGONER, J. (eds) *Sea-level Changes: An Integrated Approach*. Society of Economic Paleontologists and Mineralogists, Special Publications, **42**, 71–108.

HESSELBO, S. P. & JENKYNS, H. C. 1995. A comparison of the Hettangian to Bajocian successions of Dorset and Yorkshire. *In*: TAYLOR, P. D. (ed.) *Field Geology of the British Jurassic*. Geological Society, London, 105–150.

HOLLAWAY, S. 1985a. Triassic: Mercia Mudstone and Penarth Groups. *In*: WHITTAKER, A. (ed.) *Atlas of Onshore Sedimentary Basins in England and Wales: Post-Carboniferous Tectonics and Stratigraphy*. Blackie, Glasgow, 34–36.

—— 1985b. Lower Jurassic: the Lias. *In*: WHITTAKER, A. (ed.) *Atlas of Onshore Sedimentary Basins in England and Wales: Post-Carboniferous Tectonics and Stratigraphy*. Blackie, Glasgow, 37–40.

—— 1985c. Middle Jurassic: the Inferior Oolite and Ravenscar Groups. *In*: WHITTAKER, A. (ed.) *Atlas of Onshore Sedimentary Basins in England and Wales: Post-Carboniferous Tectonics and Stratigraphy*. Blackie, Glasgow, 41–43.

—— 1985d. Middle Jurassic: the Great Oolite Group. *In*: WHITTAKER, A. (ed.) *Atlas of Onshore Sedimentary Basins in England and Wales: Post-Carboniferous Tectonics and Stratigraphy*. Blackie, Glasgow, 44–46.

HOLLINGWORTH, N. J. J. & WIGNALL, P. B. 1992. The Callovian–Oxfordian boundary in Oxfordshire based on two temporary sections. *Proceedings of the Geologists' Association*, **103**, 15–30.

HOUSE, M. 1985. A new approach to an absolute timescale from measurements of orbital cycles and sedimentary microrhythms. *Nature*, **316**, 721–725.

—— 1987. Are Jurassic sedimentary micro-rhythms due to orbital forcing? *Proceedings of the Ussher Society*, **6**, 299–311.

—— 1993. *Geology of the Dorset Coast. Geologists' Association Guide 22*. Holywell Press, Oxford.

HOWARTH, M. K. 1980. The Toarcian age of the upper part of the Marlstone Rock Bed of England. *Palaeontology*, **23**, 637–656.

HUDSON, J. D. & MARTILL, D. M. 1991. The Lower Oxford Clay: production and preservation of organic matter in the Callovian (Jurassic) of central England. *In*: TYSON, R. V. & PEARSON, T. M. (eds) *Modern and Ancient Continental Shelf Anoxia*. Geological Society, London, Special Publications, **58**, 363–379.

JENKYNS, H. C. & SENIOR, J. R. 1991. Geological evidence for intra-Jurassic faulting in the Wessex Basin and its margins. *Journal of the Geological Society of London*, **148**, 245–260.

KARNER, G. D., LAKE, S. D. & DEWEY, J. F. 1987. The thermal and mechanical development of the Wessex Basin, southern England. *In*: COWARD, M. P., DEWEY, J. F. & HANCOCK, P. L. (eds) *Continental Extensional Tectonics*. Geological Society, London, Special Publications, **28**, 517–537.

KENT, P. E. 1949. A structure contour map of the surface of the buried pre-Permian rocks of England and Wales. *Proceedings of the Geologists' Association*, **60**, 87–104.

—— 1980. *Eastern England from the Tees to the Wash*. HMSO, London.

KELLING, G. & MOSHRIF, M. A. 1977. The orientation of fossil bivalves in a pene-littoral sequence (the Rhaetian of South Wales). *Journal of Sedimentary Petrology*, **47**, 1342–1346.

KNOX, R. W. 1982. The petrology of the Penarth Group (Rhaetian) of the Winterborne Kingston borehole, Dorset. *In:* RHYS, G. H., LOTT, G. K. & CALVER, M. A. (eds) *The Winterborne Kingston Borehole, Dorset, England*. Report of the Institute of Geological Sciences No. 81, 127–134.

——, MORTON, A. C. & LOTT, G. K. 1982. Petrology of the Bridport Sands in the Winterborne Kingston borehole, Dorset. *In*: RHYS, G. H., LOTT, G. K. & CALVER, M. A. (eds) *The Winterborne Kingston Borehole, Dorset, England*. Report of the Institute of Geological Sciences No. 81, 107–121.

LAKE, S. D. & KARNER, G. D. 1987. The structure and evolution of the Wessex Basin, southern England: an example of inversion tectonics. *Tectonophysics*, **137**, 347–378.

MARTIN, A. J. 1967. Bathonian sedimentation in southern England. *Proceedings of the Geologists' Association*, **78**, 473–488.

MELVILLE, R. V. & FRESHNEY, E. C. 1982. *British Regional Geology: the Hampshire Basin and Adjoining Areas*, 4th edition. HMSO for the Institute of Geological Sciences, London.

MELYNK, D. H., ATHERSUCH, J. & SMITH, D. G. 1992. Estimating the dispersion of biostratigraphic events in the subsurface by graphic correlation: an example from the Late Jurassic of the Wessex Basin, UK. *Marine and Petroleum Geology*, **9**, 602–607.

——, SMITH, D. G. & AMIRI-GARROUSSI, K. 1994. Filtering and frequency mapping as tools in subsurface cyclostratigraphy, with examples from the Wessex Basin, UK. *In*: DE BOER, P. L. & SMITH, D. G. (eds) *Orbital Forcing and Cyclic Sedimentary Sequences. International Association of Sedimentologists*, Special Publications, **19**, 35–46.

MORRIS, K. A. 1980. Comparison of major sequences of organic-rich mud deposition in the British Jurassic. *Journal of the Geological Society, London*, **137**, 157–170.

MORTER, A. A. 1984. Purbeck–Wealden Mollusca and their relationship to ostracod biostratigraphy, stratigraphical correlation and palaeoecology in the Weald and adjacent areas. *Proceedings of the Geologists' Association*, **95**, 217–234.

OSCHMANN, W. 1988. Kimmeridge Clay sedimentation – a new cyclic model. *Palaeogeography, Palaeoclimatology, Palaeoecology*, **65**, 217–251.

—— 1991. Distribution, dynamics and paleocology of Kimmeridgian (Upper Jurassic) shelf anoxia in western Europe. *In*: TYSON, R. V. & PEARSON, T. M. (eds) *Modern and Ancient Continental Shelf Anoxia*. Geological Society, London, Special Publications, **58**, 381–395.

PAGE, K. N. 1989. A stratigraphic revision for the English Lower Callovian. *Proceedings of the Geologists' Association*, **100**, 363–382.

PARKINSON, D. N. 1996. Gamma-ray spectrometry as a tool for stratigraphical interpretation: examples from western European Lower Jurassic. *In*: HESSELBO, S. P. & PARKINSON, D. N. (eds) *Sequence Stratigraphy in British Geology*, Geological Society, London, Special Publications, **103**, 231–256.

PENN, I. E. 1982. Middle Jurassic stratigraphy and correlation of the Winterborne Kingston borehole. *In*: RHYS, G. H., LOTT, G. K. & CALVER, M. A. (eds) *The Winterborne Kingston Borehole, Dorset, England*. Report of the Institute of Geological Sciences No. 81, 53–76.

——, DINGWALL, R. G. & KNOX, R. W. O'B. 1980. *The Inferior Oolite (Bajocian) Sequence from a Borehole in Lyme Bay, Dorset*. Report of the Institute of Geological Sciences No. 79, 1–27.

POOLE, E. G. 1978. The stratigraphy of the Withycombe Farm Borehole, near Banbury, Oxfordshire. *Bulletin of the Geological Survey of Great Britain*, **68**, 1–63.

RAWSON, P. F. & RILEY, L. A. 1982. Latest Jurassic–Early Cretaceous events and 'Late Cimmerian Unconformity' in the North Sea area. *Bulletin of the American Association of Petroleum Geologists*, **66**, 2628–2648.

RHYS, G. H., LOTT, G. K. & CALVER, M. A. (eds) 1982. *The Winterborne Kingston Borehole, Dorset, England*. Report of the Institute of Geological Sciences No. 81.

RIOULT, M., DUGUE, O., JAN, DU CHENE, R., PONSOL, C., FILY, G., MORON, J. M. & VAIL, P. R. 1991. Outcrop sequence stratigraphy of the Anglo-Paris Basin, Middle to Upper Jurassic (Normandy, Maine, Dorset). *Bulletin Centres de Recherches Exploration–Production Elf-Aquitaine*, **15**, 101–194.

SELLWOOD, B. W. 1978. Shallow water carbonate environments. *In*: READING, H. G. (ed.) *Sedimentary Environments and Facies*. Blackwell Scientific Publications, Oxford, 259–313.

—— & EVANS, R. 1987. Middle Jurassic Great Oolite of southern England, facies development and regional significance of 'dedolomitization' during basin evolution. *Bulletin of the American Association of Petroleum Geologists*, **71**, 612.

—— & JENKYNS, H. C. 1975. Basins and swells and the evolution of an epeiric sea (Pliensbachian–Bajocian of Great Britain) *Journal of the Geological Society, London*, **131**, 373–388.

——, SCOTT, J., MIKKELSEN, P. & AKROYD, P. 1985. Stratigraphy and sedimentology of the Great Oolite Group in the Humbly Grove Oilfield, Hampshire. *Marine and Petroleum Geology*, **2**, 44–55.

—— & WILSON, R. C. L. 1990. *Jurassic Sedimentary Environments of the Wessex Basin*. 13th International Sedimentological Congress, Nottingham, UK. Field Guide No. 7.

STONELEY, R. 1982. The structural development of the Wessex Basin. *Journal of the Geological Society, London*, **139**, 545–552.

—— & SELLEY, R. C. 1983. *A Field Guide to the Petroleum Geology of the Wessex Basin*. The World Petroleum Jubilee Congress Guide. R. C. Selley & Co. Ltd.

SUN, S. Q. 1989. A new interpretation of the Corallian (Upper Jurassic) cycles of the Dorset Coast, southern England. *Geological Journal*, **24**, 139–158.

—— 1990. Facies-related diagenesis in a cyclic shallow marine sequence; the Corallian Group (Upper Jurassic) of the Dorset coast, southern England. *Journal of Sedimentary Petrology*, **60**, 42–52.

TAITT, A. H. & KENT, P. E. 1958. *Deep Boreholes at Portsdown (Hants) and Henfield (Sussex)*. British Petroleum, London.

TALBOT, M. R. 1973. Major sedimentary cycles in the Corallian Beds (Oxfordian) of southern England. *Palaeogeography, Palaeoclimatology, Palaeoecology*, **14**, 293–313.

THOMAS, L. P. & HOLLIDAY, D. W. 1982. *Southampton No. 1 (Western Esplanade) Geothermal Well: Geological Well Completion Report*. Report of the Deep Geology Unit, Institute of Geological Sciences No. 82.

TOWNSON, W. G. 1975. Lithostratigraphy and deposition of the type Portlandian. *Journal of the Geological Society, London*, **131**, 619–668.

TYSON, R. V. 1996. Sequence stratigraphical interpretation of organic facies variation in marine siliciclastic systems: general principles and application to the onshore Kimmeridge Clay Formation, UK. *In*: HESSELBO, S. P. & PARKINSON, D. N. (eds) *Sequence Stratigraphy in British Geology*. Geological Society, London, Special Publications, **103**, 75–96.

——, WILSON, R. L. & DOWNIE, C. 1979. A stratified water column environmental model for the Kimmeridge Clay. *Nature*, **277**, 377–380.

UNDERHILL, J. R. & PATERSON, S. 1998. Genesis of tectonic inversion structures: seismic evidence for the development of key structures along the Purbeck–Isle of Wight Disturbance. *Journal of the Geological Society, London*, in press.

—— & STONELEY, R. 1982. Introduction to the development, evolution and petroleum geology of the Wessex Basin. *This volume*.

WARRINGTON, G., AUDLEY-CHARLES, M. G., ELLIOT, R. E., EVANS, W. B., IVIMEY-COOK, H. C., KENT, P. E., ROBINSON, P. L., SHOTTON, F. W. & TAYLOR, F. M. 1980. A correlation of Triassic rocks in the British Isles. Special Report of the Geological Society, London, No. 13.

—— & IVIMEY-COOK, H. C. 1992. Triassic. *In*: COPE, J. C. W., INGHAM, J. K. & RAWSON, P. F. (eds) *Atlas of Palaeogeography and Lithofacies*. Memoir of the Geological Society, London, No. 13, 97–106.

WEEDON, G. P. 1986. Hemipelagic shelf sedimentation and climatic cycles; the basal Jurassic (Blue Lias) of south Britain. *Earth and Planetary Science Letters*, **76**, 321–335.

—— & JENKYNS, H. C. 1990. Regular and irregular climatic cycles and the Belemnite Marls (Pliensbachian, Lower Jurassic, Wessex Basin). *Journal of the Geological Society, London*, **147**, 915–918.

WEST, I. M. 1975. Evaporite and associated sediments of the basal Purbeck Formation (Upper Jurassic) of Dorset. *Proceedings of the Geologists' Association*, **86**, 205–225.

WESTHEAD, R. K. & MATHER, A. E. 1996. An updated lithostratigraphy for the Purbeck Limestone Group I the Dorset type-area. *Proceedings of the Geologists' Association*, **107**, 117–128.

WHATLEY, R. C. 1965. *Callovian and Oxfordian Ostracoda from England and Scotland*. PhD thesis, University of Hull.

WHITTAKER, A. (ed.) 1985. *Atlas of Onshore Sedimentary Basins in England and Wales: Post-Carboniferous Tectonics and Stratigraphy*. Blackie, Glasgow.

—— & GREEN, G. W. 1983. *Geology of the Area around Weston-super-Mare*. Memoir of the Geological Survey of Great Britain Sheet 279 with parts of 263 and 295.

——, HOLLIDAY, D. W. & PENN, I. E. 1985. Geophysical Logs in British Stratigraphy. Special Report of the Geological Society, London, No. 18.

WIGNALL, P. B. 1989. Sedimentary dynamics of the Kimmeridge Clay; tempests and earthquakes. *Journal of the Geological Society, London*, **146**, 273–284.

—— 1991. Test of the concepts of sequence stratigraphy in the Kimmeridgian (Late Jurassic) of England and northern France. *Marine and Petroleum Geology*, **8**, 430–441.

—— & HALLAM, A. 1991. Biofacies, stratigraphic distribution and depositional models of British onshore Jurassic black shales. *In*: TYSON, R. V. & PEARSON, T. H. (eds) *Modern and Ancient Continental Shelf Anoxia*. Geological Society, London, Special Publications, **58**, 291–309.

—— & RUFFELL, A. H. 1990. The influence of a sudden climatic change on marine deposition in the Kimmeridgian of Northwest Europe. *Journal of the Geological Society, London*, **147**, 365–371.

WILSON, R. C. L. 1968. Upper Oxfordian palaeogeography of southern England. *Palaeogeography, Palaeoclimatology, Palaeoecology*, **4**, 5–28.

WILSON, V., WELCH, F. B. A., ROBBIE, J. A. & GREEN, G. W. 1958. *The Geology of the Country Around Bridport and Yeovil*. Memoir of the Geological Survey of Great Britain.

WIMBLEDON, W. A. & HUNT, C. O. 1983. The Portland–Purbeck junction (Portlandian–Berriasian) in the Weald, and correlation of latest Jurassic–Early Cretaceous rocks in southern England. *Geological Magazine*, **120**, 267–280.

WRIGHT, J. K. 1986a. A new look at the stratigraphy, sedimentology and ammonite fauna of the Corallian Group (Oxfordian) of South Dorset. *Proceedings of the Geologist's Association*, **97**, 1–22.

—— 1986b. The Upper Oxford Clay at Furzy Cliff, Dorset: stratigraphy, palaeoenvironment and ammonite fauna. *Proceedings of the Geologists' Association*, **97**, 221–228.

YOUNG, B. & LAKE, R. D. 1988. *Geology of the Country Around Brighton and Worthing*. Memoir of the Geological Survey of Great Britain Sheets 318 and 333.

ZEIGLER, P. A. 1990. *Geological Atlas of Western and Central Europe*. Shell Internationale Petroleum Maatschappij, B.V., Amsterdam.

Use of palynofacies analysis to define Lower Jurassic (Sinemurian to Pliensbachian) genetic stratigraphic sequences in the Wessex Basin, England

DAVID C. COLE & IAN C. HARDING

Department of Geology, University of Southampton, Southampton Oceanography Centre, Empress Dock, Southampton SO14 3ZH, UK (e-mail: ich@mail.soc.soton.ac.uk)

Abstract: Quantitative palynological and palynofacies analyses allow the Lower Jurassic succession of the Dorset coast (Wessex Basin, southern England, UK) to be placed within a genetic sequence stratigraphic framework. The multifaceted approach employed utilizes various palynological criteria adopted from previous publications (e.g. both relative and absolute particle abundance data, allochthonous versus autochthonous palynomorphs ratios, phytoclast type, etc.), and suggests new parameters which may be studied in this context (e.g. ratios of acritarch spine length). Interpretation of palynofacies trends within the studied succession enables the identification of eight genetic sequences informally described as Genetic Stratigraphic Sequences I–VIII. Genetic stratigraphic sequence boundaries are placed at the following horizons: Mongrel, Grey Ledge, Pavior, the Coinstone, Hummocky, the Belemnite Stone, Day's Shell Bed and the *margaritatus* Stone. These results compare well with sequence stratigraphic frameworks for published for the British Lower Jurassic interval both in Dorset and elsewhere, suggesting that the maximum flooding events recognized are likely to have been of inter-regional extent.

Insoluble organic matter, when acid-macerated from a sediment, may be used to interpret the sedimentary environment prevailing at the time of deposition (Müller 1959, Combaz 1964). The study of the particulate organic matter contained within a sediment is usually referred to as palynofacies analysis, a technique recently defined by Tyson (1995, p. 4) as 'the palynological study of depositional environments..., based upon the total assemblage of particulate organic matter'. This new definition broadens the somewhat more restricted definitions proposed elsewhere (e.g. Traverse 1988, Prauss 1993), which have attempted to delimit depositional environments solely from distinctive palynomorph assemblages (dinocysts, spores, pollen, etc.) with little attention being paid to the plant-derived detritus (phytoclasts).

Qualitative and quantitative studies of the particulate organic matter (POM) contained within a sediment allow the determination of a number of characteristics of the depositional environment (Tyson 1995). These characteristics include, for example, such parameters as variations in proximal–distal location of the depocentre with respect to sedimentary source, regressive–transgressive trends within a stratigraphic sequence and the nature of the environment of deposition (such as water column oxygenation, salinity, etc.), all of which have great potential to aid sequence stratigraphic interpretations.

Analyses of POM have only recently been utilized as a tool to aid sequence stratigraphic interpretations (e.g. Prauss 1993; Steffen & Gorin 1993). Although studies concerned with the relationships between palynofacies and sequence stratigraphy in several different depositional settings had been published earlier (Pasley & Hazel 1990, Gregory & Hart 1992; Powell *et al.* 1992; Blondel *et al.* 1993; Jones *et al.* 1993). The term 'sequence palynology' was introduced by Prauss (1993). The study and subsequent interpretation of the palynofacies response to relative sea-level chnge forms the foundations of sequence palynology. Such studies rely on the assumption proposed by Posamentier *et al.* (1988) that relative sea-level change controls accommodation space (*sensu* Jervey 1988) which is the principle control on sequence stratigraphic architecture.

Principles and methodology of sequence stratigraphy as applied to sequence palynology

Sequence stratigraphy provides a method for the examination of sedimentary successions within a chronostratigraphic framework. The two main sequence stratigraphic models developed allow for the correlation of syndepositional facies as opposed to the more traditional correlation of like lithologies (lithostratigraphy). It is well established that sedimentary successions can be subdivided into a series of depositional episodes or cycles.

The principle controlling mechanism on such cyclicity is relative sea-level change (Posamentier et al. 1988). The repetitive, cyclic nature of such change is the product of the ongoing complex interplay between sediment supply, basin subsidence (and uplift) and eustatic sea-level fluctuations. Workers from the Exxon school (see Vail et al. 1977) developed the interpretations made regarding such cyclicity a stage further and defined the principles underlying current sequence stratigraphic studies, integrating many of the seismic stratigraphic concepts concerned with sequence architecture. The basic ('slug') model developed by the Exxon group was further honed in a series of important conceptual papers (e.g. Jervey 1988; Posamentier et al. 1988; van Wagoner et al. 1988) into the depositional model in usage today.

An alternative, genetic sequence stratigraphy has, however, also been developed (Galloway 1989) and is widely employed within certain depositional settings such as the North Sea (Partington et al. 1993). Central to genetic sequence stratigraphic interpretation is the recognition of the key bounding surfaces to each individual sequence, which form at the point of maximum shelfal accommodation, i.e. maximum transgression. Such levels within a marine stratigraphic succession are termed maximum flooding surfaces (MFSs). The boundaries to genetic sequences differ from those utilized in the depositional or Exxon model, where sequence boundaries are formed at relative sea-level minima. Further conceptual differences between the two models are outlined in Galloway (1989).

Geological setting

The Dorset Coast between Pinhay Bay (SY 318908) to the west and Eype Mouth (SY 447910) to the east (Fig. 1) exposes an almost complete succession of Lower Jurassic strata (see House 1989). The studied section comprises some 260 m of predominantly fine-grained, marine siliciclastic sediments with occasional limestone horizons. Lithostratigraphic nomenclature for the Lower Jurassic Dorset section follows that of Cope et al. (1980), based on almost two centuries of work by various authors (de la Beche 1826, Day 1863, Woodward 1893, Lang 1914, etc.) (see Fig. 2).

Sediment grain size is seen to increase progressively up through the section (House 1989, Hesselbo & Jenkyns 1995): the Lower Lias consisting of predominantly fine-grained shales, marls and limestones, whereas the Middle and Upper Lias comprise coarser silts, sands and limestones.

At the time of deposition, the Dorset area lay within the Wessex Basin, bounded to the north and east by the London–Brabant Massif, to the south by the Armorican Massif, to the west by the Cornubian Massif and to the northwest by the Welsh Massif (Bradshaw et al. 1992). The basin itself can be divided into a number of sub-basins and associated topographic highs, controlled principally by east–west-orientated normal faults (Holloway 1985). The studied section was deposited in the area of the

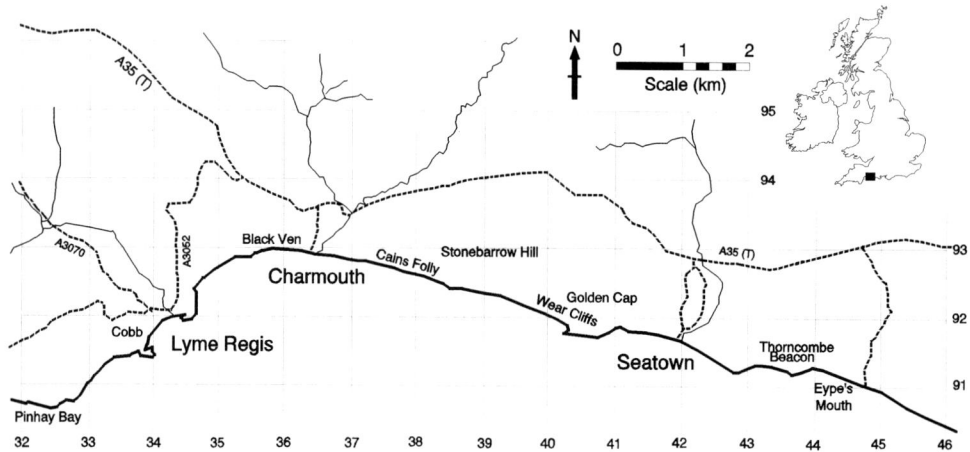

Fig. 1. Location map of study area, highlighting main localities referred to in text (inset map of British Isles shows location of Dorset coast succession in southern England).

STAGE	ZONE	SUBZONE	DORSET COAST	BED No.
PLIENSBACHIAN	spinatum	hawskerense	JUNCTION BED (sensu lato)	Px+P
PLIENSBACHIAN	spinatum	apyrenum	JUNCTION BED (sensu lato)	R
PLIENSBACHIAN	margaritatus	gibbosus	thorncombiensis BED	32-33
PLIENSBACHIAN	margaritatus	subnodosus	THORNCOMBE SANDS margaritatus STONE	24-31
PLIENSBACHIAN	margaritatus	stokesi	DOWNCLIFF SANDS EYPE CLAY THREE TIERS	6-23
PLIENSBACHIAN	davoei	figulinum	GREEN AMMONITE BEDS	127-130
PLIENSBACHIAN	davoei	capricornus	GREEN AMMONITE BEDS	122f-126
PLIENSBACHIAN	davoei	maculatum	GREEN AMMONITE BEDS	122a-122e
PLIENSBACHIAN	ibex	luridum	BELEMNITE STONE	121
PLIENSBACHIAN	ibex	valdani	BELEMNITE MARLS	118d-120
PLIENSBACHIAN	ibex	masseanum	BELEMNITE MARLS	118c
PLIENSBACHIAN	jamesoni	jamesoni	BELEMNITE MARLS	115-118b
PLIENSBACHIAN	jamesoni	brevispina	BELEMNITE MARLS	110-114
PLIENSBACHIAN	jamesoni	polymorphus		
PLIENSBACHIAN	jamesoni	taylori	HUMMOCKY	103-109
SINEMURIAN	raricostatum	aplanatum		
SINEMURIAN	raricostatum	macdonnelli		
SINEMURIAN	raricostatum	raricostatoides	BLACK VEN MARLS	99-102
SINEMURIAN	raricostatum	densinodulum	BLACK VEN MARLS	90-98
SINEMURIAN	oxynotum	oxynotum		
SINEMURIAN	oxynotum	simpsoni		
SINEMURIAN	obtusum	denotatus		
SINEMURIAN	obtusum	stellare		85-89
SINEMURIAN	obtusum	obtusum	BLACK VEN MARLS	83f-84
SINEMURIAN	turneri	birchi	SHALES WITH BEEF	76-83e / 74g-75
SINEMURIAN	turneri	brooki	SHALES WITH BEEF	73-74f
SINEMURIAN	semicostatum	resupinatum	SHALES WITH BEEF	54-72
SINEMURIAN	semicostatum	scipionanum	BLUE LIAS	50-53
SINEMURIAN	semicostatum	lyra	BLUE LIAS	47-49
SINEMURIAN	bucklandi	bucklandi	BLUE LIAS	41-46
SINEMURIAN	bucklandi	rotiforme	BLUE LIAS	30-40
SINEMURIAN	bucklandi	conybeari	BLUE LIAS	21-29

Fig. 2. Ammonite biozonation and lithostratigraphic nomenclature for the Sinemurian and Pliensbachian strata of the Dorset coast (after Cope *et al.* 1980; for references to bed numbers see text).

Mid-Dorset High, an upstanding fault block, bounded to the north by the Bere Regis Fault and to the south by the Abbotsbury–Ridgeway Fault (Holloway 1985) (see Fig. 3). Numerous publications have dealt with both the formation of and structural controls on the Wessex Basin (Sellwood & Jenkyns 1975; Stoneley 1982; Chadwick 1986; Lake & Karner 1987; Hesselbo & Jenkyns 1995, 1997; Underhill & Paterson 1998; Underhil & Stoneley this volume).

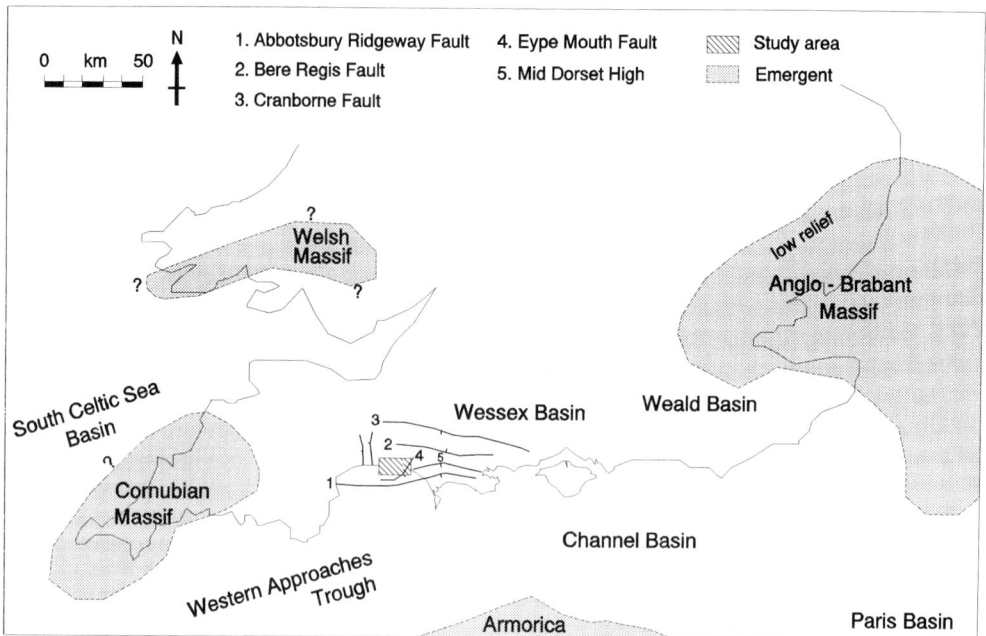

Fig. 3. Simplified palaeogeographic reconstruction of southern Britain in mid to late Pliensbachian times (after Bradshaw *et al.* 1992), with major Mesozoic structural features round the study area (after Chadwick 1986; Lake & Karner 1987).

From a period of general regression and associated continental sedimentation in Permo-Triassic times, the Lower Jurassic period saw a worldwide marine transgression, with northwest Europe being covered by a large, generally shallow, epeiric sea (Hallam 1975). This first-order transgressive cycle was punctuated and locally modified by a number of smaller-scale second- and third-order changes in sea level (Vail *et al.* 1984, Haq *et al.* 1988). Owing to the relatively subdued Early Jurassic topography within the Wessex Basin these lower-order fluctuations caused quite marked sedimentary variations within the localized depositional environments. These variations are considered to be controlled by the interplay between eustatic change and localized epeirogenic movement (Sellwood & Jenkyns 1975; Chadwick 1986; Jenkyns & Senior 1991).

The role of relative sea-level cyclicity in controlling the characteristics of sedimentary deposition has been investigated at several different scales within the Lower Jurassic succession of the Wessex Basin (see Arkell 1933; Sellwood & Jenkyns 1975; Hallam 1988). These have varied from the Milankovitch-scale cyclicity discussed by House (1985) and Weedon (1986) for the Blue Lias and Weedon & Jenkyns (1990) for the Belemnite Marls, to the major second-order (*sensu* Duval *et al.* 1992) transgressive–regressive cycles and third-order eustatic cycles proposed by Haq *et al.* (1988).

Most sequence stratigraphic interpretations, of both genetic and depositional nature, are based around the third-order sequence cycles (0.5–3 Ma) which relate to changes in shelfal accommodation. Hesselbo & Jenkyns (1997) propose a sequence stratigraphic framework for the British Lower Jurassic succession based principally on lithological facies evidence, utilizing outcrop information from the Wessex, Bristol Channel, Cleveland and Hebrides basins. This study recorded four large-scale (second-order) lithological cycles, with durations of 3–10 Ma that 'appear to be synchronously developed in all onshore British basins' (Hesselbo & Jenkyns in press). In addition, numerous smaller sequence cycles (0.5–3 Ma). were also proposed.

Materials and methods

One hundred samples were taken through the section, using an average sampling interval of between 2 and 3 m (Fig. 7). Each sample consisted of approximately 250 g of sediment, of which an accurately weighed 15 g was processed according to standard acid-maceration techniques (Phipps & Playford 1984; Traverse 1988), with the following modifications made to enhance the recovery of POM:

(i) structured debris was concentrated by tuned ultrasonic treatment of residues for a standard 15 s and subsequent sieving at 6 mm to remove as much of the finely disseminated amorphous organic matter (AOM) as possible. Tests were carried out to ensure that none of the structured debris was destroyed by this process;
(ii) a known volume of a pre-calibrated spike of the exotic modern spore *Lycopodium clavatum* was added as a suspension to all samples prior to the addition of hydrofluoric acid. This was undertaken to allow absolute quantitative counts of the POM components;
(iii) small volumes of concentrated nitric acid were added for a period of not more than 5 min to remove any diagenetic pyrite from the residues. Tests showed that this limited oxidation period did not significantly alter POM assemblages.

Investigation of the samples was undertaken using an Olympus BH-2 stereoscopic binocular microscope following the strew-mounting of residues in Elvacite on glass slides. Palynofacies analyses were conducted utilizing a Swift Model F 12-channel point-counter. Counts were made of 400 particles per sample residue, a separate running total of *Lycopodium* spike also being recorded. Absolute abundances of each POM category per gramme of sediment were obtained by ratioing abundance data with the recorded abundance of the exotic spike.

The definition of palynofacies analysis by Tyson (1995, p. 4) encompasses only structured particulate organic matter (POM) and not the total insoluble organic matter fraction, i.e. it excludes unstructured, amorphous organic matter (AOM). The non-particulate nature of AOM renders the collection of comparable abundance data impossible in palynofacies counts, but it may be assessed separately either by visual estimation of AOM yields in kerogen slides or quantitatively by integrating palynofacies and total organic carbon (TOC) data.

Classification of particulate organic matter

Since the work of Combaz (1964) many schemes have been proposed to classify insoluble sedimentary organic particles, such diversity being a reflection of the different requirements of each specific study (see for examples Whitaker 1984; Boulter & Riddick 1986; van der Zwan 1990). A universally accepted, all encompassing, working classification is still some way from fruition, although attempts to this end have been made (Open Workshop on Organic Matter Classification, Amsterdam 1991; Boulter 1995; Tyson 1995).

The POM classification scheme used in this work (Fig. 4) is a composite based around the schemes proposed by Whitaker (1984) and Boulter & Riddick (1986). Essentially POM can be divided into two categories: autochthonous and allochthonous. The autochthonous component is composed principally of marine algae – dinocysts, acritarchs and prasinophyte phycomata. Most allochthonous material is derived from higher-order plants (with the exception of fungal spores and freshwater algae) and comprises terrestrial sporomorphs (pollen and spores) and phytoclasts. In this study the phytoclasts may be further subdivided according to shape, size and preservational state:

- well preserved/brown wood – large (25 mm), angular woody fragments, largely unoxidized and still displaying evidence of cell structure, tracheids, etc;
- equidimensional black debris – small (15 mm), black, well rounded, highly oxidized material, thought to be of woody origin;
- blade-shaped black debris – black laths, variable in length (15–40 mm) of highly oxidized woody material.

For a detailed description of the POM categories utilized the reader is referred to Whitaker (1984), Boulter & Riddick (1986), Boulter (1995) and Tyson (1995).

Palynofacies and sequence stratigraphy

The relative sea-level interpretations proposed in this study are based on a multifaceted palynofacies approach utilizing relative and absolute abundance data, as well as environmentally controlled assemblage and morphological variations in microplankton taxa. To test published palynofacies responses to relative sea-level change only limited sedimentological information has been utilized to aid in interpretation. The results will principally be compared to the sequence stratigraphic interpretations of Hesselbo & Jenkyns (1997) for the same Lower Jurassic Dorset section (based on sedimentological facies evidence) and also to published sequence stratigraphic frameworks developed

APPROX. COAL MACERAL EQUIVALENT	MAIN CATEGORIES OF POM			
		Whitaker (1984)	Boulter & Riddick (1986)	This Study
VITRINITE	ALLOCHTHONOUS	PALYNOMACERAL 1	DEGRADED DEBRIS	DEGRADED DEBRIS
			BROWN WOOD	BROWN WOOD
		PALYNOMACERAL 2	WELL PRESERVED WOOD PARENCHYMA	WELL PRESERVED WOOD
CUTINITE		PALYNOMACERAL 3	LEAF CUTICLE, UNSTRUCTURED DEBRIS	LEAF CUTICLE
INERTINITE		PALYNOMACERAL 4	BLACK DEBRIS	BLACK DEBRIS / BLADE-SHAPED / EQUIDIMENSIONAL
SPORINITE		NON-SACCATE SPOROMORPHS	BISACCATES AND SPORES (PRO PARTE)	SPORES
		FUNGAL SPORES	FUNGI	FUNGAL SPORES
LIPTINITE		FRESHWATER ALGAE	ALGAE (PRO PARTE)	FRESHWATER ALGAE
		SACCATE SPOROMORPHS	BISACCATES AND SPORES (PRO PARTE)	POLLEN
LIPTINITE	AUTOCHTHONOUS	DINOCYSTS	DINOFLAGELLATE CYSTS	DINOCYSTS
		ACRITARCHS	ALGAE (PRO PARTE)	ACRITARCHS
		MARINE ALGAE		MARINE ALGAE
TECTIN		FORAM TEST LININGS	FORAM TEST LININGS	FORAM TEST LININGS
		STRUCTURELESS ORGANIC MATTER	AMORPHOUS MATTER, SPECKS	

Fig. 4. Comparison of the classifications of organic matter adopted by Whitaker (1984), Boulter & Riddick (1986) and the scheme used in this study.

for the Lower Jurassic strata of other British depositional basins.

Sedimentological sequence stratigraphic studies differ from sequence palynological analyses in that they are able to examine continuous vertical sedimentary sequences. Each palynological residue represents a single discrete horizon which may only be compared to other spot samples directly above or beneath that level. Candidate stratigraphic surfaces bounding sequences may be discerned from fluctuations in the palynofacies signatures. The bounding surfaces most readily recognized by palynological means are maximum flooding surfaces (MFSs). These bound genetic stratigraphic sequences or depositional episodes (*sensu* Galloway 1989). MFSs and their correlative basinal condensed sequences (*sensu* Loutit *et al.* 1988) display the highest abundances and diversities of microplankton and thus are horizons of major importance for biostratigraphic correlation both within and between basins (Gorin & Steffen 1991, Habib *et al.* 1992, Powell 1992). It is often possible for approximations to be made as to the probable location of MFSs from reconnaissance, low-resolution palynological studies in successions which might then be resampled in more detail in order to delimit the precise location of the surface.

Palynofacies response to relative sea-level cyclicity at a number of different hierarchical scales has been reported. Cycles of a lower (first and second) order may overprint their own long-term palynofacies characteristics onto the third-order sequence-level variations. However, at the scale of this study, third-order sequence palynological responses may be readily distinguished within the more long-term, lower-order relative sea-level cycles (Steffen & Gorin 1993). In a similar manner, the palynofacies response to higher orders of cyclicity has also been noted (e.g. Waterhouse 1995). For sediments deposited in shallow-marine settings the effects of higher-order sea-level oscillations on palynofacies response can be discerned. Where a high-sampling resolution is employed within a section

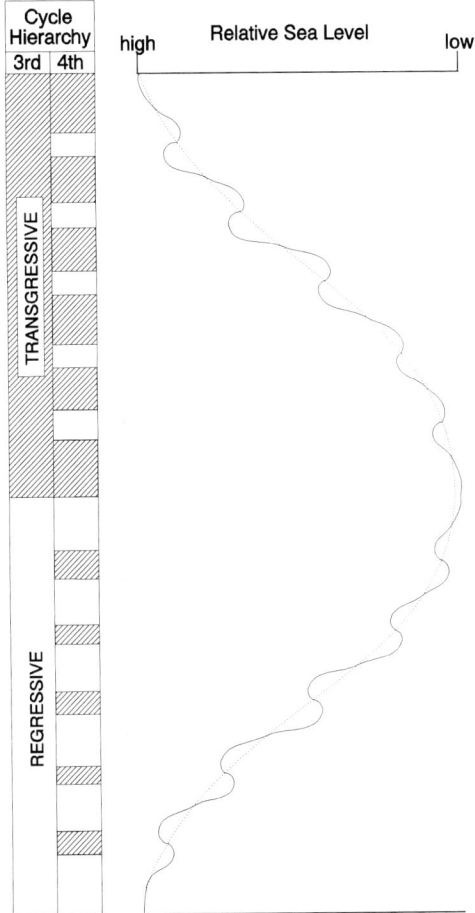

Fig. 5. Schematic diagram to illustrate the component nature of cycle hirachy with respect to palynological sampling resolution.

a higher-order cyclic palynofacies response is observed. Such variations, however, can be considered to be a component of the third-order cycle (see Fig. 5).

Palynofacies response to relative sea-level change

The responses of a number of different palynological criteria have been used in previous studies to infer changes in relative sea level. Garnered from published research, a review of predicted trends in palynofacies characteristics through a third-order siliciclastic sequence was given by Tyson (1995). However, as definitions of many of the parameters used in Tyson's work and the present study are scattered throughout the literature, a detailed explanation of the various palynological analyses and expected trends is considered justified in the present study (see Fig. 6). Three different approaches have been utilized to interpret relative sea-level change:

(i) the relative abundance variations of different POM categories throughout the study section;
(ii) the fluctuations in absolute abundances of the different POM categories;
(iii) variations in assemblage composition and species morphology within the microplankton fraction.

Where appropriate, within such a framework, information has been provided regarding the behaviour of individual POM categories, as deduced in the present study.

Variations in the relative abundance of POM

Terrigenous vs. marine palynomorphs. One method in common usage is to compare the relative abundances of allochthonous terrigenous sporomorphs and autochthonous marine palynomorphs (Gregory & Hart 1992; Prauss 1993), the so-called Terrestrial/Marine index (T/M index). The relative abundance of the autochthonous fraction increases through transgressive intervals as less allochthonous material is transported to the site of deposition. This index has only a limited application to the study of the Dorset succession (Fig. 7) because *Classopollis* pollen dominates the sporomorph fraction (~90%). As the percentage abundance of this cheirolepidiaceous pollen has been shown to increase distally in European Jurassic sediments due to its size and buoyancy (Frederiksen 1980), the T/M index can be artificially depressed.

Autochthonous POM response. The distribution of autochthonous POM is controlled by physical, biological and ecological parameters. Prauss (1993) studied the autochthonous response to relative sea-level change through a marine depositional cycle from Cenomanian–Turonian sediments in Germany, determining that in transgressive periods prasinophyte phycomata were usually absent and the relative proportions of dinoflagellates increased whilst that of acritarchs diminished. However, in oxygen-deficient conditions a prasinophyte maxima may occur in early transgressive sediments (Leckie *et al.* 1992). The MFS correlative in distal environments, the condensed section (marine condensed horizon), corresponds to the point of maximum planktonic palynomorph

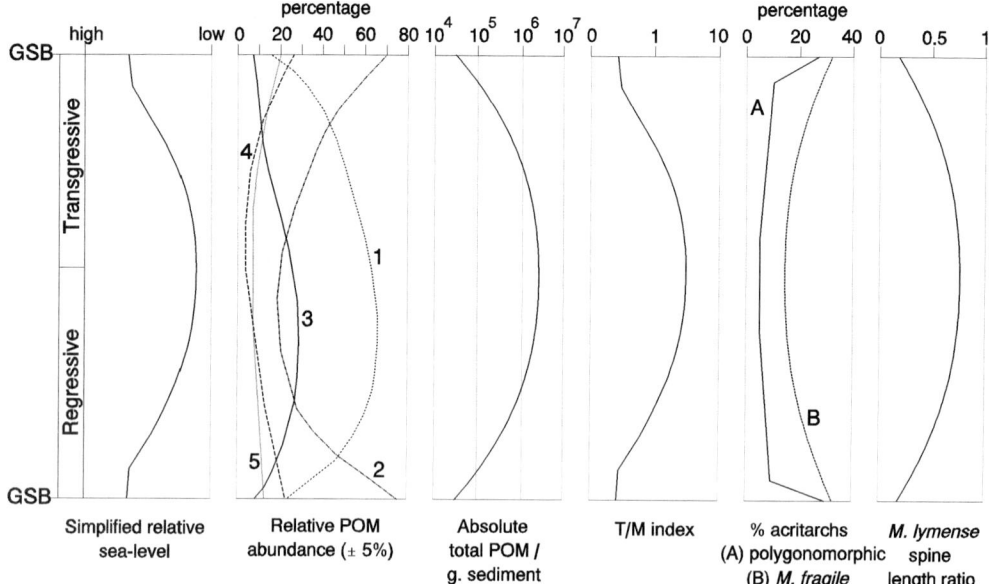

Fig. 6. Predicted palynological and palynofacies responses through one simplified genetic stratigraphic sequence cycle. 1, Cheirolepidiaceous pollen; 2, equidimensional black debris; 3, well-preserved brown wood; 4, blade-shaped black debris; and 5, acritarchs.

diversity (Powell 1992; Blondel *et al.* 1993; Steffen & Gorin 1993). Foram test linings also display a peak in relative abundance associated with MFSs. Regressive phases of relative sea-level change show a diminishing proportion of dinoflagellates relative to acritarchs, a trend that continues until the next relative sea-level rise occurs (Prauss 1993). Reworking of autochthonous palynomorphs is also seen to increase through early regression to a maximum frequency at the onset of lowstand conditions (Gregory & Hart 1992).

Allochthonous POM response. Phytoclasts: during periods of increases in relative sea-level, the ensuing coastal onlap implies that the site of deposition becomes relatively more distal and as a result the relative proportions of brown/well preserved wood progressively diminishes. From deep-sea carbonates Steffen & Gorin (1993) recorded the relative abundance of blade-shaped black debris as increasing during relative sea-level rise. However, under the more marginal, higher-energy conditions intermittently present in the early Jurassic Wessex Basin, much of the phytoclast fraction was liable to oxidation and fragmentation through transportation and hence peaks in the relative abundance of blade-shaped black debris are diluted as this fraction becomes broken down into smaller, more equant, fragments. Thus, the relative proportion of small, equidimensional black debris gradually increases with increasing distance from source.

The POM responses to a fall in relative sea level are the inverse of those seen through transgressive conditions: the proportion of blade-shaped black debris diminishes quite rapidly above the MFS, and the size and quality of preservation of the phytoclast fraction increases through highstand conditions.

Allochthonous POM response. Sporomorphs: studies of Recent sediments have shown most sporomorphs to be introduced into a basin by river transport (Heusser 1988) and as such this fraction can be treated as any other allochthonous fragment. However, due to their relatively high degree of buoyancy, sporomorphs may be transported varying distances from shore before being sedimented out and as such may show a peak in relative abundance with respect to the phytoclasts some distance offshore, as the latter are sedimented out more rapidly.

Variations in the absolute abundance of POM

The absolute abundance of POM per gramme of sediment is controlled by the interplay of three important variables: the proximity to

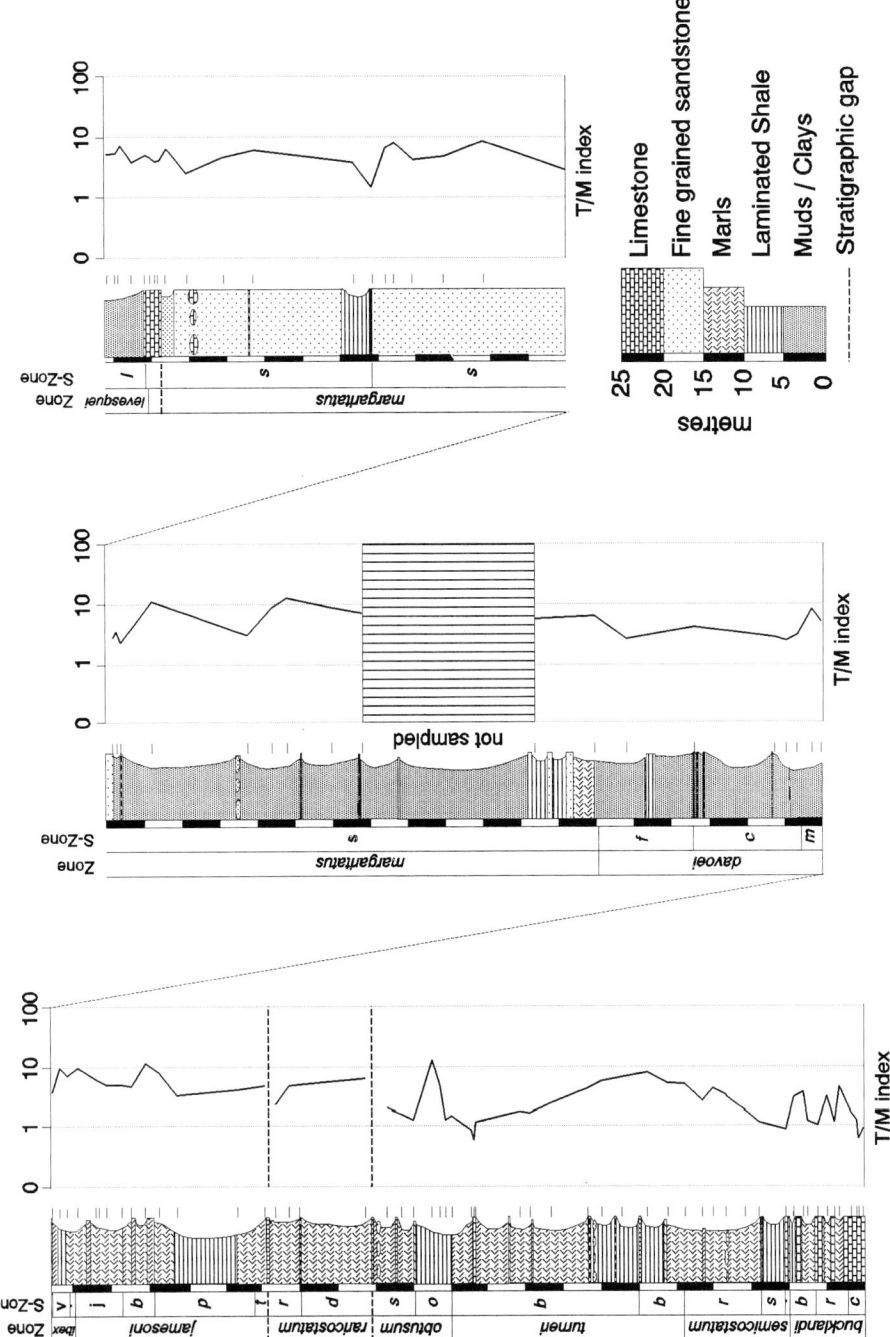

Fig. 7. Terrestrial/Marine index (T/M index) for the Sinemurian and Pliensbachian succession, Dorset coast.

source of supply (for allochthonous POM); the degree of sediment dilution (both siliciclastic and carbonate), and the oxygen regime within the basin.

Under these controls, the vertical trends in *absolute* abundance for the different POM categories appear to parallel one another, irrespective of their constituent *relative* abundance variations. The absolute POM abundance decreases away from source as POM becomes progressively sedimented out. In this study the absolute abundance of POM in the dominantly fine-grained Sinemurian and Lower–Middle Pliensbachian sediments does not appear to have been influenced by sediment dilution. However, the Upper Pliensbachian Down Cliff and Thorncombe Sands show a progressive upwards increase in mean grain size (Hesselbo & Jenkyns 1995) which can be seen to result in the progressive dilution of POM (see Fig. 9).

The degree of oxygenation prevalent at the time of deposition may also have a control over absolute total POM abundances (Tyson 1993). The MFS correlative condensed section, for example, under oxic conditions may produce a hardground or surface of non-deposition with low absolute total POM counts; however, under oxygen deficient conditions the organic matter will continue to accumulate and be preserved (Creaney & Passey 1993).

The degree of oxygenation at the time of deposition can also be assessed, irrespective of relative sea level. High relative abundances of allochthonous pollen (and high absolute POM abundances) are associated with more oxygen-deficient conditions (suggested by the increased abundance of AOM): this results in suppression in productivity of the autochthonous fraction and an increased probability of organic preservation under such conditions. In such sediments cheirolepidiaceous pollen dominates the total palynofacies assemblage (~78%).

Hesselbo & Jenkyns (1997) interpret most condensed successions in the Wessex Basin as corresponding to relative sea-level rise (this report provides corroborative evidence for this intepretation). Sedimentological, ichnological and palynofacies observations indicate that these successions were deposited under relatively well-oxygenated conditions. Conversely, expanded parts of the succession are regarded as having been associated with relative sea-level fall. It is in these successions that the markers for oxygen deficiency occur: a high abundance of AOM and an increase in the relative abundance of prasinophyte phycomata, etc.

Variations in the microplankton assemblage and microplankton morphology

Wall (1965), in a study of the Lower Jurassic Dorset succession, showed that the distribution of microplankton was environmentally controlled. In this study, it has been found that Lower Jurassic sediments from the Dorset Coast contain microplankton assemblages in which the relative abundance of the acritarch component is ~95%. By studying the ratio of polygonomorphic to acanthomorphic acritarchs it can be shown that acanthomorphic acritarchs dominated in proximal environments, whilst the proportion of polygonomorphs increased in more distal settings. This supports the supposition of Wall (1965).

Wall (1965) also determined that long-spined acritarchs were associated with low-energy, distal deposition and with increased water depth, whilst the short-spined forms were tolerant of proximal, more turbulent environments. The present study has utilized a ratio between the relative abundances of subspecies of a single acritarch taxon, *Micrhystridium lymense*, to identify proximal–distal environmental fluctuations (Fig. 6). The relative abundance of *M. lymense rigidum* (spine length 15–33% of body diameter) is compared to the combined relative abundances of *M. lymense gliscum* (spine length 33–66% of body diameter) and *M. lymense lymense* (spine length 66–100% of body diameter). A low value for this ratio is considered indicative of more distal settings. Significant fluctuations in the relative abundance of many common, long-ranging, acritarch taxa (e.g. *Micrhystridium fragile*) were also recorded. These variations are considered to be environmentally constrained.

Results

Sinemurian

The base of the Sinemurian stage in Dorset is taken as being the Second Mongrel horizon, Bed 21 (Lang 1924), and 78 m above this the base of the Hummocky Limestone, Bed 103 (Lang & Spath 1926), marks the top of the stage. The results of the palynological study of the Sinemurian stage are summarized in Fig. 8. Four distinct genetic stratigraphic sequences, two complete and two truncated by stratigraphic breaks, have been identified in the succession each bounded by candidate flooding surfaces, as follows.

Genetic Sequence I – Mongrel (Bed 23, Lang 1924) to Grey Ledge (Bed 49, Lang 1924), *conybeari–lyra* ammonite subzones.

Genetic Sequence II – Grey Ledge to Pavior (Bed 82, Lang & Spath 1926), *scipionanum*–mid *birchi* ammonite subzones.

Genetic Sequence III – Pavior to Coinstone (Bed 89, Lang & Spath 1926), upper *birch*, *obtusum* and *stellare* ammonite subzones.

Genetic Sequence IV – Coinstone to Hummocky (Bed 103, Lang & Spath 1926), *densinodulum* and *raricostatoides* ammonite subzones.

Genetic Sequence I, interpreted from Mongrel to Grey Ledge, is characterized by a relatively long regressive phase through the *bucklandi* zone which was followed by a rapid transgression (*lyra* subzone). The alternating Milankovitch-controlled limestones and mudstones of this cycle (House 1985, 1989; Weedon 1986) appear to exert a strong control over the palynofacies assemblages, hindering interpretation. The Mongrel horizon contains high relative abundances of equidimensional black debris (40%) and autochthonous components (27.9%) and has a T/M index of <1. Low absolute total POM abundances (150 000 particles/g sediment) relative to the rest of the Sinemurian stage characterizes a well-oxygenated, relatively distal depositional setting.

Sediments lying stratigraphically above the Mongrel horizon, through the *bucklandi* zone, show generalized regressive characteristics, for example an increasing T/M index: 0.59 in the Mongrel horizon up to the Venty Shales, Bed 42i (mid *bucklandi* subzone) where the index peaks at 3.69. Absolute total POM abundances also increase from the Mongrel horizon (150 000 particles/g sediment) up to the Glass Bottle Shales, Bed 46 (mid *bucklandi* subzone) (3 700 000 particles/g sediment). The autochthonous fraction decreases from 27.9% in the Mongrel to 8.7% in the Glass Bottle Shales. Within this autochthonous fraction the average acritarch spine length becomes reduced and the proportion of polygonomorphic acritarchs decreases.

This general regression was punctuated by a short-lived, minor transgressive event associated with the Second and Best Beds (Beds 39 and 41, Lang 1924), a conspicuous limestone horizon at the base of the *bucklandi* subzone. This horizon shows many of the palynofacies characteristics of a possible flooding surface: low absolute abundances, low T/M index, a low relative proportion of sporomorphs, a high relative proportion of equidimensional black debris, etc.

However, the overall regressive characteristics outlined above continue until the upper *bucklandi* subzone prior to what is interpreted as a rapid rise in relative sea level which resulted in the formation of the much condensed *lyra* subzone at the base of the *semicostatum* zone. The horizon of maximum flooding is taken at Grey Ledge, a discontinuous tabular limestone, above which a stratigraphic gap has been postulated (Hallam 1960). The POM assemblage characteristics are similar to those described for the Mongrel horizon, for example low absolute total POM abundance (64 000 particles/g sediment) and a high relative percentage of polygonomorphic acritarchs (27%).

Genetic Sequence II begins at Grey Ledge in the *lyra* subzone at the base of the *semicostatum* zone. Deposition throughout the *semicostatum* and lower *turneri* zones appears to have occurred under falling relative sea levels which progressively introduced more proximal characteristics to the POM assemblages. Palynofacies data indicate that this relative sea-level fall terminated in the upper *brooki* to lower *birchi* subzones. Absolute total POM abundance peaks (1 140 000 particles/g sediment) in the mid *brooki* subzone as does the relative abundance of pollen (46.8%), and the T/M index is also high (7.76). However, the abundance maxima of sporomorph tetrads/clusters, generally a good indicator of proximity to source (i.e. maximum regression Tyson 1993), occurs in the *Birchi* Nodular and *Birchi* Tabular, Beds 75a-76a (Lang *et al.* 1923). This peak is considered to be representative of a low-energy, oxygen-deficient depositional environment with consequently a reduced disaggregation of sporomorph tetrads/clusters. Evidence for such an environment is threefold: the presence of organic-rich paper shales in this interval (Wignall & Hallam 1991); a high relative abundance of the dysaerobia-tolerant prasinophyte *Tasmanites* sp.; and the well-preserved fossil insect remains recovered from these beds (Whalley 1985).

The succeeding palynofacies assemblages reflect a subsequent relative sea-level rise which persists up to the Pavior or Upper Cement Bed horizon: a decreasing T/M index from 7.76 (mid *brooki* subzone) to 0.56 in the Pavior, and an increase in relative abundance of the equidimensional black debris fraction from 28.1% (lower *birchi* subzone) to 42.3% (Pavior). The Pavior horizon can be interpreted as a flooding surface on similar palynological criteria to those previously discussed (see Fig. 6). Further evidence for such a designation is the small stratal gap recorded directly above this horizon, thought to be the result of sediment starvation (Hesselbo & Jenkyns 1995).

The proposed Genetic Sequence III is incomplete, truncated by the major stratigraphic gap represented by the Coinstone horizon. Above

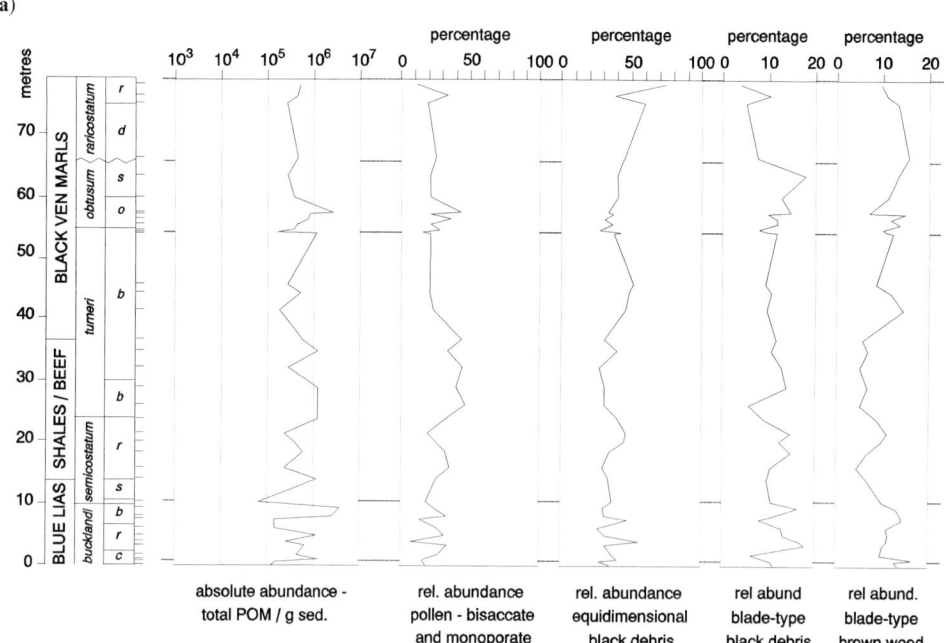

Fig. 8. Palynofacies characteristics of the Sinemurian succession, Dorset coast. FAD, first appearance datum; LAD, last appearance datum.

the Pavior, in the upper *birchi* subzone, the POM assemblages from the base of this succession show regressive characteristics. However, the interpretation of the POM signals within the organic-rich, high TOC *Obtusum* Shales above is more difficult. The increasing T/M index (0.59 in the Pavior to 12.00 in the *Obtusum* Shales), rising absolute total POM abundances (155 000 to 2 500 000 particles/g sediment) and the falling relative abundances of acritarchs (24.0 to 2.2%) would be expected during relative sea-level fall. However, the relative abundance of well-preserved brown wood also declines through the *obtusum* subzone from 14.7% in Bed 83c (Lang & Spath 1926) to 7.0% at the top of the shales. Also in the *Obtusum* Shales, the acritarch assemblages show an increase in relative abundance of polygonomorphic forms (up to 25%) and a sharp decrease in the *M. lymense* spine length ratio: both regarded as indicative of increasing water depth. The POM assemblages extracted from sediments immediately below the Coinstone are indicative of a continuing fall in relative sea level. Interpretation of this part of the sequence is therefore ambiguous. Major stratigraphic gaps are recorded both above (*macdonelli* and *aplanatum* subzones) and beneath (upper *stellare*, *denotatus*, *simpsoni* and *oxynotum* subzone) the Coinstone horizon (Lang & Spath 1926) preventing a continuous palynological sequence being recorded. This study can support the presence of such a stratigraphic break as the dinoflagellate *Liasidium variabile*, restricted to the *oxynotum* zone elsewhere around Britain (Riding 1987), is completely absent from the Dorset Coast sediments. This is not considered to be the result of ecological exclusion as previous studies have recorded *Liasidium variabile* from a number of different depositional environments (Morbey 1975; Woollam & Riding 1983; Davies 1985; etc.). Hesselbo & Palmer (1992) provide a detailed discussion of the formation of the Coinstone, but no palynological residues were obtained from the Coinstone itself prohibiting corroboration of any of the interpretations presented by these authors.

Genetic Sequence IV appears to span the lower two subzones of the *raricostatum* zone. This sequence is incomplete being bounded by two unconformities, one associated with the above-mentioned Coinstone and the erosional upper discontinuity known to occur beneath the Hummocky limestone. From the ammonite record this latter discontinuity represents the *macdonelli* and *aplanatum* subzones. Palynofacies indicators through that part of the *raricostatum* zone that is developed indicate a

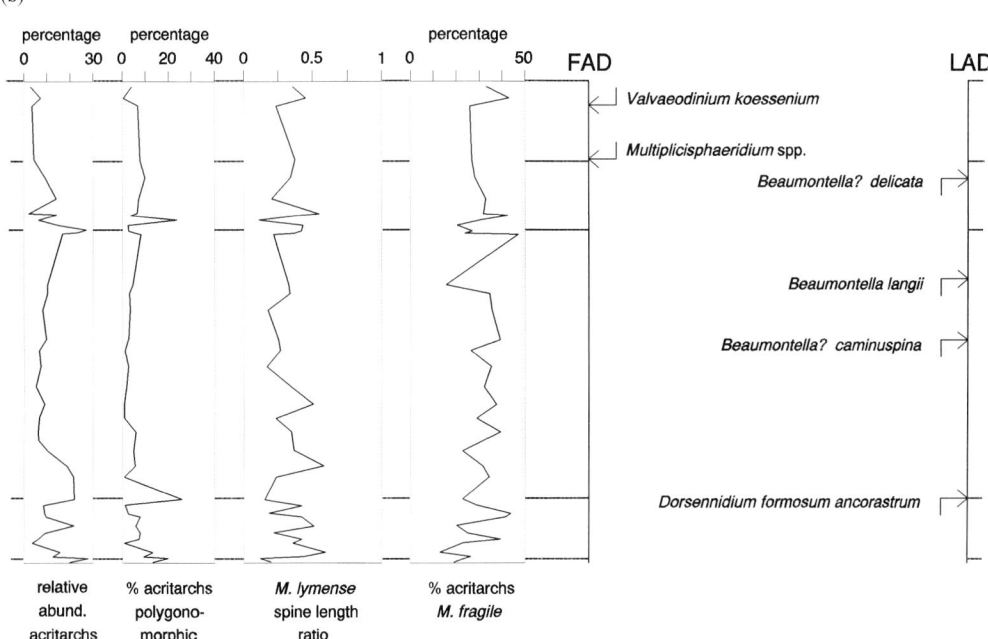

transgressive character to sediments moving towards the Sinemurian–Pliensbachian boundary: increasing relative abundances of equidimensional black debris (45.5 to 74.1%), a fall in the relative abundance of the pollen (22.3 to 8.1%) and well-preserved brown wood fractions (15.5 to 9.7%). A declining T/M index (5.87 to 2.29) is also re-corded. The *raricostatum* zone contains the only occurrence of the acritarch genus *Multiplicisphaeridium* through the section. This distinct polygonomorphic acritarch has previously only been recorded in the Jurassic period from sediments deposited under transgressive conditions from the Lias ζ in Germany (Prauss *et al.* 1991).

Pliensbachian

The base of the Pliensbachian stage in Dorset is taken as the base of the Hummocky limestone (Bed 103) and its top at the base of the Junction Bed (*sensu lato*). The Junction Bed and overlying Down Cliff Clays are of Toarcian age. Palynological results for the Dorset Pleinsbachian are summarized in Fig. 9.

The Pliensbachian stage of Dorset represents an almost complete succession of predominantly siliciclastic sediments, 176 m in thickness, with only the upper *subnodosus* and *gibbosus* subzones missing. Within this succession four genetic stratigraphic sequences may be distinguished.

Genetic Sequence V – Hummocky (Bed 103, Lang & Spath 1926) to the Belemnite Stone (Bed 121, Lang 1928), *jamesoni* and *ibex* ammonite zones.

Genetic Sequence VI – Belemnite Stone to Day's Shell Bed (Bed 20, Howarth 1957), *davoei* zone and lower *stokesi* subzone.

Genetic Sequence VII – Day's Shell Bed to the *margaritatus* Stone (Bed 24, Howarth 1957), upper *stokesi* subzone.

Genetic Sequence VIII – *margaritatus* Stone to the base of the Junction Bed (*sensu lato*) (Beds 34–35, Howarth 1957), lower *subnodosus* subzone.

The Belemnite Marls are considered to represent one complete sedimentary cycle (Hesselbo & Jenkyns 1995). This view is supported by palynological evidence. Within Genetic Sequence V two minor, yet distinct, regressive–transgressive phases may be discerned.

Following the major Upper Sinemurian–Lower Pliensbachian transgression responsible for the erosional discontinuity and the formation of Hummocky, the *polymorphus* and *brevispina* subzones represent the first regressive–transgressive phase. A minor flooding surface may be postulated from palynofacies evidence at the Middle Pale Band (Bed 113, Lang *et al.* 1928), for

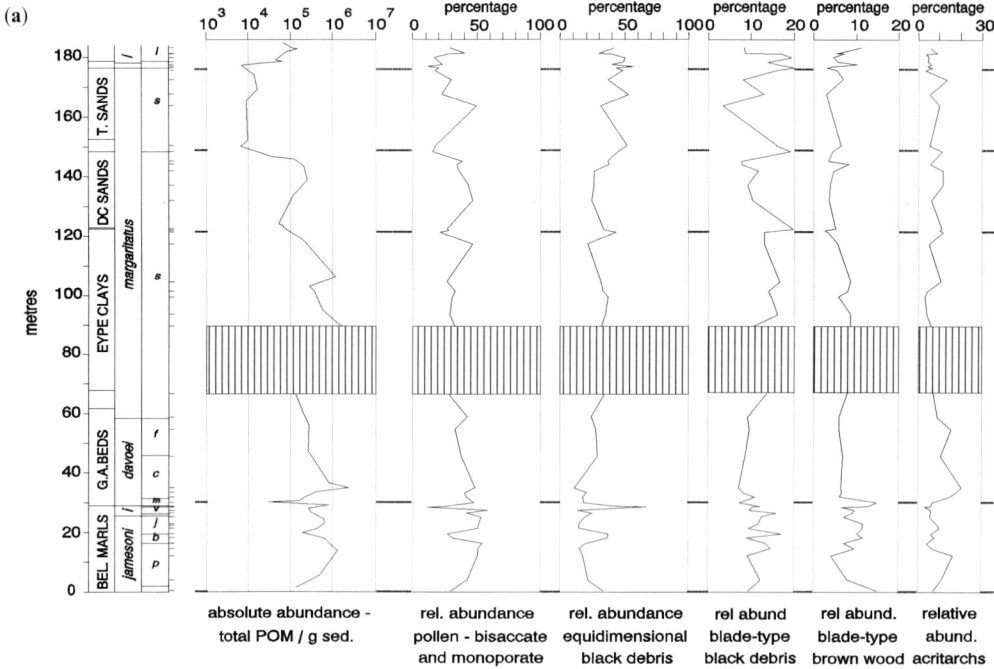

Fig. 9. Palynofacies characteristics of the Pliensbachian succession, Dorset coast. FAD, first appearance datum; LAD, Last appearance datum.

example, the absolute total POM abundance peak in the *polymorphus* subzone of 1 250 000 particles/g sediment (Bed 110a, Lang *et al.* 1928) declines to 192 000 particles/g sediment in the *brevispina* subzone (Bed 113, Lang *et al.* 1928), whilst across the same interval the relative abundance of equidimensional black debris increases from 14.9 to 37.0%, and that of sporomorphs decreases from 54.6 to 27.6%. High-resolution natural gamma-ray spectrometry further supports the presence of a flooding surface at this level (Bessa & Hesselbo 1997), although such an interpretation was not made for this interval in Dorset by Parkinson (1996), based on a lower-resolution study. The overlying *jamesoni* subzone and *ibex* zone represent the second regressive–transgressive phase. Palynofacies assemblages through this fourth-order cycle display regressive characteristics from the Middle Pale Band through to the Upper Dark Band (Bed 115, Lang 1928) in the mid-*jamesoni* subzone: the absolute total abundance of POM rises from 192 000 to 623 000 particles/g sediment, whilst the relative abundance of equidimensional black debris falls (37.4–15.1%). The ensuing transgressive phase occurs from the upper *jamesoni* subzone to the Belemnite Stone (Bed 121, Lang *et al.* 1928), a conspicuous condensed limestone representing the *luridum* subzone. This horizon is characterized by a low T/M index relative to the preceding *jamesoni* and lower *ibex* zones (3.54), very high relative abundances of equidimensional black debris (65%), low relative abundance of well-preserved brown wood (6.7%) and a very low absolute total POM abundance (29 800 particles/g sediment). The relative abundance of polygonomorphic acritarchs within the Belemnite Stone was also high (28% of the total acritarch assemblage) suggesting more open marine conditions. Ongoing high-resolution studies suggest that regressive–transgressive pulses of even shorter duration may be recognizable within the *ibex* zone.

Genetic Sequence VI comprises the Green Ammonite Beds, Three Tiers and the Eype Clays, and is bounded by the Belemnite Stone at the base with Day's Shell Bed as the upper bounding surface. Callomon & Oates (1993) report a minor stratigraphic gap directly above the Belemnite Stone. The overlying Green Ammonite Beds represent a fully marine succession of mudrocks which generally becomes more silty upwards towards the Three Tiers. The base of the *margaritatus* zone is recognized as a second-order depositional sequence boundary (*sensu* Vail *et al.* 1984) throughout Britain and the North Sea (Partington *et al.* 1993). The gradual increase in grain size through this part of the

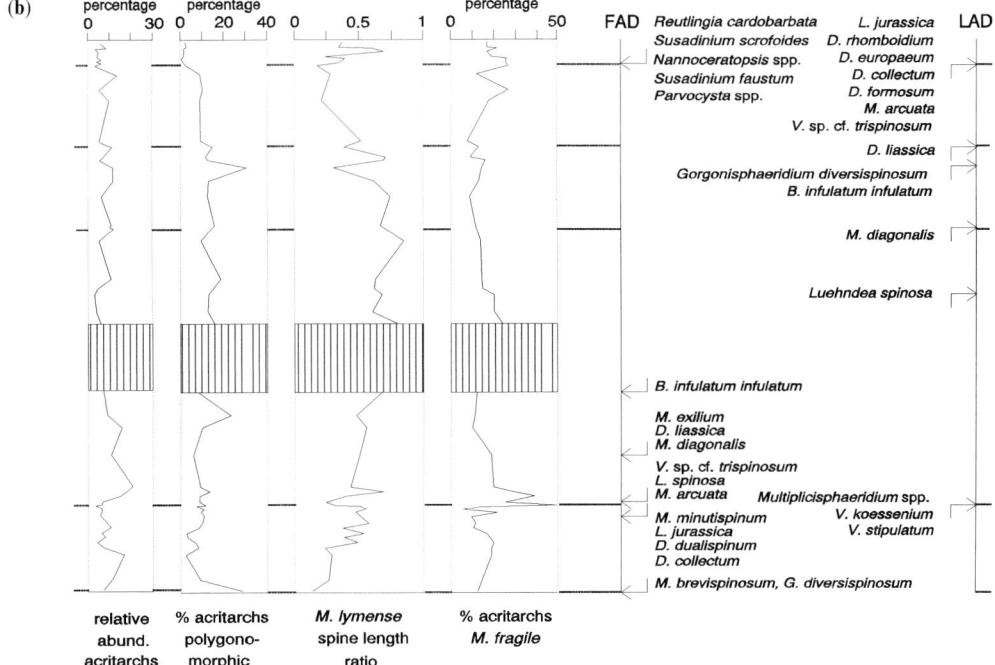

succession influences the relative abundance of POM due to the effects of hydrodynamic sorting. The absolute total POM abundance declines upwards towards the Three Tiers from 2 400 000 particles/g sediment at the base of the *capricornus* subzone to 138 000 particles/g sediment in Three Tiers. In most situations this would be interpreted as indicative of transgressive conditions. However, this interpretation is contradicted by the increase in the *M. lymense* spine length ratio from 0.25 at the Belemnite Stone through the Green Ammonite Beds to a maximum in the Three Tiers (0.70) and Eype Clays (0.83): such high figures being considered indicative of a proximal depositional setting, arguing against a transgressive depositional regime. Taking into account the high *M. lymense* spine length ratio and the increasing sediment grain size through the section, the decrease in absolute total POM abundance is interpreted as due to the effects of sediment dilution in a more proximal setting. The progressive increase in grain size through the *davoei* zone is considered to have negated the typical proximal–distal trends in absolute total POM abundance response, and to have overprinted its own response characteristic.

Following the formation of the Three Tiers, sea level is interpreted as having risen, enabling accommodation space to be created to allow for the rapid deposition of the stratigraphically expanded Eype Clays. The top of the cycle is marked by Day's Shell Bed. Palmer (1966) interpreted the Shell Bed as a winnowed shell accumulate. However, the palynological results suggest another interpretation. The relative abundance of both equidimensional and blade-type black debris are considered high relative to the adjacent sediments, 41.0% and 13.2% respectively, whilst the pollen (20%) and brown-wood fractions (8%) are low. A trough in the T/M index (2.2) is also associated with this horizon and the absolute abundance of total POM falls to 67 000 particles/g sediment. All these parameters may be interpreted as being characteristic of a mid-cycle shell bed (*sensu* Vail *et al.* 1984), thus enabling Day's Shell Bed to be interpreted as a maximum flooding surface.

The upper *stokesi* subzone represents one complete regressive–transgressive cycle, Genetic Sequence VII, bounded by Day's Shell Bed and the *margaritatus* Stone. Above the Shell Bed 1 m of shaley sands occur which in turn are overlain by the coarser Down Cliff Sands. Grain size also coarsens upwards through this cycle (pers. obs.). Absolute total POM abundance varies from 67 000 particles/g sediment in the Shell Bed to a maximum of 240 000 particles/g sediment 15 m above the Starfish Bed and then gradually falls to 6 500 particles/g sediment in the *margaritatus* Stone. The relative abundance of pollen displays much the same pattern, from 21.3% in the Shell Bed to a maximum of 47.2% 9 m above the Starfish Bed and then gradually declining through the Down Cliff Sands to 15.7% in the *margaritatus* Stone. An inverse pattern in the relative abundances of both equidimensional and blade-type black debris is recorded.

Genetic Sequence VIII, represented by the interval from the *margaritatus* Stone to the base of the Junction Bed (*sensu lato*), is incomplete. The *margaritatus* Stone has herein been interpreted as a flooding surface. Stratigraphically above this 2 m of the *margaritatus* Clay occurs, which is in turn overlain by the fine siliciclastics of the Thorncombe Sands. The succession is believed to represent the regressive phase of a genetic sequence cycle truncated by the unconformity immediately below the Junction Bed (*sensu lato*). Absolute total POM abundances do not exceed 17 500 particles/g sediment throughout the Thorncombe Sands due to the effects of increased sediment dilution and winnowing, the result of clastic deposition close to source. Despite this, variations in the absolute abundance of POM are still characteristic of a regressive phase: 6500 particles/g sediment in the *margaritatus* Stone (Bed 24, Howarth 1957), rising to 17 300 particles/g sediment in Bed 28 (Howarth 1957).

Discussion

Solely from palynological evidence it has been possible to discern eight discrete genetic stratigraphic sequence cycles (or depositional episodes) for the Sinemurian and Pliensbachian stages of the Dorset Coast. The palynofacies response for a complete genetic sequence cycle is characterized by a typical regressive–transgressive trend (see Fig. 6).

In order to test the validity of this independent study the proposed genetic sequence stratigraphic framework is compared here with sequence frameworks for Lower Jurassic sections both within the Wessex Basin and elsewhere in Britain developed principally from sedimentary facies studies (Haq *et al.* 1988; Partington *et al.* 1993; Hesselbo & Jenkyns 1997). Hesselbo & Jenkyns (1997) analysed sedimentary facies changes in the Dorset succession allowing the construction of a sequence stratigraphic framework for the Lower Jurassic of Britain. Partington *et al.* (1993) utilized biostratigraphy to calibrate Jurassic genetic stratigraphic sequences within the North Sea area, whilst Haq *et al.* (1988) incorporated the Dorset and Yorkshire Lower Jurassic sections in the construction of a worldwide Mesozoic eustatic sea-level curve. The principal results of these studies are reproduced in Fig. 10.

Most of the major candidate flooding surfaces proposed by Hesselbo & Jenkyns (1997) in their study of the Dorset section have been recorded here. The upper bounding flooding surface of Genetic Sequence I is taken to be Grey Ledge within the *lyra* subzone, Hesselbo & Jenkyns (1997) likewise propose this horizon to represent a candidate MFS. The lower boundary for proposed Genetic Sequence V the upper *raricostatum*–lower *jamesoni* zonal interval, in Dorset represented by a stratigraphic break followed by the condensed Hummocky limestone, is recorded by Partington *et al.* (1993), Hesselbo & Jenkyns (1997) and the present authors as a MFS.

Despite the general trend of palynofacies assemblages showing progressively more proximal characteristics throughout much of the upper *jamesoni* and *ibex* zones, the palynofacies assemblages recovered immediately below and at the Belemnite Stone within the stratigraphically condensed *luridium* subzone suggest that this subzone represents a short-lived period of relative sea-level rise, culminating in the formation of the condensed Belemnite Stone. Hesselbo & Jenkyns (1997) suggest the condensation of the *luridium* subzone to be related to sediment starvation and hence deepening, as opposed to the sediment by-pass (associated with shallowing) proposed for condensation within the underlying *valdani* subzone. This interpretation supports the present proposal that the Belemnite Stone represents a flooding surface correlative. It should be noted that Haq *et al.* (1988) provided a conflicting interpretation for the mid-upper *ibex* zone, postulating a sequence boundary at the *valdani–luridum* transition. Such an interpretation is not supported by high-resolution palynofacies analyses currently being undertaken on sample material from the Dorset succession.

The *margaritatus* Stone (the lower bounding surface of Genetic Sequence VIII) is proposed as a major candidate MFS by Hesselbo & Jenkyns (1997), time equivalent to the upper *stokesi* and lower *subnodosus* subzones. This is supported by the results of the palynological study. However, despite the fact that Hesselbo & Jenkyns (1997) recorded this candidate MFS from a number of basins within the British Isles, neither Partington *et al.* (1993) nor Haq *et al.* (1988) record a

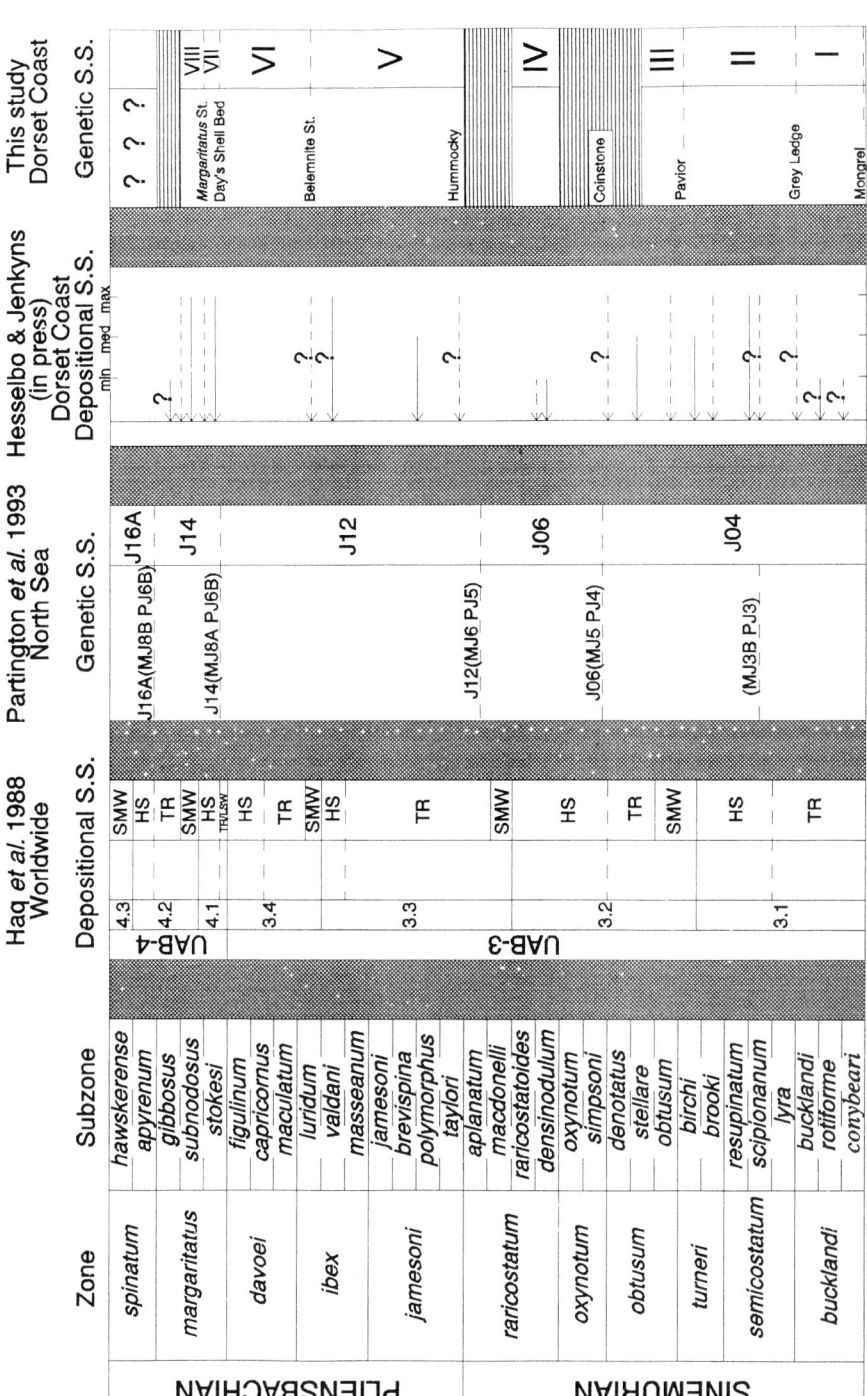

Fig. 10. Comparison of the genetic sequence framework proposed in this study with those of Partington *et al.* (1993), Hesselbo & Jenkyns (1997) and the depositional sequence stratigraphic model of Haq *et al.* (1988).

flooding surface as present at the *stokesi–subnodosus* subzonal boundary horizon. Day's Shell Bed (mid-*stokesi* subzone) is herein proposed as a genetic sequence boundary. Similar time-equivalent flooding surfaces have been recorded both by Haq *et al.* (1988, the Cycle 4.1 flooding surface) and by Partington *et al.* (1993, the J14 (MJ8A PJ6B) genetic sequence boundary at the base of genetic sequence J14). Hesselbo & Jenkyns (1997) do, however, record the presence of several fourth-order coarsening-upwards cycles within the *margaritatus* zone. Genetic Sequence VII proposed herein may represent one such fourth-order cycle (see Fig. 10), the base of which, Day's Shell Bed, might be correlated to the flooding surfaces recorded elsewhere for the mid-*stokesi* subzone.

Despite the clear corroboration of the palynological data by sedimentological studies for certain horizons in the section, the following examples indicate that discrepancies between the proposed palynological and sedimentological interpretations do occur. Hesselbo & Jenkyns (1997) recognize a major candidate MFS in the *scipionanum* subzone (also recorded by Partington *et al.* 1993 as the MJ3B PJ3 level). However, at the time this study was undertaken this part of the succession was not available for palynological sampling, hence this candidate MFS cannot be confirmed palynologically. Subsequent higher-resolution palynological studies might elucidate this interval further.

The placement of the lower bounding flooding surface for Genetic Sequence III also differs between studies. Hesselbo & Jenkyns (1997) regard the strong sedimentary facies change at the base of the *obtusum* Shale to represent the point of maximal flooding; however, palynological evidence contradicts this lithological interpretation, suggesting that the Pavior horizon represents the surface of maximum flooding. The *obtusum* Shales comprise sediment deposited under oxygen-deficient conditions (cf. high relative abundance of both AOM and the prasinophyte *Tasmanites*) which it is thought may have unduly biased the palynofacies assemblages recovered (see the discussion in the section on '*Variations in absolute abundance of POM*').

Hesselbo & Jenkyns (1997) interpret the Coinstone as representing a MFS, however, no palynological data were obtained from this horizon. Natural gamma-ray spectrometry further supports the presence of a flooding surface at this level (J. Bessa, pers. comm.). Partington *et al.* (1993) record their '*oxynotum*' J06 (MJ5 PJ4) maximum flooding surface from the stratigraphically more complete Upper Sinemurian sections in the North Sea, whilst Haq *et al.* (1988) similarly document their Cycle 3.2 flooding surface at the base of the *oxynotum* zone.

Conclusions

Operating within sample preparation/observational constraints, this study demonstrates that a detailed palynofacies investigation can aid the construction of a sequence stratigraphic framework for a sedimentary succession. For the Dorset Coast Lower Jurassic succession it has been possible to recognize eight discrete palynologically based genetic sequences for the Sinemurian and Pliensbachian stages, informally designated Genetic Stratigraphic Sequences I–VIII. The placement of the bounding flooding surfaces of these sequences (by palynological interpretation) can be corroborated by comparison with the sedimentological facies studies of Hesselbo & Jenkyns (1997) in an independent study of the same section, as well as by comparison with existing published frameworks for the Lower Jurassic (Haq *et al.* 1988; Partington *et al.* 1993). Such comparisons enhance the validity of a sequence palynological approach to sequence stratigraphic interpretation.

D. C. Cole would like to acknowledge the receipt of a NERC postgraduate studentship, GT4/92/260/G. The authors are grateful to the following individuals for their assistance and helpful discussion on matters concerning the Dorset Lower Jurassic succession: M. R. House (Southampton), S. Hesselbo (Oxford) and J. Bessa (Shell Venezuela).

References

ARKELL, W. J. 1933. *The Jurassic System in Great Britain*. Oxford University Press, Oxford.

BESSA, J. L. & HESSLEBO, S. P. 1997. Gamma-ray character and correlation of the Lower Lias, southwest Britain. *Proceedings of the Geologists' Association*, **108**, 113–129.

BLONDEL, T. G. A., GORIN, G. E. & JAN DU CHÊNE, R. 1993. *Sequence Stratigraphy in Coastal Environments: Sedimentology and Palynofacies of the Miocene in C. Tunisia*. International Association of Sedimentologists, Special Publications, **1**, 161–179.

BOULTER, M. C. 1995. An approach to a standard terminology for palynodebris. *In*: TRAVERSE, A. (ed.) *Sedimentation of Organic Particles*. Cambridge University Press, Cambridge, 199–216.

—— & RIDDICK, A. 1986. Classification and analysis of palynodebris from the Paleocene sediments of the Forties Field. *Sedimentology*, **33**, 871–886.

BRADSHAW, M. J., COPE, J. C. W., CRIPPS, D. W., DONOVAN, D. T., HOWARTH, M. K., RAWSON, P. F., WEST, I. M. & WIMBLEDON, W. A. 1992.

Jurassic. *In*: COPE, J. C. W., INGRAM, J. K. & RAWSON, P. F. (eds) *Atlas of Palaeogeography and Lithofacies*. Geological Society of London Memoir No. 13, 107–129.

CHADWICK, R. A. 1986. Extension tectonics in the Wessex Basin, Southern England. *Journal of the Geological Society, London*, **143**, 465–488.

CALLOMON, J. H. & OATES, M. J. 1993. Jurassic of Oxfordshire and Cotswolds. *In*: *An excursion guide for the Arkell Symposium on Jurassic Geology*.

COMBAZ, A. 1964. Les palynofaciès. *Revues de Micropaléontologie*, **7**, 205–218.

COPE, J. C. W., GETTY, T. A., HOWARTH, M. K., MORTON, & N. & TORRENS, H. S. 1980. *A Correlation of Jurassic Rocks in the British Isles. Part one: Introduction and Lower Jurassic*. Special Report of the Geological Society, London No. 14, 73.

CREANEY, S. & PASSEY, Q. R. 1993. Recurring patterns of total organic carbon and source rock quality within a sequence stratigraphic framework. *Bulletin of the American Association of Petroleum Geologists*, **77**, 386–401.

DAVIES, E. H. 1985. The miospore and dinoflagellate cyst Oppel – zonation of the Lias of Portugal. *Palynology*, **9**, 105–132.

DAY, E. G. H. 1863. On the Middle and Upper Lias of the Dorsetshire coast. *Quarterly Journal of the Geological Society, London*, **19**, 278–97.

DE LA BECHE, H. T. 1826. On the Lias of the coast, in the vicinity of Lyme Regis, Dorset. *Transactions of the Geological Society, London*, **2**, 21–30.

DUVAL, B., CRAMEZ, C. & VAIL, P. R. 1992. Types and hierarchy of stratigraphic cycles. *Abstracts, International symposium on Mesozoic and Cenozoic Sequence Stratigraphy of European Basins, Dijon*, p. 44.

FREDERIKSEN, N. O. 1980. Significance of monosulcate pollen abundance in Mesozoic sediments. *Lethaia*, **13**, 1–20.

GALLOWAY, W. E. 1989. Genetic stratigraphic sequences in basin analysis I: architecture and genesis of flooding – Surface bounded depositional units. *Bulletin of the American Association of Petroleum Geologists*, **73**, 125–142.

GORIN, G. E. & STEFFEN, D. 1991. Organic facies as a tool for recording eustatic variations in marine fine-grained carbonates – example of the Berriasian stratotype at Berrias (Ardèche, SE France). *Palaeogeography, Palaeoclimatology, Palaeoecology*, **85**, 303–320.

GREGORY, W. A. & HART, G. F. 1992. Towards a predictive model for the palynofacies response to sea-level changes. *Palaios*, **7**, 3–33.

HABIB, D., MOSHKOVITZ, S. & KRAMER, C. 1992. Dinoflagellate and calcareous nannofossil response to sea-level change in Cretaceous–Tertiary boundary sections. *Geology*, **20**, 165–8.

HALLAM, A. 1960. A sedimentary and faunal study of the Blue Lias of Dorset and Glamorgan. *Philosophical Transactions of the Royal Society, London*, **B234**, 1–44.

—— 1975. *Jurassic Environments*. Cambridge University Press, Cambridge.

—— 1988. A re-evaluation of the Jurassic eustasy in the light of new data and the revised Exxon curve. *In*: WILGUS, C. K., HASTINGS, B. S., KENDELL, G. ST. C., POSAMENTIER, H. W., ROSS, C. A. & VAN WAGONER, J. C. (eds) *Sea-level Changes – An Integrated Approach*. Society of Economic Paleontologists and Mineralogists, Special Publications, **42**, 261–273.

HAQ, B. U., HARDENBOL, J. & VAIL, P. R. 1988. Mesozoic and Cenozoic chronostratigraphy and eustatic cycles. *In*: WILGUS, C. K., HASTINGS, B. S., KENDELL, G. ST. C., POSAMENTIER, H. W. ROSS, C. A. & VAN WAGONER, J. C. (eds) *Sea-level Changes – An Integrated Approach*. Society of Economic Paleontologists and Mineralogists, Special Publications, **42**, 71–108.

HESSELBO, S. P. & JENKYNS, H. C. 1995. A comparison of the Hettangian to Bajocian successions of Dorset and Yorkshire. *In*: TAYLOR, P. D. (ed.) *Field Geology of the British Jurassic*. Geological Society, London, 105–150.

—— & PALMER, T. J. 1992. Reworked early diagenetic concretions and the bioerosional origin of a regional discontinuity within British Jurassic marine mudrocks. *Sedimentology*, **39**, 145–165.

—— & —— 1998. Sequence stratigraphy of the British Lower Jurassic. *In*: DE GRACIANSKY, P. C., HARDENBOL, J., JACQUIN, T., FARLEY, M. B. & VAIL, P. R. (eds) *Mesozoic and Cenozoic Sequence Stratigraphy of Western European Basins*. Society of Economic Mineralogists and Paleontologists, Special Publications, in press.

HEUSSER, L. E. 1988. Pollen distribution in marine sediments on the continental margin off northern California. *Marine Geology*, **80**, 131–147.

HOLLOWAY, S. 1985. Lower Jurassic: the Lias. *In*: WHITTAKER, A. (ed.) *Atlas of Onshore Sedimentary Basins in England and Wales – Post-Carboniferous Tectonics and Stratigraphy*. Blackie, Glasgow, 37–40.

HOUSE, M. R. 1985. A new approach to an absolute timescale from measurements of orbital cycles and sedimentary microrhythms. *Nature*, **315**, 721–725.

—— 1989. *Geology of the Dorset Coast*. Geologists Association Guide.

HOWARTH, M. K. 1957. The Middle Lias of the Dorset Coast. *Quarterly Journal of the Geological Society, London*, **113**, 185–204.

JENKYNS, H. C. & SENIOR, J. R. 1991. Geological evidence for intra-Jurassic faulting in the Wessex Basin and its margins. *Journal of the Geological Society, London*, **148**, 245–260.

JERVEY, M. T. 1988. Quantitative Geological Modelling of siliciclastic rock sequences and their seismic expression. *In*: WILGUS, C. K., HASTINGS, B. S., KENDELL, G. ST. C., POSAMENTIER, H. W. ROSS, C. A. & VAN WAGONER, J. C. (eds) *Sea-level Changes – An Integrated Approach*. Society of Economic Paleontologists and Mineralogists, Special Publications, **42**, 47–69.

JONES, R. W., VENTRIS, P. A., WONDERS, A. A. H., LOWE, S., RUTHERFORD, H. M., SIMMONS, M. D., VARNEY, T. D., ATHERSUCH, J., STURROCK, S. J., BOYD, R. & BRENNER, W. 1993. Sequence stratigraphy of Barrow Group (Berriasian–Valanginian siliciclastics, N.W. Shelf, Australia), with emphasis on the sedimentological and palaeontological characterization of systems tracts. *In*: JENKINS, D. G. (ed.) *Applied Micropalaeontology*. Kluwer, Dordrecht 193–223.

LAKE, S. D. & KARNER, G. D. 1987. The structure and evolution of the Wessex Basin, southern England: an example of inversion tectonics. *Tectonophysics*, **137**, 347–378.

LANG, W. D. 1914. The geology of the Charmouth cliffs, beach and foreshore. *Proceedings of the Geologists Association, London*, **25**, 294–359.

—— 1924. The Blue Lias of the Devon and Dorset Coasts. *Proceedings of the Geologists Association, London*, **35**, 169–185.

—— & SPATH, L. F. 1926. The Black Marl of Black Ven and Stonebarrow, in the Lias of the Dorset Coast. With notes on the Lamellibranchia by L. R. Cox; on the Brachiopoda by H. M. Muir-Wood; on certain Echioceratidae by A. E. Trueman and D. M. Williams. *Quarterly Journal of the Geological Society, London*, **82**, 144–187.

——, ——, COX, L. R. & MUIR-WOOD, H. M. 1928. The Belemnite Marls of Charmouth, a series in the Lias of the Dorset Coast. *Quarterly Journal of the Geological Society, London*, **84**, 179–257.

——, —— & RICHARDSON, W. A. 1923. Shales with Beef, a sequence in the lower Lias of the Dorset Coast. *Quarterly Journal of the Geological Society, London*, **79**, 47–99.

LECKIE, D. A., SINGH, C., BLOCH, J. WILSON, M. & WALL, J. 1992. An anoxic event at the Albian-Cenomanian boundary: the Fish Scale Marker Bed, northern Alberta, Canada. *Palaeogeography, Palaeoclimatology, Palaeoecology*, **92**, 139–66.

LOUTIT, T. S., HARDENBOL, J., VAIL, P. R. & BAUM, G. R. 1988. Condensed sections: the key to age dating and correlation of continental margin sequences. *In*: WILGUS, C. K. *et al.* (eds) *Sea-level changes, an integrated approach*. Special Publications of the Society of Economic Paleontologists and Mineralogists, **42**, 183–216.

MORBEY, S. J. 1975. A palynostratigraphy of the Rhaetian Stage, Upper Triassic in the Kendelbachgraben, Austria. *Palaeontographica Abt. B*, **152**, 1–75.

—— 1978. Late Triassic and Early Jurassic subsurface palynostratigraphy in Northwestern Europe. *Palinologia*, **1** (Numéro Extraordinario), 355–365.

MÜLLER, J. 1959. Palynology of Recent Orinoco Delta and shelf sediments: reports of the Orinoco Shelf expedition; volume 5. *Micropalaeontology*, **5**, 1–2.

PALMER, C. P. 1966. The fauna of Day's Shell Bed in the Middle Lias of the Dorset Coast. *Dorset Natural History and Archaeological Society*, **87**, 69–80.

PARKINSON, D. N. 1996. Gamma-ray spectrometry as a tool for stratigraphical interpretation: examples from the western European Lower Jurassic. *In*: HESSELBO, S. P. & PARKINSON, D. N. (eds) *Sequence Stratigraphy in British Geology*. Geological Society, London, Special Publication **103**, 231–255.

PARTINGTON, M. A., COPESTAKE, P., MITCHENER, B. C. & UNDERHILL, J. R. 1993. Biostratigraphic calibration of genetic sequence stratigraphic sequences in the Jurassic–Lowermost Cretaceous (Hettangian–Ryazanian) of the North Sea and adjacent areas. *In*: PARKER, J. R. (ed.) *Petroleum Geology of Northwest Europe: Proceedings of the 4th Conference*. Geological Society, London, **1**, 371–386.

PASLEY, M. A. & HAZEL, J. E. 1990. Use of organic petrology and graphic correlation of biostratigraphic data in sequence stratigraphic interpretation. An example from the Eocene–Oligocene boundary section, St. Stephens Quarry, Alabama. *Transactions of the Gulf Coast Association of Geological Sciences*, **40**, 661–683

PHIPPS, D. & PLAYFORD, G. 1984. Laboratory techniques for the extraction of palynomorphs from sediments. *Papers of the Department of Geology, University of Queensland*, **11**, 1–23.

POSAMENTIER, H. W., JERVEY, M. T. & VAIL, P. R. 1988. Eustatic controls on clastic deposition I – conceptual framework. *In*: WILGUS, C. K., HASTINGS, B. S., KENDELL, G. ST. C., POSAMENTIER, H. W. ROSS, C. A. & VAN WAGONER, J. C. (eds) *Sea-Level Changes – An Integrated Approach*. Society of Economic Paleontologists and Mineralogists, Special Publications, **42**, 109–124.

POWELL, A. J. 1992. Making the most of microfossils. *Geoscientist*, **2**, 12–16.

PRAUSS, M. 1993. Sequence palynology – Evidence from Mesozoic sections and conceptual framework. *Neues Jahrbuch für Geologie und Paläontologie Abhandlungen*, **190**, 143–163.

——, LIGOUIS, B. & LUTERBACHER, H. 1991. Organic matter and palynomorphs in the 'Posidonienschiefer' (Toarcian, Lower Jurassic) of southern Germany. *In*: TYSON, R. V. & PEARSON, T. H. (eds) *Modern and Ancient Continental Shelf Anoxia*. Geological Society, London, Special Publications, **58**, 335–351.

RIDING, J. B. 1987. Dinoflagellate cyst stratigraphy of the Nettleton Bottom Borehole (Jurassic: Hettangian to Kimmeridgian), Lincs, England. *Proceedings of the Yorkshire Geological Society*, **46**, 231–266.

SELLWOOD, B. W. & JENKYNS, H. C. 1975. Basins and swells and the evolution of an epeiric sea (Pliensbachian–Bajocian of Great Britain). *Journal of the Geological Society, London*, **131**, 373–388.

STEFFEN, D. & GORIN, G. E. (1993) Sedimentology of organic matter in Upper Tithonian – Berriasian deep sea carbonates of, S.E. France: evidence of eustatic control. *In*: KANTZ, B. & PRATT, L. (eds) *Source Rocks in a Sequence Stratigraphic Framework*. American Association of Petroleum Geologists Studies in Geology, **37**, 49–65.

STONELEY, R. 1982. The structural development of the Wessex Basin: implications for exploration. *Proceedings of the Ussher Society*, **8**, 1–6.

TRAVERSE, A. 1988. *Paleopalynology*. Unwin Hyman, Boston.

TYSON, R. V. 1993. Palynofacies analysis. *In*: JENKINS, D. J. (ed.) *Applied Micropalaeontology*. Kluwer, Dordrecht, 153–191.

—— 1995. *Sedimentary Organic Matter – Organic Facies and Palynofacies*. Chapman & Hall, London.

UNDERHILL, J. R. & PATERSON, S. 1998. Genesis of tectonic inversion structures: seismic evidence for the development of key structures along the Purbeck–Isle of Wight Disturbance. *Journal of the Geological Society, London*, in press.

—— & STONELEY, R. 1982. Introduction to the development, evolution and petroleum geology of the Wessex Basin. *This volume*.

VAIL, P. R., HARDENBOL, J. & TODD, R. G. 1984. Jurassic unconformities, chronostratigraphy and sea level changes from seismic stratigraphy and biostratigraphy. *AAPG Memoirs*, **36**, 129–144.

——, MITCHUM, R. H. JR, THOMPSON, M. III, TODD, R. G., SANGREE, J. B., WIDMIER, J., BUBB, N. N. & NATELID, W. G. 1977. Seismic stratigraphy and sea level charges. *AAPG Memoirs*, **26**, 49–212.

VAN DER ZWAN, C. J. 1990. Palynostratigraphy and palynofacies reconstruction of the Upper Jurassic to Lowermost Cretaceous of the Draugen Field, offshore Mid-Norway. *Review of Palaeobotany and Palynology*, **62**, 157–187.

VAN WAGONER, J. C., POSAMENTIER, H. W., MITCHUM, R. M. JR, VAIL, P. R., SARG, J. F., LOUTIT, T. S. & HARDENBOL, J. 1988. et al. 1988. An overview of the fundamentals of sequence stratigraphy and key definitions. *In*: WILGUS, C. K. et al. (eds) *Sea-level changes, an integrated approach*. Special Publication of the Society of Economic Paleontologists and Mineralogists, **42**, 39–46.

WALL, D. 1965. Microplankton, pollen and spores from the Lower Jurassic in Britain. *Micropaleontology*, **11**, 151–190.

WATERHOUSE, H. K. 1995. High-resolution palynofacies investigation of Kimmeridgian sedimentary cycles. *In*: HOUSE, M. R. & GALE, A. S. (eds) *Orbital Forcing and Cyclostratigraphy*. Geological Society, London, Special Publications, **85**, 75–114.

WEEDON, G. P. 1986. Hemipelagic shelf sedimentation and climatic cycles: the basal Jurassic (Blue Lias) of South Britain. *Earth and Planetary Science Letters*, **76**, 321–335.

—— & JENKYNS, H. C. 1990. Regular and irregular climatic cycles in the Belemnite Marls (Pliensbachian, Lower Jurassic, Wessex Basin). *Journal of the Geological Society, London*, **147**, 915–918.

WHALLEY, P. E. S. 1985. The systematics and palaeogeography of the Lower Jurassic insects of Dorset, England. *Bulletin British Museum (Natural History) Geology*, **39**, 107–189.

WHITAKER, M. F. 1984. *The Usage of Palynology in Definition of Troll Field Geology*, 6th Offshore Northern Seas Conference and Exhibition, Stavanger, 1984. Paper G6.

WIGNALL, P. B. & HALLAM, A. 1991. Biofacies, stratigraphic distribution and depositional models of British onshore Jurassic black shales. *In*: TYSON, R. V. & PEARSON, T. H. (eds) *Modern and Ancient Continental Shelf Anoxia*. Geological Society, London, Special Publications, **58**, 291–309.

WOODWARD, H. B. 1893. *The Jurassic Rocks of Britain. III. The Lias of England and Wales (Yorkshire Excepted)*. Memoir of the Geological Survey.

WOOLLAM, R. & RIDING, J. B. 1983. *Dinoflagellate cyst zonation of the English Jurassic*. Institute of Geological Sciences Report No. 83/2.

Regional uplift in the English Channel: quantification using sonic velocity

ADAM LAW

BG Exploration and Production Ltd, 100 Thames Valley Park Drive, Reading, Berkshire RG6 1PT, UK
Present address: Amerada Hess Ltd, 33 Grosvenor Place, London SW1X 7HY, UK
(e-mail: a.law@amhess.co.uk)

Abstract: Analysis of the substantial well database held by British Gas Exploration and Production Ltd enables regional estimates of uplift for the English Channel area to be derived from anomalous measurements of downhole sonic velocity. Estimates of section uplift, or apparent erosion, can be derived from variations in sonic velocity/travel-time between wells in a basin relative to a function defining the velocity/travel-time and depth relationship for the basin; the normal compaction relationship. These estimates are, however, extremely sensitive to errors in the normal compaction relationship, and to variations in sedimentary facies and/or diagenesis between wells. In this study, these problems have been minimized using regionally persistent, homogeneous thick facies such as shales and by carefully constraining the normal compaction relationship by 'backstripping' apparent erosion values to determine 'residual apparent erosion'.

Apparent erosion has been determined for three intervals from the Mid- and Upper Jurassic, broadly corresponding to the Fuller's Earth, Oxford Clay and Kimmeridge Clay Formations. Averaging these values and constraining them as above results in robust estimates of total uplift across the English Channel Basin, with uncertainties of around 20%. The results of this study suggest kilometre-scale uplift within the area, with maxima over the Purbeck–Isle of Wight Fault Zone and the Mid Channel High inversion axes. To the north of the Purbeck–Isle of Wight Fault Zone the apparent erosion calculations estimate uplift of 500 m or less, suggesting that this area may have remained relatively stable during basin inversion. When compared to equivalent estimates from AFTA/VR (apatite fission track/vitrinite reflectance) palaeotemperature data, the sonic-derived uplift is similar within errors in inverted areas. However, in more stable areas such as to the north of the Purbeck–Isle of Wight Fault Zone, calculated apparent erosion indicates that little uplift has occurred, whereas uplift estimates derived from AFTA/VR models are still significant.

The tectonically complex history of the English Channel Basin has been the major control on the success of hydrocarbon exploration within the area (see Buchanan 1997 this volume). In general, the magnitude of the well-documented Late Cretaceous–Tertiary basin inversion episode is poorly understood. One method of quantifying uplift and/or erosion of section within a semi-mature basin, such as the English Channel area, is to study variations in the interval transit-time/velocity between wells relative to a derived regional velocity–depth relationship, the *normal compaction relationship*, to derive estimates of uplift or *apparent erosion* for that well (e.g. Hillis et al. 1994; Menpes & Hillis 1995).

Using the extensive well database held by BG Exploration and Production, estimates of uplift/erosion have been determined for the Portland–Wight Basin (after Smith & Curry 1975) and the southern Wessex Basin using this method. Over 45 wells have been used in this study, 26 of which are in the public domain and are presented here (Table 1). Owing to the lack of Cretaceous facies within many of the wells used estimates are of total uplift of section within the basin, and are derived using the regionally homogeneous Upper and Mid-Jurassic shales and clays of the Kimmeridge Clay Formation, the Oxford Clay Formation and the Fuller's Earth Formation. The Lower Jurassic lithofacies are too heterogeneous to provide meaningful uplift estimates using this method.

Estimating uplift using sonic velocity–apparent erosion

Lateral variations in sonic velocity or travel-time within regionally homogeneous facies have been widely used to estimate the amount of erosion or missing section within sedimentary basins. Shales have traditionally been used (Marie 1975; Cope 1986; Bulat & Stoker 1987), although recently workers have argued that other facies can also

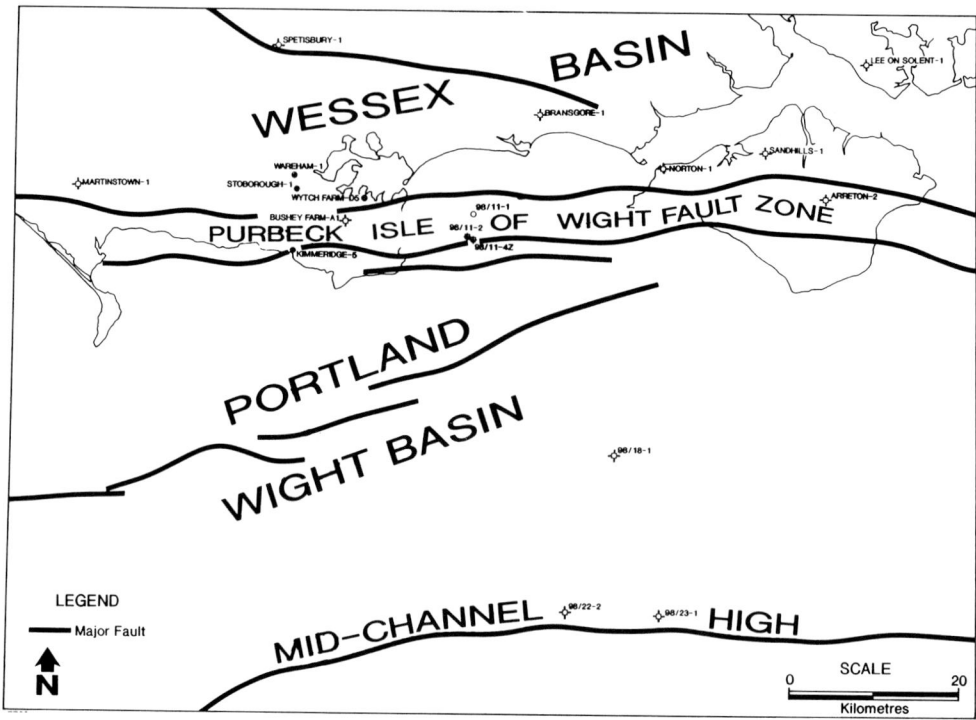

Fig. 1. Location of study area.

give valid results, for instance the Cretaceous Chalk of the United Kingdom Continental Shelf (Hillis *et al.* 1994; Menpes & Hillis 1995).

For a given study interval, a relationship can be derived relating the effect of burial on interval transit-time/velocity: *the normal compaction relationship*. Although the loss of porosity with depth has been shown to be logarithmic (Magara 1976), a linear relationship between interval velocity and depth is found to yield equivalent results over the formation mid-point depth range commonly used in these studies (Bulat & Stoker 1987). Theoretically, overcompaction or uplift of a particular formation is therefore represented by interval velocities that show anomalously high values relative to this normal compaction relationship.

Methodology

A method similar to Menpes & Hillis (1995) has been used in this study to estimate apparent erosion using wells from the southern Wessex Basin and the UK sector of the Central English Channel Basin: the Portland–Wight Basin (Fig. 1). As discussed above, the Cretaceous and Tertiary sections are poorly preserved or missing from most of the wells used. The study was therefore focused on the regionally persistent shale–clay-dominated facies of the Jurassic section. Each well was first subdivided on the basis of the main formation tops of the Jurassic used in the basin; from Top Kimmeridge Clay to Top Corallian, corresponding broadly to the Kimmeridge Clay Formation, the Top Corallian to Top Cornbrash, corresponding to the Oxford Clay Formation, and the Top Cornbrash to Top Inferior Oolite, corresponding to the Fuller's Earth Formation. The Lower Jurassic (Top Inferior Oolite to Top Rhaetic) interval was also used to examine the effect of large-scale facies variations on apparent erosion estimates, but could not be used to determine meaningful estimates of uplift/erosion within errors. These results are, however, presented here for completeness (Table 1).

For each of these intervals, the average sonic velocity was computed and plotted against

Table 1. *Mid-point depths, interval velocities, apparent erosion estimates and normal compaction relationships for the public domain wells used in the study.**

Well	Top Kimmeridge Clay to Top Corallian			Top Corallian to Top Cornbrash			Top Cornbrash to Top Inferior Oolite			Top Inferior Oolite to Top Rhaetic			Cumulative E_a (m)	Jurassic Error (m)	Missing section (m)
	Mid-point depth (m)	Interval velocity (m s^{-1})	E_a (m)	Mid-point depth (m)	Interval velocity (m s^{-1})	E_a (m)	Mid-point depth (m)	Interval velocity (m s^{-1})	E_a (m)	Mid-point depth (m)	Interval velocity (m s^{-1})	E_a (m)			
98/11-1	826	2659	1042	964	3029	1083	1088	3396	663	1287	3574	1008	930	189	808
98/11-2				671	2859	1074	908	3614	1253	1231	3805	1656	1163	89	1676
98/11-3	822	2361	440							1247	3319	397	440		807
98/11-4z	835	2655	1026				1006	2980		1337	4101	2307	1026		747
98/16-1	877	2999	1687	1022	3150	1241	1254	3862	1372	1665	4110	2005	1306	66	1494
98/18-1	787	2575	912	1027	2914	815	1268	3487	654	1534	3964	1762	794	106	880
98/22-2				86	1928	0	286	3093	899	543	3566	1733	899		1450
98/23-1	56	1771	0	221	2990	1756	462	3714	1885	747	4263	3314	1821	65	1661
99/12-1				810	2496	287							287		
99/16-1	1045	2913	1343	1271	3104	909	1473	3802	1040	1665	4493	2986	974	66	896
99/18-1	807	2631	1005	930	3034	1125	1045	3464	835				988	119	1082
Arreton-2	950	3382	2396	1229	3604	1841	1405	4498	2412	1706	4777	3672	2216	265	1220
Bushey Farm-A1	615			721	2446	287	910	3162	404	1188	3440	765	345	58	
Bransgore-1	766	2655	1095	872	2468	177	1032	3392	713	1195	3327	469	445	268	
Kimmeridge-5				246			653	3180	695	1393	4381	2969	695		1200
Hurn-1	637	2056	0	744	2407	195	918	3063	210	1148	3126	0	202	7	
Lee-on-Solent-1	816	2619	972	994	2817	674	1157	3740	1239	1399	3661	1119	962	231	
Martinstown-1				21			168	2742	358	571	2901	0	358		
Norton-1	918	3023		1030	2458	0	1165	3584	939	1280	3648	1205	469	103	
Sandhills-1				1016	2873	753	1105	3562	958	1190	3658	1321	856	54	365
Spetisbury-1	296			446	2389	460	666	3120	568	1052	3286	505	514	259	
Stoborough-1				610	2365	254	827	3314	772				513		
Waddock Cross-1				363	2590	902	506	2731		815	2861		902		
Wareham-1				579	2345	250	758	2972	229	1034	3195	291	240	9	
Wytch Farm-D5				576	2506	539	740	2890	64				302	237	
Winterbourne Kingston-1	288	2696		477	2230	147	722	2919	136	1176	3410	699	141	6	420
Normal compaction gradient (s^{-1})	0.4897			0.5613			0.5340			0.3904					
Surface intercept QC (m s^{-1})	1743			1880			2447			2678					

* Results are shown as mid-point depth, interval velocity and estimated apparent erosion (E_a) for the three study intervals (where present): Top Kimmeridge Clay to Top Corallian, Top Corallian to Top Cornbrash; and Top Cornbrash to Top Inferior Oolite. The average of these results (cumulative Jurassic E_a) provides the results discussed in the text. The results for the interval between Top Inferior Oolite and Top Rhaetic are included for completeness. A column containing estimates of missing section is also presented for comparison.

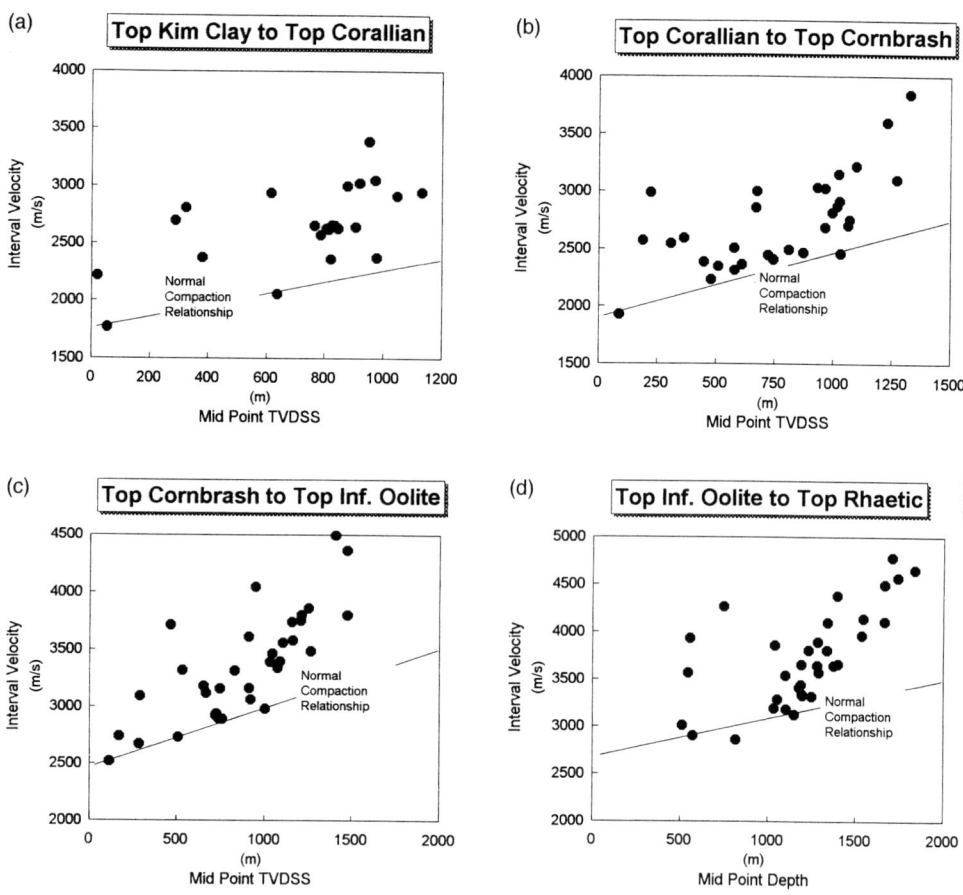

Fig. 2. Interval velocity (m s^{-1}) vs. mid-point depth (m TVDSS) for study wells. (**a**) Top Kimmeridge Clay to Top Corallian; (**b**) Top Corallian to Top Cornbrash; (**c**) Top Cornbash to Top Inferior Oolite; (**d**) Top Inferior Oolite to Top Rhaetic.

mid-point depth (Fig. 2). A normal compaction relationship was derived for each interval (Table 1) by finding the best-fit linear function that passed through those wells that showed the lowest interval velocity for a given burial depth – *the reference wells*. The amount of apparent erosion (E_a) could be determined within any given well relative to this normal compaction relationship, and therefore relative to the reference wells, using the equation:

$$E_a = \frac{1}{m}(V_i - V_o) - z_i,$$

where m is the gradient of normal compaction line; V_o is the surface intercept of the normal compaction relationship; V_i is the interval velocity of the interval under consideration; and z_i is the mid-point depth of interval under consideration.

Limitations and error constraints

Apparent erosion studies can be subject to large and potentially unmanageable errors, which must be understood and constrained as much as possible for such a study to be successful. The three main sources of error in such estimates result from the reference wells being at some elevation above their maximum burial depth, from variations in diagenesis or facies type between wells, and from statistical errors in the derivation of the normal compaction relationship.

Reference wells and maximum burial depth

Apparent erosion calculated in this manner is determined relative to the normal compaction relationship derived from a series of reference wells. If these reference wells are not at their maximum burial depth, then the calculations will underestimate the amount of uplift/erosion. To provide an absolute calibration of apparent erosion estimates, uplift/erosion of the reference wells can be examined using estimates of missing section derived from regional geological knowledge, or by calibrating the results using other estimates, for instance using AFTA or vitrinite reflectance (see Bray et al. this volume). The implications of reference well uplift to this study are discussed later.

Normal compaction relationship

In this and other studies, the intercept velocity of the normal compaction relationship, that is, the velocity predicted by the normal compaction relationship at zero depth, has been constrained between around 1500 and 3000 m s^{-1}. An error in the gradient of the normal compaction relationship of around ±10% results in an equivalent error in apparent erosion values of between ±5% and ±20%. This, however, only results in a small error in the intercept or surface velocity, well within the limits imposed on the intercept velocity discussed above. Figure 3 shows the effect this variation has on the values of apparent erosion calculated for the Fuller's Earth Formation. Here, an error of 10% has been built into the normal compaction relationship, resulting in an average error in apparent erosion of around 12%. However, the error induced in the intercept velocity is only around 25 m s^{-1}, or around 1%. Thus, the use of intercept velocities to constrain the normal compaction relationship is a relatively insensitive control.

The normal compaction relationship may be better constrained by using residual apparent erosion. In general, geologically older intervals should not show a decrease in apparent erosion relative to formations above. Thus, if apparent erosion values from younger intervals are sub-tracted, the residual apparent erosion should not be negative (Cope 1986). Figure 3 shows an example of how this can be used. In this study, two normal compaction relationships can be derived for the Lower Jurassic interval. Both give valid intercept velocities within the constraints used here (Fig. 4a), however, the difference in the estimated apparent erosion values calculated relative to the two possible normal compaction relationships are around 50%. Calculating the residual apparent erosion for this interval relative to the Upper and Mid Jurassic average results in a considerable number of negative residuals if well 'A' is used as a reference well, implying that well 'B' is a better choice as reference well (Fig. 4b).

Using this concept to constrain the normal compaction relationship for each interval used in the study results in an average statistical error of around 20% in the estimated apparent erosion values, with a resolution limit of around 200 m

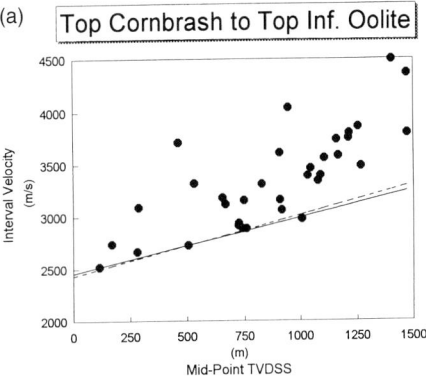

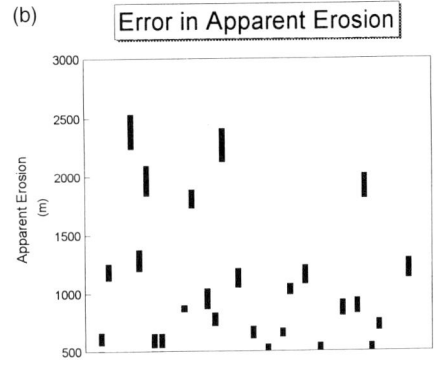

Fig. 3. The effect of intercept velocity on apparent erosion estimates. (**a**) Derivation of the normal compaction relationship. An error has been introduced in the gradient of the compaction relationship to give a 10% error in intercept velocity. The original compaction relationship (solid line) has an intercept velocity of 2460 m s^{-1}, whereas the erroneous relationship (broken line) has an intercept velocity of 2435 m s^{-1}, both within the range of intercept velocities used as a constraint. (**b**) The error effects the resulting apparent erosion estimates for this interval.

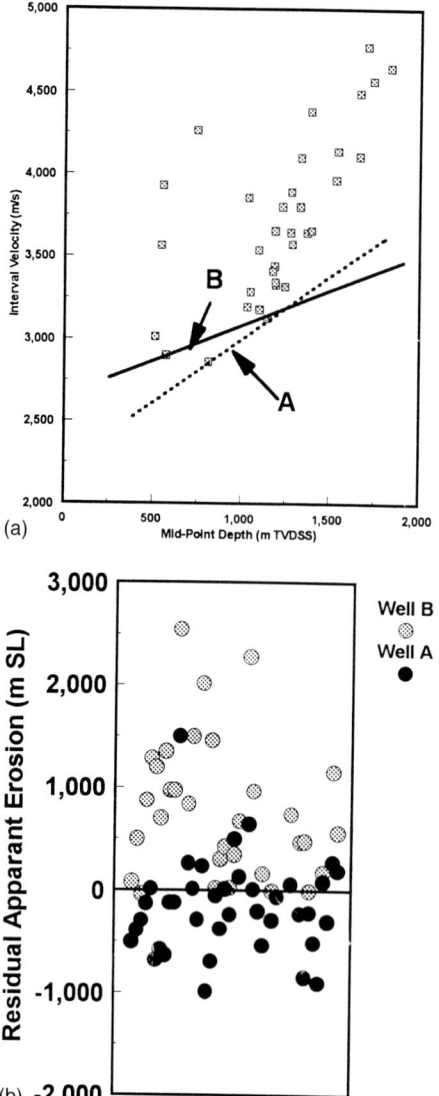

Fig. 4. Using residual apparent erosion as a control on the normal compaction relationship. (**a**) Two possible normal compaction relationships for the Lower Jurassic, derived using the well 'A' (broken line) and the well 'B' (solid line). Their respective normal compaction relationships are shown in boxes. Each has an equally valid intercept velocity within the constraints used in this study (1500–3000 m s^{-1}). (**b**) Residual apparent erosion between the Lower and Upper to Mid-Jurassic intervals for each well. Using the well 'A' to define the normal compaction relationship for the Lower Jurassic results in negative residuals for most wells. However, the normal compaction relationship derived using well 'B' results in few negative residual. Well 'B' is therefore a better choice as a reference well to define the normal compaction relationship for this interval.

below which the associated velocity anomalies are less than the error in the normal compaction relationship.

The effects of facies and diagenetic variation

The effect of sedimentary facies variation on apparent erosion estimates can also be examined by calculating residual apparent erosion values. By removing the apparent erosion calculated for the Upper to Mid-Jurassic interval, the Lower Jurassic interval shows large residual apparent erosion values where wells contain a high proportion of Bridport Sandstone Equivalent limestone facies (solid circles, Fig. 6), and smaller but significant values where the Bridport Sandstone is thick (solid squares, Fig. 6). Excluding these values, residual apparent erosion within the Lower Jurassic, that is, uplift before the deposition of the Inferior Oolite or equivalent, appears to be less than 700 m.

But does this have a significant effect on the Upper and Mid-Jurassic intervals used, where facies were assumed to be regionally homogeneous and persistent? This can be studied using cross-plots of apparent erosion estimates from geological intervals within each of the wells used (Fig. 5). Where facies are regionally persistent, adjacent intervals should have a high degree of correlation within a well if apparent erosion has been calculated successfully, assuming there are no major erosive episodes present between intervals. In this study, cross-plots of the three intervals within the Upper to Mid-Jurassic give regression coefficients of around 0.7. However, the variation in facies type within the Lower Jurassic interval results in a regression coefficient of 0.5 when correlated with apparent erosion values derived from the Upper to Mid-Jurassic.

Results

The apparent erosion values calculated here provide estimates of the *total* present-day uplift within the southern Wessex Basin and the Portland–Wight Basin, due to the lack of regionally extensive and homogeneous Cretaceous and Tertiary facies within many of the study wells used. These estimates are relative to the burial depth of the reference wells used to define the normal compaction relationships for each interval, and are therefore relative to any uplift or erosion that has occurred in the area of the

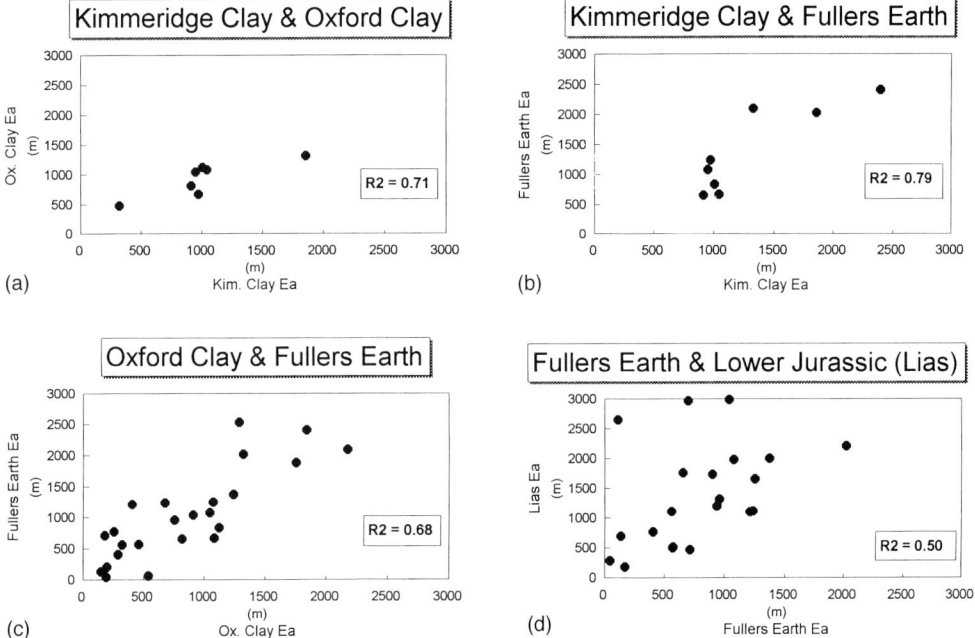

Fig. 5. Cross-plots of apparent erosion for Jurassic intervals. Regression coefficients are shown in boxes. (**a**) Kimmeridge Clay Formation (Top Kimmeridge Clay to Top Corallian) vs. Oxford Clay Formation (Top Corallian to Top Cornbrash); (**b**) Kimmeridge Clay Formation (Top Kimmeridge Clay to Top Corallian) vs. Fuller's Earth Formation (Top Cornbrash to Top Inferior Oolite); (**c**) Oxford Clay Formation (Top Corallian to Top Cornbrash) vs. Fuller's Earth Formation (Top Cornbrash to Top Inferior Oolite); (**d**) Fuller's Earth Formation (Top Cornbrash to Top Inferior Oolite) and Lower Jurassic (Lias).

reference wells. To increase the robustness of the results and statistically quantify errors, the three main Upper and Mid-Jurassic intervals used have been averaged to provide one combined estimate of apparent erosion (Table 1). This further reduces the effect of facies variation. As discussed previously, the results from the Lower Jurassic interval have not been used in this averaging, due to the heterogeneous nature of the facies within this interval.

Table 1 summarizes the results from the Upper and Mid-Jurassic sections for the released wells used in the study.

These results have been combined with results from 19 unreleased wells to produce a contour map of the estimated total present day uplift (Fig. 7). To the north of the Purbeck–Isle of Wight Fault Zone, here termed the 'Wytch Farm area', the combined apparent erosion estimates give values of 500 m or less, approaching the resolution limits of the method, suggesting that this area has undergone little inversion. However, many of the reference wells used for the study are located within this area, indicating that this method will not be able to resolve any inversion that the area has undergone. Estimates of missing section from the composite logs where derived (Table 1) imply around 500 m or less has been eroded from the wells within the 'Wytch Farm area', suggesting that the results may be underestimating by around 200–300 m. The total estimated uplift increases from this area eastwards to around 1000 m at Lee-on-Solent-1.

The maximum apparent erosion estimates come from wells within the Purbeck–Isle of Wight Fault Zone, drilled on the main inversion axis of the basin. Estimates of total present-day uplift range from 600 m at Bushey Farm-A1, increasing rapidly eastwards to approximately 1100 m in the 98/11-2 well, and reaching a maximum of around 2000 m at Arreton-2, although estimates of missing section indicate that this may be an overestimate.

Well information is sparse within the Portland Wight Basin, to the south of the Purbeck–Isle of Wight Fault Zone. However, regional estimates of apparent erosion can still be

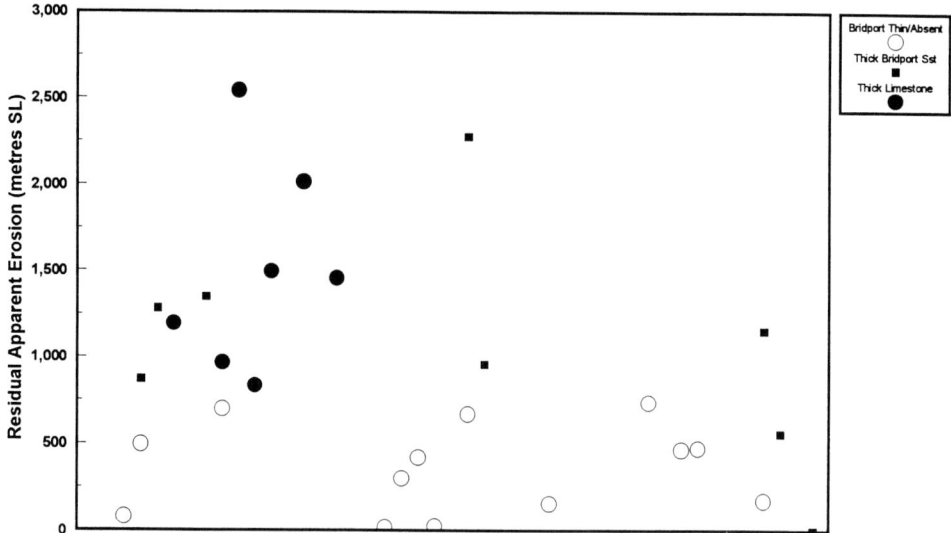

Fig. 6. The effects of facies variation on apparent erosion – Lower Jurassic residual apparent erosion. Significant positive residuals result if residual apparent erosion is calculated for the Lower Jurassic interval. However, the anomalous values are primarily the result of facies variations. Large values result where the Bridport Sandstone Formation equivalent limestone is present within the well (solid circles), and smaller but still significant anomalous values where the Bridport Sandstone is thick (solid squares). Removing these points shows that the residual apparent erosion within the Lower Jurassic is less than 700 m.

derived. Immediately south of the Purbeck–Isle of Wight Fault Zone, the estimated total present-day uplift drops to less than 1000 m at Kimmeridge-5 and the 98/11 wells to the south of the fault system. Uplift/erosion appears to remain constant southward to 98/18-1 (900 m), and increases to a maximum of over 1000 m at the Mid Channel High (98/22-2 and 98/23-1).

Comparison with other estimates of uplift and erosion

Several other methods also exist for the estimation of uplift–erosion within a basin, including the estimates of missing section discussed above. Using data from 10 of the wells in this study area, Geotrack International Pty Ltd (see Bray *et al.* 1997 this volume) have determined absolute values of uplift using AFTA/VR data. Uplift/erosion of missing section can be estimated from palaeotemperature profiles derived from AFTA/VR data. An estimate of uplift is determined by extrapolating the derived palaeogeothermal gradient from the well to an assumed palaeosurface intercept, assuming a linear relationship between palaeotemperature and burial depth. The maximum resolution of palaeotemperature for such studies is commonly 10°C, giving a corresponding maximum resolution of around ±300 m for typical geothermal gradients. The Geotrack study was limited by the lack of suitable lithologies within wells for the collection of apatite samples (commonly restricted to Bridport Sandstone Formation and the Sherwood Sandstone Group) and the possible geochemical suppression of vitrinite reflectance within organic-rich shales. However, AFTA/VR results can provide an absolute value of uplift and in this case can also subdivide uplift episodes in geological time.

The estimates of total present-day uplift derived using AFTA/VR data are generally greater than the apparent erosion estimates derived in this study (Fig. 8), again suggesting that the reference wells used to derive normal compaction relationships may not be at their maximum burial depth. This difference is less than 500 m over most of the study area, with the estimates from wells 98/11-2 and 98/23-1 agreeing within errors.

Wells within the 'Wytch Farm area' (Kimmeridge-5, Wytch Farm-D5 (AFTA/VR is a Wytch Farm composite) and Winterbourne Kingston-1) have a much larger difference

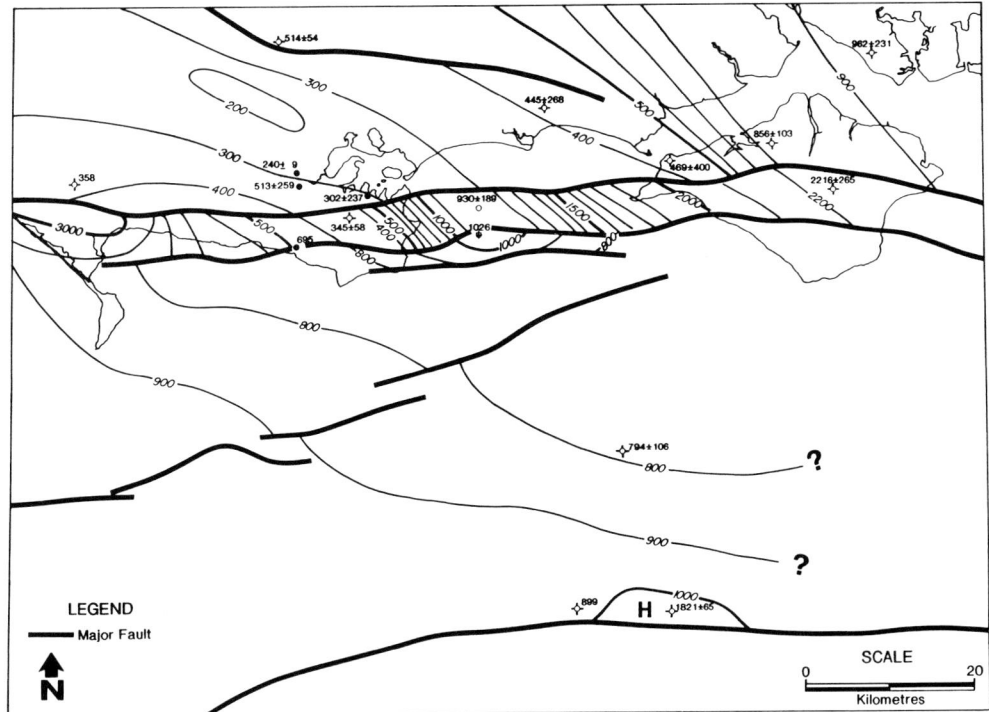

Fig. 7. Map of Upper and Mid-Jurassic apparent erosion for the study area. Generalized major faults have been used to constrain contouring, along with regional seismic interpretation where available. Data points are from wells in Fig. 1.

between AFTA/VR and sonic velocity derived estimates of uplift and erosion, on average 1200 m. The AFTA/VR derived estimates suggest that 1000–2000 m of uplift and erosion has occurred in an area which seems to have remained relatively stable through geological time. Estimates of missing section for Kimmeridge-5 and Winterbourne Kingston-1 (Table 1) agree more with the apparent erosion estimates from sonic velocity, at around 1200 and 400 m, respectively. Although the wells in the Wytch Farm area are used as reference wells for the apparent erosion estimates presented here, and are therefore assumed to have undergone little or no uplift, it is apparent that the AFTA/VR estimates are significantly greater than both estimates of missing section and of apparent erosion from sonic velocities throughout the basin. The AFTA/VR estimates from this area could possibly be reflecting a component of non-burial related heating, such as hot fluid flow through the sampled Bridport Sandstone Formation and the Sherwood Sandstone Group, thus be significantly overestimating uplift/erosion (Bray et al. 1997 this volume). This could also be a possible explanation for he disparity between estimates for the 98/18-1 well.

Conclusions

Borehole measured sonic velocity has been used here to derive estimates of uplift/erosion for wells within the southern Wessex Basin and the Portland–Wight Basin, using the concept of apparent erosion. By using residual apparent erosion as well as surface intercept velocity to constrain the normal compaction relationship, and by statistically averaging the results over several intervals of regionally homogeneous facies, statistical errors are limited to an average of ±20%. The minimum resolution limit is around 200 m, below which it is impossible to resolve velocity anomalies relative to errors in the normal compaction relationship. The estimates are derived relative to the burial depth of the reference wells that define the normal

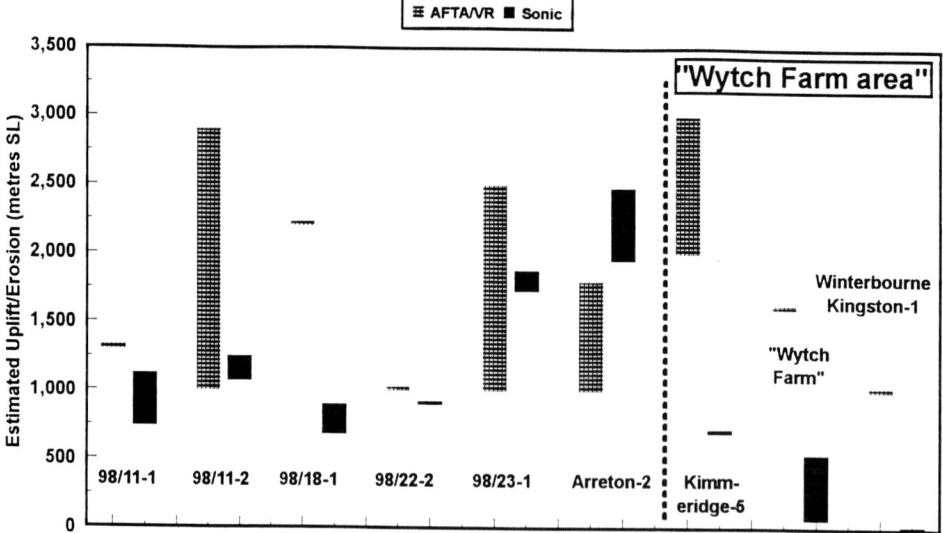

Fig. 8. Comparison of AFTA/VR and apparent erosion derived uplift–erosion estimates for study wells (AFTA/VR data courtesy of Geotrack International Pty Ltd). Results are shown in pairs with error ranges for each selected well. In all cases, the AFTA/VR estimates form the left-hand column, with the sonic velocity derived estimates in the right-hand column.

compaction relationship for each interval. The apparent erosion calculations will therefore underestimate uplift/erosion should these wells not be at their maximum burial depth within the basin.

Owing to the lack of Cretaceous and Tertiary facies within many of the study wells, estimates of the total uplift/erosion have been derived using this method for the Kimmeridge Clay Formation, Oxford Clay Formation and Fuller's Earth Formation. These were then averaged to provide one combined estimate of total present-day uplift/erosion per well. The results show that the area to the north of the Purbeck–Isle of Wight Fault Zone may have remained a relatively stable block, with estimates of total present-day uplift of 500 m or less. The largest estimates of inversion are from within the Purbeck–Isle of Wight Fault Zone itself, and vary from 1000–2000 m. Around 800–900 m of uplift is estimated from within the Portland–Wight Basin where wells are available, increasing to the north to over 1000 m along the Mid Channel High inversion axis.

The apparent erosion estimates compare well with estimates of missing section derived for the wells, although in the 'Wytch Farm area' missing section estimates imply that the apparent erosion calculations are underestimating uplift/erosion by around 200–300 m, suggesting that the reference wells may not be at their maximum burial depth. However, the estimates of uplift/erosion derived from AFTA palaeo-temperature estimates are significantly greater than the apparent erosion estimates, by on average 1200 m in the 'Wytch Farm area' within errors. One possibility is that the reference wells used in the apparent erosion calculations may not be at maximum burial. However, another explanation for these significant differences may be that the AFTA palaeotemperature estimates in the 'Wytch Farm area' show a component of non-burial related heating, possibly as a result of hot fluid flow through the Bridport Sandstone Formation and the Sherwood Sandstone Group sample intervals. When combined, all the estimates are consistent with kilometre-scale uplift/erosion within the area.

Apparent erosion is only one parameter that can be used to describe basin evolution. There is scope for these values of minimum erosion to be tied to and checked against the more absolute values derived using AFTA/VR data and other independent methods. As sonic information is generally more widely available for drilled wells, a few well chosen AFTA or VR points should be enough to calibrate apparent erosion estimates to an absolute depth, enabling erosion or maximum burial depths to be quickly and cheaply extrapolated across a basin.

The author would like to thank BG Exploration and Production for permission to publish the results of this study, and also the many colleagues and friends at BG who assisted in the preparation and review of this manuscript. Geotrack International Pty Ltd also provided considerable assistance, helpful discussion and also gave permission to publish their results of their work in the English Channel area, for which the author would like to express his thanks. However, the author takes full responsibility for any errors and inconsistencies within these data.

References

BRAY, R. J., DUDDY, I. R. & GREEN, P. F. 1997. Multiple heating episodes in the Wessex Basin: implications for geological evolution and hydrocarbon generation. *This volume*.

BUCHANAN, J. B. 1997. The exploration history and controls on hydrocarbon prospectivity in the Wessex basins, southern England, UK. *This volume*.

BULAT, J. & STOKER, S. J. 1987. Uplift determination from interval velocity studies. *In:* BROOKS, J. & GLENNIE, K. (eds) *Petroleum Geology of North West Europe*. Graham and Trotman, London, 293–305.

COPE, M. J. 1986. An interpretation of vitrinite reflectance data from the Southern North Sea Basin. *In:* BROOKS, J., GOFF, J. & VAN HOORNE, B. (eds) *Habitat of Palaeozoic Gas in, N.W. Europe*. Geological Society, London, Special Publications, **23**, 85–100.

HILLIS, R. R., THOMPSON, K. & UNDERHILL, J. R. 1994. Quantification of Tertiary erosion in the Inner Moray Firth using sonic velocity data from the Chalk and the Kimmeridge Clay. *Marine and Petroleum Geology*, **11**, 283–293.

MAGARA, K. 1976. Thickness of removed sedimentary rocks, palaeopore pressure, and palaeotemperature, southwestern part of the Western Canada Basin. *Bulletin of the American Association of Petroleum Geologists*, **60**, 554–565.

MARIE, J. P. P. 1975. Rotliegendes stratigraphy and diagenesis. *In:* WOODLAND, A. (ed.) *Petroleum and the Continental Shelf of NW Europe*. Institute of Petroleum, London, 205–210.

MENPES, R. J. & HILLIS, R. R. 1995. Quantification of Tertiary exhumation from sonic velocity data, Celtic Sea/South-Western Approaches. *In:* BUCHANAN, J. G. & BUCHANAN, P. J. (eds) *Basin Inversion*. Geological Society, London, Special Publications, **88**, 191–207.

SMITH, A. J. & CURRY, D. 1975. The structure and evolution of the English Channel. *Philosophical Transactions of the Royal Society of London*, **A279**, 3–20.

Multiple heating episodes in the Wessex Basin: implications for geological evolution and hydrocarbon generation

RICHARD J. BRAY,[1] IAN R. DUDDY[2] & PAUL F. GREEN[2]

[1] *Geotrack International UK Office, 5 The Linen Yard South Street, Crewkerne, Somerset TA18 8AB, UK*
[2] *Geotrack International Pty Ltd, 37 Melville Road, Brunswick West, Victoria 3055, Australia*

Abstract: The timing of maturity development and hydrocarbon generation with respect to structuring is a primary factor controlling oil and gas accumulation. Hydrocarbon generation is largely a temperature-dependent process and in order to predict accumulation it is necessary to reconstruct thermal history. In this study of the Wessex Basin and adjacent areas, direct thermal history data are obtained from Apatite Fission Track Analysis (AFTA) on samples from wells, shallow boreholes and outcrops, enabling the key facets of a complex thermal and tectonic history to be reconstructed. Several distinct styles of thermal history are recognized, with clear differences across the region in the time of cooling from maximum palaeotemperatures. Areas of east Devon show evidence of an early thermal episode, sometime after the mid-Triassic and before ~170 Ma. The cause remains uncertain at present. In the area of the 'Wytch Farm block' maximum palaeotemperatures were reached prior to cooling in the early Cretaceous, in the range ~140–100 Ma, due mainly to uplift and erosion. This area remained relatively high through the rest of its history compared to the areas to the north and south. The Lias (early Jurassic) was not heated sufficiently to cause significant generation and any that occurred effectively ceased in the early Cretaceous. South of the Portland–Isle of Wight faults, maximum palaeotemperatures were reached prior to cooling due to uplift and erosion in the mid- and late Tertiary, at ~40 Ma and ~20 Ma, respectively. Tertiary uplift and erosion affected a wide region. Maximum erosion occurred immediately to the south of the Purbeck structure where maturation of Lias source rocks continued through peak oil and into the gas generation phase until terminated by the Tertiary cooling. In most wells, AFTA and vitrinite reflectance data define low or non-linear palaeogeothermal gradients indicative of an element of heating due to fluid movement in addition to heating due to burial, making quantitative estimation of erosion difficult, but any reasonable model requires kilometre-scale uplift and erosion over a wide region. West of Portland in Lyme Bay, a key question is whether maturation history is dominated by early Cretaceous or by mid- to late Tertiary events. At present there are not sufficient data to extrapolate confidently from the area of well control east of Portland. No parts of the basin are currently generating hydrocarbons, implying that surface seepages are from previously reservoired accumulations.

The rocks of the Wessex Basin have been subject to detailed study since the 18th century. The coastal outcrop provides exposure of all the main stratigraphic divisions from Palaeozoic basement to Oligocene. This, combined with the more recent stratigraphic data from exploration wells and seismic from both onshore and offshore, has allowed numerous models of basin subsidence, uplift and geological evolution to be constructed (e.g. Buchanan 1997 this volume; McMahon & Turner 1997 this volume). The clear success of some of these models is illustrated by the history of exploration for oil at Wytch Farm (Colter & Harvard 1981; Underhill & Stoneley this volume). The problems of accurately predicting hydrocarbon accumulation in the basin are equally well demonstrated by the number of dry holes, most located on surface features, drilled in the period leading up to the Wytch Farm discovery. A relatively small number of exploration wells have been drilled since then without notable success. At the time of writing, despite the numerous natural seepages of oil and gas, the Wytch Farm field remains the only large discovery in the basin.

The timing of the development of maturity and hydrocarbon generation with respect to the time of structuring and trap formation is a primary factor controlling oil and gas accumulation in the Wessex Basin. Hydrocarbon generation is largely a temperature-dependent process and in order to predict whether hydrocarbon accumulations will exist in potential traps it is

BRAY, R. J., DUDDY, I. R. & GREEN, P. F. 1998. Multiple heating episodes in the Wessex Basin: implications for geological evolution and hydrocarbon generation. *In*: UNDERHILL, J. R. (ed.) *Development, Evolution and Petroleum Geology of the Wessex Basin*, Geological Society, London, Special Publications, **133**, 199–213.

necessary to reconstruct thermal history as accurately and precisely as possible.

It is now well established that the Wessex Basin has undergone several episodes of subsidence and tectonism resulting in uplift and erosion. In basins with such complex histories it is very difficult, if not impossible, to confidently reconstruct the development of palaeotemperature and maturity by the traditional methods of burial history reconstruction based on the preserved stratigraphic section combined with theoretical models of palaeo-heat flow. In such settings direct data on thermal history can, however, be obtained from Apatite Fission Track Analysis (AFTA).

In sediments that have been hotter in the past AFTA can identify the timing and magnitude of thermal maxima, and provide qualitative determination of maximum and subsequent peak palaeotemperatures, thus allowing the key facets of thermal history to be reconstructed. Palaeotemperature estimates from several AFTA data points in a vertical sequence from a well can define a palaeotemperature profile which gives unique insight to the causes of heating and cooling in discrete palaeothermal episodes. For example, palaeotemperature profiles can be inspected to provide estimates of palaeogeothermal gradients and the amount of section removed by uplift and erosion, or alternatively to recognize episodes of high heat flow and heating due to fluid movement or igneous activity (Bray *et al.* 1991, Duddy *et al.* 1994).

Vitrinite reflectance (VR) can also provide quantitative estimates of maximum palaeotemperatures. Once the timing of maximum palaeotemperatures is determined from AFTA, VR values can be used to give independent estimates of maximum palaeotemperature, providing the potential to greatly increase the number of data points in a vertical sequence. Combination of AFTA with VR thus brings a number of advantages in thermal history reconstruction.

This paper presents a review of new thermal history data from the Wessex Basin. These comprise AFTA and VR measurements from onshore and offshore wells, shallow offshore boreholes and outcrop samples from south Devon and Dorset (Fig. 1). These data have enabled the main features of the thermal history to be reconstructed. Several episodes of heating and cooling through the basin's history are revealed, some of which can be linked to the tectonic events recognized or speculated upon by previous workers.

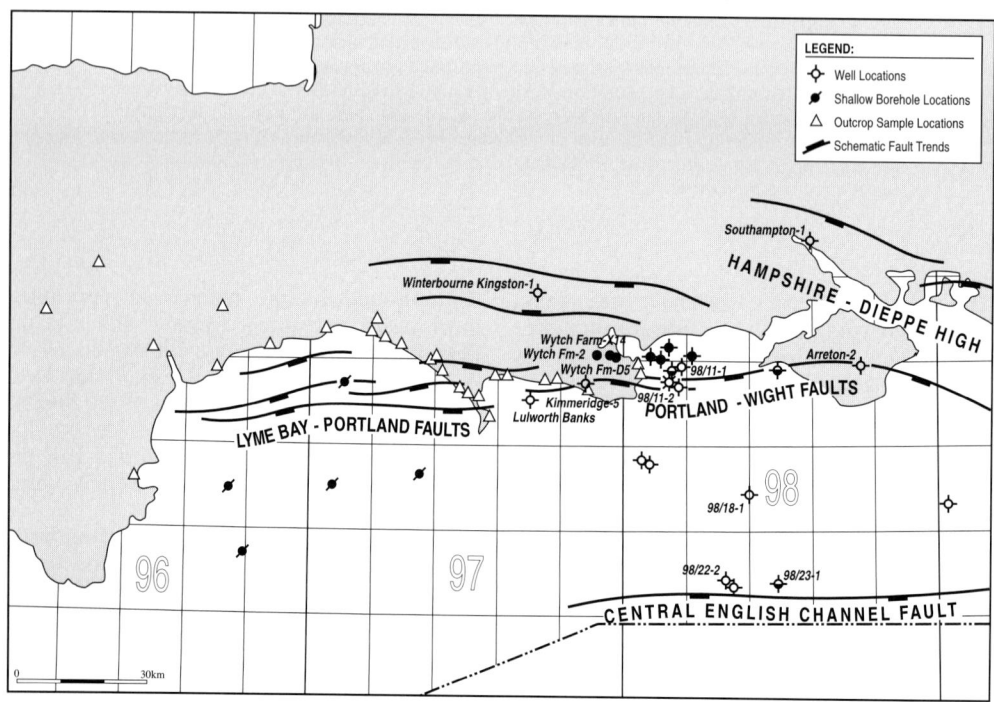

Fig. 1. Location map.

The timing of the key palaeothermal episodes affecting the Wessex Basin

The AFTA and VR data show that all samples examined from wells, shallow boreholes and outcrop have experienced significantly higher temperatures at some time in the geological past. Several styles of thermal history can be recognized, with clear differences across the region in the time at which the Mesozoic section began to cool from maximum palaeotemperatures.

Evidence for five distinct episodes of heating and cooling emerge from regional synthesis of the data: mid-Triassic–early Jurassic; early Cretaceous; late Cretaceous–early Tertiary; mid-Tertiary; and late Tertiary. For each of these, the range of geological time in which the cooling commenced is constrained directly by AFTA.

It is important to note that any particular point in the study area could have experienced each of the above heating episodes, with the timing of cooling from *maximum* palaeotemperatures varying across the region. The kinetic responses of thermal indicators such as AFTA and VR are dominated by the effects of maximum temperature, so that earlier, lesser temperature peaks are obscured by subsequent maxima. For example, the early Cretaceous episode could to some extent have affected all the samples that show *maximum* palaeotemperatures in the Tertiary.

Conversely, AFTA can identify palaeotemperature peaks after the maximum. For example, over the Wytch Farm field where maximum palaeotemperatures occurred in the early Cretaceous, a later palaeotemperature peak in the Tertiary can also be recognized and measured.

The early event (?mid-Triassic–early Jurassic)

An early episode of cooling from maximum palaeotemperatures in the range 65–95°C is recognized in outcropping Permian and Triassic rocks at the western end of study area, from south and southeast Devon (Fig. 2). Direct evidence for this early event comes from the Permian Breccias just west of Crediton and from the Triassic Sherwood Sandstone at Ladram Bay. AFTA data from these rocks show that cooling from maximum palaeotemperatures began at some time before ~170 Ma.

Samples of the Sherwood Sandstone from Ottery St. Mary and the Permian Breccias from the vicinity of Exeter and Paignton show cooling

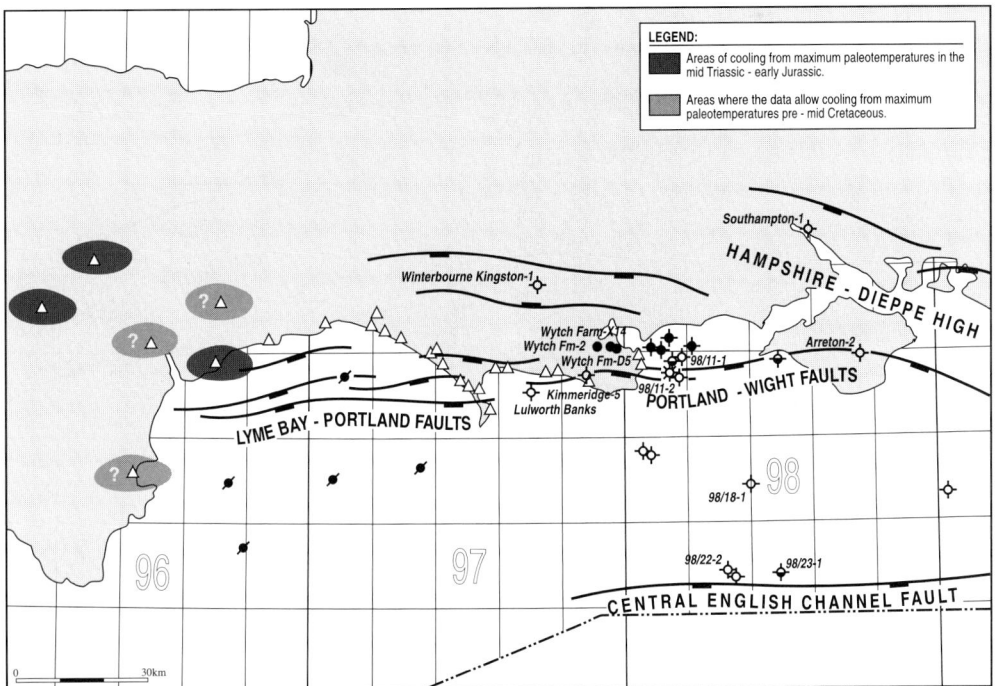

Fig. 2. Areas showing cooling from maximum palaeotemperatures in the mid-Triassic to early Jurassic event.

from maximum palaeotemperatures of >110°C began prior to 110, 100 and 120 Ma, respectively. These could, therefore, have commenced cooling from their maximum palaeotemperatures during the early episode, or some later episode in the Jurassic or early Cretaceous, and indeed the data show that these samples were still at relatively high palaeotemperatures in the Tertiary. However, as the data are consistent with cooling during the early episode revealed by the Crediton and Ladram Bay samples, it is concluded that the Sherwood Sandstone and older samples from the western margin of the basin cooled from their maximum palaeotemperatures during an early episode post-mid-Triassic and pre ~170 Ma (early Jurassic).

The cause of this early episode remains uncertain at present. Regional uplift and erosion, a regional hydrothermal episode or elevated heat flow combined with uplift and erosion associated with Triassic rifting are possible causes. Palaeotemperature profiles from vertical sequences of samples would enable further insight but these are not available in the study area. A similar mid-Triassic to early Jurassic timing is recognized in AFTA studies elsewhere in the UK and Northwest European Continental Shelf, raising the possibility that these data are a manifestation of a widespread regional heating event.

Warrington et al. (pers. com.) point to evidence of east–west extension which caused a north–south cross-cutting vein system in Cornwall and east Devon, with vein mineralization during the Triassic. Sandeman et al. (1995) present $^{40}Ar/^{39}Ar$ data from the Lizard Complex in Cornwall which suggest a low-temperature thermal overprint at ~220 Ma. They note that this is in good agreement with the age estimated for an episode of hydrothermal activity affecting the Lizard Complex reported by Halliday & Mitchell (1976) and suggest that this was caused by fluxing of hydrothermal fluids related to rifting of the North Atlantic in the late Triassic.

The early Cretaceous episode

The AFTA data from wells on the 'Wytch Farm block' to the north of the main Portland–Isle of Wight Fault Zone show cooling from maximum palaeotemperatures at some time in the early Cretaceous (Fig. 3). Taking all the available data constraints into account, the cooling commenced

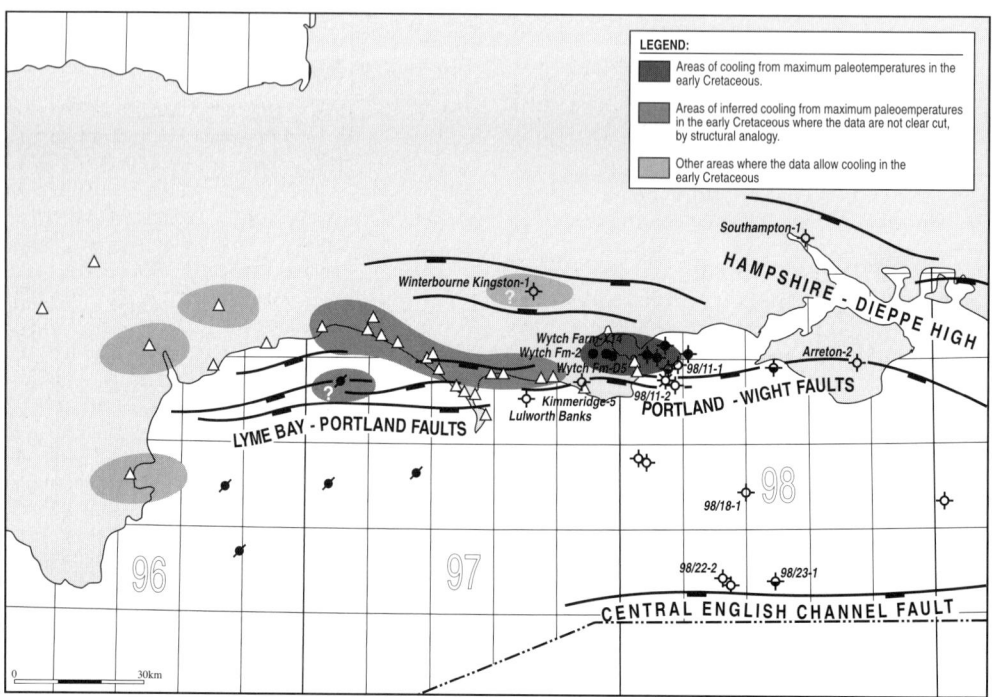

Fig. 3. Areas showing cooling from maximum palaeotemperatures in the early Cretaceous event.

some time in the range 140–100 Ma. This places this episode in the early Cretaceous – possibly as early as the latest Jurassic Ryazanian to Valanginian, possibly as late as the Albian. The data also show that the area stayed relatively cool through the rest of its history, although there is evidence for a distinct reheating followed by cooling during the Tertiary. Thus, the Wytch Farm block cooled in the early Cretaceous and apparently remained as a relatively cool, structurally high, block through to the present day.

AFTA data from outcrop samples extending westwards from Lulworth into Lyme Bay, to the north of the Lyme Bay–Portland Fault System, also show some evidence of early Cretaceous cooling. In most cases the maximum palaeotemperatures experienced by these currently outcropping rocks was in the range 65–80°C and most samples show evidence of later phases of cooling in the Tertiary. These two factors – the relatively low palaeotemperatures and the subsequent heating–cooling episodes – make precise observations on an early Cretaceous cooling difficult, but it seems likely that the area affected by early Cretaceous cooling recognized in the Wytch Farm block can be extended westward into Lyme Bay. Evidence for early Cretaceous cooling is also seen over the Cornwall granites from fission track work by Chen et al. (1996).

The timing of this early Cretaceous episode recognized by AFTA coincides with a major unconformity (e.g. McMahon & Underhill 1995; MaMahon & Turner 1997 this volume). There is a clear break in the section at the base of the Aptian–Albian, which rests locally on various stages of the Jurassic or Wealden Beds. Over the Wytch Farm field the Albian rests on the Callovian Oxford Clay, therefore much of the Upper Jurassic and Lower Cretaceous is missing. However, the correlatable unconformity to this event from the most basinal succession occurs in the lowermost Wealden beds and is of Berriasian–Valanginian in age (McMahon & Underhill 1995).

The area where the AFTA data clearly recognize early Cretaceous cooling coincides with the area where Wealden sediments are absent, traditionally considered not to have been deposited. The presently available data do not have sufficient resolution to say whether the early Cretaceous event was pre- or post-Wealden. However, it is easier to explain the palaeotemperatures attained prior to the cooling if burial by a Wealden section is allowed, i.e. if the erosion is post-Wealden. This interpretation would involve some re-evaluation of present concepts of Wealden palaeogeography (e.g. Hamblin et al. 1992).

The late Cretaceous–early Tertiary episode

A regional base Tertiary unconformity exists between the Upper Cretaceous Chalk and the Tertiary rocks of east Dorset and Hampshire. Variable amounts of Chalk section are missing beneath the Tertiary and the oldest preserved Tertiary sediments are of late Palaeocene age. A similar break in the stratigraphic record exists in other parts of Britain and offshore on the UK Continental Shelf. Widespread cooling commencing in the latest Cretaceous–early Tertiary associated with uplift and erosion has been revealed by AFTA data over much of northern and eastern England and adjacent offshore areas (Green 1986, 1989; Bray et al. 1991; Green et al. 1993; Green et al. 1997). It might be expected therefore that an early Tertiary timing would dominate AFTA data from most of the Dorset and Hampshire basins, but this appears not to be the case.

The AFTA data in Winterbourne Kingston-1 do allow an episode of cooling from maximum palaeotemperatures some time in the range 80–60 Ma, but the data from all the other wells showing cooling from maximum palaeotemperatures in the Tertiary are dominated by mid or late Tertiary effects. It is therefore concluded from the present data that early Tertiary erosion is of only minor significance in this region.

The mid- and late Tertiary cooling episodes

The AFTA data from wells from areas immediately south of the Portland–Wight Fault Zone, from the Portland–Wight Basin and the Central Channel Basin, and from one sample from the Southampton-1 well show clear evidence of a significant cooling episode in the mid-Tertiary, with cooling from maximum palaeotemperatures at ~40 Ma (late Eocene) (Figs 4 and 5). The exception to this is a single AFTA sample from Lulworth Banks-1 which shows that cooling commenced somewhat later at ~20 Ma (Miocene). These episodes relate to the major phases of mid-Tertiary structuring along the Portland–Wight monocline, inversion of the channel basin, and possible uplift of regions to the north of the 'Wytch Farm block' as suggested by the Southampton-1 data. The Lulworth Banks data suggest that the latest phase of inversion commenced in the Miocene, causing uplift of the area around the well on the Weymouth Bay Anticline.

The literature has long speculated a Miocene timing for the main phase of inversion on the Purbeck–Isle of Wight monocline. The late

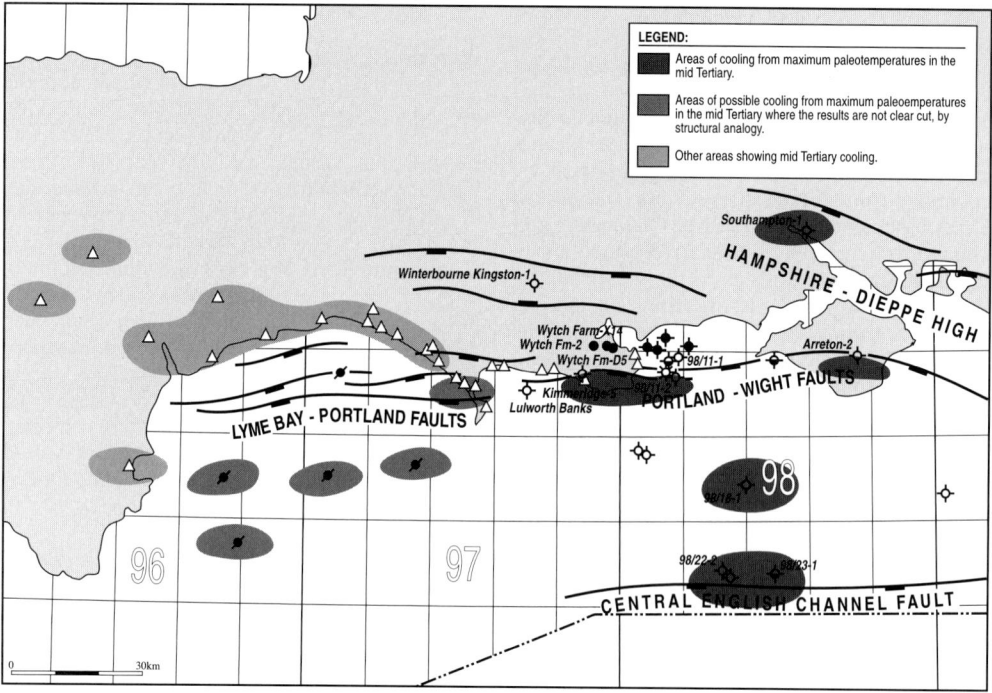

Fig. 4. Areas showing cooling in the mid-Tertiary.

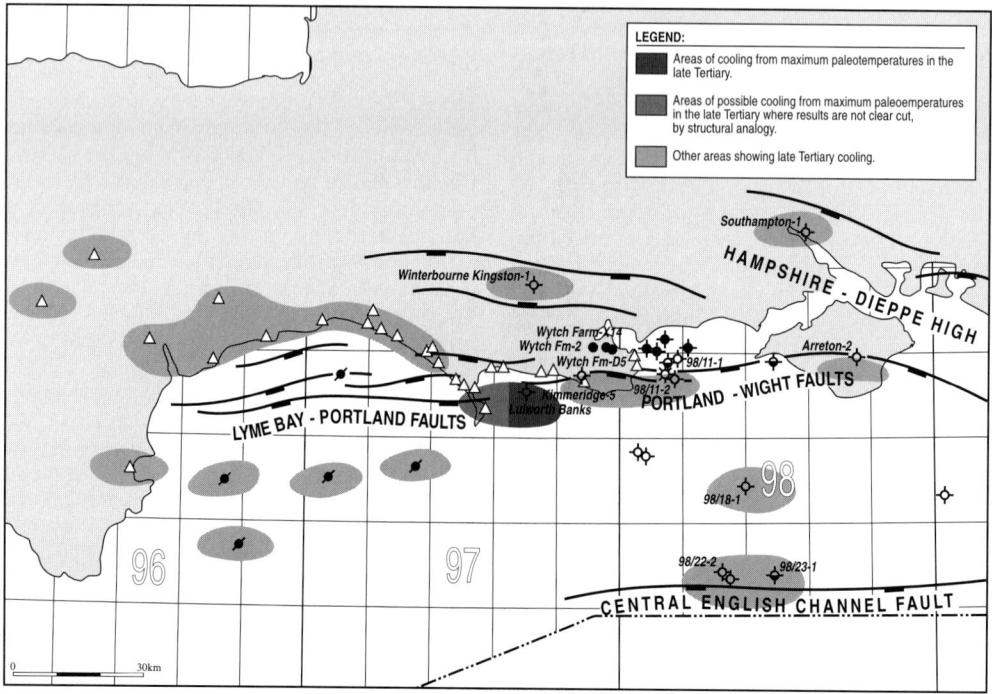

Fig. 5. Areas showing cooling in the late Tertiary.

Palaeocene–Eocene sediments of the Hampshire Basin were deformed during the inversion tectonics. Early Oligocene sediments lie unconformably on the Eocene in the Isle of Wight, and post-early Oligocene sediments are absent from the study area. Post- or late Eocene and post-Oligocene phases of movement can therefore be deduced from the stratigraphic record. The timing of the Tertiary episodes derived from the AFTA data therefore fit well with the geological evidence and with the 'Laramide' inversions of Northern Europe and the Helvetic and Pyrrenean orogenic events.

As mentioned above, cooling from elevated palaeotemperatures during the Tertiary is recognized by AFTA over a wide region. High degrees of cooling are seen on the classical inversion structures, but samples away from obvious inversions to the west of study area also show a significant degree of cooling. It is therefore concluded that the Tertiary cooling is regional in its effects and not restricted to the main inversion axes. Similar conclusions concerning the regional extent of Tertiary cooling have been drawn from AFTA studies elsewhere in Britain, e.g. the Sole Pit axis and the East Midlands Shelf (Bray *et al.* 1992) and the Irish Sea region (Green *et al.* 1997), and seem to be characteristic of inverted basins (Green *et al.* 1995).

Palaeotemperature profiles, palaeogeothermal gradients and estimates of uplift and erosion

Palaeotemperature estimation, the construction of palaeotemperature profiles and measurement of palaeogeothermal gradients are important aspects of thermal history reconstruction. Inspection of palaeotemperature profiles allows interpretation of the mechanisms of heating and cooling, and allows recognition of lateral heat transfer in the past, for example by igneous intrusion or the passage of hot fluids. Measurement of palaeogeothermal gradients can provide constraints on viable heat flow models and in some cases enables estimation of the amount of section removed by uplift and erosion, as described below.

Integration of AFTA and VR data

Both AFTA and VR record maximum palaeotemperatures. AFTA gives direct estimates of palaeotemperatures in °C. VR values can be converted to maximum temperature estimates using an appropriate kinetic model relating temperature–time history to reflectance (e.g. Burnham & Sweeney 1989).

Once the timing of maximum palaeotemperature has been established from AFTA, percentage reflectance values can be converted to temperature in °C. Where both the AFTA and VR parameters were set by the same heating episode, palaeotemperature estimates from the two techniques are generally highly consistent. This consistency is not surprising because the independently derived kinetic descriptions of fission track annealing (Laslett *et al.* 1987) and of VR (Burnham & Sweeney 1989) are very similar in their response to temperature, irrespective of heating rate. This implies that a particular degree of fission track annealing should always associate with a particular level of VR. This is borne out by observation from a variety of basins and geological settings (e.g. Kamp & Green, 1990; Bray *et al.* 1992; Green *et al.* 1995).

Thus, AFTA- and VR-derived palaeotemperature estimates from a well can be plotted on the same axes to provide palaeotemperature–depth profiles. The integration of AFTA and VR data in this way has big advantages. For example, independent estimates of maximum palaeotemperature from AFTA and VR can be compared in the same or in adjacent samples; AFTA and VR are run in different lithologies, enabling greater vertical stratigraphic coverage and therefore better constraint on gradient; AFTA can be used to recognize geochemical suppression or other problems in VR data; and, although it does not apply in this study, combined AFTA and VR can be used to identify histories involving multiple heating episodes.

Estimating amounts of section removed during uplift and erosion from palaeothermal data

The primary information derived from AFTA is in the form of time–temperature history, which must be known before hydrocarbon generation can be predicted. However, for a number of reasons it is also important to reconstruct burial history, which in inverted basins involves determining the timing and amount of uplift and erosion.

Inspection of palaeotemperature profiles as described above can help determine whether heating and cooling was due to burial followed by uplift and erosion, or to other causes. In the case of cooling due to uplift and erosion, the time of cooling derived from AFTA gives the time of uplift. An estimate of the amount of section removed by uplift and erosion can be made by extrapolating the palaeogeothermal gradient in the preserved section through the removed section to a palaeosurface temperature intercept. This approach is illustrated in Fig. 6.

Certain important assumptions are made in applying this method. It must be assumed that the palaeogeothermal gradient in the preserved section is linear and can be extrapolated as a straight line to the palaeosurface temperature, i.e. that the gradient in the removed section is the same as in the preserved section. This implies that the conductivity and therefore lithological mix in the removed section is similar to that of the preserved section. We must also assume that we know the palaeosurface temperature at the time of uplift.

The linearity of the palaeogeothermal gradient in the preserved section can be evaluated from the data. It is more difficult to assess whether the assumption of linearity in the removed section is valid. It is important to realize that these assumptions must be made in order to use any sort of palaeothermal data to estimate uplift and erosion, regardless of whether we use a heat flow modelling approach or the palaeogeothermal gradient approach described above. The key point is that although this assumption is supported by a range of evidence (Green *et al.* 1995) we cannot be certain whether the assumptions are valid, and the resulting estimates of uplift and erosion should be viewed with this in mind.

Because of the uncertainties involved, estimates of section removed by uplift and erosion from palaeothermal data should be used in conjunction with estimates from non-thermal techniques such as the sonic velocity method. However this method is also subject to a range of uncertainties and assumptions (see Law this volume).

Estimates of uplift and erosion in the Wessex Basin

With the presently available dataset from the Wessex Basin there are several difficulties in relating palaeotemperatures to depth of burial, with consequent problems in using the palaeotemperature data to estimate the amount of section removed by uplift and erosion. First, there are difficulties in acquiring well-constrained palaeotemperature profiles. In several wells good AFTA data are only available from two or three samples per well, due to restrictions on access to samples and because of a tendency for good apatite yields to be limited to the Bridport Sands and the Permo-Triassic section. There is generally no problem in collecting samples for VR from the Jurassic and Cretaceous shales but there are distinct problems in interpreting VR data from the Wessex Basin, as discussed below. The

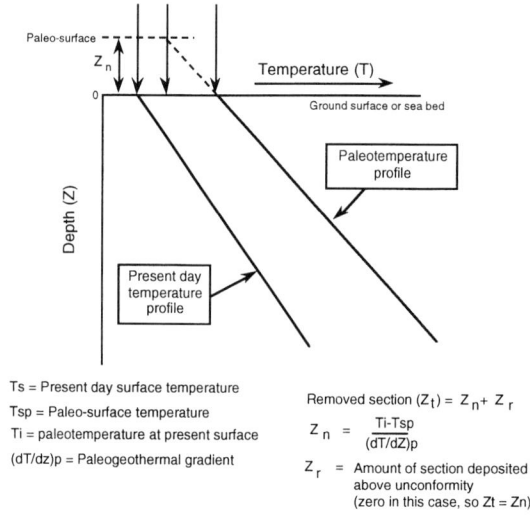

Fig. 6. Method of estimating section removed by uplift and erosion from palaeotemperature data.

data often form a wide scatter or are anomalously low due to suppression. The second problem is that where palaeotemperature profiles can be established from AFTA and VR data, there are widespread suggestions of a component of localized heating in the section possibly due to fluid movement through the permeable aquifer or reservoir horizons.

Lateral transfer of heat due to fluid movement gives rise to perturbed, non-linear palaeotemperature profiles or low palaeogeothermal gradients (Ziagos & Blackwell 1986; Duddy *et al.* 1994). A non-linear temperature profile implies a non-linear relationship between temperature and depth of burial. A low palaeogeothermal gradient measured in the preserved section in a well may be indicative of high palaeogeothermal gradient in the removed section. If non-linear palaeotemperature profiles or low palaeogeothermal gradients are evident, then we cannot be confident that the assumptions implicit in estimation of uplift from palaeotemperature data are valid. In the Wessex Basin, the complex form of the palaeotemperature profiles make it difficult to estimate of the amounts of section removed by uplift and erosion in individual wells. Examples from selected wells are discussed below.

The palaeotemperatures derived from AFTA and VR in the Lulworth Banks-1 well form a good linear profile, typical of a section that has been cooled by uplift and erosion with a 'normal' geothermal gradient (Fig. 7). These data can be explained by ~1800 m of section removed by Tertiary uplift. The resulting burial history is illustrated in Fig. 7.

In Arreton-2 (Fig. 8) the AFTA- and VR-derived palaeotemperatures do not form a linear profile and suggest a reduction in the gradient in the lower part of the section. Estimates of removed section depend on how the gradient is extrapolated. Between 1 and 2 km can be accommodated within the data which seems reasonable as there is ~1200 m of preserved Chalk and Tertiary section nearby and both of these units are incomplete. The timing of the uplift is well constrained to the mid-Tertiary by the AFTA data.

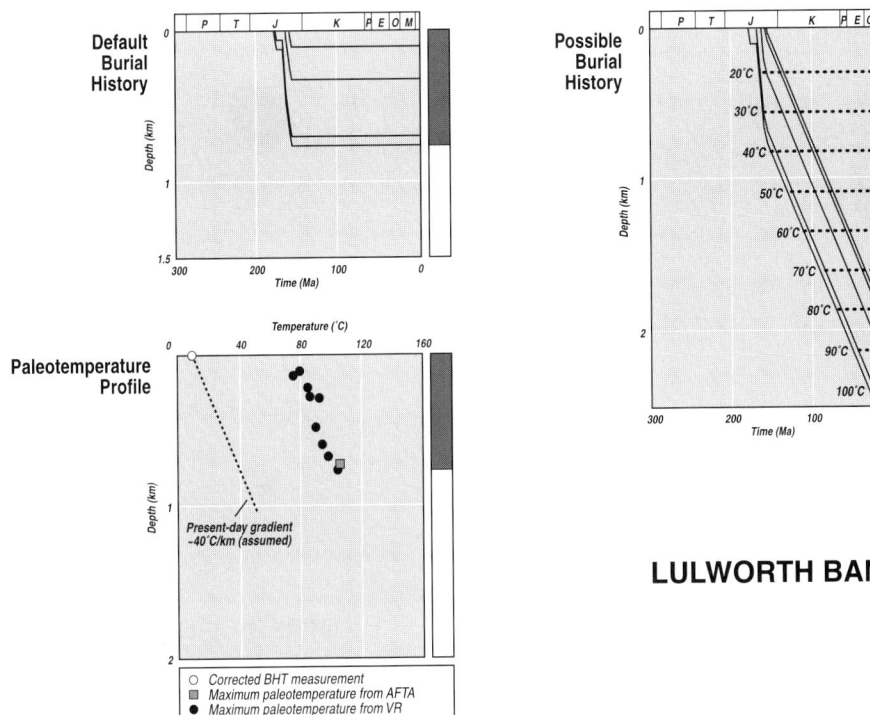

Fig. 7. Palaeotemperature profile and reconstructed burial history from AFTA and vitrinite reflectance (VR) data from the Lulworth Banks-1 well. The 'Default' burial history are constructed based on the preserved stratigraphy alone (i.e. by assuming no uplift and erosion at uncomformities).

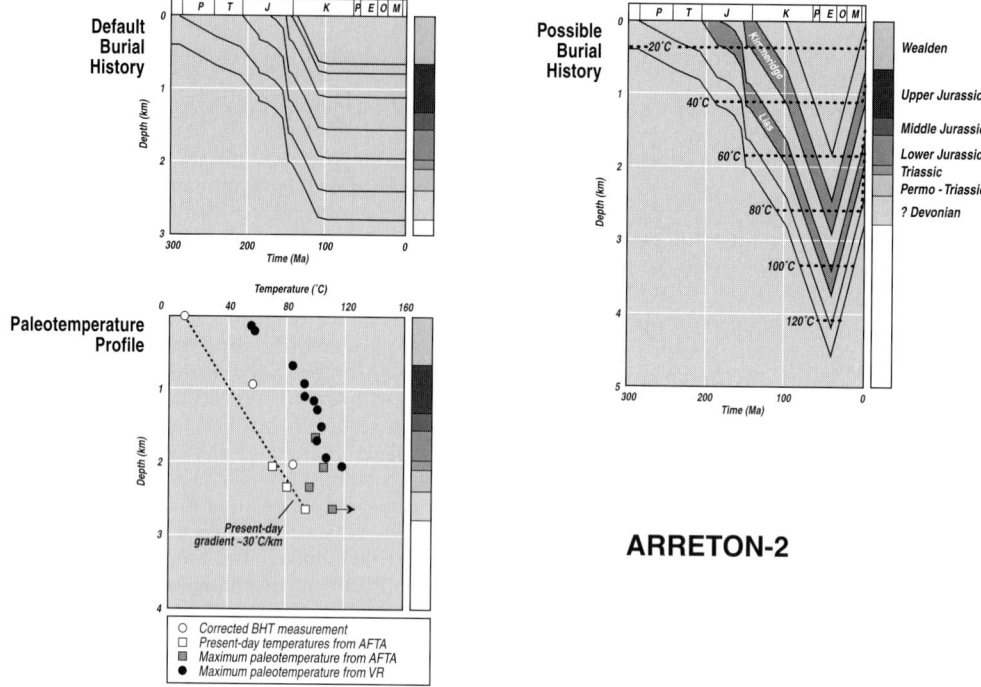

Fig. 8. Palaeotemperature profile and reconstructed burial history from AFTA and VR data from the Arreton-2 well. See caption to Fig. 7 for further details.

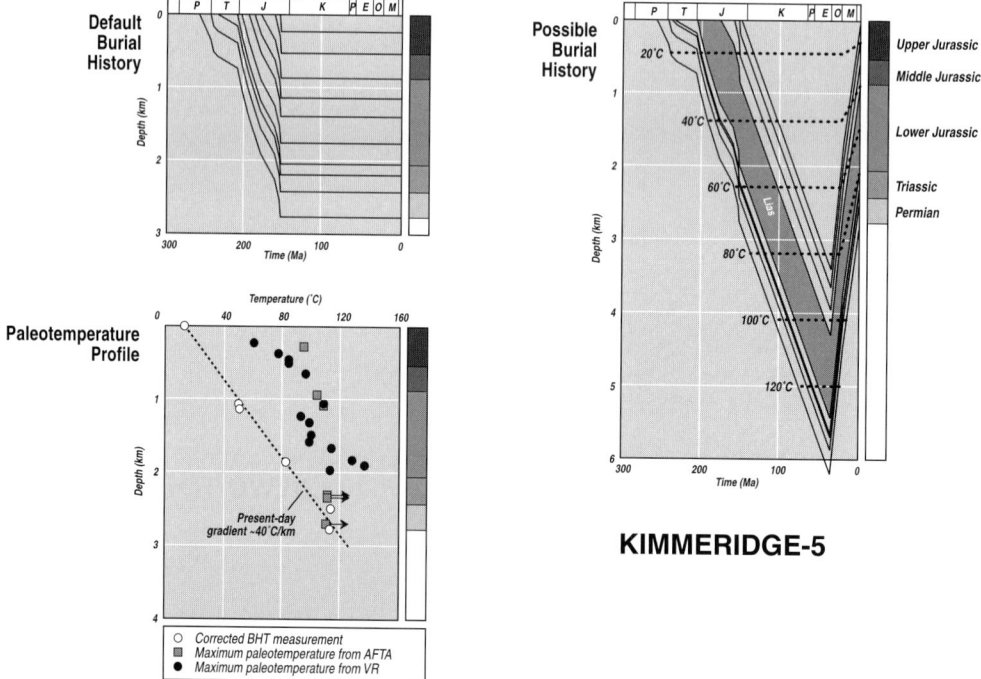

Fig. 9. Palaeotemperature profile and reconstructed burial history from AFTA and VR data from the Kimmeridge-5 well. See caption to Fig. 7 for further details.

In Kimmeridge-5 (Fig. 9) the uplift also commenced in the mid-Tertiary. The data show a similar palaeotemperature profile to the Arreton-2 well but here we have AFTA samples from the shallower section. These define a relatively low palaeogeothermal gradient, whereas the VR data are quite scattered. There is a strong possibility that the low VR values in the Lias section are suppressed and not providing a true measure of palaeotemperature or maturity, as discussed below. As with the Arreton-2 well, an estimate of removed section depends on how the gradient is fitted and could range from 2 to 3.4 km. The missing section ranges from Kimmeridgian to mid-Tertiary so this range does not seem unreasonable.

Figure 10 shows a composite palaeotemperature plot made up of data from three Wytch Farm wells. Here maximum palaeotemperatures in the pre-Cretaceous section were attained in the early Cretaceous. The AFTA data reveal a distinct palaeotemperature peak prior to a second phase of cooling commencing in the mid-Tertiary. Again the palaeotemperature profile appears to be non-linear and consequently the amount of uplift is very difficult to constrain. However, up to 2.3 km of erosion prior to reburial in the late Cretaceous and early Tertiary is compatible with the data.

Law (this volume) presents the results of a study of uplift in the Wessex Basin based on sonic velocity data. The sonic velocity and palaeothermal data are consistent in that they both show that significant uplift and erosion has taken place, but there is a wide range in the differences between the estimates derived from the two techniques. This range in differences reflects the problems in applying the palaeothermal and sonic velocity techniques in this basin, as described above and in detail in Law's paper, respectively. The estimates derived from the AFTA and VR data are generally significantly higher than those from the sonic velocities. As discussed by Law, this could be explained if the reference wells (i.e. those with the lowest interval velocity for a given depth of burial) have themselves undergone a degree of uplift and erosion. The AFTA and VR data show that none of the wells studied in the Wessex Basin are currently at their maximum depth of burial and the outcrop data suggest that the heating was regional, therefore the sonic velocity-derived estimates indeed reflect the difference in uplift between a given well and a reference well which has itself been uplifted.

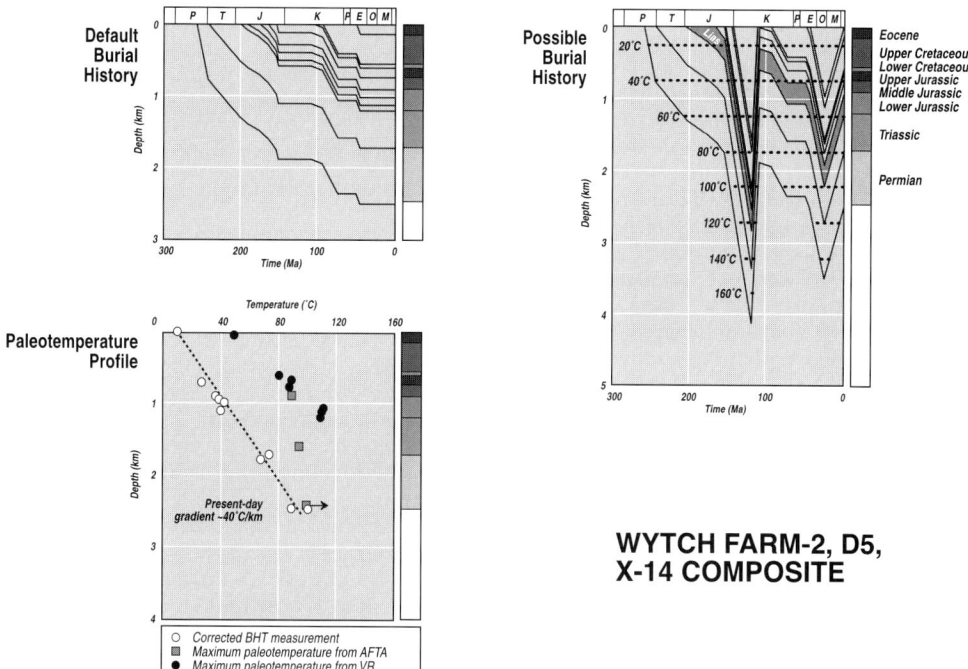

WYTCH FARM-2, D5, X-14 COMPOSITE

Fig. 10. Palaeotemperature profile and reconstructed burial history from AFTA and VR data from the Wytch Farm wells. See caption to Fig. 7 for further details.

Clearly both the palaeothermal and sonic velocity approaches suffer from problems in this region for the reasons discussed above. However, because of the high palaeotemperature estimates from relatively shallow in the section, any reasonable model attempting to explain the AFTA and VR data requires regional, kilometre-scale uplift and erosion.

A note on vitrinite reflectance data from the Wessex Basin

VR values need to be measured as accurately as possible for the purpose of palaeotemperature estimation. The difficulties of measuring and interpreting reflectances in good source rock sections such as the Kimmeridge Clay and Lias of the Wessex Basin are well known. Because of this, new vitrinite reflectance data were acquired to a high standard for this study using the exacting $R_{V_{max}}$ method of measurement on whole-rock sample mounts (Cook unpublished data). Measurement on whole-rock samples enables the organic particles to be petrographically observed and measured *in situ*. This greatly assists in the identification of true vitrinite and allows assessment of liptinite fluorescence, petrographic observation of the type and mix of organic macerals present in the sample, estimation of their relative abundances, etc.

It was noted that many of the VR values are lower than would be expected on the basis of progressive heating to present-day temperatures, or on the basis of the interpretation of the AFTA data. In some wells a large scatter in the VR data is evident where the AFTA palaeotemperatures form relatively linear profiles. The reasons for this are not immediately clear but there are strong suggestions that 'geochemical suppression' of VR values occurs in parts of the source rock sections. Factors known to cause suppression, such as the presence of oils, bitumens, sulphur and liptinite macerals, are common in some of the samples. The inclusion of measurements made on bituminite, often very difficult to distinguish from vitrinite, could also be a factor in lowering mean reflectance values.

The integration of AFTA with VR data forms a basis for investigating suppression effects. This is not pursued here but is a topic for further study.

Evidence for fluid flow

Natural seepages of oil and gas along the Purbeck–Isle of Wight Fault Zone are well known and these faults are evidently conduits for fluid flow at the present day. It is interesting therefore that there is some evidence in the palaeotemperature data for an episode of fluid movement in the mid- to late Tertiary.

The low interval palaeogeothermal gradients in the Tertiary through parts of the section in Kimmeridge-5, Arreton-2 and the Wytch Farm wells are also recognized in other wells along the trend of the Portland–Isle of Wight faults analysed in this study. These gradients may be due partially to episodes of hot fluid flow through reservoirs or other permeable zones, introducing heat laterally into shallower parts of the section.

In Wytch Farm, for example, the deeper AFTA sample has cooled by a maximum of $\sim$10–20°C since the late Tertiary, whereas the the shallower samples have cooled by $\sim$30–40°C during the same period. If all the cooling in the deeper sample was due to uplift and erosion, this leaves a significant degree of cooling in the shallower samples that could have been due to cessation of hot fluid flow. A similar line of reasoning can be taken to explain the patterns of palaeotemperatures in other wells, although more detailed sampling for AFTA is needed before firm conclusions can be drawn.

The similarity in timing between the late phase of uplift and erosion seen in the Lulworth Banks well and the apparent cessation of fluid flow north of the Purbeck structure suggests some link, and this episode could mark the termination of oil migration into the Wytch Farm structure.

Implications for source rock maturity and hydrocarbon accumulation

Hydrocarbon generation effectively ceases as the source rock begins to cool from maximum palaeotemperature. Important conclusions on oil and gas generation and migration can therefore be drawn from the thermal histories reconstructed in this study (Fig. 11). The time of onset of generation in the basin is more difficult to constrain as, in all wells studied, generation commenced during burial by sediments now removed by uplift and erosion. In addition, to determine the temperature of the onset of generation, the kinetic parameters for kerogen in specific source rocks must be known. Nevertheless, the AFTA thermal history reconstructions allows some insight.

Putting all the data together, a picture emerges of early Cretaceous structuring causing significant erosion on the Wytch Farm block, which remained relatively high compared to the areas to the north and south through the rest of its history. The palaeotemperature data suggest that the Lias in the Wytch Farm block did not reach sufficient maturity for significant generation *at any time*. It is probable that the maturity levels in this block were a response to burial in the Jurassic and early Cretaceous, and that no further hydrocarbon generation occurred after cooling due to uplift and erosion commencing in the early Cretaceous.

Burial during the late Cretaceous and early Tertiary probably occurred throughout the region and in the north, around the Winterbourne Kingston well, post-Chalk uplift and erosion may have caused cooling, terminating any hydrocarbon generation there. Subsequent reburial during the Tertiary was limited in this area and not sufficient to re-initiate generation from Lias source rocks.

The most widespread phase of structuring commenced in the mid-Tertiary (late Eocene–Oligocene) with uplift and erosion over the whole of the Wessex Basin area covered by this study. This culminated in major uplift and erosion immediately to the south of the Purbeck structure in the late Tertiary.

In Kimmeridge-5, generation from Lias source rocks may have commenced in the early Cretaceous and continued through peak and late generation phases until the mid-Tertiary cooling episode. At Arreton-2, the Lias section is much less mature, probably only becoming early mature in the mid-Cretaceous prior to cooling in the mid-Tertiary. This period of cooling effectively terminated hydrocarbon generation over a wide region, although it is possible that active generation continued into the late Tertiary around the Lulworth Banks well on the Weymouth Bay Anticline until uplift of this area in the Miocene at ~20 Ma. There is also some evidence from the AFTA that fluid migration into the Wytch Farm reservoirs also ceased at this time, as noted previously.

All samples analysed over the study area have cooled from significantly higher temperatures in the past and therefore no parts of the basin are currently generating hydrocarbons. This implies that the numerous active oil and gas seepages along the Dorset coast are from leakage of previously reservoired accumulations, or re-migration due to the late Tertiary structuring, and are not due to generation at the present day.

Conclusions

The Wessex Basin has undergone a complex thermal history and several phases of uplift and erosion. An understanding of this history is critical to the prediction of maturity development and hydrocarbon generation. AFTA can provide data on the timing of significant thermal episodes and periods of uplift and erosion which are difficult or impossible to obtain from other methods. Combination of AFTA with VR data to construct palaeotemperature profiles is a useful approach. These provide constraints on the amount of uplift and erosion and recognition of non-burial related heating such as the effects of fluid movement. The main conclusions of the study are as follows.

- Four thermal episodes have been recognized, occurring in the mid-Triassic to early Jurassic, the early Cretaceous, the mid-Tertiary and the late Tertiary, each affecting different parts of the basin to different degrees. The recognition of a mid-Triassic to early Jurassic episode in the western part of the study area is important to the understanding of the early history of the basin.

- The 'Wytch Farm block' and the area of Bournemouth Bay to the north of the Portland–Isle of Wight Fault Zone commenced cooling from maximum palaeotemperature during uplift and erosion in the early Cretaceous. The Lias source rock section here did not reach sufficient maturity to generate significant hydrocarbons. Any generation that did occur would have ceased during the early Cretaceous.

- The area south of the Portland–Isle of Wight Fault Zone experienced a major thermal episode prior to cooling in the mid-Tertiary caused by regional uplift and erosion. Uplift culminated along the recognized inversion structures, with maximum amounts immediately south of the Purbeck disturbance. A second distinct phase of Tertiary erosion caused cooling from maximum palaeotemperatures during the Miocene in the Lulworth Banks well. Hydrocarbon generation ceased during Tertiary uplift with the latest phase of generation occurring prior to the Miocene cooling around Lulworth Banks. Evidence exists for cessation of hot fluid movement in the Wytch Farm block at this time, which could represent the end of oil migration into the reservoirs.

- Uncertainty exists over the timing of maximum palaeotemperatures and maturity development in Lyme Bay. Existing AFTA data here are consistent with cooling in either the early Cretaceous or Tertiary episodes recognized to the east. Resolution of this timing question is critical for exploration in Lyme Bay and could be resolved by AFTA in future wells.

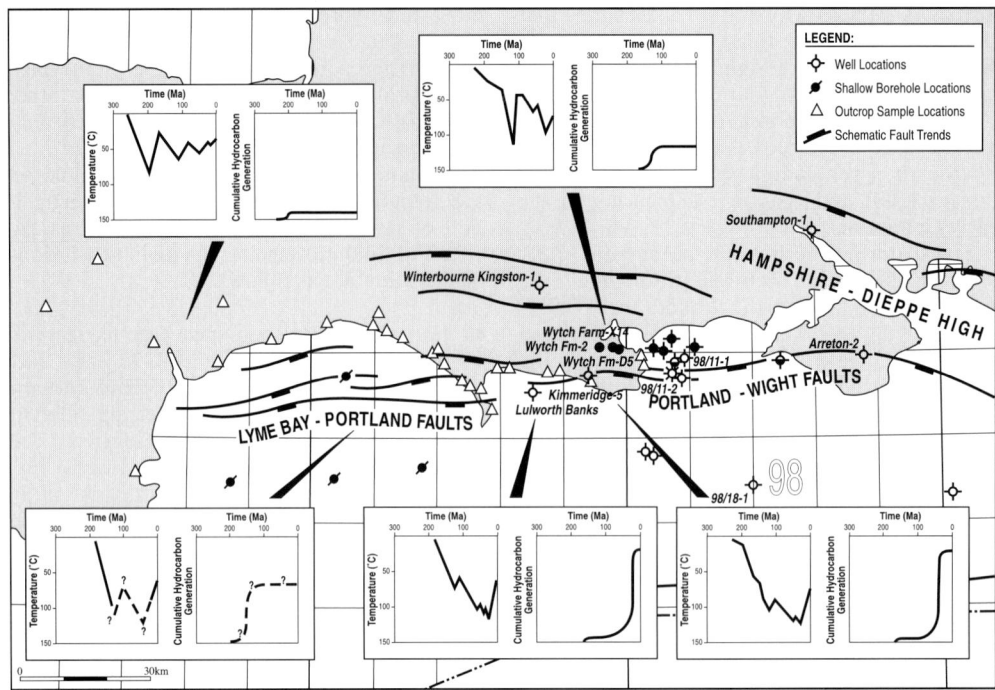

Fig. 11. Schematic summary of the thermal history styles recognized in this study. Corresponding schematic cumulative hydrocarbon generation curves are also shown to illustrate the variation in the style of generation histories. Note that hydrocarbon generation ceases on cooling from maximum temperature. The data from the Wytch Farm block suggest that the Lias source rock probably did not attain significant maturity here, but any generation would have ceased during the early Cretaceous. South of the Portland–Isle of Wight Fault Zone, in the English Channel Basin, peak hydrocarbon generation was caused by maximum palaeotemperatures in the mid-Tertiary, prior to cooling at ~40 Ma. The data suggest that around the Lulworth Banks well generation continued into the late Tertiary until cooling at ~20 Ma. In the western part of the study area, maximum palaeotemperatures occurred early in the history, prior to ~170 Ma. The data show that this area was still at elevated palaeotemperatures in the Tertiary and it is likely that this region experienced all of the episodes recognized to the east to some degree, but with an overall cooling trend from the late Triassic–early Jurassic to the present day.

- The data show that Tertiary uplift was regional and not restricted to the inversion of the area south of the Portland–Isle of Wight Fault Zone. Significant cooling occurred in west Dorset and east Devon, and in the Wytch Farm, Winterbourne Kingston and Southampton wells.
- Estimates of the amount of uplift and erosion are hampered by the low or non-linear palaeogeothermal gradients defined by the AFTA and the scattered, sometimes anomalous, VR data. Lateral transfer of heat by fluid movement may have had significant effects and implies non-linear relationships between palaeoburial depth and palaeotemperature. However, kilometre-scale uplift and erosion during both the early Cretaceous and Tertiary is required to explain the data.
- Anomalously low VR values in parts of the section have been recognized by a combination of AFTA and VR data. The possibility of widespread VR 'suppression' has significant implications for correct measurement of source-rock maturity. In these circumstances AFTA provides an independent check on maturity levels.

The authors would like to thank BP Exploration Operating Co Ltd for allowing information from the Lulworth Banks-1 well to be published. We would also like to express our thanks Dr A. Law at British Gas Exploration and Production Ltd for helpful discussions concerning estimates of uplift and erosion from their sonic velocity data, and to many friends and colleagues who have assisted in sample collection and analytical work during the course of this study. AFTA is a registered trademark of Geotrack International Pty Ltd.

References

BRAY, R. J., GREEN, P. F. & DUDDY, I. R. 1992. Thermal history reconstruction in sedimentary basins using apatite fission track analysis and vitrinite reflectance: a case study from the east Midlands of England and the Southern North Sea. *In*: *Exploration Britain: Geological Insights for the Next Decade*. Geological Society, London, Special Publications, **67**, 3–25.

BURNHAM, A. K. & SWEENEY, J. J. 1989. A chemical kinetic model of vitrinite reflectance maturation. *Geochimica et Cosmochimica Acta*, **53**, 2649–2657.

BUCHANAN, J. G. 1997. The exploration history and controls on hydrocarbon prospectivity in the Wessex basins, southern England, UK. *This volume*.

CHEN, Y., ZENTILLI, M. A., CLARK, A. H., FARRAR, E., GRIST, A. M. & WILLIS-RICHARDS, J. 1996. Geochronological evidence for post-Variscan cooling and uplift of the Carnmenellis granite, SW England. *Journal of the Geological Society, London*, **153**, 191–195.

COLTER, F. S. & HAVARD, D. J. 1981. The Wytch Farm Oil Field, Dorset. *In*: ILLING, L. V. & HOBSON, D. G. (eds) *Petroleum Geology of the Continental Shelf of NW Europe*. Hayden & Sons, London, 494–503.

DUDDY, I. R., GREEN, P. F, BRAY, R. J. & HEGARTY, K. A. 1994. Recognition of the thermal effects of fluid flow in sedimentary basins. *In:* PARNELL, J. (ed.) *Geofluids: Origin, Migration and Evolution of Fluids in Sedimentary Basins*. Geological Society, London, Special Publications, **78**, 325–345.

GREEN, P. F. 1986. On the thermo-tectonic evolution of Northern England: evidence from fission track analysis. *Geological Magazine*, **123**, 493–506.

—— 1989. Thermal and tectonic history of the East Midlands shelf (onshore UK) and surrounding regions assessed by apatite fission track analysis. *Journal of the Geological Society, London*, **146**, 755–733.

——, DUDDY, I. R. & BRAY, R. J. 1995. Applications of thermal history reconstructions in inverted basins. *In*: BUCHANAN, J. G. & BUCHANAN, P. G. (eds) *Basin Inversion*. Geological Society, London, Special Publications, **88**, 149–166.

——, —— & —— 1997. Variation in thermal history styles around the Irish Sea and adjacent areas: implications for hydrocarbon occurrence and tectonic evolution. *In*: TRUEBLOOD, S. P. & MEADOWS, N. (eds) *Petroleum Geology of the Irish Sea Basins*. Geological Society, London, Special Publications, **124**, 73–93.

——, ——, —— & LEWIS, C. L. E. 1993. Elevated paleotemperatures prior to early Tertiary cooling throughout the UK region: implications for hydrocarbon generation. *In:* PARKER, J. R. (ed.) *Petroleum Geology of Northwest Europe: Proceedings of the 4th Conference*. Barbican, London, 1067–1074.

HALLIDAY, A. N. & MITCHELL, J. G. 1976. Structural K–Ar and $^{40}Ar/^{39}Ar$ age studies of adularia K-feldspars from the Lizard Complex, England. *Earth and Planetary Science Letters*, **29**, 227–237.

HAMBLIN, R. J. O., CROSBY, A., BALSON, P. S., JONES, S. M., CHADWICK, A. R., PENN, I. E. & ARTHUR, M. J. 1992. *United Kingdom Offshore Regional Report: The Geology of the English Channel*. HMSO, London.

KAMP, P. J. J. & GREEN, P. F. 1990. Thermal and tectonic history of selected Taranaki Basin (New Zealand) wells assessed by apatite fission track analysis. *Bulletin of the American Association of Petroleum Geologists*, **74**, 1401–1419.

LASLETT, G. M., GREEN, P. F., DUDDY, I. R. & GLEADOW, A. J. W. 1987. Thermal annealing of fission tracks in apatite 2. A quantitative analysis. *Chemical Geology (Isotope Geoscience Section)*, **65**, 1–3.

LAW, A. 1997. Regional uplift in the English Channel: quantification using sonic velocity data. *This volume*.

McMAHON, N. A. & TURNER, A. 1998. The documentation of a latest Jurassic–earliest Cretaceous uplift throughout southern England and adjacent offshore areas. *This volume*.

—— & UNDERHILL, J. R. 1995. The regional stratigraphy of the southwest United Kingdom and adjacent offshore areas with particular reference to the major intra-Cretaceous unconformity. *In*: CROCKER, P. F. & SHANNON, P. M. (eds) *The Petroleum Geology of Ireland's Offshore Basins*. Geological Society, London, Special Publications, **93**, 323–325.

SANDEMAN, H. A., CHEN, Y., CLARK, A. H. & FARRAR, E. 1995. Constraints on the P–T conditions and age of emplacement of the Lizard Ophiolite, Cornwall: plagioclase thermobarometry and $^{40}Ar/^{39}Ar$ geochronology of basal amphibolites. *Journal of Earth Sciences*, **32**, 261–272.

UNDERHILL, J. R. & PATERSON, S. 1998. Genesis of tectonic inversion structures: seismic evidence for the development of key structures along the Purbeck–Isle of Wight Disturbance. *Journal of the Geological Society, London*, in press.

—— & STONELEY, R. 1982. Introduction to the development, evolution and petroleum geology of the Wessex Basin. *This volume*.

ZIAGOS, J. P. & BLACKWELL, D. D. 1986. A model for the transient temperature geothermal systems. *Journal of Volcanology and Geothermal Research*, **27**, 371–397.

The documentation of a latest Jurassic–earliest Cretaceous uplift throughout southern England and adjacent offshore areas

NEIL A. McMAHON[1] & JONATHAN TURNER[2]

[1] *Arthur D. Little, Berkeley Square House, Berkeley Square, London W1X 6EY, UK*
[2] *Department of Geology and Geophysics, University of Edinburgh, West Mains Road, Edinburgh EH9 3JW, UK*

Abstract: The development of the Wessex Basin, Celtic Sea Basins and the Western Approaches Trough was initiated during the break-up of Pangea with active subsidence continuing throughout the Mesozoic until the basins were inverted in the Tertiary. The Mesozoic stratigraphy in these basins is characterized by a number of unconformities. In the Early Cretaceous two major erosional events can be identified, one in the Berriasian and the second in the Aptian. These unconformities have a significant bearing on both the Cretaceous stratigraphy and the hydrocarbon potential in the region.

Stratigraphic data from onshore exposures in the Wessex Basin, from over 100 boreholes, and seismic data document the evolution and erosion pattern of both unconformities. The Aptian unconformity has been well documented by previous authors, and may be related to uplift associated with commencement of sea-floor spreading in the Bay of Biscay. However, an older unconformity can be identified as Berriasian in age on the basis of tracing its correlative conformity using vertical and lateral facies changes and a rapid lateral increase in the thickness of the Wealden Group. This older unconformity is herein interpreted to have been formed by a latest Jurassic–earliest Cretaceous uplift event and can be recognized in all basins surrounding the Cornubian Platform. The areal extent of the unconformity implies that the interpreted uplift was of regional significance, centred on the Cornubian Platform, but unlike the younger Aptian event was unrelated to any specific intra-basinal extensional event or the Apto-Albian sea-floor spreading in the nearby Bay of Biscay. Its effects dominated Early Cretaceous sedimentation, the results of which are demonstrated in subsidence patterns within these basins.

The thermal maturity of the Lias (Early Jurassic), the main hydrocarbon source rock in the area, has also been affected by the latest Jurassic–earliest Cretaceous uplift, as well as Tertiary uplift and inversion. The latest Jurassic–earliest Cretaceous event resulted in the erosion of much of the Jurassic succession in the basins closest to the Cornubian Platform, and in many areas uplifted the Lias from the top of the oil window. Only on the flanks of the uplift, where high sedimentation rates occurred during the deposition of the Purbeck and Wealden Beds, were source rocks buried to depths sufficient to generate hydrocarbons. Only by understanding the tectonic evolution of the latest Jurassic–earliest Cretaceous uplift can accurate play risk modelling and prediction of the hydrocarbon fairway take place.

The Celtic Sea, Wessex Basin and the Western Approaches Trough ('the Basins') are situated on- and offshore between the land masses of southern Ireland, southwest Britain and northwest France (Fig. 1). These Basins actively subsided during the Mesozoic but were all offset from the main axis of North Atlantic rifting. Their preserved stratigraphy records a number of inter- and intra-basinal unconformities during the Middle Jurassic to 'Mid'-Cretaceous, two of which, identified in the earliest Cretaceous (e.g. McMahon & Underhill 1995) and in the Aptian (e.g. Ruffell 1992), appear to be the most significant. This paper concentrates on the erosion associated with both events (but in particular the earliest Cretaceous unconformity), the possible causes for their formation and reasons for separating these two unconformities which are commonly grouped together and assigned an Apto-Albian age.

Information from onshore exposures, over 100 exploration wells (Fig. 2), additional research boreholes, regional seismic lines and gravity data were integrated using a detailed biostratigraphic framework to describe the Late Jurassic–Early Cretaceous evolution of the Basins. Where possible, wells from the basin depocentres, where the most complete stratigraphic succession would be preserved, were used. Subsidence analysis was also undertaken on 27 wells calibrated using apatite fission track (AFTA) and vitrinite reflectance data with a view to understanding the tectonic evolution of the area. The results document the erosion, subsidence and sedimentation patterns that developed during the Mesozoic in the basins surrounding

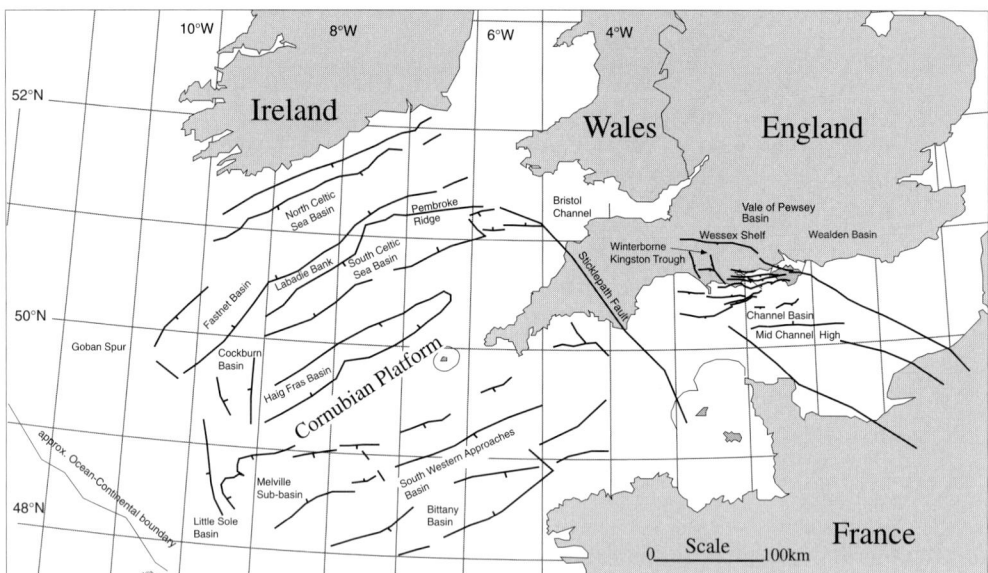

Fig. 1. The location of the main basins discussed in the text. The structure was complied from Lake & Karner (1987), Ziegler (1987a–c) and Petrie et al. (1989). (All subsequent maps in this paper use the same structural data.)

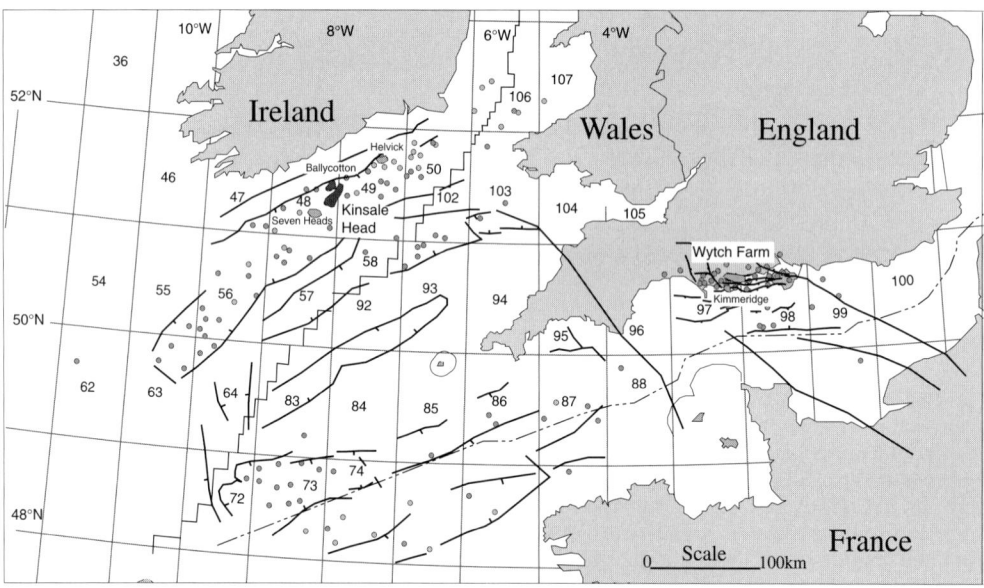

Fig. 2. The location of the wells used in this study and the main oil and gas fields in the area of interest.

the Cornubian Platform. This allows an evaluation of the likely mechanism for the development of the major intra-Cretaceous unconformities and a critical assessment of the likely contribution of intra-basinal extension and Apto-Albian sea-floor spreading in the Bay of Biscay.

The Early Cretaceous was also the time of initial hydrocarbon generation in the study area (Stoneley & Selley 1986). However, apart from the discovery of the producing Wytch Farm oilfield, and the Ballycotton–Kinsale Head gas fields (Fig. 2), and despite containing all the elements of a hydrocarbon fairway (source, reservoir, trap) the discoveries in this area have been disappointing compared to the neighbouring North Sea oil and gas province. This paper attempts to show the effects of Early Cretaceous tectonics on the maturity of the main hydrocarbon source rocks and, therefore the hydrocarbon prospectivity of the area.

Geological setting

The Basins are subdivided into two areas for this discussion, the Wessex Basin and the Celtic Sea Basins–Western Approaches Trough, as they display different late Mesozoic depositional environments. However, on the whole, there are gross similarities in the depositional environments of the Triassic–Middle Jurassic section and the Late Cretaceous–Tertiary section.

The Wessex Basin

The Wessex Basin is composed of a number of sub-basins namely; the Channel Basin, the Winterborne Kingston Trough, the Vale of Pewsey Basin and the Weald Basin (Lake & Karner 1987) which will be described collectively, except where used as specific examples.

Throughout the latest Jurassic–earliest Cretaceous the sedimentary sequence in the Wessex Basin reflects a gradual relative fall in sea level, climaxing in a widespread unconformity surface (the intra-Cretaceous unconformity of McMahon & Underhill 1995). Other unconformities can be recognized during this period particularly in the Portlandian (e.g. Coe 1996) where erosion appears to increase in a westerly direction. This drop in relative sea level was initiated during the Late Jurassic rift phase and resulted in the transition from the open marine Kimmeridge Clay, through the shallow to marginal marine Portland and Purbeck Group to the non-marine deposits of the Wealden (Fig. 3). However, by the Barremian times a gradual increase in relative sea level was initiated with marginal marine incursions in the Weald Basin (Allen 1975) and in the Channel Basin (Ruffell 1992) which increased in magnitude and spread westwards eventually leading to the deposition of the Aptian–Cenomanian Greensand Group. During this time renewed uplift, volcanic activity and fault movements, possibly associated with the opening of the Bay of Biscay, generated a number of unconformities the most significant of which was a Mid-Aptian event (Ruffell 1992). A further rise in sea level led to the deposition of the Chalk facies during the Late Cretaceous after which time the initiation of basin inversion commenced. Throughout the Early Tertiary, Late Palaeozoic–Mesozoic depocentres were up-lifted and became sediment source areas while former Mesozoic highs became major Tertiary basins (Lake & Karner 1987).

The Celtic Sea area basins

The Late Jurassic deposits of the Celtic Sea Basins and the Western Approaches Trough display distinct differences from the classical exposures documented onshore in the Wessex Basin (Millson 1987). The preserved Late Jurassic deposits throughout the Celtic Sea Basins are mostly composed of Oxfordian–Mid Kimmeridgian non-marine sediments (Millson 1987), although marine conditions existed to the south and west (Petrie et al. 1989). Major rifting and subsidence occurred during the Oxfordian with the formation of restricted basins (Millson 1987; Tucker & Arter 1987; Petrie et al. 1989; Shannon 1991a) and rotated fault blocks (Kamerling 1979; Roberts et al. 1981; Robinson et al. 1981). A brief transgression occurred in the Late Kimmeridgian and Early Portlandian but, as with the Wessex Basin, by Purbeckian times the area (apart from the Fastnet Basin, Biscay Margin and the Western Approaches Trough) was once again in a marginal marine to non-marine setting (Millson 1987; Petrie et al. 1989) before the deposition of the wholly continental lower Wealden Group in the Berriasian–Hauterivian period.

Throughout the Celtic Sea Basins and the Western Approaches Trough the pattern of Early Cretaceous sedimentation was dominated by the occurrence of a regional, earliest Cretaceous aged unconformity described by McMahon & Underhill (1995) amongst others (e.g. Ziegler 1982; Millson, 1987). This unconformity caused considerable erosion of basin margins and centres producing a significant gap in the Mesozoic succession. The unconformity occurred towards the end of the Late Jurassic–Early Cretaceous regressive sequence, and was followed by a transgressive sequence characterized by marine incursions into the non-marine Wealden in the latest Hauterivian (McMahon & Underhill 1995). Throughout the Late Hauterivian to the Albian a series of transgressions from the southwest, building up in intensity and magnitude, punctuated by regressions and unconformities, penetrated the Early Cretaceous terrestrial basins up to the Apto-Albian when a period of erosion is identified (Colley et al. 1981; Ainsworth et al.

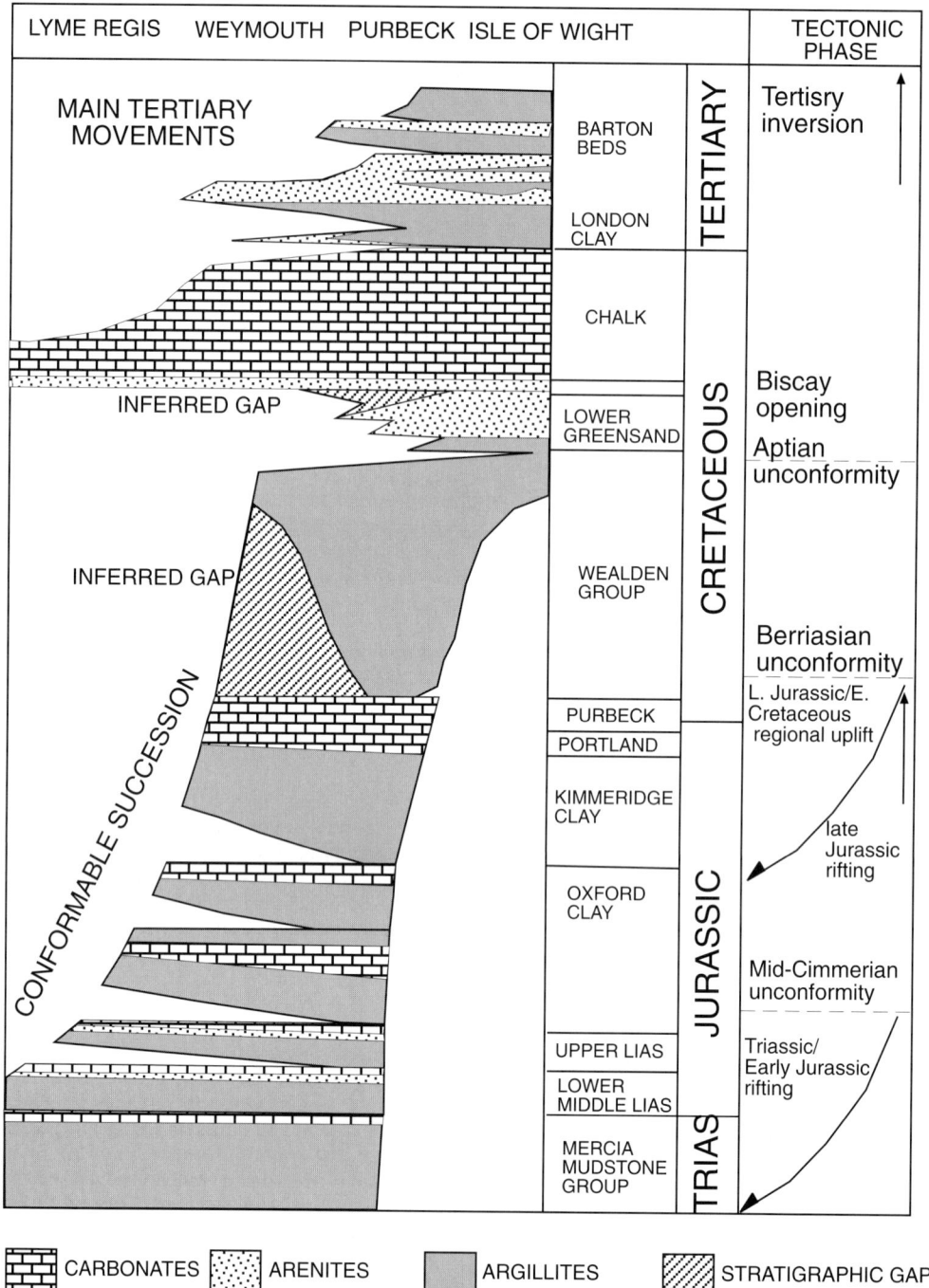

Fig. 3. The sedimentological and structural history of the Wessex basins showing the stratigraphic position of the main formations and groups named in the text. Adapted after House (1989).

1987; Evans 1990; Tappin et al. 1994). As with similarly aged events in the Wessex Basin, discussed above, these unconformities are thought to be the result of renewed tectonic activity (Ziegler 1987b) which was associated with volcanic intrusions in the Western Approaches Basin (Bennet et al. 1985; Ziegler 1987b) and the emplacement of oceanic crust first on the northern Biscay margin in the Early Aptian (Montadert et al. 1979; Masson & Miller 1984) then west of Goban Spur in the Middle Albian (Scrutton 1985), and with intensive rifting in the Bay of Biscay.

Intra-Cretaceous unconformities

Evidence for major discrete Berriasian and Aptian unconformities

Many workers suggest that erosion and deformation during the late Mesozoic did not occur at one discrete point in time but over a period from the Late Jurassic to the Aptian (e.g. Roberts et al. 1981; van Hoorn 1987). However, this paper aims to show that discrete erosional events within the Early Cretaceous were responsible for the majority of the erosion associated with the large stratigraphic gap observed in the Mesozoic succession.

The Late Jurassic–Early Cretaceous stratigraphy of the Wessex Basin is dominated by two significantly different depositional environments; the marine strata of the Late Jurassic and the Aptian–Albian; and the non-marine to marginal marine strata of the Valanginian–Aptian period. This stratigraphic sequence appears to correspond to a large-scale regressive–transgressive cycle and therefore one would expect a major unconformity at the regressive–transgressive boundary. This boundary has been given a Berriasian–Valanginian age by McMahon & Underhill (1995) and occurs at the top of a large-scale coarsening-upwards sequence from the Kimmeridgian–Portlandian marine clays to the non-marine Wealden (Fyfe et al. 1981).

Unconformities in the Wealden Group have also been recognized with a similar age to that of the maximum extent of the latest Jurassic–earliest Cretaceous regression. In the Channel Basin, wells 98/11-1 (Ruffell 1991) and 99/12-1 (Hamblin et al. 1992) record a pronounced unconformity, as do onshore exposures along the Dorset coast (Hesselbo & Allen 1991). These erosive surfaces lie in the lowermost Wealden Group (possibly Berriasian) and appear to die out in an easterly direction in the Wessex Basin where the Wealden succession thickens rapidly. Other evidence for widespread erosion during this time are the exotic minerals from the Cornubian Massif (Allen 1975) and Early Jurassic fossils which occur within the Wealden sediments. More tentative observations for Early Cretaceous erosion come from the interpreted palaeovalleys in northern France (Juignet et al. 1973) and the northern margin of the Wessex Basin (Ruffell & Wignall 1990).

However, in order to establish the exact age of the regressive–transgressive boundary and associated unconformities the first marine incursions (the initiation of the transgressive cycle) into the non-marine setting (the Wealden) must first be traced.

The sediments of the Wealden Group are generally represented by non- to marginal marine facies throughout the Basins. This characteristic poses a major problem when defining the age of any event which lies in the earliest Cretaceous. However, detailed analysis of the shale–mudstone successions from the Celtic Sea Basins has identified a number of transgressive and maximum flooding surfaces defined by biostratigraphic data (McMahon 1995). These surfaces appear to mark the initiation or peaks of transgressions into the Wealden Group, a trend which would continue throughout the Early Cretaceous ending with the deposition of the Greensand and Chalk Groups. Exact dating of these surfaces is difficult but it appears that many of the events herein identified in the Celtic Sea Basins are synchronous with (Early Cretaceous) transgressions already documented in the Wessex, Wealden and the Paris basins (e.g. Pomeral et al. 1989; Stewart et al. 1991; Ruffell 1992). The base of these shale–mudstone units is in many places erosional, often entraining pebble-sized clasts of the underlying sand unit (observed in the pebble beds of the Wealden Group of the Wealden Basin) and may represent ravinement surfaces (sensu Nummedal & Swift 1987).

The transgressions into the non-marine Wealden of the North Celtic Sea Basin (and later in the South Celtic Sea Basin) appear to have advanced from the southwest and increased in magnitude throughout the Early Cretaceous. The main events identified, dated using confidential oil industry biostratigraphic reports, can be found in the latest Hauterivian–earliest Barremian, Mid- to Late Barremian, latest Barremian–earliest Aptian; Mid- to Late Aptian; latest Aptian–earliest Albian; latest Albian and in the Mid-Cenomanian. These transgressions suggest that the reappearance of marine conditions in this region occurred well before the deposition of the Greensand Group in the Apto-Albian, thereby dating the Early Cretaceous regression–transgression boundary as pre-Hauterivian in age. This characteristic together with the observation that Wealden sediments lie directly on the unconformity surface (further described in the next section) suggest that a Berriasian age could be ascribed to this earliest Cretaceous unconformity.

The argument for a major Apto-Albian aged unconformity has been widely published and is best summarized by Ruffell (1992) who describes the unconformity as varying in age from Early Aptian to Early Albian depending on the location and the first transgressive unit. This event can be clearly identified using data from the basin palaeodeeps, where the maximum section is preserved, the amount of missing stratigraphy is seen at many localities where it erodes into the Wealden Group. However, it is important to separate this event from the older Berriasian-aged unconformity, as the latter may have caused the majority of erosion previously attributed to the Apto-Albian event.

The stratigraphic gap associated with the unconformity

A major stratigraphic gap in the Mesozoic succession can be recognized in the Wessex Basin and is best observed in the coastal outcrops of Devon and Dorset where Cretaceous sediments lie unconformably on Jurassic or older strata. The amount of missing section decreases in an easterly direction from the Cornubian Platform, where Permian and older stratigraphy subcrop the unconformity surface, to the east, where erosion is limited to the Berriasian section of the lowermost Wealden Group (see McMahon & Underhill 1995, fig. 1) and also in the Aptian (Ruffell 1992). A stratigraphic gap is also recognized along south–north transects from the basin centre in the Channel Basin increasing towards the Wessex Shelf to the north. However, this gap is exaggerated due to the basin margin location of most of this section which was extensively eroded during Wealden deposition. Whilst it is possible to recognize two separate unconformities within the Early Cretaceous, it is unclear from the stratigraphic record in areas which have undergone substantial erosion whether both, or only one, of the unconformities was responsible. However, where discrete events can be identified, erosion at the base of the Wealden is profound in many places (described below), whereas erosion associated with the Aptian unconformity is less significant and only appears to penetrate the uppermost Wealden succession even in the basin centres. Therefore, it is possible that the Berriasian unconformity developed parallel to the main structural grain of the Wessex Basin and was responsible for the large stratigraphic gap which exists in western Devon and on the Cornubian Platform.

The stratigraphic gap is again recognized in the South Celtic Sea Basin and Fastnet Basin where over 500 m of Wealden sediments sit directly on Bathonian or older strata (Tappin et al. 1994). Wealden deposits of Hauterivian–Aptian age also rest unconformably on Triassic or Lower Jurassic rocks in the Western Approaches Basin (Chapman 1989; Evans 1990), and these deposits are also thought to lie unconformably on older stratigraphy in the Hurd Deep north of the Channel Islands (Curry et al. 1970) and on the margins of the Plymouth Bay Basin (Lott et al. 1980) and in the Bristol Channel Basin (Fyfe et al. 1981). However, in the central parts of the North Celtic Sea Berriasian deposits lie conformably on Lower Purbeck (Colin et al. 1981; Naylor & Shannon 1982), as is true in the Brittany Basin, Western Approaches Trough (Evans 1990) and interpreted on the Biscay Margin (Masson & Roberts 1981) and the Goban Spur Basin (Colin et al. 1992). In these areas it is thought that a correlative conformity of Berriasian age exists (Fig. 4), nevertheless, any basin margin or basin centre highs present at this time is expected to have been eroded. The interpretation of the chronostratigraphic diagram is also mirrored on regional seismic data in the same orientation (Fig. 5). Line MPCR-17 (composite) shows the progressive subcrop of older stratigraphy beneath the Berriasian unconformity from northwest to southeast. Outside the area of erosion the Wealden deposits change character with distance from the Cornubian Platform (Ruffell 1995) (Fig. 6) with a typical non-marine succession passing southwest into marine clays and eventually into shelf limestones on the continental margin (Masson & Roberts 1981).

Using the above information, together with published data and proprietary oil industry reports, a subcrop map can be constructed from the most basinal setting (Fig. 7). This map reveals the progressive subcrop of older stratigraphy, Late Jurassic–Permian or older, towards the Cornubian Platform below Wealden or younger Cretaceous strata. Outside the edges of this elliptical pattern lies the correlative conformity, where a complete stratigraphy exists commonly represented by 500–1000 m of non-marine Wealden or marginal to marine 'Wealden equivalent' deposits. From the depositional environments of the formations surrounding

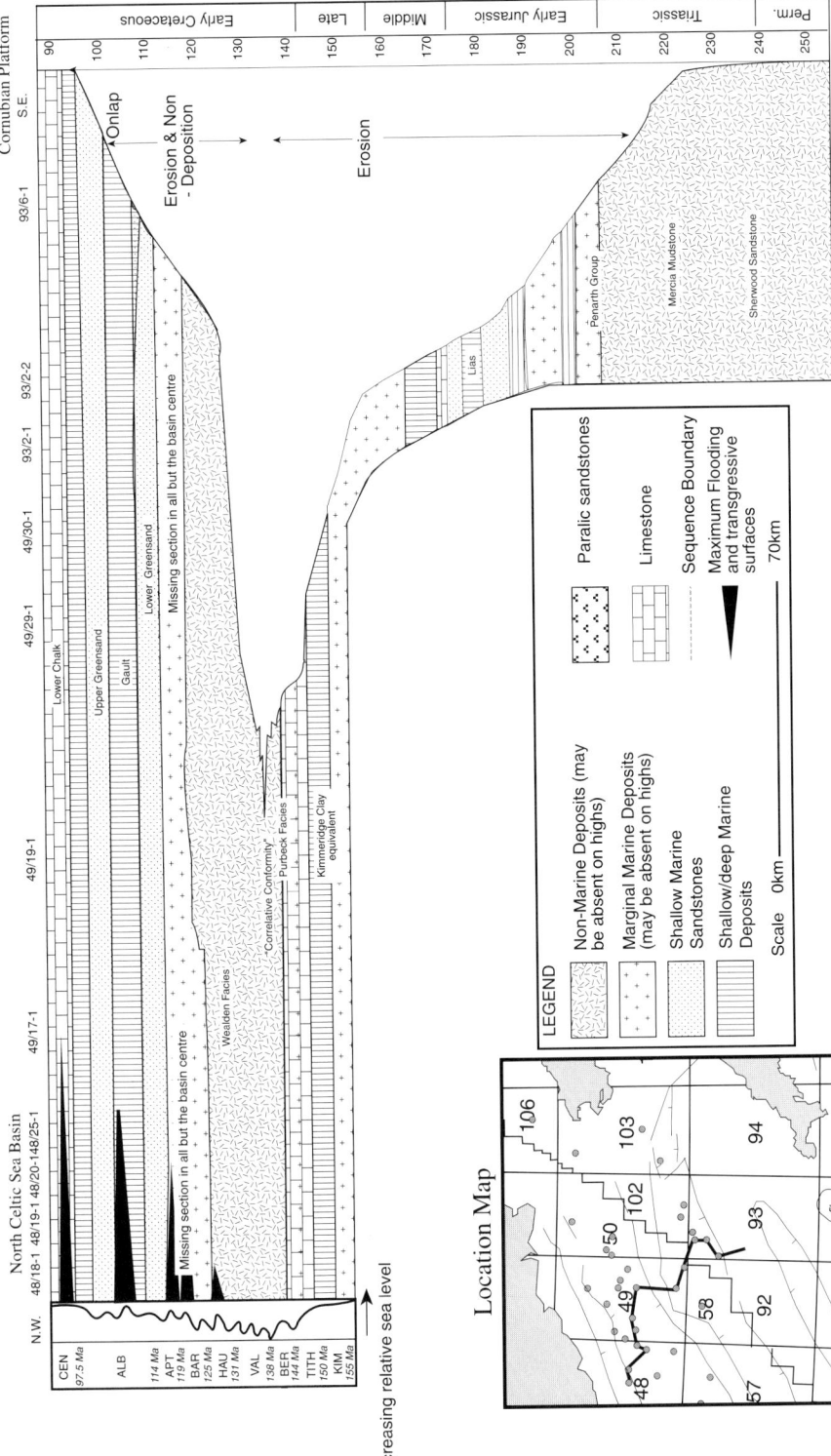

Fig. 4. A chronostratigraphic diagram from the Cornubian Massif into the North Celtic Sea Basin, avoiding footwall high locations where possible, showing the stratigraphic gap associated with the Berriasian unconformity. This gap is observed to widen towards the Cornubian Platform from the North Celtic Sea Basin.

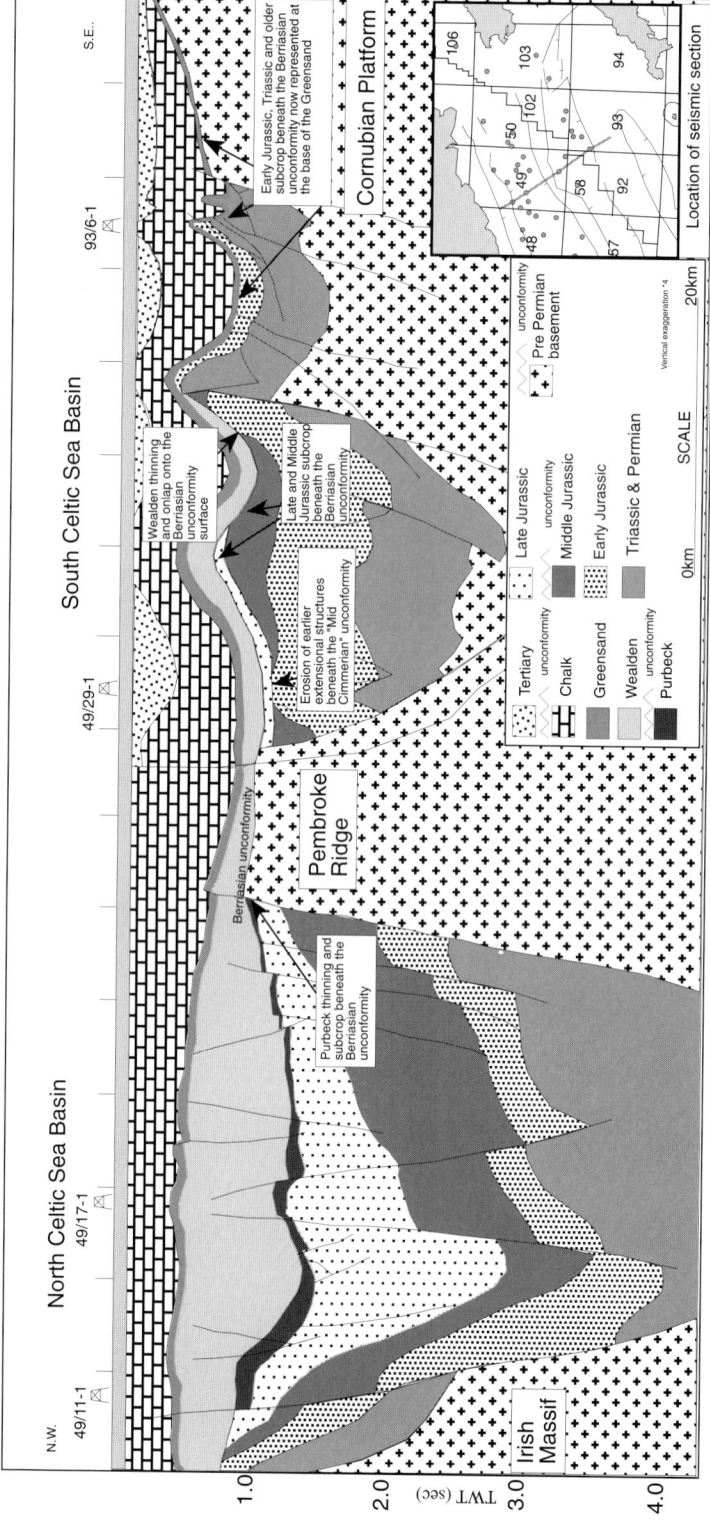

Fig. 5. Interpretation of the composite seismic line MPCR-17. This northwest-southeast-orientated line shows the southeastwards truncation of older stratigraphy beneath the Berriasian unconformity. It also shows the existence of an earlier 'Mid-Cimmerian' unconformity. The tops of the lower units, such as the Early Jurassic, Triassic and basement, are tentative in the North Celtic Sea Basin due to poor seismic quality and the lack of well penetration. The North Celtic Sea Basin is also interpreted to have experienced a higher degree of inversion during the Tertiary than the South Celtic Sea Basin as few deposits of this age have been observed in this area.

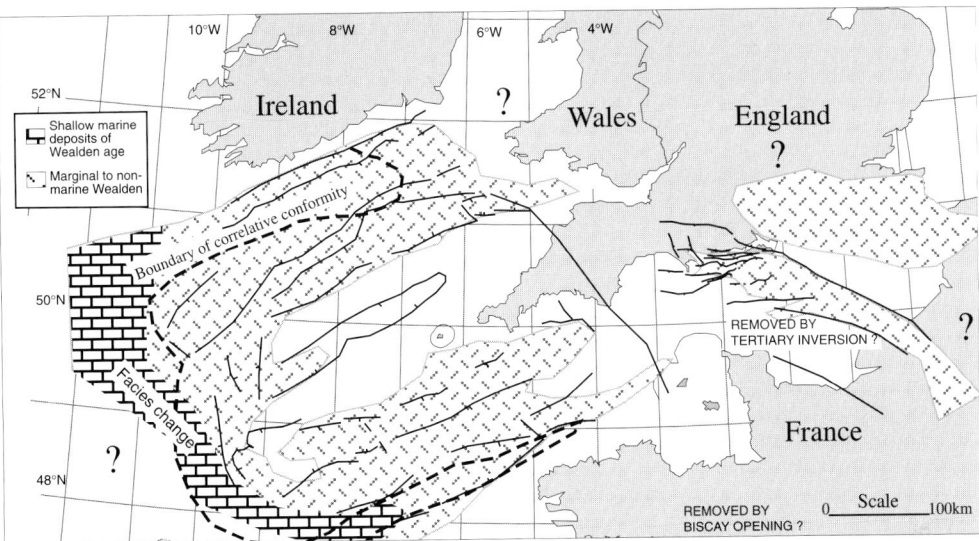

Fig. 6. The distribution of Wealden sediments, from well penetrations, seismic data, regional interpretations and the published work of Lott *et al.* (1980), Evans (1990), Tappin *et al.* (1994). Some areas have been affected by severe Tertiary erosion which has removed any possible Wealden strata which may have been present. The change in pattern denotes a facies change from non- to marginal marine deposits to shales and limestones.

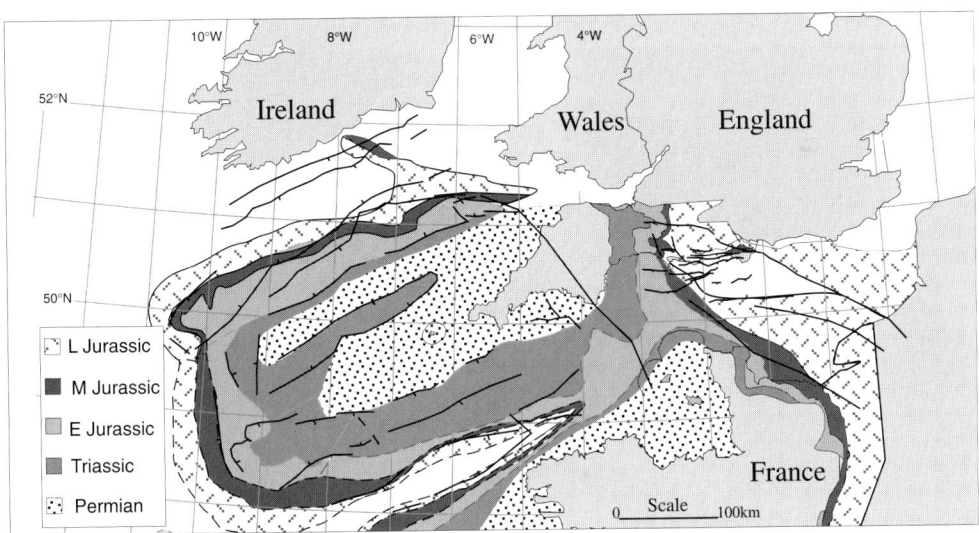

Fig. 7. The subcrop map beneath the Berriasian and Aptian unconformity. White areas are interpreted to have suffered no or minimal erosion during Berriasian and in these locations a correlative conformity to this event exists. Basement highs and ridges which may have experienced many periods of erosion throughout the Mesozoic have not been included. Erosion associated with the Berriasian unconformity increased towards the Cornubian Platform where Permian or older strata exists beneath Late Cretaceous sediments. Dashed lines indicate boundaries which are poorly constrained.

the eroded area, a relationship is evident between the most basinal shift in facies (the peak of the latest Jurassic–earliest Cretaceous regression) and the timing of the maximum areal extent of erosion which appears to be Berriasian in age. After this time the Wealden deposits became increasingly more marine through a series of transgressions and regressions (described in the last section) in an overall transgressive setting until the fully marine conditions of the Gault, Greensand and Chalk Groups are established.

It might be tempting to assume that such an elliptical subcrop pattern purely results from existing topography and the buoyant nature of the Hercynian granites in the Cornubian Platform and the Armorican Massif. However, the area of erosion appears to extend well past the termination of the granites, interpreted from gravity data to end in UKCS quadrant 84, indicating that the Cornubian Platform may only have played a part in the latest Jurassic–earliest Cretaceous uplift. Independent uplift studies using the apatite fission track analysis method (AFTA) from Palaeozoic sediments from Devon and Cornwall, near the centre of the Cornubian Platform, reveal that the most significant uplift event experienced by these rocks occurred in the earliest Cretaceous (Chen et al. 1994 unpublished; Bray et al. this volume). Before this time the Cornubian Platform and the Armorican Massif were probably not significant highs as the Lias was deposited in a deep-marine setting in west Dorset, close to the edge of the Cornubian Platform, and shows no signs of shallowing in a westerly direction. The subcrop pattern about the Cornubian Platform is, however, distorted by the Armorican Massif and the Biscay Margin (Fig. 7). One explanation for this is the uplift and erosional effect of the Aptian unconformity mentioned earlier, possibly a result of the emplacement of oceanic crust rifting in the Bay of Biscay. Again, although a large stratigraphic gap in the Mesozoic succession is obvious it is difficult to differentiate between the erosion associated with the Berriasian and Aptian unconformities. Only the record of a much larger stratigraphic gap beneath the Wealden Group of the Western Approaches Basin in wells 87/12-1a, 86/18-1 and 73/7-1 (Evans 1990) than beneath the Aptian–Albian section suggests that the Berriasian event was the most significant.

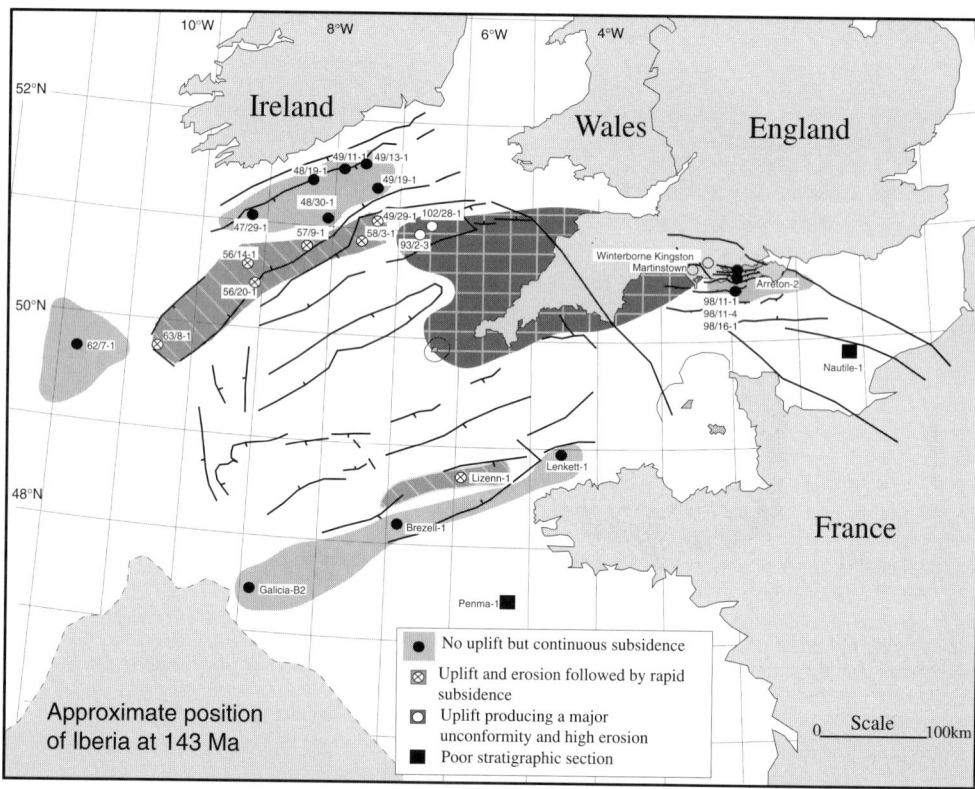

Fig. 8. The location of wells used in the subsidence study. The wells record three patterns of sedimentation and erosion associated with the latest Jurassic–earliest Cretaceous uplift: continued subsidence with no erosion; uplift during the Berriasian followed by rapid subsidence and uplift throughout the Late Jurassic and the Early Cretaceous.

Subsidence analysis, methodology and database

Twenty-seven wells were chosen for subsidence analysis from the Basins (Fig. 8) in order to quantify the tectonic response of these basins during the Early Cretaceous. This study was designed to cover a wider area than previous workers (e.g. Karner *et al.* 1987; Hillis 1988) who concentrated specifically on the Wessex Basin or the Western Approaches Trough. By using this regional approach it was possible to formulate a model for the pattern of Early Cretaceous tectonics from the preserved stratigraphy.

Subsidence analysis, or backstripping, was carried out using the method of Steckler &

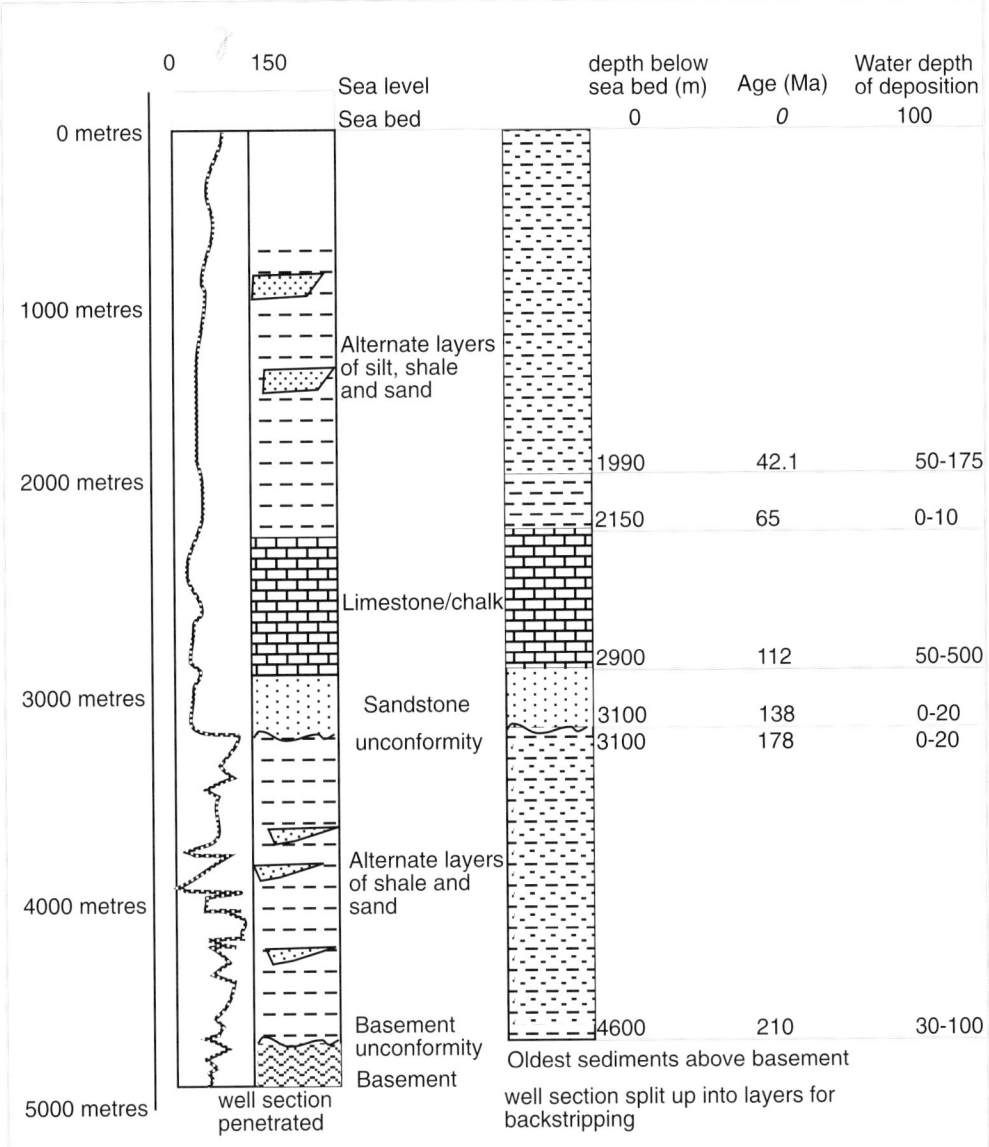

Fig. 9. Data preparation for the subsidence calculation. The succession found in each well is divided into layers the age, depth below sea level, water depth of deposition and bulk lithology of which are assigned values. The age, depth below sea bed and water depth of deposition of the oldest sediments sitting on basement are also required.

Watts (1982) and sedimentation rates were calculated from the simple equation of van Hinte (1978). The wells used were chosen because as they had the most complete stratigraphy and/or had penetrated the basement The successions in these wells were split into a number of layers, the tops of which could be assigned an age (Fig. 9). Each layer was assigned a bulk lithology and the water depth of deposition estimated for both its top and bottom. Unconformities were treated as periods of non-deposition except where information was available for the amount of missing section. This was the case for the Tertiary unconformity throughout much of the study area for which the amount of erosion was corrected using the apparent erosion estimates of Hillis (1995). The philosophy followed in this study has been to concentrate on the accuracy of the stratigraphic data and make only very simple corrections for sediment compaction. More compli-cated corrections are difficult or impossible to verify and are not included, therefore minimizing the risk of introducing artefacts from the data processing into the subsidence results. The porosity–depth relationships established by Sclater & Christie (1980) for North Sea lithologies were used for decompaction; the curves were plotted against the time scale of Harland et al. (1990).

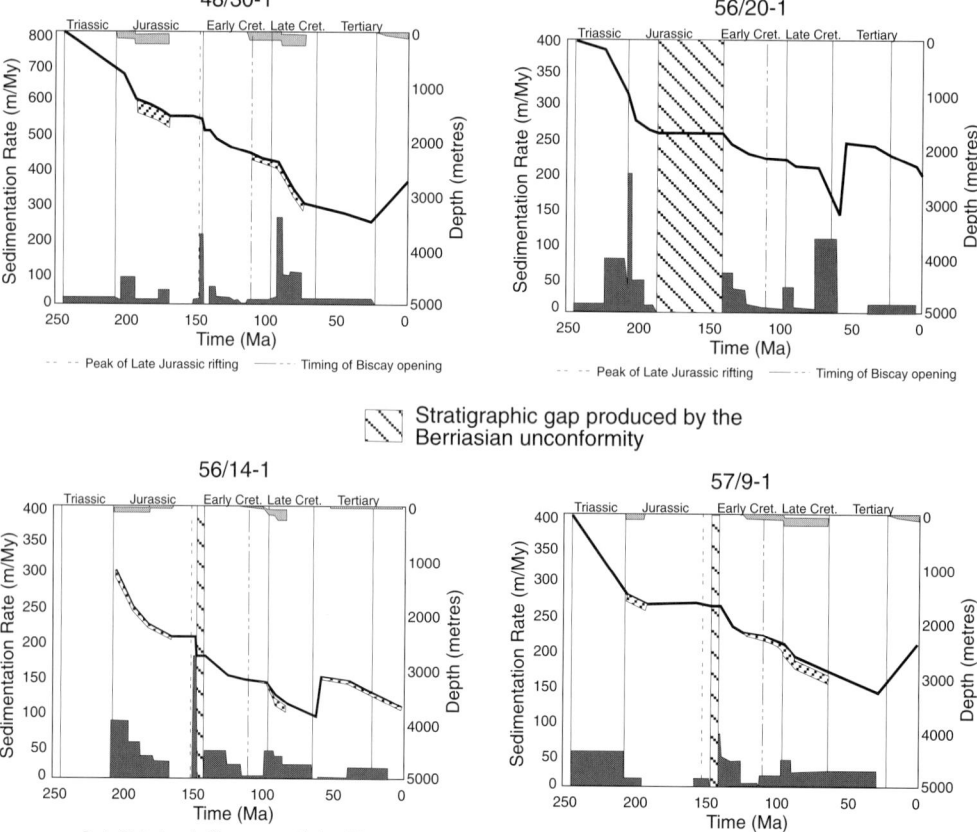

Fig. 10. Subsidence plots from the North Celtic Sea Basin. Sedimentation rates were fairly high during the Early Cretaceous with no obvious erosion associated with the Berriasian unconformity apart from 57/9-1. The locations of all these wells can be found in Fig. 8. The shaded area beneath the upper horizontal axis indicates water depths of deposition in metres. The solid line is depth to basement through time (in metres) with minimum and maximum water depth of deposition corrections indicated by the cross-hatched area beneath. Sedimentation rates (in metres per million years) are plotted along the bottom horizontal axis as a dark shaded area.

Although this simple approach introduces some errors and uncertainty into the subsidence calculations these are insignificant compared to the differences in the subsidence history between different wells, sub-basins and basins. The subsidence plots presented (Figs 10–13) show total basement subsidence, no correction has been made for sediment loading. Water depth information is also illustrated (Figs 10–13) where it was available.

Interpretation of the subsidence curves

North Celtic Sea Basin

Wells from the North Celtic Sea Basin (Fig. 10) show three major phases of high subsidence and sedimentation rates: the Late Triassic–Early Jurassic; the Late Jurassic; and the Late Cretaceous, followed by a period of inversion which started either in the Palaeocene or the Oligocene (depending on author and/or location). The first two phases of high subsidence and sedimentation rates are related to rifting events which started in the Norian and the Callovian, respectively. It is interesting that high sedimentation rates are also recorded from the Turonian to the Campanian. These high values, which are seen elsewhere (Turner unpublished data), may relate to the cooling of the Aptian- to Cenomanian-aged oceanic crust west of the Goban Spur–Biscay margin.

Only three wells record an unconformity which could be dated as earliest Cretaceous in age (56/14-1, 56/20-1 and 57/9-1), with 56/20-1 showing a major stratigraphic gap associated with the amalgamation of this unconformity and the older mid-Cimmerian event. It can be seen that sedimentation rates increase rapidly after the Late Barremian in these wells following a period of non-deposition, a feature which

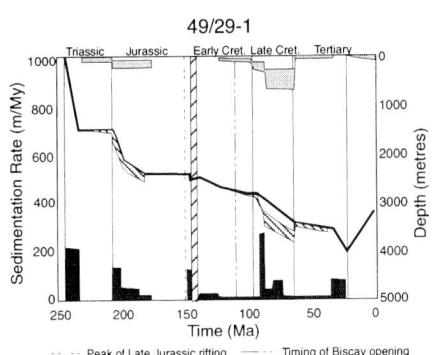
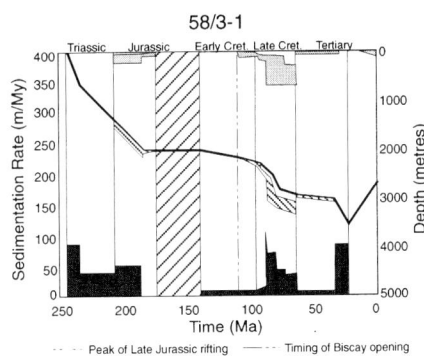
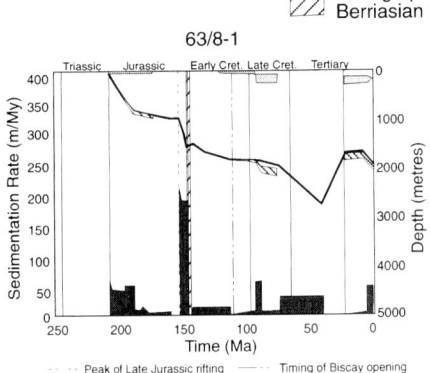
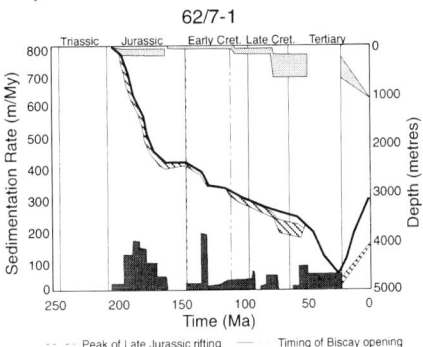

Fig. 11. Subsidence plots from the South Celtic Sea and Fastnet Basin (63/8-1) illustrating the erosion and shallowing associated with the Berriasian unconformity. Wells 93/2-3 and 102/28-1, not shown, show extreme erosion associated with Early Cretaceous tectonics. Subsidence on the Goban Spur margin (62/7-1) appears to be unaffected by Berriasian erosion. See Fig. 10 for key.

can also be seen in other wells in this area (e.g. 48/30-1). The increase in sedimentation rates also follows the climax of the latest Jurassic–earliest Cretaceous regression observed in the plot of water depth for all wells in the North Celtic Sea Basin. Following the increased sedimentation during the Early Cretaceous, sedimentation rates again slowed in the Aptian. An unconformity of Middle–Late Jurassic age is also recognized in wells 48/30-1, 56/14-1, 56/20-1 and 57/9-1 which possibly represents the mid-Cimmerian unconformity and may be associated with the eruption of the Fastnet volcanics (Caston *et al.* 1981).

South Celtic Sea, Goban Spur and Fastnet basins

Wells from the South Celtic Sea, Goban Spur and Fastnet basins (Fig. 11) do not show the same simple subsidence pattern that can be seen in the North Celtic Sea Basin. An earlier extensional phase, which began in the Early Triassic, is evident in all wells, apart from 63/8-1 and 62/7-1 which show a Late Triassic to Middle Jurassic extensional period of moderate subsidence. Subsidence data for the Late Jurassic are limited, apart from wells 63/8-1 and 49/29-1, as much of this section was removed by Early Cretaceous erosion and it is unclear whether an extensional phase took place in these basins at this time. The evidence from 63/8-1 and 49/29-1 suggests that a Late Jurassic rift phase did exist, which began possibly in the Late Oxfordian. As with the North Celtic Sea Basin, subsidence and sedimentation rates increased again during the Late Cretaceous, especially from the Turonian to the Campanian period, prior to Tertiary inversion in the Eocene–Oligocene.

A number of wells (63/8-1, 62/7-1 and 49/29-1) record the development of the mid-Cimmerian unconformity (Early to Mid-Jurassic). This event may also exist in wells 58/3-1, 93/2-3 and 102/28-1, but is undetectable due to younger Early Cretaceous erosion. A period of non-deposition and/or erosion associated with either or both the Berriasian and Aptian unconformities is evident in all wells, apart from 62/7-1 in which a correlative conformity is proposed to exist. Moderate sedimentation rates developed immediately after the Mid–Late Berriasian in wells 62/7-1, 49/29-1, 63/8-1 and 58/3-1. This increase in sedimentation may be either due to a short period of extension or the resumption in deposition in an area which had suffered erosion during the previous Valanginian and Berriasian period. Owing to the amount of missing section in many of these wells it is hard to comment on the effect that the Berriasian unconformity had on the water depths of deposition, although it seems to have decreased throughout the interval of erosion.

Western Approaches Trough and Brittany Basin

Subsidence plots from wells located in the Western Approaches and Brittany Basins (Fig. 12) show an almost complete stratigraphic record from Permian to Early Tertiary times. Three phases of high subsidence and sedimentation rates are observed on these curves. The first occurred in the Late Triassic–Early Jurassic, the second in the Late Jurassic and the third in the Early Cretaceous. However, sedimentation slows during the Berriasian with an unconformity being recorded in Lizenn-1 which shows minor erosion of the Purbeck strata. In Brezell-1 and Lennket-1 no obvious unconformity is identified and a correlative conformity, representing a complete Early Cretaceous section, is presumed to exist. Following the Berriasian, sedimentation rates again increase during the Hauterivian–Barremian with all wells recording a decrease in subsidence during the Mid–Late Aptian. Sedimentation rates again increased slightly during the Late Cretaceous before Tertiary inversion in the Oligocene.

Wessex Basin

Karner *et al.* (1987) suggested that subsidence in the Wessex Basin showed general background thermal subsidence through time punctuated by a number of rapid, finite and renewed subsidence events reflecting polyphase extension. Each stretching event was predicted to initiate in a major period of clay deposition and contain two phases: an active rift or mechanical phase (from 10 to 50 Ma in length) and a passive thermal phase (from 80 to 200 Ma in length) (Karner *et al.* 1987). This study shows the timing of these rapid rifting events and agrees with the overall passive thermal subsidence through time which has been supported by independent research (K. Thompson unpublished data). However, this work goes further than the study of Karner *et al.* (1987) by identifying the periods of rifting and uplift in basin formation.

All of the wells studied show a period (Fig. 13) of moderate subsidence and sedimentation rates from the Early Triassic to the Early Jurassic

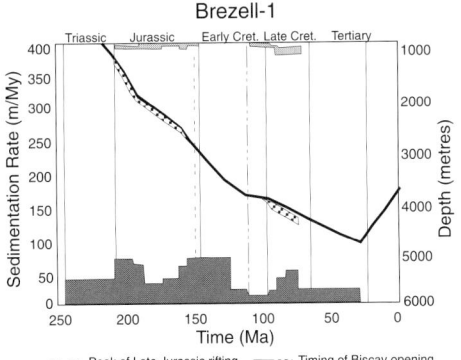
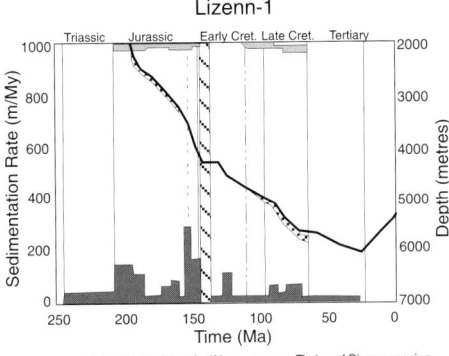
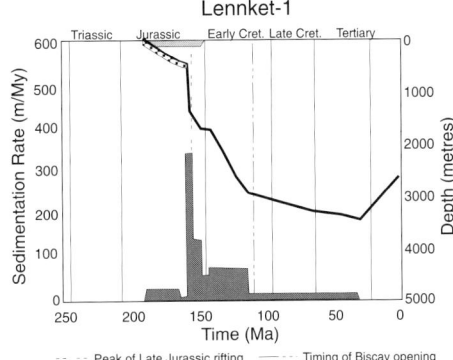

Fig. 12. Subsidence plots from the Western Approaches and Brittany Basins illustrating rapid subsidence in the Late Jurassic and high sedimentation rates after the climax of the latest Jurassic–earliest Cretaceous uplift. Erosion associated with the Berriasian unconformity is only recorded in Lizenn-1 and a complete section and correlative conformity are present in the other wells. The depth scale remains the same on all these plots so that they can be easily compared to other curves. See Fig. 10 for key.

associated with active extension. A period of Middle Jurassic extension is also evident in wells 98/11-1, 98/11-4 and 98/16-1, although in Nautile-1 and Martinstown this time is represented by low sedimentation rates. A Late Jurassic phase of high subsidence and sedimentation rates is evident in all wells relating to a rifting event which began in the Oxfordian and appears to have been terminated during the Valanginian. For the Early Cretaceous the subsidence curves from the wells fall into three groups: wells suffering no erosion associated with the Berriasian unconformity (i.e. 98/11-4, 98/16-1 and Arreton 2); wells which have minor erosion associated with the Berriasian unconformity (i.e. 98/11-1 and Nautile 1); and wells which have major erosion associated with both the Berriasian and Apto-Albian unconformities (i.e. Winterborne Kingston and Martinstown).

Moderate sedimentation rates are associated with Early Cretaceous sedimentation, especially in basinal locations such as in 98/16-1 and Arreton 2 whereas to the west and in basin margins to the north sedimentation did not resume until the Aptian or Albian (e.g. Martinstown and Winterbourne Kingston). All wells show either erosion or low sedimentation rates during the Mid–Late Aptian. This period of erosion and low subsidence is more pronounced in this area than in any other studied location indicating that separate or more pronounced regional tectonic forces affected the Wessex, Channel and surrounding basins at this time. Some wells (e.g. Winterbourne Kingston and 98/11-1) also show moderately high sedimentation during Chalk deposition in the Turonian–Campanian before Tertiary inversion which took place from the Eocene to the Oligocene.

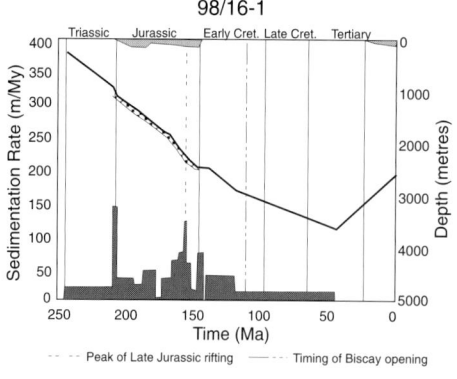
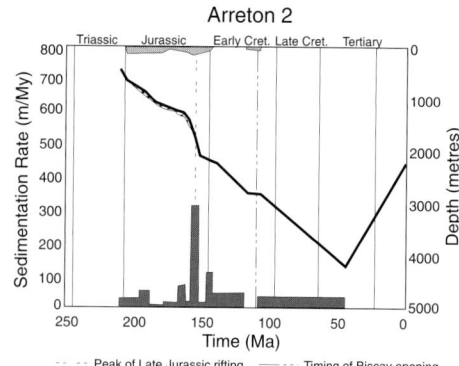
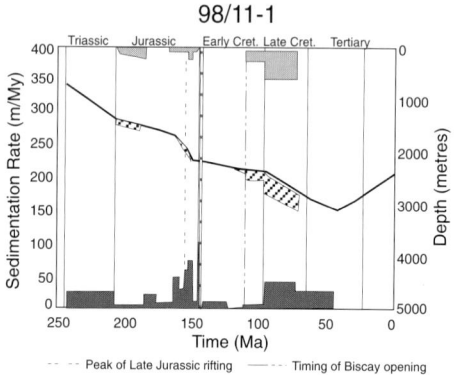
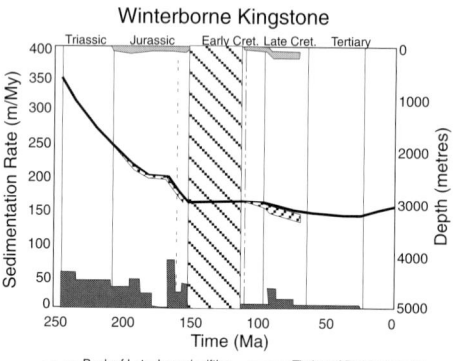

Fig. 13. Subsidence plots from the Wessex Basin. Three of the wells (98/16-1, 98/11-1 and Arreton-2) record no or very minor erosion associated with the Berriasian unconformity. Rapid subsidence associated with Late Jurassic rifting is observed to slow during the Berriasian and Valanginian. Winterborne Kingston shows major erosion associated with the Berriasian unconformity. This erosion is not a result of being located on a palaeo-high as these wells record continued subsidence from the Triassic. See Fig. 10 for key.

Summary of the subsidence results

A number of common features have been illustrated by the subsidence analysis. There appears to have been two main extension phases which occurred in all areas: the Triassic–Early Jurassic; and the Late Jurassic. A minor extensional phase may also have occurred in the Middle Jurassic. This may be related to the mid-Cimmerian unconformity which is best developed in the area surrounding the Fastnet Basin. A period of moderate to high sedimentation rates is also recognized locally during the Early Cretaceous (Hauterivian–Mid-Aptian). This could be interpreted to be associated with a minor extensional phase, although there is little evidence of extension prior to a minor phase in the Early Aptian (Ziegler 1987b) from seismic data. High sedimentation rates returned to the area during the Late Aptian–Albian and Late Cretaceous times after a period of low sedimentation or erosion associated with the Apto-Albian unconformity. These high rates may be related to either the infilling of new accommodation space formed by increasing sea level or an increase in regional subsidence due to sea-floor spreading in the North Atlantic and are probably not the result of active faulting.

The Berriasian unconformity exists in many wells as an erosional feature. Three distinct groupings of wells show different aspects of the unconformity's character:

(a) no erosion and the development of a correlative conformity;
(b) minor erosion and moderate to high sedimentation rates immediately after the unconformity;

(c) major erosion with sedimentation only resuming from the Barremian to the Albian.

It is also interesting that the rifting event in the Late Jurassic appears to have been terminated in the earliest Cretaceous. The view that rifting ended around the time of the Berriasian unconformity has also been suggested by a number of authors (e.g. Robinson *et al.* 1981; Cook 1987; Tucker & Arter 1987; Shannon 1991*b*) and implies that most of the Early Cretaceous succession qualifies as post-rift, although it has long being regarded as syn-rift by many workers (e.g. van Hoorn 1987; Ziegler 1987*a*; Evans 1990). The increase in sedimentation rates during the Hauterivian–Aptian could also be associated with a renewed phase of rifting, as discussed above, although if extension took place it was minor in comparison with that of the Early and Late Jurassic.

The effects of the latest Jurassic–earliest Cretaceous regional uplift on subsidence curves

When the latest Jurassic–earliest Cretaceous regression is coupled with the existence of an earliest Cretaceous unconformity, which has been attributed a Berriasian age, one could draw the conclusion that an uplift event was responsible for the formation of both events. This uplift can also be recognized in the decrease in water depths of deposition from the Late Jurassic to the Berriasian and the transgressions onto the unconformity surface as well as the regional subsidence pattern up to the Apto-Albian. Figure 14 summarizes the results of the subsidence analysis together with the correlative conformity–unconformity boundary and the Permian or older subcrop boundary beneath the Berriasian unconformity. These results delineate three areas, each with a distinct subsidence history from the time of Berriasian unconformity development to the Apto-Albian.

Constant sedimentation and the correlative conformity

The first area represents the region in which the Berriasian unconformity does not result in any erosion of underlying strata in the basin centre (Fig. 14). During the Berriasian subsidence slowed, suggesting that the Early Cretaceous uplift had terminated the active faulting in the region. Sedimentation rates were generally low

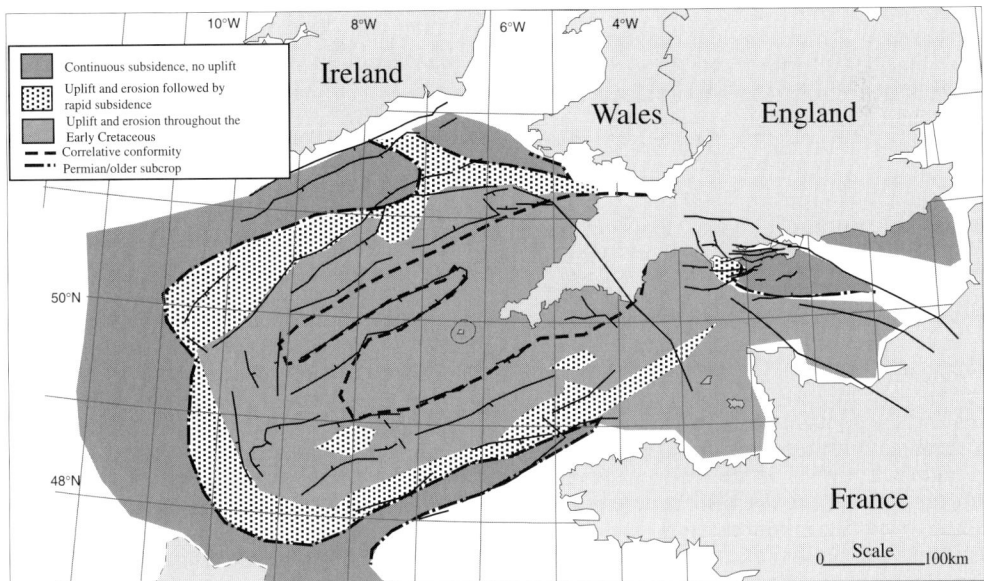

Fig. 14. Three zones which can be defined from the subsidence curves relating to the sedimentation patterns associated with the latest Jurassic–earliest Cretaceous uplift. Zone 1 exists outboard of the correlative conformity boundary. Zone 2 occurs between the Berriasian correlative conformity boundary and the Cornubian Platform. Zone 3 exists from the area surrounding the Cornubian Platform (indicated by the Permian subcrop boundary beneath the Berriasian and Aptian) onto the Cornubian Platform itself.

during this period although in some places they increased slightly immediately after the time of maximum uplift. This can be seen in the deposition of the Wealden Group which, in this area contains the erosion products from the region surrounding and including the Cornubian Platform. A facies change, described earlier, also exists in this region from continental deposits of the Wealden Group to their shallow marine carbonate equivalents. The above factors demonstrate that the effects of the latest Jurassic–earliest Cretaceous uplift in this zone were minor, resulting in erosion on highs and a regional sea-level fall which can be observed in the Late Jurassic–earliest Cretaceous regression and the formation of the Wealden.

Minor erosion, rapid post-uplift sedimentation

The second zone lies between the correlative conformity–unconformity boundary and the margins of the Cornubian Platform, although this zone suffers from the exact character of the poor data coverage. Wells within this zone (e.g. 57/9-1, 56/20-1 and Lizenn 1) show minor erosion associated with the Berriasian unconformity followed by rapid sedimentation for approximately 10 Ma after the unconformity climax until the Barremian. This zone is interpreted to represent the area which experienced immediate deflation of the uplift after its maximum areal extent. What was once an uplifted area during the latest Jurassic to the Early Berriasian, supplying sediment towards the area outside the correlative conformity boundary was, during the Valanginian–Barremian, itself on the margins of the uplift receiving sediment from the Cornubian Platform. Typically wells within this zone found pre-Barremian non-marine Wealden deposits resting unconformably on older stratigraphy.

Major erosion

The third zone represents the area immediately adjacent to and including the Cornubian Platform, which itself is delineated by the Permian subcrop boundary. Wells in this zone (e.g. 93/2-3, 102/28-1 and Winterbourne Kingston) all record major erosion associated with both the Berriasian and Aptian unconformities followed by sedimentation in the Barremian or Aptian–Turonian period. These wells do not simply correspond to highs which have experienced erosion during every drop in sea level. Subsidence curves show that their locations have subsided since the Early Triassic. This area, where the latest Jurassic–earliest Cretaceous uplift was most long-lived, was flooded by marine transgressive deposits in the Albian relating to the general eustatic sea-level rise of the Late Cretaceous. This is supported by research boreholes that have found Wealden deposits of Aptian and Albian age on the margins of the Plymouth Basin (Lott *et al.* 1980), next to the Cornubian Platform, suggesting that if an area of erosion existed during this time it was very small indeed.

Evans (1990) attributed the subsidence in the northern basins of the Western Approaches Trough and the Cornubian Platform to cooling of the lithosphere following a thermally driven uplift in the Late Jurassic. Although the evidence presented herein may suggest a slightly younger timing for this uplift (climaxing in the Berriasian) one might argue that the subsidence patterns in the three zones mentioned above reflect the waning of an uplift throughout the latest Jurassic–earliest Cretaceous. However, it is difficult to see how this uplift could have been sustained long after rifting had ended, without some permanent change to the structure or thickness of the crust.

The timing of the Bay of Biscay opening and the development of the Apto-Albian unconformity

Many workers (e.g. van Hoorn 1987; Ziegler 1987*a;* Evans 1990) have suggested that the rifting in the Celtic Sea Area Basins has been controlled by the Aptian crustal separation of the Bay of Biscay (Olivet *et al.* 1984). However, it has been shown, in this and other studies, that the rifting which initiated in the Late Jurassic probably terminated in the Berriasian–Valanginian with a renewal in extension during the Early Aptian. This suggestion implies that the Berriasian unconformity formed after the initiation of Late Jurassic rifting and before the crustal separation of the Bay of Biscay and, therefore, was unrelated to both tectonic events.

To qualify the last statement a knowledge of the timing of rifting and sea-floor spreading in the Bay of Biscay must be established. Montadert *et al.* (1979) and Olivet *et al.* (1984) suggested that sea-floor spreading initiated west of the north margin of the Biscay Margin in the Aptian which corresponded with a period of faulting and erosion (Ziegler 1982, 1987*c*). This tectonic period possibly corresponds to the slight deformation and erosion seen as far away as the

Wessex and Weald basins during the Aptian (Ruffell pers. comm)

Subsidence analysis from the Aquitaine Basin, at the southeastern end of the Bay of Biscay, suggests that a Late Jurassic–Early Cretaceous rift phase was followed by rapid subsidence in the Late Aptian and Albian related to sea-floor spreading in the Bay of Biscay (Curnelle & Dubois 1986; Desegaulx and Brunet 1990). This is supported by the subsidence analysis of two wells from the Bay of Biscay, Galicia B-2 and Penma-1 (Fig. 15). These wells date the timing and nature of the opening of the Bay of Biscay.

A Late Jurassic–Early Cretaceous rifting event is observed in Galicia-B2 with the transition from non-marine Late Jurassic clastics into shallow marine limestones of Early Cretaceous age (Fig. 16). This equates well with the timing of rifting in both the Aquitaine Basin (Desegaulx & Brunet, 1990) and the Galicia Margin (Boillot et al. 1989). No unconformity is observed in the Late Berriasian and therefore it is assumed that this well was outside the area of influence of the Berriasian unconformity. Deposition of shallow-marine limestones continued in Galicia-B2 until a phase of non-deposition or erosion in the Aptian, which marks a change in deposition from carbonate-dominated to clastic-dominated facies of the Albian. The interpreted erosion has resulted in the absence of the much of the Aptian section in Galicia-B2 and may relate to a period of uplift. The Aptian–Albian continental deposits of Penma-1, which rest unconformably on basement, attest to this uplift which soon declined with the initiation of widespread subsidence (seen in Fig. 15) and the formation of oceanic crust in the Bay of Biscay. Both Penma-1 and Galicia-B2 show continued subsidence throughout the Late Cretaceous and Early Tertiary before inversion in the Oligocene (Fig. 15).

The uplift associated with the crustal separation in the Bay of Biscay in the Aptian affected a wide area particularly on the rift flanks, possibly due to rising asthenosphere. Such an uplift probably led to the renewal of continental sedimentation from the Armorican Massif, Aptian–Albian unconformities in the Wessex and channel Basins, and the anomalous subcrop pattern of northwest France associated with the Berriasian unconformity, described earlier. This uplift event was not on the same scale as the earlier Berriasian event as it only caused major erosion on the flanks of the Bay of Biscay and minor erosion elsewhere.

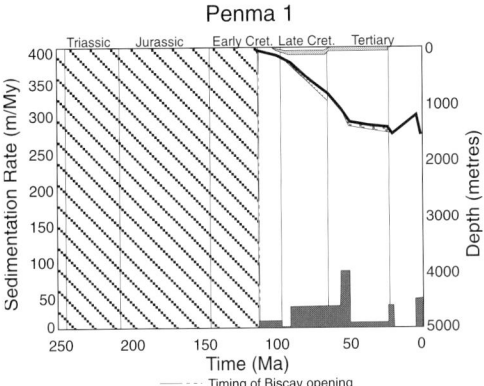

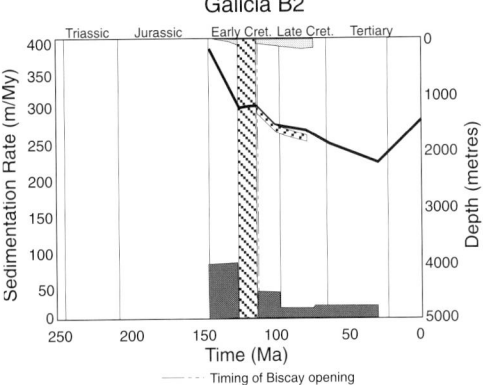

Fig. 15. Subsidence plots from the Bay of Biscay region showing erosion during the early Cretaceous. Note the rapid thermal subsidence in both wells in the Late Albian and Cenomanian associated with sea-floor spreading in the Bay of Biscay. Galicia B2 also shows rapid subsidence during the Early Cretaceous when areas to the northeast were experiencing uplift. See Fig. 10 for key.

The cause of the latest Jurassic–earliest Cretaceous uplift

The amount of Late Jurassic and Early Cretaceous stratigraphy preserved in the Basins decreases in direction towards the Cornubian Platform. This indicates that erosion and or non-deposition were greatest in the Cornubian Platform area during the latest Jurassic and earliest Cretaceous, suggesting that the area may have been actively uplifted at this time.

Roberts (1989) suggested that this uplift was the result of Late Jurassic–Early Cretaceous extension. However, many workers regard the

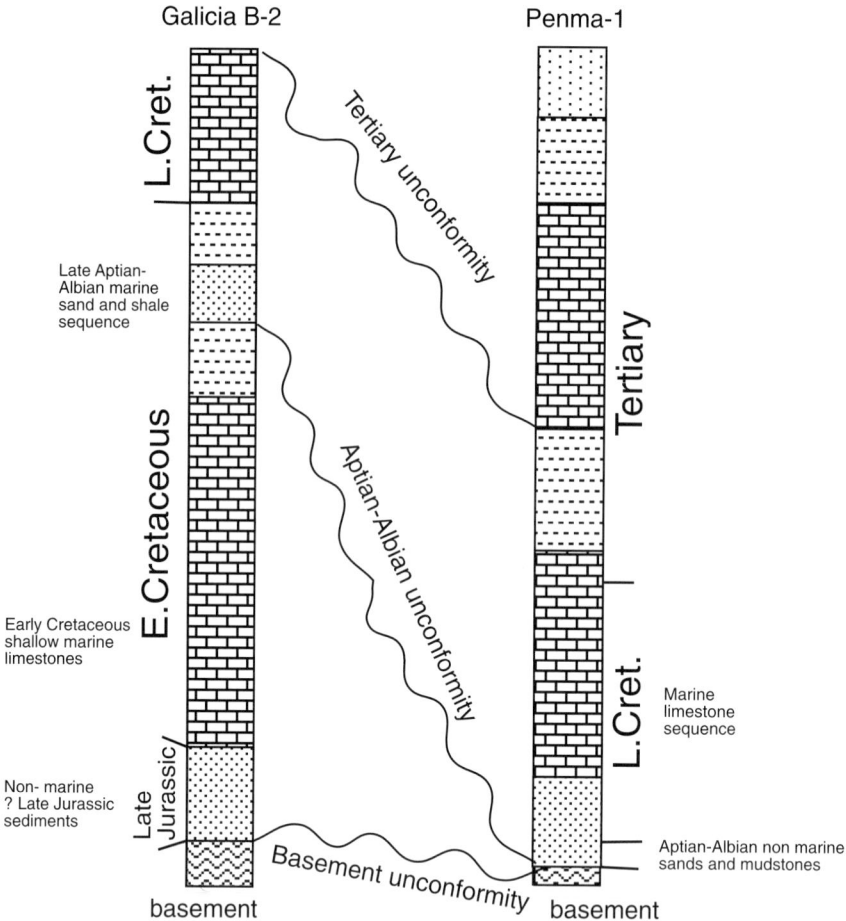

Fig. 16. Two wells from the Bay of Biscay region (for location see Fig. 8). Both wells show an Apto-Albian unconformity associated with Biscay opening in their successions and subsidence curves. Stratigraphic logs represent 1743 m and 1640 m of section, respectively, although they are not drawn to scale.

Berriasian unconformity to represent the cessation of Late Jurassic extension (e.g. Tucker & Arter 1987; Chapman 1989), with any extension in the Early Cretaceous being minor in comparison. Hillis (1988) also noted that extension, which may have started in the Callovian–Oxfordian, took place in a marine setting in many areas without substantial doming, only becoming non-marine in the Early Cretaceous long after rifting had initiated. The uplift is also unlikely to be associated with either North Atlantic ocean-floor spreading, as the Atlantic at the time of the uplift maximum (Late Berriasian) was only actively spreading below the Azores–Gibraltar Fracture Zone, or sea-floor spreading in the Bay of Biscay which began in the Aptian.

The subcrop map indicates that the size of this uplifted feature is at least 750 × 500 km across, which is somewhat smaller in size than uplifts predicted by mantle blobs (1500 km across for the North Sea; Underhill & Partington 1993, 1994) and for mantle plumes (>2000 km; White & McKenzie, 1989). However, some 2000 m of Jurassic sediment is estimated to have been removed from the northern shoulder of the Western Approaches Basin (Hillis 1988; Chapman 1989) and over 3300 m are thought to have been removed from the South Celtic Sea Basin (Tappin et al. 1994) during this event. This erosion was not simply due to footwall uplift or erosion specifically located on basinal highs, as even basin centres (e.g. the South Celtic Sea Basin) suffered from erosion. Such significant erosion can either be the result of a prolonged period of erosion (similar to North Sea Dome described by Underhill & Partington 1993, 1994) or by a significant vertical uplift which, on this scale, would suggest a thermal mechanism.

A thermal mechanism for the generation of the Early Cretaceous uplift was postulated by McMahon & Underhill (1995) based on the occurrence of Early Cretaceous volcanics in the vicinity of the Cornubian Platform which were offset from the main area of North Atlantic volcanism and the fact that the volcanics pre-dated the opening of the Bay of Biscay. Chapman (1989) also suggested that the interpreted doming and crustal extension during this time relates to mantle processes. Intrusive rocks of Early Cretaceous age (Hauterivian–Barremian) do occur in the Goban Spur, Brittany Basin (Ziegler 1987b), and on the Cornubian Platform at Wolf Rock and Epsom Shoal. During the Hauterivian–Barremian active faulting had probably ceased and had only a minor effect on sedimentation in the region (e.g. Tucker & Arter 1987; Chapman 1989) suggesting that the volcanics are unlikely to be related to active rifting although a further phase of volcanism occurred in the Apto-Albian in the Western Approaches Basin (Bennet et al. 1985) prior to and at the same time as the emplacement of oceanic crust west of Goban Spur and the Biscay Margin (Montadert et al. 1979; Masson & Miller 1984; Scrutton 1985; Cook 1987).

Implications for source rock maturity and hydrocarbon prospectivity

To date the number of hydrocarbon discoveries in the study area has been disappointing despite the region displaying optimum hydrocarbon generating and trapping mechanisms. This low success rate is generally regarded as being due to the lack or low maturity of source rocks, although no model has been generated to be of use predictively. The model developed here indicates that the absence or low maturity of the main known source rock, the Lower Jurassic Lias deposits, can be explained by uplift and erosion during the Late Jurassic and Early Cretaceous. The subcrop map (Fig. 7) can be used as a first step in assessing the potential source rock maturity of an area.

The subcrop pattern observed at Lyme Regis, in the west of the Wessex Basin, shows that in this area the Albian Gault and Greensand rest unconformably on the Lias. Using the results of the subsidence analysis it is possible to suggest that Middle–Late Jurassic sediments, deposited prior to the Early Cretaceous event, were uplifted and eroded. It is not surprising, therefore, that the maturity of the Lias in these locations is reduced as a result of this uplift and the sediments are today immature for oil generation (McMahon 1995). However, to the east, in an area south of the Wytch Farm oil field where the subcrop pattern indicates no erosion beneath the Berriasian unconformity, the Lias is known to have matured at least to the early gas generation phase.

Late Jurassic source rock maturity evolution in the Wessex Basin

The results from subsidence analysis indicate a phase of intense rifting in the Late Jurassic which was certainly region-wide during the Kimmeridgian (Fig. 17a) but more localized during the Portlandian. It is possible that the first signs of the Late Jurassic–Early Cretaceous uplift are recorded in the shallowing in the stratigraphic record from the Late Kimmeridgian into the Early Cretaceous. The effects of this uplift were probably minor in the east of the Channel Basin which continued to subside resulting in the Lias being buried deep enough to initiate oil generation (McMahon 1995). To the west, the uplift was more significant resulting in the still immature Lias deposits being exposed and eroded at the surface (Fig. 17b).

Early Cretaceous source rock maturity evolution of the Wessex Basin

The development of the latest Jurassic–earliest Cretaceous uplift is interpreted to have reached its peak in the Late Berriasian. Rifting had slowed or stopped by this stage and the entire area experienced the effects of uplift resulting either in erosion or in the shallowing of the stratigraphic record to a continental facies depositional style. Erosion occurred over the uplifted area to the west (Fig. 17c), whereas to the east only the highest areas, such as footwall high locations, were eroded. Basinal areas in the Channel Basin continued to receive sediment from the uplifted area and were slowly subsiding, further burying the Lias source rock which continued to generate oil.

Although both the east and west of the Channel Basin continued to subside throughout the Late Cretaceous subsequent to the uplift event, the Lias in the Lyme Regis area did not reach sufficient depths to mature further.

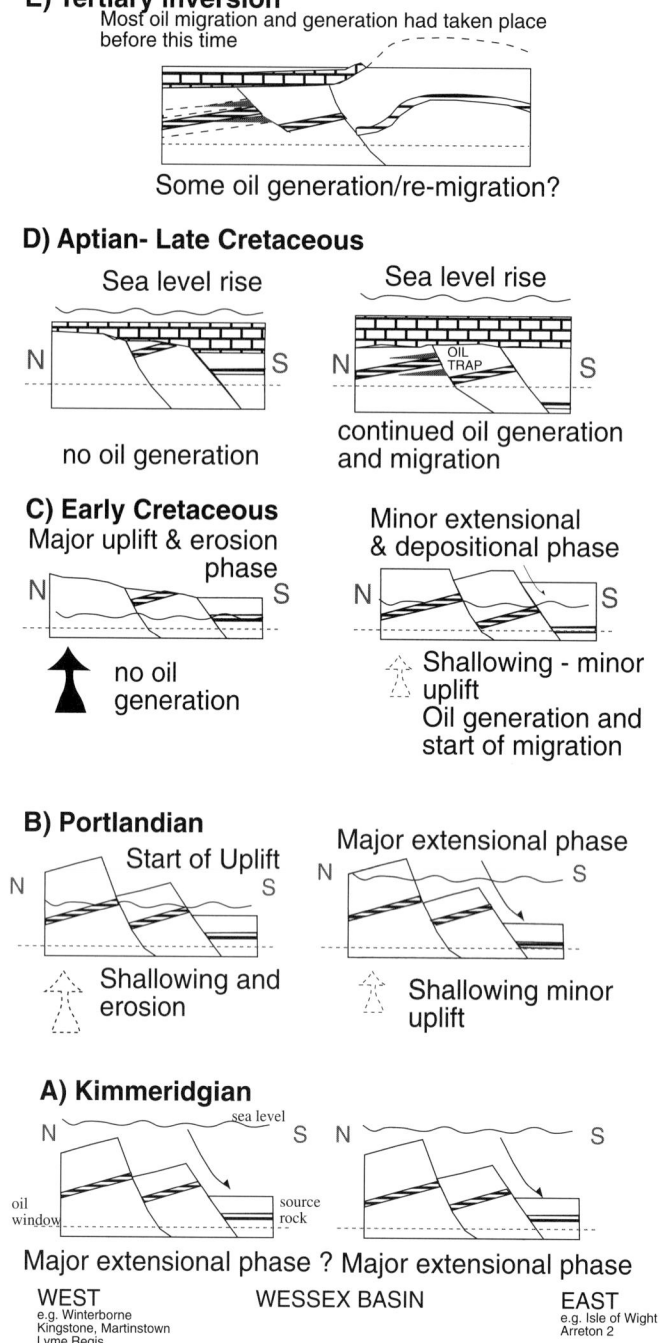

Fig. 17. Cartoons depicting the possible evolution of the Wessex Basin before, during and after the latest Jurassic–earliest Cretaceous uplift event. Clearly areas in the west Wessex Basin experienced erosion during the uplift during which time Lias source rock were exhumed. In the east Wessex basin sedimentation continued during the uplift and the Lias source rocks reached the depth of maturation. Such a scenario suggests that hydrocarbon prospectivity is limited to the east Wessex Basin where the only commercial hydrocarbon discoveries have occurred.

Late Cretaceous–Tertiary source rock maturity evolution of the Wessex Basin

The regional subsidence and rise in sea level throughout the Late Cretaceous led to the deposition of the Greensand and the Chalk facies over the areas which had experienced Early Cretaceous erosion. Renewed sedimentation continued the burial of the Lias succession and the juxtaposition of Late Cretaceous sediments on older units towards the west. In the east of the Channel Basin in the most basinal settings the Lias had entered the early gas phase (McMahon 1995) and by this time oil is likely to have migrated into the tilted fault block traps of the Wytch Farm oilfield (Fig. 17d). Oil and gas generation continued in the east until the Tertiary when widespread inversion uplifted even the most basinal areas (Fig. 17e). By this time oil generation and migration had taken place so that the Tertiary inversion event simply terminated hydrocarbon generation in a basin where the Lias was in the gas window.

Implications for hydrocarbon prospectivity

The above account illustrates that the hydrocarbon potential of the basin is governed by the erosion associated with the Berriasian unconformity, itself controlled by the distance from the Cornubian Platform on which it was centred. The Tertiary inversion event, contrary to generally held views, is of negligible importance in controlling the distribution of source rock maturity. This study therefore emphasizes the fundamental importance of including the effects of the Late Jurassic–Early Cretaceous uplift and the Early Cretaceous unconformities (particularly the Berriasian unconformity) in defining the source rock play fairway.

Conclusions

In the Wessex Basin, Celtic Sea Basins and the Western Approaches Trough a number of unconformities can be identified using outcrops, well logs and seismic data, combined with biostratigraphic analysis. By far the most significant of these events is an earliest Cretaceous unconformity, which has a correlative conformity in the Berriasian interval of the Wealden Group. The correlative conformity is dated as Berriasian from evidence in the North Celtic Sea Basin where deposits of Valanginian age overly the unconformity surface, and because the first regionally correlatable (accurately dated) transgressive event above the unconformity is of Late Hauterivian age. The Berriasian unconformity is more subtle in areas which experienced high sedimentation during the Early Cretaceous, although unconformities of this age are recorded at outcrop in the Wessex Basin and in boreholes throughout the study area. This event also corresponds to the regressive–transgressive boundary indicated by the change from the marine Late Jurassic, non-marine earliest Cretaceous to the marginal marine to shallow-marine Early Cretaceous. Evidence for an Aptian unconformity can also be observed particularly in the Wessex and Western Approaches Trough areas where the associated erosion cuts into the underlying non-marine Wealden succession.

Using all available data a subcrop map can be constructed for the study area representing erosion from both the Berriasian and Aptian unconformities. The majority of the erosion can be ascribed to the Berriasian event and shows that the stratigraphic gap associated with this unconformity increases towards the Cornubian Platform. Outside the area of erosion a correlative conformity, representing a complete stratigraphic section during the period of uplift, occurs within non-marine deposits on the uplift margins, and within shallow-marine deposits further to the southwest.

The extreme amounts of Early Cretaceous erosion surrounding the Cornubian Platform, and the shallowing facies trends recognized in the transition from marine to non-marine strata during the Late Jurassic to earliest Cretaceous, are interpreted to be the result of an uplift event. The timing of this event suggests that it initiated after the main period of Late Jurassic rifting had ceased, and before the emplacement of oceanic crust in the Bay of Biscay during the Apto-Albian. This indicates that most of the Early Cretaceous sedimentary succession post-dates the Late Jurassic extensional event and was deposited during a period of limited fault activity. With no direct tectonic mechanism for the formation of the uplift, circumstantial evidence from anomalous Early Cretaceous volcanics within the uplifted area implies that this region may have been uplifted by a thermal mechanism.

The sedimentation record of this uplift can be documented by subsidence analysis. The subsidence curves generated indicate that three distinct geographical zones can be identified, each of which have their own distinct sedimentary response to the uplift. The outermost zone (Zone 1) is characterized by the absence of erosion, with sedimentation in the Early Cretaceous. An intermediate zone (Zone 2) exhibits minor erosion associated with the Berriasian unconformity and subsequent high sedimentation rates. The inner zone (Zone 3) records a major stratigraphic gap produced by significant erosion with sedimentation not resuming until the Barremian. Subsidence analysis of these zones also indicates that the uplift was transgressed during the Late Berriasian to Late Cretaceous.

The timing and evolution of the Late Jurassic–Early Cretaceous uplift had a significant effect on the Lias, the main hydrocarbon source rock in the area. Only in the areas which received sediment during the period of uplift (Zone 1 and in places Zone 2) was the Lias buried to a sufficient depth for the generation of hydrocarbons. In other areas closer to the centre of the uplift (most of Zone 2 and Zone 3) the Lias was either eroded or elevated above the oil window. Areas where the Lias had not reached maturity by the Early Cretaceous remained immature as the region was again uplifted during the Tertiary. Any regional maturity study must, therefore, consider the effects of the Late Jurassic–Early Cretaceous uplift and Berriasian unconformity as these are fundamental in controlling the hydrocarbon potential and fairway distribution in these Basins.

This paper has benefited greatly from discussions and fieldwork with J. R. Underhill and without his input many important concepts would have gone unresearched. The authors would also like to thank B.P., British Gas E&P, Elf Enterprise Caledonia, Ranger, B.G.S. and Geco for the data used in this paper and/or permission to publish specific well and seismic information. In particular B. Lovell, A. MacGregor and R. Wrigley are thanked for helping with data acquisition. This research was undertaken at the University of Edinburgh with the financial support from B.P., PSTI and DENI. K. Thomson is thanked for discussing the results of his work in this area. The final text benefited significantly from the comments of C. Brooke and A. Payne. The authors wish to thank A. Ruffell whose input greatly improved the content of this paper.

References

AINSWORTH, N. R., O'NEILL, M., RUTHERFORD, M. M., CLAYTON, G. HORTON, N. F. & PENNY, R. A. 1987. Biostratigraphy of the Lower Cretaceous, Jurassic, and Late Triassic of the North Celtic Sea and Fastnet Basins. *In*: BROOKS, J. & GLENNIE, K. W. (eds) *Petroleum Geology of North West Europe*. Graham & Trotman, London, 611–622.
ALLEN, P. 1975. Wealden of the Weald: a new model. *Proceedings of the Geologists' Association*, **86**, 389–437.
BENNET, G., COPESTAKE, P. & HOOKER, N. P. 1985. Stratigraphy of the Britoil 72/10-1A well, Western Approaches. *Proceedings of the Geologists' Association*, **96**, 255–261.
BOILLOT, G., MOUGENOT, D., GIRARDEAU, J. & WITERER, E. L. 1989. Rifting processes on the West Galicia Margin, Spain. *In*: TANKARD, A. J. & BALKWILL, H. R. (eds) *Extensional Tectonics and Stratigraphy of the North Atlantic Margins*. American Association of Petroleum Geologists Memoir, **46**, 363–377.
BRAY, R. J. GREEN, P. F. & DUDDY, J. R. 1998. Multiple heating episodes in the Wessex Basin: Implications for geological evolution and hydrocarbon generation. *This volume*.
CASTON, V. N., DEARNLEY, R., HARRISON, R. K., RUNDLE, C. C. & STYLES, M. T. 1981. Olivine–dolerite intrusions in the Fastnet Basin. *Journal of the Geological Society, London*, **138**, 31–46.
CHAPMAN, T. J. 1989. Permian to Cretaceous structural evolution of the Western Approaches Basin (Melville sub-basin, UK). *In*: COOPER, M. A. & WILLIAMS, G. D. (eds) *Inversion Tectonics*. Geological Society, London, Special Publications, **44**, 177–200.
COE, A. L. 1996. Unconformities within the Portlandian Stage of the Wessex Basin and their sequence stratigraphical significance. *In*: HESSELBO, S. P. & PARKINSON, D. N. (eds) *Sequence Stratigraphy in British Geology*. Geological Society, London, Special Publications, **103**, 109–143.
COLIN, J. P., IOANNIDES, N. S. & VINNING, B. 1992. Mesozoic stratigraphy of the Goban Spur, offshore South-West Ireland. *Marine and Petroleum Geology*, **9**, 527–541.
——, LEHMANN, R. A. & MORGAN, B. E. 1981. Cretaceous and Late Jurassic Biostratigraphy of the North Celtic Sea Basin, Offshore Ireland. *In*: NEALE, J. W. & BRASIERS, M. D. (eds) *Microfossils From Recent and Fossil Shelf Seas*. Ellis Horwood, Chichester, 122–155.
COLLEY, M. G., MCWILLIAMS, A. S. F. & MEYERS, R. C. 1981. Geology of the Kinsale Head Gas field, Celtic Sea, Ireland. *In*: ILLING, L. V. & HOBSON, G. D. (eds) *Petroleum Geology of the Continental Shelf of North-west Europe*. Heydon and Son, London, for the Institute of Petroleum, 504–510.
COOK, D. R 1987. The Goban Spur – exploration in a deep-water frontier basin. *In*: BROOKS, J. & GLENNIE, K. W. (eds) *Petroleum Geology of North-west Europe*, Vol. 2. Graham and Trotman, London, 623–632.
CURNELLE, R. & DUBOIS, P. 1986. Evolution Mesozoique des grands bassins sedimentaires francais (Bassin de Paris, d'Aquitaine et Sud-Est). *Bulletin de la Société géologique de France, Série 8*, **2**, 529–546.
CURRY, D., HAMILTON, D. & SMITH, A. J. 1970. *Geological and Shallow Subsurface Geophysical Investigations in the Western Approaches to the English Channel*. Report of the Institute of Geological Sciences No. 70/3.

DESEGAULX, P. & BRUNET, M. 1990. Tectonic subsidence of the Aquitaine Basin since Cretaceous times. *Bulletin de la Société geologique de France, Série 8*, **2**, 295–306.

EVANS, C. D. R. 1990. *United Kingdom Offshore Regional Report: The Geology of the Western English Channel and its Western Approaches.* HMSO, London.

FYFE, J. A., ABBOTS, I. & CROSBY, A. 1981. The subcrop of the Mid-Mesozoic unconformity in the UK area. *In*: ILLING, L. V. & HOBSON, G. D. (eds) *Petroleum Geology of the Continental Shelf of North-west Europe.* Heydon and Son, London, for the Institute of Petroleum, 236–244.

HAMBLIN, R. J. O., CROSBY, A., BALSON, P. S., JONES, S. M., CHADWICK, R. A., PENN, I. E. & ARTHUR, M. J. 1992. *United Kingdom Offshore Regional Report: The Geology of the English Channel.* HMSO, London, p. 106.

HARLAND, W. B., ARMSTRONG, R. L., COX, A. V., CRAIG, L. E., SMITH, A. G. & SMITH, D. G. 1990. *A Geological Time Scale 1989.* Cambridge University Press, Cambridge.

HESSELBO, S. H. & ALLEN, P. A. 1991. Major erosional surfaces in the basal Wealden Beds, Lower Cretaceous, south Dorset. *Journal of the Geological Society, London*, **148**, 105–113.

HILLIS, R. R. 1988. *The Geology and Tectonic Evolution of the Western Approaches Trough.* PhD thesis, University of Edinburgh.

—— 1995. Regional Tertiary exhumation in and around the United Kingdom. *In*: BUCHANAN, J. G. & BUCHANAN, P. G. (eds) *Basin Inversion.* Geological Society, London, Special Publications, **88**, 167–190.

HOUSE, M. R. 1989. *Geology of the Dorset Coast, Geologist's Association Guide.* Geologists' Association, Oxford.

JUIGNET, P. M., RIOULT, M. & DESTOMBES, P. 1973. Boreal influences in the Upper Aptian–Lower Albian beds of Normandy, northwest France. *In*: CASEY, R. & RAWSON, P. F. (eds) *The Boreal Lower Cretaceous. Geology Journal*, Special Issue, **5**, 303–326.

KAMERLING, P. 1979. The geology and hydrocarbon habitat of the Bristol Channel Basin. *Journal of Petroleum Geology*, **2**, 75–93.

KARNER, G. D., LAKE, S. D. & DEWEY, J. F. 1987. The thermal and mechanical development of the Wessex Basin, southern England. *In*: COWARD, M. P., DEWEY, J. F. & HANCOCK, P. L. (eds) *Continental Extension Tectonics.* Geological Society, London, Special Publications, **28**, 517–537.

LAKE, S. D. & KARNER, G. D. 1987. The structure and evolution of the Wessex Basin, southern England: an example of inversion tectonics. *Tectonophysics*, **137**, 347–378.

LOTT, G. K., KNOX, R. W., BRIGG, P. J., DEWEY, R. J. & MORTON, A. C. 1980. *Aptian Cenomanian Stratigraphy in Boreholes From Offshore Southwest England.* Report of the Institute of Geological Sciences No. 80/8.

MASSON, D. G. & MILLER, P. R. 1984. Mesozoic sea-floor spreading between Iberia and North America. *Marine Geology*, **56**, 279–287.

—— & ROBERTS, D. G. 1981. Late Jurassic–Early Cretaceous reef trends on the continental margin S.W. of the British Isles. *Journal of the Geological Society, London*, **138**, 437–443.

MCMAHON N. A. 1995. *The Role of Uplifts in the Rifting and Sedimentation History of the N. Atlantic.* PhD thesis, University of Edinburgh.

—— & UNDERHILL, J. R. 1995. The regional stratigraphy of the southwest United Kingdom and adjacent offshore areas with particular reference to the major intra-Cretaceous unconformity. *In*: CROCKER, P. F. & SHANNON, P. M. (eds) *The Petroleum Geology of Ireland's Offshore Basins.* Geological Society, London, Special Publications, **93**, 323–325.

MILLSON, J. A. 1987. The Jurassic evolution of the Celtic Sea Basins. *In*: BROOKS, J. & GLENNIE, K. W. (eds) *Petroleum Geology of North-west Europe*, Vol. 2. Graham & Trotman, London, 599–610.

MONTADERT, L., ROBERTS, D. G., DE CHARPAL, O. & GUENNOC, P. 1979. Rifting and subsidence of the northern continental margin of the Bay of Biscay. *In*: MONTADERT, L., ROBERTS, D. G. *et al.* (eds) *Initial Reports of the Deep Sea Drilling Program*, **48**, 1025–1060.

NAYLOR, D. & SHANNON, P. M. 1982. *The Geology of Offshore Ireland and West Britain.* Graham & Trotman, London.

NUMMEDAL, D. & SMITH, D. J. P. 1987. Transgressive stratigraphy at sequence bounding unconformities: some principles derived from Holocene and Cretaceous examples. *In*: NUMMEDAL, D., PILKEY, O. H. & HOWARD, J. D. (eds) *Sea Level Fluctuation and Coastal Evolution.* Society of Economic Paleontologists and Mineralogists Special Publications, **41**, 241–260.

OLIVET, J. L., BONNIN, J., BEUZART, P. & AUZENC, J. M. 1984. *Cinematique de l'Atlantique Nord et Central.* Publications du Centre National pour l'Exploitatons des Oceans Rapports Scientifiques et Techniques, 54.

PETRIE, S. H., BROWN, J. R., GRANGER, P. J. & LOVELL, J. P. B. 1989. Mesozoic history of the Celtic Sea Basins. *In*: TANKARD, A. J. & BALKWILL, H. R. (eds) *Extensional Tectonics and Stratigraphy of the North Atlantic Margins.* American Association of Petroleum Geologists Memoir, **46**.

POMEROL, CH., DEBELMAS, J., MIROUSE, R., RAT, P. & ROUSSET, C. 1980. *The Geology of France.* Guides Geologiques Regionaux, Masson, Paris.

ROBERTS, D. G. 1989. Basin inversion in and around the British Isles. *In*: COOPER, M. A. & WILLIAMS, G. D. (eds) *Inversion Tectonics.* Geological Socity, London, Special Publications, **44**, 131–150.

——, MASSON, D. G. & MILES, P. R. 1981. Age and structure of the southern Rockall Trough – new evidence. *Earth and Planetary Science Letters*, **52**, 115–128.

ROBINSON, K. W., SHANNON, P. M. & YOUNG, D. G. G. 1981. The Fastnet Basin: an integrated analysis. *In*: ILLING, L. V. & HOBSON, G. D. (eds) *Petroleum*

Geology of the Continental Shelf of North-west Europe. Heydon and Son, London, for the Institute of Petroleum, 444–454.

RUFFELL, A. H. 1991. Geophysical correlation of the Aptian and Albian formations in the Wessex Basin of southern England. *Geological Magazine*, **128**, 67–75.

—— 1992. Early to mid-Cretaceous tectonics and unconformities of the Wessex Basin (southern England). *Journal of the Geological Society, London*, **149**, 443–454.

—— 1995. Seismic stratigraphic analysis of non-marine Lower Cretaceous strata in N.W. Europe, *Cretaceous Research*, **16**, 603–637.

—— & WIGNALL, P. B. 1990. Depositional trends in the Upper Jurassic – Lower Cretaceous of the northern margin of the Wessex Basin. *Proceedings of the Geologists' Association*, **101**, 279–88.

SCLATER, J. G. & CHRISTIE, P. A. F. 1980. Continental stretching; explanation of post Mid-Cretaceous subsidence of the Central North Sea basin. *Bulletin of the American Association of Petroleum Geologists*, **64**, 781–782.

SCRUTTON, R. A. 1985. Modelling of magnetic and gravity anomalies at Goban spur, northeastern Atlantic. *In*: DE GRACIANSKY, P. C. & POAG, C. W. et al. (eds) *Initial Reports of the Deep Sea Drilling Project*, **80** (part 2), 1141–1151.

SHANNON, P. M. 1991a. The development of Irish Offshore sedimentary basins. *Journal of the Geological Society, London*, **148**, 181–189.

—— 1991b. Tectonic framework and petroleum potential of the Celtic Sea, Ireland. *First Break*, **9**(3), 107–122.

STECKLER, M. S. & WATTS, A. B. 1982. Subsidence history and tectonic evolution of Atlantic-type continental margins. *In*: SCRUTTON, R. A. (ed.) *Dynamics of Passive Margins*. AGU, Geodynamics Series, **6**, 184–196.

STEWART, D. J., RUFFELL, A. H., WACH, G. D. & GOLDRING, R. 1991. Lagoonal sedimentation and fluctuating salinities in the Isle of Wight, Southern England. *Sedimentary Geology*, **72**, 117–134.

STONELEY, R. & SELLEY, R. C. 1986. *A Field Guide to the Petroleum Geology of the Wessex Basin*. Department of Geology, Imperial College, London.

TAPPIN, D. R., CHADWICK, R. A., JACKSON, A. A., WINGFIELD, R. T. R. & SMITH, N. J. P. 1994. *United Kingdom Offshore Regional Report: The Geology of Cardigan Bay and the Bristol Channel.* HMSO, London, 107.

TUCKER, R. M. & ARTER, G. 1987. The tectonic evolution of the North Celtic Sea and Cardigan Bay basins with special reference to basin inversion. *Tectonophysics*, **137**, 291–307

UNDERHILL, J. R. & PARTINGTON, M. A. 1993. Jurassic thermal doming and deflation in the N. Sea – Implications for sequence stratigraphic evaluations. *In*: PARKER, J. R. (ed.) *Petroleum Geology of Northwest Europe, Proceedings of the 4th Conference*. Geological Society, London, 337–345.

—— & —— 1994. Use of genetic sequence stratigraphy in defining and determining regional tectonic control on the Cimmerian unconformity – implication for N. Sea basin development and the Exxon sea level chart. *In*: POSSIMENTIER, H. & WEIMER (eds) *Siliclastic Sequence Stratigraphy.* American association of Petroleum Geologists Memoir, **58**, 449–484.

VAN HINTE, J. E. 1978. Geohistory analysis – application of micropalaeontology in exploration geology. *Bulletin of the American Association of Petroleum Geologists*, **62**, 201–222.

VAN HOORN, B. 1987. The North Celtic Sea/Bristol Channel Basin: origin, deformation and inversion history. *Tectonophysics*, **137**, 309–334.

WHITE, R. S. & MCKENZIE, D. 1989. Magmatism at rift zones: The generation of volcanic continental margins and flood basalts. *Journal of Geophysical Research*, **94**, B6, 7685–7729.

ZIEGLER, P. A. 1982. *Geological Atlas of Western and Central Europe*. Shell International Petroleum, Maatschappij BV.

—— 1987a. Late Cretaceous and Cenozoic intra-plate compressional deformations in the Alpine foreland – a geodynamic model. *Tectonophysics*, **137**, 389–420.

—— 1987b. Evolution of the Western Approaches Trough. *Tectonophysics*, **137**, 341–346.

—— 1987c. Manx-Furness Basin. *Tectonophysics*, **137**, 335–340.

Influence of salt on the structural evolution of the Channel Basin

MICHAEL J. HARVEY[1,2] & SIMON A. STEWART[3]

[1] *Department of Geology, Imperial College, London, UK*
[2] *Present address: Nederlandse Aardolie Maatschappij, Postbus 28000, 9400HH Assen, The Netherlands*
[3] *Amerada Hess Ltd, 33 Grosvenor Place, London SW1X 7HY, UK*

Abstract: The north margin of the Channel Basin is defined in Dorset and the Isle of Wight by Mesozoic extensional faults which were reversed during Tertiary contraction, causing forced folding of the Albian–Oligocene post-rift cover. The resulting contractional structures (the Purbeck and Isle of Wight Monoclines and the reversed Abbotsbury–Ridgeway Fault) have previously been used as type examples for inversion tectonics. The Jurassic–Cretaceous basin margin crosses the eastern edge of a Triassic salt basin (Dorset Halite, Upper Triassic Mercia Mudstone Group) near Swanage. Detachments in Triassic salt around Weymouth resulted in extension and inversion styles which differ from the better known basement-linked structures to the east. The Abbotsbury–Ridgeway Fault is listric and unlike the Tertiary hangingwall folds to the east the Weymouth Anticline formed as an extensional rollover, only tightened during inversion.

Seismic and well data and structure maps are used to illustrate for the first time the increasing effects of salt on structural style as the salt thickens westwards. The onset of detachment on the basin margin is placed at Ringstead, where a north-stepping shift in the margin at pre-salt level defines a relay zone. Here extension was accommodated on many basement faults and the ratio of basement fault displacement to salt thickness was reduced sufficiently to allow the post-salt section to detach throughout extension (Abbotsbury–Ridgeway Fault). Tight Tertiary compressional folds in this area may result from linkage of the pre- and post-salt sections during inversion, a consequence of salt welds developed during the last stages of extension.

In the western part of the Channel Basin (Lyme Bay) the Triassic salt is sufficiently thick to have detached the pre- and post-salt sections over a wide area. North–south shortening during basin inversion may have been accommodated in the Lyme Bay post-salt section by displacements on conjugate, northwest–southeast- and northeast–southwest-oriented, strike-slip faults such as the Mangerton Fault. The Dorset Halite is a complex sequence of halite and mudstone interbeds, different from the 'pure' halites seen in the North Sea Basin, and its influence on structural style differs accordingly. Variation in salt tectonics style between the Weymouth area (salt swells and rollers) and Lyme Bay (salt rollers only) may reflect lateral facies variation and hence rheology in the Dorset Halite Formation.

The northern margin of the Channel Basin is exposed in south Dorset and the Isle of Wight as an east–west-trending fold and fault zone. Tertiary basin inversion led to the reversal of major Mesozoic down-to-the-south normal faults and forced folding of the overlying Cretaceous–Palaeogene post-rift sediments. Progressively deeper Recent erosion towards the west has exposed a range of structural levels, from monoclinal forced folds in the Isle of Wight and Purbeck to the underlying reversed faults outcropping north and northeast of Weymouth. Commercial seismic data in Bournemouth Bay have enabled the fault zone to be traced offshore between the exposed Purbeck and Isle of Wight monoclines, but west of Weymouth in Lyme Bay the basin-bounding fault zone becomes more diffuse (Fig. 1).

The Channel Basin was identified as a distinct structural unit from early offshore seismic data by Smith & Curry (1975). The increase in hydrocarbon exploration activity in the area since 1980 has provided enough well and seismic data to allow the extension and inversion history of the basin to be constrained. Stoneley (1982), Whittaker (1985), Chadwick (1986) and Karner *et al.* (1987) described Permian–early Cretaceous extensional structures, recording several phases of approximately north–south extension. Stoneley (1982) and Selley & Stoneley (1987) used the concave-southward map geometry of the major northern basin-bounding faults onshore in Dorset to suggest that they are deep-seated (basement-linked) listric structures, whereas Chadwick (1986, 1993) proposed that these faults are planar and link to re-activated mid-crustal Variscan thrusts. The geometry of extensional, predominantly dip-slip, basement-linked cover faults has previously been illustrated with seismic lines from Bournemouth Bay (Penn *et al.*

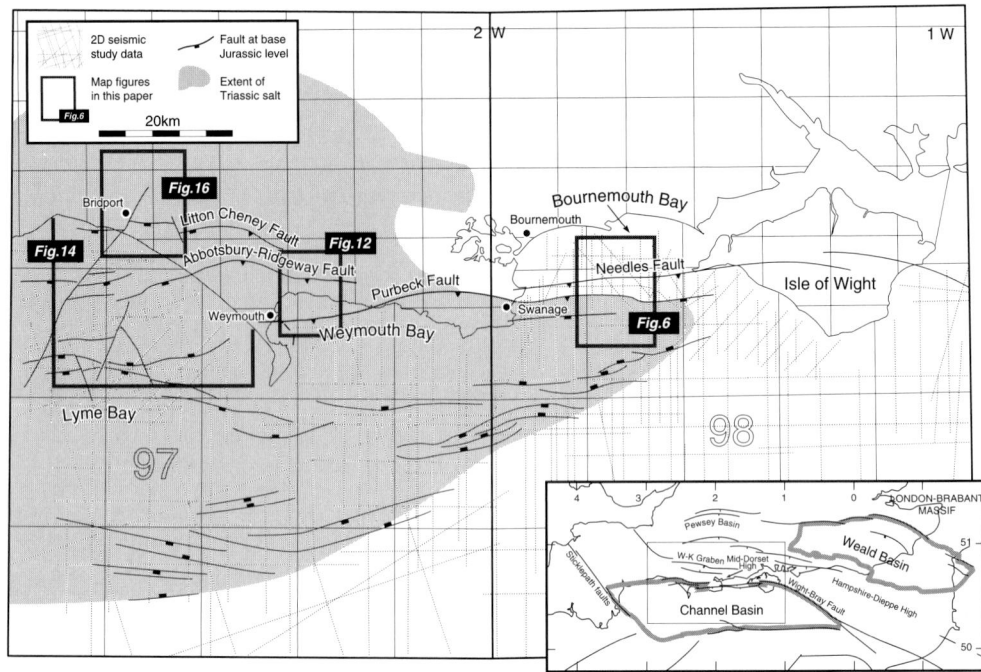

Fig. 1. Location map showing the northern margin of the Channel Basin. Main faults are mapped at base Jurassic level from the 2D seismic grid shown. Shading shows the estimated extent of Triassic salt; boxes outline detailed structure maps presented in this paper. Inset map: location of the Channel Basin relative to depocentres and highs of the Wessex Basin system.

1987; Hamblin *et al.* 1992; Chadwick 1993) and the Isle of Wight (Chadwick 1993; Underhill & Paterson 1998; Underhill & Stoneley this volume).

The geometry of the basin margin further west is less well documented. The possibility that Permian-Triassic salt influenced the structural evolution of the west of the basin was first proposed by Lees & Cox (1937) for the Tertiary periclines south of the Abbotsbury–Ridgeway Fault and by Falcon & Kent (1960) for the Compton Valence structure to the north. Jenkyns & Senior (1991) used outcrop data to give a detailed account of the timing of extension across basin margin faults between Bridport and Weymouth, but no geometrical analysis of the onshore faults has been published. Stewart *et al.* (1996) used seismic data to illustrate extensional detachments on salt horizons of the Triassic Mercia Mudstone Group in Lyme Bay. That work is expanded here, using cross-sections and maps based on interpretation of 9000 km of offshore two-dimensional seismic data covering the whole basin (Fig. 1) and published onshore maps. The aim of this paper is to describe the variation in structural style along the northern margin of the basin, focusing on the role of Triassic salt. The distribution of Triassic salt is described first, followed by detailed descriptions of the structure of the basin margin at several points along an east–west traverse, concluding with a discussion of the salt-related structures observed.

Distribution of Triassic salt in and around the Channel Basin

The Triassic Mercia Mudstone Group crops out on the shores of Lyme Bay in south Devon as a monotonous sequence of terriginous red–brown silty mudstones containing thin interbeds of anhydrite or gypsum. A regionally correlatable Carnian sandy unit lying in the middle of the sequence, the Weston Mouth Sandstone Member – WMSM (Warrington & Scrivener 1980; Whittaker 1985), overlies mudstones which contain pseudomorphs of halite at outcrop in Devon (Lott *et al.* 1982). The dolomitic lateral equivalent to the WMSM can be correlated between onshore wells to the north of the Channel Basin (Lott *et al.* 1982). The underlying saliferous unit, the Dorset Halite, thickens along the axis of the Winterborne–Kingston Graben in

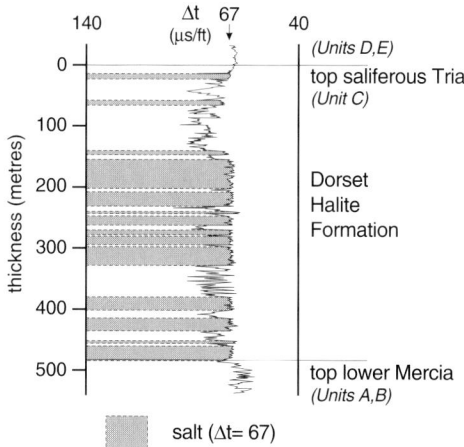

Fig. 2. Sonic log over the Triassic saliferous interval in Chickerell-1 (Δt: interval transit time in μs/foot). The whole salt interval (unit C of Lott et al. 1982) contains significant interbeds of mudstone.

the Seaborough-1, Nettlecombe-1 and Marshwood-1 wells, and is correlated with 150 m of salt cored in Winterborne Kingston-1 (Colter & Havard 1981). The released exploration well Chickerell-1, drilled south of the basin margin on the Weymouth Anticline, proved a much thicker salt-bearing sequence with many interbeds of mudstone and anhydrite (Fig. 2). The lateral equivalent of the WMSM in the Central Somerset Basin (Ruffell 1990) overlies another Carnian salt sequence (the Somerset Halite). Both Carnian salts have been correlated with evaporites in the Worcester and Cheshire basins (Whittaker 1985).

The probable limits of the Triassic salt sequence are shown in Fig. 1. Previous estimates of the minimum extent of Triassic salt (Stoneley 1982) have been restricted to the area between the Nettlecombe and Winterbourne Kingston wells, entirely to the north of the Channel Basin. More recent wells drilled within the basinal succession (e.g. Chickerell and Southard Quarry) have proved much thicker Triassic salt sequences than to the north (Fig. 3). The limit of the Dorset Halite in the south and southwest of the Channel Basin, where no well data are available, has been estimated here by mapping the extent of detached overburden faulting from offshore seismic data.

The southern margin of the salt basin is defined by a sinuous zone of symmetrical graben within the Jurassic–Cretaceous section which detach within the Triassic sequence and strike ENE–WSW from south of Swanage (block 98/16) to south of Lyme Bay (block 97/22, Fig. 3). Along most of its length this fault zone coincides with a break in basement slope, with steeper basement dips to the north, towards the basin-margin half-graben faults (see below). Such a hinge zone may have controlled the position of the edge of salt deposition and similar examples of the spatial association of cover fault zones and inflections at top basement level have been described from the Campos salt basin, Brazil (Cobbold & Szatmari 1991). The salt margin further east must lie in Bournemouth Bay, as the Arreton-2 well encountered no salt within the basin sequence on the Isle of Wight. Salt is also absent on the Mid-Dorset High around Wytch Farm and its presence immediately to the south infers that Triassic movement on the Purbeck Fault may have controlled salt distribution or preservation in Purbeck. The western limit of the Dorset Halite is controlled by the Mercia Mudstone outcrop in Lyme Bay, west of the study area (Quad 9b, Hamblin et al. 1992). The most westerly available seismic data (in Lyme Bay, 3°W; Fig. 1) image a thick Triassic interval, which is interpreted as saliferous on the basis of tectonic style, less than 20 km east of the WMSM outcrop near Sidmouth. Dissolution of the salt during early Cretaceous and Tertiary uplift may account for the absence of thick evaporite layers at outcrop, although westward thinning of the Triassic towards the Cornubian Massif may have been accompanied by lateral facies changes, limiting the original extent of the salt. The distribution of the salt-bearing Triassic sequence shown in Fig. 3 is consistent with thickness trends of the entire Permo-Triassic section offshore (Hamblin et al. 1992).

Basement-linked fault zone in Bournemouth Bay

The northern margin of the Channel Basin between the Isle of Wight and Swanage conforms well to existing models of inverted thick-skinned extensional fault zones (e.g. Williams et al. 1989). A north–south-oriented seismic section running across the basin margin in Bournemouth Bay (Fig. 4) shows that the Mesozoic basin section to the south has been uplifted during Tertiary inversion. The Purbeck–Isle of Wight Fault Zone defines a terraced basin margin. This fault terrace is bounded in the south by the Purbeck Fault and to the north by what is termed here the Needles Fault. Although seismic imaging below the Jurassic section on the basin margin is very

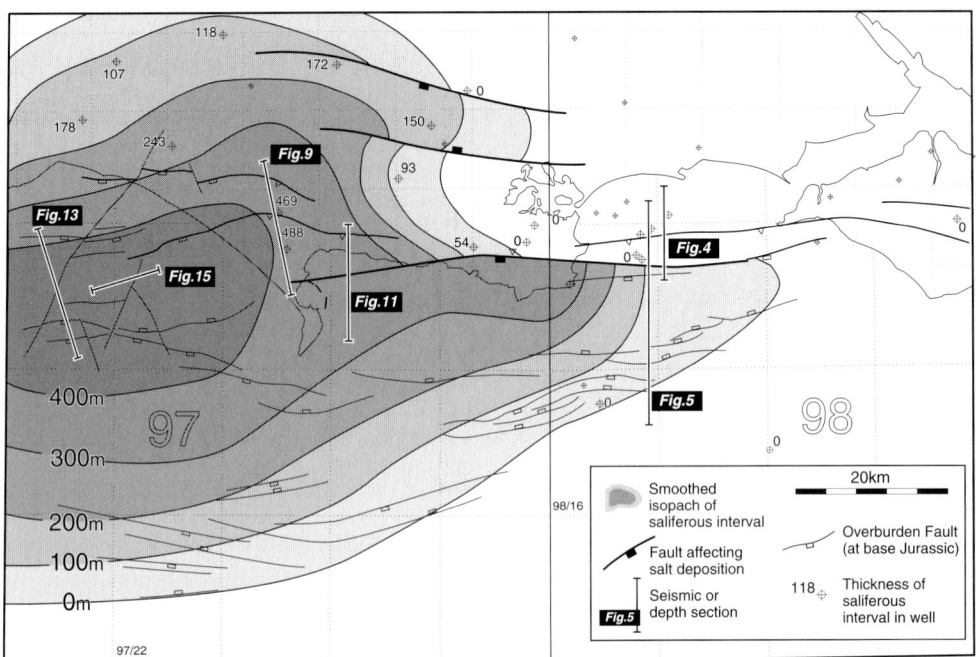

Fig. 3. Distribution of Triassic salt from well and offshore seismic data. Present-day isopachs are for the total saliferous interval, including interbeds, and are smoothed to remove local thickness changes due to overburden faulting and folding. Onshore wells show depositional control by the Purbeck Fault and the Winterbourne–Kingston Graben to the north. Figure numbers are for seismic examples and depth sections presented in text.

poor, most faults can be traced down to lowermost Triassic and top Permian levels. Even after considerable Tertiary reversal on the Purbeck Fault, Lower Jurassic rocks in its hanging wall are still juxtaposed with the Permian section in the footwall.

The basin-bounding faults appear curved on seismic data (Fig. 4) but become planar after depth conversion. The depth-converted north–south section in Fig. 5 (5 km to the west of Fig. 4) is constrained by wells on the northern footwall, the intermediate fault terrace and within the basin proper. The regional depth section shows the relationship between a break in basement slope and the Portland–Wight Graben, an overburden fault zone thought to define the southern edge of the Triassic salt (Fig. 5a, southern end). The possibility that north-dipping basement faults underlie the symmetrical overburden graben to the south cannot be ruled out with the available seismic data. The depth to top basement shown in the half-graben is a minimum estimate as no thickening of the Permo-Triassic section across the Purbeck Fault has been assumed and the Permo-Triassic thickness proved in the footwall is used. Thickness changes across the Purbeck Fault (Fig. 5b) show that it was active in extension throughout the Jurassic, with accelerated movement during separate early Jurassic and late Jurassic–early Cretaceous regional rift events (cf. Chadwick 1986; Jenkyns & Senior 1991). Upper Jurassic and Lower Cretaceous syn-rift sediments are only preserved beneath the Albian unconformity on the intermediate terrace block and within the basin proper. The post-Albian cover sediments are folded above the basin margin in two northward-facing monoclines, whose location indicates that Tertiary reversal occurred preferentially on those faults with greatest pre-Albian extensional displacement. Despite the high degree of deformation in the post-rift section related to fault reversal, the blocks between the inverted faults appear relatively undeformed and display only moderate northward dips. The footwall high to the north, which was uplifted and eroded during the last rift episode, was the site of renewed subsidence during early inversion, and deposition of Tertiary sediments of the Hampshire Basin was probably influenced by the uplift of the Channel Basin to the south (Plint 1982; Lake & Karner 1987).

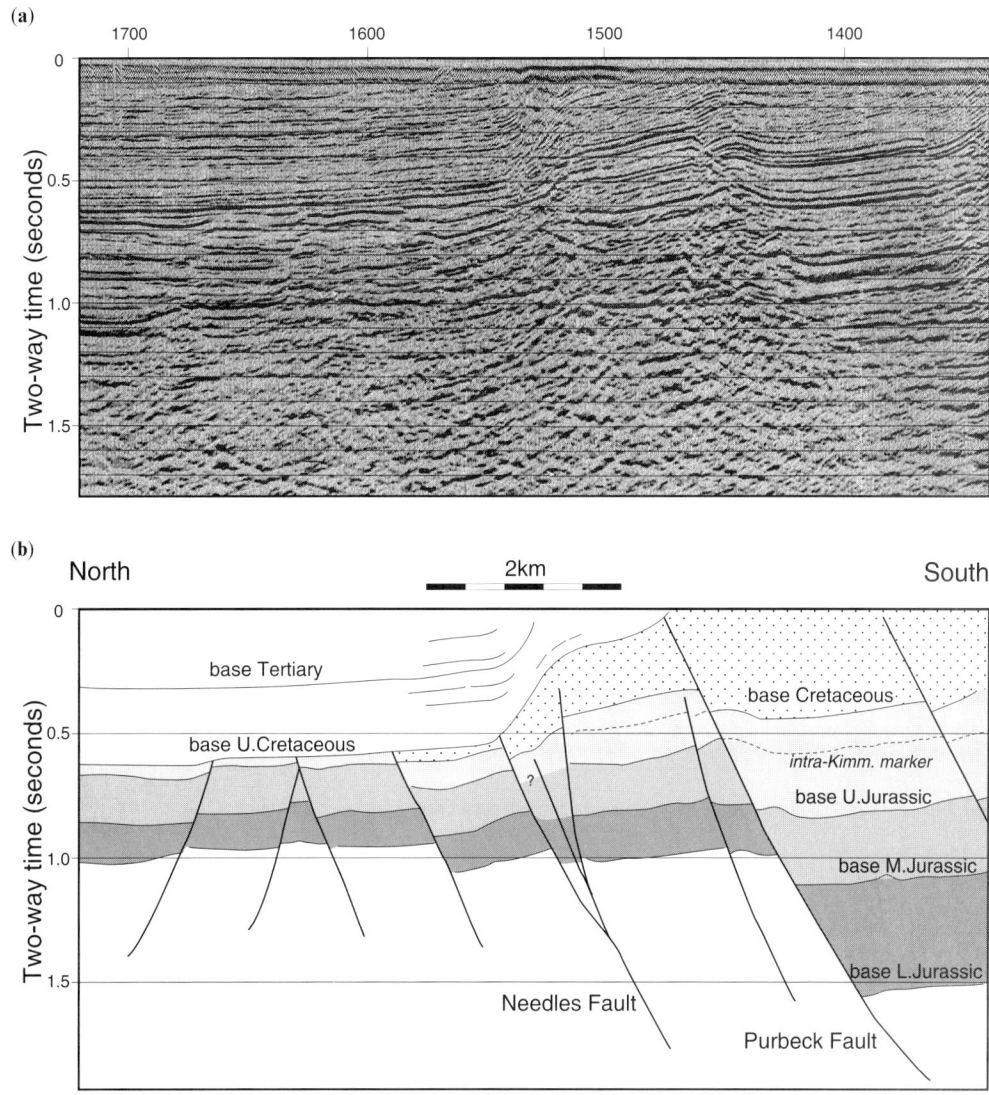

Fig. 4. North–south-trending seismic line across the inverted basin margin in Bournemouth Bay (all seismic locations shown in Fig. 3). (**a**) Time section. (**b**) Geoseismic. The Purbeck Fault to the south juxtaposes Jurassic against Permian in the footwall even after Tertiary reversal. The Needles Fault bounds an intermediate pre-Albian fault block to the north. All faults appear to link directly downwards to the pre-Triassic section.

On most inverted faults within the Purbeck–Isle of Wight Fault Zone (Fig. 5) the displacement null point lies within the uppermost syn-rift section. The Purbeck Fault displays extensional displacements for all Jurassic and older markers, with a null point within the Wealden section, close to the surface. To the north the inverted faults have null points within the Upper Jurassic sequence and in places below the Corallian Beds (base Upper Jurassic). This spatial trend infers a higher degree of reversal relative to initial extensional displacement (inversion ratio of Williams et al. 1989) for the northern faults. Figure 6 shows a structure map in two-way time drawn at the base Upper Jurassic level. The largest preserved normal displacement is clearly across the Purbeck Fault, and the Needles Fault preserves little displacement at base Upper Jurassic level. Minor short-wavelength folds in the hanging wall of the Purbeck Fault may

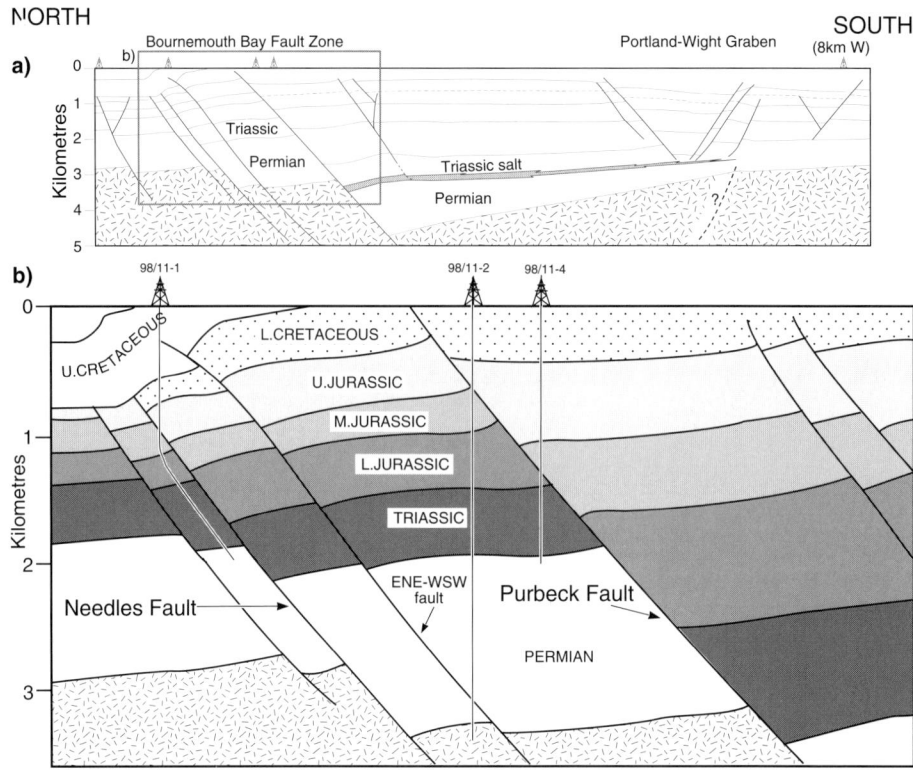

Fig. 5. True-depth section across the northern Channel Basin from depth converted seismic data tied to exploration wells. (**a**) Regional line showing inverted half-graben basin geometry. ENE–WSW-trending Portland–Wight Graben probably marks the southern edge of Triassic salt, drilled in the hanging wall of the Purbeck Fault 10 km to the west of the line. Probable intra-Triassic salt reflectors (near top Permian) are seen in the half-graben, but cannot be traced south of the Portland–Wight Graben. (**b**) Bournemouth Bay Fault Zone. All wells proved an absence of Triassic salt north of the Purbeck Fault; basement and overburden faults were hard-linked during extension and inversion. Despite 1 km of differential uplift of the basin during reversal of fault zone all faults preserve net extension at the Middle Jurassic level.

indicate that compressional detachment folding occurred at the far eastern edge of the Triassic salt basin.

Oblique movement across the Bournemouth Bay Fault Zone

The intermediate block between the Purbeck and Needles faults is affected by reversed ENE–WSW trending faults (Fig. 6) striking obliquely to the main faults by up to 25°. Faults of this orientation may indicate sinistral oblique-slip movement across the east–west-trending basin-bounding faults during extension. Similar geometries have been reproduced in physical models of cover sediment deformation during oblique extension across a basement step (Richard *et al.* 1995, fig. 1) and oblique extension across a rift zone (McClay & White 1995, fig. 5c). Although similar geometries can be produced in cover sediments by oblique shortening across a basement step, with the same shear sense as the extensional models above (Richard *et al.* 1995, fig. 10) a Jurassic–early Cretaceous transtensional origin is favoured here (Fig. 7), as faults within this set display net extension in cover sequence and some show clear evidence of pre-Albian syn-sedimentary movement. The orientations of subsidiary faults suggest that, at least during the late Jurassic–early Cretaceous, this part of the basin margin was subject to northwest–southeast or NNW–SSE extension.

The orientations of major basement-linked structures in the Channel Basin have previously been linked to a postulated east–west Variscan

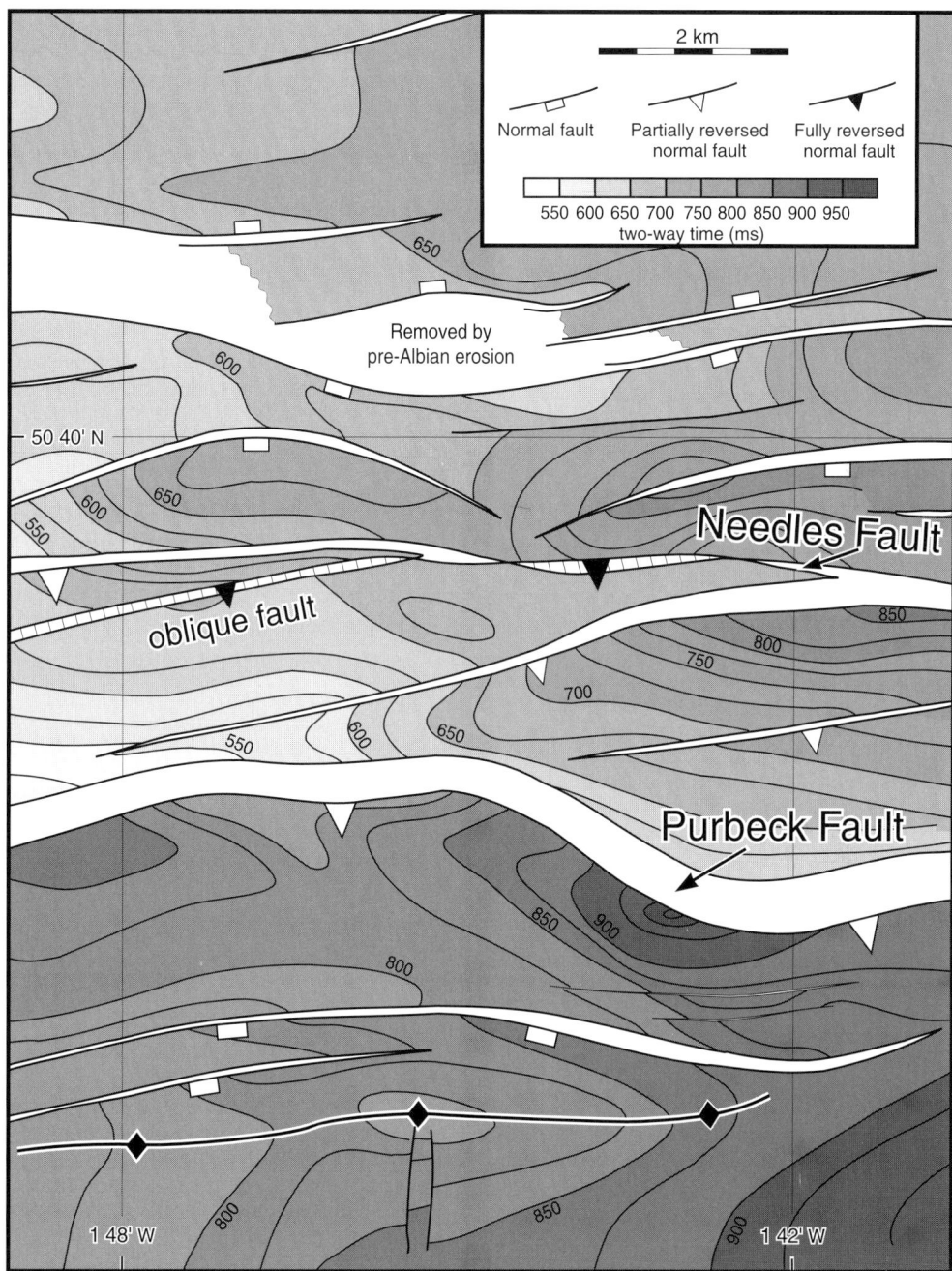

Fig. 6. Structure map of central part of Bournemouth Bay Fault Zone (blocks 98/7 and 98/12), from seismic grid shown in Fig. 1. Contours are in two-way time to the base Upper Jurassic horizon. Minor faults in the intermediate block between the Purbeck and Needles faults show evidence of late Jurassic–early Cretaceous extension and Tertiary contraction. Tertiary folds south of the Purbeck Fault may indicate detachment in the hanging wall at the eastern edge of the salt basin.

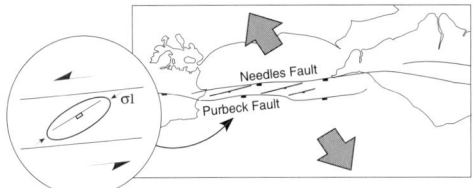

Fig. 7. Cartoon showing possible Jurassic–early Cretaceous sinistral-oblique extension across the east–west-trending basement-linked fault zone in Bournemouth Bay, leading to obliquely trending subsidiary faults.

thrust fabric within the basement, re-activated during Mesozoic extension (Stoneley 1982; Chadwick 1986). Basement fabric control on the east–west-trending basin-bounding faults may have resulted in the persistence of this orientation of structure despite an oblique extension direction. Other Variscan fabric orientations have been proposed as having controlled the extensional geometry of Wessex Basin depocentres, notably the northwest–southeast or WNW–ESE Variscan wrench trends (Drummond 1970; Stoneley 1982; Lake & Karner 1987) picked out by the Sticklepath, Quantock, Pays de Bray and Rouen-Senelly fault zones. The Sticklepath and Pays de Bray faults define the western and eastern margins of the Channel Basin (Fig. 1) but within the eastern part of the basin there is little evidence for basement-controlled faults with this orientation. Minor north–south-trending faults in Bournemouth Bay (Fig. 6) and to the east in Purbeck and Weymouth Bay are associated with east–west-trending Tertiary compressional folds.

The salt-detached basin margin north of Weymouth

Location of major structures

To the west of Bournemouth Bay the northern margin of the Channel Basin is marked by the Purbeck Monocline, a forced fold overlying the reversed Purbeck Fault which trends from Swanage westwards to Lulworth Cove in south Dorset. The monocline runs offshore between Lulworth Cove and Ringstead (Fig. 8), and the Purbeck Fault can be traced on seismic data westwards in Weymouth Bay until it dies out onshore beneath Weymouth. During extension the westward decrease in displacement on the Purbeck Fault was compensated by increasing displacement across the Abbotsbury–Ridgeway Fault, exposed onshore north of Weymouth. This northward step in the extensional basin margin, as defined by preservation of the late Jurassic–early Cretaceous syn-rift section beneath the Albian unconformity, defines a relay zone where the two major south-dipping faults overlap northeast of Weymouth. During Tertiary inversion contractional displacement was also transferred northwestwards from the Purbeck Fault to the Abbotsbury–Ridgeway Fault.

At outcrop between the villages of Abbotsbury and Poxwell the Abbotsbury–Ridgeway Fault shows net reverse displacement (Fig. 8), with the Upper Jurassic–Lower Cretaceous syn-rift section to the south upthrown against post-rift Cretaceous and Tertiary rocks to the north. The short-wavelength periclines in the hanging wall are clearly related to contraction across the fault as the pre-Albian subcrop shows that a syncline was present here at the end of the extension (Ridd 1973). Where the footwall to the fault is exposed in inliers west of Poxwell no Upper Jurassic–Lower Cretaceous syn-rift sediments are preserved and the Albian unconformity cuts down to Middle Jurassic level (Arkell 1947), indicating that the fault acted as a major basin-bounding structure during the final extension event.

The Litton Cheney Fault lies 5 km north of, and parallel to, the Abbotsbury–Ridgeway Fault and also shows evidence of movement during the last rift event. A narrow syncline cored by Upper Jurassic Kimmeridge Clay subcrops the post-rift section on the north-dipping hanging wall, close to the fault (Fig. 8). Similar pre-Albian folds are present in the relay zone between the Purbeck and Abbotsbury–Ridgeway faults between Poxwell and Ringstead where the pre-Albian subcrop pattern shows the Upton Syncline in the hanging wall, close to the Abbotsbury–Ridgeway Fault and an anticline to the south (Ridd 1973; Selley & Stoneley 1987). House (1961) used the continuity of the south limb of the pre-Albian Upwey Syncline and the north limb of the Weymouth Anticline to infer that the Weymouth Anticline was, at least in part, an early Cretaceous structure. The contrast in geometry between the nearly symmetrical Weymouth structures and the north-facing inversion-related folds on the Isle of Wight and Purbeck also supports a more complex mode of formation than simple hanging-wall folding during inversion. Stoneley (1982) deduced from the cuspate map trace of the Abbotsbury–Ridgeway Fault that it has a listric geometry, and placed the detachment level within the basement. Recent well and seismic data clearly show, however, that the fault detaches within the Triassic salt series.

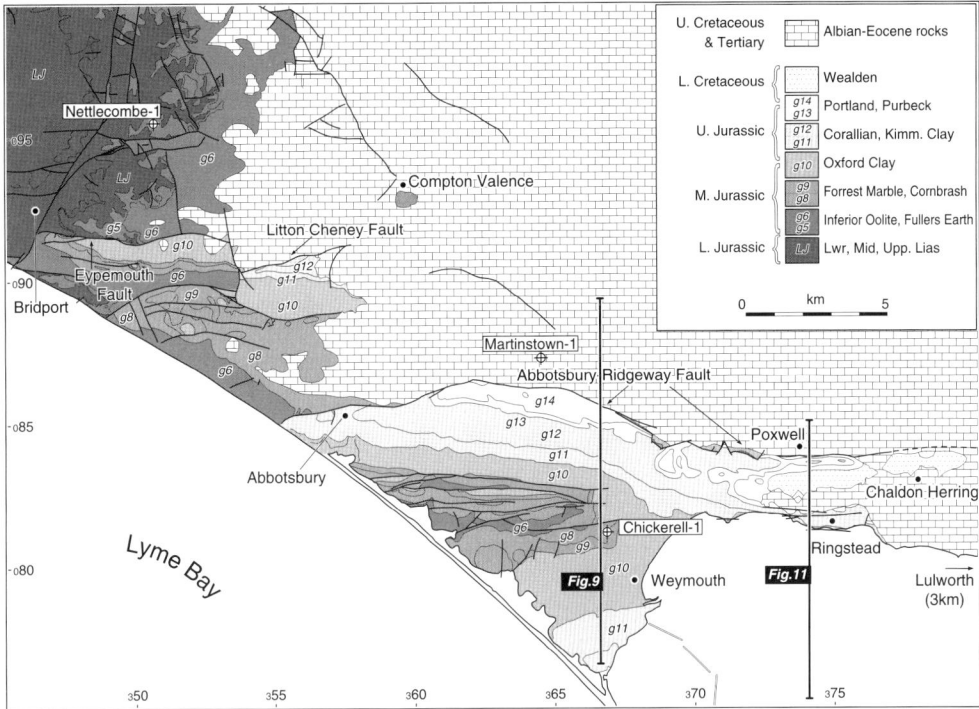

Fig. 8. Simplified geology of the basin margin north of Weymouth. After BGS 1:50 000 sheets 327, 328, 341 and 342, except around Weymouth (House 1961) and Ringstead (Arkell 1951). The Jurassic–early Cretaceous basin-bounding faults step northwest towards the footwall. The Purbeck Fault steps across a relay zone at Ringstead to the Abbotsbury–Ridgeway Fault. The Litton Cheney and Eypemouth Faults define the margin to the west.

The Weymouth Anticline

Recently released exploration wells Martinstown-1 and Chickerell-1 (Figs 1 and 2) were drilled, respectively, on the immediate footwall and in the hanging wall of the Abbotsbury–Ridgeway Fault and both wells proved over 450 m of Triassic saliferous beds correlated with the Dorset Halite. Nettlecombe-1 showed that the Triassic salt series exceeds 240 m in thickness to the north of the Litton Cheney Fault. High-quality two-dimensional seismic data shot by Brabant Petroleum in 1992 image the full Triassic section clearly for the first time onshore and enable confident identification of reflectors at the base of the salt sequence (Butler this volume). Figure 9 shows a true-scale depth section running north–south across the Litton Cheney and Abbotsbury–Ridgeway faults and the Weymouth Anticline, produced by depth-converting the Brabant line B92-40 using a velocity model derived from all nearby wells. The true-depth geometries of both major south-dipping faults are clearly listric in the Jurassic section. The lower part of the Abbotsbury–Ridgeway Fault is especially well constrained by clear fault-plane reflections in the seismic data and can be seen to sole out close to the base of the Triassic salt sequence (Butler this volume). The post-salt Jurassic and uppermost Triassic section displays a markedly different structural style to that picked out in the pre-salt section by bright reflectors at base salt and base Triassic levels. Numerous normal faults at pre-salt level constitute a relatively complex basin margin which still preserves a slope towards the basin. There are fewer, but larger, normal faults at base Jurassic level and the post-salt section is gently folded over steps at pre-salt level. The Triassic salt series acted to separate the uppermost Triassic and Jurassic section (cover) from pre-salt level structures. The pre-salt lower Mercia Mudstone, Sherwood Sandstone, Permian and Pre-Permian section are or for simplicity referred to here as 'basement'.

The Weymouth Anticline is cored by a thickened salt sequence and there is a similar swell in the hanging wall of the Litton Cheney

Fault. The salt swell beneath Weymouth formed initially during extension by movement of salt away from the descending hanging wall of the Abbotsbury–Ridgeway Fault to passively fill the core of the rollover anticline and form a salt roller in the footwall of the fault (following the terminology of Jackson & Talbot 1986). The amount of Tertiary reversal on the Abbotsbury–Ridgeway Fault can be estimated by assuming a thickness of 150 m for the missing pre-Albian section in the immediate hanging wall (from Lower Cretaceous outcrops at Bincombe cutting, 1 km E). Restoring the base Albian level in the hanging wall reveals 350 m of reversal on the fault. Slip-line restorations of this reversal show that the pre-inversion displacement on the fault exceeded 1300 m at base Jurassic level. None of the basement faults show present-day displacements of this magnitude and even allowing for Tertiary reversal on the pre-salt faults, the difference in displacement, location and angle of the faults above and below the salt infers complete detachment of the overburden during extension.

Thickness changes in the Lower Jurassic section across the Litton Cheney Fault (Fig. 9) show that, as in Bournemouth Bay, active extension occurred during the early Jurassic. Depositional facies changes and the presence of neptunian sills and dykes at outcrop (Jenkyns & Senior 1991) record early Jurassic extension on the Bride and Eypemouth faults, along strike to the west. Although the Lower Jurassic is only 9 m thicker in Chickerell-1 than in Martinstown-1 to the north of the Abbotsbury–Ridgeway Fault, the depth section illustrates rapid thickening of this interval northwards from Chickerell-1 towards the fault. This evidence demonstrates that the Weymouth Anticline began to form as a rollover feature at this time, and Lower Jurassic sediments also thicken southwards on the southern flank of the fold into the basin. Both major south-dipping faults were also active during late Jurassic–early Cretaceous extension, although the Abbotsbury–Ridgeway Fault accommodated by far the most displacement. During Tertiary inversion the hanging wall of the Abbotsbury–Ridgeway Fault was uplifted and eroded so that the entire Cretaceous and Tertiary sequence is now absent over the southern part of the depth section. Removal of the effects of this uplift at pre-salt level indicates that the basinward slope across the length of the depth section (Fig. 9) was greater prior to inversion than at the present day. Even if most of the Tertiary uplift was generated by reversal on a single basement fault (e.g. the south-dipping fault 1 km south of the axis of the Weymouth Anticline), the restored Jurassic–early Cretaceous basement dip between the Purbeck and Abbotsbury–Ridgeway faults would still have been towards the basin, as evidenced by southward thickening of the detached Jurassic section.

This evidence leads to the conclusion that extensional deformation in the overburden was

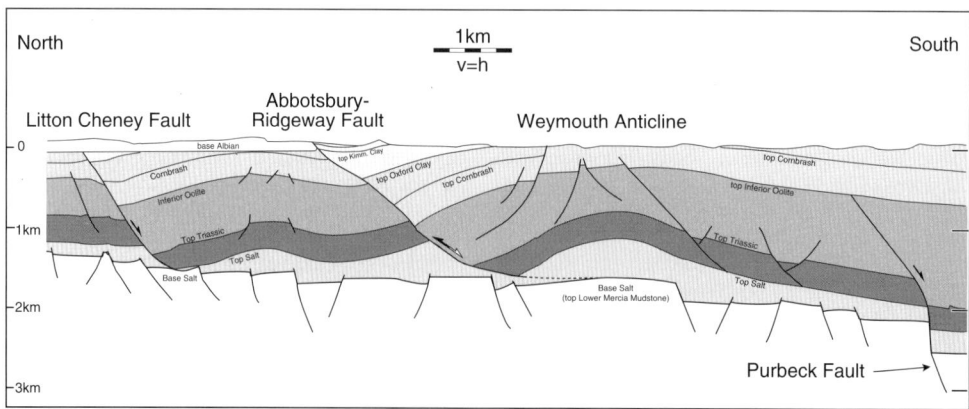

Fig. 9. True-scale depth section across the Litton Cheney and Abbotsbury–Ridgeway faults and the Weymouth Anticline. Depth converted from 1992 Brabant onshore seismic (Butler, this volume) using a velocity model derived from all adjacent wells. Triassic salt detaches the Upper Triassic–Lower Cretaceous overburden from sub-salt structures. The true-depth section shows the listric geometry of detached overburden faults: the Weymouth Anticline is a pre-Albian rollover structure tightened during Tertiary reversal on the Abbotsbury–Ridgeway Fault. Slope-driven overburden extension occurred above a basinward-dipping part of the margin, in the relay zone between major pre-salt faults (see text).

closer in nature to thin-skinned tectonics than to extension directly linked to basement displacements. The thickness of the salt sequence appears to have been sufficient for complete detachment of the overburden from basement fault movements. The location of the Abbotsbury–Ridgeway Fault may, however, have been influenced by basinward fault steps at pre-salt level (e.g. Gaullier et al. 1993), which would have been larger prior to Tertiary inversion. Basinwards migration of cover extension during thin-skinned rifting has been demonstrated on the East African margin by Lundin (1992) who linked the end of extension on up-slope cover faults to grounding of the hanging-wall block and creation of a new pin (salt weld) between the overburden and pre-salt sections (Fig. 10). This process may explain the greater late Jurassic–early Cretaceous extension across the Abbotsbury–Ridgeway Fault compared to the Litton Cheney Fault, as there is a salt weld where the hanging wall of the latter fault has grounded (Fig. 9). The more prolonged extension across the southern fault may, of course, have been a response solely to shifts in the locus of basement extension. The overall geometry of the Channel Basin margin in this area reflects synchronous down-dip extension of the cover and active basement rifting – this temporal association is unlikely to be a coincidence and displacements on the basement faults probably balance cover faulting via detachment in

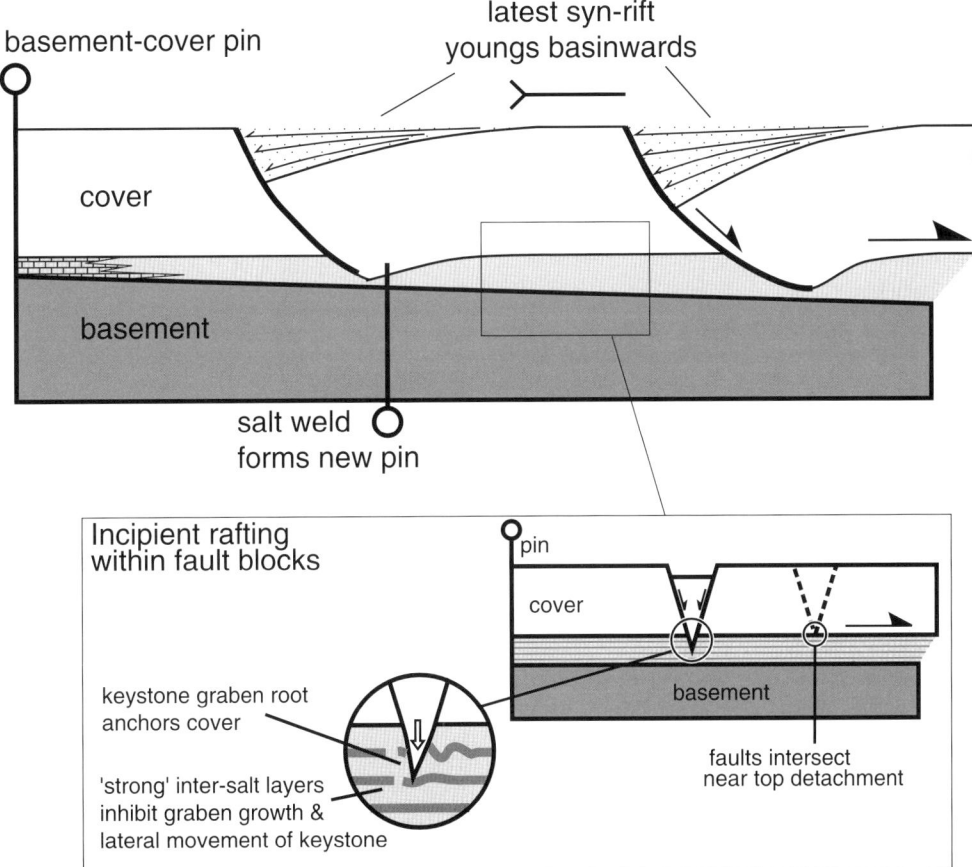

Fig. 10. Simple model of basinward migration of overburden rifting during slope-driven extension (cf. East African margin, Lundin et al. 1992). Grounding of hanging-wall blocks results in the formation of salt welds which pin the overburden to the pre-salt section, halting fault growth up-dip. This process may explain greater early Cretaceous extension on the Abbotsbury–Ridgeway Fault than that on the Litton Cheney Fault. Inset: possible grounding of narrow overburden graben within the saliferous section; stronger mudstone layers may pin overburden during incipient raft tectonics (see text, Fig. 13).

the Triassic salt. In other words, each cover fault represents the upwards continuation of basement fault(s), offset via a 'flat' in the salt. Such ramp–flat geometry is characteristic of extensional faults in multilayers (Nalpas & Brun 1993; Jackson & Vendeville 1994).

Lateral onset of detachment faulting: the Ringstead relay zone

The Chickerell and Martinstown wells and recent data demonstrate that salt detachments have had a significant influence on the extensional geometry of the basin margin around Weymouth. As cover faulting in Bournemouth Bay is clearly basement-linked a question arises as to where the lateral transition lies between the two geometrical styles. Key parameters that may affect whether extensional detachments develop include the thickness of salt and overburden sequences and the magnitude and rate of basement fault displacement (Koyi et al. 1993). Two of the these parameters are described by the displacement ratio D_r, defined as the thickness of the viscous layer T_v divided by the basement fault displacement D (Koyi et al. 1993). Stewart et al. (1996) suggested that for values of $D_r > 1$ salt detachments might be expected to develop, but where basement fault displacement exceeds salt thickness ($D_r < 1$) a single, hard-linked fault is likely to propagate from basement to cover. Such classifications are complicated in the Channel Basin by the impure nature of the Triassic salt sequence (Fig. 2) as the effective thickness of the viscous layer may be much smaller than the thickness of the entire salt sequence. Along the Jurassic–Cretaceous basin margin salt is only present in the hanging wall of the Purbeck Fault, inhibiting the usual horizontal offset between cover and basement elements of fault strands (cf. Nalpas & Brun 1993; Jackson & Vendeville 1994).

The thickness of the Triassic salt sequence in the Purbeck Fault footwall increases rapidly westwards from the point at which the salt oversteps onto the footwall near Kimmeridge (Fig. 3). Between this point and the emergence of the Purbeck Fault in Weymouth Bay, however, extensional displacement was far greater than the salt thickness and a single, planar fault can be seen to pass from the pre-salt section into the cover (Fig. 11). Even where the salt was thick enough to detach the Abbotsbury–Ridgeway Fault, the Purbeck Fault had sufficient displacement to completely displace the salt (Fig. 9) and link directly to the cover section, indicating that $D_r < 1$ along much of the length of the Purbeck Fault strand. To the north of the Purbeck Fault in the relay zone between Ringstead and Poxwell, extension at pre-salt level appears to have been partitioned between many small faults. Each basement fault has a small displacement compared to the probable thickness of Triassic salt (i.e. $D_r < 1$) enabling faults in the Jurassic–Cretaceous section to detach at Triassic level (Fig. 11).

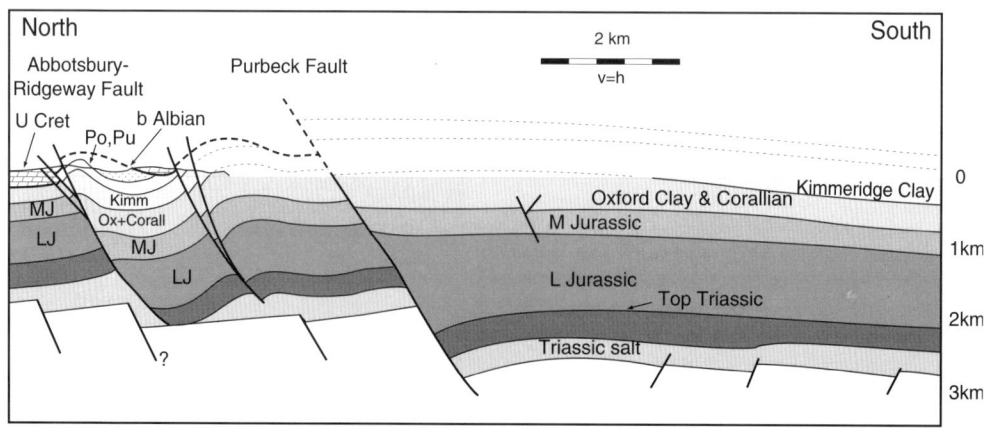

Fig. 11. Depth section across the Abbotsbury–Ridgeway and Purbeck faults. The southern part from depth conversion of offshore seismic data using separate interval velocities for the footwall and (inverted) hanging wall sequences from adjacent wells. Northern part modified in the shallow section from Mottram (1949) and Stoneley (1982), redrawn in deep section assuming extensional salt detachments and using thickness estimates based on data from Chickerell, Creech and Bushey Farm wells.

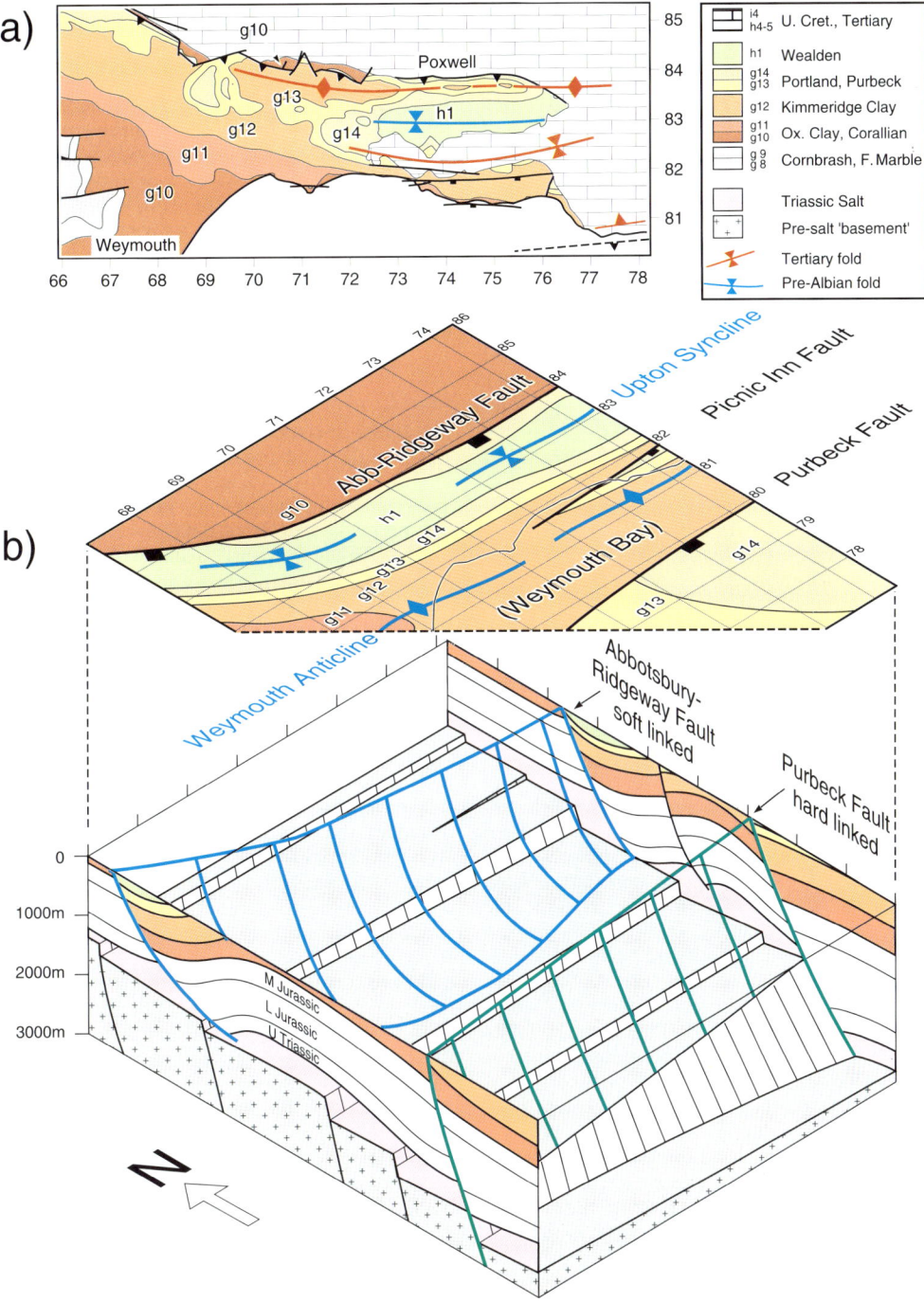

Fig. 12. Geometry of the Ringstead extensional relay zone. (**a**) Outcrop geology of the area between the Abbotsbury–Ridgeway and Purbeck faults, from BGS sheet 341. (**b**) Three-dimensional pre-Albian restoration (i.e. with the effects of Tertiary inversion removed). The easstern depth section was constructed using the mapped subcrop beneath the Albian post-rift unconformity (pre-salt structure north of the Purbeck Fault is conjectural). The western depth section is from fault-to-bed chevron constructions of the rollover geometry on the Abbotsbury–Ridgeway Fault north of Weymouth, using heaves corrected for Tertiary inversion and a fault profile constrained by seismic.

It is difficult to estimate the eastward extent of the Abbotsbury–Ridgeway Fault, as it becomes obscured within the Chalk east of the Poxwell Pericline (Fig. 8) (Mottram 1949). Reversal of the fault during Tertiary inversion resulted in the formation of this tight fold and presumably also the Chaldon Herring Pericline to the east. A broad anticline north of Lulworth is picked out by the base Tertiary outcrop and by mapping of Chalk zones (Taitt *in* Arkell 1947). This was probably caused by inversion of the eastward extension of the Abbotsbury–Ridgeway Fault, as the base Tertiary contact describes a tight syncline directly along strike from the fault's easternmost outcrop. It is unlikely that significant salt-detached fault zones developed to the north of the Purbeck Fault much further east of here, as the thickness of the salt series reduces very rapidly (Fig. 3), being less than 60 m at Creech and absent on the Wytch Farm block (e.g. in Bushey Farm-1 in the immediate footwall of the Purbeck Fault).

The cropping-out Tertiary periclines are superimposed on the pre-Albian Upton Syncline which probably formed as a hanging-wall syncline during extension on the Abbotsbury–Ridgeway Fault. The overburden structure prior to Tertiary inversion can be reconstructed using the pre-Albian subcrop pattern. Figure 12 is a three-dimensional representation of the geometry of the relay zone between the Purbeck and Abbotsbury–Ridgeway faults at the end of extension (early Cretaceous). The top and eastern side panels of the block were constructed using the subcrop pattern. The western side panel is constrained by fault-to-bed chevron constructions, which model the rollover geometry in the hanging wall of the Abbotsbury–Ridgeway Fault for known detachment shape and pre-inversion heaves. The pre-salt faults have been modelled using the assumption that they strike roughly eastwards from the line of Fig. 9. Their positions beneath the eastern edge of the block diagram and in Fig. 11b are therefore conjectural, although they conform to the structural style observed further west (Fig. 9) and on seismic data across the Poxwell Pericline (Butler this volume). The more pronounced inversion-related folding at the eastern end of the Abbotsbury–Ridgeway Fault (compare Figs 9 and 11) may indicate that there was a higher degree of linkage between pre- and post-salt level faults in the east, at least during the Tertiary. This hard linkage may have arisen from grounding of the hanging wall of the Abbotsbury–Ridgeway Fault during extension where the salt sequence thins to the east. A salt weld in this area may have led to the development of the extensional Spring Bottom Fault downthrowing the crest of the rollover anticline to the south (Fig. 12). The westwards thickening of salt explains a sharp northward swing in strike of the Abbotsbury–Ridgeway Fault (Fig. 8, [GR 369 084]). The overburden fault is more fully detached in the west where the salt is thicker, and is offset further towards the footwall of the main basement steps, conforming to analogue models of cover deformation above basement faults with varying thicknesses of an intervening ductile layer (e.g. Larroque 1993).

Detached basement and cover faulting in Lyme Bay

The relationship between pre- and post-salt faults can be studied in more detail where the basin margin strikes westwards from Weymouth into Lyme Bay. Figure 13 shows an offshore seismic line from Lyme Bay, running southeast from near Lyme Regis, and displays high-amplitude reflections at top Triassic and top pre-salt (lower Mercia Mudstone) levels. These events, along with top Permian (base Budleigh Salterton Pebble Beds), top Triassic salt and intra-Jurassic markers, have been correlated with seismic data tied to the Chickerell-1 well. Although the pre-salt reflectors are difficult to map over long distances, the data illustrate a similar structural style to that already described in the Weymouth area, with relatively flat-lying basement fault blocks stepping down into the basin. Four major south-dipping normal faults displace the cover sequence (at shot-points 1700, 1200, 800 and 600) and have far greater heaves than those at pre-salt level. All faults in the overburden display listric geometries on the time data (Fig 13a and b) and in the hanging wall of the northernmost fault an anticline in the Jurassic section is cut by crestal collapse faults like those on the Weymouth Anticline. After depth conversion (Fig. 13c), using interval velocities from Chickerell-1, the overburden faults appear more planar than the Abbotsbury–Ridgeway Fault, and they separate rotated fault blocks which are less folded but more faulted than those further east.

The four major south-dipping overburden faults in Fig. 13 each have a displacement of >600 m at top Triassic level, whereas none of the present-day pre-salt level displacements exceeds 350 m. The overburden faults in the centre and at the north end of the section show a clear spatial relationship with steps at pre-salt level, in contrast to the Abbotsbury–Ridgeway Fault which developed away from any large basement

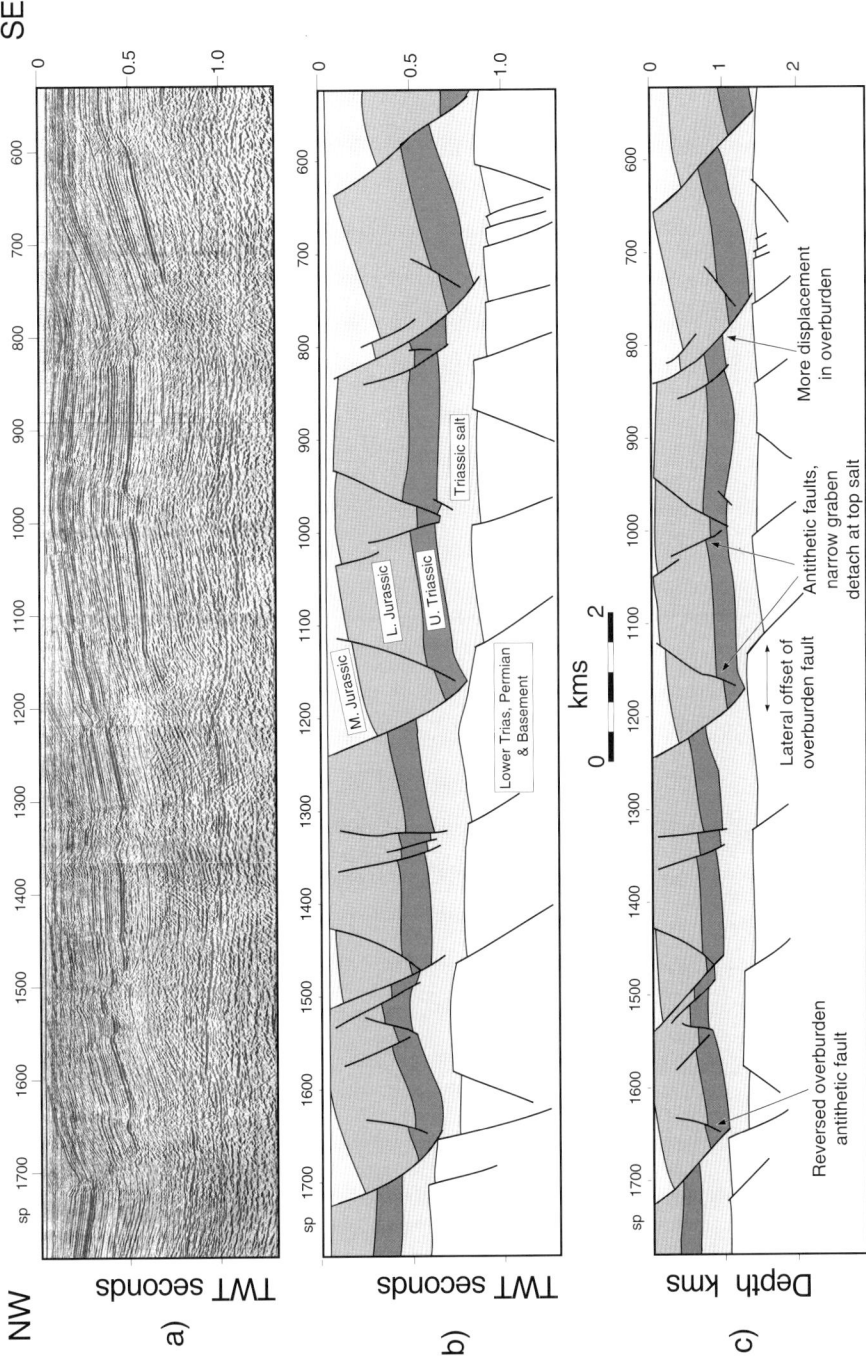

Fig. 13. (a) NNW–SSE seismic line from Lyme Bay (location in Fig. 14). (b) Geoseismic section. (c) Depth section, depth converted using a simple interval velocity model from Chickerell-1 well. Curved overburden faults in time become more planar in the depth section and define rotated overburden blocks: cf. listric faults, salt swells and overburden folds to east (Fig. 9). Lateral offsets and displacement disparities between basement and overburden faults indicate detachment tectonics. (Data courtesy of Kerr McGee Oil Ltd.)

structure. The difference in displacement between the faults at top Triassic level and those in the pre-salt section, along with the significant lateral offsets between overburden faults and basement steps, suggests however that they are not 'hard'-linked. The two southernmost overburden faults overlie an area of poor seismic imaging beneath the salt, but top Permian reflectors appear to show that there are no corresponding basement faults of this magnitude. Antithetic faults in the hanging wall of each major south-dipping fault intersect the major faults close to the top of the salt sequence (cf. Childs *et al.* 1993) as do small symmetrical grabens within the main overburden fault blocks (see below), showing that detachment tectonics did occur in the area. The current easterly tilt of the basin means that the Middle and Lower Jurassic section crops out in Lyme Bay, making it difficult to estimate the timing of Jurassic–Cretaceous movement on overburden faults. Slight thickening of the Lower Jurassic sequence is apparent across the major south-dipping fault in the centre of the depth section (Fig. 13c, shotpoint 1200) but not on those at the southern end, and early Jurassic growth appears to have been less pronounced than on the Abbotsbury–Ridgeway and Litton Cheney faults. Most displacement on the overburden faults must have occurred after the early Jurassic, as is the case onshore to the east. As Albian outcrops overstep east–west normal faults onshore to the north, the major extension on this fault set is interpreted as late Jurassic–early Cretaceous.

Narrow, salt-detached grabens

The lateral continuity of overburden structures at top Triassic level has been mapped in Lyme Bay (Fig. 14) using a 1.5–2 km spaced grid of two-dimensional seismic data. In addition to the major south-dipping cover faults, several smaller symmetrical grabens can be traced laterally for

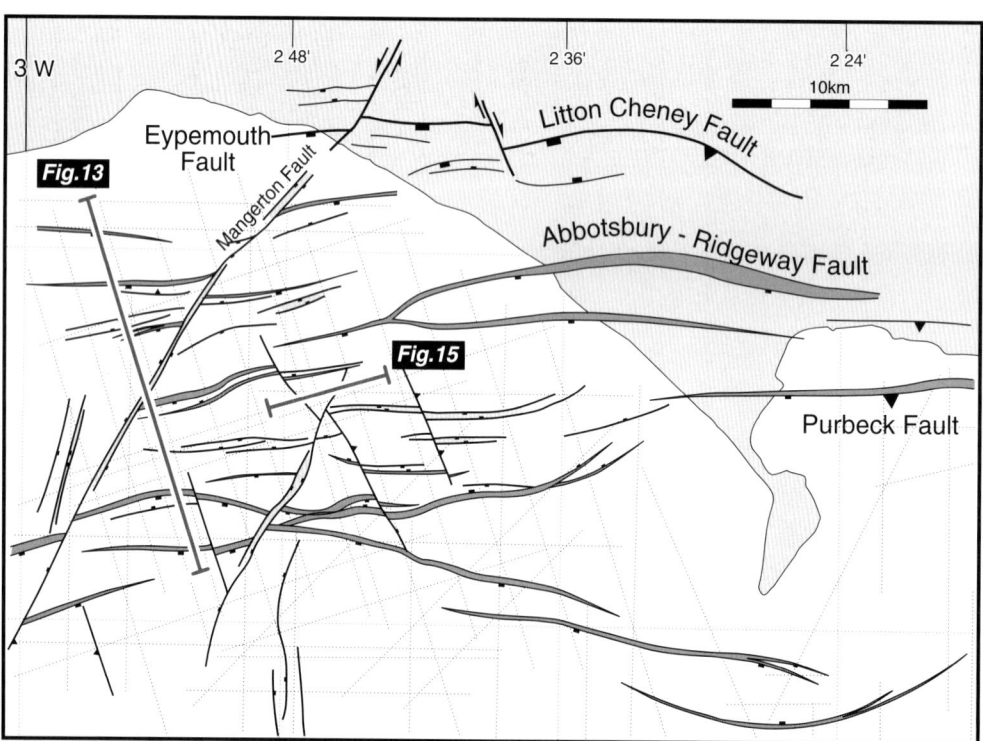

Fig. 14. Summary map of overburden structures in the Lyme Bay area, mapped at base Jurassic level. All east–west faults, apart from Purbeck Fault, detach within the Triassic salt series. A narrow sinuous graben in Lyme Bay may indicate raft tectonics within the main overburden fault blocks. Northwest–southeast and northeast–southwest fault zones form a conjugate sets and appear as steep upward-branching structures on seismic.

up to 10 km. The hanging-wall cut-off spacing of around 300 m at top Triassic level is constant along each graben, further indicating that they share a common keystone level at the top of the salt sequence. These grabens result from minor amounts of gravity-driven down-dip slip of the cover sequence, accommodated by formation of planar faults detaching at top salt. After the first increments of displacement on such ramp–flat systems, hanging-wall rollover is accommodated by formation of an antithetic fault which roots onto the ramp–flat junction (at top detachment) in the synthetic strand (cf. Faure & Chermette 1989). With further increments of down-dip cover slip, collapse of the keystone grabens formed in this way may lead to a local anchoring

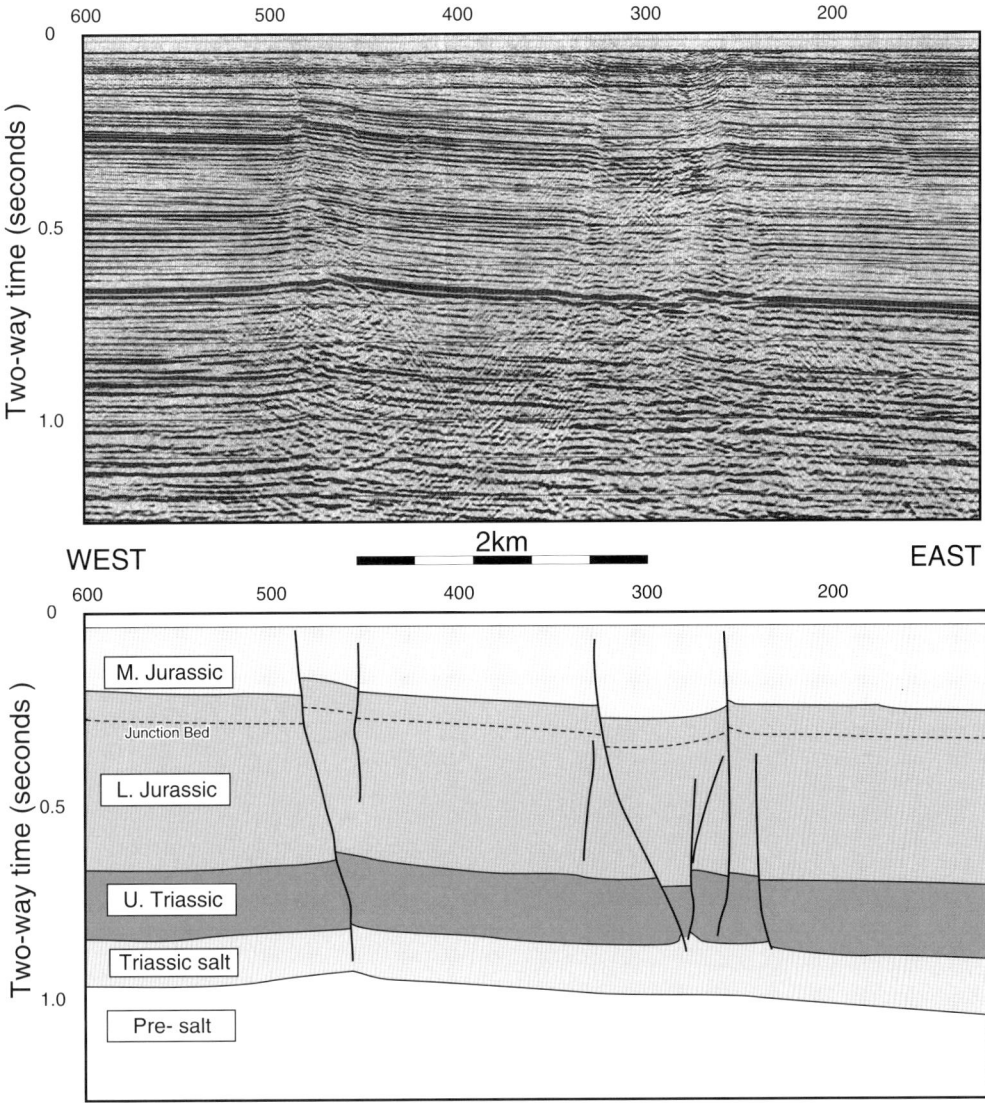

Fig. 15. ENE–WSW seismic (a) and geoseismic (b) across northwest–southeast and northeast–southwest fault zones in Lyme Bay. Steepness and the anastamosing geometry of the fault zones is similar to the 'flower' geometry of strike-slip faults. No clear continuations of the structures are detectable within the pre-salt section. (data courtesy of Kerr McGee Oil Ltd.).

effect liable to inhibit further extension across the graben in favour of nucleation of new faults down-dip (Fig. 10). These narrow grabens may represent a nascent raft tectonics system in this part of the basin (cf. Smith & Hatton this volume), geometrically similar to distributed graben systems related to low-angle detachments which have been described, for example, in Canyonlands, Utah (Trudgill & Cartwright 1994).

Conjugate strike-slip fault zones

A striking feature of the map pattern of faulting in Lyme Bay is the presence of northwest–southeast- and northeast–southwest-trending fault zones which intersect and apparently displace the east–west-trending normal faults and narrow grabens. These fault zones are most easily detected on strike lines which run subparallel to the east–west-trending extensional structures. Figure 15 shows a strike line that images two steep fault zones, one trending northeast–southwest and the other northwest–southeast. Each zone contains several fault strands with both normal and reverse displacements within the Lower Jurassic section. In cross-section the fault strands tend to splay upwards at rheological interfaces (Fig. 15) giving 'flower' geometries – a structural style which has been reproduced in analogue models of strike-slip faults (Richard et al. 1995, fig. 9). Seismic resolution of the pre-salt horizons beneath the fault zones is too poor to be able to trace the faults downwards with confidence and it is unclear whether the fault zones are basement-linked or are restricted to the overburden. A similar northeast–southwest-trending fault zone crosses the line of section of Fig. 13 at shot-point 1350.

The relationship between the northwest–southeast and northeast–southwest fault zones, which form a conjugate fault set in map view, and the east–west-trending normal faults which they appear to truncate can be examined in more detail onshore to the north. The Mangerton Fault trends northeast–southwest from Mangerton in Dorset to the coast south of Bridport (Fig. 16) and is a continuation of the steep fault zone shown in Fig. 13 (shot-point 1350). Like those mapped offshore, the Mangerton Fault appears to displace east–west trending normal faults. The Eypemouth Fault runs eastwards from the coast, where it is well exposed at Watton Cliff, until it meets the Mangerton Fault at West Bay and is offset sinistrally by 1 km to Bothenhampton, from where it strikes eastwards towards Litton Cheney. The fault forms the northern boundary of a half-graben, which is cut 1.5 km to the south by a north-dipping antithetic fault (Fig. 16). The spacing between synthetic and antithetic strands at the level of the Inferior Oolite outcrop is similar to that on the major south-dipping faults in Fig. 13 which suggests that the Eypemouth Fault has a similar geometry to those in Lyme Bay, detaching within the salt.

Kinematic indicators on the surface of the Eypemouth Fault at Watton Cliff infer pure dip-slip movement during the last displacement event. Jenkyns & Senior (1991) demonstrated several early Jurassic periods of fault activity but the last extension must have been during the late Jurassic or early Cretaceous as the top Middle Jurassic sequence is downthrown in the hanging wall. Although the Mangerton Fault truncates the Eypemouth Fault and the narrow graben to the north in map view, the possibility that the two fault sets developed simultaneously must be considered, as the northeast–southwest- and northwest–southeast-oriented faults might have acted as transfer zones during extension on the east–west structures. The dip-slip movement on the Eypemouth Fault suggests, however, that it could not have moved simultaneously with the Mangerton Fault, which from recent onshore data (Butler pers. comm.) and offshore seismic data is near vertical. A space problem would prevent movement of the hanging wall of the extensional fault in the west (Fig. 17) and a pull-apart would have developed in the transfer zone. The geometrical character of the steep northeast–southwest and northwest–southeast fault zones, along with outcrop evidence that they post-date extensional faulting, suggests that they formed as a conjugate strike-slip set during late Cretaceous–Tertiary compression. Faults of each set are shown on BGS maps (e.g. Sheet 327, Bridport) to post-date the Cretaceous post-rift sequence, including those cutting the Chalk north of Compton Valence and several northwest–southeast faults separating the Eypemouth and Litton Cheney Faults (Figs 8 and 16). The development of strike-slip faults in the Lyme Bay post-salt section during inversion implies that it was easier for the overburden to accommodate north–south shortening in this area by an element of east–west expulsion than by uplift alone. Partitioning of some inversion-related shortening into lateral expulsion in the western part of the Channel Basin may be related to the distribution of Triassic salt, which is the obvious candidate for facilitating such lateral movement.

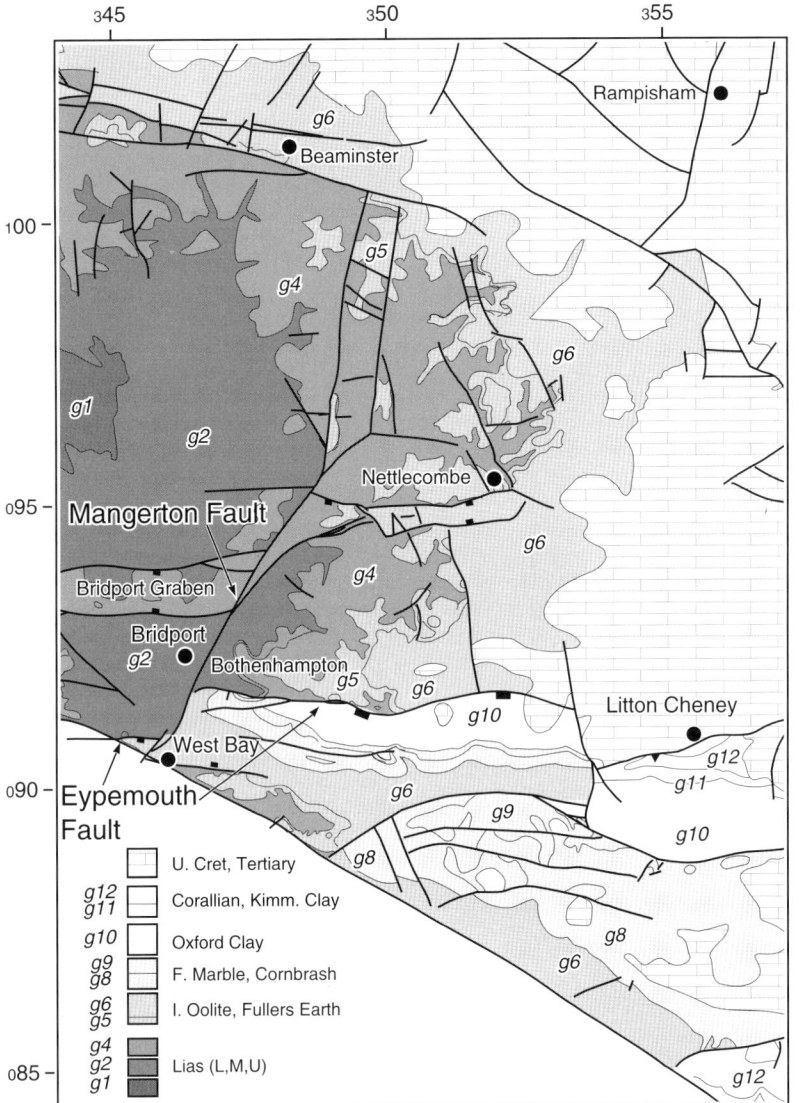

Fig. 16. Geological map of West Dorset showing relationship between east–west extensional faults and conjugate northwest–southeast and northeast–southwest fault set (after BGS Sheet 327). East–west-trending Eypemouth Fault is displaced sinistrally by the Mangerton Fault and is separated from the Litton Cheney Fault by dextral displacement on a northwest–southeast structure. The Litton Cheney Fault detaches in Triassic salt (Fig. 9). Northwest–southeast and northeast–southwest faults also affect the Cretaceous post-rift section at Compton Valence and to the north (Fig. 8).

Summary of salt tectonics in the Channel Basin

Seismic and well data demonstrate the presence and influence of significant Triassic salt detachments on the extensional basin margin in the west, contrasting with the basement-linked deformation style in the east. Within the region of detachment tectonics several different extensional fault and fold patterns exist, ranging from low-angle listric faults, with associated salt rollers and hanging wall anticlines,

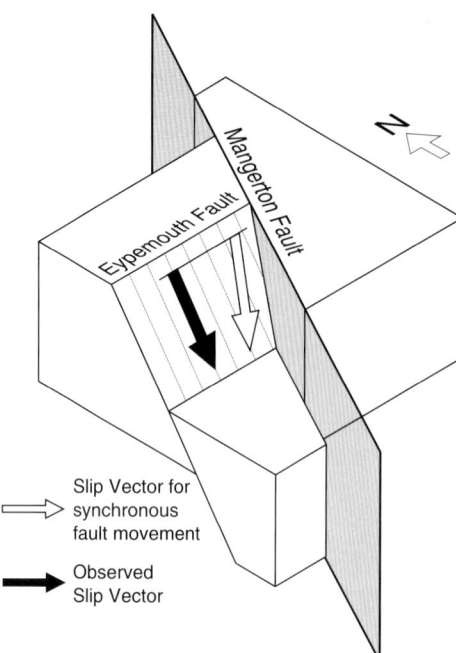

Fig. 17. Cartoon of relationship between the Eypemouth and Mangerton faults. The Mangerton Fault must post-date the final extension on the Eypemouth Fault as kinematic indicators on the latter indicate pure dip-slip.

to more rigid rotated blocks bounded by steeper, planar faults. Slope-driven overburden extension around Weymouth may have been localized to a pre-salt relay zone where the basement dipped basinwards. The variation in style suggests that the halite is more mobile in some parts of the basin (e.g. Weymouth area) than in others (Lyme Bay). The mobility of the Channel Basin 'salt' section is related to the stratigraphy of the saliferous succession which, unlike the North Sea Permian salt, contains many non-evaporitic interbeds which are variably developed across the basin. Although sufficiently mobile to redistribute locally into swells and rollers, the salt would seem to be reinforced by less mobile, interbedded units and could not achieve the degree of lateral movement required for diapiric intrusion of cover faults, despite cover extension exceeding that which characterizes the onset of reactive diapirism ($\beta = 1.2$; cf. Vendeville & Jackson 1992).

Analogue models of extension in overburden sequences separated from basement faults by a ductile layer have shown that where the ductile layer is thick enough to detach the overburden, faulting in the cover will be offset towards the footwall (Richard 1991; Larroque 1993; Vendeville et al. 1995). Spatial variation in such ramp–flat geometries within the Channel Basin can be considered in terms of the viscous layer thickness to fault displacement ratio D_r (Koyi et al. 1993; Stewart et al. 1996). Departures from these simple models occur within the Channel Basin and may reflect stratigraphically controlled variations in rheology of the salt sequence. The variation in basement–cover fault linkage as a function of salt layer thickness in the Channel Basin is summarized in Fig. 18, which also attempts to portray the impact on inversion style of salt redistribution during extension. The end members of hard-linked and unlinked overburden faulting can be easily distinguished within the Channel Basin. The intermediate cases of soft-linked and 'firm-linked' overburden faulting have been observed along the north margin of the basin and differ in their relationship to basement faults in terms of lateral offset and differences in displacement. Firm-linked overburden faults developed during extension around Ringstead (Fig. 12). Extensional folding of the Abbotsbury–Ridgeway Fault hanging wall shows that some detachment between cover and basement occurred but the location of the overburden faults was probably closely controlled by basement faulting. Grounding of the hanging wall during extension allowed it to become hard-linked with the basement and facilitated cover fault reversal during inversion. Two categories of soft-linked overburden faulting can be distinguished (Fig. 18). Around Weymouth the overburden deformed by down-slope extension on listric normal faults, which show no close spatial association with basement steps. Displacement on large cover faults may have been soft-linked to many basement faults via a master detachment within the salt. In Lyme Bay the major overburden faults are either com-pletely separated or substantially offset from basement fault steps. In the latter case present-day displacement on the cover fault strands far exceeds that beneath the salt, either due to reversal at basement level that is not transferred to cover faults or to linkage of overburden displacement to many basement faults via a shared detachment. The salt appears to have been less mobile than around Weymouth resulting in rotation of rigid overburden blocks rather than folding above swells and rollers.

There is also evidence for temporal variation in basement–cover fault linkage during fault growth. The early formation of a rollover fold in

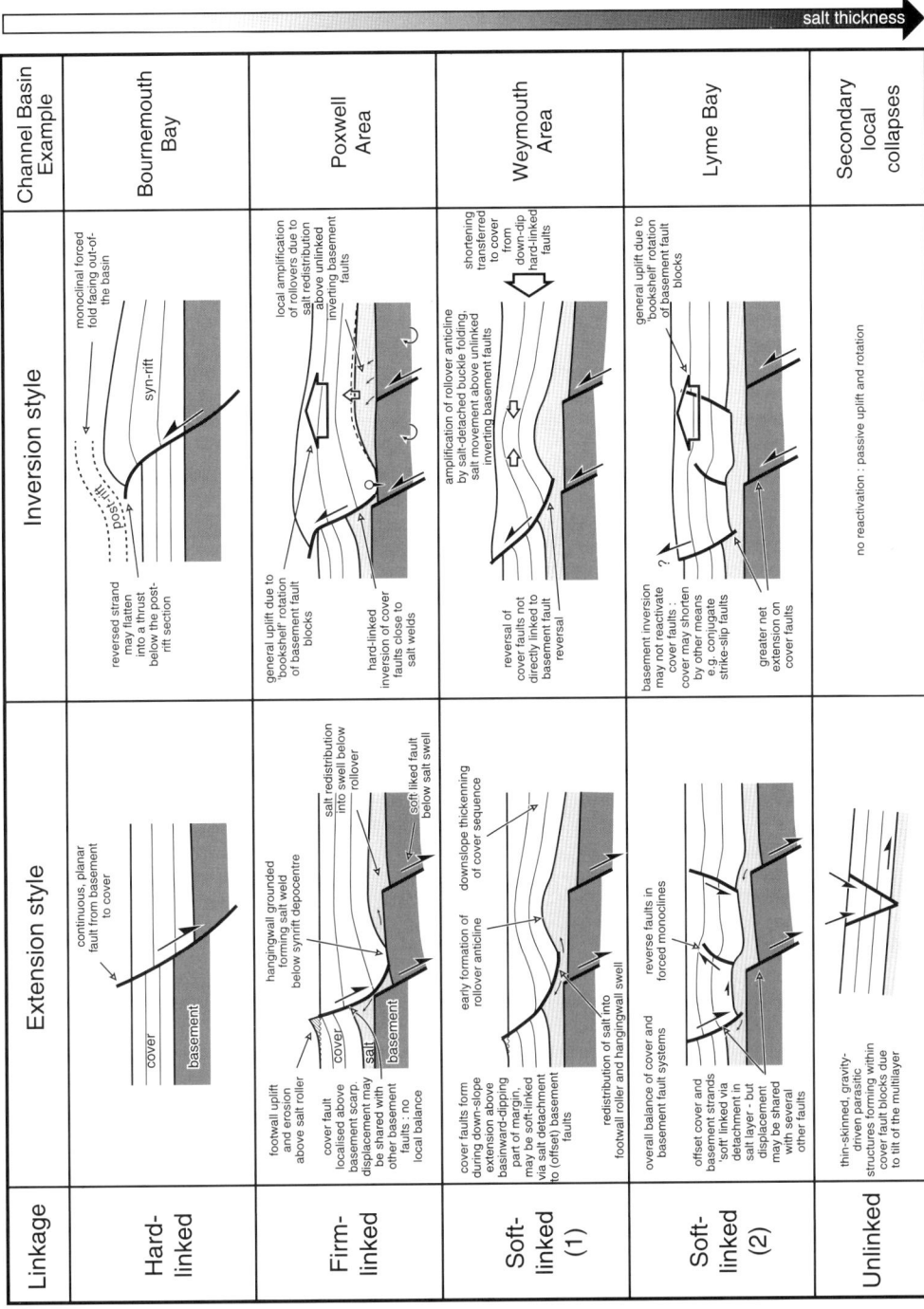

Fig. 18. Summary of the variation in linkage between the overburden and basement faults with increasing salt thickness. Differences in soft-linked fault geometries, with similar displacement ratios (e.g. between Weymouth and Lyme Bay) may reflect variations in rheology of the salt interval or changes in extensional basement slope along strike. Channel Basin examples are discussed in text.

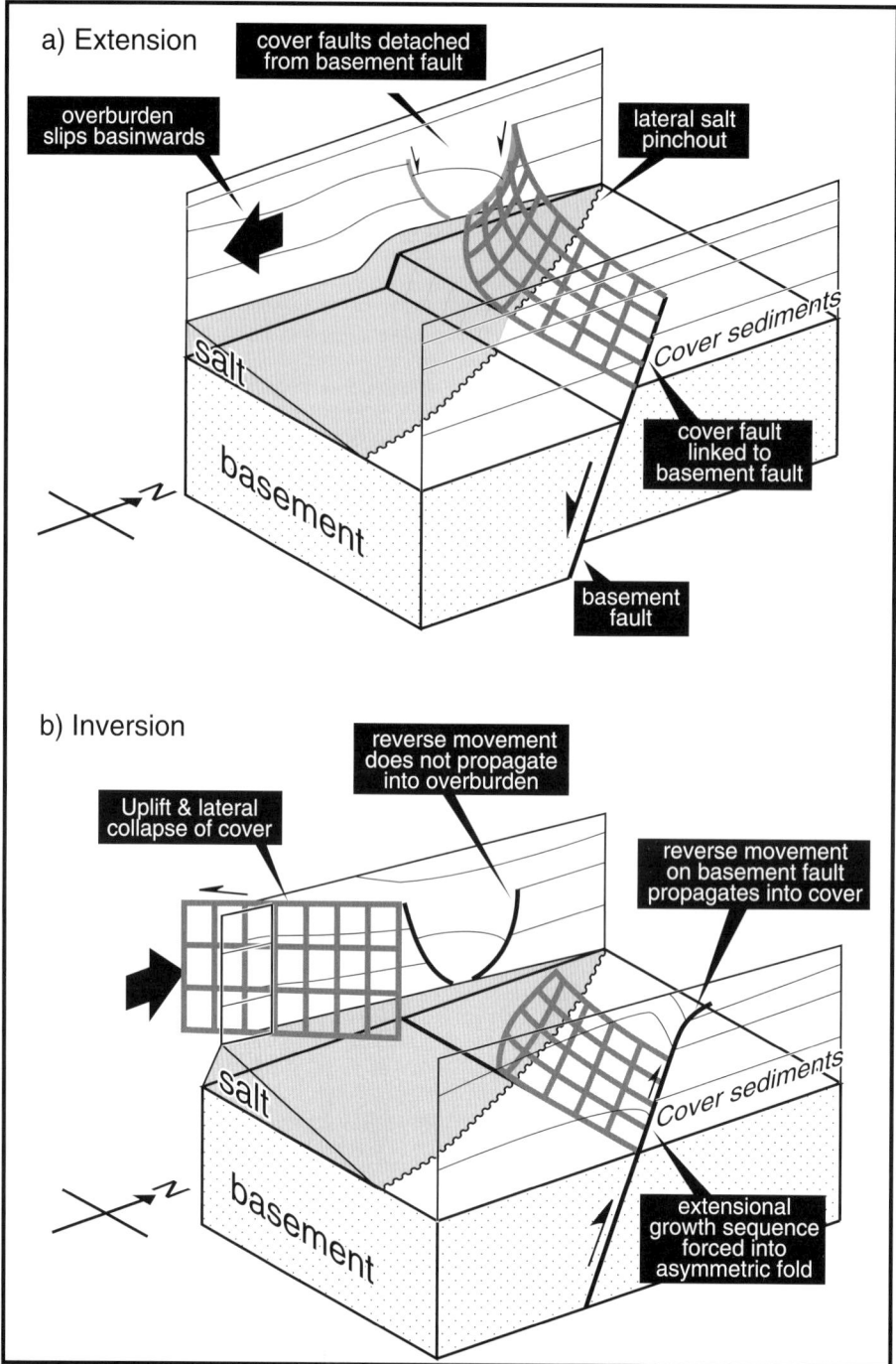

Fig. 19. Summary block diagrams showing a westward increase in the influence of salt on the evolution of the north margin of the Channel Basin. (**a**) During extension salt detachments cause the step-back of overburden margin faults towards the footwall (e.g. from the Purbeck to Litton Cheney Faults). (**b**) During inversion salt detachments inhibit propagation of basement fault reversal into the overburden, and other shortening mechanisms may be utilized (e.g. strike-slip faulting).

the hanging wall of Purbeck Fault can be inferred from rapid northward thickening of the Lower Jurassic isopach close to the fault (Fig. 11). The less pronounced thickening of the overlying Jurassic intervals towards the fault indicates that by mid-Jurassic times, after further displacement, a hard link between basement and cover fault strands had been established. Figure 13 shows a reverse fault within the Lower Jurassic section in Lyme Bay (shot-point 1650) that, in its geometry and relationship to the basement fault, is very similar to reverse faults produced during monoclinal forced folding above the tip of an underlying extensional fault (Richard 1991, fig. 6; Couples *et al.* 1994; Vendeville *et al.* 1995). This suggests that early draping of the cover occurred above the basement fault with relatively small displacement and a link between basement and cover faults (via a 'flat' within the salt interval) only developed after further extension.

An additional control on the current geometry of the margin is the degree to which extensional structures in the basement and cover were reactivated during basin inversion (Fig. 18). The presence of conjugate strike-slip fault zones in the Lyme Bay area, which apparently accommodate some Tertiary north–south contraction, may represent an unusual partitioning of shortening styles between basement and cover (Fig. 19). However, there remains a possibility that the cover strike-slip faults are linked to unimaged basement equivalents as major basement structures parallel to both of the conjugate trends are present around the Channel Basin (Lefort & Max 1992, fig. 1). The effects of inversion on east–west structures included reduction of extensional displacement on cover faults and modification of existing extensional folds, with fold amplification (e.g. Weymouth Anticline) or superposition of shorter wavelength folds (Poxwell Pericline superposed on the Upton Syncline). The likelihood of reverse displacement propagating from basement to cover fault strands depends on the thickness of the intervening viscous layer, a parameter which will have been altered by salt redistribution during extension (Fig. 18). For example, basement faults close to salt welds will be more likely to pass into the overlying cover than an inverting fault below a salt swell, which is likely to generate further redistribution and fold amplification. The style of inversion is therefore controlled by salt distribution and rheology directly, and indirectly following tectonic modification of salt thickness during extension.

The authors acknowledge Amerada Hess Limited, Brabant Petroleum Limited, Kerr-McGee Oil (UK) Limited and OMV (UK) Limited for permission to use the seismic examples. The views expressed here are those of the authors and not necessarily those of Amerada Hess Limited, Brabant Petroleum Limited, Kerr-McGee Oil (UK) Limited or OMV (UK) Limited.

References

ARKELL, W. J. 1947. *The Geology of the Country around Weymouth, Swanage, Corfe and Lulworth.* Memoir of the Geological Survey of Great Britain.
—— 1951. The structure of Spring Bottom Ridge, and the origin of the mud-slides, Osmington, Dorset. *Proceedings of the Geologists' Association*, **62**, 21–30.
BUTLER, M. 1998. The geological history of the Wessex Basin – a review of new information from oil exploration. *This volume.*
CHADWICK, R. A. 1986. Extension tectonics in the Wessex Basin, southern England. *Journal of the Geological Society, London*, **143**, 465–488.
—— 1993. Aspects of basin inversion in southern Britain. *Journal of the Geological Society, London*, **150**, 311–322.
CHILDS, C., EASTON, S. J., VENDEVILLE, B. C., JACKSON, M. P. A., LIN, S. T., WALSH, J. J. & WATTERSON, J. 1993. Kinematic analysis of faults in a physical model of growth faulting above a viscous salt analogue. *Tectonophysics*, **228**, 313–329.
COBBOLD, P. R. & SZATMARI, P. 1991. Radial gravitational gliding on passive margins. *Tectonophysics*, **188**, 249–289.
COLTER, V. S. & HAVARD, D. J. 1981. The Wytch Farm Oil Field, Dorset. *In*: ILLING, L. V. & HOBSON, G. D. (eds) *Petroleum Geology of the Continental Shelf of North-west Europe.* Heyden, London, 494–503.
COUPLES, G. D., STEARNS, D. W. & HANDIN, J. W. 1994. Kinematics of experimental forced folds and their relevance to cross-section balancing. *Tectonophysics*, **233** 193–213.
DRUMMOND, P. V. O. 1970. The Mid-Dorset Swell. Evidence of Albian–Cenomanian movements in Wessex. *Proceedings of the Geologists' Association*, **81**, 679–714.
FALCON, N. L. & KENT, P. E. 1960. *Geological Results of Petroleum Exploration in Britain 1945–1957.* Memoir of the Geological Society of London No. 2, 1–56.
FAURE, J.-L. & CHERMETTE, J.-C. 1989. Deformation of tilted blocks, consequences on block geometry and extension measurements. *Bulletin de la Société géologique de Francais*, **5**, 461–476.
GAULLIER, V., BRUN, J. P., GUERIN, G. & LECANU, H. 1993. Raft tectonics: the effects of residual topography below a salt decollement. *Tectonophysics*, **228**, 363–381.
HAMBLIN, R. J. O., CROSBY, A., BALSON, P. S., JONES, S. M., CHADWICK, R. A., PENN, I. E. & ARTHUR, M. J. 1992. *United Kingdom Offshore Regional*

Report: the Geology of the English Channel. HMSO, London, for the British Geological Survey.

HOUSE, M. R. 1961. The structure of the Weymouth Anticline. *Proceedings of the Geologists' Association*, **72**, 221–238.

JACKSON, M. P. A. & TALBOT, C. J. 1986. External shapes, strain rates and dynamics of salt structures. *Bulletin of the Geological Society of America*, **97**, 305–323.

—— & VENDEVILLE, B. C. 1994. Regional extension as a geologic trigger for diapirism. *Bulletin of the Geological Society of America*, **106**, 57–73.

JENKYNS, H. C. & SENIOR, J. R. 1991. Geological evidence for intra-Jurassic faulting in the Wessex Basin and its margins. *Journal of the Geological Society, London*, **148**, 245–260.

KARNER, G. D., LAKE, S. D. & DEWEY, J. F. 1987. The thermal and mechanical development of the Wessex Basin, southern England. *In*: COWARD, M. P., HANCOCK, P. L. & DEWEY, J. F. (eds) *Continental Extensional Tectonics.* Geological Society, London, Special Publications, **28**, 517–536.

KOYI, H., JENYON, M. K. & PETERSON, K. 1993. The effect of basement faulting on diapirism. *Journal of Petroleum Geology*, **16**, 285–312.

LAKE, S. D. & KARNER, G. D. 1987. The structure and evolution of the Wessex Basin, southern England: an example of inversion tectonics. *Tectonophysics*, **137**, 347–378.

LARROQUE, J. M. 1993. Decoupling of basement and overburden deformation by a viscous layer (salt): normal faulting and reactivation. *In*: Abstracts, American Association of Petroleum Geologists Hedburg Research Conference on Salt Tectonics, Bath, UK.

LEES, G. M. & COX, P. T. 1937. The geological basis for the present search for oil in Great Britain by the D'Arcy Exploration Company, Ltd. *Quarterly Journal of the Geological Society, London*, **93**, 156–194.

LEFORT, J. P. & MAX, M. D. 1992. Structure of the Variscan belt beneath the British and Armorican overstep sequences. *Geology Magazine*, **20**, 979–982.

LOTT, G. K., SOBEY, R. A., WARRINGTON, G. & WHITTAKER, A. 1982. The Mercia Mudstone Group (Triassic) in the western Wessex Basin. *Proceedings of the Ussher Society*, **5**, 340–346.

LUNDIN, E. R. 1992. Thin-skinned extensional tectonics on a salt detachment, northern Kwanza Basin, Angola. *Marine and Petroleum Geology*, **9**, 405–411.

MCCLAY, K. R. & WHITE, M. J. 1995. Analogue modelling of orthogonal and oblique rifting. *Marine and Petroleum Geology*, **12**, 137–151.

MOTTRAM, B. H. 1949. Notes on the structure of the Poxwell Pericline, and the Ridgeway Fault at Bincombe Tunnel, Dorset. *Proceedings of the Dorset Natural History and Archaeological Society*, **71**, 175–183.

—— & HOUSE, M. R. 1956. The structure of the northern margin of the Poxwell Pericline. *Proceedings of the Dorset Natural History and Archaeological Society*, **76**, 129–135.

NALPAS, T. & BRUN, J. P. 1993. Salt flow and diapirism related to extension at crustal scale. *Tectonophysics*, **228**, 349–362.

PENN, I. E., CHADWICK, R. A., ROBERTS, G., PHARAOH, T. C., ALLSOP, J. M., HULBERT, A. G. & BURNS, I. M. 1987. Principal features of the hydrocarbon prospectivity of the Wessex–Channel Basin, U.K. *In*: BROOKS, J. & GLENNIE, K. (eds) *Petroleum Geology of North West Europe.* Graham & Trotman, London, 109–118.

PLINT, A. G. 1982. Eocene sedimentation and tectonics in the Hampshire Basin. *Journal of the Geological Society, London*, **139**, 249–254.

RICHARD, P. D. 1991. Experiments on faulting in a two-layer cover sequence overlying a reactivated basement fault with oblique-slip. *Journal of Structural Geology*, **13**, 459–469.

——, NAYLOR, M. A. & KOOPMAN, A. 1995. Experimental models of strike-slip tectonics. *Petroleum Geoscience*, **1**, 71–80.

RIDD, M. F. 1973. The Sutton Poyntz, Poxwell and Chaldon Herring Anticlines, southern England: a reinterpretation. *Proceedings of the Geologists' Association*, **84**, 1–8.

RUFFELL, A. 1990. Stratigraphy and structure of the Mercia Mudstone Group (Triassic) in the western part of the Wessex Basin. *Proceedings of the Ussher Society*, **7**, 263–267.

SELLEY, R. C. & STONELEY, R. 1987. Petroleum habitat in south Dorset. *In*: BROOKS, J. & GLENNIE, K. (eds) *Petroleum Geology of North West Europe.* Graham & Trotman, London, 139–148.

SMITH, A. J. & CURRY, D. 1975. The structure and geological evolution of the English Channel. *Philosphical Transactions of the Royal Society, London*, **A279**, 3–20.

SMITH, C. & HATTON, I. R. 1998. Inversion tectonics in the Lyme Bay–West Dorset area. *This volume.*

STEWART, S. A., HARVEY, M. J., OTTO, S. C. & WESTON, P. J. 1996. Influence of salt on fault geometry: examples from the UK basins. *In*: ALSOP, G. I., BLUNDELL, D. J. & DAVISON, I. (eds) *Salt Tectonics.* Geological Society, London, Special Publications, **100**, 175–202.

STONELEY, R. 1982. The structural development of the Wessex Basin. *Journal of the Geological Society, London*, **139**, 543–554.

TRUDGILL, B. D. & CARTWRIGHT, J. A. 1994. Relay ramp forms and normal-fault linkages, Canyonlands National Park, Utah. *Bulletin of the Geological Society of America*, **106**, 1143–1157.

UNDERHILL, J. R. & PATERSON, S. 1998. Genesis of tectonic inversion structures: seismic evidence for the development of key structures along the Purbeck–Isle of Wight Disturbance. *Journal of the Geological Society, London*, in press.

—— & STONELEY, R. 1982. Introduction to the development, evolution and petroleum geology of the Wessex Basin. *This volume.*

VENDEVILLE, B. C., GE, H. & JACKSON, M. P. A. 1995. Scale models of salt tectonics during basement-involved extension. *Petroleum Geoscience*, **1**, 179–183.

—— & JACKSON, M. P. A. 1992. The rise of diapirs during thin-skinned extension. *Marine and Petroleum Geology*, **9**, 331–353.

WARRINGTON, G. & SCRIVENER, R. C. 1980. The Lyme-Regis (1901) Borehole succession and its relationship to the Triassic sequence of the east Devon coast. *Proceedings of the Ussher Society*, **5**, 24–32.

WHITTAKER, A. (ed.) 1985. *Atlas of Onshore Sedimentary Basins in England and Wales*. Blackie, Glasgow.

WILLIAMS, G. D., POWELL, C. M. & COOPER, M. A. 1989. Geometry and kinematics of inversion tectonics. *In*: COOPER, M. A. & WILLIAMS, G. D. (eds) *Inversion Tectonics*. Geological Society, London, Special Publications, **44**, 3–15.

Inversion tectonics in the Lyme Bay–West Dorset area of the Wessex Basin, UK

C. SMITH[1] & I. R. HATTON[2]

[1] *Department of Geology, Imperial College, Prince Consort Road, London SW7 2BP, UK*
[2] *Kerr-McGee Oil (UK) plc, 75 Davies Street, London W1Y 1FA, UK*

Abstract: Inversion in Lyme Bay and West Dorset was characterized by the reversal of motion on major Jurassic faults, and regional uplift which resulted in the widespread loss of the mid-Cretaceous to Tertiary post-rift cover throughout the area. North–south contraction during the Tertiary resulted in the development of a major anticline in the hanging wall of the Lyme–Portland faults, which was accompanied by widespread folding within the footwall of these faults. Northeast–southwest sinistral and northwest–southeast dextral strike-slip faults were also developed throughout the Lyme Bay–West Dorset area and accommodated much of the north–south contraction. These features are considered to be a consequence of the mechanical effects of an extensive salt interval within the Triassic Mercia Mudstone Group of this area. Major Early Jurassic faults appear to have been directly linked with basement through the salt interval, although smaller faults within the footwall apparently detached on this horizon. The major faults were inverted in a complex manner during the Tertiary, commencing with simple reversal of motion on these faults, followed by detachment of the post-salt carapace as the footwall and hanging wall salt layers were brought into close proximity, thereby resulting in the widespread deformation observed in the footwall of these faults.

West Dorset and the adjacent offshore area of Lyme Bay are located towards the western margin of the Wessex Basin (Fig. 1). In this part of the basin rocks of Triassic and Jurassic age are exposed at the surface together with a thin, patchy cover of mainly Late Cretaceous Chalk (Hamblin *et al.* 1992). A major unconformity of mid-Cretaceous age exists in this area which increases in magnitude from east to west with the result that the Late Cretaceous progressively oversteps the Jurassic formations to overlie the Triassic towards the western margin

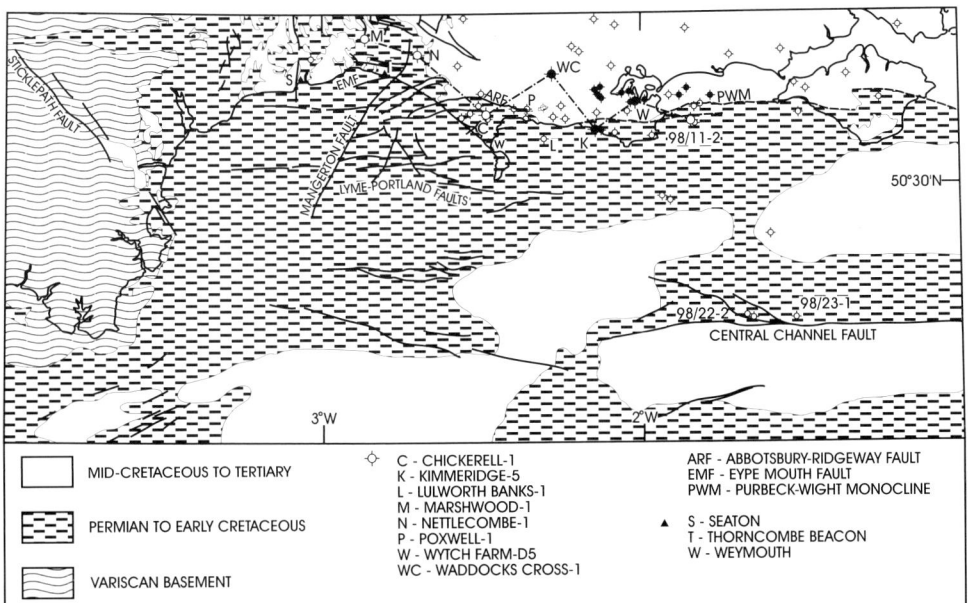

Fig. 1. Simplified geological map of the western English Channel.

SMITH, C. & HATTON, I. R. 1998. Inversion tectonics in the Lyme Bay–West Dorset area of the Wessex Basin, UK. *In*: UNDERHILL, J. R. (ed.) *Development, Evolution and Petroleum Geology of the Wessex Basin*, Geological Society, London, Special Publications, **133**, 267–281.

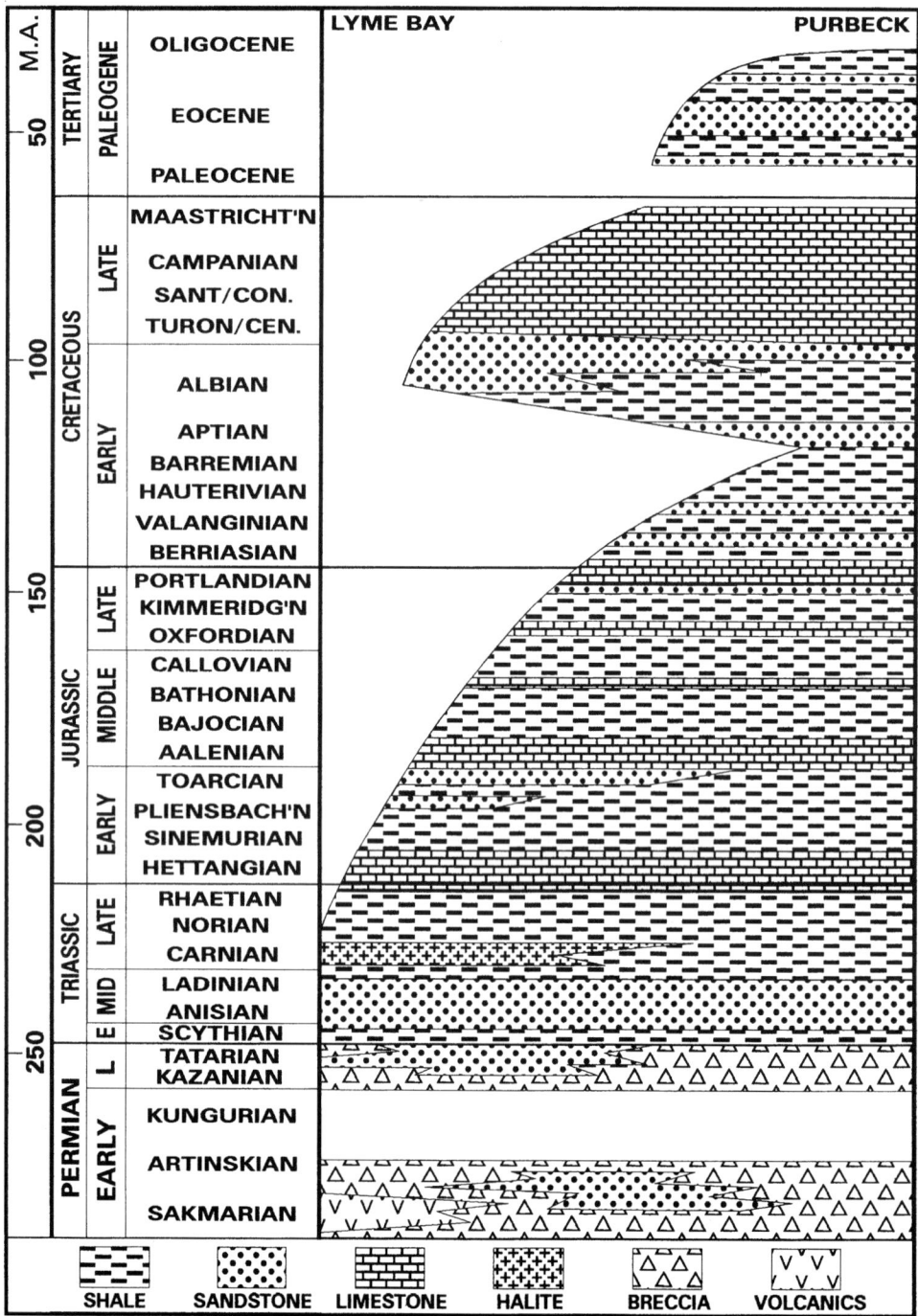

Fig. 2. Generalized stratigraphy of the Wessex Basin in the western English Channel area.

of the basin (McMahon & Underhill 1995). The stratigraphy of the area is illustrated in Fig. 2 and is generally well documented from the excellent exposures along the Dorset coast (Melville & Freshney 1982), supplemented by a number of onshore exploration wells which were drilled between 1937 (the Poxwell deep borehole, Taitt & Kent 1939) and the present day. These wells (Fig. 3) reveal dramatic variations in the thickness of the Sherwood Sandstone and Mercia Mudstone Groups of Triassic age, and the development of a substantial evaporite interval within the Mercia Mudstone Group. From downhole electric logs it would appear that this unit consists largely of halite, although significant quantities of interbedded claystone are present throughout the section which attains a gross thickness of approximately 300 m in the Chickerell-1 well.

Inversion styles in the Wessex Basin

The inversion of pre-existing extensional structures has been recognized throughout the Wessex Basin (Stoneley 1982; Whittaker 1985; Lake & Karner 1987; Chadwick 1993; Underhill & Paterson 1998; Underhill & Stoneley this volume). In the east of the area the inversion is mainly focused on the Purbeck–Isle of Wight faults, and has resulted in the development of major monoclinal flexures within the Late Cretaceous and Tertiary cover immediately overlying the inverted faults. Late Jurassic–Early Cretaceous rocks are currently exposed in the hanging wall of these major faults, while the post-Aptian succession has been largely preserved in the footwall of the same faults. However, in the west of the basin the latter interval is severely restricted in distribution, suggesting that the inversion in this area has been more diffuse in nature. In part this reflects regional west to east tilting of the British Isles, although broad doming is also evident in West Dorset where the mid-Cretaceous unconformity occurs at an elevation of approximately 125 m at Thorncombe Beacon on the Dorset coast (Melville & Freshney 1982), descending westwards (in opposition to the regional dip) to sea level immediately to the west of Seaton in East Devon (Edmonds et al. 1975).

Further insight into the nature of inversion in the Channel Basin may be gained from consideration of the residual Bouguer anomaly map of the area (Fig. 4). The inversions associated with the Purbeck–Isle of Wight faults are evident as areas of positive anomalies, reflecting the uplift of more compacted and therefore denser

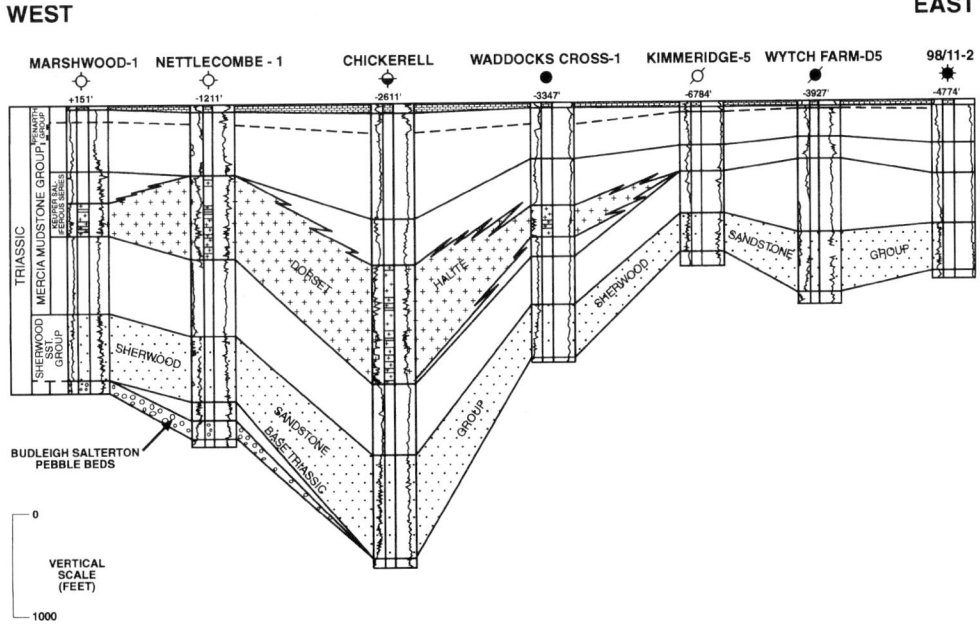

Fig. 3. East–west correlation of the Triassic Sherwood Sandstone and Mercia Mudstone Groups. Ornament as shown in Fig. 2 except shales are unshaded.

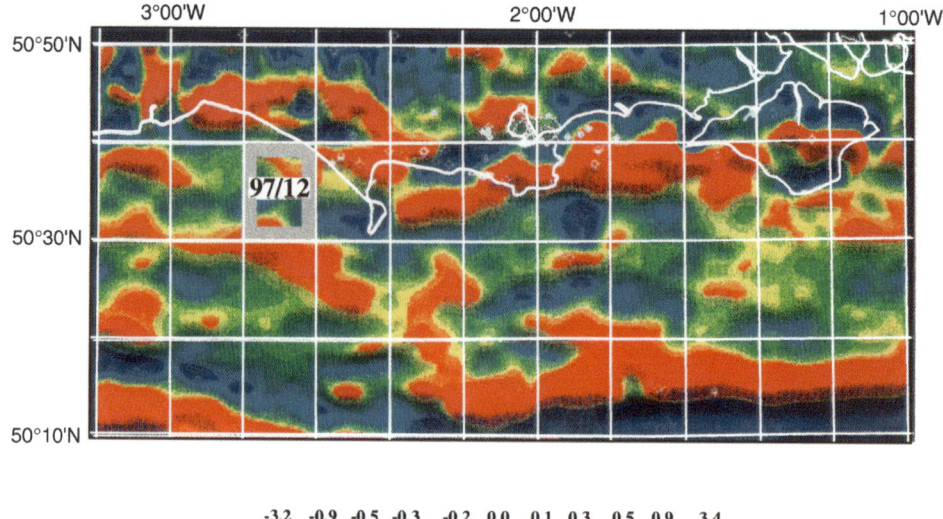

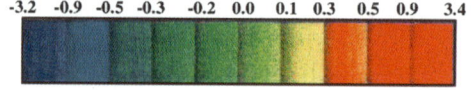

Fig. 4. Residual Bouguer gravity map of the western English Channel.

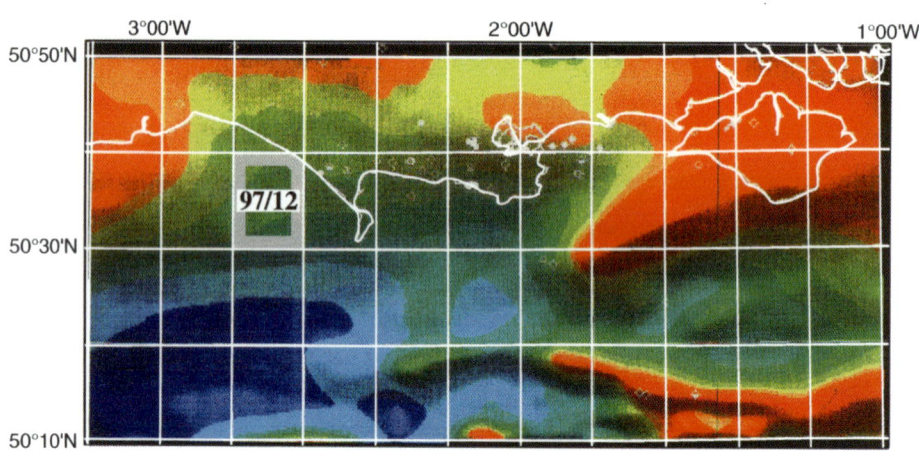

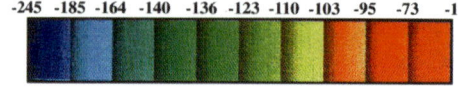

Fig. 5. Aeromagnetic anomaly map (reduced to pole) of the western English Channel.

material in the hanging walls of these faults, and details such as the sinistral offset in the Isle of Wight Monocline (Stoneley 1982) are apparent (cf. Underhill & Paterson 1998). In the east of the area the anomalies parallel the edge of the Chalk outcrop, although west of Purbeck the anomaly heads offshore before terminating to the east of Weymouth, while the Chalk outcrop swings inland to parallel the Ridgeway–Abbotsbury Fault. The gravity anomaly appears to be associated with a prominent anticline upon which the Lulworth Banks-1 well was located, but not exactly coincident with the structure, although this may be an artefact of gaps in data coverage between the onshore and marine datasets from which the map was produced.

The major east–west anomaly close to the southern boundary of the map corresponds to the inversion of the Central Channel Fault, structural culminations along which have been unsuccessfully tested by wells 98/22-2 and 98/23-1. This fault has been previously considered not to have undergone significant reversal (Hamblin *et al.* 1992), which is difficult to reconcile with the magnitude of the observed gravity anomaly, although this may partly be accounted for by the presence of more magnetic material (typically basic in nature and therefore denser) along the fault zone (Fig. 5). The correlation of seismic reflections across this fault is severely hampered by the lack of nearby wells to the south of the structure, although recent high-quality regional seismic lines clearly show the development of a south-facing monocline above a fault which dips steeply to the north, rather than to the south as previously reported by Hamblin *et al.* (1992). Assuming the structural maps published by these authors to be otherwise correct in that the Mesozoic section is 'downthrown' on the southern side of the Central Channel Fault, this implies that the structure is now in net contraction and has therefore undergone much more severe inversion than that observed in the Isle of Wight–Purbeck area.

Within Lyme Bay a positive residual Bouguer anomaly of comparable magnitude is located in the hanging wall of a series of major faults with southerly throw (the Lyme–Portland faults of Hamblin *et al.* 1992). The aeromagnetic anomaly map (Fig. 5) in this part of the area is relatively featureless, suggesting that the observed gravity anomaly is also probably the consequence of structural inversion. It would therefore appear that in addition to the widespread inversion observed in this area, the reversal of motion on major faults has also occurred in this part of the Wessex Basin.

Lyme Bay

The eastern half of Lyme Bay is dissected into numerous fault blocks (Fig. 6), typically of 3–5 km in width, by a series of east–west- to ENE–WSW-oriented faults with throws frequently in excess of 100 m. The central part of this area is characterized by the development

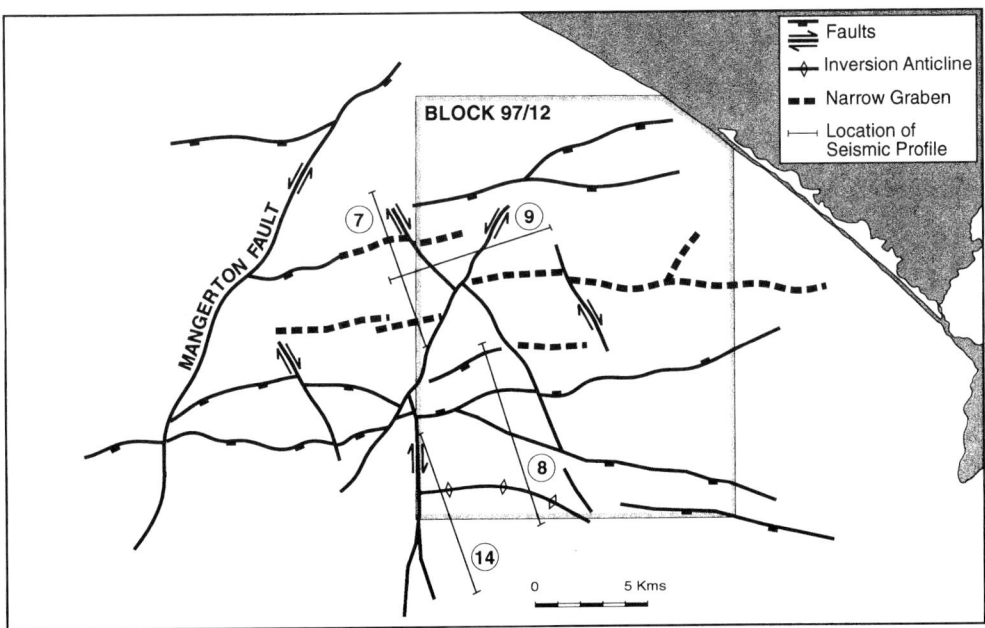

Fig. 6. Structural features of the Lyme Bay area.

of several narrow graben (Figs 6 and 7) which persist laterally for distances commonly in excess of 5 km, despite having relatively small throws. The bounding faults of these graben invariably intersect close to the top of the Dorset Halite and are usually unrelated to the sub-salt structure. These features are considered to be the result of incipient raft tectonics under conditions of relatively small amounts of extension in this part of the area. The grabens exhibit growth mainly during the Upper Lias, which in the example shown (Fig. 7) increases in thickness by some 30% relative to the footwall succession. The base of this interval is marked by a prominent onlap surface which corresponds to the Middle–Upper Lias Junction Bed exposed onshore. Onlapping reflections are observed to be repeated across the faults, indicating that these faults extended up to the sea floor, an observation which is consistent with the syn-tectonic interpretation of fissure-fills associated with the Eype Mouth Fault (Jenkyns & Senior 1977).

A series of major faults with southerly throws occurs immediately to the west of the Isle of Portland. These faults exhibit considerable growth throughout the Lower Jurassic succession (Fig. 8) and are associated with a broad hanging wall anticline, the crest of which is located approximately 3 km to the south. Figure 8 also demonstrates that deformation is not restricted to the hanging wall of this major fault, as the footwall horst immediately to the northwest has been folded into a syncline, and an anticline has developed in the adjacent graben. This style of deformation is also evident further to the north (Fig. 7), although the folding is greatly reduced in amplitude.

The hanging wall fold clearly affects pre-rift as well as syn-rift sequences and is therefore not a classical inversion anticline (Williams *et al.* 1989), although it is similar in the respect that the limb which faces the fault is significantly steeper than the gently dipping southern limb. The elevation of the hanging wall anticline (with respect to datum) is comparable to that of the footwall (Fig. 8), despite a southerly throw of approximately 500 m across the fault, and increases westwards until the structure is effectively terminated by a north–south- to NNE–SSW-trending fault. This is also evident from the residual

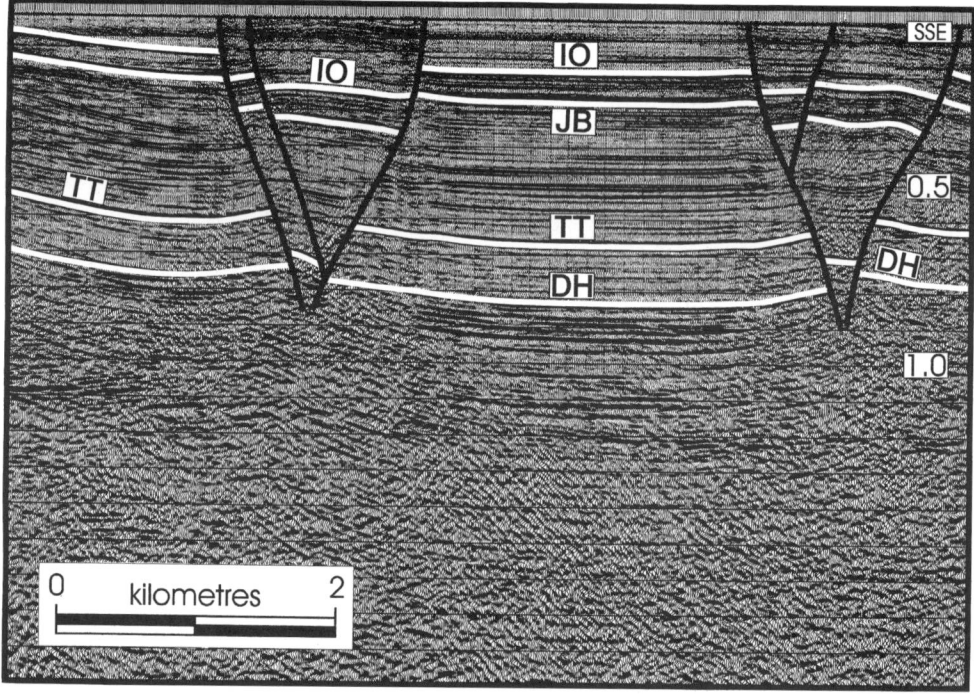

Fig. 7. Seismic profile from the centre of Lyme Bay illustrating narrow grabens, thought to represent incipient raft tectonics. IO, Inferior Oolite; JB, Junction Bed; TT, near top Triassic; DH, near top Dorset Halite. Data courtesy of Kerr-McGee Oil (UK) plc.

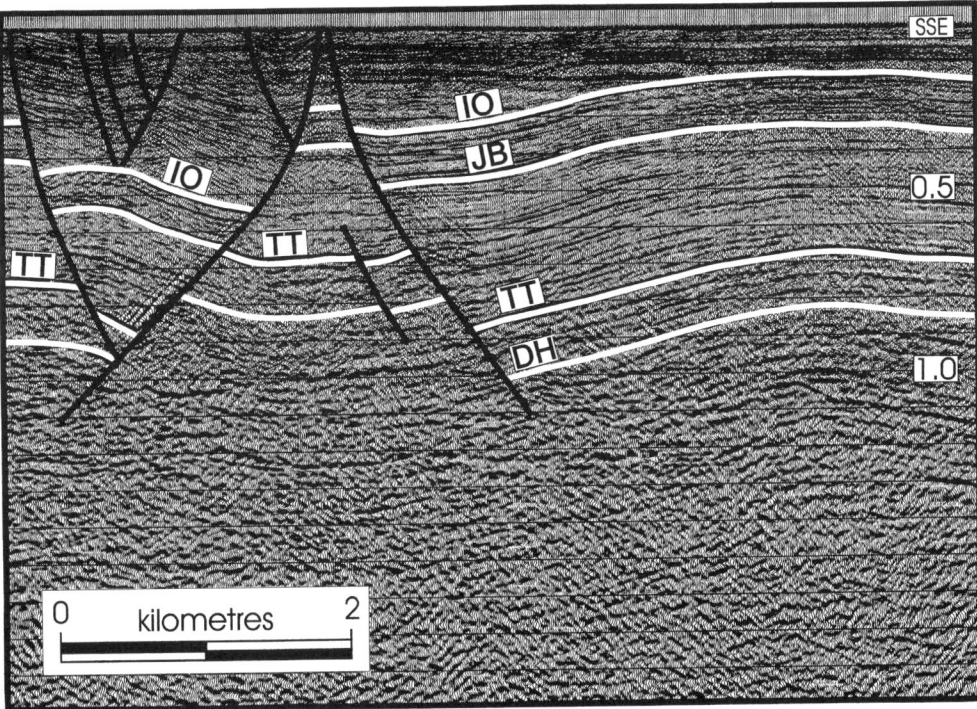

Fig. 8. Seismic profile showing a major Early Jurassic growth fault with a prominent hangingwall anticline. For legend see Fig. 7. Data courtesy of Kerr-McGee Oil (UK) plc.

Bouguer gravity (Fig. 4) where the positive anomaly referred to in the preceding section is greatly reduced in magnitude to the west of 2°48′. The throw on the major fault substantially exceeds the interpreted thickness of the Dorset Halite in this area, and for this reason it is considered likely that the fault is directly linked through the salt interval to a major basement fault (Stewart et al. 1996).

The west of the area (Fig. 6) is more complex structurally with a series of sinistral northeast–southwest faults, including a probable extension of the onshore Mangerton Fault, and dextral northwest–southeast faults forming conjugate strike-slip arrays which testify to widespread north–south contraction throughout the Lyme Bay area. The northwest–southeast faults invariably exhibit a small reverse component, as is the case with the northwest–southeast Watchet Fault in Somerset (Whittaker 1972), while the northeast–southwest faults show a degree of extension and commonly form small negative flower structures (Fig. 9). The latter set appear to form longer and more continuous faults than the northwest–southeast dextral set, which may reflect the dominance of NNE–SSW structures in the basement of the western half of Lyme Bay, as revealed by the trends of magnetic anomalies in that area and further evidenced by the orientations of faults and outcrop patterns in East Devon. However, the degree to which the northeast–southwest sinistral faults are linked with the basement structures remains conjectural as a result of the lack of sufficient good-quality seismic reflection data in the west of the area. This also precludes any assessment of the possible role of the Mangerton Fault during the Mesozoic development of the Lyme Bay area. Jenkyns & Senior (1991) suggested that this fault was active during the Bajocian from observations of facies distribution within the Inferior Oolite Group, although no other evidence of pre-Tertiary displacement has been reported. However, significant changes in the thickness of Cenomanian strata in East Devon (Smith 1961) appear to be associated with north–south, to NNE–SSW-trending faults in the vicinity of Seaton (Fig. 1), suggesting that structures of this orientation in Lyme Bay may well have been re-activated during the Mesozoic.

North–south to NNE–SSW structures are unique to this part of the Wessex Basin, within

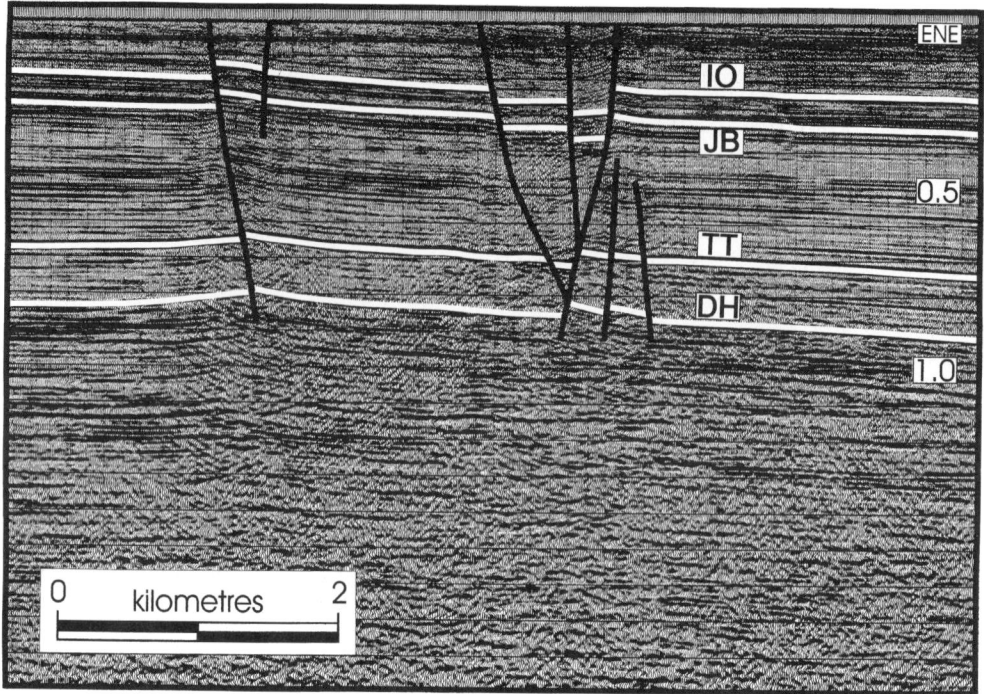

Fig. 9. Seismic profile illustrating the characteristics of a conjugate strike-slip fault-set from the Lyme Bay area. For legend see Fig. 7. Data courtesy of Kerr-McGee Oil (UK) plc.

which northwest–southeast basement trends are more commonly observed and may have exerted a fundamental control on the evolution of the Channel Basin (Lake & Karner 1987). However, the re-activation of northwest–southeast basement faults in the basins to the south and southwest of Britain is distinctly patchy, with major structures such as the Scilly Wrench Zone (Hillis & Chapman 1992) playing no role in the Mesozoic–Cenozoic evolution of the Western Approaches Basin (Smith 1995). This patchiness is also evident from the northwest margin of the Wessex Basin where the Cothelstone Fault (Webby 1965b), which offsets Variscan basement structures dextrally by 14–16 km (Miliorizos & Ruffell this volume), has undergone demonstrable post-Variscan displacement (Whittaker 1972) while the smaller northwest–southeast Timberscombe and Monksilver faults (Webby 1965a) immediately to the west appear not to have been re-activated. It is reasonable to assume that similar considerations should apply to the north–south to NNE–SSW structures of the Lyme Bay area, and that although the Mangerton Fault clearly could have been active during the Mesozoic, there is no reason to conclude that this must have necessarily been the case.

Structural development

Inversion tectonics in the Lyme Bay area are characterized by the development of a broad anticline in the hanging wall of a major Liassic growth fault, and folding of the footwall post-salt carapace accompanied by conjugate strike-slip fault sets. The widespread nature of this deformation is to be expected, considering the presence of a relatively thick salt succession in the Triassic which can effectively decouple contraction in the carapace from that within the pre-salt succession, and it was noted in the previous section that many of the Jurassic–?Early Cretaceous extensional faults in the Lyme Bay area detach on this layer. In fact, the nature of the inversion structures in South Dorset led Lees & Cox (1937) to postulate the existence of an extensive salt layer in the area before the drilling of any deep exploration wells took place. However, the influence of this salt interval on inversion tectonics in South and West Dorset has otherwise been ignored, with the exception of Chadwick (1993).

A range of possible mechanisms can be envisaged depending upon the degree to which the major faults are linked through the salt interval with basement. This depends on a number of factors, including the thickness of the salt layer relative to the fault displacement, the thickness of the overburden (Koyi *et al.* 1993) and the amount of interbedded claystone in the salt succession (Stewart *et al.* 1996).

The simplest mechanism is illustrated in Fig. 10, which is considered to be the most likely scenario where the salt is free of interbedded claystones and is relatively thick compared with the fault displacement. During the extensional phase, faulting in the carapace is detached from that in the pre-salt succession and results in the development of a syn-rift hanging wall anticline. Subsequent contraction then results in the amplification of this pre-existing fold, together with more widespread deformation in the carapace. However, an inverted major fault within the pre-salt succession is still required to explain the magnitude of the positive residual Bouguer gravity actually observed.

This mechanism has been suggested for the periclinal structures associated with the Ridgeway Fault (Harvey & Stewart this volume) where there is clear evidence of pre-Aptian folds in the hanging wall of the fault, and the Purbeck Monocline (Stoneley 1982), although in the latter example the Purbeck Fault is considered to be directly linked with basement, but listric in geometry rather than planar.

A key prediction from this model is that the anticline is an early structure, and should be evident in the isopachs of formations in the hanging wall of the fault. An isochore (vertical isopach) in two-way time of the gross interval between the Inferior Oolite and the top of the Triassic in the vicinity of the Lyme–Portland faults is illustrated in Fig. 11. This reveals that there is substantial thickening of the Lias to the north of the anticlinal axis, in the direction of the faults, but minimal thickening on the southern flank which disappears entirely if allowance is made for the effect of structural dip on the isochore. This does not preclude the development of a hanging wall anticline during a later phase of extension, (e.g. Late Jurassic or Early Cretaceous), although this is thought to be unlikely as the displacement on the principal fault at the end of the Lias already exceeded the interpreted thickness of the salt interval and would therefore have been hard-linked to basement by this time.

An alternative mechanism is shown in Fig. 12, which is thought to be favoured in situations where the salt is thin relative to the fault displacement and contains substantial interbedded claystones. During the extensional phase a planar normal fault develops which is directly linked with a basement fault through the salt-bearing interval. Assuming the fault throw is sufficient, this will enable hydrocarbons to migrate from Lower Lias source rocks in the hanging wall of the fault into the Sherwood Sandstone Group reservoir in the footwall at the time of maximum burial. During inversion the initial response is simply the reversal of motion on the fault, presumably accompanied by a classical inversion anticline within the syn- to

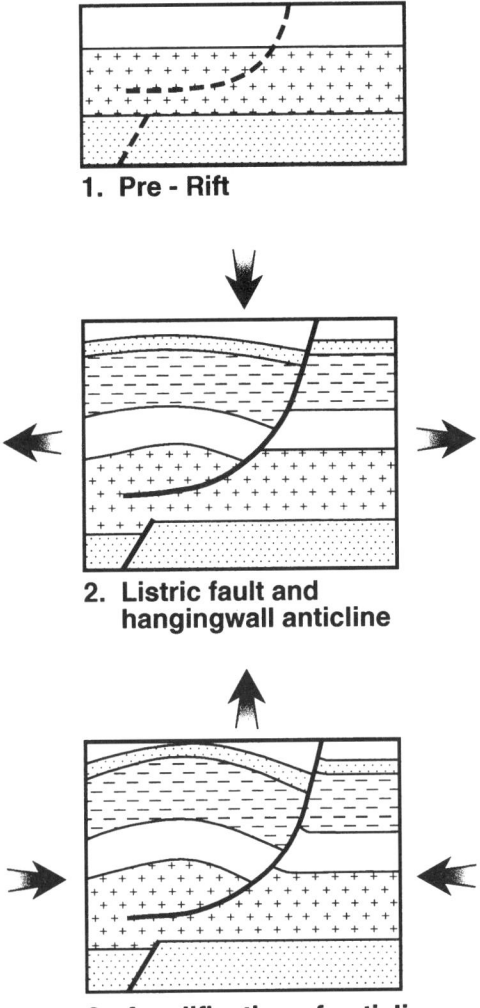

Fig. 10. Cartoon illustrating the development of a hanging wall anticline by the amplification of a pre-existing rollover developed above a listric detachment.

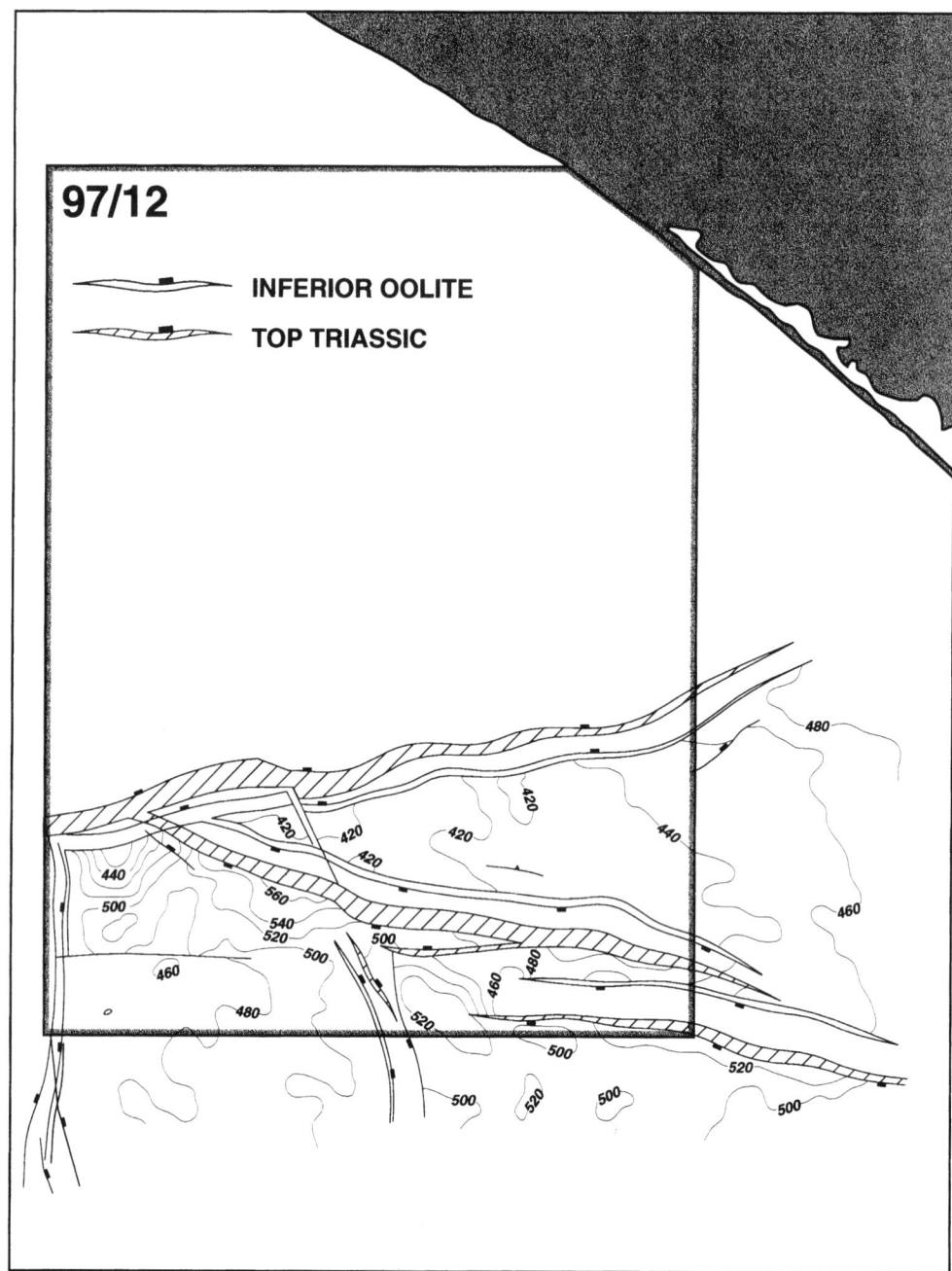

Fig. 11. Gross isochore (vertical isopach) in two-way time between the Inferior Oolite and the top of the Triassic.

post-rift interval of the hanging wall. However, as inversion proceeds a critical point is reached at which the footwall and hanging wall salt intervals begin to be juxtaposed across the fault, at which point the post-salt carapace is free to decouple from the pre-salt succession. Inversion of the carapace fault ceases at this point, and the carapace is transported northwards relative to the pre-salt succession with the development of folds and conjugate strike-slip faults within the

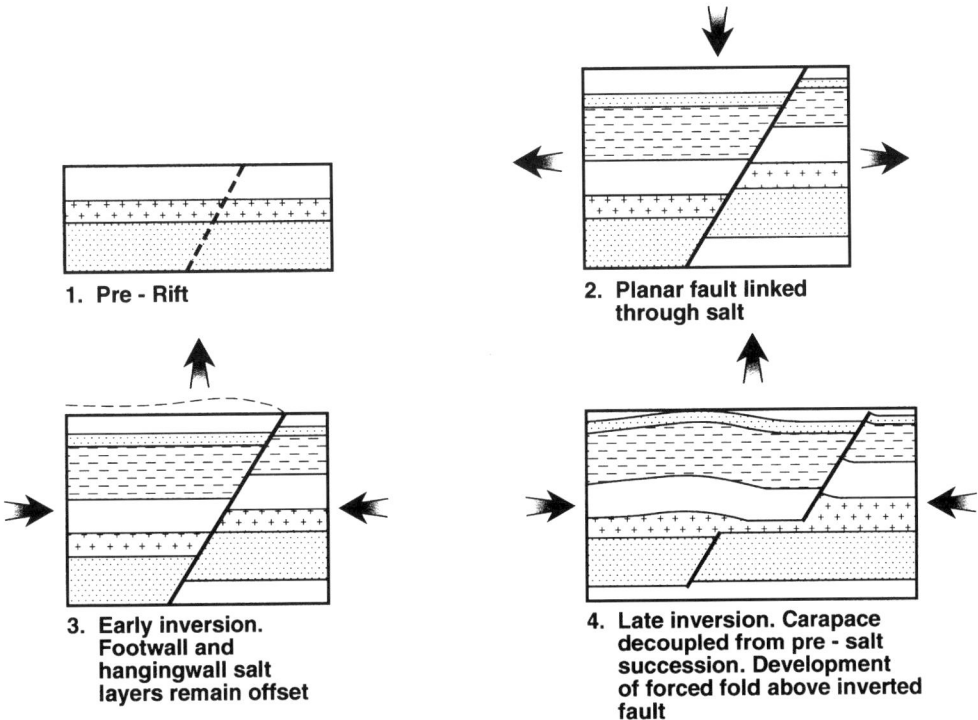

Fig. 12. Cartoon illustrating the inversion of a planar fault with detachment above a thin salt interval.

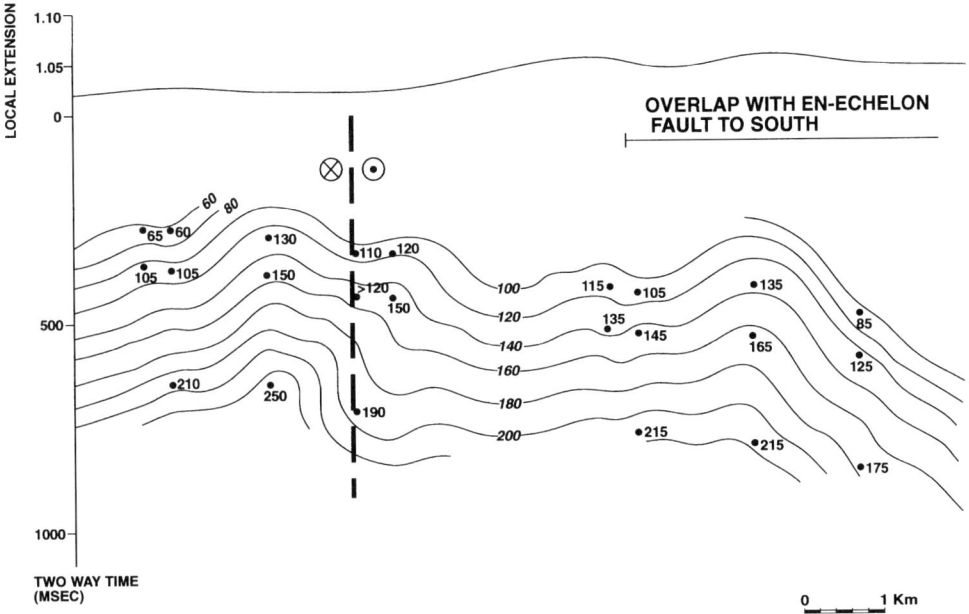

Fig. 13. Displacement (in MS TWT) and extension associated with the major fault shown in Fig. 8.

post-salt succession. However, reverse movement continues on the pre-salt fault, with the development of a forced fold in the overlying carapace, initially supported by the competent claystones within the salt interval.

A consequence of this model is that the strike-slip displacement post-dates the cessation of fault reversal within the carapace, and that the hanging wall anticline is the last inversion structure to form. It is interesting to note that apatite fission track analysis (Bray *et al.* this volume) from the Lulworth Banks anticline, which has many features in common with the anticline in Lyme Bay, indicates that the uplift of this fold post-dates all of the other structures sampled in the basin, and is clearly consistent with the sequential inversion mechanism described above.

The displacement on the principal Lyme Bay Fault is illustrated in Fig. 13, along with the extension determined from reflection segment lengths within a 10 km zone containing the fault (the 'local' extension). Little variation is observed in displacement along the length of the fault, except towards the eastern end where extension is transferred to an en echelon fault to the south via a lateral relay ramp. Displacement is maintained despite a significant decrease in the overall extension from east to west, which clearly demonstrates that shortening in the carapace is being accommodated by folding. It should be noted that the estimate of carapace shortening derived from this analysis is a minimum value, as the conjugate strike-slip arrays also contribute to the overall north–south contraction.

The throw on this fault is in fact remarkably constant, as systematic changes in displacement would normally be anticipated on a fault of almost 10 km lateral extent (Walsh & Watterson 1991). However, this is precisely what would be expected as a consequence of the proposed inversion mechanism, as the net displacement on the partially inverted carapace fault is directly related (and ideally equivalent) to the thickness of the salt interval. Unfortunately, the observed displacement on the fault at the top of the Triassic (the reflection nearest to the top of the salt which can be identified with confidence), appears to be about twice the thickness of the

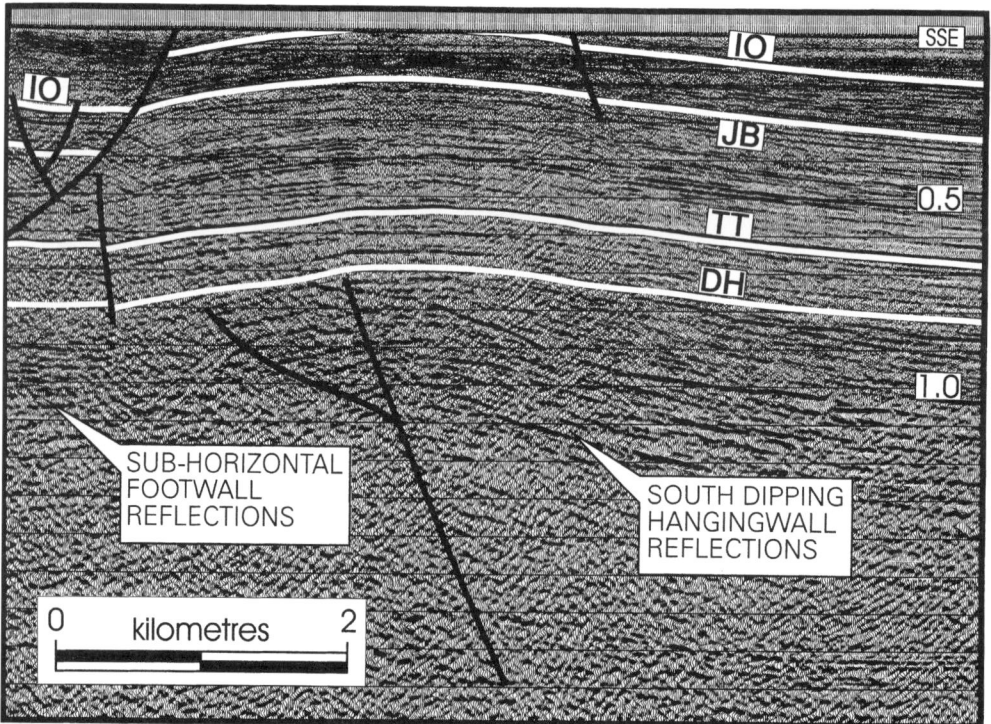

Fig. 14. Seismic profile from the crest of the inversion anticline showing possible evidence of failure of the footwall in the pre-salt succession during inversion. For legend see Fig. 7. Data courtesy of Kerr-McGee Oil (UK) plc.

salt interval interpreted from the seismic data. This estimate is derived from observations of extensional faults which detach within the salt interval and may therefore represent only the central zone of the salt where the salt/claystone ratio is highest. The effective thickness of the salt from a mechanical point of view may therefore be different in contraction relative to extension. Another possibility is that footwall collapse of the normal fault may occur before the footwall and hanging wall salt successions are juxtaposed, with a segment of pre-salt material being thrust northwards over the footwall block. Deformation of this type has been described from inverted half-grabens on the margins of the Belledonne and Pelvoux massifs in the western Alps (Coward et al. 1991). This is interpreted to be the situation with the seismic profile shown in Fig. 14, although considering the quality of the data and in particular the problems of distinguishing multiple reflections from real data in the pre-salt section this is clearly not definitive. The interpretation of this profile, which is located at the culmination of the hanging wall anticline, also suggests that the pre-salt fault is now in net contraction with the carapace anticline being directly supported by the uplifted block with elimination of the salt presumably by lateral flowage.

An important consequence of the proposed inversion mechanism for the Lyme Bay area is illustrated by Fig. 15. It is assumed that the northward displacement of the carapace is related to the amount of inversion of the fault in the pre-salt interval, and that as a consequence increases from zero at the fault tips to a maximum somewhere along the length of the fault. The trace of the carapace fault therefore bows outward to the north, which inherently requires an east–west increase in line length, with the result that north–south contraction must be accompanied by east–west extension of the footwall carapace as a necessary consequence of the geometry of the inversion. This mechanism therefore naturally results in the formation of the observed conjugate strike-slip fault-sets, whereas the alternative model results in simple north–south compression which should favour the development of folds and thrusts above such a shallow detachment. The situation in Lyme Bay is clearly more complex than this as a consequence of the control exerted on the inversion by NNE–SSW basement structures, but it is significant that there is a marked change in the strike of the Lyme–Portland faults from east–west to WNW–ESE immediately to the west of the first strike-slip carapace fault.

Conclusions

Inversion tectonics in the Lyme Bay area have been strongly influenced by the presence of a salt interval within the Mercia Mudstone Group

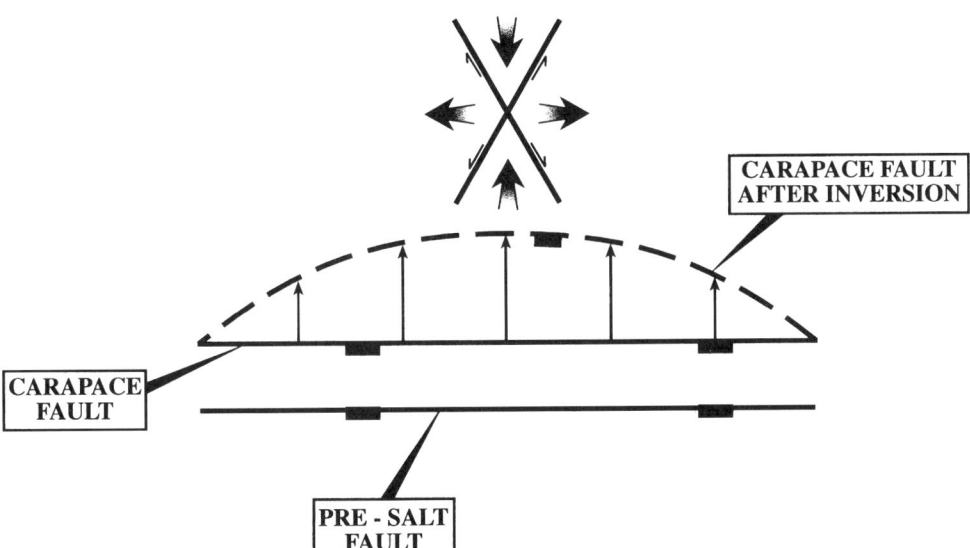

Fig. 15. Cartoon illustrating the geometrical consequences of inversion-related contraction of the post-salt carapace in plan view.

which has effectively decoupled contraction in the post-salt carapace from that in the pre-salt succession. Carapace extensional faults have consequently been only partially inverted, if at all, with north–south contraction being accommodated by the development of folds and conjugate strike-slip fault sets over a wide area, probably including West Dorset where a broad dome is evident from the elevation of the Aptian unconformity surface. Inversion in the pre-salt succession appears to be focused mainly on the Lyme–Portland faults, and it would appear from seismic observations and the average elevations of hanging wall and footwall that little net extension remains, with net contraction locally within the culmination of the inversion. Inversion in Lyme Bay therefore appears to be much more severe than in the Isle of Wight–Purbeck area.

It is considered that the severity of the inversion in this area, in combination with the delocalization of deformation as a consequence of decoupling on the Triassic Dorset Halite, has resulted in the loss of the Cretaceous and Tertiary cover throughout much of the Lyme Bay–West Dorset area and a structural style which is unique to this part of the Wessex Basin.

Much of the material presented in this paper is the result of work carried out on behalf of the partnership in UK block 97/12, comprising Kerr-McGee Oil (UK) plc, Amerada Hess Limited and OMV (UK) Limited. The permission of these companies to publish this material is gratefully acknowledged. Figures 4 and 5 are reproduced from work carried out by ARK Geophysics Ltd on behalf of the 97/12 group.

References

BRAY, R. J., DUDDY, I. R. & GREEN, P. F. 1998. Multiple heating episodes in the Wessex Basin: implications for geological evolution and hydrocarbon generation. *This volume.*

CHADWICK, R. A. 1993. Aspects of basin inversion in southern Britain. *Journal of the Geological Society*, London, **150**, 311–322.

COWARD, M. P., GILLCHRIST, R. & TRUDGILL, B. 1991. Extensional structures and their tectonic inversion in the Western Alps. *In*: ROBERTS, A. M., YIELDING, G. & FREEMAN, B. (eds) *The Geometry of Normal Faults*. Geological Society, London, Special Publications, **56** 193–203.

EDMONDS, E. A, MCKEOWN, M. C. & WILLIAMS, M. 1975. *British Regional Geology: South-west England*, 4th edition. HMSO, London, for the Institute of Geological Sciences.

HAMBLIN, R. J. O., CROSBY, A., BALSON, P. S., JONES, S. M., CHADWICK, R. A., PENN, I. E. & ARTHUR, M. J. 1992. *United Kingdom Offshore Regional Report: The Geology of the English Channel*. HMSO, London, for the British Geological Survey.

HARVEY, M. J. & STEWART, S. A. 1998. Influence of salt on the structural evolution of the Channel Basin. *This volume.*

HILLIS, R. R. & CHAPMAN, T. J. 1992. Variscan structure and its influence on post-Carboniferous basin development, Western Approaches Basin, SW UK Continental Shelf. *Journal of the Geological Society*, London, **149**, 413–417.

JENKYNS, H. C. & SENIOR, J. R. 1977. A Liassic palaeofault from Dorset. *Geological Magazine*, **114**, 47–52.

—— & —— 1991. Geological evidence for intra-Jurassic faulting in the Wessex Basin and its margins. *Journal of the Geological Society*, London, **148**, 245–260.

KOYI, H., JENYON, M. K. & PETERSEN, K. 1993. The effect of basement faulting on diapirism. *Journal of Petroleum Geology*, **16**, 285–312.

LAKE, S. D. & KARNER, G. D. 1987. The structure and evolution of the Wessex Basin, southern England. an example of inversion tectonics. *Tectonophysics*, **137**, 347–378.

LEES, G. M. & COX, P. T. 1937. The geological basis for the present search for oil in Great Britain by the D'Arcy Exploration Company Ltd. *Quarterly Journal of the Geological Society*, London, **93**, 156–194.

MCMAHON, N. A. & UNDERHILL, J. R. 1995. The regional stratigraphy of the Southwest United Kingdom and adjacent offshore areas with particular reference to the major intra-Cretaceous unconformity. *In*: CROKER, P. F. & SHANNON, P. M. (eds) *The Petroleum Geology of Ireland's Offshore Basins*. Geological Society, London, Special Publications, **93**, 323–325.

MELVILLE, R. V. & FRESHNEY, E. C. 1982. *British Regional Geology: The Hampshire Basin and Adjoining Areas*, 4th edition. HMSO, London, for the Institute of Geological Sciences.

MILIORIZOS, M. & RUFFELL, A. 1998. Kinematics and geometry of the Watchet–Cothelstone–Hatch Fault System: implications for the fault history of the Wessex Basin and adjacent areas. *This volume.*

SMITH, C. 1995. Evolution of the Cockburn Basin: implications for the structural development of the Celtic Sea basins. *In*: CROKER, P. F. & SHANNON, P. M. (eds) *The Petroleum Geology of Ireland's Offshore Basins*. Geological Society, London, Special Publications, **93**, 279–295.

SMITH, W. E. 1961. The Cenomanian deposits of south-east Devonshire. *Proceedings of the Geologists' Association*, **72**, 91–134.

STEWART, S. A., HARVEY, M. J., OTTO, S. C. & WESTON, P. J. 1996. Influence of salt on fault geometry: examples from the UK salt basins. *In*: ALSOP, G. I., BLUNDELL, D. J. & DAVISON, I. (eds) *Salt Tectonics*. Geological Society, London, Special Publications, **100**, 175–202.

STONELEY, R. 1982. The structural development of the Wessex Basin. *Journal of the Geological Society*, London; **139**, 543–554.

TAITT, A. H. & KENT, P. E. 1939. Note on an examination of the Poxwell Anticline, Dorset. *Geological Magazine*, **76**, 173–181.

UNDERHILL, J. R. & PATERSON, S. 1998. Genesis of tectonic inversion structures: seismic evidence for the development of key structures along the Purbeck–Isle of Wight Disturbance. *Journal of the Geological Society, London*, in press.

—— & STONELEY, R. 1982. Introduction to the development, evolution and petroleum geology of the Wessex Basin. *This volume*.

WALSH, J. J. & WATTERSON, J. 1991. Geometric and kinematic coherence and scale effects in normal fault systems. *In*: ROBERTS, A. M., YIELDING, G. & FREEMAN, B. (eds) *The Geometry of Normal Faults*. Geological Society, London, Special Publications, **56** 193–203.

WEBBY, B. D. 1965a. The stratigraphy and structure of the Devonian rocks in the Brendon Hills, West Somerset. *Proceedings of the Geologists' Association*, **76**, 39–59.

——1965b. The stratigraphy and structure of the Devonian rocks in the Quantock Hills, West Somerset. *Proceedings of the Geologists' Association*, **76**, 321–343.

WHITTAKER, A. 1972. The Watchet Fault – a post-Liassic transcurrent reverse fault. *Bulletin of the Geological Survey*, **41**, 75–80.

—— (ed.) 1985. *Atlas of Onshore Sedimentary Basins in England and Wales: Post-Carboniferous Tectonics and Stratigraphy*. Blackie, Glasgow.

WILLIAMS, G. D., POWELL, C. M. & COOPER, M. A. 1989. Geometry and kinematics of inversion tectonics. *In*: COOPER, M. A. & WILLIAMS, G. D. (eds) *Inversion Tectonics*. Geological Society, London, Special Publications, **44**, 3–15.

The structural development of the Central English Channel High – constraints from section restoration

H. S. BEELEY[1] & M. G. NORTON[2]

[1] *Phillips Petroleum Company, 6330 West Loop South, Bellaire, TX 77401, USA (e-mail: hsb@ppco.com)*
[2] *103 Coleraine Road, London SE3 7NZ, UK*

Abstract: The Central English Channel High is an east–west-trending asymmetric, south-facing, anticlinal structure which forms the southern boundary of the Central Channel Basin. Upper Jurassic rocks are exposed at seabed in the core of the structure, while Tertiary strata are preserved immediately to the south. Elsewhere in the Wessex–Channel basins structures of this overall form are interpreted as linear inversion structures formed during the mid-Tertiary by the reactivation of Mesozoic extensional faults, themselves formed by the partial reactivation of Variscan thrust ramps. Projection of the eroded sequence over the high using the geometry of the preserved section indicates the original presence of an approximately symmetrical Upper Jurassic–Lower Cretaceous sub-basin formed in the hanging wall of a north-dipping Mesozoic extensional structure, the Central Channel Fault. Seismic velocities from wells over the high are consistent with the implied post-Cretaceous relative uplift of up to 1500 m. The deeper geometry of the Central Channel Fault has been determined using the restored hanging wall geometry and the main aspects of the inversion structure have been recreated by forward modelling. The internal structure of the Central Channel High is consistent with a moderately steep fault near surface which passes relatively abruptly into a lower-angle fault. This is interpreted to indicate the presence of a Variscan backthrust at depth which has been partially reactivated during both Mesozoic extension and mid-Tertiary inversion.

The Central Channel High is an east–west-trending high that forms the southern margin of the Central Channel Basin (Fig. 1). It extends over 120 km from the Pays de Bray Fault in the east to a point due south of the Isle of Purbeck in the west where it becomes poorly defined. In the central part of the high, where fault displacement is at a maximum, the sequence is eroded down sufficiently for Upper Jurassic strata to outcrop at seabed. In the same part of the structure the downthrown side to the south preserves Tertiary sediments.

The section preserved over the high itself is known from hydrocarbon exploration wells 98/22-1, 98/22-2, and 98/23-1. No wells have been drilled on the downthrown side of the structure, apart from the Nautile-1 well in the French sector *c.* 50 km to the east-southeast, which has a very condensed sequence. Consequently, the interpretation of the sequence relies entirely on seismic jump correlations from the relatively well-known Central Channel Basin sequence. Hamblin *et al.* (1992) suggest that the structure is associated with a considerable southwards thickening (≥ 800 m) of the Triassic sequence.

The shallow structure appears to be a south-facing monocline, superficially a mirror-image of the Portland–Wight structure. The 1:250 000 Wight solid geology map (IGS, 1977) shows southward dips of up to 70° within the Upper Jurassic.

Previous published studies of the Wessex and Channel basins have tended to mention the Central Channel High only in passing. It has been identified either as a large down-to-the-south extensional fault (Stoneley & Selley 1992; Hamblin *et al.* 1992) or as an inversion-related structure (Beach 1987; Gibbs 1987; Simpson *et al.* 1989). To quote Hamblin *et al.* (1992) 'The role of the Central English Channel Fault System during basin inversion is enigmatic; it does not appear to have suffered significant reversal, for the sense of throw of both the Tertiary monocline and the underlying Mesozoic normal fault is down-to-the-south... The reason for this apparent lack of involvement with the basin inversion is unclear'.

The aim of this study is to understand the development of the Central Channel High in the context of the regional tectonic history of southern England from the Variscan to the Tertiary.

Regional setting

Variscan basement structure

The concealed Variscan structure of central southern England remains poorly understood. Most of the direct information comes from exposed Devonian and Carboniferous rocks to the west. In the area around Bristol the structure is consistent with a northward directed relatively

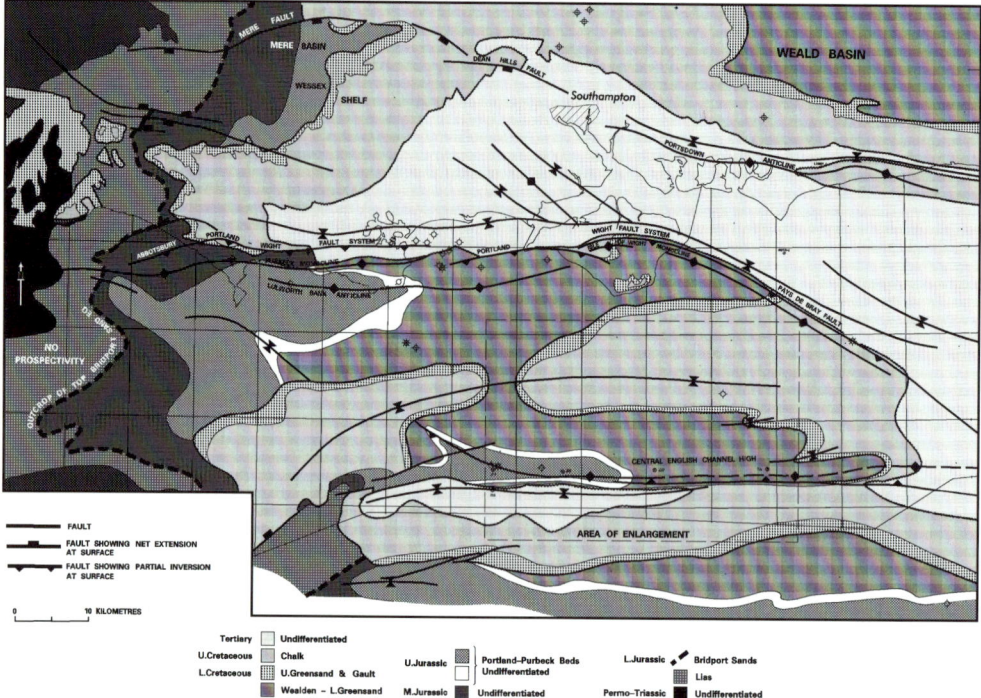

Fig. 1. Location map showing the regional setting of the Central Channel Basin. (modified from the 1:250 000 Portland and Wight Sheets. BGS 1983; IGS 1977).

thin-skinned thrust system (Williams & Chapman 1986). Seismic reflection data from c. 20 km to the east indicate the presence of two c. 25°SSE-dipping reflectors (interpreted as thrusts – T_3 and T_4; Chadwick et al. 1983, and subsequently identified as the Variscan Front Thrust and the Wardour Thrust, respectively; Chadwick 1986) which may represent the main ramp structures of this system.

The northernmost of these, probably the Variscan main frontal thrust, correlates with the location of the Vale of Pewsey faults which are interpreted to have formed by its reactivation during Mesozoic extension. It has been proposed that the thrust structure extends eastward as a relatively continuous structure along the line of the Vale of Pewsey and London Platform faults, beneath the Weald Basin (Chadwick 1986), eventually linking with the northwest–southeast-trending Faille du Midi which forms the Variscan main frontal thrust in northeastern France (Pinet et al. 1987).

The southern structure, the Wardour Thrust, appears to have been extensionally reactivated during the Mesozoic, in a similar fashion as the Mere Fault. This extensional structure is traceable to the east as the Wardour–Portsdown Fault System (Chadwick 1993) possibly linking into the northwest–southeast-trending Pays de Bray Fault which is thought to have been active as a major dextral oblique-slip zone during the Variscan (Holder & Leveridge 1986).

The next major structure to the south is the proposed Portland–Wight Thrust (Chadwick 1986). This south-dipping Variscan thrust is shown as coincident with the later Portland–Wight extensional faults, which form the northern boundary to the Mesozoic Channel Basin, and it is interpreted to link directly into the Pays de Bray Fault. The basis of the interpretation of this thrust is unknown and it remains possible that both the Wardour–Portsdown and the Portland–Wight extensional faults represent reactivation of a single Variscan structure.

The Variscan structure beneath the Central Channel Basin is unknown. Holder & Leveridge (1986) and Pinet et al. (1987) suggest that the thrust structures known from southern Cornwall and Devon are likely to continue eastwards into the study area. Such thrusts are, however, likely to be offset to the south by the major northwest–southeast-trending structures, the Lustleigh–Sticklepath and Watchet–Cothelstone faults, and their precise locations are unknown.

The northward-propagating thrust system described above formed in response to northerly directed compression and appears to be of different origin (and perhaps timing) to the east–west late Carboniferous compression which is characteristic of central/northern England and Scotland (Coward 1994). There is no indication that this latter episode produced any significant effect on the structure of southernmost Britain.

Mesozoic basin structure

The area has undergone two main periods of extension. The Wessex and Central Channel basins form part of a complex north–south-trending Triassic rift system which includes the East Irish Sea, and the Cheshire and Worcester basins. The extension was apparently ENE–WSW directed, and has been interpreted to have caused dextral-normal oblique reactivation of the Variscan thrust structures (Chadwick & Evans 1995). The location of these faults seems to have been governed by the position of the main thrust ramps, forming as steeper short-cuts in their hanging walls. The Vale of Pewsey, Wardour–Portsdown and Portland–Wight systems appear to have been initiated at this time.

The Triassic sequence on the Central Channel High is thin as proved by the 98/22-2 and 98/23-1 wells, whereas Hamblin *et al.* (1992) have interpreted up to a 1000 m of Triassic strata to the south of the high implying the existence of an intervening major south-dipping extensional fault.

The second period of basin formation was during the Jurassic–Cretaceous. This was related to an approximately north–south directed extension (Stoneley 1982) which not only reactivated the east–west-trending Triassic growth faults as essentially dip-slip faults but also activated the northwest–southeast-trending Pays de Bray Fault in an oblique normal-sinistral sense. The area affected by extension was significantly greater than during the Triassic.

No changes in thickness of the Jurassic–Lower Cretaceous sequence across the southern margin of the Central Channel High have been noted in the literature which would support active faulting down to either the north or south.

The rift sequence was blanketed by the Upper Cretaceous Chalk and some early Tertiary strata.

Mid-Tertiary basin inversion

There is some evidence for minor inversion affecting the Wessex and Central Channel basins during the late Cretaceous–early Tertiary, probably related to early phases of Alpine deformation (Chadwick 1993). The main inversion event, caused as a far-field effect of the main continental collision between Africa and Europe, occurred during the Miocene.

This inversion was in response to northerly directed Alpine compression, causing reverse reactivation of most of the Mesozoic extensional faults and uplift of the basinal areas within the Wessex and Central Channel basins. This includes the Variscan structures, described above, which were extensionally reactivated during the Mesozoic and later subjected to Tertiary compression, such as the Vale of Pewsey, Wardour–Portsdown and Portland–Wight fault systems (Underhill & Paterson, 1998; Underhill & Stoneley this volume). Chadwick (1993) has identified two principal types of inversion structure, regional upwarps (e.g. the Wealden Anticline) and linear zones of en echelon faulted monoclinal or anticlinal flexures (e.g. the Portland–Wight structures).

Both the northern and southern margins of the Central Channel Basin have the character of linear inversion structures. It can be argued that the uplift of the southern margin, the Central Channel High, is even greater than that over the Portland–Wight structure. The principal argument against an inversion origin for the high is its close spatial association with an inferred large-scale Mesozoic down-to-the-south extensional fault.

Summary of tectonic evolution

The structural development of the Wessex and Central Channel basins appears to have been almost exclusively governed by the repeated partial reactivation of Variscan thrust structures. There appears to be an almost one-to-one relationship between major Variscan thrust ramps, Mesozoic extensional faults and mid-Tertiary linear inversion structures. This can be taken to imply that the Central Channel High is likely to be sited over a north-dipping Mesozoic extensional fault which reactivates a Variscan backthrust. This possibility is explored in this paper.

Dataset used

Seismic data

This reinterpretation is based principally on a JEBCO Seismic Ltd speculative seismic survey

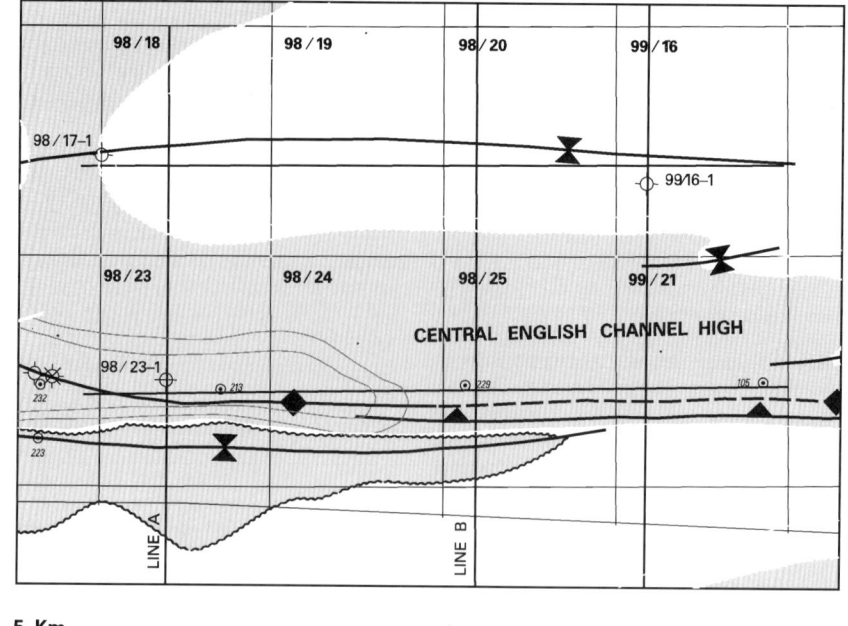

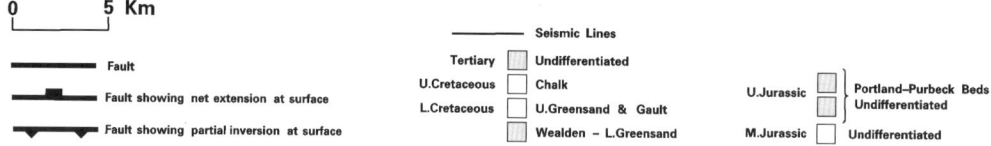

Fig. 2. Location map showing the seabed geology, dip data, borehole locations and seismic reflection profiles (modified from the 1:250 000 Portland and Wight Sheets. BGS 1983; IGS 1977).

recorded and processed in 1993 (Fig. 2). The survey consists of a grid of seven north–south, three east–west, seven northwest–southeast and three northwest–southwest lines. They were acquired using a relatively small sleeve airgun array source which produced generally very good resolution in the shallow section. This has been particularly important for measuring stratal dip within the Wealden beds to allow the construction of eroded section.

Well data

Five wells were used to aid the interpretation, Nautile-1, 98/22-2, 98/23-1, 98/18-1 and 99/16-1. Composite logs were available for all the wells and calibrated sonic logs for all but Nautile-1.

Solid geology and surface dip data

The Wight 1:250 000 series solid geology map (IGS 1977) shows the outcrop of the main geological boundaries at seabed as determined from shallow boreholes and the interpretation of shallow analogue seismic data. This has been particularly useful in identifying the top of the Wealden Group and the base of the Chalk Group on the digital seismic data used in this study. The map also shows dips of strata near seabed estimated from the analogue seismic data. This information has been incorporated into the building of cross-sections for restoration.

Stratigraphy

The stratigraphy of the Wessex Basin has been studied extensively by many authors (e.g. Arkell 1935; Callomon & Cope 1995) and is summarized in Fig. 3. Detailed understanding of the stratigraphy is possible because of the excellent outcrop, particularly along the Dorset–South Devon coastal sections and is supported by both onshore and offshore well penetrations.

The majority of basin fill comprises thick Permian–Early Cretaceous sediments, resting

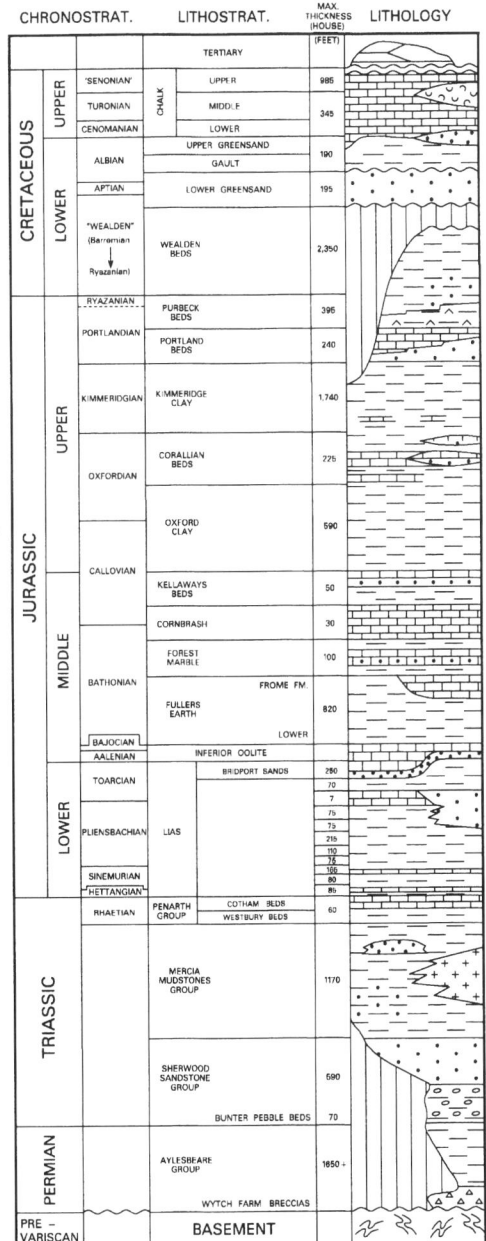

Fig. 3. Generalized stratigraphy for the Central English Channel Basin.

Sherwood Sands, deposited in a complex fluvial–alluvial setting. These provide the primary reservoir for the Wytch Farm oilfield (Bowman et al. 1993). Widespread playa lake sediments of the Mercia Mudstone Group form a thick regional seal to the underlying Sherwood Sands.

Sediments of the latest Triassic Penarth Group mark the onset of marine conditions within the basin, culminating in the deposition of the Lower Lias organic-rich shales which are the principal mature source rocks. These shales are overlain by the Downcliff Sand. This entire succession is the first of several upward coarsening cycles identified within the Jurassic (e.g. House 1989). Each cycle is characterized by a basal marine shale facies which gradually coarsens into either a shallow-marine sand (e.g. Bridport Sand), a high-energy carbonate (Inferior Oolite) or lagoonal–lacustrine facies (Purbeck Beds). In some cases these facies present minor reservoir potential, such as the Bridport Sand, which at Wytch Farm is the secondary reservoir (Penn et al 1987).

The last sequence within the Upper Jurassic includes the fully marine Kimmeridge Clay and the shallow-marine to lagoonal–lacustrine Portland and Purbeck beds. Although the Kimmeridge Clay is of source quality it is not mature at the present day and will not contribute hydrocarbons.

The lacustrine–freshwater environments which developed during the deposition of the Purbeck Beds continued with terrestrial sedimentation in the Early Cretaceous, Wealden times. These sediments consist of sandstones and grits, interpreted as fluvial–alluvial deposits, associated with estuarine and lagoonal environments (House 1989). Following the development of an intra-Early Cretaceous unconformity, the area returned to marine conditions marked by a series of transgressions which deposited the Upper and Lower Greensands, the Gault Clay and culminating in the deposition of the Chalk Group during the Late Cretaceous.

A minor unconformity is present at the top of the Chalk Group and most of the Lower Palaeocene is missing. The remaining Palaeocene–Oligocene sediments are characterized by variegated freshwater to shallow-marine sands and clays.

unconformably on deformed Variscan basement. The earliest sediments, of Permo-Triassic age, consist of dominantly continental red-beds deposited in a semi-arid environment. The most important sediments in this succession are the

Seismic interpretation

Ties from the four wells within the Central Channel Basin allowed unequivocal identification of most of the observed reflectors. The most clearly identified are the Purbeck–Portland beds, the top of the Corallian, the top of the Cornbrash and the top of the Inferior Oolite. The top of the Wealden Beds and the base of the Chalk were picked at changes in seismic character tied to the wells and the seabed outcrop. The Permo-Triassic and Lower Jurassic sequence is less clearly imaged and the confidence of the picking reduces quickly away from well control.

To the south of the Central Channel High good jump correlations were found to the Central Channel Basin sequence, particularly where the latter has a relatively thin Upper Jurassic–Lower Cretaceous section. Some uncertainty remains due to the lack of well control but there appears to be no obvious alternative to that proposed here. Contrary to Hamblin *et al.* (1992) there is no evidence for a thick Permo-Triassic sequence to the south of the high.

The detailed description presented below is based on the final interpretation.

Line A

This line shows a similar overall geometry to that seen in Line B except that the Upper Jurassic–Lower Cretaceous within the Channel Basin are everywhere thicker (Fig. 4). In contrast, the sequence to the south of the high is significantly thinner.

The line stops well short of the Portland–Wight structures but the sequence at the northern end of the line can be seen gradually thickening northwards although, unlike Line B, it is affected by a series of down-to-the-north extensional faults. A central zone of relatively close-spaced uninverted extensional faults is again present, characterized by the thinnest sequence of Upper Jurassic to Lower Cretaceous strata.

The sub-basin is broadly symmetrical, although a slightly thicker Lower Cretaceous appears to be developed in minor grabens at its northern end. The Central Channel High inversion structure is distinctly asymmetric on this profile and is associated with a much greater uplift than seen on Line B. It retains an abrupt kink-like northern boundary and monoclinal southern boundary. Additional low-amplitude monoclines are also formed above the inverted boundaries of the minor grabens.

The Central Channel Fault is not imaged, although it is presumed to be north-dipping. The associated growth fault in its footwall has been partially inverted.

Line B

The Lower and Middle Jurassic sequences show relatively little change in thickness from the 99/16-1 well, which lies 14 km to the east, southwards across the Central Channel High (Fig. 5). There also appears to be no change in the Permo-Triassic thickness.

The interval including the Kimmeridge Clay and Wealden Beds shows the greatest variation in thickness, with a significant sub-basin developed between the southern side of the Central Channel High and just south of the projected position of the 99/16-1 well.

The structure seen on this line can be subdivided into three. At the northern end of the line the dip within the Cretaceous is very low and the structure is dominated by Mesozoic extension with no indication of inversion. The Lower Cretaceous and Upper Jurassic sequence shows gradual northward growth towards the Pays De Bray Fault with very little faulting.

The 99/16-1 well was drilled on a narrow horst block within a zone of uninverted extensional faulting in the central part of the Central Channel Basin. The Upper Jurassic–Lower Cretaceous sequence is at its thinnest here. Just to the south of this well is the abrupt edge of the Tertiary inversion structure which has a kink-like form, with the bedding in the Wealden showing significant divergence from that in the underlying Portland–Purbeck beds.

The Upper Jurassic–Lower Cretaceous sub-basin is broadly symmetrical with the Upper Jurassic sequence being approximately 50% thicker than that to both north and south. Thickness variations in the Lower Cretaceous are not directly imaged due to erosion of the upper part of the sequence. The inversion structure seems to be composed of two parts, each showing their greatest uplift to the south. The northernmost has the appearance of a Lower Cretaceous graben within the sub-basin that has 'popped-up' during the inversion. The inversion structure shows a very close spatial relationship to the Mesozoic sub-basin.

Although the Central Channel Fault is not imaged on this section, a northward dip is implied by the direction of growth during the Lower Cretaceous–Upper Jurassic and this is supported by unequivocally northward dipping growth faults in its footwall.

Restoration methodology

Several different interpretations of the time data were carried out during the study using different

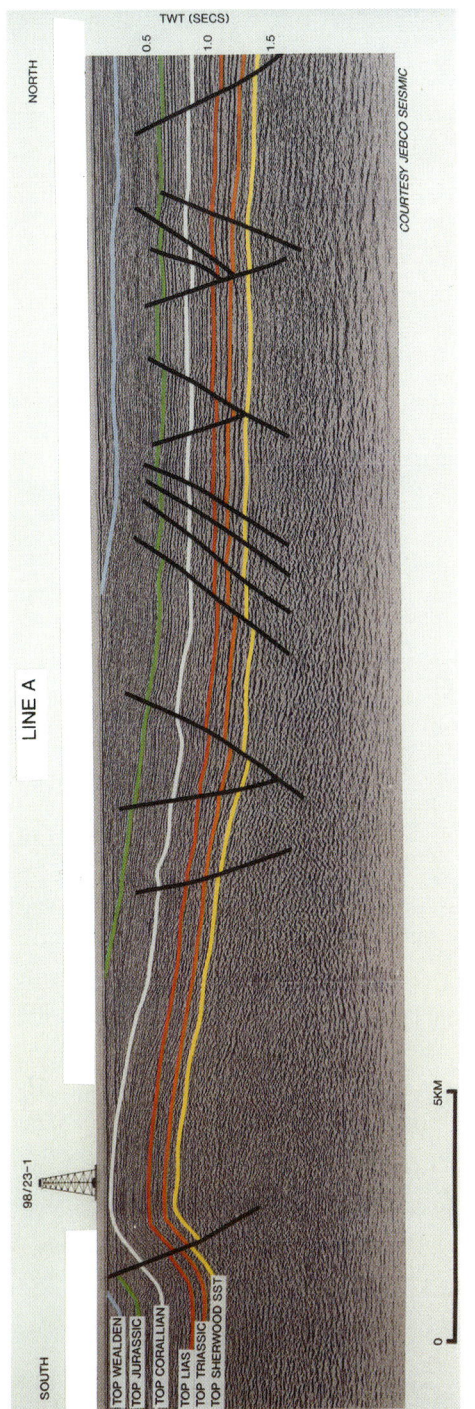

Fig. 4. Final interpretation of the northern part of seismic Line A.

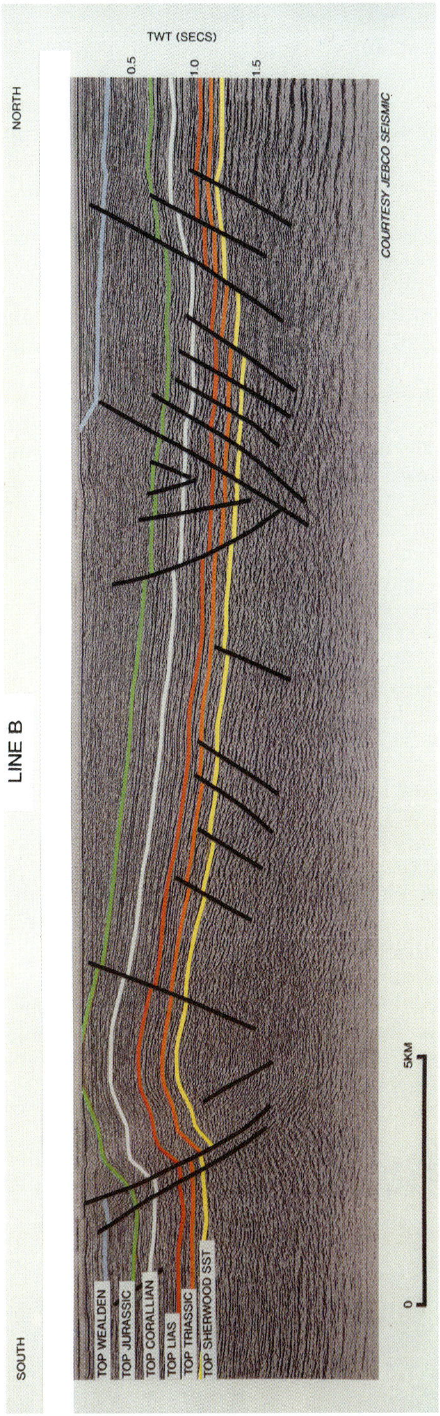

Fig. 5. Final interpretation of the northern part of seismic Line B.

assumptions. The time sections were input to GEOSEC and converted to depth using constant seismic interval velocities. The assumption of constant interval velocities is considered reasonable within the Central Channel Basin itself due to the similarities in velocities for each unit for the four wells used (see below). The depth section was then restored using the appropriate deformation mechanisms, vertical or inclined shear for extension and flexural-slip for contractional inversion.

Eroded parts of the section were constructed using the projection tool in GEOSEC using either its constant thickness or gradient options as appropriate. This allows the geometry of the preserved strata to be used to complete the missing section in a non-arbitrary way. The projection method uses dip domain boundaries, defined by an observed horizon shape, to define areas of constant dip. This is an appropriate technique for bedded sequences that have deformed by a flexural-slip mechanism.

Restored horizon geometries were used to predict the form of the controlling extensional faults making assumptions about the deformation mechanism. Fault shapes can be predicted for any angle of simple shear. Where there is a clear dominant polarity of minor faults within the hanging wall their dip is normally used to govern the angle of shear used.

The fault geometries derived in this way were then used in forward modelling using the GEOSEC Fault-Bend Fold module which assumes a flexural-slip mechanism (using the rules for progressive development outlined by Suppe 1983). It also assumes that the initial part of the fault is horizontal and it is necessary to add a horizontal segment at the base such that the algorithm treats the entire fault surface as part of a complex ramp. There is some degree of instability in the solution produced and reasonable geometries are only formed for particular applied displacements along the fault. Displacements were chosen which approximately matched the estimated relative uplift during inversion.

Results

At an early stage in this study, an initial interpretation of the data was carried out with reference to that published by Hamblin *et al.* (1992) (Fig. 6) and only considered the Permian–Jurassic section. This assumed the presence of a large down-to-the-south extensional fault. Preliminary analysis of interpretations based on this model showed that restoration using vertical simple shear produced invalid results and this model was therefore discarded.

It was then decided to look in more detail at the shallower part of the data, particularly the Wealden. This indicated that the Central Channel High was a sub-basinal area during the Upper Jurassic–Lower Cretaceous periods with a significantly thicker Upper Jurassic–Lower Cretaceous sequence than that to the south.

Figure 7 shows variations in velocity with depth of argillaceous strata for the five wells used in this study. They have been combined with those published by Chadwick (1993) for the Sandhills borehole from the northern Isle of Wight, in the footwall of the Portland–Wight structure and the Detention borehole from the central part of the Wealden Anticline, a regional mid-Tertiary upwarp.

The data show considerable scatter due to variations in lithology and only overall trends can be derived from such information. It should be noted that the velocities from Nautile-1 are uncalibrated by check-shot data.

Data from Nautile 1, 99/16-1 and 98/18-1 plot in a similar field as those from Sandhills, apart from shallow levels where they have higher velocities. This coincides with the Tertiary section in Sandhills which is lithologically distinct from the Jurassic–Lower Cretaceous rocks in the other wells. This suggests that the central part of the Central Channel Basin, which has the thinnest Upper Jurassic–Lower Cretaceous sequence, has suffered relatively minor uplift in the Tertiary. Although the number of data points is small, 98/18-1 appears to show

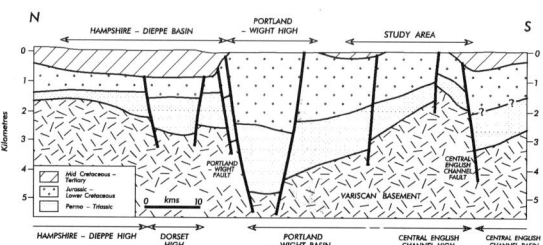

Fig. 6. An example of previous interpretations across the Central Channel Basin showing the Central Channel High controlled by a major down-to-the-south extensional fault (from Hamblin *et al.* 1992).

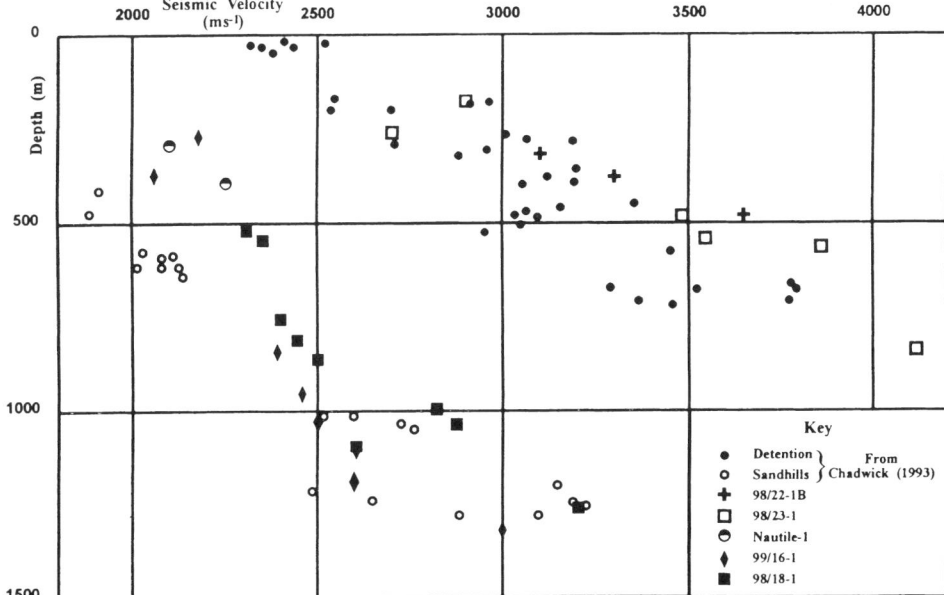

Fig. 7. Variations of seismic interval velocity with depth for wells Nautile-1, 98/18-1, 99/16-1, 98/22-1C and 98/23-1 compared to data from Chadwick (1993) for the Sandhills and Detention boreholes.

slightly greater uplift than 99/16-1, consistent with the erosion of the upper Cretaceous from the former.

Data from 98/22-1C and 98/23-1 plot in a similar field as that from Detention, overall showing slightly higher velocities for the same depth. This indicates that the Central Channel High has been uplifted, probably well over 1000 m at its maximum.

Comparisons of velocities from equivalent stratigraphic intervals between wells from within the Central Channel Basin show remarkable little variation. Overall those wells from the Central Channel High have slightly higher velocities than those from the centre of the basin. This difference is, however, less than would be expected from the interpreted thickness variations in the Upper Jurassic and Lower Cretaceous. This might indicate that there was a significant Tertiary section developed between the uplifting Central Channel High and Portland–Wight inversion structures since removed by erosion. Chalk in the 99/16-1 well has velocities of up to 3400 m s^{-1} at 150 m below seabed, suggesting maximum burial of between several hundred and a thousand metres (cf. Menpes & Hillis 1995).

Initially each section was restored using vertical simple shear as the mechanism. The geometry in both the immediate footwall and hanging wall of the Central Channel Fault is unrealistic (e.g. Fig. 8) with mismatches and unrestored folding. The section was then successfully restored using a flexural-slip mechanism for those fault blocks involved in the inversion. All other fault blocks were still restored using a vertical simple-shear mechanism.

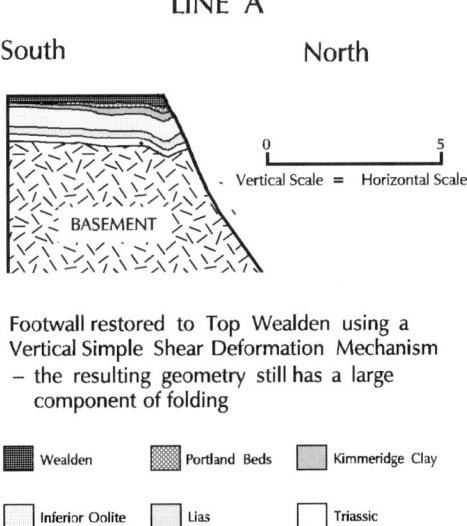

Fig. 8. Effect of simple shear restoration on the folded footwall of Line A. Scale is km.

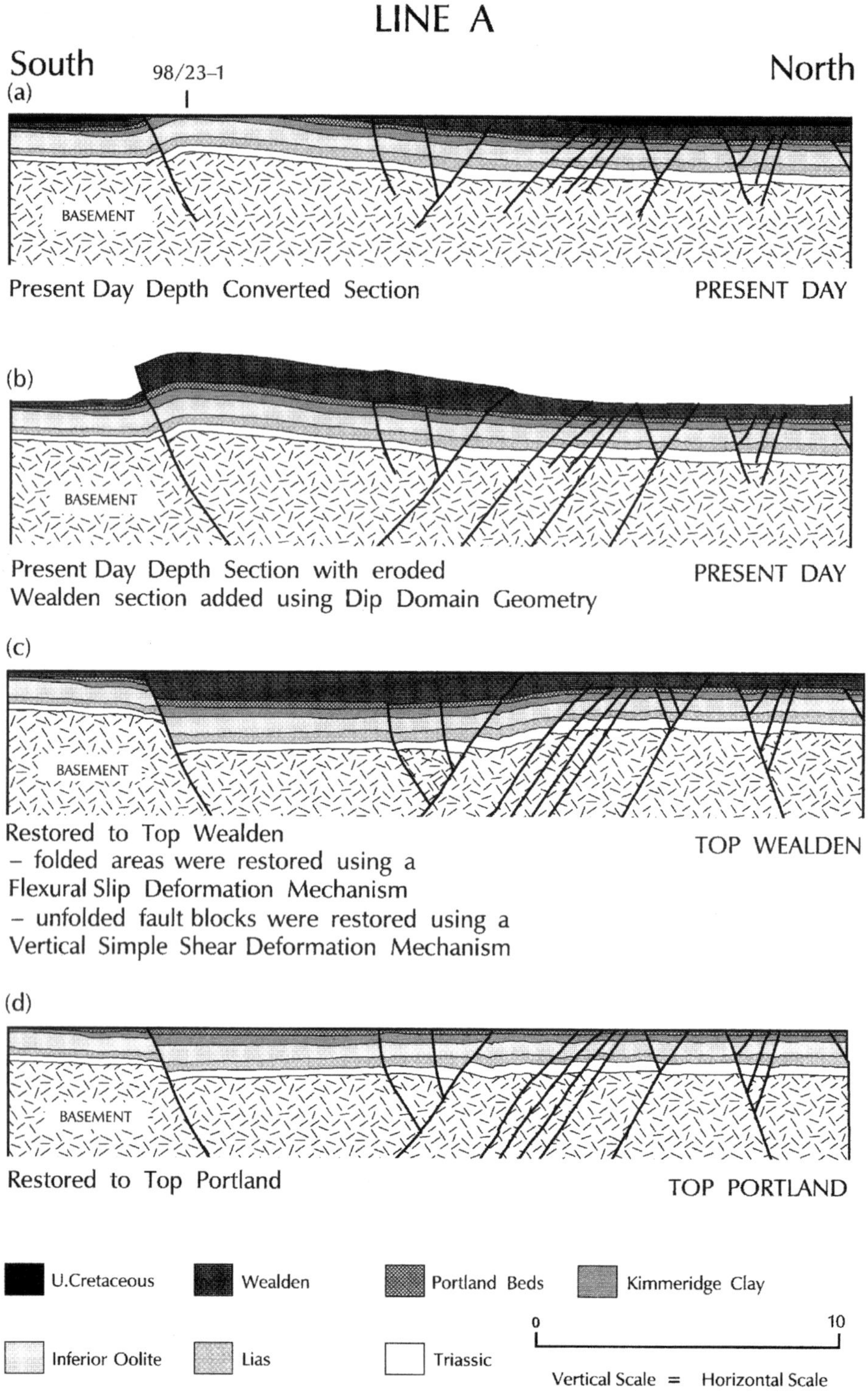

Fig. 9. Sequential restoration of section along seismic Line A: (**a**) original depth section: (**b**) section completed to the top of the Wealden Beds: (**c**) section restored to the top of the Wealden Beds; and (**d**) section restored to the top of the Portland Beds.

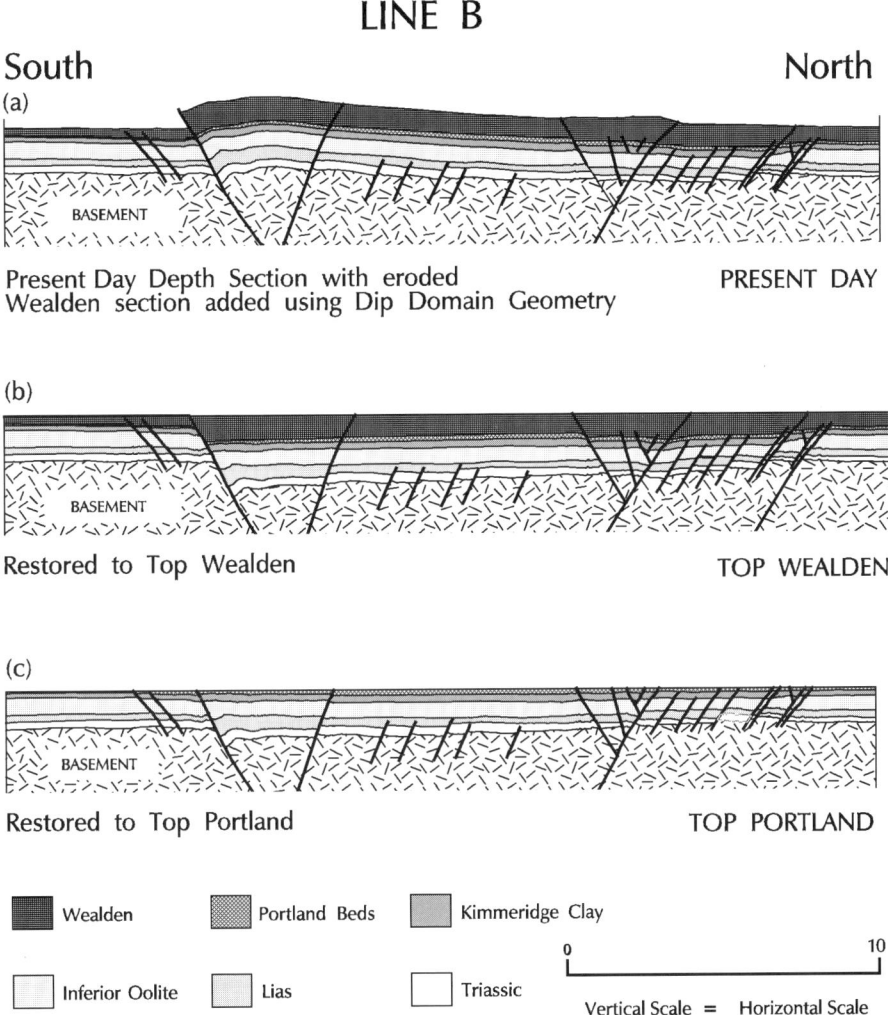

Fig. 10. Sequential restoration of section along seismic Line B: (**a**) original depth section completed to the top of the Wealden Beds: (**b**) section restored to the top of the Wealden Beds; and (**c**) section restored to the top of the Portland Beds. Scale is km.

The revised interpretation for Line A is shown in Fig. 9a, and the depth section completed to top Wealden in Fig. 9b. The constructed Wealden thickness in this case reaches over 950 m and the uplift over the high is implied to exceed 1500 m relative to the footwall. Figure 9c and 9d shows the restoration to top Wealden and top Portland and shows the geometry of the Upper Jurassic–Lower Cretaceous sub-basin.

Figure 10a shows the revised depth interpretation for Line B with the section completed to the top of the Wealden Beds using the geometry of the underlying Jurassic strata. The thickness of the Wealden reaches a maximum of 800 m which implies a minimum 1100 m uplift relative to the footwall of the Central Channel Fault. Figure 10b and c shows the restoration to the top of the Wealden and top Portland, again showing the sub-basin geometry.

Hydrocarbon prospectivity

Hydrocarbon prospectivity of this, and similar structures, is problematic. Although the structure itself is valid, questions arise concerning the timing of the structure with respect to hydrocarbon charge and the nature of the potential reservoir (governed by basinal position). The Sherwood Sand reservoir has the additional problem of hydrocarbon charge from source rocks that are stratigraphically higher in the section. These issues need to be addressed further before any definitive predictions regarding prospectivity can be made.

Discussion

The results of the seismic interpretation backed up by section restoration and well velocity variations provide clear evidence that the Central Channel High was caused by the inversion of a late Mesozoic sub-basin developed in the hanging wall of a major north-dipping extensional fault, the Central Channel Fault.

Elsewhere in the Wessex–Channel basins such structures have been interpreted as being formed by the partial reactivation of Variscan thrusts. In this case, however, the structure responsible would be a backthrust for which no other form of evidence is available.

In this study the geometry of both the observed inversion structure and the restored

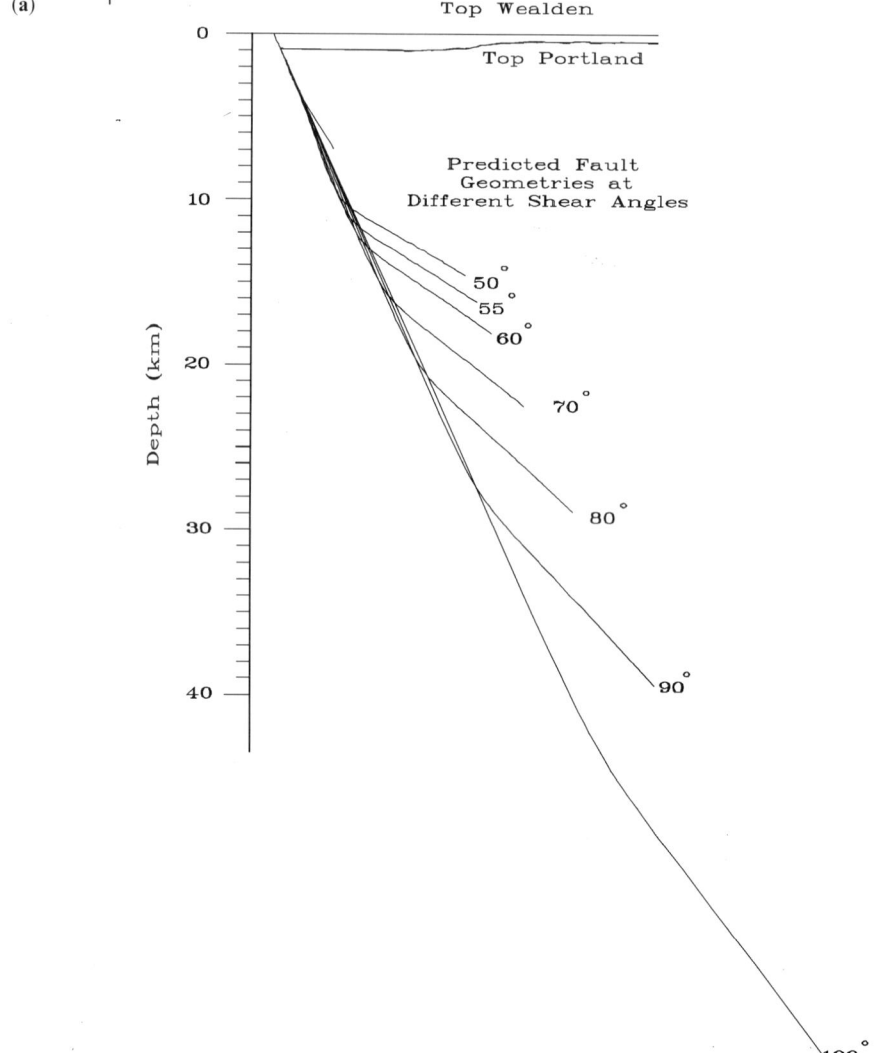

Fig. 11. Range of fault geometries predicted for generalized hanging wall horizon shapes for varying angles of inclined simple shear for: (**a**) Line A; and (**b**) Line B.

extensional sub-basin have been used to determine the underlying fault geometry. The first stage used the fault prediction capability in GEOSEC in which the fault geometry is derived from the shape of horizons in the hanging wall.

Figure 11a and b shows the range of fault geometries derived from smoothed hanging wall geometries in the restored sections of Lines A and B for varying angles of simple shear. For Line A the predicted fault shows a relatively abrupt downward dip change. The fault derived from Line B shows a similar change in dip, although the upper part of the fault is more irregular and of lower dip. The preferred geometry is that for 55° antithetic shear, which matches the dip of the south-dipping faults within the hanging wall on both lines. In both cases the low-angle portion of the fault dips at about 30° and the change in dip occurs at approximately 12.5 km depth. This is taken to imply that the Central Channel Fault formed by reactivation of the deeper part of an approximately 30° north-dipping structure which propagated upwards as a relatively planar fault dipping 55–65°.

The preferred geometries from the two profiles were used in forward modelling inversion structures for comparison with those observed. Figure 12a and b shows the results of the forward modelling carried out using the GEOSEC Fault-Bend Fold module. In each case generalized versions of the restored hanging wall structure have been transferred into the constructed top Lower Cretaceous geometry using flexural slip. The modelled structure for Line B reproduces most aspects of the observed inversion structure, except that the reverse kink nucleated at the abrupt change in fault dip does not coincide with the edge of the sub-basin but is located somewhat further to the south. The result for Line A is less convincing with regard to the shape of the inversion anticline but still provides a broad match to the observed geometry.

Differences between the observed and modelled inversion geometries could have a number

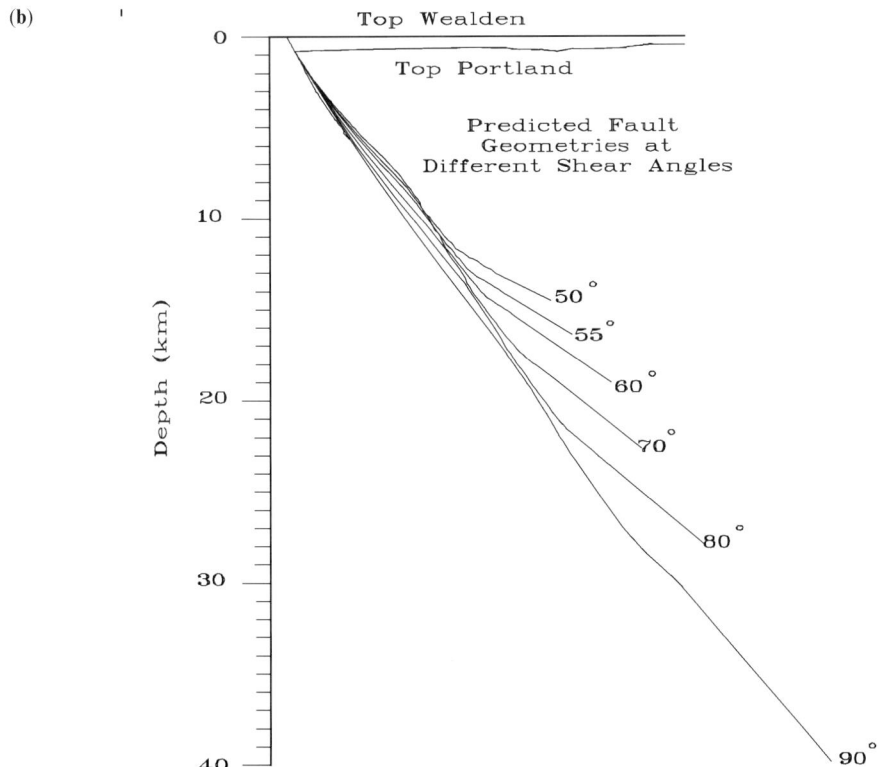

Fig. 11. (*continued*)

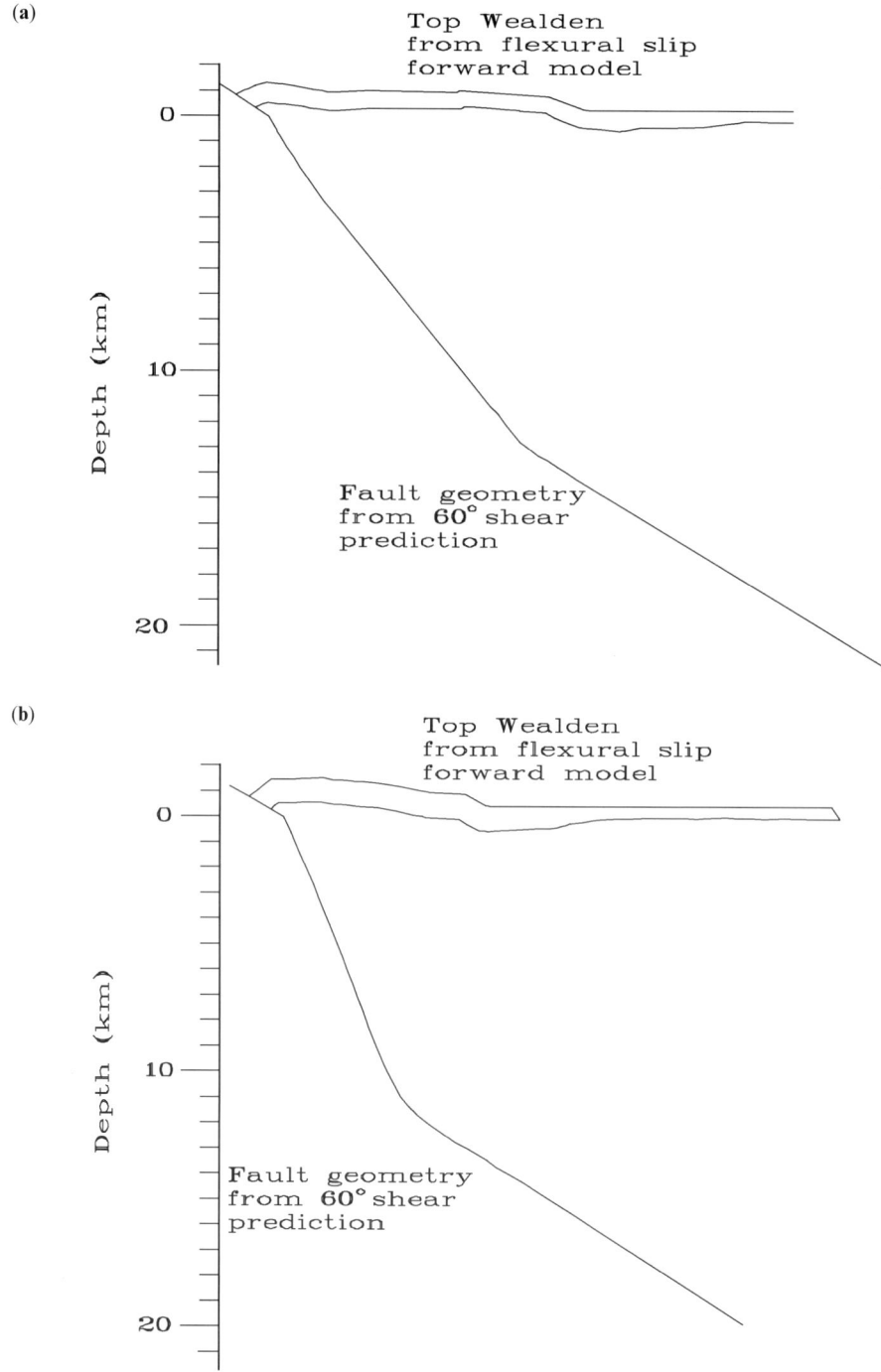

Fig. 12. Results of forward modelling using the GEOSEC Fault-Bend Fold module: (**a**) Line A; and (**b**) Line B.

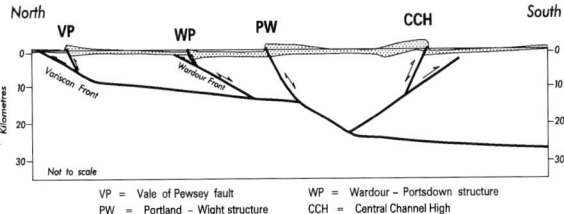

Fig. 13. Schematic cross-section across the Wessex and Central Channel basins showing the proposed deep fault geometry.

of origins. The original projection of the top Lower Cretaceous may have failed to take into account structures which have failed to induce dip changes within the Wealden. This unit contains no correlatable internal reflectors which could be used to verify this interpretation. Second, the Fault-Bend Fold algorithm assumes the hanging wall consists of a sequence of parallel-bedded strata for its full thickness. Most of the section in this case is Variscan basement. Lastly, the algorithm also assumes that the hanging wall is unfaulted. It may be that the orientation of the reverse kink is controlled by the angle of the antithetic faults rather than that predicted by the theory of Suppe (1983).

Considering the potential problems with the methodology employed – going from the observed geometry via surface projection using dip domains, restoration using flexural slip, extensional fault prediction using inclined simple shear and Fault-Bend Fold forward modelling – the results are sufficiently internally consistent to suggest that the interpreted deep fault geometry is close to that responsible for generating the structure of the Central Channel High.

The interpreted geometry of the Central Channel Fault has implications for the deeper structure of the Portland–Wight, Wardour–Portsdown and Vale of Pewsey fault systems. It is very unlikely that the Portland–Wight structure represents a direct reactivation of a pre-existing Portland–Wight Thrust. The geometry of both the late Mesozoic sub-basin and the superimposed inversion structure suggest that the extensional fault is mildly listric to a considerable depth (≥ 15 km). There is no sign of the abrupt dip change that would be expected if it passed at shallow levels into a low-angle fault.

Figure 13 shows a schematic north–south cross-section across the Wessex–Channel basins indicating possible deep fault geometries. The Portland–Wight extensional fault is shown as passing into a ramp in the basal Variscan thrust at depth.

Conclusions

(1) The form of the Central Channel High is best explained as being the result of the reversal of a major down-to-the-north later Mesozoic extensional structure, the Central Channel Fault, during mid-Tertiary inversion; a mirror image of the Portland–Wight structure.

(2) The Central Channel Fault was active in extension mainly during the Upper Jurassic–Lower Cretaceous. There is no indication that it significantly affects the thickness of Permo-Triassic to Middle Jurassic strata. No evidence has been found to support the presence of a large down-to-the-south Mesozoic extensional fault on the southern side of the Central Channel High.

(3) The geometry of both the Upper Jurassic–Lower Cretaceous sub-basin and the superimposed Tertiary inversion structure provides evidence for control by a relatively steep north-dipping fault which joins relatively abruptly with a lower-angle fault at depth. This geometry probably formed by the extensional reactivation of a Variscan backthrust.

(4) The Portland–Wight Mesozoic extensional faults are unlikely to pass downwards into a low-angle fault at a shallow level suggesting that there is no evidence to support a Variscan Portland–Wight Thrust. It is considered more likely that it joins with a ramp on the basal Variscan thrust at depths of ≥ 15 km.

The authors would like to thank Phillips Petroleum UK Ltd for permission to publish this paper. JEBCO Seismic Ltd are gratefully acknowledged for allowing their seismic lines to be used and shown. A. Chadwick and J. Gutmanis gave additional information on the Detention well and the Pays de Bray Fault, respectively. Finally we are indebted to T. Rickards and R. Diver for the draughting work.

References

ARKELL, W. J. 1935. The Portland beds of the Dorset mainland (England). *Proceedings of the Geologists' Association*, **46**, 301–347.

BEACH, A. 1987. A regional model for linked tectonics in north-west Europe. *In*: BROOKS, J. & GLENIE, K. (eds) *Petroleum Geology of North West Europe*. Graham & Trotman, London, 43–48.

BOWMAN, M. B. J., MCCLURE, N. M. & WILKINSON, D. W. 1993. Wytch Farm oilfield: deterministic reservoir description of the Sherwood Sandstone. *In*: PARKER, J. R. (ed.) *Petroleum Geology of North West Europe: Proceedings of the 4th Conference*. Geological Society, London, 1513–1518.

BGS. 1983. *1:250 000 Portland Sheet*. British Geological Society.

CALLOMON, J. H. & COPE, J. C. W. 1995. The Jurassic Geology of Dorset. *In*: TAYLOR, P. D. (ed.) *Field Geology of the British Jurassic*. Geology Society, London, 51–104.

CHADWICK, R. A. 1986. Extension tectonics in the Wessex Basin, southern England. *Journal of the Geological Society, London*, **143**, 456–488.

—— 1993. Aspects of basin inversion in southern Britain. *Journal of the Geological Society, London*, **150**, 311–322.

—— & EVANS, D. J. 1995. The timing and direction of Permo-Triassic extension in southern Britain. *In*: BOLDY, S. A. R. (ed.) *Permian and Triassic Rifting in Northwest Europe*. Geological Society, London, Special Publications, **91**, 161–192.

——, KENOLTY, N. & WHITTAKER, A. 1983. Crustal structure beneath southern England from deep seismic reflection profiles. *Journal of the Geological Society, London*, **140**, 893–912.

COWARD, M. P. 1994. Inversion Tectonics. *In*: HANCOCK, P. L. (ed.) *Continental deformation*. Pergamon Press, 289–304.

GIBBS, A. D. 1987. Basin development, examples from the United Kingdom and comments on hydrocarbon prospectivity. *Tectonophysics*, **133**, 189–198.

HAMBLIN, R. J. O., CROSBY, A., BALSON, P. S., JONES, S. M., CHADWICK, R. A., PENN, I. E. & ARTHUR, M. J. 1992. *United Kingdom Offshore Regional Report: The Geology of the English Channel*. HMSO, London.

HOLDER, M. T. & LEVERIDGE, B. E. 1986. Correlation of the Rhenohercynian Variscides. *Journal of the Geological Society, London*, **143**, 141–148.

HOUSE, M. 1989. *Geology of the Dorset Coast*. Geologists' Association Guides, **22**.

IGS. 1977. *1:250 000 Wight Sheet*. Insitute of Geological Sciences.

MENPES, R. J. & HILLIS, R. R. 1995. Quantification of Tertiary exhumation from sonic velocity data, Celtic Sea/southwestern approaches. *In*: BUCHANAN, P. G. (eds) *Basin Inversion*. Geological Society, Special Publications, **88**, 191–207.

PENN, I., CHADWICK, R. A., HOLLOWAY, S., ROBERTS, G., PHARAOH, T. C., ALLSOP, J. M., HULBERT, A. G. & BURNS, I. M. 1987. Principal features of the hydrocarbon prospectivity of the Wessex–Channel Basin. In: BROOKS, J. & GLENNIE, K. (eds) *Petroleum Geology of North West Europe*. Graham & Trotman, London, 109–118.

PINET, B., MONTADERT, L., MASCLE, A., CAZES, M. & BOIS, C. 1987. New insights on the structure and the formation of sedimentary basins from deep seismic profiling in Western Europe. *In*: BROOKS, J. & GLENNIE, K. (eds) *Petroleum Geology of North West Europe*. Graham & Trotman, London, 11–31.

SIMPSON, I. R., GRAVESTOCK, M., HAIN, D., LEACH, H. & THOMPSON, S. D. 1989. *In*: COOPER, M. A. & WILLIAMS, G. D. (eds) Inversion Tectonics. Geological Society, London, Special Publications, **44**, 123–129.

STONELEY, R. 1982. The structural development of the Wessex Basin. *Journal of the Geological Society, London*, **139**, 545–554.

STONELEY, R. & SELLEY, R. C. 1992. *Field Guide to the Petroleum Geology of the Wessex Basin*, 3rd edition. Department of Geology, Imperial College of Science and Technology, University of London.

UNDERHILL, J. R. & PATERSON, S. 1998. Genesis of tectonic inversion structures: seismic evidence for the development of key structures along the Purbeck–Isle of Wight Disturbance. *Journal of the Geological Society, London*, in press.

—— & STONELEY, R. 1982. Introduction to the development, evolution and petroleum geology of the Wessex Basin. *This volume*.

SUPPE, J. 1983. Geometry and kinematics of fault-bend folding. *American Journal of Science*, **283**, 684–271.

WILLIAMS, G. D & CHAPMAN, T. J. 1986. The Bristol–Mendip foreland thrust belt. *Journal of the Geological Society, London*, **143**, 63–74.

Fault size distribution analysis – an example from Kimmeridge Bay, Dorset, UK

R. HUNSDALE* & D. J. SANDERSON

Geomechanics Research Group, Department of Geology, University of Southampton, Southampton Oceanography Centre, Empress Dock, European Way, Southampton SO14 3ZH, UK
* Present address: Phillips Petroleum Norway, PO Box 220, Tanager, N-4056 Norway

Abstract: Fault displacement data were measured over four orders of magnitude for a fault set cross-cutting Upper Jurassic rocks exposed along the Dorset coast. Fault data were subdivided into three data-sets, based on data source and field character. Distribution analysis showed that these faults conform to a power law. The scaling relationship is, however, not constant over the entire displacement range of the faults. Faults with displacement >2 m are characterized by a negative power law having an exponent of ≈ 0.96, while faults with displacement <1 m are related by a negative exponent of ≈ 0.7. The change at the small scale is interpreted as being a product of the influence of lithological heterogeneity on fracture initiation and growth.

Many geological features display scale-invariant properties. Although this is not a new concept (Mandelbrot 1977), renewed interest has been generated in geology through the application of the power law distribution model to quantify the relationship between phenomena of a similar type over large scale-ranges. A power law is defined by

$$N = cU^{-D} \quad (1)$$

or in log form

$$\log N = \log c - D \log U, \quad (2)$$

where N is the number of objects with size $\geq U$, D is the power law exponent of the distribution and c is a constant. Unlike other distribution models a power law theoretically extends over an infinite scale-range. As a result, a scale-limited sample from a population that conforms to a power law distribution can be used to predict the number and magnitude of similar features, at other scales.

Fracture characteristics such as displacement (Kakimi 1980; Childs et al. 1990; Jackson & Sanderson 1992; Yielding et al. 1992; Pickering et al. 1994), trace length (Heffer & Bevan 1990; Scholz & Cowie 1990) and aperture (Barton & Zoback 1992) have been shown to obey a power law. The applicability of the power law approach can be examined by looking at fault displacements across a basin as represented on seismic sections. On a commercial seismic section faults with displacements of less than 30–50 m are not resolved. Small-scale faults are, however, important in terms of reservoir characterization and for estimating total strain (Marrett & Allmendinger 1991). If displacement data above the limit of seismic resolution can be shown to follow a power law, then sub-seismic fault frequencies can be predicted by extrapolating the power law distribution of the known population to the small size-range. This additional information, although only an estimate, results in both a more accurate prediction of strain within a basin and also a more realistic idea of how fractured potential reservoirs might be.

In North Sea basins fault displacement data have been shown to be power law over many orders of magnitude (Marrett & Allmendinger 1991; Walsh et al. 1991; Yielding et al. 1992; Pickering et al. 1994). On the south coast of England, at Kimmeridge Bay in Dorset (Fig. 1), a fault set has been identified which allows assessment of fault displacement (throw) over four orders of magnitude (1 mm–11 m). Data collection and methods of analysis will be discussed and data presented in several forms (after Pickering et al. 1995). The applicability of the power law model to fault displacement at Kimmeridge Bay will be discussed and the results compared with similar data sets from the Inner Moray Firth, NE Scotland (Pickering et al. 1994, fig. 1) and from Nash Point, South Wales (Fig. 1). In all data-sets, displacement was measured as the vertical component, or throw of a fault.

HUNSDALE, R. & SANDERSON, D. J. 1998. Fault size distribution analysis – an example from Kimmeridge Bay, Dorset, UK. *In:* UNDERHILL, J. R. (ed.) *Development, Evolution and Petroleum Geology of the Wessex Basin*, Geological Society, London, Special Publications, **133**, 299–310.

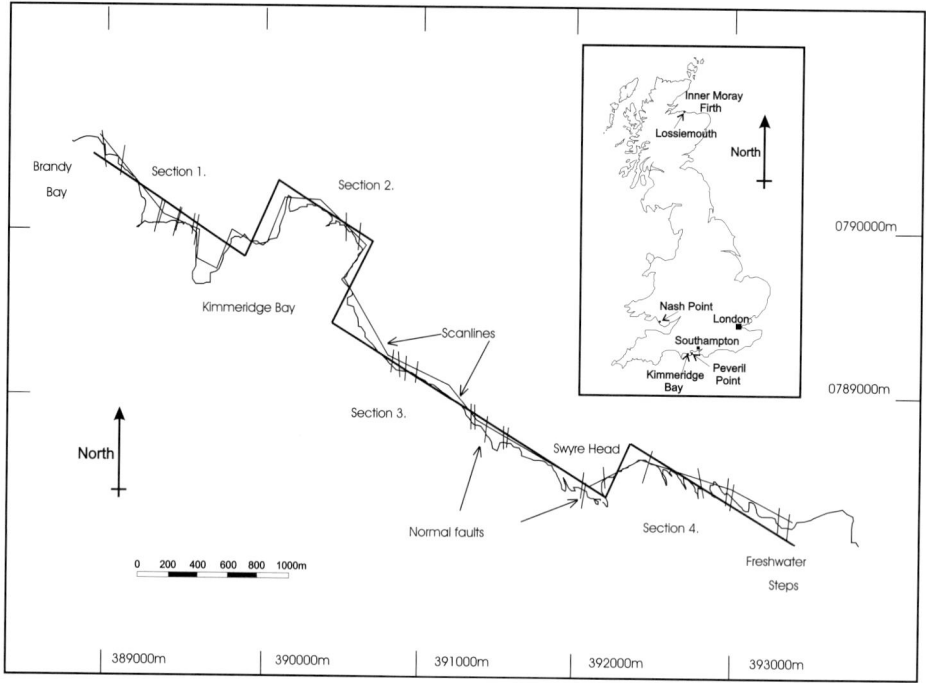

Fig. 1. Locality map showing the nature of the Dorset coast around Kimmeridge Bay and the position of prominent faults, scanline traverses and normalized sections. Inset shows position of Nash Point and the Inner Moray Firth in relation to Kimmeridge Bay.

Lithology and fault geometry

Along the Dorset coast (Fig. 1) rocks of Upper Jurassic age (Kimmeridgian through to Purbeck) are cross-cut by a conjugate set of normal faults (BGS Sheet No. 343, Swanage). These faults have a mean strike of 010° and a mean dip of 56° and are locally restricted to the area south of the Purbeck Monocline. In the adjacent offshore area similar faults can be seen cross-cutting the core of the Purbeck Anticline (Donovan & Stride 1961).

A 5.6 km continuous coastal section, through the Upper Kimmeridgian, was identified (Fig. 1). The succession is dominated by mudstones, unlaminated and laminated shales, within which thin carbonate layers are developed (Cox & Gallois 1981). Laminated shales are more brittle than unlaminated shales, having a high proportion of carbonate cement (Cox & Gallois 1981). Mudstone and shale layers vary in thickness from hundreds of millimetres to tens of metres (Cox & Gallois 1981). Carbonate layers vary from organic-rich cocco-lithic limestones (House 1989) to crystalline diagentic dolostone beds (Irwin et al. 1977).

Carbonate beds, although thin (0.5–2 m), are laterally persistent (Arkell 1947). The coastal section is centred on a shallow ENE-plunging, open upright anticline, which may represent the onshore expression of the offshore Purbeck Anticline to the west (Donovan & Stride 1961; Ramsay 1992).

Fault morphology varies as faults cross-cut the different lithological units of the Kimmeridge Formation. Within mudstone and unlaminated shale units faults comprise a single slip-plane, which often has thin zones of extensional veining associated. Where faults cross-cut laminated shales and carbonate bands the simple slip-plane morphology is lost and a more complex pattern of fracturing is seen. Within laminated shales, faults can be seen to splay and branch, before coalescing into a single slip plane as they cross the next substantial mudstone or unlaminated shale unit. Carbonate layers show an identifiable zone of damage associated with faults which cross cut them. Dolostones display narrow zones of intense hydro-brecciation and extensional veining, whereas some limestones show well-developed zones of extensional veining and faulting. In the latter case, small size faulting

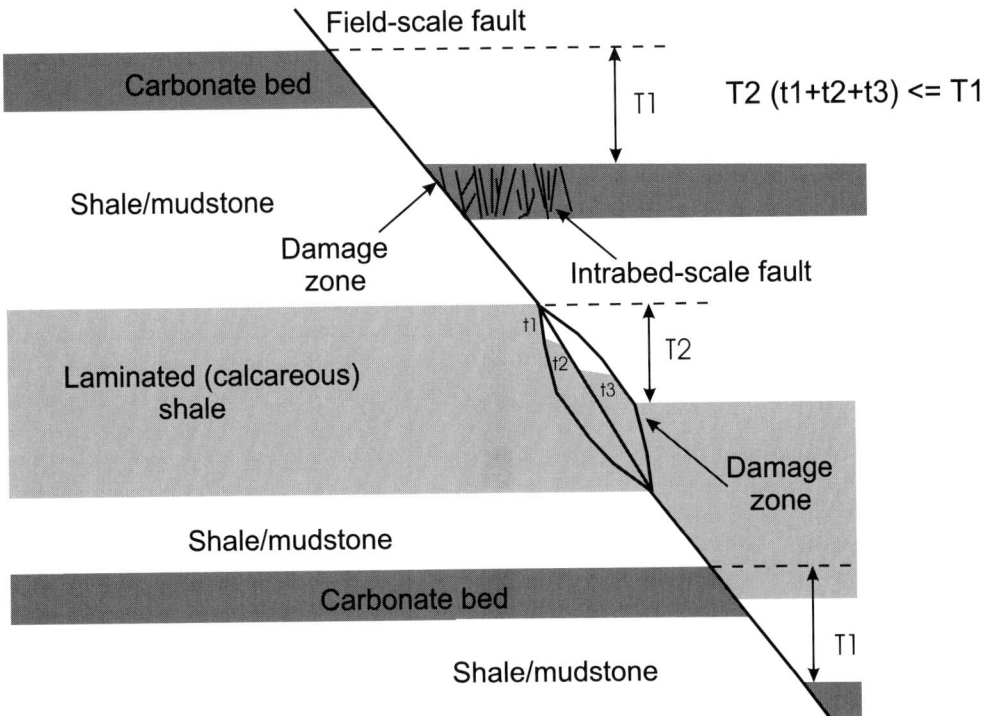

Fig. 2. Schematic representation of fault morphology related to lithology in the Kimmeridge Bay sections. Damage zone development is seen in carbonate layers, while faults may be manifest as single slip planes or as splayed and branched multiple slip planes in laminated shales. The latter arrangement is another expression of a damage zone. Where these occur, the displacement can be distributed over all branches. The total displacement across a fault can be decreased where many branches exist.

is often restricted to the limestone bed. Figure 2 summarizes the change in fault morphology seen at Kimmeridge Bay.

One consistent feature of fault geometry is that zones of damage (extensional veining, brecciation and small-scale normal faulting) are concentrated in the hanging wall of all faults seen cross-cutting the Kimmeridgian. Such a strong association indicates that as faults propagated through Kimmeridgian rocks, tensile stresses were asymmetrically distributed and concentrated to the hanging wall side of faults. A similar distribution of tensile stress was noted by Reches & Lockner (1994). In their experimental modelling of fracture nucleation and propagation they showed that zones of tensile stress developed and were asymmetrically distributed around fault tips (Reches & Lockner 1994, fig. 11). Fault propagation, by the linkage of tensile microcracks, is preceded by a zone of tensile stress, which migrates through the rock in front of the advancing tip. In the case of normal fault geometry the zone of tensile stress would be asymmetrically distributed in front of and to the hanging wall, at the upward-propagating tip of the fault, and in front of the fracture tip and to the footwall, at the downward-propagating tip.

The geometric relationship of extensional veining, small-scale faulting in limestones and brecciation in dolostones at Kimmeridge could be accounted for by such a process of fault tip migration. The association of damage with hanging walls of exposed faults indicates that all faults propagated upwards and that the current coastal section represents a high tectonic window.

A good example of this is seen where faults cross-cut the White Stone Bands, which are a series of three coccolithic limestone marker horizons in the Upper Kimmeridge (Cox & Gallois 1981). The measured section is centred on a broad open anticline and the White Stone Bands are exposed both at the extreme west (Brandy Bay, Fig. 1) and near the eastern end (between Swyre Head and Freshwater Steps, Fig. 1) of the section. Where faults cross-cut the White Stone Bands, discrete zones of small-scale faults and veins occur in the hanging wall (Fig. 2). These faults are developed in the coccolithic limestone and rarely penetrate the over- and under-lying shales. The majority of these faults are strike parallel to the main throughgoing fault and have normal displacements. It should be noted, however, that not all small-size faults are related to damage zones in the hanging wall of larger-size faults. Examples of isolated small-size (1 to tens of millimetres throw) faults can be found along the entire section.

Although tip propagation processes could be invoked to explain localized damage in the hanging wall of larger size faults, it does not account for the more subtle pattern of damage seen as faults cross-cut mudstone–unlaminated shale and laminated shale beds. Within the former lithologies faults are characteristically single slip planes. As faults approach laminated shale horizons the fault plane may refract slightly, becoming steeper, and often become splayed or branched (Fig. 2). This occurs on scales of several metres to several tens of metres and is related to the thickness of the lithologic units the faults cross-cut. Where branching can be identified and throws measured on all branches, the cumulative throw is often less than the throw on the single slip plane beyond the branched fault segment (Fig. 2). At large scales it can be difficult to distinguish branched segments from single slip planes due to the constraints of the coastal section. Branched fault segments as seen at Kimmeridge Bay do not show the geometry of extensional or contractional overstepped fault segments (Peacock & Xhang 1995). They are considered to have formed as a type of damage resulting from the lithological variation seen in the Kimmeridge Formation

Data collection

Data were collected both from map sources and in the field. British Geological Survey Sheet No. 343, the 1:50 000 (Swanage Sheet) was the basis for a map section. A tape and compass method was used to construct 15 scanline traverses (La Pointe & Hudson 1985) along an irregular 5.7 km coastal section (Fig. 1). Data from these traverses were projected onto four sections, orientated at N116°E, that best fit local coastal variation. These sections provide a continuous section through the Kimmeridgian rocks. An orientational weighting method was applied to correct for frequency, due to fault strike variation in a one-dimensional sample (Terzaghi 1965; Peacock & Sanderson 1994). The faults strike, almost orthogonal to the normalized section trend (mean fault strike of 010° and normalized section trend of 116°), resulted in insignificant variation between uncorrected and corrected data.

Combining data from different sources or from different fault sets can disguise the underlying distribution pattern of a sample. To avoid this, data taken from map sources and that gathered in the field were considered separately. In addition, field data were subdivided into faults which clearly cross-cut bedding and those which have an 'intrabed' nature. Although all faults measured on the coastal traverse are geometrically and temporally related, a clear distinction can be made between faults seen cross-cutting the Kimmeridgian stratigraphy and those which have resulted from fault propagation through the White Stone Bands. Three data-sets have been established, as follows.

(i) Map-scale faults

(Displacement range $= 0.3$–11 m, $n = 37$.) Fault displacement data are taken from the 1:50 000 scale Geological Survey Sheet No. 343 (Swanage Sheet). Thirty-seven faults outcrop in a 14.2 km coastal section and cross-cut rocks of the Upper Kimmeridge Clay, west of Kimmeridge Bay, through to the Purbeck Beds, at Peveril Point in the east (Fig. 1).

(ii) Field-scale faults

(Displacement range $= 10$ mm–8 m, $n = 85$.) These faults were measured in a 5.7 km coastal section centred on Kimmeridge Bay (Fig. 1). All faults cross-cut rocks of the Upper Kimmeridge Clay. While collecting field-scale data it proved impossible to distinguish between isolated faults and all branched faults, except at scales where faults branched and coalesced within the vertical cliff section. All fault planes were treated as single entities and a throw recorded for each discontinuity where a displacement could be measured.

(iii) Intrabed-scale faults

(Displacement range $= 1$–800 mm, $n = 145$.) These faults represent damage zones restricted to the hanging wall of six faults from the field-scale fault data-set. All faults are restricted to the White Stone Bands and do not cross-cut the over- and under-lying shales. A lower displacement limit of 1 mm was imposed as measurement of smaller displacements could not be achieved with any confidence in field conditions.

Analysis of data

Cumulative frequency graphs

The most common form of graphical representation used to measure D values is the cumulative frequency graph (Childs *et al.* 1990; Heffer & Bevan 1990; Scholz & Cowie 1990; Walsh *et al.* 1991; Barton & Zoback 1992; Jackson & Sanderson 1992; Yielding *et al.* 1992; Gillespie *et al.* 1994; Pickering *et al.* 1994, 1995). From equation (2), a plot of $\log N$ against $\log U$ should give a straight line with slope $-D$. Natural data-sets that follow a power law distribution approximate a straight line on such graphs.

Data derived from map or field section sources are scale-limited, so degradation is common at the size limits of data-sets. Many authors have noted this (Walsh *et al.* 1991; Barton & Zoback 1992; Jackson & Sanderson 1992) but have not always taken it into consideration when estimating the D value of the sample. At lower sizes

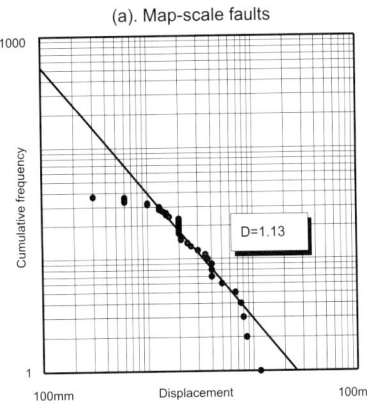

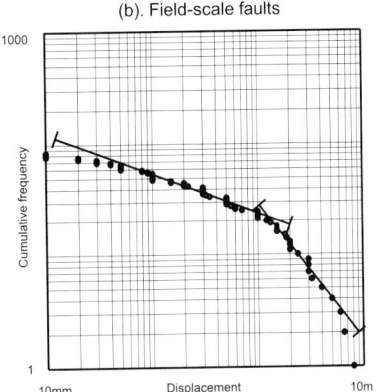

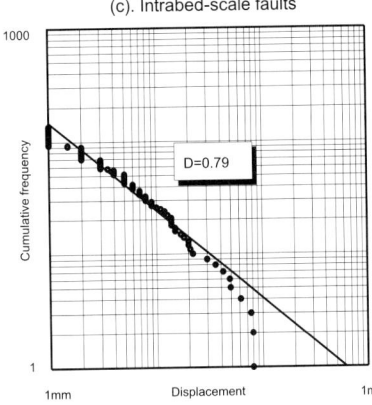

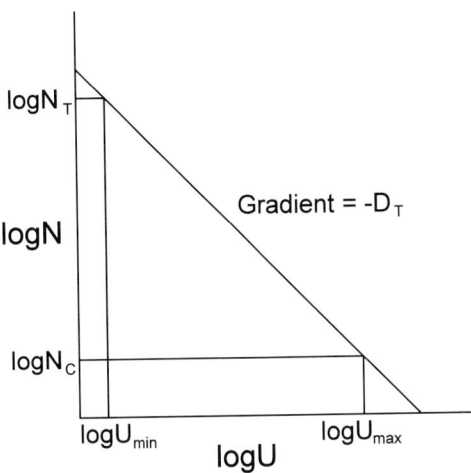

Fig. 3. Cumulative frequency graph of a self-similar sample from a power law distribution with exponent D_T (after Pickering *et al.* 1995) U is in arbitrary units and N is the cumulative number. If a sub-sample from U_{min} to U_{max} is taken, then N must take values from N_C to N_T, but in practice the data will be plotted from 1 to $N_T - N_C$. The geometry of the plot may be used to estimate N_C.

Fig. 4. Cumulative frequency graphs for three subsamples of fault population in Dorset. (**a**) Map-scale faults. (**b**) Field-scale faults. (**c**) Intrabed-scale faults. The map- and intrabed-scale faults indicate power law distributions, with D values of 1.13 and 0.79, respectively. Field-scale faults do not show power law behaviour. This distribution may represent two straight-line segments, which intersect at 1.5 m displacement. No meaningful line can be fitted to this distribution.

there is often a loss of data associated with the lower resolution limit of the sampling technique. For example, in field conditions it is highly unlikely that all faults with throws of <100 mm will be sampled. This deviation is referred to as *truncation*. At the upper size limit of a data-set, degradation can be produced in several ways. In exposure, the hanging wall or footwall cut-off of a large fault may not be visible, leading to a minimum estimate of the throw, or on seismic sections basin-bounding faults may lie beyond the survey limits. Alternatively, as large faults are few in number they have less likelihood of being sampled. Perhaps the most important reason for degradation at the upper size limit is that natural data-sets tend to have such a limit, whereas an ideal power law has no such limit. Deviation at the upper size limit of data has been referred to as the *finite range effect* by Pickering *et al.* (1994, 1995), a term which is employed here. Truncation and finite range effects which modify the shape of a cumulative frequency graph (see Fig. 4a) can influence the estimate of D.

Pickering *et al.* (1995) use computer simulations to quantify the effect of scale limitation on D values calculated on cumulative frequency graphs. Their work shows that truncation has no effect, as long as data below the point of deviation are not included in the calculation of D. However, random samples of a power law distribution, of known D value, over a finite range produces cumulative frequency plots with slopes which systematically overestimate D.

Finite range corrected cumulative frequency graphs

The finite range effect can be corrected using an iterative process that accounts for the under-sampled or 'missing' faults at the upper scale limit of a data-set. Pickering *et al.* (1995) have outlined the full methodology and derivation of the process outlined here. Consider a line representing an exact power law distribution on a log-scale cumulative frequency graph (Fig. 3). The slope of the line is $-D_T$ and the scale range of a sample taken from the distribution is U_{min} to U_{max}. For this sample there should be N_C faults with displacement $>U_{max}$, which may be omitted or beyond the finite range of the population. If U_{max} is plotted with rank N_C rather than rank 1, then the finite range effect should be removed. To make the sample self-similar to the ideal distribution, $(N_C - 1)$ extra faults of displacement $>U_{max}$ need to be added, where N_C is the cumulative number assigned to U_{max} in the original distribution. From Fig. 3

$$\log N_T - \log N_C = -D_T(\log U_{min} - \log U_{max}), \quad (3)$$

where N_T is the sample size plus the correction $(N_C - 1)$. Therefore

$$\log N_C = \log N_T - D_T(\log U_{max} - \log U_{min}). \quad (4)$$

For real data-sets, only U_{min}, U_{max} and N, the number of data points, are known. Pickering *et al.* (1995) argue that if $N_C \approx N_T$ then $N_T \approx N$ and by using an initial estimate D_E for D_T then equation (4) allows an estimate of N_C

$$\log N_C = \log N - D_E(\log U_{max} - \log U_{min}). \quad (5)$$

Finite range correction (FRC) has been successfully used to remove the finite range effect in simulations, where the derived D value can be compared to the known D value, D_T (Pickering *et al.* 1995). The correction N_C is significant if $D < 1$ and the size range of the data is less than two orders of magnitude, conditions which generally apply to field and seismic data. The iterative process outlined in Pickering *et al.* (1995) is used to correct for finite range effect here.

Log-interval graphs

By using a discrete frequency graphing technique the finite range effect can be avoided, as no assumption is made about the nature of the data beyond the upper limit of the range. Determination of a D value, using a linear discrete frequency method, is dependent on the distribution being continuous. It is unlikely that this criterion can be fully satisfied by fault displacement data. Using a log rather than a linear class interval reduces this problem. Log class intervals of 0.1 were used in this analysis. In data-sets where the sample number is small, the log-interval graph can break down if class intervals with zero frequency occur. D cannot be estimated in this case. A full explanation and methodology for this technique is given by Pickering *et al.* (1995).

Experimental work by Pickering *et al.* (1995) shows that cumulative frequency plots tend to overestimate the D value of a sample, while finite range corrected cumulative frequency graphs and log-interval graphs give a more accurate estimation of the true D value. Data-sets from Kimmeridge Bay are presented in all three forms for comparison.

Line-fitting to estimate D

In all three graphing techniques D is estimated by fitting a line to the distribution pattern on the relevant graph. The least-squares method can be used to fit a line to a cumulative frequency graph. This method assumes that any deviations of points from a line have a normal distribution, implying the probability of a point lying a long way from a line is low. Should points lie significantly far from the line, a least-squares line-fitting algorithm will move the line towards those points. If data show a good approximation to a straight-line a least-squares algorithm can be used to estimate D. Where data are erratic a more robust line-fitting technique is needed. Pickering (1995) uses a robust method, which minimizes the absolute deviation of a set of points from a line. This algorithm ignores outlying points as long as most of the data fall close to a linear trend. This method of line-fitting is utilized for log-interval graphs and cumulative frequency graphs with erratic distribution traces.

Estimation of D values

Figure 4 shows the distribution pattern produced by map, field and intrabed-scale fault data-sets on cumulative frequency graphs. The approximate straight-line trend produced by both map-scale (Fig. 4a) and intrabed-scale (Fig. 4c) faults suggests that these data-sets behave in a power-law manner. A D value of 1.13 was obtained from the map-scale data-set. The distribution trace shows a significant amount of truncation below a displacement of 1.5 m (Fig. 4a). An element of finite range effect can also be seen. Intrabed-scale faults produce a D value of 0.79 when plotted on a cumulative frequency graph. The distribution trace shows little truncation, but clearly shows finite range effect (Fig. 4c). In comparison, the distribution pattern produced by the field-scale data (Fig. 4b) shows little resemblance to a straight line. The distribution is almost a convex-up curve, but more closely resembles two roughly straight segments intersecting at 1.5 m. No meaningful estimation of D can be achieved from this graph.

Finite range corrections (FRC) were applied to both the map-scale and intrabed-scale data-sets and they were replotted on cumulative frequency graphs (Fig. 5). The iterative process outlined in Pickering et al. (1995) was used to corrected for finite range effect. For both map- and intrabed-scale data-sets $N_C = 4$ proved optimal. A D value of 0.93 was obtained for map-scale faults (Fig. 5a), while $D = 0.71$ for

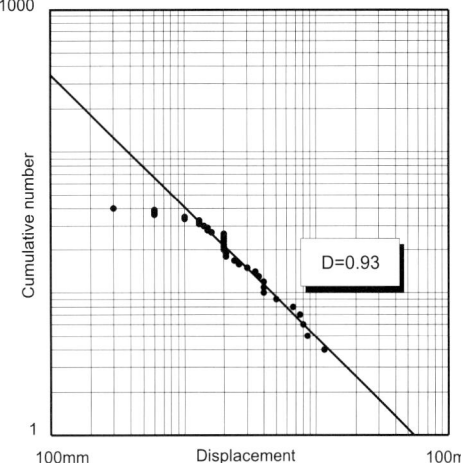

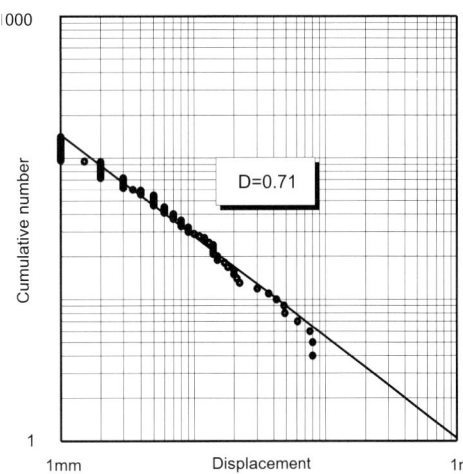

Fig. 5. Finite range corrected cumulative frequency graphs (after Pickering et al. 1995). Applying a finite range correction to both map-scale (**a**) and intrabed-scale (**b**) data reduced the D values of both data-sets. D values of 0.93 for map-scale faults, and 0.71 for field-scale faults were obtained. These are considered more accurate estimates of the power law exponents of the underlying distributions.

intrabed-scale faults (Fig. 5b). The D values of both data-sets are lower than the corresponding uncorrected cumulative frequency graphs (cf. Fig. 4).

The three data-sets were also plotted on log-interval graphs (Fig. 6). Map-scale faults (Fig. 6a) produce a distribution pattern with a point of inflection at 1.5 m, the point at which truncation was observed on the cumulative frequency graph (Fig. 4a). Data points below 1.5 m are therefore not representative of the distribution and were not included in the determination of D, which is 0.94 for the distribution

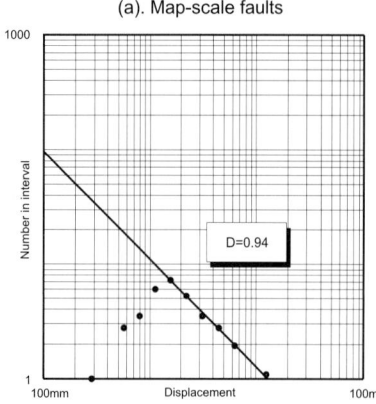

(a). Map-scale faults

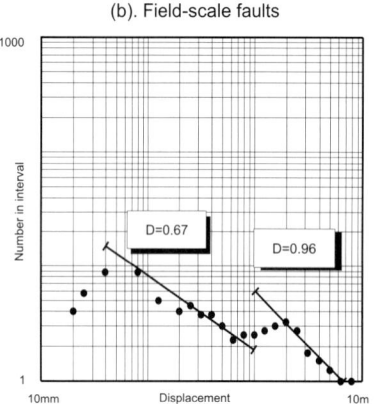

(b). Field-scale faults

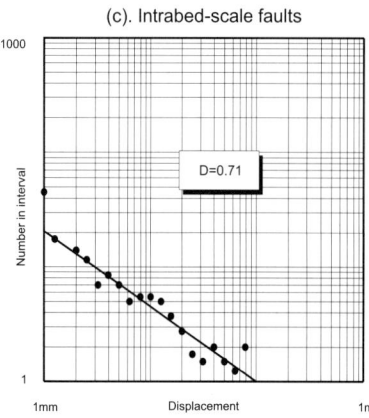

(c). Intrabed-scale faults

points above 1.5 m throw. Intrabed-scale faults also approximate a straight line on a log-interval graph (Fig. 6c), indicating power law behaviour. A D value of 0.71 was obtained. Plotting field-scale data on a log-interval graph (Fig. 6b) produces two straight-line segments. Between 2 and 8 m of displacement the distribution produces a D value of 0.96, while between 80 and 750 mm of displacement the distribution suggests a D value of 0.67. The estimated D values for the various data-sets, using the three graphing techniques, are summarized in Table 1 and are discussed below.

Discussion of the variation in D values

Distribution analysis of fault displacement data for the three data-sets, from Kimmeridge Bay in Dorset, suggest that these faults have displacements that conform to a power law distribution. Faults from the three data-sets are geometrically and temporally related, and all faults formed as a result of the same set of far-field stress conditions. The distributions displayed by the three data-sets however, produce different estimates of D. A cumulative frequency plot of map-scale data produces a D value of 1.13, but a more accurate estimate of ≈ 0.94 is obtained from finite range corrected cumulative frequency and log-interval graphs. These data show a significant amount of truncation because there is a limit to the size of faults that can be represented on a 1:50 000 scale map. Intrabed-scale faults, when plotted on a cumulative frequency graph, also appear to follow a power law, producing a D value of 0.79. This is corrected to 0.71 using the finite range corrected cumulative frequency and log interval graphs.

Fig. 6. Log-interval graphs for the three data-sets at Kimmeridge. Similar D values to those obtained from finite range corrected cumulative frequency graphs are produced for the map-scale (**a**) and intrabed-scale (**c**) data, these being 0.94 and 0.71, respectively. The field-scale data is resolved into two displacement ranges which approximate power-law distributions with different exponents (**b**). Between 2 and 8 m of displacement a D value of 0.96 is obtained, while for the displacement range 80–750 mm, $D = 0.67$. The D value of the upper displacement range is in good agreement with that produced by map-scale data, while the D value for the smaller-scale displacements is similar to that obtained from the intrabed-scale data. (**b**) shows that predictions of the number of faults with displacement 60 mm from the power law relationship of faults with displacement 2–8 m would give an estimate almost one order of magnitude greater than the actual number.

Table 1. *D value determination for Dorset fault sets using cumulative frequency, corrected cumulative frequency and log-interval graphing techniques*

Data-sets from Kimmeridge Bay			D values		
Sample	Scale range	Sample No.	Cumulative frequency graph	Corrected cumulative frequency graph*	Log-interval graph
Map-scale faults	300 mm–11 mm	37	1.13	0.93	0.94
Field-scale faults	2 m–8 m	85	N/A	N/A	0.96
	10 mm–1.5 m		N/A	N/A	0.67
Intrabed-scale faults	1–80 mm	145	0.79	0.71	0.71

*Using finite range correction of $N_C = 4$.

Unlike the map-scale data, intrabed-scale data do not appear to be truncated. This results from the lower scale limit of 1 mm displacement, imposed during sampling. Map-scale and intrabed-scale faults, although having formed under the same stress regime, represent different size ranges. The gap in the size range is bridged by the field-scale data-set. Field-scale fault data when plotted on a cumulative frequency graph do not appear to follow a simple power law relationship, but when plotted on a log-interval graph the data resolves into two straight-line segments (Fig. 6b). Faults with displacements >2 m have a D value of 0.96, while faults with displacement <1 m have a D value of 0.67. A change in the scaling relationship of fault displacement is confirmed.

Some indication of why this change in D occurs may be found by looking at the field relationships of these faults. There are two possible explanations. First, faults in the measured sections at Kimmeridge Bay cross-cut a heterogeneous lithological sequence. Fault initiation and growth are intimately related to the mechanical properties of the rocks. In an isotropic solid fractures at all scales would form and grow in a similar manner. The lithological contrast produced by the layered mudstones, shales and carbonates of the Upper Kimmeridge Clay may, however, influence the fracture process, at least for faults of small size. If this were the case the scaling relationship of the faults would depend on the lithological properties of individual beds within the sequence, up to a point where displacement was large enough for a fault to become independent of such controls. The change in the fault displacement–scaling relationship reflects a change from faults controlled by individual beds to those controlled by the formation as a whole.

Second, the change in D value could relate to the occurrence of damage zones. Two forms of damage zone have been noted at Kimmeridge Bay; both related to specific lithologies. Intrabed-scale faults represent the damage caused by the migration of larger faults through the White Stone Bands, while faults of the field-scale data-set are seen to branch in laminated shale units. The D value for intrabed-scale faults ($D = 0.71$) is significantly lower than that of map-scale faults ($D = 0.94$). While collecting field-scale data isolated faults and branched faults were included in the same data-set. If the scaling of faults is different within damage zones, as suggested by the intrabed-scale data-set, then the inclusion of faults in branched damage zones with isolated faults in the field-scale data-set could produce the distribution pattern seen in Fig. 4b. The similarity between the D value of the intrabed-scale faults and field-scale faults with displacement <1 m ($D = 0.71$ and 0.67, respectively) suggests that the scaling relationships of faults within damage zones is different than that of the through-going, causative faults. Thus, the faults at Kimmeridge Bay behave in a power law manner and are characterized by a D value of 0.96, but small size faults produced in response to the 'damage' caused by these faults are characterized by a different D value of ≈ 0.7.

To assess which of these mechanisms is more probable, comparison is made with other data-sets. Pickering *et al.* (1994) described the fractal behaviour of fault displacements over seven orders of magnitude for a fault set in the Inner Moray Firth. At large displacements, data were taken from seismic sections and a good power law relationship was achieved with a D value of 0.8. Displacement data were measured along the top reflector of a massive Triassic sandstone unit. The same fault set can be seen cutting the Triassic sandstones in the onshore area near Lossiemouth. Faults occur as discrete fractures and in complex zones of granulation seams containing many small-scale displacements (Pickering *et al.* 1994). These field-scale faults also produce a good power law relationship with a D value of 0.8. It would appear that even though the field-scale data comprise both discrete faults and faults within damage zones this arrangement does not effect the scaling relationship of the fault displacements. The important difference between these faults and those at Kimmeridge Bay is that, in the Inner Moray Firth, all fault displacements are measured within a fairly isotropic sandstone, whereas those at Kimmeridge are in a mixed mudstone–shale–carbonate sequence. This suggests that it is lithological variation, rather than incorporation in damage zones, that cause a change in the scaling relationship in small faults.

If lithological variation is indeed responsible for the behaviour described here, then faults that cross-cut similar mixed lithological sequences should also show a change in the displacement–scaling relationship. To test this hypothesis, measurements were made of a set of normal faults, which cross-cut inter-bedded limestone and shale units of the Lower Lias, at Nash Point, South Wales. Here 0.5–2 m thick limestones are inter-bedded with 1–6 m shale units. The resultant distribution pattern is presented on cumulative frequency and log-interval graphs in Fig. 7. On a cumulative frequency graph these data do not conform to a simple power law (Fig. 7a). The distribution trace shows two straight-line segments, intersecting at 0.8 m displacement. This is similar to, though more pronounced than, the pattern produced by the field-scale data at Kimmeridge Bay (cf. Fig. 4b). When plotted on a log-interval graph these data form two straight lines over 0.8–2 m and 80–500 mm displacement ranges (Fig. 7b). The larger size faults (0.8–2 m) produce a D value of 1 while the smaller size faults (80–500 mm) have a D value of 0.67. These values are similar to those produced by the Kimmeridge Bay fault data-sets. The displacement size at which the change in scaling relationship occurs is ≈ 0.8 m at Nash Point and ≈ 1.5 m at Kimmeridge Bay. These data provide strong evidence to suggest that the scaling relationship of small displacement faults is controlled to some extent by variations in lithology. The change in scaling relationship occurs at a lower throw value at Nash Point, where the overall bed thickness is smaller than at Kimmeridge Bay. It could be tentatively suggested that the change in scaling will occur at lower throw values in more thinly bedded sequences, but more data-sets are required to fully establish any empirical relationship between fault throw scaling and bed thickness.

This observation, nevertheless, has important implications when considering the value of power law analysis as a predictive tool in geology. The D value of small-size faults is lower than that for large-size faults which formed under the same far-field stress field. Using the D value of the larger size faults to predict the number of small-size faults in the population would result in an overestimate of the number

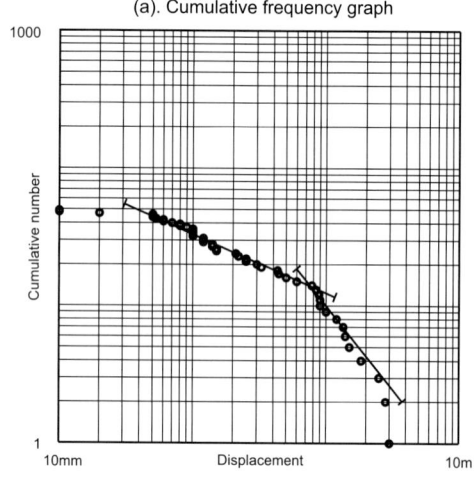

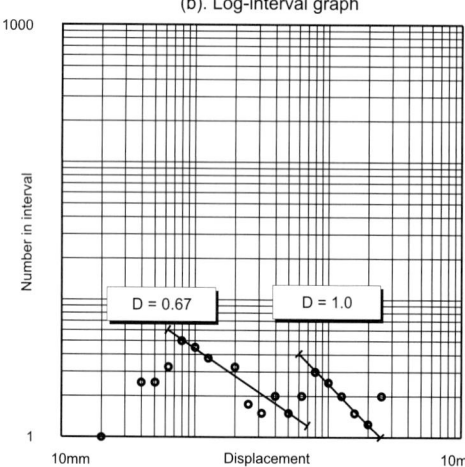

Fig. 7. Nash Point data-set. Forty-nine normal faults were collected from a 2.2 km coastal section in a Lower Liassic interbedded limestone–shale sequence at Nash Point, South Wales. (**a**) Cumulative frequency graph, which suggests that the data do not represent a simple power law distribution. Two straight-line segments occur, intersecting at 0.8 m displacement (cf. Fig. 5b). (**b**) Log-interval graph which resolves the distribution into two displacement scale ranges which approximate power laws with different exponents. Between 0.8 and 2 m a D value of 1.0 is obtained, while between 80 and 500 mm of displacement $D = 0.67$.

of small-size faults. In the case of the faults at Kimmeridge Bay there is an order of magnitude less faults with displacement ≥ 50 mm, than would be predicted by down-scaling the faults with displacement >2 m (Fig. 6b). Consequently, if using this information to assess strain, the resulting estimate would be too high. This observation is, however, only relevant for faults in sequences of heterogeneous lithology. Data for such as those which cut uniform Triassic sandstone in the Inner Moray Firth area maintain a constant scaling relationship for displacement over many orders of magnitude. Changes in scaling relationship also highlight that care is needed when undertaking fault sampling procedures. Scanlines (one-dimensional sample traverses) of limited length, if placed across a damage zone, could produce a D value which was not representative of all faults from the distribution.

Conclusions

Several important conclusions can be drawn from the data presented here.

(1) Fault displacement data from Kimmeridge Bay, Dorset, behave in a power law manner.

(2) The scaling relationship does not remain constant over the size range represented by the data. Faults with displacements >2 m form a power law characterized by $D \approx 0.96$, while faults with displacements <1 m are related by a D value of ≈ 0.7.

(3) The change in the scaling relationship of fault displacement at small size is influenced by lithological variation within the fracturing rock mass. Small-size fault initiation and growth are, to some extent, controlled by the lithological character of the fracturing rock mass. Once a fault grows to a certain displacement size it becomes independent of such controls and conforms to a different scaling relationship.

(4) When looking at faults in mixed lithological sequences, estimates of basinal strain and fracture density, based on the extrapolation of a power law relationship from larger sizes, will be erroneous, because of this change in scaling relationship.

(5) Care must be taken in the placement of one-dimensional (or scanline) samples, especially if the one-dimensional distributions are to later be extrapolated to a three-dimensional volume.

The authors would like to thank G. Pickering for the use of his software and graphing techniques and, together with D. Peacock, for much useful discussion. N. H. Dawers is also thanked for critical review of the initial manuscript. This research was funded by Phillips Petroleum UK Ltd.

References

ARKELL, W. J. 1947. *The Geology of the country around Weymouth, Swanage, Corfe and Lulworth.* Memoir Geological Survey of Great Britain.

BARTON, C. A. & ZOBACK, M. D. 1992. Self-similar distribution and properties of macroscopic fractures at depth in crystalline rock in the Cajon Pass scientific drill hole. *Journal of Geophysical Research*, **97**, 5181–5200.

CHILDS, C., WALSH, J. J. & WATTERSON, J. 1990. A method for estimation of the density of fault displacements below the limits of seismic resolution in reservoir formations. *In:* BULLER, A. T. (ed.) *North Sea Oil and Gas Reservoirs II.* Graham & Trotman, London, 309–318.

COX, B. M. & GALLOIS, R. W. 1981. *The Stratigraphy of the Kimmeridge Clay of the Dorset Type Area and its Correlation With Some Other Kimmeridgian Sequences.* Report of the Institute of Geological Sciences, No. 80/4.

DONOVAN, D. T. & STRIDE, A. H. 1961. An acoustic survey of the sea floor south of Dorset and its geological interpretation. *Philosophical Transactions of the Royal Society of London*, **244**, 299–330.

GILLESPIE, P. A., HOWARD, C. B., WALSH, J. J. & WATTERSON, J. 1994. Measurement and characterization of spatial distributions of fractures. *Tectonophysics*, **226**, 113–141.

HEFFER, K. & BEVAN, T. 1990. Scaling relationships in natural fractures – data, theory and applications. *Proceedings of the European Petroleum Conference*, vol. 2 367–376 (SPE paper No. 20981).

HOUSE, M. 1989. *Geology of the Dorset Coast.* Geologists' Association Guide. 2nd edition.

IRWIN, H. CURTIS, C. & COLEMAN, M. 1977. Isotopic evidence for source of diagenetic carbonates formed during burial of organic-rich sediments. *Nature*, **269**, 209–213.

JACKSON, P. & SANDERSON, D. J. 1992. Scaling of fault displacements from the Badajoz–Cordoba shear zone, SW Spain. *Tectonophysics*, **210**, 179–190.

KAKIMI, T. 1980. Magnitude–frequency relation for displacement of minor faults and its significance in crustal deformation. *Bulletin of the Geological Survey of Japan*, **31**, 486–487.

LA POINTE, P. R. & HUDSON, J. A. 1985. *Characterization and Interpretation of Rock Mass Joint Patterns.* Geological Society of America Special Paper **199**.

MANDELBROT, B. B. 1977. *Fractals: Form, Chance and Dimension.* Freeman, San Francisco.

MARRETT, R. & ALLMENDINGER, R. W. 1991. Estimates of strain due to brittle faulting: sampling of fault populations. *Journal of Structural Geology*, **13**, 735–738.

PEACOCK, D. C. P. & SANDERSON, D. J. 1994. Strain and scaling of faults in the chalk at Flamborough Head, U.K. *Journal of Structural Geology*, **16**, 97–108.

—— & ZHANG, X. 1994. Field examples and numerical modelling of oversteps and bends along normal faults in cross-section. *Tectonophysics*, **234**, 147–167.

PICKERING, G. 1995. *An Analysis of Fault Scaling in the North Sea*. PhD thesis, University of Southampton.

——, BULL, J. M. & SANDERSON, D. J. 1995. Sampling power-law distributions. *Tectonophysics*, **248**, 1–20.

——, ——, —— & HARRISON, P. V. 1994. Fractal fault displacements: a case study from the Moray Firth, Scotland. *In*: KRUHL, J. H. (ed.) *Fractals and Dynamics Systems in Geosciences*. Springer, Frankfurt.

RAMSAY, J. G. 1992. Some geometric problems of ramp-flat thrust models. *In*: MCCLAY, K. (ed.) *Thrust Tectonics*. Chapman-Hall, London, 191–200.

RECHES, Z. & LOCKNER, D. A. 1994. Nucleation and growth of faults in brittle rocks. *Journal of Geophysical Research*, **99**, 159–173.

SCHOLZ, C. H. & COWIE, P. A. 1990. Determination of total strain from faulting using slip measurements. *Nature*, **346**, 837–839.

TERZAGHI, R. D. 1965. Sources of errors in joint trace surveys. *Geotechniques*, **15**, 287–304.

WALSH, J. J., WATTERSON, J. & YIELDING, G. 1991. The importance of small scale faulting in regional extension. *Nature*, **351**, 391–393.

YIELDING, G., WALSH, J. J. & WATTERSON, J. 1992. The prediction of small-scale faulting in reservoirs. *First Break*, **10**, 449–460.

Kinematics of the Watchet–Cothelstone–Hatch Fault System: implications for the fault history of the Wessex Basin and adjacent areas

M. MILIORIZOS[1] & A. RUFFELL[2]

[1] *Department of Geology, University of Wales, College of Cardiff, PO Box 914, Cardiff CF1 3YE, UK*
[2] *Department of Geology, School of Geosciences, Queen's University, Belfast, Northern Ireland, UK*

Abstract: The Watchet–Cothelstone–Hatch Fault (WCHF) comprises a system of northwest–southeast-trending basement and cover faults that are traceable for at least 40 km from the Bristol Channel, southeast into the western Wessex Basin. The west-dipping WCHF displays outcrop and seismic evidence of a complex movement history involving early (Variscan) and late (?Cretaceous or Tertiary) strike-slip, separated by phases of normal extension. In the Variscan basement of North Somerset, the WCHF shows a 14–16 km dextral offset to Devonian markers and Variscan fold axes, and can be linked to similar trending faults in South Wales. Offsets to early Permian (and possibly late Carboniferous) rifts indicate a dextral transfer of north–south extension from the Crediton and Tiverton troughs, across the WCHF and into the Wessex Basin. In Mesozoic cover, the WCHF swings to become part of the east–west Central Bristol Channel Fault Zone in the north and Lopen Fault Belt (western Wessex Basin) in the south. This geometry reflects a present-day preservation of a later sinistral movement. The structure of juxtaposed Palaeozoic and Mesozoic successions along the WCHF provides an analogue for what might be expected at the basement–cover contact beneath the Wessex Basin. The existence of Mesozoic cover to the WCHF most likely explains why no Cenozoic pull-apart basins were formed, unlike the Sticklepath Fault (in granites and metamorphic rocks) to the southwest.

The Northwest European plate is dissected by a number of northwest–southeast strike-slip faults (references below). These structures are known to have accommodated regional compression during both the Variscan (Carboniferous–Permian) and Alpine (late Cenozoic) mountain-building phases. Variscan strike-slip was dextral in origin, resulting from the collision of the Palaeozoic African and European plates (Badham 1982). Early Cenozoic compression along the Lustleigh–Sticklepath Fault system in Devon is suggested by Holloway & Chadwick (1986) to have been dominated by sinistral motion, followed by Alpine (Miocene) dextral reactivation. In this study we aim to compare the kinematic history of the WCHF to the Lustleigh–Sticklepath Fault and to the extension–inversion history of the Wessex Basin, through examination of the effects on Permian and Mesozoic strata.

Extensional reactivation of many northwest–southeast faults occurred during the Mesozoic (Coward 1990). Consequently, complex structures and continuous phases of movement are known (Whittaker 1972, Holloway & Chadwick 1986, Smithurst 1990). The above authors cite strike-slip as the dominant fault movement, whilst Shearman (1967) attempted integration of the known strike-slip history with north–south extension, resulting in a model of transtension across northwest–southeast faults. Although numerous northwest–southeast faults are documented from basement terrains across southwest England, relatively few are exposed that effect both the Variscan basement and overlying Mesozoic cover. As the WCHF affects both Upper Palaeozoic and Mesozoic rocks at its present level of exposure, it provides a unique insight into the possible geometry and kinematics of other northwest–southeast faults in the basement to the Wessex Basin. The Lustleigh–Sticklepath Fault system in South Devon and North Cornwall displaces a deformed and metamorphosed Devonian–Carboniferous succession and was also the site of Cenozoic pull-apart basin formation (Bovey, Petrockstow and Stanley Bank basins: Arthur 1989). The Watchet–Cothelstone–Hatch system of faults elucidated by Whittaker (1972) displaces deformed Devonian–Carboniferous strata. Conversely, ?Permian, Triassic and Jurassic strata are also affected in a dextral sense (from outcrop displacements), yet no Cenozoic pull-apart basins are preserved along the line of the WCHF. The known history of the WCHF includes post-Carboniferous (Webby 1965a) and post-Jurassic (Whittaker 1972) dextral strike-slip. Mesozoic extension in the Wessex Basin is known to be north–south (Chadwick 1985). The conclusion may be drawn that the WCHF was active as a strike-slip fault at discrete times only. Outcrop evidence from the Quantock area (BGS sheets 295 and 211), shows extension across the fault in addition to the effects of strike-slip (Fig. 1). Further to comparing the WCHF to other northwest–southeast zones of transfer in the Wessex Basin, we aim in this paper to summarize to main phases of movement by integrating outcrop, shallow borehole and seismic data. The results have direct significance for structures that underlie the Wessex Basin to the east.

Previous work

This section is intended as a reappraisal and synthesis of relevant previous work describing the WCHF and possible tectonically linked faults in the Bristol Channel and Western Wessex Basin (Somerset).

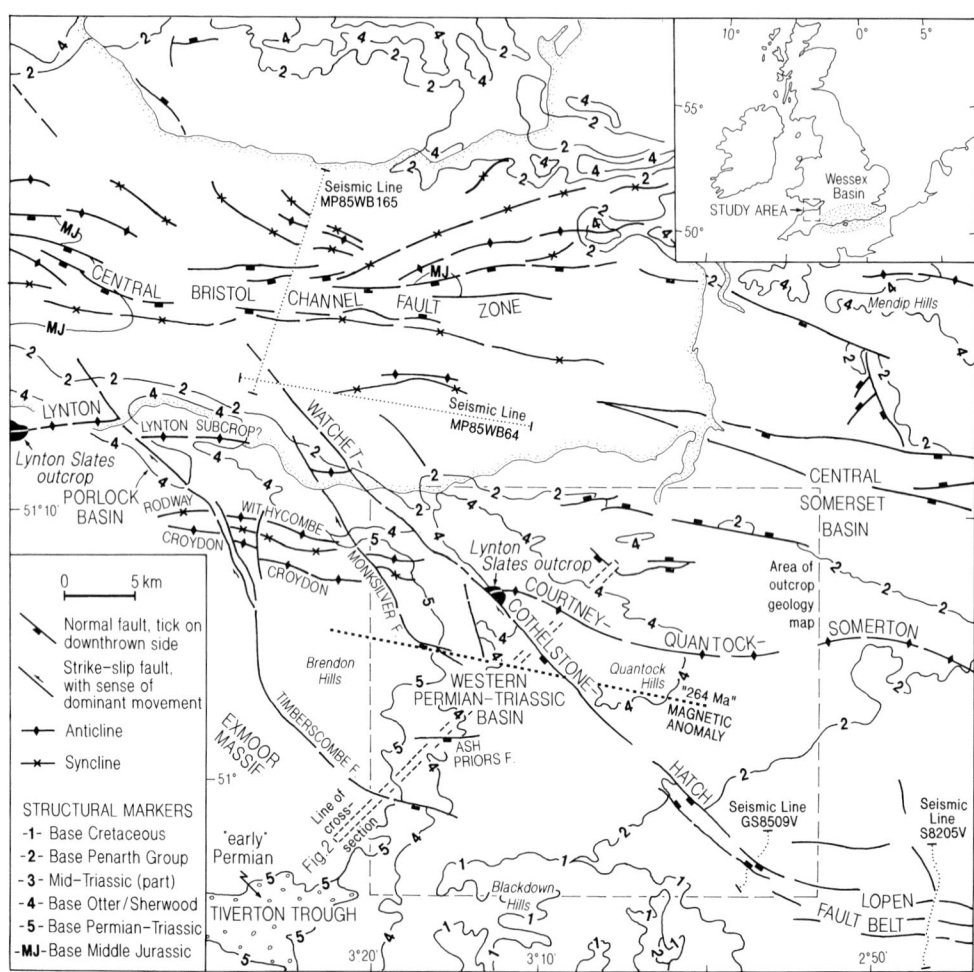

Fig. 1. Present-day surface structure map along the Watchet–Cothelstone–Hatch Fault (WCHF) Zone, as defined by the deformation observed between the Central Bristol Channel Fault Zone to the north and Lopen Fault Belt to the south. The outcrop and offset of key structural markers and location of figured seismic lines are shown. The top 'early' Permian of the Tiverton Trough (stipple) is taken arbitrarily at the base of the Thorverton Sandstones (Scrivener & Edwards 1990). Data from: British Geological Survey 1:250 000 Series Continental Shelf maps; Edmonds & Williams (1985a, b) and Webby (1965a, b). Area of Fig. 2 indicated; Structural marker (3: mid-Triassic) is included in the key for reference to other diagrams.

Western Wessex Basin (Somerset)

The northwest–southeast-oriented boundary between deformed Devonian rocks of the Quantock Hills and Permian–Triassic sediments to the southwest in North Somerset is abrupt and was suspected as representing a faulted contact during early geological studies of the area (De la Beche 1839; Etheridge 1867; Woodward 1893). This structural interpretation was confirmed through mapping by Ussher (1908). Earlier in 1906 Ussher had observed a similar northwest–southeast-trending contact between the Lias and Keuper Marl (= Mercia Mudstone Group), to the southeast and on-strike with the WCHF of the Quantock Hills (Ussher 1906). Here, Ussher noted associated deformation of the Lias around Hatch Beauchamp, thus the fault here was later to become known as the Hatch Fault (Whittaker 1972).

The structure of the Quantock Hills was analysed by Webby, who described the southwestern Devonian–Permian–Triassic faulted contact as the Cothelstone Fault (Webby 1965a, p. 328). Webby regarded the fault as a complex feature, displaying both normal and strike-slip senses of movement. This is critical to the present study, as Webby's (1965a) estimates of horizontal displacement have been cited subsequently, with no critical appraisal (such as is offered here). Webby's (1965a) method was to imply that the Cothelstone Fault exhibited similar horizontal to vertical senses of movement to the Timberscombe Fault System in the Brendon Hills, also studied by Webby (1965b). Using the offset to the Ilfracombe Beds (= Slates)–Morte Slates contact on the eastern side of the WCHF (estimated to dip 25° south), Webby (1965a, p. 339) suggested an "actual dextral displacement of just over three miles" (approximately 4.8 km). Webby's (1965a, b) works indicate that the WCHF is one of a series of northwest–southeast-trending Variscan strike-slip faults in the Quantock–Exmoor Upper Palaeozoic succession. The three main faults (as shown on Fig. 1) may be summarized in order of their importance to this study. (1) The WCHF displays the greatest Variscan strike-slip component (at least 4.8 km), can be traced furthest (at least 40 km) and is adjacent to the most areally extensive Permian–Triassic basin (the Western Permo-Triassic Basin). (2) The Timberscombe Fault (Fig. 1) is the next largest of the three: it shows a Variscan strike-slip movement of 2 km and has a small (7 km^2) Permian–Triassic basin (the Porlock Basin, adjoining the Bristol Channel) associated with it. The easterly-dipping Permian–Triassic of the Porlock Basin gives it a small yet similar half-graben geometry to the Western Permian–Triassic Basin and WCHF to the east. The Timberscombe Fault swings into an east–west orientation as its southern strand passes into the Permian and Mesozoic of the western Wessex Basin. (3) The smallest strike-slip fault, the Monksilver Fault (Fig. 1), shows minimal displacement of Devonian marker beds and only preserves a Permian–Triassic outcrop 300 m in diameter (Wiveliscombe Sandstone exposed at Elworthy on Exmoor at ST 086347).

The above observations were all made by Webby (1965a, b). Their synthesis here allows us to conclude that the larger the fault (length, throw, horizontal displacement) then the thicker and more areally extensive the associated Permian–Triassic basin. This analogy between fault 'size' (see above) and Permian–Triassic basin development may be useful in predicting the kinematics of the WCHF fault system at depth. We may observe in Fig. 1 the pronounced swing of the Timberscombe Fault from northwest–southeast to east–west in its southern strand. If a minor fault of the tripartite northwest–southeast-trending array in this area shows such a swing, we might anticipate the WCHF or Monsilver Fault to behave similarly. As the Monksilver Fault does swing to become the Ash Priors Fault of the Western Permian–Triassic Basin, perhaps a similar geometry occurs in the WCHF beneath the Mesozoic cover of the western Wessex Basin. Below, we examine the intersection of the WCHF with both the east–west Lopen Fault Belt of the Wessex Basin and the Central Bristol Channel Fault Belt to the north in light of this predicted tectonic swing (Fig. 1).

The effect of the Cothelstone Fault on the cover succession (Triassic–Jurassic) of North Somerset was discussed by Whittaker (1972). Northwest of the Quantock Hills, Whittaker (1972) located a similar fault at Watchet, on the North Somerset coast. Distinction of this feature from the Cothelstone Fault led Whittaker to term it the Watchet Fault, using displacements between the Lias and uppermost Triassic strata to estimate both reversed and strike-slip movement. The reversed geometry of the Watchet Fault was summarized by Whittaker (1972) as dipping 55° to the southwest, with horizontal (post-Liassic strike-slip) displacement of 275 m. Whittaker (1972) discussed the southeasterly continuation of this fault along the Quantock Hills, under the Vale of Taunton Deane to Ussher's (1906) mapped offset to the Lias–Mercia Mudstone Group at Hatch (Fig. 1). Whittaker (1972) noted the difficulty in tracing the WCHF through alluvium-covered Mercia Mudstone Group around Taunton. This problem is confirmed by making a comparison of the different traces of the fault made by Edmonds & Williams (1985a), who place a distinct curve around Taunton, to Ruffell (1990) who preferred a straight trace, with an antithetic normal fault to the southwest. Edmonds & Williams (1985a) based their fault line on outcrop mapping, whilst Ruffell (1990) used the location of borehole displacements in estimating the line of the fault.

Tracing the WCHF in basement beneath Mesozoic cover of the Wessex Basin to the south is also problematic. Whittaker (1972) speculated on a link to the southeast with the southern margin of Drummond's (1970) 'Mid-Dorset Swell', a northwest-trending axis of mid-Cretaceous uplift. Workers on the Variscan basement structure of southern Britain similarly appear to favour the continuation southeast of the WCHF to the Dorset Coast. Thus, Ziegler (1982), Edmonds & Williams (1985b), Beach (1987) and Coward (1990) depict the Variscan Northwest European plate as being dissected by northwest–southeast zones of transfer, including the WCHF, which is shown by all those cited above as continuing into the English Channel. Views on the shallow expression of the WCHF vary. Whilst Whittaker (1972) linked Cretaceous structures on-strike to the WCHF to the south-east, Shearman (1967) described (from theoretical modelling) how northwest–southeast basement strike-slip faults could accommodate east–west extension in the cover above. This view of the WCHF was supported by Chadwick (1985) who depicted a northwest–southeast basement structure, en echelon to the east–west Cranborne–Fordingbridge High in the Mesozoic–Cenozoic cover. Such a 'splay' from northwest–southeast basement to east–west cover was suggested by Ruffell (1990) to occur within the Triassic section southeast of Taunton, using evidence from the seismic data presented here.

Bristol Channel Basin

The probable outcrop of Mesozoic strata on the floor of the Bristol Channel was suggested by Ussher (1908) from extrapolation of onshore successions. This was confirmed by later dredging and shallow coring (Lloyd et al. 1973). The structure of both the Mesozoic and Palaeozoic rocks beneath the Bristol Channel has required geophysical experiments (Bott et al. 1958; Brooks & James 1975; Brooks & Al-Saadi 1977) and more detailed geophysical surveys (Kammerling 1979; BIRPS & ECORS 1986;

Brooks et al. 1988). The results of these extensive investigations include the likelihood that the Palaeozoic rocks are deformed by major thrusts that show extensional reactivation in the Mesozoic. Most significantly for this study, Brooks et al. (1988) show the extensive network of northwest–southeast-trending strike-slip faults dominating Mesozoic structure in the Bristol Channel. Tracing the path of the WCHF system through the Bristol Channel proves difficult in Mesozoic strata due to the intersection of the east–west Central Bristol Channel Fault Zone (CBCFZ; after Brooks et al. 1988). In Carboniferous–Devonian basement, northwest–southeast-trending strike-slip faults are common throughout the South Wales Coalfield (Gayer & Jones 1989; Hathaway & Gayer 1994). Similar faults occur on the coast of South Wales, including the Nash Point Fault (Glamorgan) which lies on strike with the WCHF at its last onshore outcrop in North Somerset (Fig. 1).

The intersection of the WCHF with the CBCFZ in the Bristol Channel and with the Lopen Fault Belt in the Wessex Basin suggests a common origin for all three features: Shearman's (1967) model of extensional transfer might apply equally to the Bristol Channel as to the Wessex Basin. In summary, similar northwest-trending faults occur on both sides of the Bristol Channel: the Watchet Fault of the WCHF system to the south, and the Nash Point Fault to the north. As the entire WCHF seems unlikely to lose displacement completely beneath the Bristol Channel, what is the relationship between the fault and Mesozoic cover? As there is only limited exposure of Mesozoic rocks in the Bristol Channel borderlands, an answer must be sought in the seismic database across the area. Thus, the outcrop geology of the WCHF and its link with the seismic expression of the fault onshore to the south can be used in comparison with the offshore continuation to the north. The geographic focus of this onshore history is where the fault is exposed – in North Somerset.

Onshore outcrop geology of the WCHF

The structure of the North Somerset area is shown in Fig. 1 with names of individual features derived largely from the works of Webby (1965a, b), Edmonds & Williams (1985a) and Donato (1988).

Devonian–Carboniferous

Webby (1965a, b) defined a series of northwest–southeast faults dissecting the Devonian massifs of the Brendon and Quantock Hills. These all displayed a dextral strike-slip sense of movement with respect to Devonian marker surfaces (fold-axial traces; individual beds). Webby (1965b) conducted a detailed analysis of the Timberscombe Fault System in the Brendon Hills, using the offsets to localized fold axes to assess horizontal displacements. Webby (1965a) used this previous analysis of horizontal v. vertical components in comparison with the WCHF, considering a dextral movement on the WCHF of 4.8 km likely. This displacement was used by Edmonds & Williams (1985a, fig. 14) to project the offset of the Quantock (= Courtway of Webby, 1965a) Anticline 5 km to the northwest, beneath Mesozoic cover in the area north of Bicknoller (ST 105 400).

Edmonds & Williams projection (1985a) implies a match of the Quantock Anticline with Webby's (1965b) Withycombe Anticline of the Brendon Hills Devonian. In this study we have used a comparison of the Quantock–Brendon Hills Devonian displayed on both the BGS 1:250 000 Bristol Channel Sheet, and 1:50 000 sheets of the area. These suggest that correlation of the Quantock Anticline with the Withycombe Anticline is unlikely: the east-plunging Quantock Anticline brings Lynton Slates to surface in its core around Triscombe Quarry (ST 156 354), yet the same stratigraphic horizon is only brought to surface west of the WCHF in the Devon Brendon Hills by the east-plunging Lynton Anticline. Analysis of the variations in dip across these fold-axial traces also suggests that the Quantock–Lynton Anticline is the dominant fold of the area (= anticlinorium), with such folds as the Rodway Syncline and Withycombe–Croydon anticlines being parasitic to the major structure. The Quantock Anticline of the Devonian plunges east beneath Mesozoic cover in the Central Somerset Basin (Fig. 1) where a low-amplitude fold, the Somerton Anticline, can be found deforming Mesozoic strata. The Lynton Anticline also plunges east in North Devon–Somerset.

Correlation of the Quantock with the Lynton Anticlines (*sensu stricto*) indicates a dextral offset between the Devonian structural locations of Lynton and Triscombe in excess of 18 km. A similar correlation was suggested by Donato (1988). This is a cumulative offset produced by movement on the Timberscombe, Monksilver and WCH faults. Horizontal offsets in the Timberscombe and Monksilver Faults can be deduced from the locally displaced fold-axial traces (the Rodway, Withycombe and Croydon, used by Webby 1965b). These minor strike-slip faults account for between 2 and 4 km of dextral displacement. Removal of these offsets allows a sub-Mesozoic projection of the Lynton Anticline beneath the Watchet area. This leaves a residual 14–16 km dextral displacement to the Variscan Lynton–Quantock anticlines' axial traces along the WCHF.

Webby (1965a) assumed a vertical and horizontal displacement on the WCHF to produce his 4.8 km offset to the projected (sub-Mesozoic) Ilfracombe (= Leighland)–Morte Slates junction. Measurement of the total horizontal offset caused to this junction by the WCHF can be made from near Cothelstone in the Quantocks (ST 188 315) to Stogumber in the Brendon Hills (ST 099 379). With no account made for vertical displacement of this dipping contact, the horizontal displacement may be conservatively estimated at 10–12 km. The latter figure is only 2 km less than that derived from the Quantock–Lynton anticlines' offset. If we may conjecture that the sub-cropping Lynton Anticline (beneath the North Somerset coast) bends in a southerly direction, an exact match of the two measured offsets to the major fold-axial traces may be possible. These two measurements (displaced major fold axes and stratigraphic boundaries), compared to the semi-quantitative estimate of Webby (1965a), may be taken as evidence in favour of a much greater post-Devonian dextral strike-slip offset to the WCHF than previously thought. The Lynton–Quantock anticlines most likely formed synchronously to dextral strike-slip, although their upright attitude implies limited overturning.

In the southern part of the Quantock Hills ENE–WSW-trending lamprophyre dykes intrude Devonian metasediments: they are considered to be Variscan in age (264 ± 36 Ma; Edmonds & Williams 1985a,b). Unfortunately, no similar dykes have been observed west of the WCHF, thus no direct assessment of post-intrusive movement can be made. However, Edmonds & Williams (1985a) report a ENE–WSW linear magnetic anomaly coincident with the lamprophyre intrusions in the Quantocks and mineral veins in the Brendon Hills, west of the WCHF (Figs 1 and 11). This linear feature shows no apparent offset across the WCHF; certainly no post-intrusive strike-slip occurred on a similar scale to that affecting the Quantock–Lynton axial trace (14–16 km). Most of the Variscan strike-slip movement on the WCHF took place between the formation of folds in the Devonian ('early Variscan' and intrusion of the linear magnetic dykes and mineral veins ('late Variscan').

Permian and Mesozoic

Variscan deformation of Devonian–Carboniferous strata exposed in the Brendon and Quantock Hills resulted in the folding and strike-slip tectonics discussed above. This may have been accompanied by thrusting from the south (see the Bristol Channel section, above). Late Carboniferous–early Permian extension in the Crediton and Tiverton Troughs allowed fault-controlled accommodation of red-beds that continued into the Triassic. The lowest Permian–Triassic sediments preserved in the study area are the breccias and sandstones of the east–west-trending Crediton and Tiverton Troughs (Fig. 1). These strata are partially underlain by, and interbedded with, igneous rocks of the Exeter Volcanic Series (estimated to be 291 ± 6 Ma by Forster & Warrington 1985). The infill of the Crediton–Tiverton troughs is onlapped by the Wiveliscombe Sandstones; beds that rest unconformably on the Devonian of the Brendon Hills and thus representing the earliest sedimentation of the Western Permian–Triassic Basin. This relationship (Fig. 1) indicates that the east–west Crediton and Tiverton troughs formed earlier than the northwest–southeast Western Permian Triassic Basin. Biostratigraphic control of all Permian–Triassic strata in the study area is limited, and the timing of associated movements on the WCHF is imprecise. This may not be critical, as by Permian–Triassic times the Devonian massif of the Quantock Hills represented a fault-controlled high (the Taunton High of Donato 1988). East–west cross-sections through the Quantock Hills (Edmonds & Williams 1985a; this study, Fig. 2) confirm the two-dimensional geometry as a simple half-graben. The western infill of this half-graben was termed the 'Western Permo-Triassic Basin' by Edmonds & Williams (1985a). This basin was largely infilled by late Triassic–early Jurassic times, as is evident from the onlap of Sherwood Sandstone (and equivalents) and Mercia Mudstone onto Upper Palaeozoic basement. Whittaker (1975) recognized continued Liassic thinning over the northern extension of the Quantock Massif, indicating the persistence of this positive structure into early Jurassic ('post-rift') times. The Pre-*planorbis* Beds (Triassic–Jurassic) exposed at the type section of St. Audries Bay (3 km east of Watchet) lie in the core of the Western Permian–Triassic Basin: this might be why these beds attain their maximum thickness here, compared to the rest of the British Isles (Cope *et al.* 1980).

Present-day mapped offsets to Mesozoic marker beds can be estimated from the BGS Taunton (295) and Wellington (211) 1:50 000 geological maps (Fig. 3), allowing comparison to Devonian offsets derived from areas to the north. These may be summarized as the following present-day, purely horizontal, offsets (i.e. no strike-slip is yet implied).

(1)	Base Cretaceous Blackdown Greensand (Albian–Cenomanian)	not observed
(2)	Base Penarth Group (approx. Triassic–Jurassic boundary)	2.6 km offset

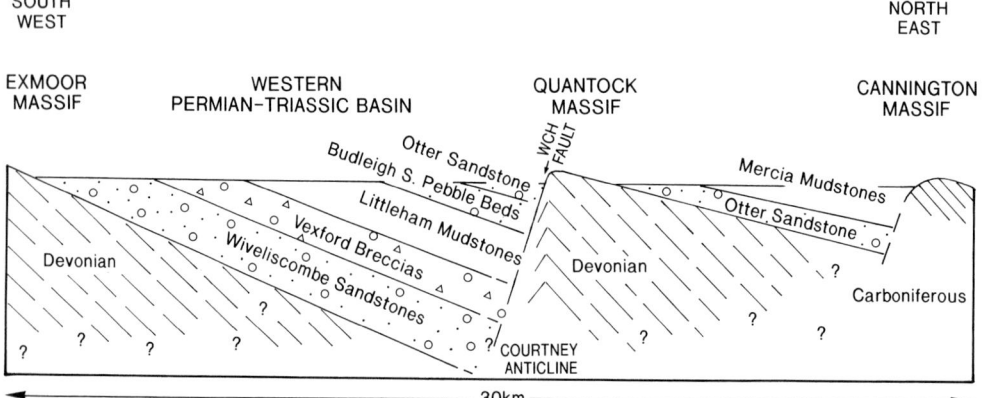

Fig. 2. Southwest–northeast cross-section (derived from outcrop alone) across the WCHF (location on Fig. 1) showing Permian–early Triassic syn-rift type deposition in the hanging wall (West Somerset or Western Permo-Triassic Basin) to the WCHF. The steep nature of the WCHF in this location is indicated by its straight outcrop in the slopes of the SW Quantock Hills. Minor faults are omitted. Grossly exaggerated vertical scale.

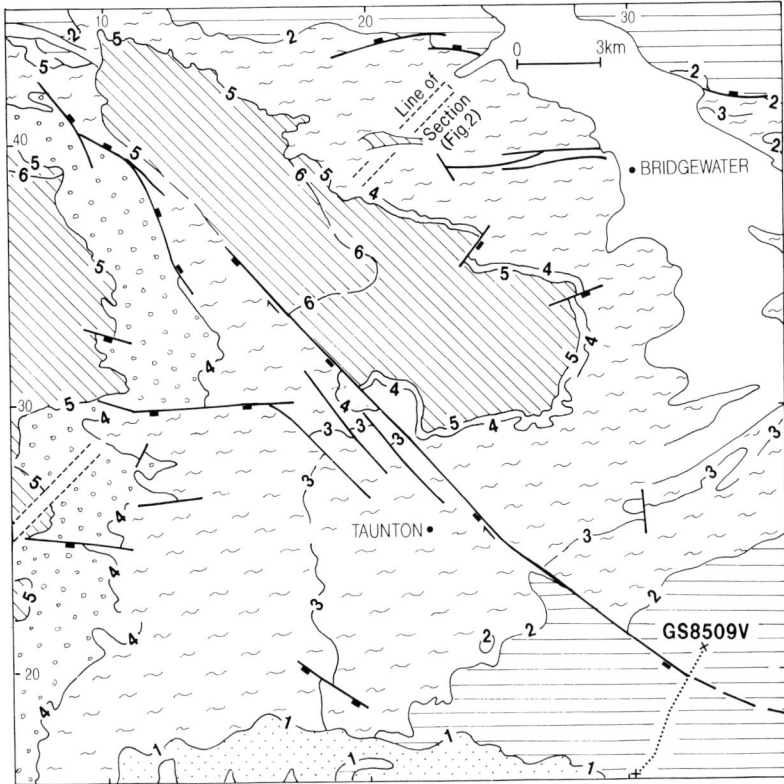

Fig. 3. Sketch outcrop geology map of the central (focus) strand of the WCHF zone in north Somerset. Areas of Mesozoic to the northwest and southeast are known largely from seismic data only (see Fig. 9). Kinematic marker beds discussed in text also serve to split the stratigraphy: 1, base Blackdown Greensand (Albian–Cenomanian); 2, base Penarth Group (near base Jurassic); 3, base North Curry Sandstone (a Carnian (late Triassic) intra-Mercia Mudstone Group arenaceous member); 4, base Mercia Mudstone Group (intra-Triassic); 5, base Permian–Triassic (Wiveliscombe or Otter Sandstones, ?early Triassic); 6, Morte–Lynton Slates boundary (Givetian–Frasnian, Devonian). Note the wide outcrop of the late Triassic (interval 2–3) in the hanging wall of the WCHF around Taunton.

(3)	Base North Curry Sandstone Member (Carnian: early–late Triassic)	6.2 km offset
(4)	Base Otter Sandstone (?early Triassic Sherwood equivalent)	variable 5.5–6 km offset
(5a)	Base Wiveliscombe Sandstone (?early Triassic)	unknown
(5b)	Linear magnetic anomalies (264 ± 36 Ma)	no offset
(6a)	Morte Slates–Lynton Slates junction (Givetian–Frasnian: Devonian)	9.5–10.5 km offset
(6b)	Lynton (projected)–Quantock/Somerton anticlines	14–16 km offset
(6b)	Cumulative Lynton–Quantock/Somerton anticlines, including effects of Timberscombe and Monksilver Faults	18–20 km offset

Many of the observed offsets in Mesozoic marker horizons can be accounted for through normal faulting of inclined strata or complex dip-slip movements. More specifically, the different offsets to the Otter and North Curry sandstones and Penarth Group in the Taunton area could be derived purely by syn-sedimentary normal movement on the WCHF from the Carnian (North Curry Sandstone) to the early Jurassic. The only check available is the comparative offset to the linear magnetic anomalies and fold-axial trace to the Lynton–Quantock/Somerton anticlines. The latter show considerable horizontal offsets, indicating that in the more steeply dipping Devonian strata, normal faulting alone cannot account for the displacement. No vertical markers are known from the Mesozoic with which to check the offsets. Thus, the following problems still remain in attempting structural restoration of the WCHF.

(i) Early Triassic (Otter Sandstone) strata display generally low dips west of the WCHF at the southwestern end of the Quantock Hills. This has resulted in a complex outcrop pattern (BGS Sheet 295) that allows a variety of mapped offsets to be matched across the fault, allowing horizontal displacement of the base Otter Sandstone by between 0 and 3 km. The greatest estimate of the match shows no more displacement to the Otter Sandstone than that observed in the later Triassic (Carnian: North Curry Sandstone Member) above. This indicates that only limited normal, oblique or strike-slip fault movement occurred in the time between the mid- and late Triassic. Jones (1991) records palaeocurrent flow toward the WCHF in Permian–Triassic (rather than parallel to it), suggesting syn-sedimentary growth.

(ii) Very limited late Triassic movement of the WCHF is indicated by the different displacements to the base North Curry Sandstone Member (Carnian: Upper Triassic; 6.2 km) and Penarth Group (approximately Triassic–Jurassic boundary; 2.6 km). A variation in displacement could also be produced by thickening of post-North Curry Sandstone sediments across the WCHF, or shallow dips in the late Triassic southwest of the fault and steeper dips to the northeast. The former mechanism requires uplift or decreased relative subsidence of the northeastern side of the WCHF (relative to the southwest) in late Triassic times. Beds in the latest Triassic (post-Carnian North Curry Sandstone) succession can be traced across the fault, and show similar displacement to the intra-Mercia Mudstone marker beds below (Ruffell 1990), constraining any possible fault movement to close to the Triassic–Jurassic boundary. No significant clastic influx southwest of the fault is recorded, and thus a strike-slip movement that did not drastically alter topography may explain the different offsets to Triassic and Jurassic marker beds. Some minor uplift along both sides of the WCHF may nonetheless have occurred: Ruffell (1990) noted the existence of a mappable hardened surface (most likely a palaeosol: P. Wright pers. comm.) in the Mercia Mudstone Group found only in the geographic vicinity of the WCHF. If we make a simple comparison of the present-day horizontal offsets to Triassic marker horizons, we may note that early late Triassic offsets (Otter Sandstone) are 0–3 km. The Carnian North Curry Sandstone Member (mid–late Triassic: Fig. 3) shows a 6.2 km offset and the latest Triassic Penarth Group is offset by 2.6 km. This increase in offset is intriguing, suggesting differential slip along the WCHF or a syn-sedimentary increase in movement through the Triassic. Either explanation (above) requires variable dips across the trace of the WCHF in late Triassic successions compared to more even structure in the Jurassic. In the area of steeper dips, a structural unconformity between late Triassic and early Jurassic should be encountered: only very scant evidence in favour of this is available (Ruffell, 1990). As far as is known, the Triassic appears conformable with the Jurassic in most areas of Southwest England, with few changes in dip across the boundary. This study indicates that the most likely structural change between Triassic and Jurassic is a phase of faulting. This may be a localized expression of renewed late Triassic–early Jurassic extension–transtension (Musgrove et al. 1995). The geometry of the WCHF observed on seismic data shot on Mesozoic strata may be the only indication of later kinematics.

Seismic data

Additional evidence in the kinematic analysis of the WCHF exists in the form of high-quality seismic data across the northern and southern strands of the northwest–southeast-oriented part of the fault system. In 1982 and 1985 Goal Petroleum plc carried out land seismic surveys to the southeast of Taunton in Somerset: some details of this data are included in Donato (1988) and Ruffell (1990). The data now belong to Talisman Energy UK Ltd. In 1985 Merlin Geophysical conducted marine seismic surveys of the Bristol Channel, two lines of which are utilized here. Previous accounts of this data can be found in Brooks et al. (1988).

Somerset

Two lines selected from the Talisman Energy UK Ltd dataset demonstrate how the WCHF is expressed on seismic in this area (Figs 4 and 5). The northernmost line, S85–09V displays a Permian–Triassic half-graben similar to that deduced from outcrop studies across the WCHF to the north (see above). The fault appears vertical (or with a steep westerly dip) in basement, splitting into two or more synthetic–antithetic normal faults in Mesozoic cover. The expression of the faulted basement–cover contact as a normal fault with a southwestern hanging wall side, mirrors the inferred geometry from outcrop (Fig. 2). Whittaker (1972) observed a reverse fault at outcrop in the Triassic–Jurassic of North Somerset (Watchet), noting a southwest

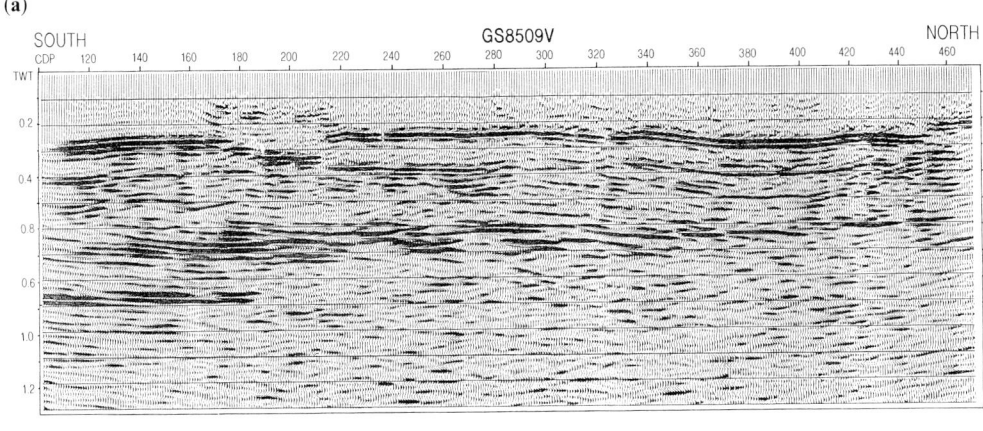

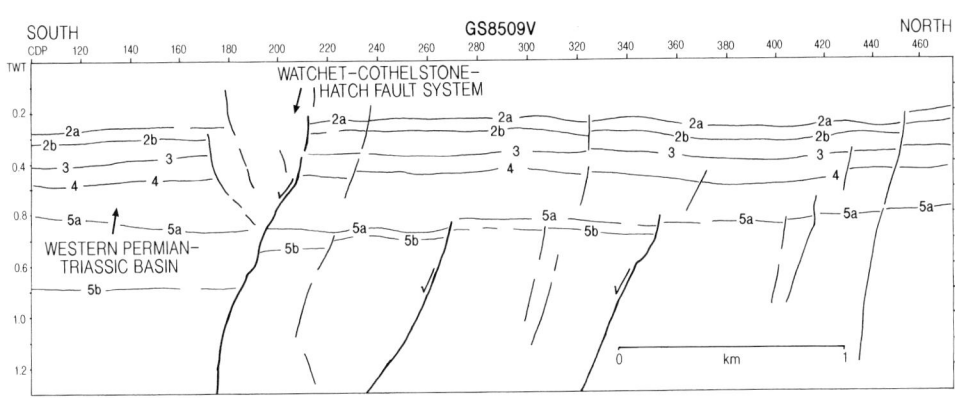

Fig. 4. Seismic line GS8509V (location on Figs 1 and 2) and interpretation. Note that the WCHF shows no obvious vertical offset to reflectors above the interpreted base Otter Sandstone (5a, early Triassic) in contrast to those below. This is paralleled in the outcrop pattern of the preserved Wiveliscombe to base Otter Sandstone in the West Somerset half-graben, which shows negligible faulted offset. Note also the appearance of a negative flower structure along the WCHF zone. 2a, base Lias; 2b, near base Penarth; 3, intra-Somerset Halite Formation (below the North Curry Sandstone marker of Figs 1 and 3); 4, base Mercia Mudstone Group; 5a, base Otter (or Sherwood equivalent) Sandstone; 5b, base Permian–Triassic unconformity (on basement). Data courtesy of Talisman Energy UK Ltd.

dip to the fault plane. Similar southwest-dipping (although normal) faults occur as minor synthetic normal faults on the seismic data. No similar thickness variations in later Triassic and Jurassic strata are observed across the WCHF on seismic line 09, although this is also dependant on seismic resolution: minor thickening across the WCHF in the late Triassic Mercia Mudstone Group may be interpreted. The data show clearly how an early (?Permian) rift phase caused syn-sedimentary fault-accommodated accumulation of sediments during extensional reactivation of the WCHF (Fig. 6). More regional subsidence began in mid- to late Triassic times, reflected in the Sherwood Sandstone onlap of the Quantock Massif (at outcrop) and in the Penarth Group marine transgression. The absence of significant thickness variations in the Triassic and Jurassic belies the obvious post-early Jurassic deformation that these strata have suffered. Antithetic and synthetic splay faults emanating from the basement–Mesozoic cover levels of the WCHF form a synthetic–antithetic fault pair that may be a primitive flower structure or master with two antithetic faults (Fig. 4).

A seismic line to the southeast (S82-05V) shows the continuation of splayed faults from the WCHF at depth as the structure swings into

320 M. MILIORIZOS & A. RUFFELL

(a)
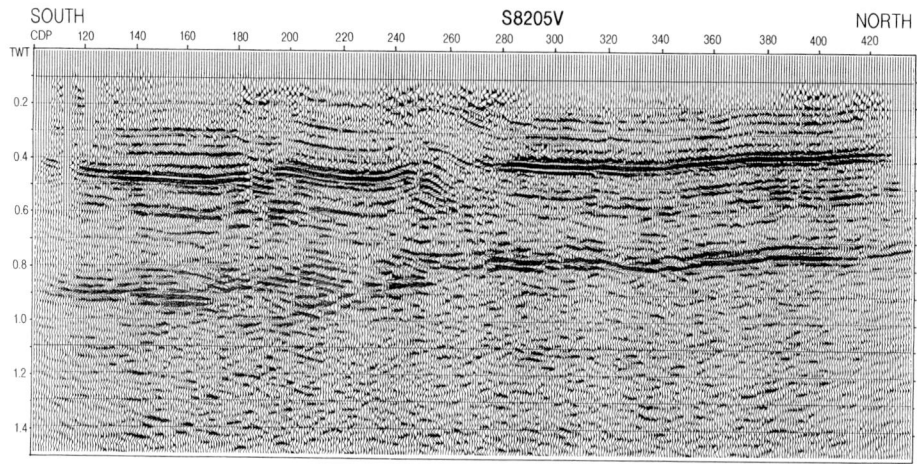

(b)
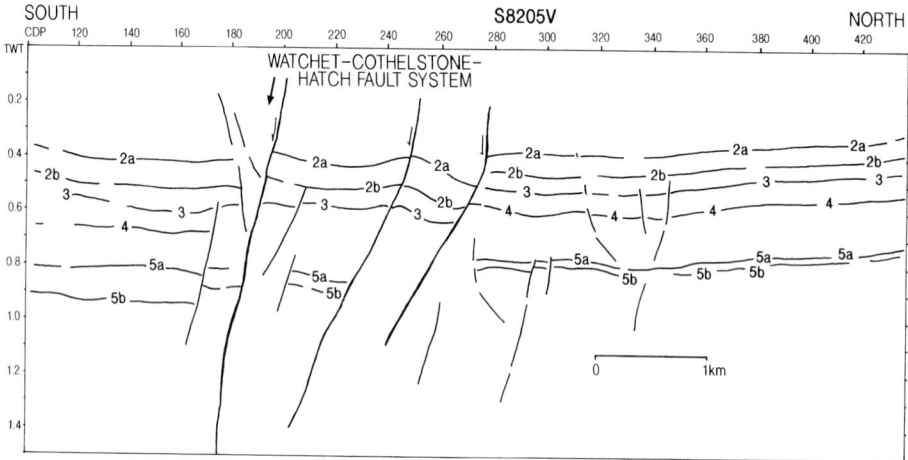

Fig. 5. Seismic line S8205V, located 13 km southeast of line 09 (Fig. 4) along the strike of the WCHF. Similar hanging wall preservation of Permian–Triassic sediments and limited vertical displacement to late Triassic–early Jurassic reflectors are observed. Note the less obvious negative flower structure in this location across the Lopen Fault Belt. 2a, base Lias; 2b, near base Penarth; 3, intra-Somerset Halite Formation (below the North Curry Sandstone marker of Figs 1 and 3); 4, base Mercia Mudstone Group; 5a, base Otter (or Sherwood equivalent) Sandstone; 5b, base Permian–Triassic unconformity (on basement). Data courtesy of Talisman Energy UK Ltd.

an east–west trend. The half-graben geometry of the Permian–early Triassic section is less obvious in this orientation, yet the continuation of splayed faults most likely explains the appearance of the flower-type structure observed on seismic. This suggests that maximum apparent Permian extension (across the northwest–southeast WCHF) in the western Wessex Basin was northeast–southwest, as compared to both earlier and later north–south extension found to the west (this study) and east (Chadwick 1985). Comparison of seismic data to that derived from outcrop shows similarities and some apparent paradoxes. In outcrop structural restoration, difficulty was experienced in matching the Otter Sandstone–Mercia Mudstone Group boundary across the WCHF. This is interpreted as the product of predominantly normal movement

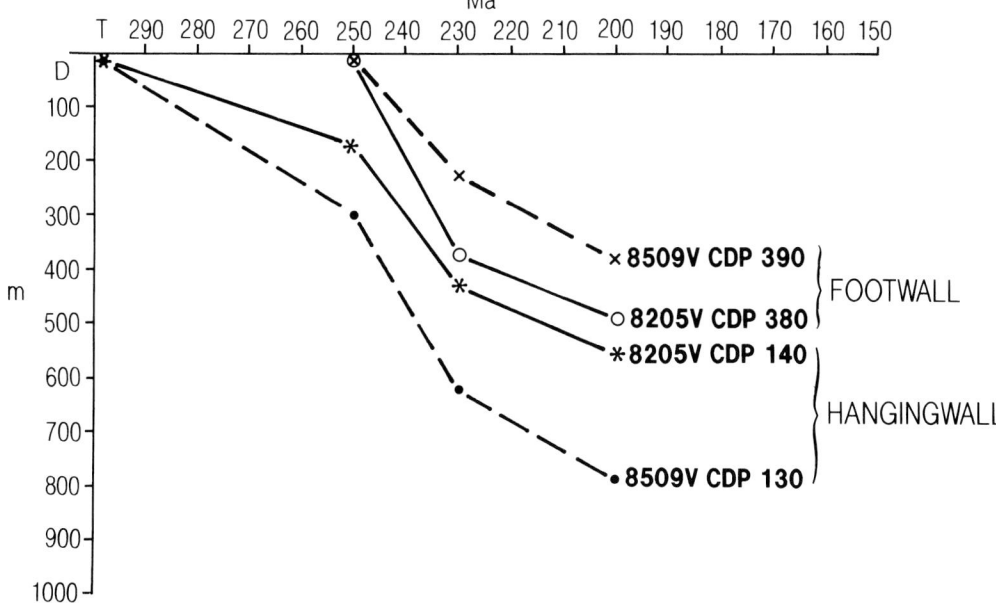

Fig. 6. Time v. depth plots from depth-converted points along the Talisman Energy UK Ltd seismic lines shown in Figs 4 and 5. Depth conversion was achieved using typical interval velocities (from Kearey & Brooks 1991) compared to the stratigraphy recorded in the Seaborough borehole (Donato 1988). The plots confirm the early Permian–early Triassic syn-sedimentary movement of the WCHF, preserving sediments in the hanging wall. Similar subsidence across the WCHF is recorded in Triassic–Jurassic times.

during extension, confirmed by the half-graben geometry observed on seismic data. The large dextral offsets to Devonian fold axes are presumed to have formed through late Carboniferous–early Permian (Variscan) strike-slip deformation, yet the next kinematic event observed in the Western Permian–Triassic Basin is extension. We may surmise that either Variscan strike-slip compression was followed immediately by northeast–southwest extension, or that the Western Permo-Triassic Basin formed during pull-apart (mimicking an extensional basin in two dimensions). The early Permian sediments of the east–west Crediton–Tiverton troughs (the Exeter Volcanic Traps, Crediton Breccias, etc.) are not preserved in the early fill of the Western Permian–Triassic Basin, suggesting that a considerable time gap occurred between Variscan deformation and the first Permian–Triassic deposition in the hanging wall to the WCHF. The fault splays observed on seismic data, plus Whittaker's (1972) strike-slip of the North Somerset coast, were formed in post-early Jurassic times. No evidence is found for successive pull-apart basins so this strike-slip could be the result of any later tectonism. Possible compressional events include Jurassic–Cretaceous inversion ('late Cimmerian'), mid-Cretaceous (Austrian), Cretaceous–Tertiary (early Alpine) or Miocene (Alpine) tectonic episodes: extensional episodes of strike-slip may also have occurred between compression. Evidence from the mapped geometry of the fault in the Bristol Channel and Wessex Basin suggests that late, sinistral strike-slip has occurred, although the remnant offset to Jurassic markers south of Taunton is still dextral.

The swing of the WCHF from northwest–southeast to east–west into the Lopen Fault Belt occurs in Mesozoic cover. The question raised then, is what happens to the fault in the basement? Data acquired by British Gas to the south of the study area show no obvious basement structures: this may be due to acquisition problems in the Blackdown Hills and on Chalk cover where basement may not have been imaged. Splay faults observed on the Talisman seismic data occur in the Triassic section, where the change from northwest–southeast to east–west faults also occurs. To demonstrate this, a structure contour map on a prominent intra-Triassic reflection horizon ('Keuper Evaporites') was constructed (Fig. 7). This clearly shows a swing in the WCHF at this level, via a series of

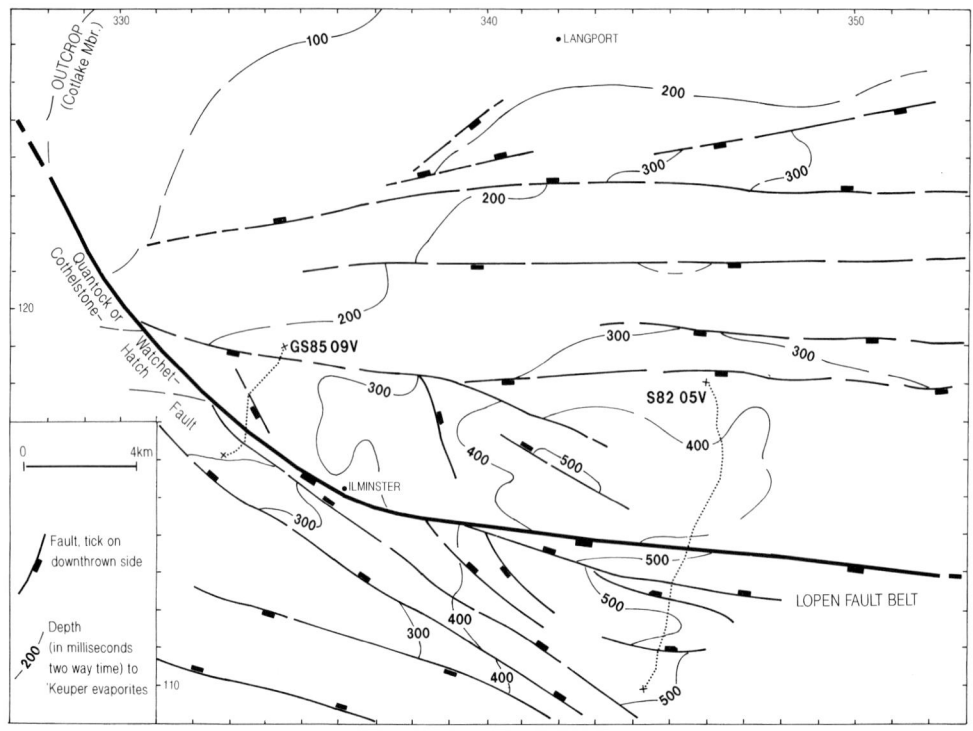

Fig. 7. Time structure map derived from seismic data acquired by Goal Petroleum plc and owned by Talisman Energy UK Ltd in South Somerset (location in Fig. 10). This dataset includes the lines shown in Figs 4 and 5; other line locations are confidential. The map has been constructed on an informal reflection horizon tied to the 'Keuper Evaporites' (intra-upper Mercia Mudstone Group) identified in the Seaborough borehole (Donato, 1988). This interval displays both the swing from northwest–southeast to east–west in the WCHF seen in late Triassic and early Jurassic strata, as well as the continued northwest–southeast trend observed in early Triassic and ?basement.

horsetail-like splays along the Lopen Fault Belt, all of which probably accommodated some post-early Jurassic movement. As suggested in the section on 'Previous work', Webby's (1965a, b) maps of the Exmoor and Quantock Upper Palaeozoic may aid our understanding of the WCHF to the east. The swing observed in the traces of the Timberscombe and Monksilver faults may be a valuable additional clue in interpreting the geometry of the WCHF at depth. The change in orientation to the smaller faults may be mirrored in the intersection of the WCHF with the east–west Lopen Fault Belt, indicating that this area has inherited structure from the Palaeozoic rocks at depth.

Bristol Channel

To the north of the Taunton (–Quantock) High a similar basement–cover relationship to that observed to the south occurs in the Bristol Channel. Here, the northwest–southeast WCHF intersects the east–west Central Bristol Channel Fault Zone, where it may splay (Brooks et al. 1988). In (Carboniferous and ?older) basement, the WCHF is thought by Kammerling (1979) to continue northeast and appear onshore at Merthyr Mawr in South Wales. The east–west and northwest–southeast-oriented seismic surveys conducted by Merlin Geophysics in the Bristol Channel (including Figs 8 and 9 of this study) intersect the WCHF and other northwest–southeast-trending faults obliquely and thus provide more limited information about the fault than the Talisman Energy lines in South Somerset. Nonetheless, in the South Bristol Channel (Fig. 8) the steep-faulted half-graben geometry known from outcrop in the Quantock Hills and from seismic in South Somerset can also be seen. In addition, line MP85WB64 (Fig. 8) shows the relationship between the West–Central Somerset basins and the Quantock–Mendip

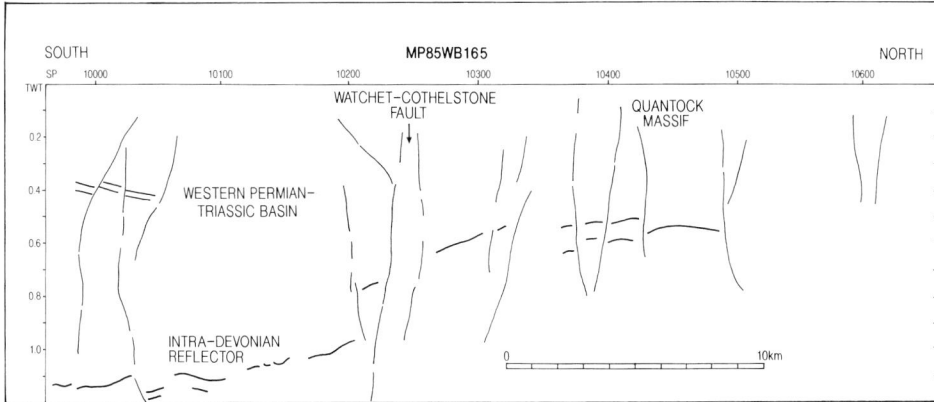
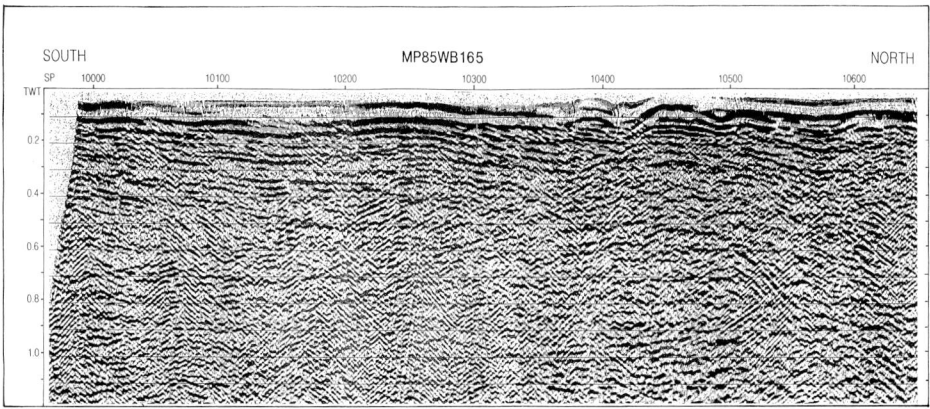

Fig. 8. This southern portion of line MP85WB165 intersects the WCHF obliquely (shown on speculative line interpretation). The half-graben geometry observed at outcrop to the south (Western Permian–Triassic Basin) is interpreted in the inclined Permian–Triassic reflectors. A deep intra-Devonian reflector appears similar to the south-dipping Variscan thrusts interpreted by Brooks *et al.* (1988) to occur below the Bristol Channel Basin. 5a, ?base Otter–Sherwood Sandstone; 5b, base Permian–Triassic unconformity (no close well-tie allows better correlation).

massifs. The Central Somerset Basin was known to Ussher (1911) and named by Whittaker (1975) who suggested from gravity and borehole data that a rift graben model be applied. In general terms this model is confirmed by the seismic data; the basin is roughly symmetrical with evidence of Permian–Triassic synsedimentary faults and late Triassic onlap on to the Quantock Massif (Fig. 8). Whittaker (1975) suggested that the Permian–Triassic of the Central Somerset Basin had a flat base: a feature also observed on seismic data (Fig. 8). There is no reason to doubt that this basin is an onshore continuation of the Bristol Channel Basin.

Kinematic history of the WCHF

We may summarize the data accumulated (so far) on the WCHF by integrating outcrop and seismic data from the Bristol Channel to the western Wessex Basin.

Devonian and Carboniferous

Late Carboniferous–early Permian Variscan tectonics form the major Quantock and Lynton anticlines and cause dextral displacement of Devonian stratigraphy/structure of at least 14 km, and at most 16 km, along the WCHF and associated fault systems. East–west-trending

(a)

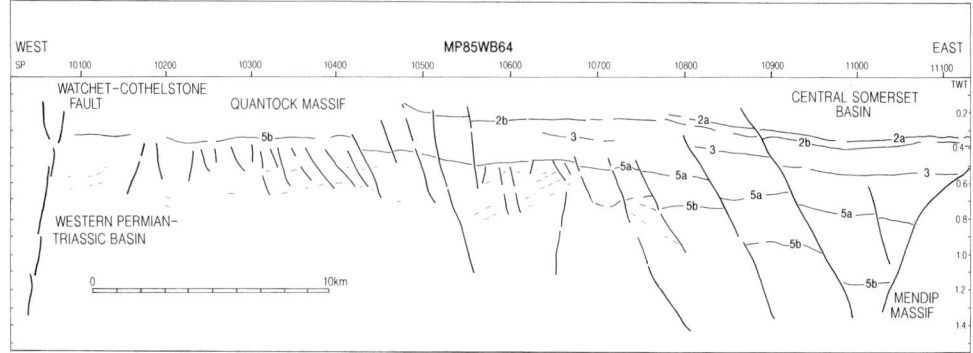

(b)

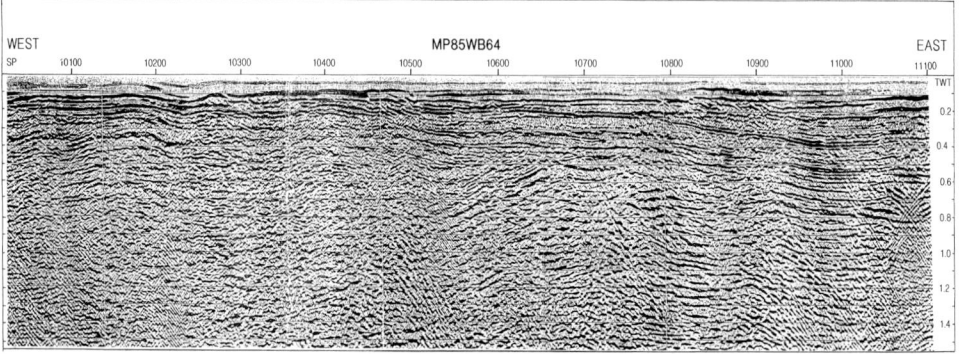

Fig. 9. Seismic line MP85WB64 shows the nature of the WCHF, Quantock Massif and Central Somerset Basin. The WCHF appears similar in geometry to that observed on Talisman Seismic Line 05 to the southeast (Fig. 5). The Central Somerset Basin appears as a narrow flat-based graben similar to that described by Whittaker (1975).

late Carboniferous–early Permian igneous intrusions and mineralization *may* have accompanied the late formation of the east–west Crediton and Tiverton rifts as dextral pull-apart basins (early Permian Cadbury Breccias to Belfield Sandstone after Scrivener & Edwards, 1990). A similar, but less radical interpretation merely uses the WCHF as a transfer of north–south extension from the Crediton–Tiverton troughs to the Winterborne Trough, with neither syn- or post-sedimentary strike-slip implied.

Late Permian–early Triassic

Northeast–southwest extension causes half-graben development across the WCHF. This is unlikely to represent strike-slip pull-apart as previously emplaced late Carboniferous–early Permian linear magnetic anomalies in the south Quantock–south Exmoor (Brendon Hills) tract were unaffected. During deposition of the Otter Sandstone (mid-Triassic) onlap of the Quantock footwall to the WCHF occurs. Infill of earlier Crediton and Tiverton Troughs is largely complete.

Early to mid-Triassic

No strike-slip fault movement in the earlier Triassic is indicated by similar offsets to the base Otter Sandstone and base North Curry Sandstone around Taunton. The variations in offset that do occur may be explained by late Triassic extension. Onlap of the margins to the Western Permian–Triassic, Tiverton, Crediton and Central Somerset troughs/basins.

Late Triassic–earliest Jurassic

Unquantified minor strike-slip or normal movement (not observed on seismic) is indicated by different mid–late Triassic to early Jurassic offsets. This forms the anomalously wide Mercia Mudstone outcrop around Taunton (Fig. 3). Late Triassic hardened surfaces (possibly palaeosols) both side of the WCHF indicate minor uplift.

Post Jurassic

Dextral (North Somerset coast: Whittaker 1972) and sinistral (Fig. 10) displacement occurs. This may reflect early–mid-Jurassic and late Jurassic–early Cretaceous movements. The cumulative effect of these has been minor dextral offsets, formation of primitive flower structures and present-day sinistral splays to the WCHF. As no major offset to the late Carboniferous–early Permian magnetic anomaly occurs, such sinistral strike-slip either resulted in only limited horizontal movement or merely restored earlier dextral motion. Unpublished isopach mapping by one of us (M. Miliorizos) suggests a Cretaceous age for Middle and Upper Jurassic sinistral displacements in the Bristol Channel.

Recent

Probably negligible net movement, yet significant enough to cause historical earthquakes in Taunton (Davison 1924). Hugh Prudden (pers. comm.) has documented the influence of northwest–southeast structures on landforms both in the Devonian of Exmoor and in the Jurassic around Ilminster.

Summary – comparison of seismic data to outcrop

The Devonian outcrop of the Quantock Hills is the product of a fault-controlled Permian–Triassic high, the Taunton High of Donato (1988). Bounding this high, the WCHF shows a predominant northwest–southeast orientation. To the south and north, the orientation of the fault at different levels is less clear. Ziegler

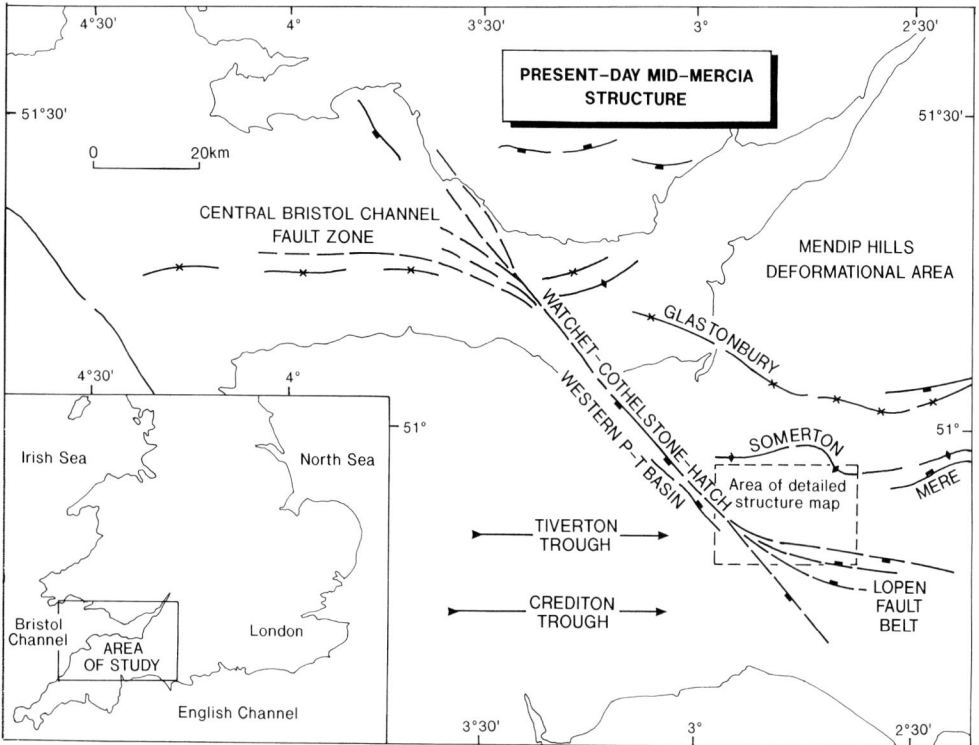

Fig. 10. Present-day structural map of the western Central Bristol Channel Fault Zone, WCHF and Lopen Fault Belt, indicating a common genetic origin through limited late sinistral strike-slip on the WCHF.

(1982), Edmonds & Williams (1985b) and Beach (1987) all indicate the continuation of the fault in a northwest–southeast orientation in the basement succession to the south. Stoneley (1982), Chadwick (1985) and Karner et al. (1987) show the northwest–southeast orientation intersecting with the predominant east–west fault pattern in the Mesozoic–Cenozoic of the Wessex Basin, largely in accord with Shearman's (1967) model. Mapping of seismic reflection data on the southeast margin to the Quantock High (data courtesy of Talisman Energy UK Ltd) and comparison with the Central Bristol Channel Fault Zone from Brooks et al. (1988) shows that at the critical level (late Triassic), where the northwest–southeast WCHF of the basement and Permian–Triassic passes up into the east–west Lopen Fault Belt of the Jurassic, a series of splay faults occur (Fig 10).

The evidence for both strike-slip and dip-slip extension on the WCHF is convincing. In the former case, laterally offset fold axes (Donato 1988) and marker beds (Webby 1965a; Whittaker 1972); and in the latter, the half-graben geometry of the WCHF indicate that both strike- and dip-slip must have occurred. Permian–Triassic stratigraphy differs markedly and shows thickening across the fault, suggesting synsedimentary movement. The east–west Crediton and Tiverton troughs contain both volcanic rocks and an early Permian stratigraphy that pre-date the formation of the Western Permian–Triassic Basin. The western Wessex Basin and Bristol Channel probably also formed in the Permian, the WCHF acting as a zone of transfer between southern and northern (possibly syn-orogenic) extension. This structural division is likely to continue well into the subsurface of the western Wessex Basin as Lee et al. (1993) clearly show the Quantock Massif extending 25 km eastwards from its last outcrop.

The association of northwest–southeast-trending strike-slip faults with a major offset Variscan anticline (the Quantock–Lynton trace) suggests a genetic relationship. The presence of east–west-striking basement thrusts on all the seismic datasets examined in this study (Somerset, Donato 1988; southern Bristol Channel, this study, Fig. 8; Central Bristol Channel, Brooks et al. 1988) may further be linked to the northwest–southeast faults by explaining the latter as lateral ramps to the thrusts. Such a kinematic link in the Variscan is reflected in the Permian–Triassic extensional reactivation of the Timberscombe–Monksilver–WCHF array as all three preserve later basins. The largest Permian–Triassic basin (judged by thickness and lateral extent) is in the hanging wall to the largest fault (the Western Permian–Triassic Basin and the WCHF, respectively). Similarly, the Porlock Basin, with its easterly-dipping Permian–Triassic conglomerates in the hanging wall of the Timberscombe Fault and the minor basin at Yarford on the Monksilver Fault all imply a simple relationship: the bigger the fault, the bigger the associated Permian–Triassic basin. This model may have repercussions for hydrocarbon prospectivity: major northwest–southeast faults may preserve significant Permian–Triassic clastic reservoir units in their hanging walls.

Comparison of the WCHF with the Sticklepath Fault

The WCHF is essentially similar to the Sticklepath Fault in Devon (Holloway & Chadwick 1986), in that both are northwest–southeast faults displaying extensive Variscan and later movements. Both show evidence of dextral offsets to basement fabrics. Holloway & Chadwick (1986) suggested that the Sticklepath Fault preserved an early Tertiary sinistral component, resulting from lateral motion of around 6 km. As no Tertiary strata are preserved along the length of the WCHF, no direct evidence of a similar Tertiary sinistral displacement is observed: the presence of a regional sinistral deformation to the WCHF at the present-day obviously links the two. The Sticklepath Fault and WCHF are similar in another way: both Webby (1965a) and Holloway & Chadwick (1986) made outcrop estimates of offset that increased considerably when regional geology and geophysical data were utilized (Arthur 1989 and this study, respectively). Thus, Arthur (1989) suggests sinistral movement of 28–40 km along the Sticklepath Fault. Conversely, no east–west or northwest–southeast-trending Permian–Triassic basins are associated with the Sticklepath Fault: again, uplift and erosion in the Mesozoic or early Cenozoic could have removed such material.

The preservation (through strike-slip pull-apart) of Eocene–Oligocene sediments in the Bovey–Petrockstow–Lundy basins along the Sticklepath Fault is not observed in the WCHF zone. This does not preclude such basins having been formed, and now not observed through uplift and erosion. There are theoretical reasons why this is unlikely: Holloway & Chadwick (1986) concluded that the Bovey and Petrockstow basins developed as pull-aparts developed between en echelon relays in the Sticklepath Fault; few such relays occur in the WCHF system of faults close to surface (and thus detectable from mapping and geophysics). This would have effectively precluded the opening of pull-aparts and the creation of sedimentary accommodation space in Cenozoic fault-bounded basins. En echelon relays may have developed more readily along the Sticklepath Fault where the brittle but inhomogenous Crackington Formation and similarly brittle yet homogenous Dartmoor Granite forms basement to the Tertiary Bovey and Petrockstow basins. Strike-slip deformation in the Mercia Mudstone Group and Lias along the WCHF achieved deformation of more homogenous and initially more ductile sediments. The Devonian–Carboniferous of the Brendon–Quantock Hills is more brittle than its cover, and thus Permian or Tertiary pull-apart basins are more likely to have developed: the Crediton and Tiverton troughs may be such basins. Evidence for minor late Carboniferous–early Permian and Tertiary basins originating as east–west pull-aparts may be sought adjacent to the Crediton and Tiverton troughs and in relays along the northwest–southeast faults of the area.

Similarly, Tertiary strike-slip may also have aided preservation of Triassic sediments in down-faulted areas of the Timberscombe or Monsilver faults. Arthur (1989) proposed that movement on the Sticklepath Fault aided the emplacement of the Lundy Granite through brittle graben and horst faulting. This is considered unlikely in the softer Mesozoic cover of the Wessex Basin, but highly likely in more brittle rocks at depth. Regional gravity maps of southern England acquired by JEBCO Geophysical (P. Baxter pers. comm.) indicate a likely Tertiary granite in the footwall of the Pays de Bray Fault (southeast of the Isle of Wight) in the English Channel. This is one of the predicted zones of northwest–southeast strike-slip faults from Karner *et al.* (1987) and this work.

The model of extensional transfer presented in Fig. 11 may explain why the '264 Ma' lamprophyre dykes are restricted to the eastern (Quantock Hills) side of the WCHF. In this model, Permian–Triassic north–south extension

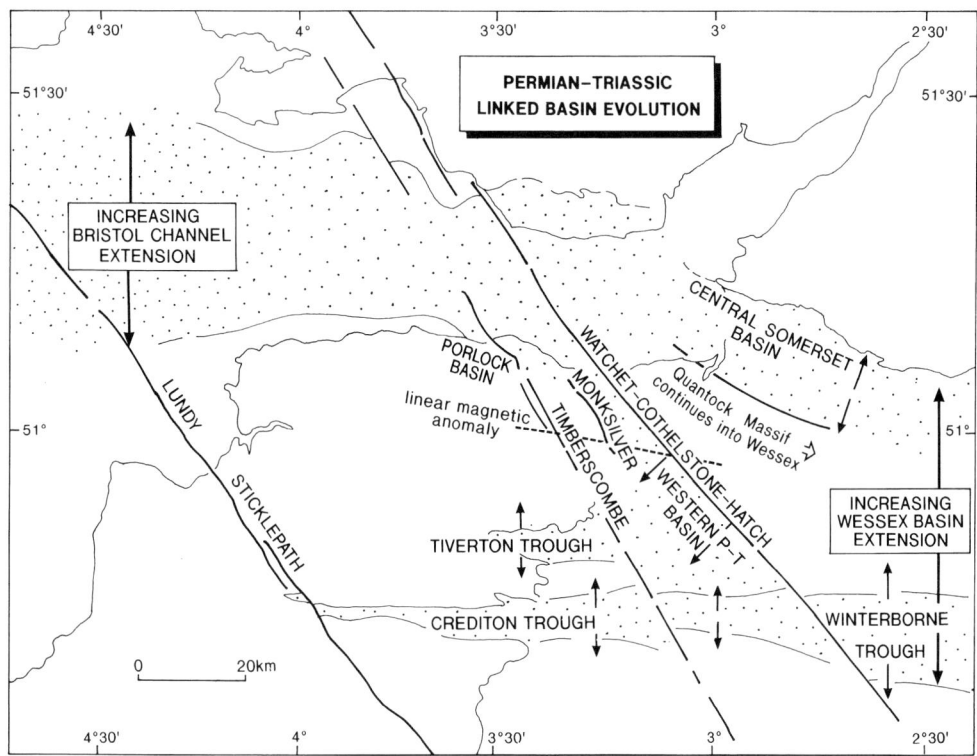

Fig. 11. Permian–Triassic rift model for the western Wessex Basin, in which the WCHF acts in a dextral sense in Tiverton–Crediton times (ealiest Permian), accommodates northeast–southwest extension through the late Permian and Triassic, and preserves increasing amounts of Mesozoic in the Wessex and Bristol Channel by later sinistral motion. Basins and inferred extension from Kammerling (1979), Chadwick (1985), Donato (pers. comm.) and Wall (pers. comm.).

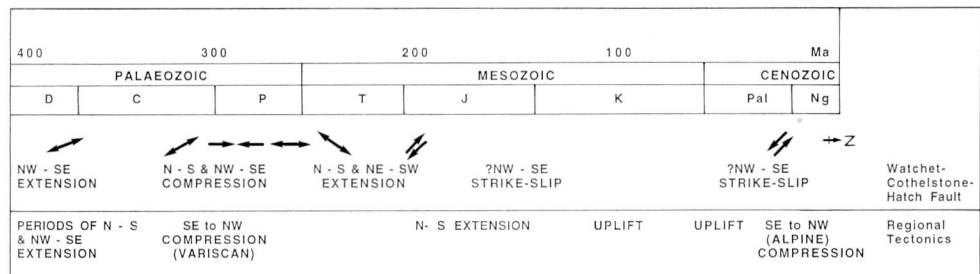

Fig. 12. Summary of the main kinematic events to have affected the WCHF and associated basins. Extension, divergent arrows; compression, convergent arrows; dextral and sinistral strike-slip arrows.

increased eastwards into the Wessex Basin and westwards into the Bristol Channel (and possibly the South Celtic Sea Basin). Thus, basement on the eastern side of the WCHF in northern Wessex was extended more than to the west, creating more tension and allowing igneous intrusions to penetrate. Permian igneous rocks might be expected near the basement–cover contact in the western Wessex Basin. Yet earlier extension could have controlled the intrusion of the '291 Ma' Exeter Volcanics to the south: supporting evidence may be found in the distribution of the Exeter Volcanics, which are only very obvious (from the published maps of the Geological Survey) on the northern margins of the Crediton–Tiverton troughs. The above statement should not be read as meaning that Exeter Volcanics are restricted to these areas, they are widely distributed but appear as extensive and continuous outcrops to the north of the troughs. To the west of the WCHF and Timberscombe Fault (in the Brendon Hills) Devonian–Carboniferous basement may still have been under dextral compression (or, at least, subject to less extension than in the east) when the lamprophyres intruded, possibly precluding dyke formation and allowing only mineralized vein-filled fractures to form.

Consideration of the early Permian basin kinematics of this part of the Western Wessex Basin produces a paradox. Interpreting the northern margins of the Crediton and Tiverton troughs as faults (exploited by igneous activity) implies an east–west-trending half-graben geometry which in this orientation would be most easily explained by north–south-directed extension. This may not be a regional stress field, the east–west orientation being a product of extensional reactivation of Variscan thrusts at depth. North–south extension produces a sinistral early Permian motion on the WCHF which is at odds with the last known Variscan dextral movement. A realistic approach, in line with radiometric dates, would be to have the *origin* of the Tiverton and Crediton troughs as late Variscan (dextral) basins, formed mostly before the '264' magnetic anomaly. The *preservation* of Permian–Triassic and Jurassic strata in the Winterborne Trough (Western Wessex) and Bristol Channel Basin would then require a sinistral transfer in later Jurassic or Cretaceous times. Minor dextral and sinistral motion could also occur at any time (Whittaker 1972).

The Sticklepath Fault and WCHF are approximately 65 km apart. It is interesting to speculate whether similar strike-slip faults occur beneath the Wessex Basin to the east, as suggested by Karner *et al.* (1987). Given an homogenous Variscan crust, similar strike-slip faults might be predicted to occur across the English Variscides. Similarly spaced northwest–southeast faults are depicted by Ziegler (1982), Beach (1987) and Coward (1990). Thus Sticklepath and/or WCHF type structures are predicted to occur: (i) beneath the Isle of Wight, running from the Pays de Bray structure up into the Forest of Dean; (ii) from Brighton, through Reading to Oxford; and (iii) from Folkestone to Maidstone, under London to the South Midlands. Each of these faults may have the variable Permian–Triassic stratigraphy and kinematics discussed here.

The helpful advice of M. Coward, J. Donato, R. Gayer, M. Harvey, G. Leslie, H. Prudden, S. Stewart, M. Storey, G. Wall, G. Warrington and A. Whittaker greatly enabled this work. The views expressed here are the authors sole responsibility. Data provided by Merlin Geophysics (with the assistance of T. Kilenyi and B.P. Structural/Western Margins Groups) and Talisman Energy UK Ltd (with the assistance of M. Trees) formed a core element to this work and is gratefully acknowledged.

References

ARTHUR, M. J. 1989. The Cenozoic evolution of the Lundy pull-apart basin into the Lundy rhomb horst. *Geological Magazine*, **126**, 187–198.

BADHAM, J. P. N. 1982. Strike-slip orogens – an explanation for the Hercynides. *Journal of the Geological Society, London,* **139**, 493–504.

BEACH, A. 1987. A regional model for linked tectonics in north-west Europe. *In:* BROOKS, J. & GLENNIE, K. (eds) *The Petroleum Geology of Northwest Europe, Proceedings of the 3rd Conference.* Graham & Trotman, London, 43–48

BIRPS & ECORS 1986. Deep seismic reflection profiling between England, France and Ireland. *Journal of the Geological Society, London,* **143**, 45–52.

BOTT, M. P. H., DAY, A. A. & MASSON-SMITH, D. 1958. The geological interpretation of gravity and magnetic surveys in Devon and Cornwall. *Philosophical Transactions of the Royal Society, London,* **A251**, 161–191.

BROOKS, M. & AL-SAADI, R. H. 1977. Seismic refraction studies of geological structure in the inner part of the Bristol Channel. *Journal of the Geological Society, London,* **133**, 433–445.

—— & JAMES, D. G. 1975. The geological results of seismic refraction surveys in the Bristol Channel 1970–1973. *Journal of the Geological Society, London,* **131**, 163–182.

——, TRAYNER, P. M. & TRIMBLE, T. J. 1988. Mesozoic reactivation of Variscan thrusting in the Bristol Channel area, UK. *Journal of the Geological Society, London,* **145**, 439–444.

CHADWICK, R. A. 1985. Permian–Triassic. Mesozoic and Cenozoic structural evolution of England and Wales in relation to the principles of extension and inversion tectonics. *In:* WHITTAKER, A. (ed) *Atlas of Onshore Sedimentary Basins in England and Wales.* Blackie, Glasgow, 9–25.

COPE, J. C. W., GETTY, T. A., HOWARTH, M. K., MORTON, N. & TORRENS, H. S. 1980. *A Correlation of Jurassic Rocks in the British Isles. Part 1: Introduction and Lower Jurassic.* Special Report of the Geological Society, London No. 14.

COWARD, M. P. 1990. The Precambrian, Caledonian & Variscan framework to NW Europe. *In:* HARDMAN, R. F. P. & BROOKS, J. (eds) *Tectonic Events Responsible for Britain's Oil and Gas Reserves.* Geological Society, London, Special Publications, **55**, 1–34.

DAVISON, C. 1924. *A History of British Earthquakes.* Cambridge University Press, Cambridge.

DE LA BECHE, H. T. 1839. *Report on the Geology of Cornwall, Devon and Somerset.* Memoirs of the Geological Survey of Great Britain.

DONATO, J. A. 1988. Possible Variscan thrusting beneath the Somerton Anticline, Somerset. *Journal of the Geological Society, London,* **145**, 431–438.

DRUMMOND, P. V. O. 1970. 1970. The mid-Dorset Swell. Evidence of Albian–Cenomanian movements in Wessex. *Proceedings of the Geologists' Association,* **81**, 679–714.

EDMONDS, E. A. & WILLIAMS, B. J. 1985a. *Geology of the Country around Taunton and the Quantock Hills.* Memoirs of the British Geological Survey (Sheet 295).

—— & —— 1985b. *The Regional Geology of Southwest England.* British Geological Survey, Regional Guidebooks, Vol. 17, 138.

ETHERIDGE, R. 1867. On the physical structure of west Somerset and north Devon and on the palaeontological value of Devonian fossils. *Quarterly Journal of the Geological Society, London,* **23**, 568–698.

FORSTER, S. C. & WARRINGTON, G. 1985. Geochronology of the Carboniferous, Permian and Triassic. *In:* SNELLING, N. J. (ed.) *The Chronology of the Geological Record.* Geological Society of London, Memoir, **10**, 99–113.

GAYER, R. A. & JONES, J. A. 1989. The Variscan foreland in South Wales. *Proceedings of the Ussher Society,* **7**, 177–179.

HATHAWAY, T. M. & GAYER, R. A. 1994. Variations in the style of thrust faulting in the South Wales Coalfield and mechanisms of thrust development. *Proceedings of the Ussher Society,* **8**, 279–284.

HOLLOWAY, S. & CHADWICK, R. A. 1986. The Sticklepath–Lustleigh fault zone: Tertiary sinistral reactivation of a Variscan strike-slip fault. *Journal of the Geological Society, London,* **143**, 447–452.

JONES, D. J. 1991. *Triassic Basin Sedimentation in the U.K. and North Sea.* PhD thesis, University of London.

KAMMERLING, P. 1979. The geology and hydrocarbon habitat of the Bristol Channel Basin. *Journal of Petroleum Geology,* **2**, 75–93.

KARNER, G. D., LAKE, S. D. & DEWEY, J. F. 1987. The thermal & mechanical development of the Wessex Basin, southern England. *In:* COWARD, M. P., DEWEY, J. F. & HANCOCK, P. L. (eds) *Continental Extensional Tectonics.* Geological Society, London, Special Publications, **28**, 517–536.

KEAREY, P. & BROOKS, M. 1991. *An Introduction to Geophysical Exploration.* Blackwell Scientific Publications, Oxford.

LEE, M. K., PHARAOH, J. P., WILLIAMSON, J. P., GREEN, C. A. & DE VOS, W. 1993. Evidence on the deep structure of the Anglo-Brabant Massif from gravity and magnetic data. *Geological Magazine,* **130**, 575–582.

LLOYD, A. J., SAVAGE, R. J. G., STRIDE, A. H. & DONOVAN, D. T. 1973. The geology of the Bristol Channel floor. *Philosophical Transactions of the Royal Society, London,* **A274**, 595–626.

MUSGROVE, F. W., MURDOCH, L. M. & LENEHAN, T. 1995. The Variscan fold-thrust belt of southeast Ireland and its control on early Mesozoic extension and deposition: a method to predict the Sherwood Sandstone. *In:* CROCKER, P. F. & SHANNON, P. M. (eds) *The Petroleum Geology of Ireland's Offshore Basins.* Geological Society, London, Special Publications, **93**, 81–100.

RUFFELL, A. 1990. Stratigraphy and structure of the Mercia Mudstone Group (Triassic) in the western part of the Wessex Basin. *Proceedings of the Ussher Society,* **7**, 263–267.

SCRIVENER, R. C. & EDWARDS, R. A. 1990. Field excursion to the New Red Sandstone of the eastern Crediton Trough, 3rd January 1990. *Proceedings of the Ussher Society,* **7**, 304–305.

SHEARMAN, D. J. 1967. On Tertiary fault movements in north Devonshire. *Proceedings of the Geologists' Association,* **78**, 555–566.

SMITHURST, L. J. M. 1990. Structural remote sensing of south-west England. *Proceedings of the Ussher Society*, **7**, 236–241.

STONELEY, R. 1982. The structural development of the Wessex Basin. *Journal of the Geological Society, London*, **139**, 545–552

USSHER, W. A. E. 1906. *Geology of the Country between Wellington and Chard*. Memoirs of the Geological Survey of Great Britain.

——1908. *Geology of the Quantock Hills and of Taunton and Bridgwater*. Memoirs of the Geological Survey of Great Britain.

——1911. Excursion to Dunball, Burlescombe, Ilminster, Chard, Ham Hill and Bradford Abbas. *Proceedings of the Geologists' Association*, **22**, 246–254.

WEBBY, B. D. 1965a. The stratigraphy and structure of the Devonian rocks in the Quantock Hills, west Somerset. *Proceedings of the Geologists' Association*, **76**, 321–343.

——1965b. The stratigraphy and structure of the Devonian rocks in the Brendon Hills, west Somerset. *Proceedings of the Geologists' Association*, **76**, 39–60.

WHITTAKER, A. 1972. The Watchet Fault – a post-Liassic transcurrent reverse fault. *Bulletin of the Geological Survey*, **41**, 75–80.

——1975. A postulated post-Hercynian rift valley system in southern Britain. *Geological Magazine*, **112**, 137–149.

WOODWARD, H. B. 1893. *The Jurassic Rocks of Britain. Part 3 the Lias of England and Wales (Yorkshire Excepted)*. Memoirs of the Geological Survey of Great Britain.

ZIEGLER, P. A. 1982. *Geological Atlas of Western and Central Europe*. Elsevier, Amsterdam.

Tectonic accentuation of sequence boundaries: evidence from the Lower Cretaceous of southern England

A. RUFFELL

School of Geosciences, Queen's University, Belfast BT7 1NN, Northern Ireland, UK

Abstract: Three sequence boundaries (SB1, SB2 and SB3) are documented in the Upper Aptian *martinioides* to *jacobi* zones of the Lower Greensand Group of southern England. Correlation shows how folding affects only the strata of the lowest depositional sequence, indicating a discrete phase of tectonism in the *nutfieldiensis* zone/subzone (SB2). No similar tectonic deformation occurs at the sequence boundaries above or below. The regional features of this unconformity surface demonstrate that tectonically enhanced sequence boundaries may be discerned from those of eustatic/global tectonic origin. The tectonically enhanced sequence boundary (SB2) was initially interpreted as a transgressive surface at individual outcrops, but the regional pattern of truncation indicates that the transgressive surface is amalgamated onto a sequence boundary. Comparison of the geometry of each sequence on a regional (basin-wide) scale shows clearly the structural origin to truncation. Correlation of the three depositional sequences is made using their bounding surfaces, and the variable deposits of fuller's earth bentonite, estuarine sand and lagoonal muds between SB1 and SB2 are resolved as time-equivalent facies deposited at the turn-around from sea-level lowstand to transgression. This correlation can be compared to that achieved by two conflicting ammonite zonal/subzonal schemes: the older biostratigraphy suggests diachroneity of the deformed lower sequence, whilst a new zonation shows this same sequence to be approximately synchronous.

Previous to the publication of van Wagoner *et al.* (1990), outcrop sequence stratigraphic methods (and the tests of those methods), have been largely concerned with the analysis of vertical sedimentary profiles in terms of chronostratigraphy (Haq *et al.* 1988; Galloway 1989; Hesselbo *et al.* 1990*b*). The correlation and geometry of surfaces related to sequence stratigraphy at outcrop and in boreholes proposed by van Wagoner *et al.* (1990) marks a return to the seismic stratigraphic method (Vail 1987) in that the nature of the sedimentary surface is critical,

Table 1. *Comparison of the two ammonite biostratigraphic schemes applicable to the succession under discussion*

	Casey's (1961) scheme		Owen's (1994) scheme	
	zone	subzone	zone	subzone
Lower Albian	*tardefurcata*	*regularis*	*tardefurcata*	*regularis*
		milletioides		*acuticostata*
		farnhamensis		*schrammeni*
Upper Aptian	*jacobi*	*anglicus*	*nodosocostatum*	*jacobi*
		rubricosus		*nolani*
		nolani		
	nufieldiensis	*cunningtoni*	*martinioides*	*nutfieldiensis*
		subarcticum		
	martinioides	*buxtorfi*		*buxtorfi*
		gracile		*gracile*
		debile		*debile*

albeit at a far higher resolution. Two related features of the stratigraphic record are currently considered important in the analysis of depositional sequences. These are parasequences and important or key surfaces. For reasons outlined below, this study concentrates on the formation and correlation of surfaces. A problem still exists concerning the role of tectonics in influencing the development of sedimentary sequences and their bounding surfaces (e.g. Williams & Dobb 1993). The present work was undertaken to assess the consequences of applying an analysis of depositional sequences to a succession known to have been deposited during a discrete phase of tectonic activity (Ruffell 1992). This work was conducted using the ammonite zonal scheme of Casey (1961), whose terminology is used throughout the initial descriptive parts of this article. Recently, Owen (1996) and Ruffell & Owen (1994) have proposed and utilized (respectively) an alternative ammonite zonal scheme (Table 1). The implications for the application of either scheme in a sequence stratigraphic framework are discussed in the section 'Biostratigraphic interpretation'. Thus, three methods of correlation may be employed: two conflicting ammonite biostratigraphies and sequence stratigraphy. The latter method is outlined here and compared to the ammonite schemes. No new lithostratigraphic terminology is introduced: where members and formations have been defined (e.g. for the Isle of Wight succession) then these are adhered to. Conversely 'beds' (e.g. Bargate Beds) are commonly the last lithostratigraphic usage and are not redefined here.

The Upper Aptian (*sensu* Casey 1961) (Table 1) *nutfieldiensis* zone transgression in southern England was recognized by Casey (1961), Middlemiss (1962) and Kirkaldy (1963). The recent studies of Hesselbo *et al.* (1990*b*), Ruffell & Wach (1991) and Ruffell (1992) have suggested that the early Upper Aptian (*martinioides–nutfieldiensis* zones) was a time of significant sea-level fluctuation, the final record of which are the previously recognized transgressive deposits. Three sequence boundaries are interpreted in the mid- to late Aptian succession (Lower Greensand) of southern England (Hesselbo *et al.* 1990*a, b*; Ruffell 1992). Comparing the tectonic models proposed for the Lower Cretaceous of southern England (Whittaker, 1985; Karner & Lake 1987; Karner *et al.* 1987), Ruffell (1992) considered minor compression (basin-margin uplift) and the formation of associated unconformities in the mid-Cretaceous likely to be associated with extensional tectonics. McMahon & Turner

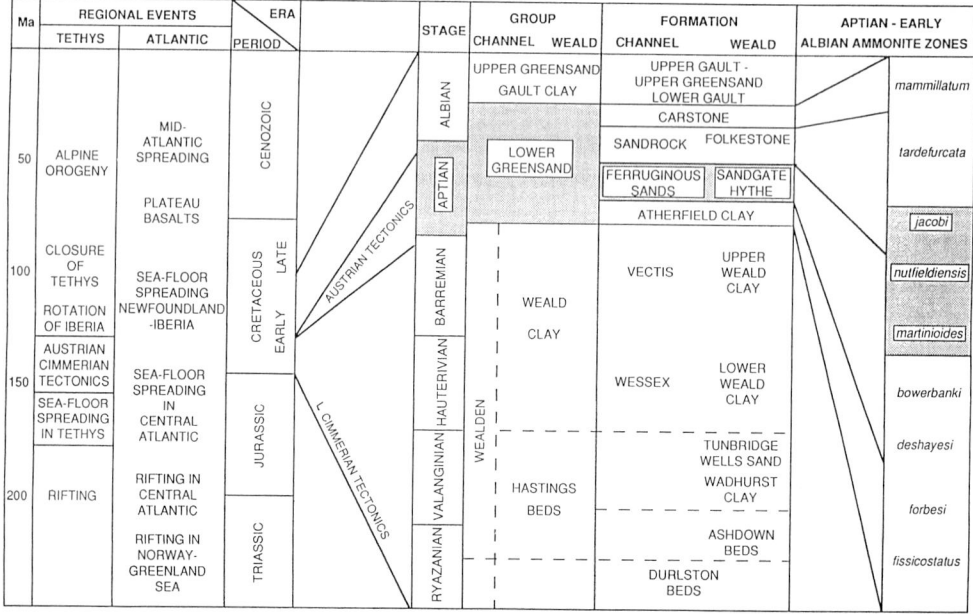

Fig. 1. Bio- and lithostratigraphy of the early to mid-Cretaceous successions of southern England. The horizons discussed in text are highlighted. Mesozoic stratigraphy and approximate ages of known tectonic changes are included in the left-hand columns. After Casey (1961) and Ruffell (1992).

(1997 this volume) consider uplift to be more widespread and generated from the opening of the Bay of Biscay. Other work on this interval includes detailed outcrop logging during a study of fuller's earth bentonite clays within the Lower Greensand Group in southern England (Moorlock & Highley 1991), from which sections showing significant lithological changes were analysed in detail (see below).

The Lower Cretaceous is characterized throughout much of the northern hemisphere by tectonic and volcanic activity (Fig. 1), along with significant changes in global sea level (Haq et al. 1988). In southern England, episodes of crustal activity thought to have affected stratigraphic successions occurred in the very earliest Cretaceous (Stoneley 1982; Whittaker 1985); in the late Ryazanian–Berriasian (McMahon & Turner 1997 this volume) and around the middle of the Aptian stage (this study). Karner et al. (1987) and Ruffell (1992) suggest that, during the Aptian, formation of a tectonically enhanced unconformity was associated with hanging wall deformation, footwall uplift–erosion and the accumulation of thick layers of volcanic ash (fuller's earth bentonites). Lower Cretaceous depocentres of the Weald and Channel basins in southern England (Fig. 2) remained around the core of the (present-day) Weald Anticline and south of the Isle of Wight Monocline, respectively, until mid-Aptian times. Beds of mid-Aptian to mid-Albian age show progressive onlap of depositional basin margins, exhibiting deposition on former massifs throughout southern England in late Albian (Gault–Upper Greensand/Chalk) times (Whittaker 1985; McMahon & Turner 1997 this volume). The northern outcrops of the 'mid'-Cretaceous in southern England extend southwest from the Wash, through Cambridge and Oxford, striking west through Wiltshire and then due south through Dorset (Fig. 2). Along this outcrop the Aptian–Albian Lower Greensand is preserved in a series of minor basins that originated as erosional features, cut during the preceding early Cretaceous uplift along lines of structural weakness (Casey 1961; Hesselbo et al. 1990a; Ruffell & Wignall 1990).

In this study key sections from basin margin and depocentre successions in the Upper Aptian Lower Greensand Group of southern England were logged in detail (Figs 3 and 4). Possible sequence stratigraphic interpretations of outcrop bounding surfaces and (where possible) internal parasequence stacking geometry (sensu van Wagoner et al. 1990) were then made prior to correlation. Thus, comparison was made without any presupposition as to whether parasequences or key surfaces would be of most use, or indeed whether the sequences were synchronous or diachronous. The ensuing correlation is presented here in an attempt to describe and document the geometry and associated facies of depositional sequences that formed under different levels of tectonic influence. Outcrop logging and correlation demonstrated the existence of estuarine sands on the Isle of Wight, fuller's earth bentonites (at various locations) and lagoonal muds and limestones around Maidstone (Kent) at similar biostratigraphic horizons to those reported by Casey (1961). The absence of thick fuller's earths in the Isle of Wight succession has previously thrown some confusion into the correlation of the Lower Greensand Group between the Channel and Weald Basins (Casey 1961; Moorlock & Highley 1991; Ruffell 1992). Correlation of sequence boundaries rather than the internal facies of each depositional sequence explains this variation where the ammonite biostratigraphy allows no better resolution, due to the absence of fossils. In this study, lithostratigraphic correlation has been constrained by the available biostratigraphy such that the three loosely dated depositional sequences recognized may be compared in geometry and facies. The upper- and lowermost sequence boundaries are defined by a basinward shift of facies, however, the

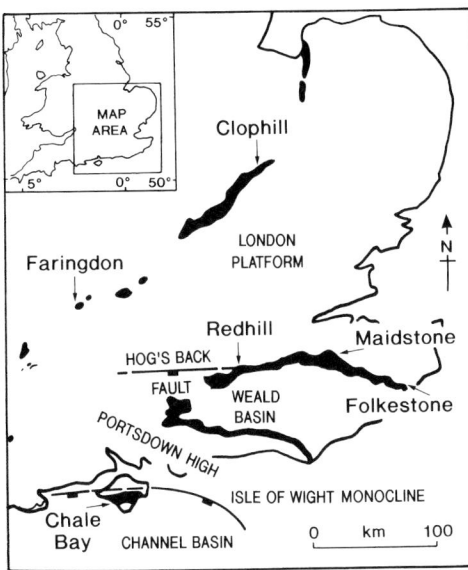

Fig. 2. Location of the study area with locations mentioned in text. Lower Greensand Group (Aptian–Albian) outcrop in black.

intervening sequence boundary is formed by a transgressive surface overlying folded sediments beneath.

Outcrop stratigraphy

Following outcrop logging of mid-Aptian exposures throughout the Weald and Channel basins, lithostratigraphic correlation was compared to the established biostratigraphy (Casey 1961) (Fig. 3). In order to provide a separate method of correlation, these sections were then measured in detail in order to elucidate parasequence stacking patterns (Fig. 4) and thus their position within systems tracts. Owing to the number of sequence boundaries within the Upper Aptian–Lower Albian succession, complete depositional sequences are rarely preserved; consequently systems tracts were difficult to define. Quite commonly the depositional sequence preserved represents part or parts of just one systems tract and thus bounding surfaces rather than parasequences are highlighted.

Clophill (Bedfordshire)

Clophill (UK National Grid: TL 100 384) is the northernmost exposure analysed in this study (Fig. 2). Here, the Aptian Lower Greensand rests unconformably on folded and faulted Ampthill Clay (Upper Jurassic), and is regarded as a basin-margin succession. A quarry opened in 1984 to the west of Clophill town extracts fuller's earth from the Lower Greensand; no previously published records exist for this location. Ammonites of the *nutfieldiensis* zone (identified by Casey pers. comm.) were found by Shephard-Thorn in the fuller's earths beds (interbedded massive bentonite clays with bioturbated glauconitic sands). The fuller's earths are eroded at their upper contact by cross-bedded, bioturbated glauconitic sands with a basal pebble bed (Fig. 4). Phosphate nodules from this horizon include the steinkerns of reworked Upper Jurassic ammonites (*Cardioceras*; various Perisphinctids; *Pavlovia*), most likely derived from the London Platform to the south: a sediment source/land area in Aptian times (Middlemiss 1962). Rip-up clasts of fuller's earth bentonite are also common in this pebble-bed, indicating the erosional reworking of bentonites deposited above seams now exposed. A second, stratigraphically higher erosion surface is present in outcrop sections to the southwest (Fox Corner sand-pit, SP 927 293 Leighton Buzzard) at the base of the Woburn Sands (Wonham & Elliot 1996). Compared to the other mid-Aptian successions (described below and in Fig. 3) the Clophill section displays both thinner individual beds and depositional sequences; this is a product of both minor internal and sequence boundary-scale erosion surfaces. The final analysis of these erosion surfaces is dependant on their correlation with other sections: the quarry outcrop exposes only 5–6 m of strata, precluding complete analysis of parasequence stacking patterns.

Redhill (Surrey)

Fuller's earth bentonite clays occur within the Lower Greensand of the Redhill area and their commercial extraction has provided numerous quarry sections. Fuller's earth seams occur intermittently within the Lower Greensand of the Weald Basin (Fig. 2) and are not always present within the mid-Aptian Sandgate Formation under consideration here. Instead, where absent, a single unconformity may be present (Sandgate–Hythe formations contact), a time-gap of one or more ammonite zones is often provable (Casey 1961). From this information it appears likely that the fuller's earth-bearing horizon has been removed during formation of the unconformity, and that the most complete ('down-dip') stratigraphic section is preserved within the fuller's earth-bearing succession. We may compare the complete succession at Redhill with a stratigraphically incomplete section 20 km west of Redhill around Albury in Surrey (Grid Ref. TQ 054 476). Here, a single mid-Aptian unconformity occurs at the contact of the Lower Aptian Hythe Formation (calcareous and siliceous fine sandstone) and Upper Aptian Bargate Beds (calcareous and fossiliferous poorly sorted pebble-rich sandstones of the basal Sandgate Formation). Lake & Shephard-Thorn (1985) suggested that the Bargate Beds formed by the reworking of both lithologies below, and derived Upper Jurassic material from the London Platform to the north (Arkell 1939). Casey's (1961) biostratigraphic analysis of the succession demonstrated that the *martinioides* zone was absent in this incomplete succession; presumably eroded prior to the *nutfieldiensis* zone transgression. Eastwards toward Redhill a more complete succession (Figs 4, 5b and 6), including the fuller's earths is preserved. Dines & Edmunds (1933) demonstrated how the fuller's earths are truncated by glauconitic, pebbly then cross-bedded sands above (Fig. 6). The succession at Redhill is thus more complete than to the west, indicating that the Hythe Formation–Bargate Beds contact (Fig. 4) represents the amalgamation of at least two erosive surfaces (Knox *et al.* 1998). The sequence stratigraphic interpretation of these surfaces is outlined below.

TECTONIC ACCENTUATION OF SEQUENCE BOUNDARIES 337

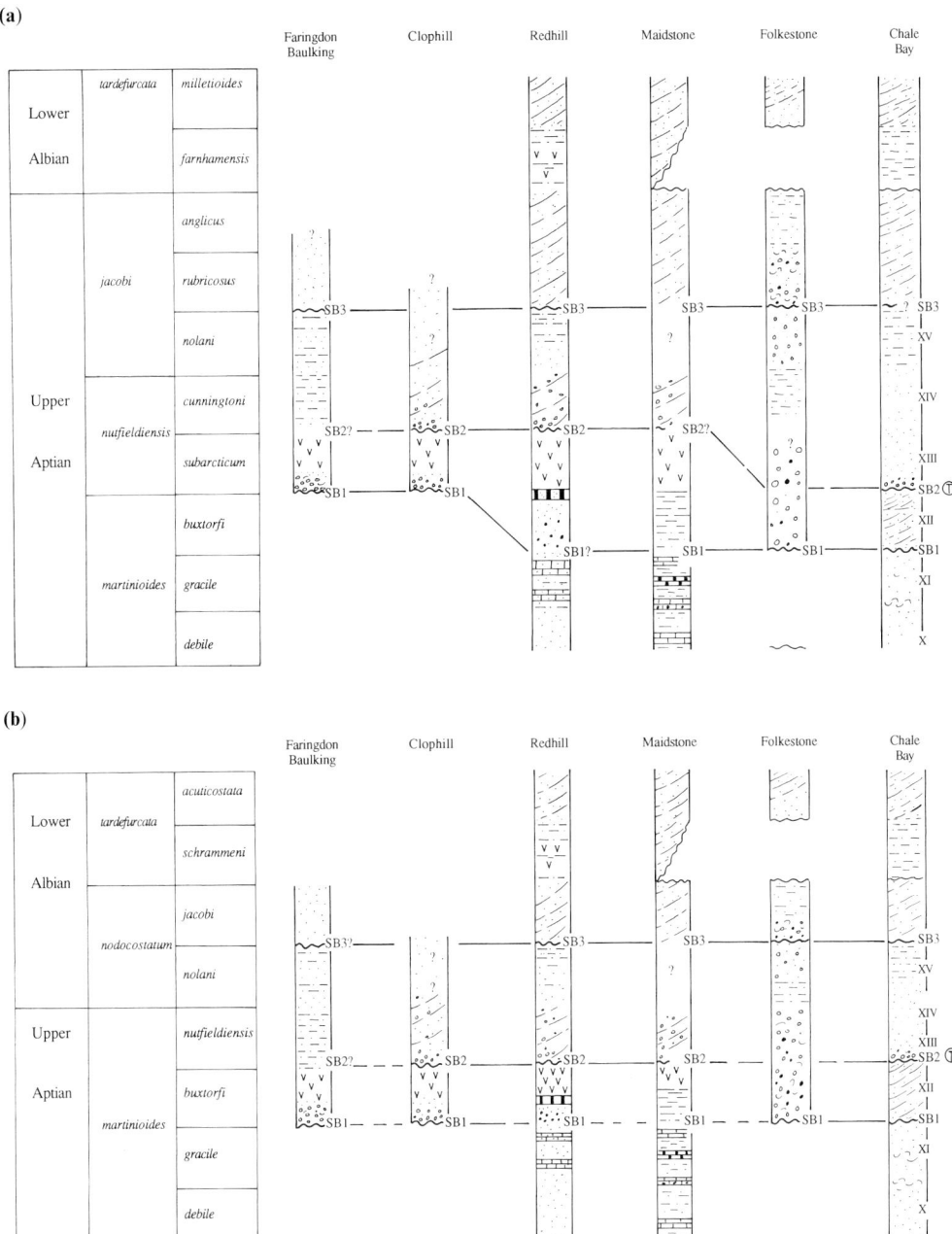

Fig. 5. (**a**) Chronostratigraphic diagram based on the Upper Aptian–Lower Albian ammonite biostratigraphy of Casey (1961) utilising the sections discussed in text. On this correlation, SB2 is diachronous between the Isle of Wight and Redhill (and possibly other locations). SB1 may also be diachronous between Faringdon–Clophill and the other sections, although no underlying Lower Greensand occurs to constrain this. T (circled) is the position of tectonically accentuated unconformity. Key as on Fig. 4. (**b**) Chronostratigraphic diagram based on the Upper Aptian–Lower Albian ammonite biostratigraphy of Owen (1996). In contrast to (**a**), neither SB1 nor SB2 appear diachronous. Key as for Fig. 4. T (circled) is the position of tectonically accentuated unconformity. Key as on Fig. 4.

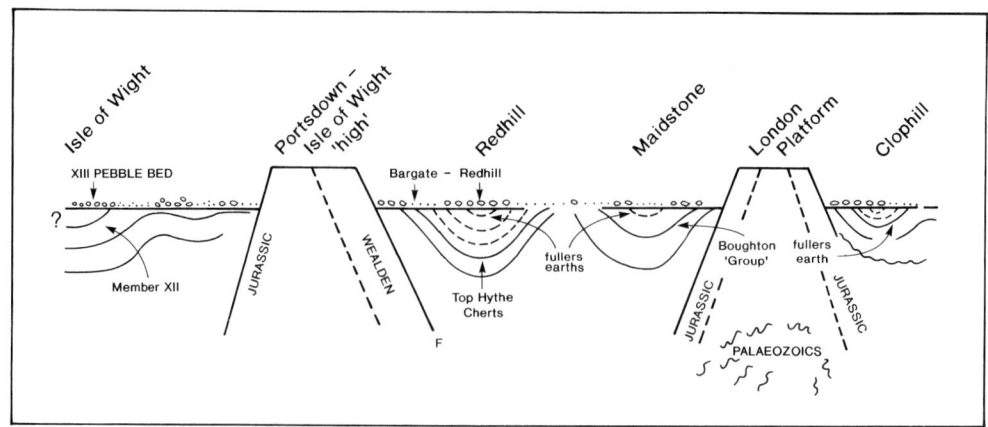

Fig. 6. Geometry of early mid-Aptian incised deposits (estuarine sands of Member XII, Isle of Wight), truncated fuller's earths (Redhill, Clophill) and lagoonal muds and limestones (Maidstone). The *nutfieldiensis* zone

The outcrop sections described here from around Redhill are situated toward the western end of the North Downs, on the northern margin of the Weald Basin (Fig. 2). The lowest part of the succession was exposed at Bletchingley (4 km east at TQ 328 504). In Cretaceous times the Palaeozoic massif of the London Platform lay to the north. This was the site of Cretaceous–Tertiary basin inversion, and now lies at depth, between 10 and 15 km north of Redhill. Fuller's earths (Ca-montmorillonite bentonites) have been worked in the Redhill ('Nutfield') area (around Grid Ref. TQ 300 510) since Roman times (Casey 1961). Over the past century the fuller's earth quarries of the area have deepened as the seams were followed down-dip to the north: thus the thickest stratigraphic succession was seen immediately prior to closure of the quarries in 1992–93. Although Redhill is 155 km south of the Clophill section and also south of the London Platform (Fig. 2), it is remarkable that both the lithofacies and depositional sequences are similar. These comprise interbedded fuller's earth bentonites and calcareous bioturbated greensands erosively overlain by pebbly cross-stratified sands. The *nutfieldiensis* zone deposits of Redhill are thicker and more fossiliferous than at Clophill, while late Aptian–early Albian sandwave bedsets are 3–4 m thick north of the London Platform and only 1–2 m thick to the south.

The stratigraphy of the fuller's earth pits (Redhill)

The last deep working quarry (Patteson Court, TQ 294 513) exposed 35 m of strata in 1992, and showed an almost complete Upper Aptian succession (Fig. 4). The fuller's earth seams were worked from between calcareous glauconitic sandstones which have a varied fauna of brachiopods, bivalves, echinoids and the rare ammonites (*Parahoplites*, *Tropaeum*) that allow the fuller's earth beds to be placed in Casey's *nutfieldiensis* zone. Other formations exposed here are not biostratigraphically dated in this immediate area. Individual fuller's earth beds contain the low angle cross-strata, ripples and fluid-escape structures noted by Hallam & Sellwood (1968), and by Jeans *et al.* (1982), as well as fossil nautiloids and *Lingula*. The latter may indicate oxygen-deficient or brackish conditions (West 1976).

A planar to low-relief erosion surface separates the fuller's earth-bearing strata below (Nutfield Beds of Knox *et al.* 1998) from coarse pebbly sands around 3 m thick with bands of white calcareous concretions. Truncation at the base of the sands (Redhill Sands of Fig. 4) is visible in the field, but best observed from former sections and boreholes along strike (Leighton 1895; Dines & Edmunds 1933, Knox *et al.* 1998). Such pebbly sands (with an occasional calcareous cement) are considered by the author to be of the same lithology and stratigraphic position to the Bargate Beds best developed 20 km to the west. Middlemiss (1962) suggested that the Bargate Beds (the likely stratigraphic equivalents to the beds above SB2) were derived from the uplands of the London Platform to the north (Fig. 2), and that the influx of this material in the *nutfieldiensis* zone deposits of the North Downs was due to uplift of this source. Dines & Edmunds (1933) observed glauconitic, cross-stratified and pebbly sands resting erosively on folded Hythe Formation and fuller's earths in the numerous shallow quarries of the Redhill area. They placed this erosion surface within their Sandgate 'Beds', describing the pebbly and occasionally calcite-cemented beds above as similar to the Bargate Beds of the western Weald. Knox *et al.* (1998) now term this horizon the Redhill Sands. During the 1980s deeper quarries were excavated north of, and adjacent to, the locations studied by Dines & Edmunds (1933). These clearly displayed the stratigraphic succession and two discrete erosion surfaces (SB2 and SB3) shown in Fig. 4 and documented by Ruffell (1990) Moorlock & Highley (1991) and Knox *et al.* (1998). The presence of abundant rip-up clasts of fuller's earth bentonite above SB2 indicates the prior erosion of the underlying succession. This erosion surface may be traced in numerous exploratory boreholes in the Redhill area and show (as in Dines & Edmunds 1933) that the post-folding erosion surface of the area lies within the Sandgate Formation at the base Redhill Sands (SB2). The combined analysis of historic work (Dines & Edmunds 1933) and more recent alteration to the stratigraphy (Knox *et al.* 1998) allows better definition of the position at which the erosive surface overlying folded beds (SB2) occurs. This truncation was of intra-*nutfieldiensis* zone age using Casey's (1961) biostratigraphy, or of *martinioides* zone age when compared to Owen (1996); see below for discussion of this assignment.

Whilst the fuller's earth quarries around Redhill have provided much information on the stratigraphy of the beds above the fuller's earth, the quarry at Bletchingley (TQ 328 504), which was formerly worked for the Mid-Hythe Sand and fuller's earth (until 1988), was unique in showing the clear transition from Hythe Formation to fuller's earth-bearing beds (or the approximate horizon of SB1). A conformable passage from chertified sands of the Hythe Formation to fuller's earth-bearing Nutfield Beds was observed with no clear break visible. The lack of a basal Sandgate Formation discontinuity in the field is equally well demonstrated by the conformable geometry of the Hythe–Sandgate formations across the whole area (Fig. 6) (Dines & Edmunds 1933, fig. 10, p. 63).

Sequence stratigraphic interpretation – Redhill

Three depositional sequences can be mapped in the quarry faces in the Redhill area. The lowest has no clearly defined base, being somewhere in the conformable (early to mid-*martinioides* zone) transition from fine-grained sands into coarse sands containing phosphate nodules (the Top Hythe Pebble Bed of Leighton 1895). This transition may be the cryptic representation of Sequence Boundary 1 (SB1), bioturbation, silica diagenesis and transitional facies changes making precise positioning of the boundary difficult. Above this bed a thin (1–4 m) succession of interbedded cherty and glauconitic sands and clays occur, the upper clay beds occasionally being minor fuller's earths. In the upper parts of this sequence chert disappears at the expense of glauconitic sands and fuller's earth seams. This depositional sequence (?SB1–SB2) is unconformably overlain by SB2 (the base of the Redhill Sands of Fig. 4), the beds above comprising a series of 1–2 m high pebbly and calcareous cross-stratified sets. Individual bedsets may coarsen-up, but overall this package shows upwards-thinning beds and a fining-up profile into soft glauconitic sands and dark clays: this pattern of retrogradation is characteristic of parasequences developed in a transgressive systems tract (*sensu* van Wagoner *et al.* 1990). These sands are overlain by the pebbly and glauconitic sediments at the base of the Folkestone Formation (SB3), comprising the third depositional sequence mapped in the area below the Folkestone Formation.

The Nutfield Beds and underlying uppermost Hythe Formation are preserved in a broad and shallow (15–20 km wide) syncline, the axis of which trends approximately north–south (Dines & Edmunds 1933). The fuller's earth-bearing Nutfield Beds contain a marine fauna with brackish elements, may overlie a sequence boundary in the Top Hythe Pebble Beds and show occasional back-stepping parasequences: these two units could thus be regarded as representing the lowstand (top Hythe) to transgressive (Nutfield Beds) systems tracts. The basal Redhill Sands would then comprise a transgressive surface formed during ravinement and the reworking of pebbles and underlying fuller's earths. The topmost Redhill Sands are characterized by dark clays and backstepping parasequences: these are interpreted as forming in the transgressive systems tract. There is no evidence for a maximum flooding surface within these clays: such an horizon may have been removed during generation of the overlying sequence boundary (SB3). This sequence stratigraphic interpretation accounts for the transgressive reworking of Nutfield Beds at the base of the Redhill Sands, but not the degree of truncation evident across the Redhill–Bletchingley outcrop: thus, it may be that a transgressive surface is here amalgamated with a sequence boundary. The folded Hythe Formation attests to truncation at this level being structural in origin (Dines & Edmunds 1933; Ruffell 1992; this study) (Fig. 6), suggesting that this surface is a tectonically enhanced unconformity. Hesselbo et al. (1990b) suggested a similar origin for this surface at Folkestone (below). No chronostratigraphic time-gap is known at Redhill (Casey 1961) and thus we presume that the depositional sequences are more fully represented here than at Folkestone or in the northern outcrops. The only other the succession as complete is Chale Bay (Isle of Wight), with which I make a comparison later.

Maidstone (Kent)

Around Maidstone the lower beds of the Lower Greensand (Hythe Formation) are quarried for building stone, early diagenesis having imparted a limestone–glauconitic clay/siltstone rhythmicity to the succession (the 'Rag and Hassock'). The upper beds are well known historically, having yielded *Iguanodon* along with many other unusual vertebrate fossils: the local correlation of this succession has been subject to debate (Worssam 1993). Overall, the basic palaeoecology of the succession records the transition from shallow offshore deposition to lagoonal conditions (Ruffell 1992). The Hythe Formation has yielded biostratigraphically diagnostic ammonites. These indicate a *martinioides* zone age (Casey 1961). Lithologically, clays developed in the upper parts of the uppermost Hythe Formation resemble the *tscherysheweni–subnodocostatum* clays of Germany (Kemper 1979) of which they may be time equivalents. The importance of this comparison is discussed in the conclusions.

Sequence stratigraphic interpretation – Maidstone

The composite log (Fig. 4) is derived from the many shallow outcrops around Maidstone (described in Ruffell 1992). This profile may be interpreted in terms of the upper Hythe Formation (Boughton 'Group') beginning deposition during a sea-level lowstand, although any possible preceding sequence boundary (SB1) is not at all obvious. The lagoonal conditions interpreted by Ruffell (1993) that persisted from this lowstand into a transgressive systems tract are reflected in a changing parasequence stacking pattern from progradation to retrogradation (Fig. 4). The basal Sandgate Formation transgressive surface truncates the lagoonal uppermost Hythe Formation sediments below in a similar manner to the base Redhill Sands (SB2) of the Redhill outcrops: no correspondingly accurate biostratigraphic data exists in the Maidstone area to enable a chronostratigraphic correlation with the Redhill Sands (if at all).

Folkestone (Kent)

Although known to be a somewhat condensed succession compared to the Isle of Wight or Redhill (Casey 1961; this study) (Fig. 4), the Lower Greensand Group at Folkestone contains three stratotype formations (Hythe, Sandgate and Folkestone) defined by Fitton (1836) as 'Beds'. The conglomeratic bed at the base of the *nutfieldiensis* zone Sandgate Formation is the product of a long period of reworking, mineralization and penecontemporaneous erosion (Casey 1961). Hesselbo et al. (1990b) have suggested that this bed is the result of a sequence boundary and transgressive surface having been amalgamated. The mid-Aptian Lower Greensand at Folkestone is characterized by the abrupt junction between the Hythe Formation below and Sandgate Formation above (Fitton 1836; Casey 1961). The exposure at Mill Point (TR 220 352) was documented by Casey (1961) and Hesselbo et al. (1990b). These studies suggested that an angular unconformity exists between the Hythe and Sandgate Formation. Casey (1961) demonstrated that the basal conglomerate of the Sandgate Formation contains a variety of clasts, some of which contain late *martinioides* zone (*buxtorfi* subzone) ammonites. This led Casey to suggest that the basal Sandgate Formation are of *martinioides* zone age: the light brown phosphatic nodules that contain these ammonites are reworked, and the first *in situ* Sandgate Formation are of *nutfieldiensis* age. This does not preclude a transgression of the preceding late *martinioides* zone. The variety of clasts present in the Mill Point 'unconformity facies' of Hesselbo et al. (1990b) support Casey's (1961) concept of a number of erosive–depositional events, and a high level of reworking. Comparison of the Folkestone section to the other locations described around the Weald and Channel Basins suggests that, further to the interpretation of Hesselbo et al. (1990b), two sequence boundaries are possibly represented in the basal pebble bed of the Sandgate Formation at this location.

Chale Bay (Isle of Wight)

The Cretaceous succession exposed on the southern coast of the Isle of Wight forms part of a Mesozoic succession developed in the Channel Basin (a Cretaceous sub-basin of the Wessex Basin complex of Stoneley 1982). Mesozoic depocentres were concentrated south of the Isle of Wight, on the subsiding hanging wall of the east–west-trending Isle of Wight Fault (now defined by a monocline formed in the Miocene). This was evident in the early Cretaceous when no sediments of pre-Albian age were preserved north of the monocline, where Albian Carstone sediments at the base of the Gault rest unconformably on Jurassic strata (Ruffell 1992). Lack of early Cretaceous deposits north of the Isle of Wight Monocline indicate that this structure must have had some effect on deposition to the south. Early Cretaceous facies of the Weald and Channel basins are thus quite different (Stewart 1981; Stewart et al. 1991).

The Lower Greensand of the Isle of Wight is known primarily from the pioneering work of Fitton (1845), Ibbetson & Forbes (1845) and White (1921); the palaeontological studies of Casey (1961) and Simpson (1985); and the sedientological work of Dike (1972) and Wach (1994). The type section of the Lower Greensand runs along the coast at Chale Bay for 8 km, where it is around 250 m thick. Most of the work by Fitton (1845) and Casey (1961) was undertaken here: this is the reference section for both the Aptian stage in Europe, and Lower Greensand Group in southern England. Fitton's divisions (known here by Roman numerals) are still in use, and the mid-Aptian succession, dated by Casey (1961), demonstrates the *martinioides–nutfieldiensis* phase of shallowing, as seen at Maidstone, but here developed as estuarine sands.

The mid-Aptian succession is shown in Fig. 4 (after Wach & Ruffell 1991; Wach 1994). The suggested (informal) stratigraphic nomenclature of these authors (numbered members) is followed here, albeit unsatisfactory. Member X is a bioturbated (*Thalassinoides*), occasionally trough cross-bedded, fossiliferous and argillaceous sandstone; Member XI is argillaceous, glauconitic and highly bioturbated with prominent siderite-cemented burrowed firmgrounds. Bivalves with estuarine affinities are recorded from the otherwise marine (glauconitic, bioturbated) Member XI (Ruffell & Wach 1991; Simpson pers. comm.). Member XII marks a more complete departure from the shelf sedimentation observed below: the unconsolidated glauconitic and white quartzose sands with clay drapes that form this member are characteristic of the estuarine cycles found in the Aptian–Albian boundary beds above (Dike 1972; Wach 1994). The base of the member can be observed cutting down into the underlying shelf sediments of Member XI (Wach & Ruffell 1991): an erosional episode thought to be late *martinioides* zone in age (Casey 1961). This horizon marks the first major departure from shelf deposition observed in the Aptian Lower Greensand of the Isle of Wight.

Comparisons may be made between the geometry of these estuarine deposits and the fuller's earth beds (Redhill)–lagoonal Boughton Member (Maidstone); all three occupy incised or synclinal depressions and are overlain by a regionally planar and erosive transgressive surface. However, on the established biostratigraphy (Casey 1961), these three successions are of different ages (Fig. 5). Member XIII rests with a marked erosion surface and pebble bed on the sands of Member XII below. The pebble lithologies include reworked pebbles and cobbles of Upper Jurassic material, some of which are up to 30 cm in diameter. This surface is analogous in age and character to the basal Redhill Sands and post-fuller's earth sands at Clophill, also tempting a lithostratigraphic correlation to SB2. Again, on the established biostratigraphy (Casey 1961), these two horizons are of different ages: base *nutfieldiensis* on the Isle of Wight and intra-*nutfieldiensis* at Redhill.

Sequence stratigraphic interpretation – Isle of Wight.

The pebble bed shown on Fig. 4 at the base of Member XIII may be interpreted as the product of ravinement associated with a transgressive surface. Erosion is apparent (and possibly real) as the underlying sands (Member XII) were (and still are) friable. The questions raised are: why were derived Jurassic clasts introduced at this time? And how did a trans-gressive shoreline transport such large clasts? In considering these questions it serves to reconsider the Redhill succession where at outcrop the erosive surface at SB2 appeared only as a transgressive surface: wider correlation proved this surface to unconformably overlie folded strata and thus also represent a sequence boundary but without a preserved basinward shift of facies. In both cases, large clasts could have come from an adjacent footwall high and/or exposed Wealden pebble-beds. In the Isle of Wight succession, however, the nearest probable fault-scarp is the inverted monocline, some 10 km to the north. Some angular clasts of Portland Sandstone around 20 cm in length have been found by the author in the pebble-bed above SB2 in the Chale Bay succession, implying strong currents and a rejuvenated sediment hinterland.

Using the Redhill succession as an analogue for the Isle of Wight, a sequence boundary coincident with a transgressive surface, or a tectonic episode at SB2, as at Redhill might be suggested. The former explanation requires the creation of two closely spaced sequence boundaries (SB1 and SB2), each with very different characteristics: the mid-*martinioides* downcutting was associated with a basinward shift of facies where estuarine sediments overlie sands of marine origin, yet no coarse detritus was introduced. The *nutfieldiensis* surface shows a landward shift of facies where a pebble-bed and shelf sediments overlie estuarine sands, yet truncation was associated with the introduction of very coarse material. A tectonic episode preceding the *nutfieldiensis* transgression has been suggested before by Casey (1961). The ramifications for sequence stratigraphy go beyond dating a period of uplift, suggesting that a tectonic episode of the duration of an ammonite zone, followed by transgression, may be indistinguishable at one outcrop section from a sequence boundary or erosive transgressive surface formed by purely eustatic or regional sea-level changes. The evidence for tectonic activity at this horizon is only evident given a regional perspective (Figs 5a, b and 6).

Faringdon–Baulking (Berkshire)

The geology of the Faringdon–Baulking area is well known through the outcrop geology and palaeontology of the Faringdon Sponge Gravels (Krantz 1972) and the geology of fuller's earth bentonites (Poole & Kelk 1971; Poole *et al.* 1971). The Sponge Gravels rest unconformably on Jurassic Corallian Limestone (C.L. on Fig. 3). The timing of early Cretaceous erosion is not known: multiple phases of incision may have occurred between the Jurassic–Cretaceous transition and the Aptian. Some incision is also possible immediately prior to deposition of the *nutfieldiensis* zone sponge gravels (in the late *martinioides* zone). The fuller's earths are stratigraphically higher than the sponge gravels, although both units are rarely present over the whole Faringdon–Baulking area. At Baulking (SU 323 912) over 20 m of silty and fossiliferous clays follow the fuller's earths, and similar successions can be examined in cores obtained by the BGS (then IGS): see Poole & Kelk (1971) and Poole *et al.* (1971). These boreholes showed the extent of the sponge gravels; fuller's earths and overlying penecontemporaneous slumped beds; erosive-based sands (see description of Furze Hill Borehole in Poole *et al.* 1971); and the general morphology of the trough in which the Sponge Gravel–fuller's earth succession was deposited. Poole & Kelk (1971) and Poole *et al.* (1971) show two erosive horizons above the *nutfieldiensis* fuller's earth beds. The lower surface is overlain by mudstones (in the Baulking area; Poole & Kelk 1971, fig. 5). The upper erosion surface is marked by the base of sands in the cores taken around Fernham (Poole *et al.* 1971, fig. 3). This sand unit oversteps Jurassic strata to the north, and is mapped at Faringdon Folly (SU 305 988), where *nutfieldiensis* zone ammonites are recorded (see Hesselbo *et al.* 1990*b*). These sands appear to onlap the Faringdon Trough margins in the same way the Redhill Sands transgress the margins of the Nutfield Beds Syncline, and although no synchroneity is proven a genetic relationship may be implied.

Sequence stratigraphic interpretation – Faringdon–Baulking

The inconclusive nature of the evidence is such that Sequence Boundaries 2 and 3 cannot be differentiated in this area. The base of the Furze Hill Borehole sands overlaps/truncates underlying strata in a similar manner to SB2 around Redhill yet these sands are lithologically akin to beds found elsewhere above SB3. With this in mind, and in advance of more precise biostratigraphic age determination, we might suppose the following correlation by pure lithostratigraphic and depositional sequence geometry. Consideration of all the surfaces in this area have attendant problems in interpretation. The base Lower Greensand Group unconformity (on Jurassic) in the Faringdon area may equate with SB1, albeit that this sequence boundary is *martinioides* age in the Isle of Wight and ?basal *nutfieldiensis* age around Redhill and Faringdon. In addition, SB1, or the mid-Cretaceous–Jurassic unconformity, represents a long time-gap in which multiple episodes of erosion may be amalgamated. The base of the erosive mudstones of the Baulking boreholes equates to SB2; the base of the Furze Hill Borehole sands may be SB3, making these a correlative of the Folkestone Formation. Placing SB1 below the fuller's earths supposes that their trough-like geometry is similar to that of the Redhill deposits and that the ash-falls are roughly synchronous. Similarly, the Maidstone uppermost Hythe Formation and the Isle of Wight estuarine succession (Member XII) also fill minor troughs, developed in post-SB1 times and overlapped by post-SB2 sediments. The latter, however, do not appear to be associated with the introduction of coarse detritus observed above SB2 at all other locations.

Sequence stratigraphic interpretation

Surfaces

There are two possible explanations for the patterns of truncation observed at the levels of SB1 and SB 2 (respectively) in southern England: the first would be to relate all the observed features to global/regional patterns of sea-level change; the second would be to invoke tectonic processes operating during the time of SB2, as folding can be demonstrated below the level of truncation. The merits of the two explanations are outlined below.

The generally non- to shallow-marine beds of the late *martinioides*–early *nutfieldiensis* zones can be interpreted as part of a lowstand or early transgressive systems tract. This is supported by the facies analysis of estuarine sands on the Isle of Wight (Wach & Ruffell 1991) and lagoonal deposits of Maidstone (Ruffell 1993), respectively. The underlying sequence boundary (SB1) may be subtle, or not observed, due to a gradual shift of facies (e.g. in the Redhill area) in a 'down-dip' correlative conformity succession: conversely on the Isle of Wight, erosion–down-cutting is more obvious (Figs 3 and 4). As SB1 represents the Jurassic–Cretaceous unconformity at Faringdon–Baulking, it alone need not have generated all the erosive topography now observed.

The individual outcrop characteristics of the overlying *nutfieldiensis* pebble beds (base Redhill Sands–Member XIII and correlative deposits of Fig. 4) suggest that this surface formed through ravinement during transgression. Such shallow-water shoreface/beach reworking could account for the pebble-lag commonly found at this level, as this transgressive surface encroached upon surrounding Jurassic (and sometimes older) highs. This sequence stratigraphic interpretation, when compared to the established biostratigraphy (Casey 1961), indicates that the SB2 surface may be diachronous (Fig. 5a). This supposition is of some significance as Galloway (1989) and Partington *et al*. (1993) suggest that sequence boundaries and transgressive–ravinement surfaces are often diachronous. In the Galloway model (1989), only the maximum flooding surface is synchronous. The 'Exxon' model, typified by Haq *et al*. (1988) is that the sequence boundary (when dated at its correlative conformity) is synchronous. In this study the age of incision appears to vary, in accordance with the Galloway (1989) model, albeit that the correlative conformity is not preserved. However, the established ammonite biostratigraphy in this case may not be accurate enough to estimate diachroneity.

Interpreting SB2 as a transgressive surface only, implies that much of the underlying sequence was eroded during ravinement and that the hydrodynamic conditions developed during this transgressive phase were strong enough to both transport boulders up to 15 cm in diameter (found in the pebble beds of the Isle of Wight) and to truncate the underlying (folded) beds at Redhill. The degree of truncation suggested from regional correlation of SB2 (Fig. 6) implies that erosion has removed the highstand, and possibly the late transgressive systems tracts of the underlying sequence. In addition, it is possible that much of the surrounding late Jurassic was removed during this episode, explaining the absence of derived clasts in SB1 and their abundance in SB2 away from the basin margins. The truncation of sustantial thicknesses of strata (estimated at Redhill to be 10–20 m) is not usual during ravinement, as accommodation space for the deposition of sediments should increase during sea-level rise. Uplift of surrounding highs and massifs (Fig. 2) some time before the deposition of SB2 would provide a sediment source for these clasts, and explain the pattern of truncation. The outcrop classification of this horizon as a transgressive surface changes to one of a sequence boundary on regional tectonic considerations (Fig. 6).

The extensive erosion observed at SB2 is aided by truncation of dipping beds within the underlying sequence (Fig. 6). This observed variation in attitude could be a product of the preceding lowstand–transgressive systems tracts being deposited in minor basins and incised valleys, the higher beds of which were truncated. An example of this process is described by Hesselbo *et al*. (1990*a*) in the Calne area of Wiltshire, where an incised valley was swept clear of sediment by a subsequent phase of erosion. Such minor basins and valleys were often developed along folds and faults that may have aided increased subsidence than in the surrounding areas where truncation was more marked (Ruffell & Wignall 1990). Thus, passive tectonic processes may have aided the preservation of sediment in minor depositional basins, and accentuated erosion–truncation along the flanks, provided that sufficient time elapsed before deposition of SB2 to allow structuring and such significant erosion. An alternative is to suggest some minor structural deformation associated with regional uplift, preceding deposition of SB2. There is no question that in the Redhill area some minor folding took place in the short time-gap represented by SB2; any doubters need only examine fig. 10 of Dines & Edmunds (1933) which shows clear evidence of the structural attenuation of fuller's earths below an unconformity. Passive tectonic processes (gradual subsidence) during SB2 require a longer time-gap (to become accentuated) than is allowed by current biostratigraphy. Shorter-lived active movements (e.g. footwall uplift or compression from the Atlantic) are considered more likely as the folding at Redhill appears to have occurred within the space of one zone.

The presence of an additional ammonite zone at this level in the German successions (the *tscheryscheweni–subnodocostatum* clays mentioned above: Kemper 1979) may be evidence of a slightly more extensive time-gap at SB2 than previously thought. Whatever the most accurate dating tells us, the time-gap at SB2 was less than one ammonite zone.

Depositional sequences

Correlation of the bounding surfaces to the three depositional sequences elucidated in this study allows comparison of the enclosed geometry and facies of each sequence. The SB1–SB2 depositional sequence comprises either a structurally or sedimentologically preserved succession below the tectonic sequence boundary (SB2). The best example of a structurally preserved sequence is that described from Redhill (Fig. 7), where

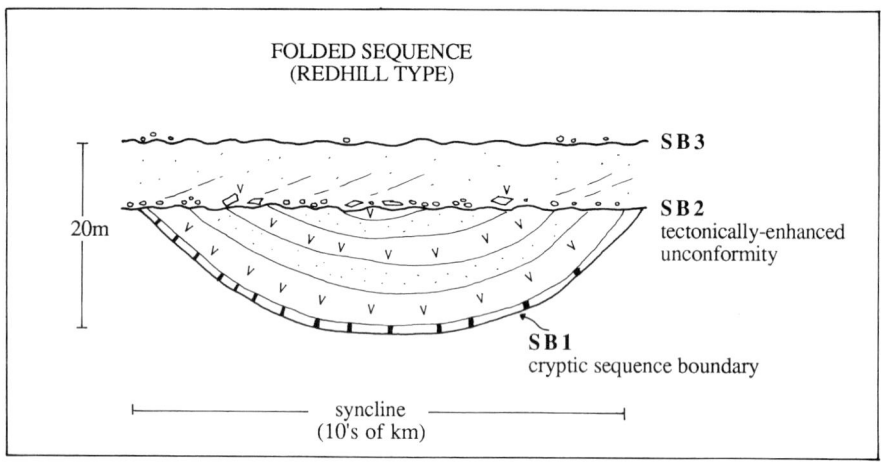

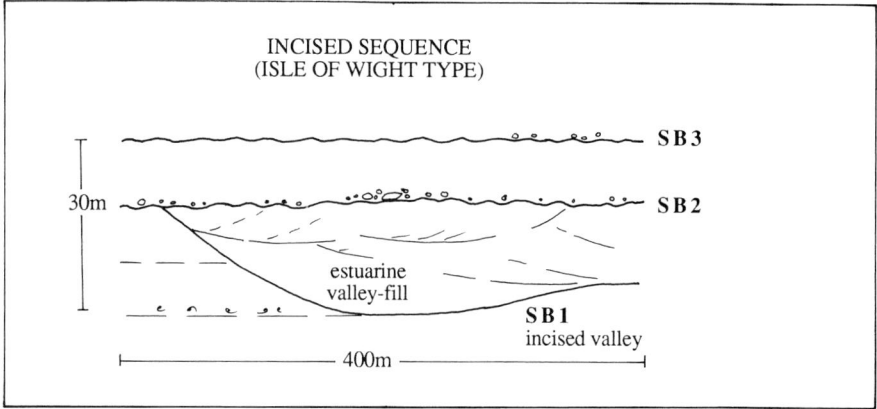

Fig. 7. Cartoon comparison of the two types of depositional sequence described here. Key as for Fig. 4.

a broad syncline is eroded prior to deposition of SB2. Similar structural basins may occur at Clophill and Faringdon, the latter succession being adjacent to an interpreted syn-sedimentary fault. The best example of a sedimentologically preserved sequence occurs in the Isle of Wight succession (Fig. 7), where a fall in base level created a valley (at least partly through fluvial incision). This incised valley was subsequently infilled with estuarine sediments before a second phase of erosion preceding SB2. Similar sedimentological preservation of the SB1–SB2 depositional sequence may occur at Maidstone, although here incision at SB1 is not observed (possibly through poor outcrop).

Tectonic influence on sequence boundary formation

The areal extent of SB2 and degree of truncation at all the mid-Aptian horizons described here is greater than in the Cretaceous succession above or below (Ruffell 1992). Thus, at outcrop purely sedimentological reasons for truncation may be invoked (e.g. erosion associated with shoreface or sandwave migration). Only regional correlation demonstrates that more erosion occurs at SB2 than at the sequence boundaries above or below. When eustacy appears to operate in a rhythmic manner (Haq et al. 1988) it is unusual to find SB1 and SB2 developed so close in time, yet with characteristics different enough to confuse their sequence stratigraphic interpretation at outcrop. Tectonic processes may have been synchronous with eustatic/regional sea-level changes during the formation of SB2, yet the level of truncation, and the tenuous or imprecise nature of the ammonite biostratigraphy (discussed below) makes a short-lived tectonic episode preceding SB2, more likely.

Biostratigraphic interpretation

When the sequence stratigraphic correlation described above is compared to the established biostratigraphy (Casey 1961), SB1 is found to be a different age on the Isle of Wight than at Redhill. Furthermore, the SB1–SB2 depositional sequence is of late *martinioides* age on the Isle of Wight and early *nutfieldiensis* age at Redhill (and possibly Faringdon). Using Casey's (1961) scheme, and trusting the direct correlation of the three late Aptian depositional sequences produces a diachronous correlation.

In the Galloway (1989) model of sequence stratigraphy, this (in itself) is not a problem. However, as an alternative, Galloway (1989) and Partington et al. (1993) suggest using the maximum flooding surface as a correlative horizon. The highly attenuated sequences described here are similar to those from the Lower Greensand of Leighton Buzzard, described by Wonham & Elliot (1996) where the maximum flooding surface is rarely preserved. Instead, we are forced to consider unconformable–transgressive surfaces in order to correlate in an up-dip location: probably not the most ideal scenario yet one that is common enough on the sort of stratigraphic highs that are cored in hydrocarbon exploration.

Recently a new ammonite zonation of the late Aptian has been published by Owen (1996). This scheme is compared to Casey's (1961) work in Table 1. Initially the subzonal divisions of Owen (1996) appear to confirm the diachroneity of the SB1–SB2 depositional sequence. However, as no ammonites have been found in this depositional sequence on the Isle of Wight, we may alternatively apply Owen's work to the sequence stratigraphic interpretation (Fig. 5b). This allows the direct correlation of SB1 and SB2 (and the intervening depositional sequence) wholly within the *martinioides* zone. An implication of this is that the dissimilar facies of estuarine sands (Isle of Wight), fuller's earths (Redhill, Faringdon and Clophill) and lagoonal clays (Maidstone) may be chronostratigraphic equivalents. Different marine and marginal marine facies are preserved by essentially the same geometry: the base marked by a minor syncline (Redhill) or incised estuarine valley (Isle of Wight), the top showing locally planar truncation (Fig. 7), usually with pebbly sands above.

Conclusions

- Three sequence boundaries in the mid-Aptian succession of southern England are identified.
- At individual outcrops, SB2 (early *nutfieldiensis* zone deposits; base Redhill Sands–Member XIII) can be interpreted as a transgressive ravinement surface.
- The degree of truncation below SB2 at outcrop (on the Isle of Wight or at Clophill), or from correlation (Redhill), allows interpretation of this surface as a sequence boundary (SB2), even though a basinward shift of facies is not observed.
- Deposits as diverse as lagoonal muds and silts, shallow-marine volcanic clays and estuarine sands are chronostratigraphic equivalents.

- The bounding surfaces of the mid-Aptian succession of southern England are of more importance in correlation than the internal facies or parasequence stacking patterns.
- On the established biostratigraphy (Casey 1961) SB2 appears to occur at different biostratigraphic horizons. This might imply that SB2 is diachronous (Fig. 5a), or that the assignation of ammonite zones is incorrect. On the more recent biostratigraphic scheme (Owen 1996) this correlation is not diachronous, SB1 and SB2 being of the same age everywhere (Fig. 5b).
- A follower of Galloway's (1989) genetic sequence stratigraphy might choose Casey's (1961) scheme to prove diachroneity, whilst a follower of the Exxon model might use Owen's (1994) scheme to prove synchroneity of sequence boundaries.
- A previously unrecognized time-gap may exist at the base of unconformity SB2 (*nutfieldiensis* zone/subzone). However, this conjectured time-gap is still not enough to allow subsidence to be accentuated during the formation of this sequence boundary.
- Truncation of folded beds at the base of SB2 occurred through short-lived tectonic uplift and minor deformation.
- Deformation below SB2 may be linked locally to the exhalation of the volcanic ash that gave rise to the fuller's earth bentonites below and regionally to North Atlantic tectonics (McMahon & Turner 1997 this volume).

Fieldwork undertaken with G. D. Wach, S. P. Hesselbo, M. Simpson, T. Bachelor R. Goldring and discussion with H. Owen and R. Casey provided much of the data and generated many of the ideas (respectively) used here. The British Geological Survey Fuller's Earth Project workers (D. Highley, R. W. O'B. Knox, B. Moorlock, R. Shephard-Thorn and R. Thurrell) are thanked for their financial and geological support. N. McMahon made many useful comments on this work. None of the views expressed here reflect those of the above. Receipt of NERC, B.P. International and British Council awards made this study possible.

References

ARKELL, W. J. 1939. Derived ammonites from the Lower Greensand of Surrey and their bearing on the tectonic history of the Hog's Back. *Proceedings of the Geologists' Association*, **50**, 22–25.

CASEY, R. 1961. The stratigraphical palaeontology of the Lower Greensand. *Palaeontology*, **3**, 487–622.

DIKE, E. F. 1972. Ophiomorpha nodosa Lundgren: environmental implications in the Lower Greensand of the Isle of Wight. *Proceedings of the Geologists Association*, **83**, 165–177.

DINES, H. G. & EDMUNDS, F. H. 1933. *The Geology of the Country Around Reigate and Dorking*. Memoir of the Geological Survey of England and Wales Sheet 285.

FITTON, W. H. 1836. Observations on some of the strata between the Chalk and the Oxford Oolite in the south-east of England. *Transactions of the Geological Society, London*, **4**, 103–390

——1845. Comparative remarks on the sections below the Chalk on the coast near Hythe, and Atherfield in the Isle of Wight. *Quarterly Journal of the Geological Society, London*, **1**, 179–189.

GALLOWAY, W. E. 1989. Genetic stratigraphic sequences in basin analysis I: architecture and genesis of flooding-surface bounded depositional units. *Bulletin of the American Association of Petroleum Geologists*, **73**, 125–142.

HALLAM, A. & SELLWOOD, B. W. 1968. Origin of fuller's earth in the Mesozoic of southern England. *Nature*, **220**, 1193–1195.

HAQ, B. U., HARDENBOL, J. & VAIL, P. R. 1988. Mesozoic and Cenozoic chronostratigraphy and cycles of sea level change. *In*: WILGUS, C. K., HASTINGS, B. S., KENDALL, G. ST. C., POSAMENTIER, H. W., ROSS, C. A. & VAN WAGONER, J. C. (eds) *Sea-level Changes: An Integrated Approach*. Society of Economic Paleontologists and Mineralogists, Special Publications, **42**, 71–108.

HESSELBO, S. P., COE, A. L., BATTEN, D. J. & WACH, G. D. 1990a. Stratigraphic relations of the Lower Greensand (Lower Cretaceous) of the Calne area, Wiltshire. *Proceedings of the Geologists Association*, **101**, 265–278.

——, —— & JENKYNS, H. C. 1990b. Recognition and documentation of depositional sequences at outcrop: an example from the Aptian and Albian on the eastern margin of the Wessex Basin. *Journal of the Geological Society, London*, **147**, 549–559.

IBBETSON, L. L. B. & FORBES, E. 1845. On the section between Blackgang Chine and Atherfield Point. *Quarterly Journal of the Geological Society, London*, **1**, 190–197.

JEANS, C. V., MERRIMAN, R. J., MITCHELL, J. G. & BLAND, D. J. 1982. Volcanic clays in the Cretaceous of southern England and Northern Ireland. *Clay Minerals*, **17**, 105–156.

KARNER, G. D. & LAKE, S. D. 1987. The structure and evolution of the Wessex Basin, southern England: an example of inversion tectonics. *Tectonophysics*, **137**, 347–378.

——, —— & DEWEY, J. F. 1987. The thermal and mechanical development of the Wessex Basin, southern England. *In*: COWARD, M. P., DEWEY, J. F. & HANCOCK, P. L. (eds) *Continental Extensional Tectonics*. Geological Society, London, Special Publications, **28**, 517–536.

KEMPER, E. 1979. Die unterkreide Nordwestdeutschlands. Ein uberlick. *Aspekte der Kreide Europas. IUGS Series*, **A6**, 1–9.

KIRKALDY, J. F. 1963. The Wealden and marine Lower Cretaceous beds of England. *Proceedings of the Geologists' Association*, **74**, 127–146.

KNOX, R. W. O'B., HIGHLEY, D. & RUFFELL, A. 1998. Stratigraphy of the Sandgate Formation (Lower Greensand; late Aptian) in the Nutfield area of southern England. *Proceedings of the Geologists' Association*, **109**, in press.

KRANTZ, R. 1972. Die Sponge Gravels von Faringdon (England). *Nues Jahrbuch Geologie und Palaontologie Abhandugen*, **140**, 207–231.

LAKE, R. D. & SHEPHARD-THORN, E. R. 1985. The stratigraphy and geological structure of the Hog's Back, Surrey, and adjoining areas. *Proceedings of the Geologists Association*, **96**, 7–21.

LEIGHTON, T. 1895. The Lower Greensand above the Atherfield Clay of East Surrey. *Quarterly Journal of the Geological Society, London*, **51**, 101–124.

MIDDLEMISS, F. A. 1962. Brachiopod ecology and Lower Greensand palaeogeography. *Palaeontology*, **5**, 253–267.

MOORLOCK, B. S. P. & HIGHLEY, D. E. 1991. *An Appraisal of Fuller's Earth Resources in England and Wales*. Technical Report of the British Geological Survey No. WA/91/75.

MCMAHON, N. A. & TURNER, J. D. 1997. The documentation of a latest Jurassic–earliest Cretaceous uplift throughout southern England. This volume.

OWEN, H. G. 1996. Boreal and Tethyan late Aptian to late Albian ammonite zonation and palaeobiogeography. *In*: SPAETH, C. (ed.) *Jost Wiedmann Memorial Volume. New Developments in Cretaceous Resaerch Topics*. Mitteilungen aus dem Geologisches–Palaontologisches Institut der Universitat Hamburg, **77**.

PARTINGTON, M.A, COPESTAKE, P., MITCHENER, B. C. & UNDERHILL, J. R. 1993. Genetic sequence stratigraphy for the North Sea Late Jurassic and Early Cretaceous: distribution and prediction of Kimmeridgian–Late Ryazanian reservoirs in the North Sea and adjacent areas. *In*: PARKER, J. R. (ed.) *Petroleum Geology of Northwest Europe*. Geological Society Proceedings, 371–386.

POOLE, E. G. & KELK, B. 1971. *Calcium Montmorillonite (Fuller's Earth) in the Lower Greensand of the Baulking Area, Berkshire*. Report of the Institute of Geological Sciences No. 71/4.

——, ——, BAIN, J. A. & MORGAN, D. J. 1971. *Calcium Montmorillonite (Fuller's Earth) in the Lower Greensand of the Fernham Area, Berkshire*. Report of the Institute of Geological Sciences No. 71/12.

RUFFELL, A. H. 1990. *The Stratigraphy and Sedimentology of the Fuller's Earth-bearing Lower Greensand (Aptian–Albian) of Southern Britain*. Report for British Geological Survey (Open-File).

—— 1992. Early to mid-Cretaceous tectonics and unconformities of the Wessex Basin (southern England). *Journal of the Geological Society, London*, **149**, 443–454.

—— 1993. Correlation of the Hythe Beds Formation (Lower Greensand Group: early–mid-Aptian), southern England. *Proceedings of the Geologists Association*, **103**, 273–291.

—— & OWEN, H. G. 1994. The Sandgate Formation of the M20 Motorway near Ashford, Kent, and its correlation. *Proceedings of the Geologists' Association*, **105**, 240–250.

—— & WACH, G. D. 1991. Sequence stratigraphic analysis of the Aptian–Albian Lower Greensand in southern England. *Marine and Petroleum Geology*, **8**, 341–353.

—— & WIGNALL, P. B. 1990. Depositional trends in the Upper Jurassic–Lower Cretaceous of the northern margin of the Wessex Basin. *Proceedings of the Geologists' Association*, **101**, 279–289.

SIMPSON, M. I. 1985. The stratigraphy of the Atherfield Clay Formation (Lower Aptian: Lower Cretaceous) at the type and other localities in southern England. *Proceedings of the Geologists' Association*, **96**, 23–45.

STEWART, D. J. 1981. A field guide to the Wealden Group of the Hastings area and Isle of Wight. *In*: ELLIOT, T. (ed.) *Field Guides to Modern and Ancient Fluvial Systems in Britain and Spain* International Fluvial Conference, Keele University.

——, RUFFELL, A. H., WACH, G. D. & GOLDRING, R. 1991. Lagoonal sedimentation and fluctuating salinities in the Vectis Formation (Wealden Group, Lower Cretaceous) of the Isle of Wight, southern England. *Sedimentary Geology*, **72**, 117–134.

STONELEY, R. 1982. The structural development of the Wessex Basin. *Journal of the Geological Society, London*, **139**, 545–552.

VAN WAGONER, J. C., MITCHUM, R. M., CAMPION, K. M. & RAHMANIAN, V. D. 1990. *Siliciclastic Sequence Stratigraphy in Well Logs, Cores and Outcrops: Concepts for High-resolution Correlation of Time and Facies*. American Association of Petroleum Geologists, Methods in Exploration Series, **7**.

VAIL, P. R. 1987. Seismic stratigraphy interpretation using sequence stratigraphy. Part 1: seismic stratigraphy interpretation procedure. *In*: BALLY, A. E. (ed.) *Atlas of Seismic Stratigraphy*, American Association of Petroleum Geologists, Studies in Geology, **27**, 1–10.

WACH, G. D. 1994. *Sedimentology and stratigraphy of the Lower Cretaceous of the Channel Basin*. DPhil thesis, University of Oxford.

—— & RUFFELL, A. H. 1991. *Sedimentology and Sequence Stratigraphy of a Lower Cretaceous Tide and Storm-dominated Clastic Succession, Isle of Wight and, S.E. England. Field Guide 4*. International Sedimentological Congress, Nottingham.

WILLIAMS, G. D. & DOBB, A. (eds) 1993. *Tectonics and Seismic Sequence Stratigraphy*. Geological Society, London, Special Publications, **71**.

WEST, R. R. 1976. Comparison of seven lingulid communities. *In*: SCOTT. R. W. & WEST, R. R. (eds) *Structure and Classification of Palaeocommunities*. Dowden, Hutchinson & Ross, 171–192.

WHITTAKER, A. 1985. *Atlas of Onshore Sedimentary Basins in England and Wales*. Blackie, Glasgow.

WHITE, H. J. O. 1921. *Geology of the Isle of Wight*. Memoirs of the Geological Survey of Great Britain.

WONHAM, J. P. & ELLIOT, T. 1996. High-resolution sequence stratigraphy of a mid-Cretaceous estuarine complex: the Woburn Sands of the Leighton Buzzard area, southern England. *In:* HESSELBO, S. P. & PARKINSON, D. N. (eds) *Sequence Stratigraphy in British Geology.* Geological Society, London, Special Publications, **103**, 9–30.

WORSSAM, B. C. 1993. Correlation of the Hythe Beds (Lower Greensand Group: early–mid-Aptian), southern England: discussion and reply. *Proceedings of the Geologists Association,* **104**, 301–307.

Basal Wealden of Mupe Bay: a new model

STEPHEN P. HESSELBO

Department of Earth Sciences, University of Oxford, Parks Road, Oxford OX1 3PR, UK

Abstract: Angular stratal relationships associated with the Mupe Bay palaeo-oilseep in the Lower Cretaceous Wealden Group of South Dorset have previously been viewed as evidence of angular unconformity. An alternative (preferred) interpretation is that the succession represents rotational channel-bank collapse at the margin of a large Early Cretaceous fluvial channel. The new interpretation explains observed bedding-plane orientations and palaeocurrent directions.

A previous paper (Hesselbo & Allen 1991) has described and interpreted the depositional environments of the Lower Cretaceous basal Wealden Beds at Mupe Bay in South Dorset (Fig. 1). In that paper angular stratal relationships at the level of the Mupe Bay 'palaeo-oilseep' (Selley & Stoneley 1987; Cornford et al. 1988; Allen 1989, p. 547) were interpreted as representing angular unconformities. Subsequently, the timing of oil migration as evidenced by the oilseep (i.e. oil-cemented clasts penecontemporaneous with deposition) has been the subject of much discussion (Miles et al. 1993, 1994; Kinghorn et al. 1994; Wimbledon et al. 1996; Parfitt & Farrimond this volume) and it has become important to re-assess our original interpretation of the sedimentological context of this horizon. In this paper it is argued that the angular relationships were generated not by tectonic tilting as initially proposed, but rather by rotational bank collapse on the margin of a large Wealden river channel.

Rotational bank collapse

Rotational bank collapse is a very common phenomenon in modern fluvial systems (Fisk 1944; Stanley et al. 1966; Turnbull et al. 1966; Laury 1971; Thorne 1982; Ullrich et al. 1986), but there have been rather few cases well documented from ancient deposits (e.g. Williams et al. 1965; Laury 1968, 1971; Alexander 1987; Guion 1987; Williams & Flint 1990). Uncommon occurrence within the geological record is ascribed by most of these authors to the unlikelihood of preservation within the active river channels of the majority of fluvial systems.

Rotational bank collapse can take place by 'base failure', 'toe failure' or 'slope failure' (Thorne 1982, and references therein). In 'base failure' the shear surface passes below the level of the thalweg of the channel; this process takes place preferentially in channels cut into clay- or silt-rich, cohesive, sediments. The preservation of rotationally collapsed banks is aided by 'base failure' rather than 'toe' or 'slope failure' because material is carried below the level of active erosion, although clearly the debris from 'toe' or 'slope failure' may be preserved, particularly if the channel is subsequently abandoned (Laury 1971; Guion 1987).

An exceptionally well-preserved and instructive example of channel-bank collapse through base failure has been described from the Miocene of the Lower Rhine Basin, Germany (Williams & Flint 1990). There, bank collapse took place in a channel at least 50 m wide and greater than 4 m deep and was accomplished by movement along a glide plane that extended below the channel floor. Both an extensional 'head' and a compressional 'tail' have been preserved; the intact blocks at the extensional end were rotated by about 30° and the initial collapse structure shows evidence of at least one phase of reactivation. This example provides an excellent analogue for interpreting the Mupe Bay palaeo-oilseep.

The Wealden Group of the southern Wessex Basin was deposited by lacustrine, lagoonal or fluvial systems which were often dominated by muddy sediment (see, for example, Allen 1975, 1981, 1989; Stewart 1981, 1983). Hence, channel fills and associated near-bank sediments commonly comprise material that would have been cohesive shortly after deposition and thus liable to collapse of the full bank height (see Laury 1971). In some respects then, it is surprising that rotational bank collapse is not a more common feature of Wealden deposits.

The Mupe Bay succession

The sedimentary succession and facies of the basal Wealden Beds of Mupe Bay and Bacon

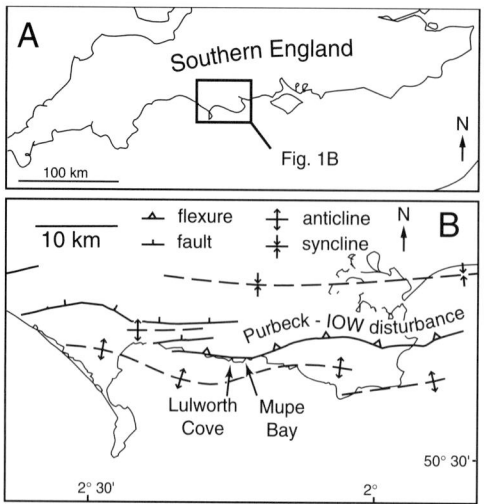

Fig. 1. Location map for the Mupe Bay Lower Cretaceous (modified from Stoneley 1982).

Hole have already described in some detail by Selley & Stoneley (1987) and Hesselbo & Allen (1991). In the following discussion reference is made to the sedimentary sequence shown in figs 2 and 3 of Hesselbo & Allen (1991); the interested reader should consult this paper for a more detailed discussion of the facies and environments of deposition. The major features of the exposure at Mupe Bay are illustrated here in Fig. 2.

The stairway to the beach (SY 844 797) is sited in a gully that undoubtedly represents a poorly resistant lithology: the upper 160 cm of this interval is a pale grey, weakly laminated mudstone. By correlation with nearby Bacon Hole, 300 m to the west (SY 841 797), the remainder of the interval probably comprises a mottled red/grey-green mudstone likely to have a fluvial overbank origin (Hesselbo & Allen 1991). At the time of writing (January 1996) 150 cm of red/grey mottled mudstone were exposed near the foot of the stairs at Mupe Bay, confirming the correlation. Above this argillaceous unit is Bed 5, a very fine- to fine-grained sandstone which is truncated westwards by an erosion surface labelled WB2 in Fig. 2. The depositional environment of this sandstone is obscure. Above erosion surface WB2, and apparently concordant with it, is Bed 6, a laminated and deformed purple-grey mudstone deposited in a quiet-water environment subject to periodic influx of coarse sand, possibly a lake or lagoon. Both the mudstone, Bed 6, and the underlying sandstone, Bed 5, are truncated westwards by another erosion surface labelled WB3 in Fig. 2. The horizon of the palaeo-oilseep immediately overlies surface WB3: the oil-cemented clasts lie within a heterolithic coarse to very coarse sandstone and mudstone. The conglomeratic interval is separated on its northern side from a fluvial, micaceous grey mudstone by an almost vertical fault of unknown throw which was could be seen in January 1996.

Discussion

Although it has been suggested before that the boulders in Bed 7 were derived through bank collapse (e.g. Miles *et al.* 1994; Wimbledon *et al.* 1996) the whole exposure has not hitherto been viewed in that context. A possible sequence of events leading to formation of the palaeo-oilseep through channel-bank collapse is shown in Fig. 3

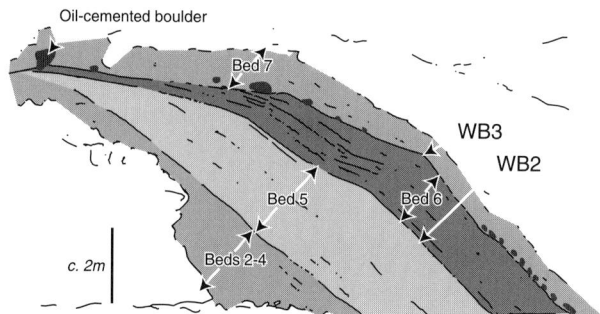

Fig. 2. Line drawing from a photograph of the Mupe Bay palaeo-oilseep showing the main erosion surfaces and beds discussed in the text (from fig. 7 in Hesselbo & Allen 1991). Bed 7 contains oil-cemented pebbles, cobbles and boulders.

(see caption for detailed commentary). Because of the potentially large scale of bank collapse features, incompletely exposed examples may commonly be mistaken for the products of tectonic deformation (Laury 1971; Guion et al. 1995). Re-interpretation of the angular relationships associated with the Mupe Bay palaeo-oilseep as due to rotational bank collapse is compatible with documented deep channelling at similar levels in the neighbouring Weald Basin and beyond (cf. Allen 1975, 1981; Ruffell 1995), and it also explains some features of the local succession otherwise unaccounted for.

The angular discordance between the palaeo-oilseep and the underlying strata is estimated to be about 5°. Because of the nature of the exposure, it has proved impossible to measure bedding orientations with sufficient accuracy to demonstrate the discordance quantitatively. Nonetheless, it is undeniable from the field observations that both surfaces WB2 and WB3 cut out progressively more of the underlying strata in a westward direction (Fig. 2). Palaeocurrent indicators in the sandstone immediately above the boulders show a northerly direction (long axis of log aligned 146°, dip 30°N; tabular(?) cross-bedding strike 078°, dip 56°N; tectonic tilt close to strike 090°, dip 40°N). Therefore, the palaeocurrent direction in the overlying channel deposit is perpendicular to the discordance with the underlying strata, which is compatible with rotational slip of the bank into a channel having a local north–south-oriented axis west of the present exposure and northward palaeoflow (cf. Laury 1971).

The co-occurrence of a rotational bank-collapse structure with the palaeo-oilseep may be viewed as connected, because bank collapse would have been associated with a topographic depression in which seeping oil could pond (either an abandoned channel or the head region of the slipped mass).

Our original hypothesis that surface WB2 and possibly surface WB3 represent angular unconformities now appears incorrect and, indeed, would have demanded a quite remarkable coincidence of critical geological relationships in a very small-scale exposure. However, the revised interpretation presented here does not exclude the possibility that these surfaces are unconformable. Comparison may be made with the study of Williams et al. (1965) who, working on the Carboniferous of western Pennsylvania, used the occurrence of anomalously thick successions within collapsed and rotated channel-bank blocks to argue that substantial unconformities occurred associated with channel cutting. In effect, they interpreted the channels as

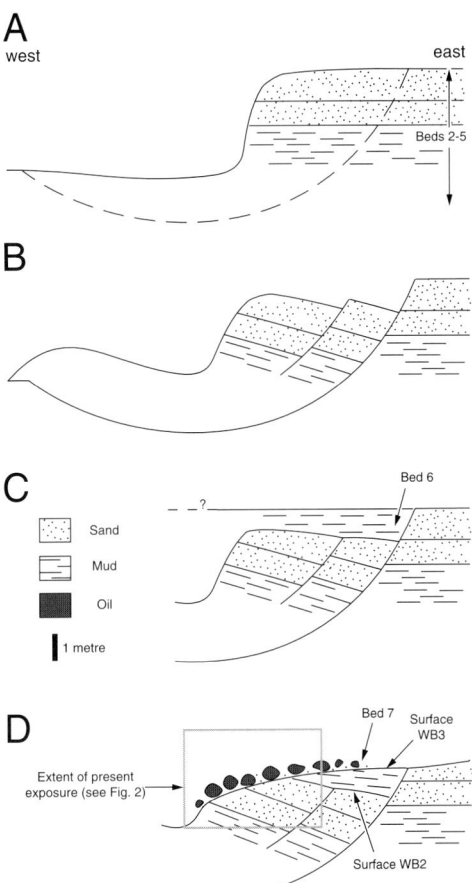

Fig. 3. Cartoon illustrating proposed stages in the genesis of the Mupe Bay palaeo-oilseep by rotational bank collapse: vertical axis exaggerated for clarity. The cartoon shows collapse by 'base failure' below the level of the channel thalweg: a more shallow failure involving only the channel wall is also a possibility. (**A**) A large fluvial channel cuts into Beds 2–5 of the Mupe Bay section. The position of the surface of bank failure is shown by a dashed line. (**B**) The channel bank collapses by rotation of blocks in the head region away from the channel axis. (**C**) Further argillaceous sedimentation occurs across eroded tops of rotated blocks (Bed 6). Sediment may have completely filled the channel, or may have been restricted to depressions between rotated blocks. At this stage seep oil could accumulate in sands within topographic depressions. (**D**) Further minor rotational failure occurs along the initial shear surface, possibly as a result of renewed fluvial activity in the channel. Deformation of the rotated blocks occurs and includes intrusion of sand dykelets within Bed 6. The crest of the rotated block is eroded and 'draped' with a remanié of oil-cemented boulders, possibly through the mechanism of 'slope failure' (Bed 7). The boulders are subsequently buried in sand transported by currents flowing in a northerly direction.

'palaeovalleys' in a sequence stratigraphic sense (see van Wagoner et al. 1990; Dalrymple et al. 1994). Exposure at Mupe Bay is not good enough to apply their criteria, and probably only refined chronostratigraphical work will resolve the unconformity issue.

Controversy over interpretation of the Mupe Bay palaeo-oilseep centres on the timing of oil migration. Most previous authors have recognized that this horizon indicates oil generation and migration (probably up fault planes associated with the Purbeck–Isle of Wight structure) during the Early Cretaceous (West 1975, p. 211; Selley & Stoneley 1987; Cornford et al. 1988). In contrast, Miles et al. (1993) claimed that the oil did not migrate until later in the Cretaceous and the Mupe Bay horizon was charged with oil substratally, a conclusion based principally on organic geochemistry and burial history analysis. This claim has been strongly disputed by Kinghorn et al. (1994). In defence of their hypothesis, Miles et al. (1994) cite the preferential oil staining in trace fossils within Bed 5 (Hesselbo & Allen 1991) as evidence that the oil charge occurred post-depositionally. This may be the case for Bed 5, but it is incredible that the sandstone boulders within Bed 7, with their unusual embayed margins, could have survived intact in an active fluvial channel if they had not been oil-cemented. Recently, in describing an oil globule with a laminated silty mudstone wrap, Wimbledon et al. (1996) have produced the most compelling evidence yet for penecontemporaneous migration of oil to the surface at the palaeo-oilseep.

Conclusions

The Mupe Bay palaeo-oilseep is re-interpreted as occurring immediately above rotated blocks of a channel-bank collapse structure. The palaeo-oilseep cannot be regarded as definitively marking the position of a major unconformity or unconformities near the base of the Wealden succession, although these possibilities, equally, cannot be excluded. Interpretation of the succession as a channel-bank collapse structure provides a possible explanation for accumulation of oil via ponding in a collapse-related depression or abandoned channel.

The author thanks J. Alexander who introduced him to the bank-collapse structures in the Middle Jurassic of the Cleveland Basin, triggering this reinterpretation. The author is also grateful to Perce Allen, Philip Allen, A. Ruffell and an anonymous referee for their helpful comments.

References

ALEXANDER, J. 1987. Syn-sedimentary and burial related deformation in the Middle Jurassic non-marine formations of the Yorkshire Basin. In: JONES, M. E. & PRESTON, R. M. F. (eds) Deformation of Sediments and Sedimentary Rocks, Geological Society, London, Special Publications, **29**, 315–324.

ALLEN, P. 1975. Wealden of the Weald: a new model. Proceedings of the Geologists' Association, **86**, 389–437.

—— 1981. Pursuit of Wealden Models. Journal of the Geological Society, London, **138**, 375–405.

—— 1989. Wealden research – ways ahead. Proceedings of the Geologists' Association, **100**, 529–564.

CORNFORD, C., CHRISTIE, O., ENDRESEN, U., JENSEN, P. & MYHR, M.-B. 1988. Source rock and seep oil maturity in Dorset, southern England. Organic Geochemistry, **13**, 399–409.

DALRYMPLE, R. W., BOYD, R. & ZAITLIN, B. A. (eds) 1994. Incised-valley Systems: Origin and Sedimentary Sequences. Society for Sedimentary Geology (SEPM), Special Publications, **51**.

FISK, H. N. 1944. Geological Investigation of the Alluvial Valley of the Lower Mississippi River. U.S. Army Corps of Engineers, Mississippi River Commission, Vicksburg, Mississippi.

GUION, P. D. 1987. Palaeochannels in mine workings in the High Hazles Coal (Westphalian B), Nottinghamshire Coalfield, England. Journal of the Geological Society, London, **144**, 471–488.

——, JONES, N. S., FULTON, I. M. & ASHTON, A. J. 1995. Effects of a Westphalian channel on coal-seam geometry: a re-appraisal of the 'Dumb Fault' of north Derbyshire. Proceedings of the Yorkshire Geological Society, **50**, 317–332.

HESSELBO, S. P. & ALLEN, P. A. 1991. Major erosion surfaces in the basal Wealden Beds, Lower Cretaceous, south Dorset. Journal of the Geological Society, London, **148**, 105–113.

KINGHORN, R. R. F, SELLEY, R. C. & STONELEY, R. 1994. The Mupe Bay oil seep demythologized? Marine and Petroleum Geology, **11**, p. 124.

LAURY, R. L. 1968. Sedimentology of the Pleasantview Sandstone, southern Iowa and western Illinois. Journal of Sedimentary Petrology, **38**, 568–599.

—— 1971. Stream bank failure and rotational slumping: preservation and significance in the geologic record. Bulletin of the Geological Society of America, **82**, 1251–1266.

MILES, J. A., DOWNES, C. J. & COOK, S. E. 1993. The fossil oil seep in Mupe Bay, Dorset: a myth investigated. Marine and Petroleum Geology, **10**, 58–70.

——, —— & —— 1994. Reply to 'The Mupe Bay oil seep demythologized?' by Kinghorn et al. Marine and Petroleum Geology, **11**, 125–126.

PARFITT, M. & FARRIMOND, P. 1998. The Mupe Bay oil seep: a detailed organic geochemical study of a controversial outcrop. This volume.

RUFFELL, A. H. 1995. Seismic stratigraphic analysis of non-marine Lower Cretaceous strata in the Wessex and North Celtic Sea Basins. Cretaceous Research, **16**, 603–637.

SELLEY, R. C. & STONELEY, R. 1987. Petroleum habitat in south Dorset. *In*: BROOKS, J. & GLENNIE, K. W. (eds) *Petroleum Geology of North West Europe*, Vol. 1. Graham & Trotman, London, 139–148.

STANLEY, D. J., KRINITZSKY, E. L. & COMPTON, J. R. 1966. Mississippi river bank failure, Fort Jackson, Louisiana. *Bulletin of the Geological Society of America*, **77**, 859–866.

STEWART, D. J. 1981. A field guide to the Hastings area and the Isle of Wight. *In*: ELLIOTT, T. (ed.) *Field Guides to Modern and Ancient Fluvial Systems in Britain and Spain*. International Fluvial Conference, Keele University, UK, 3.1–3.31.

—— 1983. Possible suspended-load channel deposits from the Wealden Group (Lower Cretaceous) of Southern England. *In*: COLLINSON, J. D. & LEWIN, J. L. (eds) *Modern and Ancient Fluvial Systems*. International Association of Sedimentologists, Special Publications, **6**, 369–384.

STONELEY, R. 1982. The structural development of the Wessex Basin. *Journal of the Geological Society, London*, **139**, 545–552.

THORNE, C. R. 1982. Processes and mechanisms of river bank erosion. *In*: HEY, R. D., BATHURST, J. C. & THORNE, C. R. (eds) *Gravel Bed Rivers*, Wiley, Chichester, 227–271.

TURNBULL, W. J., KRINITZSKY, E. L. & WEAVER, F. S. 1966. Bank erosion in soils of the Lower Mississippi Valley. *Proceedings of the American Society of Civil Engineers*, **92**, Paper 4632, *Journal of the Soil Mechanics and Foundations Division* No. SM-1, 121–136.

ULLRICH, C. R., HAGERTY, D. J. & HOLMBERG, R. W. 1986. Surficial failures of alluvial stream banks. *Canadian Geotechnical Journal*, **23**, 304–316.

VAN WAGONER, J. C., MITCHUM, R. M., CAMPION, K. M. & RAHMANIAN, V. D. 1990. *Siliciclastic Sequence Stratigraphy in Well Logs, Cores, and Outcrops*. American Association of Petroleum Geologists, Methods in Exploration Series, **7**.

WEST, I. M. 1975. Evaporites and associated sediments of the basal Purbeck Formation (Upper Jurassic) of Dorset. *Proceedings of the Geologists' Association*, **86**, 205–225.

WILLIAMS, E. G., GUBER, A. L. & JOHNSON, A. M. 1965. Rotational slumping and the recognition of disconformities. *Journal of Geology*, **72**, 534–547.

WILLIAMS, H. & FLINT, S. 1990. Anatomy of a channel-bank collapse structure in Tertiary fluvio-lacustrine sediments of the Lower Rhine Basin, Germany. *Geological Magazine*, **127**, 445–451.

WIMBLEDON, W. A., ALLEN, P. & FLEET, A. J. 1996. Penecontemporaneous oil-seep in the Wealden (early Cretaceous) at Mupe Bay, Dorset, U.K. *Sedimentary Geology*, **102**, 213–220.

east from Osmington Mills at (SY 740 814), and elsewhere in the vicinity of Weymouth, and to discuss the nature of the lower boundary with the underlying Nothe Clay and the upper boundary with the Upton Member of the Osmington Oolite Formation. The Bencliff Grit has been interpreted as open marine (Sun 1989), estuarine or intertidal (Talbot 1973b; Fürsich 1974a, b, 1975), estuarine mouth or estuarine influenced (Allen & Underhill 1989, 1990; Corbett et al. 1994) and shoreface (Grieves 1986). Interpretation has focused on the bedforms associated with the fine-grained, well-sorted sands that account for the bulk of the 5.5 m thick unit. While Sun (1989) recognized only swaley and hummocky cross-stratification, Allen & Underhill (1989, 1990) appreciated that swaley cross-stratification, produced by highly charged but essentially unidirectional flow, was dominant. Controversy followed (Allen & Underhill 1990; Hart 1990, with reply; Sun 1990): with discussion concentrating on the bedforms, how they were formed, and in what depositional setting.

Sun (1989) indicated that the Bencliff Grit displayed upward coarsening. Sellwood et al. (1990), Rioult et al. (1991), Wilson (1991) and Coe (1992, 1995) recognized the base of the member as a sequence boundary, and Rioult et al. (1991), in addition, indicated the presence of three coarsening-upwards parasequences. Our results disagree with some of the observations and inferences that have been made on the Bencliff Grit: in that (1) the member does not show a coarsening-upward trend, (2) or a shallowing-upward trend, (3) that the base does not a priori represent a sequence boundary and (4) that no parasequences (van Wagoner et al. 1990) are present. The aim of this paper is, thus, to use our own observations to test and modify existing models for the deposition of the Bencliff Grit.

J. Marshall and S. Gabbott have examined the palynofacies of the mudstone and heterolithic facies, C. D. Jenkins the foraminifera, T. R. Astin the bedforms and R. Goldring the trace fossils. The cliffs east of Osmington Mills are receding and the beach undergoes irregular aggradation and lowering. These aspects, together with cliff falls, both obscure evidence and provide new information. In 1994 the boundary with the Nothe Clay was completely obscured, but beach build-up allowed the top of the member to be more readily observed. Substantial changes in detail to Fig. 2 were evident over the period 1992–1994 but do not alter the overall facies relationships. Further information will become available in the future.

Lithofacies

Allen & Underhill (1989) carefully described the facies of the Bencliff Grit and we agree with and follow their account with minor modification and additional information (Figs 2–4 and 6–8). Each unit of the *sandstone facies* (A), unless truncated by penecontemporaneous erosion, is followed by a thin (mm–cm) unit of *mudstone–siltstone facies* (B) which passes upwards (again, unless truncated by penecontemporaneous erosion) into *heterolithic facies* (C and D). There are at least eight discrete intervals of heterolithic facies at the west end of the section. With amalgamation present in the *sandstone facies* (A) and minor breaks in sedimentation in the *heterolithic facies* the sedimentary record is far from complete. The Bencliff Grit does not coarsen upwards (Sun 1989), except over the last 0.5 m (below).

The *mudstone–siltstone facies* (B) is essentially as described by Allen & Underhill (1989). Each unit of sand (facies A) either passes upwards, or laterally as toe sets, through small-scale cross-stratification (wave and combined flow ripples) to a mm–cm thick bed of either dark blue or grey mudstone, lacking lamination and sometimes carrying comminuted plant debris set at all inclinations to the local stratification (which in some instances is related to a swale), or alternatively grades upwards to finely laminated silty mudstone. No colonization surfaces have been seen in either case.

Both types of mudstone–siltstone pass upwards, or laterally as toe sets, into a unit of flaser bedding (facies C) (grey mud and fine-grained sand) (Figs 6–8) which can be up to several centimetres thick. Trace fossils are locally present, especially higher in each unit, and appear to be attributable to *Teichichnus* isp. and *Planolites* isp. The flaser bedding is followed by facies D: heterolithic and bioturbated mudstones (Figs 6–9) (in which primary lamination is largely obscured) and cm–dm thick fine-grained and generally composite sandstones which can be traced laterally for only short distances (<10.0 m). The sandstones differ from the those of facies A in that: (1) they are generally cemented, (2) they exhibit bedforms that probably represent wave-ripple lamination or hummocky cross-stratification, (3) individual units (often composite and amalgamated, Fig. 9a) occupy shallow scours, c. 5.0 m across at outcrop, (4) the upper part is always bioturbated (although not to the extent of the typical parallel-to-bioturbated units described by Howard 1972); (5) each passes upwards into bioturbated heterolithic sandy mudstone; and (6) laminae are often strewn with pellets of the same size as those

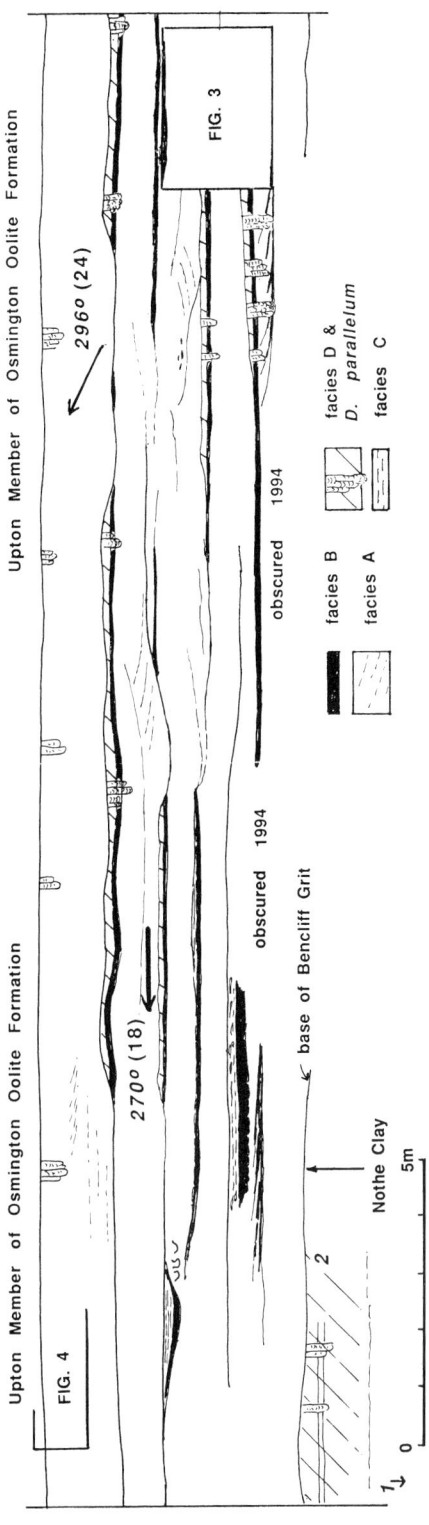

Fig. 2. Profile of a cliff at the western end of the main section of Bencliff Grit, at Osmington Mills coastal section, to show a complex cut-and-fill structure and the general succession of facies. Facies A, plain; B, blocked; C and D, parallel and diagonal ruling. Position of palynofacies and micropalaeontological sample sites 1 and 2, and vector means of palaeocurrent directions in units indicated (with number of measurements).

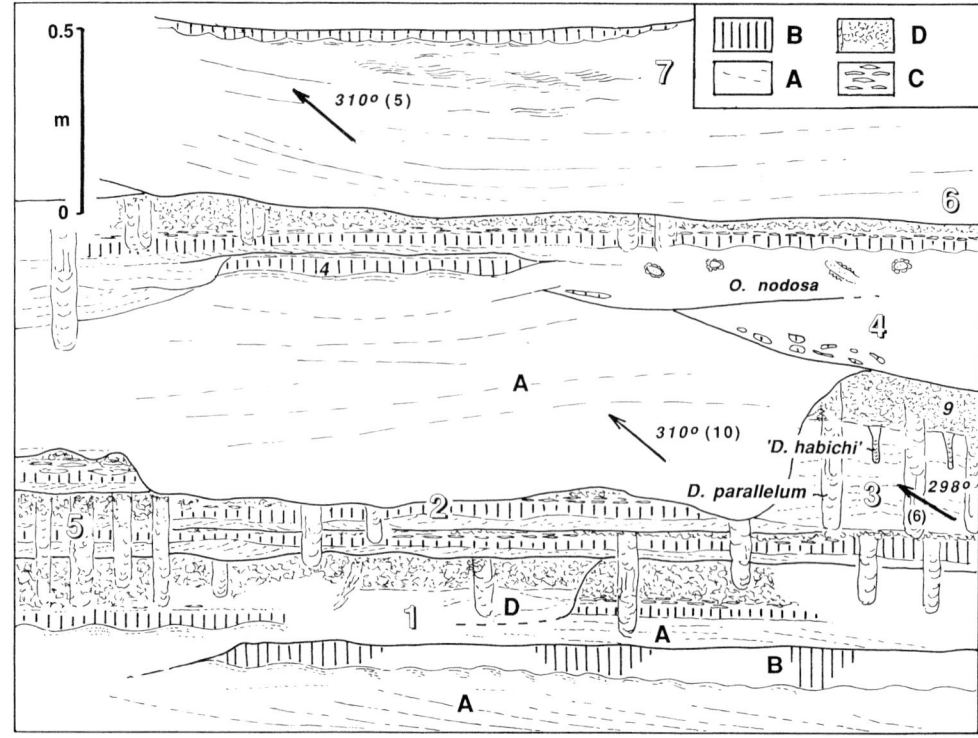

Fig. 3. Detail of the area in a box (right) in Fig. 2. Facies as indicated; major erosional boundaries with heavy lines, gradational boundaries with thinner lines. *Diplocraterion parallelum* and *D. habichi* shown diagramatically. Facies D is present in at least six levels in the area shown (shadowed numerals): 1, amalgamation of facies A with storm event bed in facies D i.e. storm bed locally erodes, channelling into other facies; 2, veneers of facies A followed by facies B and C (see Fig. 6); 3, event bed within facies D with wave ripple lamination and *D. habichi* colonizing upper surface; 4, intraformational conglomerate (clasts of facies B); 5, amalgamation of facies due to bioturbation; 6, *O. nodosa* in facies A indicates origin in an overlying unit of facies D subsequently eroded; 6, sole of bed with bivalve lag with long axes with preferred alignment 173–353°; 7, parallel lamination passing up into climbing ripple lamination (towards 275°) and then to wave ripple. Micropalaeontological and palynological sample sites 4 and 9 and vector means as for Fig. 2.

associated with *D. parallelum*, in contrast to the plant debris strews in facies A. Such sandstones show the features associated with storm event beds.

Wright (1986) recorded ammonites (*Goliathiceras microtrypa* and *Perisphinctes* sp.) from sandstone blocks of facies A at Redcliff. Aligned specimens of *Myophorella* sp. are rarely present in facies A (Fig. 3). The specimen (Wright collection DC.13) from Redcliff (facies A) with undulose and thinning plant strewn laminae contains a flattened *Perisphinctes*. The laminae are truncated and followed by small-scale undulose lamination with a convex-down impression of *Myophorella hudlestoni*. The tubercles of the *Myophorella* are relatively small (Wright pers. comm.). This and the unusually flattened ammonite suggest early shell dissolution which may explain the scarcity of bivalves in the Bencliff Grit.

While the thickness of units of facies B and C are not found to vary significantly laterally (unless reduced by penecontemporaneous erosion), the thickness of units of facies A even allowing for bed amalgamation, varies from a few millimeteres (Fig. 7) (when the following units of facies B and C are actually thicker) to almost 2.0 m (Fig. 2). The two types of sand unit (A and D) may be readily distinguished in the field by the presence or absence of a succeeding unit of mudstone unit (facies B) (Figs 6 and 7).

Corbett *et al.* (1994) recognized upward change in the nature of the laminasets (packets of sand bounded by stratal terminations) and a three-fold division of the Bencliff Grit with the stratification in the upper 1.6 m of the member

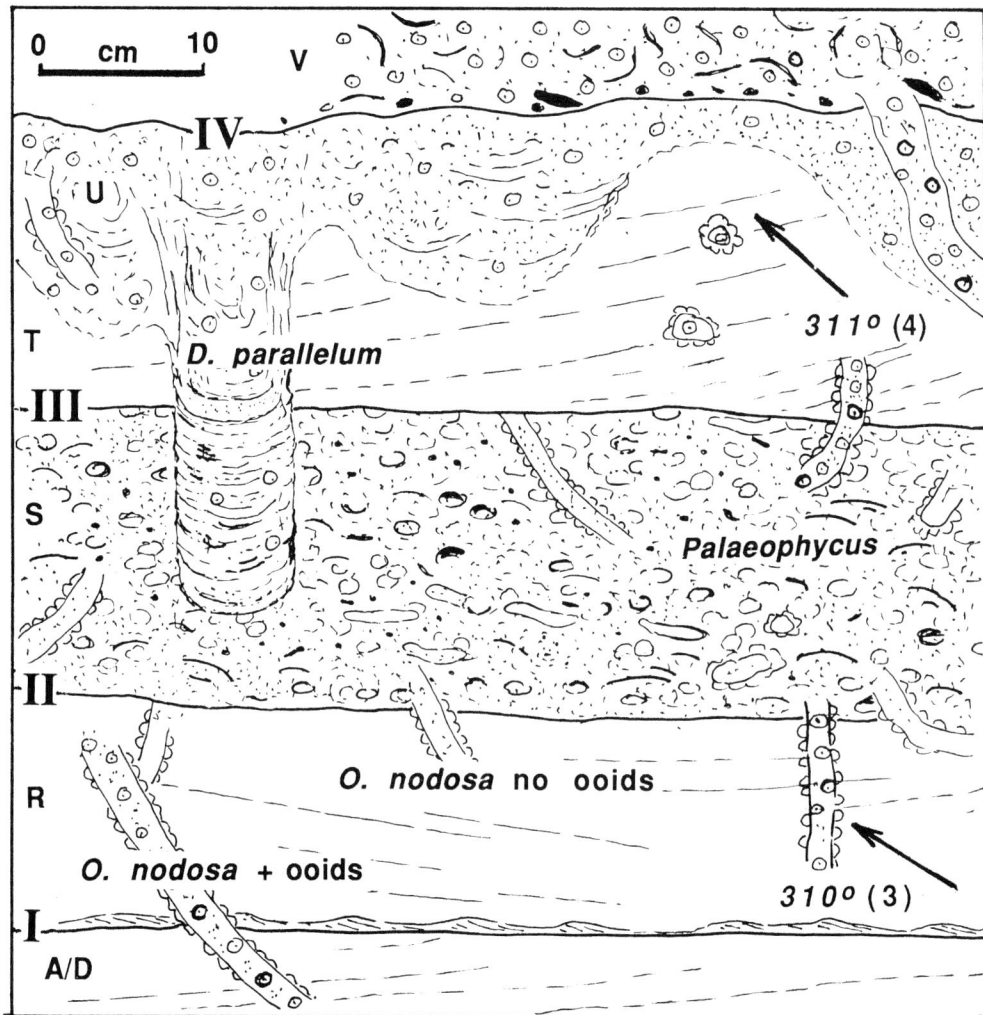

Fig. 4. Detail of the area in a box in Fig. 2 (upper left) to show the stratigraphy in the interval between the Bencliff Grit and Upton Member. The sketch is generalized but each unit R–V and the surfaces I–IV can be identified along the length of the coastal section, except locally where surface III cuts down to surface I. At the east end of the section surface I rests on facies A but on facies D at the west end of the section (area of box, Fig. 2). The cross-stratified sand (R) is followed locally by mud drapes. Unit S is bioturbated by *Palaeophycus* and *Planolites* both cut by *O. nodosa* (with sand fill, and without ooids). Unit S is truncated by surface III and followed by cross-stratified sand (without ooids) which passes upwards into very fine-grained muddy sand and bioturbated muddy oolite (U), which is truncated by surface IV. *D. parallelum* (with ooids) extends through U and into T and S, and *O. nodosa* (with ooids) extends downwards from U and V some reaching below surface I. Vector means of palaeocurrents as in Fig. 2.

laterally somewhat more continuous (less lenticular). This can probably be attributed to a relative shift in orientiation of the section in respect to the dominant unidirectional flow (Allen & Underhill 1989, 1990).

The relationships between the facies shows that the sand facies A is always erosional on facies B, C or D. Facies B shows a normal relationship with facies A and represents essentially deposition from a decelerating flow. Facies C is the least distinct of the four facies but marks a transition to facies D by increase in wave/wave-current activity and the gradual reappearance of a colonizing biota.

The results are consistent with two models. In the first, each unit of sand facies A represents

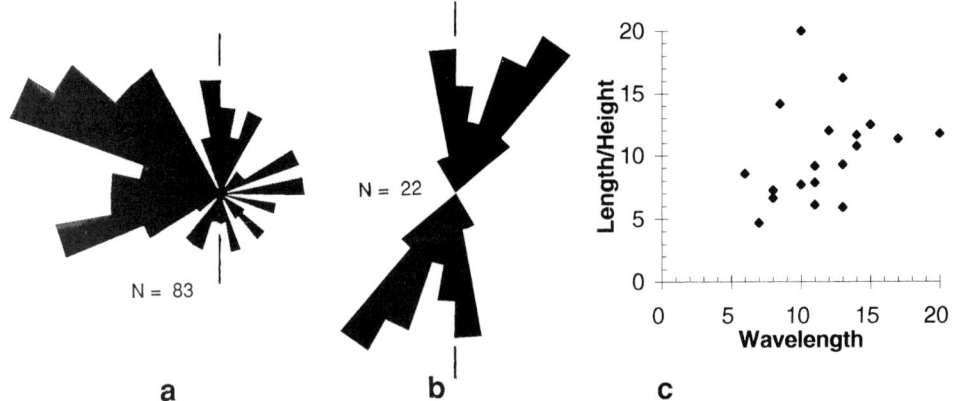

Fig. 5. Palaeocurrent data: (**a**) rose diagram of combined measurements derived from large-scale cross-stratification in facies A and D and corrected for regional dip; (**b**) rose diagram of trends of wave ripple crests in facies A and C; and (**c**) ripple form index l/h for the same data (wavelength in cm).

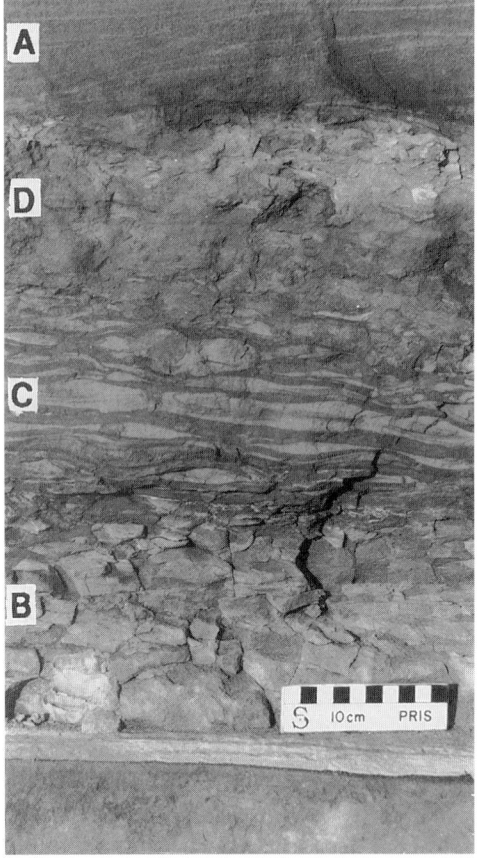

Fig. 6. Recovery succession above facies A passing successively to blocky fine muddy silt and more muddy silt (B), to flaser laminated sand and mud (C) and bioturbated muddy sand (D), truncated by succeeding unit of facies A (approximately 28 m east of area depicted in Fig. 2).

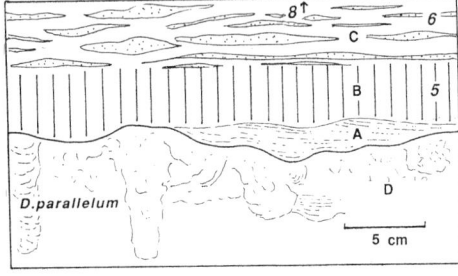

Fig. 7. Photograph and sketch of small area 5.0 m east of right margin displayed in Fig. 2. Bioturbated muddy sand (facies D) irregularly truncated and partly covered by a thin unit of cross-bedded sand (facies A), followed by laterally continuous seam of facies B (blocky mudstone) which, in turn, is succeeded by flaser-laminated sands and muds (facies C) leading to return of facies D. In turn, facies D is truncated and overlain by facies A. Palynological and micropalaeontological sample sites 5, 6 and 8.

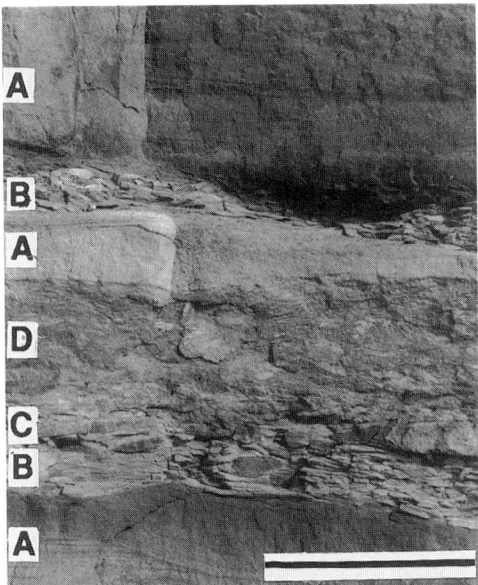

Fig. 8. Three units of facies A with intervening facies. Lower unit of facies A followed by facies B, C and D. Almost planar truncation is followed by thin units (toesets?) of facies A and B, irregularly truncated by the upper unit of facies A. Scale bar is 10 cm.

the introduction of a pulse of sediment introduced under environmental conditions different from those normally pertaining. In the second, a more Waltherian approach would suggest that facies D represents inter-dune fields over which the sand facies A migrate in a situation analogous, but clearly not identical, to the situation envisaged for the migration of large dunes in the Lower Cretaceous (Lower Greensand) of southern England (Pollard et al. 1993). Although the ichnology is consistent with the latter, though with some reservations, the micropalaeontology and palynology indicate the former as more likely. As the Bencliff Grit essentially represents 'event' rather than 'cyclic' sedimentation, neither model can correspond with upward-coarsening or upward-fining parasequences (*sensu* van Wagoner et al. 1990) where each parasequence is bounded by a sharp flooding surface.

Palaeocurrent analysis

We have remeasured the cross-stratification in facies A and D and taken measurements of wavelength and orientation of ripple crests. These (Figs 2–4 and 5) agree substantially with the data given by Allen & Underhill (1989) although, as cliff recession has taken place since their measurements were taken, these may be considered as new data. The cross-bedding (Fig. 5a) was corrected for regional dip (6° towards 019°). It is clear that when this correction is made referral of the cross-bedding to constructional hummocky cross-stratification (Sun 1989, 1990) can be dismissed.

Measurements were made for each unit of facies A and D that were available (June 1994), a cliff fall having obscured lower units available to Allen & Underhill (1989, 1990). Original dips greater than 5° (71 readings) consistently give mean directions towards the WNW ($c.\,300°$) in all units. The variance is consistent with measuring dips in a series of three-dimensional scour pits and the data demonstrate that the cross-bedding arises from the migration of typical unidirectional dune bedforms as demonstrated by Allen & Underhill (1989). Dips less than 5° (12 readings) show a mean direction towards $c.\,130°$ and represent stoss-side bedforms that have been preserved because of their strongly climbing nature.

In facies C the trend of 22 wave ripple crests (Fig. 5b) (016–196°) shows that the wave ripples originated from waves advancing in the same direction as the unidirectional currents. The wavelength of ripples in facies A and C is 6–20 cm and the ripple form index (l/h) is plotted in Fig. 5c. The graph shows that orbital ripples were present for the shorter ripples (7–13 cm) but only post-orbital ripples for the longer wavelengths. The minimum estimated water depth at which the ripples formed is 5–6 m, based on estimating minimum wave periods of 2.2 s. It may be noted that coastal water depth can increase by up to 5 m under hurricane set-up. Data for cross-stratification in the sandstone units of facies D are limited but also indicate an east to west flow.

Ichnology

The trace fossils of the Corallian Group were described and interpreted by Fürsich (1974b, 1975). Fürsich's analyses were carried out prior to the more detailed sedimentological work (listed above) and it is now clear that the trace fossils represent only colonizations of the *heterolithic facies* (C and D). Any traces associated with the *sandstone* (A) and *mudstone facies* (B) are due to downward intrusion from overlying units of facies (C) or, more especially, facies (D). *Teichichnus* isp. and *Planolites* isp. in facies (C) are uncommon. They represent initial recolonization, following each facies (A and B) event, in a marine or marine-influenced environment.

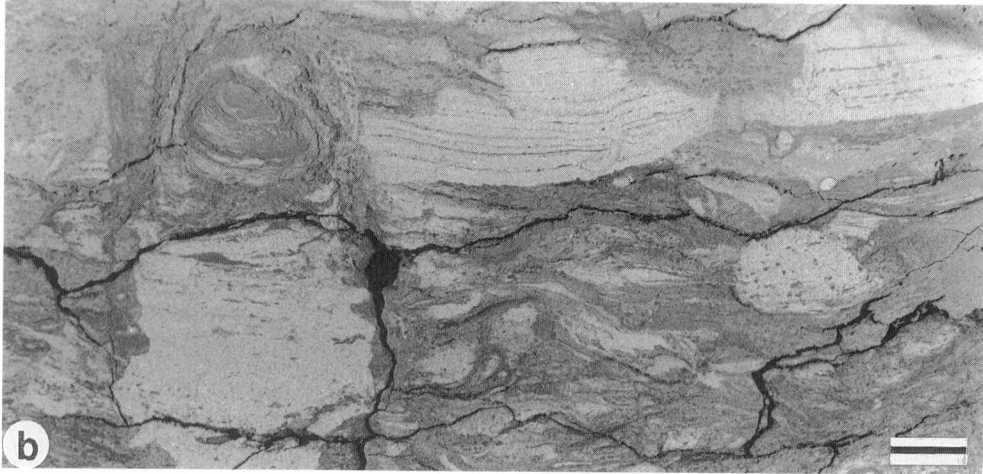

Fig. 9. (a) Three amalgamated sandstones units of facies D from just to east (right) of area depicted in Fig. 3. Upper two units with several 'D. habichi', Planolites isp., Teichichnus isp. and *Rhizocorallium* isp. (b) Roughly sawn block of bioturbated muddy sand with sandy event bed at the top. Ichnofabric constituent diagram (Fig. 10) constructed from this and slices from the same block. Constituents rapidly change in abundance, mainly due to patchy distribution of D. parallelum which cuts earlier elements. Most of spotting due to (faecal) pellets of mud released in association with D. parallelum (University of Reading, PRIS Archive S.38682). Scale bars are 1 cm.

Deeply burrowing suspension feeders relate to the ecological conditions at the sediment–water interface. Below this interface sediment consistency affects penetration and behaviour, but the sediment could have been deposited in many different environments, some of which would have been hostile to the trace makers. Two toponomic aspects are of note: (1) in the Bencliff Grit *Diplocraterion parallelum*, originating in heterolithic muds and sands (D) may penetrate up to 0.5 m into the underlying unit of *sandstone facies* (A) (e.g. Fürsich 1975, figs 9A and 12A). The maker of *D. parallelum* in traversing alternating lithologies frequently cut through mud layers into 'over-pressured' loose sand in facies A leading to extensive soft sediment deformation around the burrow and 'washing-out' of loose sand below the mud; (2) shafts of

Ophiomorpha nodosa originating on the upper surface of a sand unit of facies (D) extend downwards only a few centimetres before spreading out over the sole. The roof-lined burrows tend to turn and twist, running parallel to the stratification in a manner similar to that described for *O. irregulaire* (Frey *et al.* 1978) suggesting that downward penetration was hindered by substrate firmness.

The heterolithic facies (D) is characterized by two trace fossil associations: (1) the tops of the storm event sandstones are colonized by a small suite of distinct ichnotaxa (Bioturbation Index (B.I.) of 2, Taylor & Goldring 1993). Protrusive '*Diplocraterion habichi*' (Fürsich 1974*a*, *b*, figs 8 and 9) is common, extending down from a wave-rippled or truncated top, together with *Asterosoma* isp. (Fürsich 1974*b*, fig. 31b), *Gyrochorte comosa* (Fürsich 1974*b*, fig. 34), *Gyrophyllites* isp. (Fürsich 1974*b*, fig. 31a and c), *Ophiomorpha nodosa*, *Rhizocorallium* isp., *Scolicia* isp. (Fürsich 1974*b*, fig. 35) and *Teichichnus* isp. Laminae in the fine-grained sand units are often crowded with rod-shaped (faecal) pellets (Fig. 9) derived from the producer of *D. parallelum* (Fürsich 1974*b*, figs 12c and d). The trace fossils are too few and scattered to determine ordering.

(2) The bioturbated muddy sand has a B.I. of 3–5. *Teichichnus rectus* accounts for much of the bulk of the ichnofabric and is distinguished by the high dip of the laminar spreite. Other traces followed *T. rectus*, and are of low frequency and include: simple unlined sand- and mud-filled burrows (1.0 mm in diameter) (*Planolites montanus*); and (5.0 mm diameter) (*P. beverleyensis*), *Palaeophycus heberti*, *Rhizocorallium* isp. There is also a vague background mottling. The ichnofabric (Fig. 10) varies locally, particularly with the presence or absence of the deepest tier member, *D. parallelum*.

Discussion

The bioturbation associated with the heterolithic facies (D) is that of a typical shallow-marine assemblage, as recognized by Fürsich (1974*b*, 1975). Fürsich (1975) included the Bencliff Grit entirely within his *Diplocraterion* association which he interpreted as indicative of a high-energy environment 'very shallow subtidal or intertidal'. However, when the sedimentary context of the trace fossils is recognized, the sedimentary structures and trace fossils of the *heterolithic facies* (D) indicate a more open-marine storm-affected depositional environment.

The *D. parallelum* animal appears to have colonized the muddier intervals more or less continuously and *D. parallelum* may be considered to be over-represented. The taxon is generally regarded as a typical component of the Seilacherian *Skolithos* ichnofacies (e.g. Bromley 1990, 1996) and of the *Glossifungites* ichnofacies (e.g. Pemberton *et al.* 1992*b*). But with the latter association the taxon has taken on the role of an indicator of an erosional discontinuity with sequence stratigraphic significance in Mesozoic sediments (Pemberton *et al.* 1992*a*; Taylor & Gawthorpe 1993; Goldring 1995). This aspect of the palaeoecology of the taxon is under review. One of the main problems in understanding the trace is in identifying the sedimentary interface actually colonized. In the Bencliff Grit the main occurrences extend downwards from the lower muddier sections of facies D and are thus not as well defined as the 'discontinuity surfaces' tabled by Pemberton *et al.* (1992*b*). A possible interpretation is that facies D overall represents a somewhat restricted environment.

The two associations present indicate distinct substrate preferences: association (1) represents recolonization of the sandy upper surface of storm-event beds, and association (2) is a tiered association that colonized a more muddy substrate. The ichnotaxa of association (1) do not have the characteristics of opportunistic taxa (Frey & Goldring 1992, Goldring 1995). The

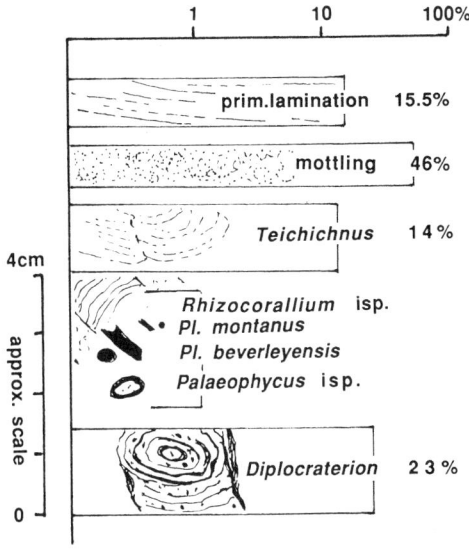

Fig. 10. Ichnofabric constituent diagram constructed from slices of block shown in Fig. 9b. On the horizontal axis the percentage area of the sedimentary structures and burrow types are plotted on a log scale. On the vertical axis, events (in order) from the initial sedimentation events (primary fabric) to subsequent modification by bioturbation (secondary fabrics).

Table 1. *Abundance of palynofacies components in the Osmington samples. The numbers are 'absolute' abundances corrected using a Lycopodium spike and expressed as numbers per gramme of rock in 100s. The palynofacies categories are similar to Waterhouse (1995). Spherricals are a frequently very abundant group of spherical palynomorphs which lack clear morphological features but are all almost certainly inaperturate pollen. Total organic carbon (TOC) data are also shown*

Sample	Facies	Acritarchs	Dinocysts	Foram tests	Bisaccate pollen	Pollen	Spores	Cuticle	Sphericals	Black wood	Brown wood	Total	TOC
1	Nothe Clay	12	77	–	530	92	61	4	284	127	350	1537	0.62
2	Nothe Clay	14	68	14	358	57	43	11	340	104	433	1430	0.42
4	B	–	220	–	1109	242	44	11	1010	472	1285	4393	1.26
5	B	–	110	29	1048	132	95	22	557	132	806	2933	1.07
7	D	–	35	13	254	62	24	8	348	30	305	1079	0.25
8	D	–	85	11	359	28	23	–	302	336	1133	2278	0.62
9	D	–	19	10	221	48	29	–	250	1038	2229	3844	1.57

storm-event sandstones cannot be followed far along the outcrop and it is likely that their cover areas were to be measured in hundreds, rather than thousands, of square metres.

The recovery of association (2) above a storm-event bed was probably associated with the rapid spread of muddy sediment, by traction and suspension fallout. It contrasts with the recovery from units of facies A where an appreciable thickness of sediment is involved with the succession of facies: A–B–C. This suggests that the factors involved in recovery from facies A were more complex than those following the storm events. A likely factor would have been salinity, which had affected an area far larger than that covered by a storm event. At small outcrops elsewhere in the stratigraphic column, the manner of biotic recovery may provide an indication of the extent of storm event beds.

Palynofacies and micropalaeontology

The samples processed for palynofacies and micropalaeontological analysis were collected from the points in the section shown in Figs 2, 3 and 7. All were fine-grained sediments and representative of the upper part of the Nothe Clay and facies B C and D as show in Tables 1 and 2.

Palynological processing of 30 g of rock was by standard procedures of HCl followed by HF. They were at this point spiked with an exotic spore *Lycopodium*. This was introduced in an aqueous suspension having been previously calibrated for particle concentration using a Coulter counter. The sample was then sieved at 20 μm before a single treatment in hot HCl to remove neoformed fluorides and mounted without oxidation on slides using Elvacite 2044.

Palynofacies counts (Table 1) were carried out using a point counter to a target of 400 (excluding *Lycopodium*) and are based on area represented rather than number of particles. The addition of the *Lycopodium* spike enables independent counts (e.g. Stockmarr 1971) to be made of each component rather than as a percentage. This permits direct comparison between samples and an assessment of the absolute quantities of each palynofacies component. To achieve this the number of each component is divided by the number of *Lycopodium* spores encountered during point counting and multiplied by the number of *Lycopodium* spores added to the sample. As each palynofacies component is present in different proportions in relation to the *Lycopodium* spores different errors in measurement are present for each of these components. In addition, there are further errors from differences in the numbers of *Lycopodium* spores added to each sample. As the summation of these errors is not easy to quantify nor applicable to each component or sample (Waterhouse 1995) only large differences (i.e. >10%) between components or samples are considered valid. In addition to reporting the data as numbers per gramme of rock the results were also normalized by recalculating as numbers per gramme of total organic carbon (TOC). These TOCs being determined (Table 1) using a Carlo Erba EA-1108 elemental analyser and fully corrected for mineral carbonate content. These TOC data shows a good correlation ($r = 0.82$) with kerogen abundance (Table 1) as determined by point counting.

The same samples were prepared for calcareous microfossils by pre-drying in an oven at 60°C for 24 h followed by soaking in calgon (sodium hexametaphosphate) for a further 24 h. The disaggregated material was then sieved through a 63-μm mesh, dried and picked. Unfortunately

Table 2. *Abundance of palynofacies components in the Osmington samples. The numbers are absolute abundances corrected using a* Lycopodium *spike and normalized using the TOC results as numbers per gramme of rock TOC in 100,000s*

Sample	Facies	Acritarchs	Dinocysts	Foram tests	Bisaccate pollen	Pollen	Spores	Cuticle	Sphericals	Black wood	Brown wood
1	Nothe Clay	2	12	–	86	15	10	1	46	20	56
2	Nothe Clay	1	16	3	85	14	10	3	81	25	103
4	B	–	17	–	88	19	3	1	80	37	102
5	B	–	10	3	98	12	9	2	52	12	75
7	D	–	14	5	101	25	10	3	139	12	122
8	D	–	14	2	58	5	4	–	49	54	183
9	D	–	1	1	14	3	2	–	16	66	142

samples 1, 2, 6 and 7 were adversely affected by subsurface dissolution with most of the hyaline foraminifera being very poorly preserved, although importantly still retaining environmentally diagnostic arenaceous and siliceous microfossils.

The quantitative palynofacies results in Table 1 show that samples from the same sedimentary facies are clearly similar in character but distinct from other depositional environments. Table 3 summarizes this combination of characteristics which allows these palynofacies to be distinguished. The one significant feature which is common to all samples is the presence of both calcareous- and organic-walled marine microfossils which shows that they were all deposited in a marine environment. In addition, more specific inferences can be drawn from these results when interpreted in the context of the sedimentological information.

Nothe Clay

Samples 1 and 2 confirm the Nothe Clay as the most marine assemblage. This is demonstrated by their high marine microfossil content including significant numbers of acritarchs, in addition

Table 3. *Summary of main facts and environmental interpretations on palynofacies, micropalaeontology and ichnology for facies B, C and D of the Bencliff Grit, and the top unit of the Nothe Clay (NC)*

Facies	Palynofacies	Micropalaeonotology	Ichnology
D	High total organic carbon (TOC) and terrestrial input but lacks nearshore terrestrial indicators, being dominated by large fragments of plant tissue	High in hyalines and dominance of *Spirillina*. Significant numbers of holothurian spicules and rhaxellid microscleres	(1) Mud rocks, sandy mudstones: *Diplocraterion parallelum*, *Teichichnus*, *Planolites*, *Palaeophycus*, *Rhizocorallium*
	Modified by differential degradation in a biologically active environment; only resistant components preserved	*Proximal to land but open unrestricted marine*	*Marine mud association*
			(2) Storm-event sands: '*D. habichi*', *Asterosoma*, *Gyrochorte*, *Gyrophyllites*, *Ophiomorpha*
			Marine sand association
C	High TOC. Dominated by large woody plant fragments	Insufficient sample	Rare *Planolites*, *Teichichnus*
	Modified by differential degradation in a biologically active environment; only resistant components preserved		*?restricted, ?brackish*
B	High TOC, high abundance of bisaccate pollen, wood, spores and dinocysts relative to other samples. All palynofacies components preserved	Diverse assemblage of agglutinated foraminifera in large numbers; *Ammobaculites*, *Trochammina*	No trace fossils
	Rapid sedimentation in a marine environment	*?Progradational, soft sediment below turbid water column, oxygen depleted*	
NC	Common and diverse marine elements, low terrestrial input, low TOC, high proportion of bisaccate pollen	No calcareous foraminifera preserved. Low abundance of agglutinates indicates a diagenetically modified assemblage. Rhaxellid microscleres preserved	Upper unit as for D
	Most marine and distant from terrestrial source	*Clear unrestricted water column, normal marine*	

to dinocysts combined with a low terrestrial input which is dominantly bisaccate pollen. This is present as a high proportion of the organic matter but a low percentage when expressed per gramme of rock. In modern environments (e.g. Heusser & Balsam 1977) bisaccate pollen increases proportionately to other terrestrial components with increasing distance from shoreline but, as with all kerogen of terrestrial origin, input rapidly declines in absolute abundance with increasing distance from the terrestrial source. These two Nothe Clay samples, by the same criteria, can also be separated by having the lowest total kerogen content both in terms of TOC and of total particle count, which in oxic marine sediments declines in absolute numbers with distance from the shoreline.

As regards foraminiferal assemblages these samples are effectively barren. The fauna found is small and mainly agglutinated with hyaline species rare and always in a poor state of preservation in showing severe diagenetic damage and are remnants of a normal marine assemblage. However, the presence of siliceous rhaxellid sponge microscleres indicates a normal marine environment, with clear unrestricted water column (Haslett 1992).

Facies B

Samples (4 and 5) from the mudstones–siltstones overlying the cross-bedded sands of facies A have a high content of both marine and terrestrially sourced organic matter, including the highest numbers of both dinoflagellate cysts and bisaccate pollen per gramme of rock and amongst the highest TOCs. They also contain a significant component of terrestrial plant debris. This shows that they are receiving and/or preserving more terrestrial input than the Nothe Clay samples. The high content of dinoflagellate cysts, although apparently anomalous, again reflects a higher organic input or preservation as the ratio of dinoflagellate cysts to bisaccate pollen is very similar to that found in the Nothe Clay samples and should not be attributed any special significance. The mudstones–siltstones of facies B occur immediately above the cross-bedded sandstones of facies A. Although individual beds of facies B are bioturbated, their upper surface is not the level from which burrowing occurs. They can thus be interpreted as units in which sedimentation has been rapid, and follow that of facies A in that they have a high input of organic matter which has been preserved.

This interpretation based on the palynofacies is complemented by the micropalaeontological assemblage. Sample 4 contains an abundant and diverse agglutinated foraminifera fauna indicating that they were dominant within the environment of deposition and, in this instance, are not a relict assemblage left after the severe dissolution of a hyaline dominated fauna. The dominant elements within this assemblage are *Trochammina* cf. *sqamata*, *Ammobaculites* cf. *agglutinans* and *A.* cf *hockleyensis* which may be indicative of restricted dysaerobic bottom conditions (Nagy *et al.* 1990*a,b*). This conclusion is supported by the relatively high TOC content. It also has been surmised that *A.* cf. *agglutinans* is indicative of a progradational environment, where there is a large input of freshwater and land-derived organic matter (Alve 1990). This environment will be characterized by soft sediment underlying a turbid water column in which there may be a degree of oxygen depletion, although the high monospecific dominance may be due to substrate control.

Facies C

Sample 6 was from a mudstone lenticle within this facies and although preserving significant amounts of organic matter proved difficult to process palynologically, as the kerogen comprised almost entirely of large three-dimensional fragments/billets of plant debris, with pollen, spores and dinoflagellate cysts being very rare. It is thus interpreted in the context of facies D as an assemblage which has undergone significant modification by degradation and differential transport within its depositional environment.

Facies D

Samples 7, 8 and 9 from the mudstone intervals of this heterolithic facies are significant in containing large amounts of detrital plant material indicative of a significant terrestrial influence. This is often in the form of large fragments with a significant proportion as opaque black wood analogous to the coal macerals of the inertinite group. The TOC, although variable, can be high. This palynofacies is notable for its combination of a high content of fragmental plant debris but lack of other nearshore material which would normally be expected to co-occur such as large spores, sculptured spores and cuticle. It is significant that it also has a reduced content of dinoflagellate cysts and bisaccate pollen. This is interpreted as resulting from differential preservation within the depositional environment with the palynofacies representing a residuum of the more resistant elements from an environment with a high terrestrial input but which has undergone significant bioturbation in oxic conditions.

The dominant presence within samples 8 and 9 of the hyaline foraminifera *Spirillina* indicates (Barnard & Shipp 1981) that the depositional environment was proximal to land. The high hyaline percentage indicates a normal unrestricted marine environment which is confirmed by the presence of large numbers of holothurian spicules.

The base of the Bencliff Grit

Coe (1992, 1995) took the base of the Bencliff Grit at the first occurrence of sand (our facies A) and explained that it follows a gradational interval from the Nothe Clay. Earlier workers (Arkell 1937; Wright 1986) included the gradational interval below facies A in the Bencliff Grit. It sees the gradual introduction of facies D and associated ichnotaxa, and includes a thin storm-event sand. This represents progradation from the underlying and muddier facies of the Nothe Clay but it is not necessarily a shallowing event. Thus, a sequence boundary at or close to the base of the Bencliff Grit cannot *a priori* be substantiated, although it would seem to represent the correlative conformity of a sequence boundary elsewhere (Coe 1995a, b).

Bencliff Grit–Upton Member contact

Although the erosive base of the Upton Member of the Osmington Oolite Formation is conspicuous in the field, where the oolitic, shelly packstone–wackestone projects above the softer units of the Bencliff Grit, there are two lower but cryptic erosion surfaces that truncate much of the face of the Bencliff Grit. The upper 0.2–0.4 m of the Bencliff Grit is somewhat variable but three units can be recognised along the section (Fig. 4), although it may be absent locally because of penecontemporaneous erosion. Unit R (0.05–0.25 m) comprises fine- to medium-grained sand (distinctly coarser than the underlying facies A) with small-scale cross-lamination and cross-stratification, followed by bioturbated muddy sand (B.I. of 5–6) with *Palaeophycus* isp. (cf. *Ophiomorpha irregulaire*) and *Planolites* isp. The former has a thin mud lining and sand fill, and the burrow is frequently somewhat collapsed, indicating passive fill. Scattered sand-filled *O. nodosa* cut *Palaeophycus*. The ichnotaxa are truncated by a sub-planar erosion surface (Fig. 4III) below cross-stratified fine-grained sand (Fig. 4T). This is irregularly broken by clusters of *D. parallelum* associated with bioturbated muddy oolite and mudstone (Fig. 4U). The clusters and top of this unit are truncated by an irregular erosion surface (IV) (amplitude 0.05 m) overlain by bioturbated muddy oolite with coarse sand and scattered siliceous granules and pebbles (Fig. 4V): the base of the Upton Member. *O. nodosa* (with ooid fill) may extend down through the whole interval and below. To the west a unit of fine-grained sand of facies D becomes amalgamated with unit U and includes '*Diplocraterion habichi*'.

The interval between the Bencliff Grit and the Upton Member thus represents at least five depositional events. The trace fossils present suggest an essentially similar marine depositional environment to facies D and the unit may be interpreted as several phases in the transgression.

Interpretation of the depositional environment

Controversy over the depositional environment of the Bencliff Grit has been due mainly to the differing hydraulic interpretations placed on the bedforms of the sand facies A. An attempt is made to incorporate other evidence: the micropalaeontology (palynofacies and foraminiferal distribution), the trace fossil assemblages, particularly the roles of *Diplocraterion parallelum* and '*D. habichi*', the distribution of the palaeocurrent data and wave ripple orientation and wavelengths, the relative lateral extents of facies A–D, and the differences in the nature of the ecological recovery in the sandstones of facies D and facies A–C.

Only the Osmington Mills section of the Bencliff Grit provides sufficient sedimentological information, as discussed by Wright (1986). The other sections to the west are almost downcurrent of the palaeoflow direction determined at Osmington Mills, so that the East Fleet section (SY 652 772) might be expected to be the most distal. But there is no outcrop. (The log of the well Hewish-1 (Fleet, SY 61 80) (Butler 1997 this volume) does not provide sufficient resolution.) East of Bowleaze Cove at Redcliff (SY 70 81), the fallen sandstone blocks of facies D give the impression that the shallow-tier trace fossils are more abundant than at Osmington Mills, and it is this section that has yielded ammonites (Wright 1986), although whether from facies A or facies D is uncertain.

Of the several interpretations hitherto advanced for the Bencliff Grit, the wholly intertidal or open-marine (shoreface) models of Sun (1989, 1990) are clearly inappropriate. None of the facies resembles in any way intertidal deposits because of the thickness and bed forms of the sand units, absence of mud drapes and the ichnology. Although the trace fossils of facies D indicate an open-marine situation, the over-representation of *D. parallelum* together with the high input of terrestrial palynological material and presence of *Spirillina*, indicate a more proximal situation. The sparse bivalves and ammonites could have been washed in, and the trace fossils (in facies C) are not environmentally specific. In any event the palynofacies from facies B, C and D are all clearly more terrestrially influenced than that of the Nothe Clay, which is generally understood to represent the deposits of a shallow shelf sea. (The sphenophyte stems referred to in Allen & Underhill (1990) are referrable to *Ophiomorpha* (Coe 1995; Brasier pers. comm.) Ecological recovery after each storm-event unit in facies D was rapid. Their apparent lateral extent appears to be less than was estimated for storm units in the Devonian (Goldring & Bridges 1973). Recovery after the intervals of facies A and B is quite different from that described from shoreface storm-event sands. The incised nature of the storm-event sands of facies D warrants comparison with the channelized storm beds of the East Texas coast (Siringan & Anderson 1994). However, the latter are probably of an order of magnitude greater in breadth but, because of the low accommodation space, appreciably less than the laterally persistent Carla bed (Hayes 1967) of the South Texas shelf.

Rather little work has been done on washover sands associated with transgressive barriers, especially their subaqueous sedimentology. Willis & Moslow (1994) had the advantage of appreciable facies differentiation in the Triassic sands they interpreted as washover fan-aprons analogous to modern examples spilling into the lagoon landward of the Chandeleur Islands, with geometry on an altogether larger scale to the washover sands described by Schwarz (1982) from the coast of North Carolina. The subaerial parts of the modern aprons are vegetated and the palaeocurrents in the submarine sediments would likely be lagoonwards. Interpreting facies A as washover sands satisfies the palaeocurrent data (i.e. suggests a temporary Corallian partial barrier to the south and southeast). Facies D would then represent mud and storm-sand sedimentation within a lagoon. This interpretation explains the appearance of sands in the Nothe Clay in a new way, as being associated with emergence of the barrier island during regression. The transgressive surface overlying the Bencliff Grit would then represent landward migration of the barrier or its elimination.

If facies A represents deposition associated with a tidal inlet (rather than washover) where the salinity was essentially normal then the question arises as to why facies A–C should be different from facies D as storm-events through tidal inlets should lead to essentially similar sedimentary units.

On balance, the estuary model of Allen & Underhill (1989, 1990) for the Bencliff Grit is supported by the abundance of plant debris in the organic matter from facies B mudstones. The high TOC and diverse assemblage of agglutinated foraminifera in facies B strongly support the hydraulic model (high-density flow) proposed by Allen & Underhill (1989, 1990). There is relatively little spread of the palaeocurrent vectors between units of facies A and the expected difference (in a distal estuarine environment) between the vectors of facies A and D was not found. The palaeocurrent data suggest a hitherto unrecognized sand source to the southeast (cf. Cope *et al.* 1992). This cannot be resolved with present information from subsurface data as the nearest data point is the well 98/16-1 (Butler 1997 this volume), situated some 14 km away SSE of Durlston Head. Only two ripple laminations with eastward migration were recorded, one is clearly within a wave ripple and is insufficient evidence for the reversing flows that would be expected in estuarine facies. Also, we know of no described modern or ancient estuarine analogue. There is no indication of any appreciable change in water depth through the Bencliff Grit, and post-orbital wave ripples in facies C indicate water depths consistent with the ichnological aspect of facies D.

Conclusions

The sedimentary structures of the Bencliff Grit and their inferred hydraulic processes, together with the micropalaeontology and ichnology, point to two distinct hydraulic and biotic sedimentary environments which are linked by the environment represented by the flaser-bedded facies C. But this, together with the palynological, micropalaeontological and ichnological evidence, does not clearly and conclusively point to a single overall depositional environment. Interpretation of the main bedform in facies A as unidirectional dune bedding indeed widens the environmental possibilities. In addition to closer micropalaeontological sampling, additional information on the regional distribution and stratigraphical context of the Bencliff Grit would be useful. But the possibilities here are limited by the local faulting and the configuration of the coast.

We support Allen & Underhill (1989, 1990) in their overall conclusions that the Bencliff Grit represents a nearshore depositional environment with units of terrestrially sourced sediment repeatedly introduced, and tentatively favour preference for a 'bay'–back barrier depositional environment.

The authors would like to acknowledge discussion and advice from P. A. Allen (Dublin), M. Butler and J. Floodpage (Brabant Petroleum), A. Coe (Open University), A. Dickson (Cambridge), A. Magyari (Budapest), M. Talbot (Bergen), J. R. Underhill (Edinburgh), G. Wang (Jiaozuo Institute of Technology, Henan, China), C. Wilson (Open University, Milton Keynes) and J. Wright (Royal Holloway, Egham) and the participation of the MSc class of 1994 in *Sedimentology and its Applications*. S. Akbari processed the palynological samples and measured the TOCs. University of Reading, PRIS Contribution No. 531.

Appendix

Systematic ichnology (RG)

Since Fürsich (1974*a*, *b*) monographed the trace fossils of the Corallian of England and Normandy a number of taxonomic changes have been made. Fürsich referred the small *Diplocraterion* associated with the event sandstones of facies D to *D. habichi* (Lisson 1904) an ichnotaxon of uncertain morphology. Fürsich's (1974*a*) revision of large cylindrical and branching burrow systems, placing these into synonymy with *Spongeliomorpha* Saporta, 1887, has not found general acceptance. Thus, his *S. nodosa* (Lungren 1891) and *S. saxonica* (Geinitz 1842) are generally referred to *Ophiomorpha nodosa* Lundgren 1891, and *S. suevica* (Reith 1932) is referred to *Thalassinoides*. In a partial revision of meniscate burrows (D'Alessandro & Bromley 1986) found *Muensteria* Sternberg, 1833 was not available as an ichnogenus and assigned the specimens figured by Fürsich (1974*a*, fig. 29a) to *Taenidium* Heer 1877. Goldring (1996) suggested that the burrows with excentric lining which Fürsich (1974*a*, figs 26 and 27) referred to *Cylindrichnus concentricus* Howard 1966 should probably be referred to another taxon.

References

ALLEN, P. A. 1990. Reply to discussion on Allen & Underhill (1989). *Journal of the Geological Society, London*, **147**, 398–400.

—— & UNDERHILL, J. R. 1989. Swaley cross-stratification produced by unidirectional flows, Bencliff Grit (Upper Jurassic), Dorset, UK. *Journal of the Geological Society, London*, **146**, 241–252.

ALVE, E. 1990 Variations in estuarine biofacies with diminishing oxygen conditions in Drammensfjord, SE Norway. *In*: HEMLEBEN, C., KAMINSKI, M. A., KUHNT, W. & SCOTT, D. B. (eds) *Paleoecology, Biostratigraphy, Paleoceanography and Taxonomy of Agglutinated Foraminifera*. Papers from the 3rd International Workshop on Agglutinated Foraminifera. Kluwer Academic, Amsterdam, NATO ASI Series, Series C Mathematical and Physical Sciences, **327**, 661–694.

ARKELL, W. J. 1933. *The Jurassic System in Great Britain*. Clarendon Press, Oxford.

BARNARD, T. & SHIPP, D. J. 1981. Kimmeridgian foraminifera from the Boulonnais. *Revue de Micropaléontologie*, **24**, 3–26.

BROMLEY, R. G. 1990. *Trace Fossils: Biology and Taphonomy*. Unwin Hyman, London.

——1996. *Trace Fossils: Biology, Taphonomy and Applications*, 2nd edition. Chapman & Hall, London.

BUTLER, M. 1997. The geological history of the Wessex Basin – a review of new information from oil exploration. *This volume*.

COE, A. L. 1992. *Unconformities Within the Upper Jurassic of the Wessex Basin, Southern England*. DPhil thesis, University of Oxford.

——1995. A comparison of the Oxfordian successions of Dorset, Oxfordshire, and Yorkshire. *In*: TAYLOR, P. D. (ed.) *Field Geology of the British Jurassic*. Geological Society, London, 151–172.

COPE, J. C. W., INGHAM, J. K. & RAWSON, P. F. 1992. *Atlas of Palaeogeography and Lithofacies*. Geological Society, London, Memoir No. 13.

CORBETT, P. W. M., STROMBERG, S. G., BRENCHLEY, P. J. & GEEHAN, G. 1994. Laminaset geometries in fine grained shallow marine sequences: core data from the Rannoch Formation (North Sea) and outcrop data from the Kennilworth Member (Utah, U.S.A.) and the Bencliff Grit (Dorset, U.K.). *Sedimentology*, **41**, 729–745.

D'ALESSANDRO, A. & BROMLEY, R. G. 1986. Meniscate trace fossils and the *Muensteria–Teanidium* problem. *Palaeontology*, **30**, 743–763.

FREY, R. W. & GOLDRING, R. 1992. Marine event beds and recolonization surfaces as revealed by trace fossil analysis. *Geological Magazine*, **129**, 325–335.

——, HOWARD, J. D. & PRYOR, W. A. 1978. *Ophiomorpha*: its morphologic, taxonomic, and environmental significance. *Palaeogeography, Palaeoclimatology, Palaeoecology*, **23**, 199–229.

FÜRSICH, F. T. 1974*a*. On *Diplocraterion* Torell 1870 and the significance of morphological features in vertical spreiten-bearing, U-shaped trace fossils. *Journal of Paleontology*, **48**, 952–962.

——1974*b*. Corallian (Upper Jurassic) trace fossils from England and Normandy. *Stuttgarter Beiträge zur Naturkunde B*, **13**, 1–52.

—— 1975. Trace fossils as environmental indicators in the Corallian of England and Normandy, *Lethaia*, **8**, 151–172.

GOLDRING, R. 1995 Organisms and the substrate: response and effect. *In*: BOSENCE, D. W. J. & ALLISON, P. A. (eds) *Fossils and Marine Palaeoenvironmental Analysis*. Geological Society, London, Special Publications, **83**, 151–180.

—— 1996. The sedimentological significance of concentrically laminated burrows from Lower Cretaceous Ca-bentonites, Oxfordshire. *Journal of the Geological Society, London*, **153**, 255–263.

—— & BRIDGES, P. 1973. Sublittoral sheet sandstones. *Journal of Sedimentary Petrology*, **43**, 736–747.

GRIEVES, I. A. T. 1986. *A Depositional Model for Part of the Corallian Sequence (Oxfordian), Dorset Coast*. MSc dissertation, University of Reading.

HART, B. S. 1990. Discussion on swaley cross-stratification produced by unidirectional flows, Bencliff Grit (Upper Jurassic), Dorset, UK. *Journal of the Geological Society, London*, **147**, 397–398.

HASLETT, S. K. 1992. Rhaxellid sponge microscleres from the Portlandian of Dorset, UK. *Geological Journal*, **27**, 339–347.

HAYES, M. O. 1967. *Hurricanes as Geological Agents: Case Studies of Hurricane Carla 1961 and Cindy 1963*. University of Texas at Austin, Bureau of Economic Geology, Report of Investigations, **61**.

HEUSSER, L. E. & BALSAM, W. L. 1977. Pollen distribution in marine sediments on the continental margin off northern California. *Quaternary Research*, **7**, 45–62.

HOWARD, J. D. 1972. Trace fossils as criteria for recognizing ancient shorelines. *In*: RIGBY, J. K. & HAMBLIN, W. K. (eds) *Recognition of Ancient Sedimentary Environments*. Society of Economic Paleontologists and Mineralogists, Special Publications, **16**, 215–225.

LISSON, C. I. 1904. Los Tigillites del Salto del Fraile y algunes Sonneratia del Morro Solar. *Boletin del Cuerpo de Ingenieros de Minas del Peru*, **17**, 1–64.

NAGY, J., LØFALDLI, M., BÄCKSTRÖM, S. A. & JOHANSEN, H. 1990*a*. Agglutinated foraminiferal stratigraphy of Middle Jurassic to basal Cretaceous shales, central Spitsbergen. *In*: HEMLEBEN, C., KAMINSKI, M. A., KUHNT, W. & SCOTT, D. B. (eds) *Paleoecology, Biostratigraphy, Paleoceanography and Taxonomy of Agglutinated Foraminifera*. Papers from the 3rd International Workshop on Agglutinated Foraminifera. Kluwer Academic, Amsterdam, NATO ASI Series, Series C, Mathematical and Physical Sciences, **327**, 969–1015.

——, PILSKOG, B. & WILHELMSEN, R. M. 1990*b*. Facies controlled distribution of foraminifera in the Jurassic North Sea Basin. *In*: HEMLEBEN, C., KAMINSKI, M. A., KUHNT, W. & SCOTT, D. B. (eds) *Paleoecology, Biostratigraphy, Paleoceanography and Taxonomy of Agglutinated Foraminifera*. Papers from the 3rd International Workshop on Agglutinated Foraminifera. kluwer Academic, Amsterdam, NATO ASI Series, Series C, Mathematical and Physical Sciences, **327**, 621–657.

PEMBERTON, S. G., FREY, R. W., RANGER, M. J. & MACEACHERN, J. A. 1992*a*. Conceptual framework of ichnology. *In*: PEMBERTON, S. G. (ed.) *Applications of Ichnology to Petroleum Exploration*. Society of Economic Paleontologists and Mineralogists, Core Workshop Guide, **17**, 1–32.

——, MACEACHERN, J. A. & FREY, R. W. 1992*b*. Trace fossil facies models: environmental and allostratigraphic significance. *In:* WALKER, R. G. & JAMES, N. P. (eds) *Facies Models, Response to Sea Level Change*. Geological Association of Canada, 47–72.

POLLARD, J. E., GOLDRING, R. & BUCK, S. G. 1993. Ichnofabrics containing *Ophiomorpha*: significance in shallow-water facies interpretation. *Journal of the Geological Society, London*, **150**, 149–164.

RIOULT, M., DUGUÉ, O., JAN DU CHÊNE, R., PONSOT, C., FILY, G., MORON, J.-M. & VAIL, P. R. 1991. Outcrop sequence stratigraphy of the Anglo-Paris Basin, Middle to Upper Jurassic (Normandy, Maine, Dorset). *Bulletin Centres Recherches Exploration–Production Elf Aquitaine*, **15**, 101–194.

SCHWARZ, R. K. 1982. Bedforms and stratification characters of some modern small-scale washover sand bodies. *Sedimentology*, **29**, 835–850.

SELLWOOD, B., WILSON, C., WEST, I. & CLITHEROE, A. 1990. *Jurassic Sedimentary Environments of the Wessex Basin*. Field Trip A16. 13th International Sedimentological Congress.

SIRINGAN, F. P. & ANDERSON, J. B. 1994. Modern shoreface and inner-shelf storm deposits off the east Texas coast, Gulf of Mexico. *Journal of Sedimentary Research*, **B64**, 99–110.

STOCKMARR, J. 1971. Tablets with spores used in absolute pollen analysis. *Pollen et Spores*, **13**, 615–621.

SUN, S. Q. 1989. A new interpretation of the Corallian (Upper Jurassic) cycles of the Dorset coast, southern England. *Geological Journal*, **24**, 139–158.

—— 1990. Discussion on swaley cross-stratification produced by unidirectional flows, Bencliff Grit (Upper Jurassic), Dorset, UK. *Journal of the Geological Society, London*, **147**, 396–397.

TALBOT, M. R. 1973*a*. *The Deposition and Diagenesis of the Corallian Beds of Southern England*. PhD thesis, University of Bristol.

—— 1973*b*. Major sedimentary cycles in the Corallian Beds (Oxfordian) of southern England. *Palaeogeography, Palaeoclimatology, Palaeoecology*, **14**, 293–317.

TAYLOR, A. M. & GAWTHORPE, R. L. 1993. Applications of sequence stratigraphy and trace fossil analysis to reservoir description: examples from the Jurassic of the North Sea. *In*: PARKER, J. R. (ed.) *Petroleum Geology of Northwest Europe: Proceedings of the 4th Conference*. Geological Society, London, 317–335.

—— & GOLDRING, R. 1993. Description and analysis of bioturbation and ichnofabric. *Journal of the Geological Society, London*, **150**, 141–148.

VAN WAGONER, J. C., MITCHUM, R. M., CAMPION, K. M. & RAHMANIAN, V. D. 1990. *Siliciclastic Sequence Stratigraphy in Well Logs, Cores, and Outcrops*. American Association of Petroleum Geologists, Methods in Exploration Series, **7**.

WATERHOUSE, H. K. 1995. High-resolution palynofacies investigation of Kimmeridgian sedimentary cycles. *In*: HOUSE, M. R. & GALE, A. S. (eds) *Orbital Forcing Timescales and Cyclostratigraphy*. Geological Society, London, Special Publications, **85**, 75–114.

WILLIS, A. J. & MOSLOW, T. F. 1994. Stratigraphic setting of transgressive barrier island reservoirs with an example from the Triassic Halfway Formation, Wembley Field, Alberta, Canada. *Bulletin of the American Association of Petroleum Geologists*, **78**, 775–791.

WILSON, R. C. L. 1968a. Upper Oxfordian palaeogeography of southern England. *Palaeogeography, Palaeoclimatology, Palaeoecology*, **4**, 5–28.

——1968b. Carbonate facies variation within the Osmington Oolite Series in southern England. *Palaeogeography, Palaeoclimatology, Palaeoecology*, **4**, 89–123.

——1991. Sequence stratigraphy: an introduction. *Geoscientist*, **1**, 13–23

WRIGHT, J. K. 1986. A new look at the stratigraphy, sedimentology and ammonite fauna of ther Corallian Group (Oxfordian) of south Dorset. *Proceedings of the Geologists' Association*, **97**, 1–21.

Biodegradation of seep oils in the Wessex Basin – a complication for correlation

M. ASHLEY BIGGE & PAUL FARRIMOND

Fossil Fuels and Environmental Geochemistry (NRG), Drummond Building, University of Newcastle upon Tyne, Newcastle upon Tyne NE1 7RU, UK

Abstract: A detailed organic geochemical investigation of seep oils from the Dorset coast has revealed notable variation in both the extent and pathways of biodegradation. All the seep samples analysed from Mupe Bay, Stair Hole (near Lulworth Cove) and Osmington Mills have had their *n*-alkanes and acyclic isoprenoid alkanes removed, but some samples are more extensively degraded, with partial loss of steranes and/or hopanes. At Mupe Bay (conglomerate matrix samples; see Parfitt & Farrimond 1997 this volume) the hopanes have been preferentially attacked, whilst at Stair Hole the steranes appear heavily degraded although there has been no alteration to the hopane distribution. 25-Norhopanes were not detected in any of the samples.

Biological marker distributions of seep oils which have suffered no hopane or sterane biodegradation are compared with those of three reservoired oils from the area (Wytch Farm and Kimmeridge fields). Molecular parameters indicate significant variation in source rock facies and maturity within the oils of the Wessex Basin.

Dorset oil seeps

Several oil seeps found along the Dorset coast (reviewed by Selley & Stoneley 1987; Selley 1992) were very important in attracting the initial interest in petroleum exploration to this area (Lees & Cox 1937). The best known of these occur in the Wealden at Mupe Bay and Lulworth Cove/Stair Hole, in the Bencliff Grit at Osmington Mills, and in the Purbeck Limestones between Durdle Door and Mupe Bay (Fig.1). Although all are separate entities in themselves, they share a common characteristic in occurring in northerly dipping beds immediately south of important faults (Selley & Stoneley 1987). Oil is currently being produced from the Wytch Farm, Wareham and Kimmeridge fields (in decreasing order of production), with characteristics indicative of a mature source rock with a vitrinite reflectance of at least 1% (Miles *et al*. 1993). Gas seeps have been reported in the sea bed at Anvil Point and small wet gas deposits discovered in drilling carried out by British Gas (Stoneley & Selley 1986).

Many burial history reconstructions have been conducted on the Liassic source rocks in Dorset (e.g. Ebukanson & Kinghorn 1986; Cornford *et al*. 1988) to estimate the timing of oil generation within the Wessex Basin. However, one widely cited (but controversial) piece of evidence for early generation of oil has been the oil staining within conglomerate clasts of the oil seep in Wealden beds at Mupe Bay (see Parfitt and Farrimond 1997 this volume). Several authors (e.g. Selley & Stoneley 1987; Cornford *et al*. 1988) have proposed that this outcrop contains two phases of oil staining: the first being at the time of original sandstone deposition in the Early Wealden, this sand being redeposited as oil-cemented clasts in a fluvial Wealden sandstone (Hesselbo & Allen 1991; Wimbledon *et al*. 1996; Hesselbo 1997 this volume) which was later stained itself. A molecular organic geochemical study by Cornford *et al*. (1988) strengthened this argument by presenting indications of a maturity difference (and thus time–burial interval) between these two phases of oil staining. However, the seep oils of the Dorset coast have all suffered extensive degradation, whereby their compositions have been dramatically altered by the action of microbes and water washing, and more recent studies of the Mupe Bay outcrop have shown that the apparent difference in maturity is actually a result of differential biodegradation (Miles *et al*. 1993; Parfitt & Farrimond 1997 this volume). The severe biodegradation of these oils has also complicated correlation between the different seeps, between seeps and reservoired oils, and between the seeps and source rocks.

Crude oil degradation

Oils undergo processes of biodegradation through the action of bacteria utilizing dissolved oxygen or sulphate to convert hydrocarbons in oil to biomass and carbon dioxide. In addition to available oxidants and nutrients, the temperature must be below 60–70°C for this process to occur

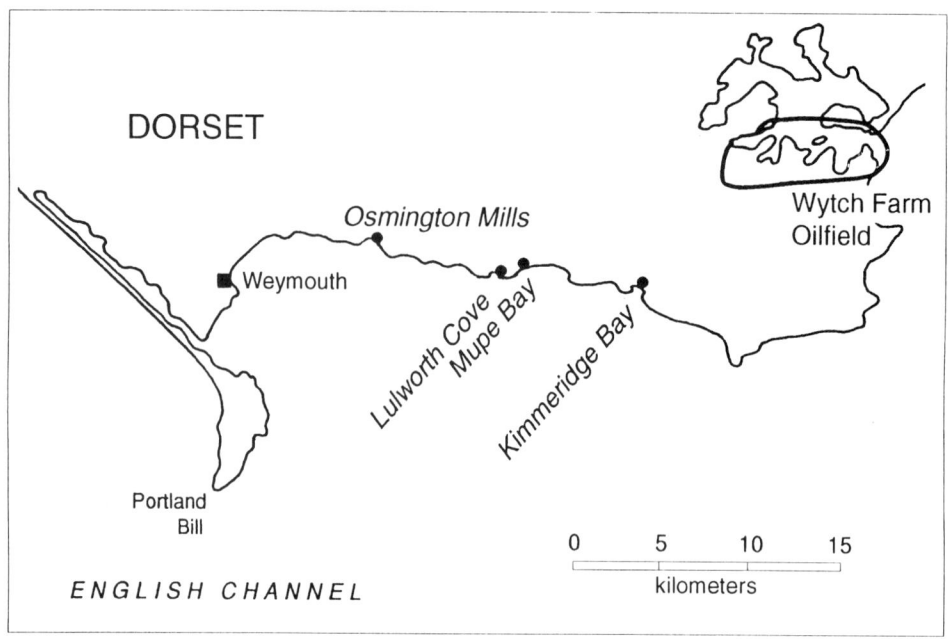

Fig. 1. Locality map of oil seeps on the Dorset Coast. (Note: Stair Hole lies on the western edge of Lulworth Cove.)

(e.g. Connan 1984). Most studies of crude oil biodegradation have addressed reservoired petroleums, where meteoric water can supply the necessary constituents to allow degradation in reservoirs shallower than around 1500 m. However, as our environmental awareness increases, the role of bacteria in degrading hydrocarbons in oil spills at the surface (bioremediation) is becoming more widely studied. Oils in natural seeps can be considered analogous to man-made spillages, and also possess suitable conditions for degradation to occur. However, there is little published data which allow comparison of the relative effects of biodegradation of oil under reservoir conditions as opposed to the very different surface environment.

Biodegradation or weathering of oil in natural seepages will differ from in reservoir biodegradation due to the different prevailing conditions and, therefore, the active bacteria. During migration and entrapment of an oil subsurface, the conditions will be of high temperature, high pressure and largely anaerobic (although meteoric waters can provide nutrients and dissolved oxygen). In contrast, for seeps at outcrop temperatures will be relatively low (and influenced by location), pressures will be atmospheric, and coastal seeps will be influenced by tidal movement and salt water washing. Oxygen is, of course, in abundance. Furthermore, the facing direction of the outcrop will influence the action of rainfall and run-off, and the intensity of incident sunlight.

The effects of biodegradation of crude oil are often observed in combination with the process of water washing, which selectively removes the more water-soluble components of the petroleum (Tissot & Welte 1984). Water washing can occur either at depth by low-salinity waters produced by compaction of smectite-rich clays, or in shallow environments by flow of meteoric waters through reservoirs. In the case of oil seeps the action of near-surface ground water, rainfall and run-off will lead to dissolution of certain components in the petroleum.

Although the processes involved are very different, both biodegradation and water washing of petroleums tend to remove components leading to an increase in the oil gravity (API°). Aliphatic hydrocarbons are generally biodegraded in preference to aromatic hydrocarbons, which are, in turn, more easily attacked than polar constituents. This results in a progressive change in the bulk geochemical composition of the oil, from a normal paraffinic crude to a heavy tar. Water washing preferentially removes the most water soluble components of the oil, principally those of low molecular weight

($<C_{15}$), and with aromatics generally being more easily dissolved (Lafargue & Barker 1988). Higher molecular weight ($>C_{15}$) aliphatic compounds are unaffected by water washing.

The processes of biodegradation do not equally attack different compounds within a single compound class. Many studies of in-reservoir biodegradation (or laboratory simulations of the same) have revealed sequential loss of *n*-alkanes, acyclic isoprenoid alkanes, followed by the cyclic biomarker compounds (hopanes and steranes), which themselves exhibit quite dramatic differences in susceptibililty to biodegradation (e.g. Seifert & Moldowan 1979; Goodwin *et al.* 1983; Connan 1984; Chosson *et al.* 1992). Steranes are typically degraded prior to the hopanes and diasteranes (Chosson *et al.* 1992). In detail, the steranes typically degrade in the order $\alpha\alpha\alpha 20R > \alpha\alpha\alpha 20S > \alpha\beta\beta 20R > \alpha\beta\beta 20S$ and $C_{27} > C_{28} > C_{29}$, although the order can vary slightly (Chosson *et al.* 1992; Peters & Moldowan 1993), $\alpha\beta\beta$ compounds sometimes being seen to degrade prior to the $\alpha\alpha\alpha 20S$. Diasteranes are particularly resistant to biodegradation, and thus increase in relative concentration. For the $\alpha\beta$ hopanes, the 22R isomers are more susceptible to biodegradation than the 22S, but the order of loss of the different carbon numbers appears to vary (Peters & Moldowan 1993). In most cases the longer chain hopanes appear more easily degraded than the shorter ones (Goodwin *et al.* 1983; Chosson *et al.* 1992), although some studies have shown preferential loss of the lower homologues (e.g. Peters & Moldowan 1991). These progressive changes in oil composition as biodegradation proceeds form the basis of using molecular geochemical parameters to rank oils with respect to their extent of degradation (Peters & Moldowan 1993). However, it seems that under varying conditions different orders of susceptibility of different compounds can occur, such that an absolute sequence of biodegradation cannot be constructed, only an outline scheme (e.g. Peters & Moldowan 1993).

For surface seeps and oil spillages, where the conditions of biodegradation are very different from those in a petroleum reservoir, the molecular effects might also be expected to differ considerably. However, weathering of oils at the surface appears to remove components in a similar manner to sub-surface biodegradation (Atlas *et al.* 1981), in the order *n*-alkanes > acyclic isoprenoid alkanes > steranes and hopanes. In a study of seep oils in Greece, Seifert *et al.* (1984) noted removal of steranes in the order $\alpha\alpha\alpha 20R > \alpha\alpha\alpha 20S > \alpha\beta\beta(20R + 20S)$ and $C_{27} > C_{28} > C_{29}$, consistent with that observed in many cases of in reservoir biodegradation. Steranes were degraded prior to attack on the hopanes, although Seifert *et al.* (1984) did note a certain amount of variability in relative susceptibilty of different biomarker components.

The molecular effects of biodegradation, whether in reservoir or at the surface, make correlation of oils, and the interpretation of biomarker distributions in terms of organic matter inputs and maturity level, very much more difficult.

This study

The aim of this study was to characterize the effects of biodegradation–weathering on the oils in three seeps in the Wessex Basin: Osmington Mills, Mupe Bay and Stair Hole (near Lulworth Cove; see Selley & Stoneley 1987; Selley 1992). Unpublished data revealed that the molecular composition of the seep oils showed some variability as a result of biodegradation, complicating the assessment of their relative maturity (see Parfitt & Farrimond 1997 this volume) and oil–oil correlations. Through the identification of these differences we hoped to optimize molecular correlation of the seep oils with locally reservoired oils. Furthermore, the molecular information provided, and the likely controls upon the variable pathways of biodegradation, is of relevance to workers using biological markers to monitor the extent of biodegradation of surface oil spills (e.g. Bragg *et al.* 1994).

Methods

Sampling

Samples were collected from seeps at three localities (Table 1): Osmington Mills, Stair Hole (near Lulworth Cove) and Mupe Bay, during several visits to Dorset (Sephton 1992; Bigge 1993; Parfitt 1993). The Osmington Mills outcrop (Corbett *et al.* 1994) comprises variably oil stained sandstone of the Bencliff Grit (5–6 m thick), and has been interpreted as an exhumed petroleum reservoir (Cornford *et al.* 1988). The sandstone is a poorly cemented, orange- or yellow-stained, well-sorted, very fine to fine quartz arenite. Thin, laterally discontinuous clay layers locally break up the sandstone into compartments. The sample from Stair Hole (Lulworth Cove) was collected from a slumped slope, where oil-stained sands (medium to coarse grain size) of Wealden age are found (Sephton 1992). Several samples were collected from the

Table 1. *Sample codes, descriptions and extractable oil content (% EOM)*

Sample	Description	EOM (%)
Osmington Mills		
OM1	Dark stained sand	8.09
OM4	Medium stained sand	2.82
OM17	Light stained sand	6.48
OM25	Light stained sand	4.75
OM27	Dark stained sand	11.8
OM38	Medium stained sand	9.65
OM(S)	Dark surface 'tarry skin' of outcrop	4.56*
OM(M)	Sample from just below the surface	4.23*
OM(D)	Sample from deeper (c. 6 cm) into the outcrop	4.31*
Stair Hole (near Lulworth Cove)		
LC1	Soft oil-cemented sand from slump	8.77
Mupe Bay		
MB-C1	Small clast from main outcrop	8.9*
MB-C7	Sample from larger clast sampled from slump	7.2
MB-M2	Very coarse poorly sorted matrix from slump	9.4*
MB-M5	Visibly weathered matrix from slump	3.2
MB3	Matrix sample from main outcrop	5.58
MB6	Matrix sample from main outcrop	5.07
MB4	Sample from underlying yellow sandstone	4.84
Reservoired Oils		
SO1	Wytch Farm (Sherwood Sandstone)	–
BO1	Wytch Farm (Bridport Sands)	–
KO1	Kimmeridge Field	–

* Average of duplicate extracts.

oil-stained conglomerate of Wealden age found in Mupe Bay (detailed by Parfitt & Farrimond 1997 this volume).

Samples were collected where possible after removal of the outer 1 cm to prevent excessive surface water and weathering contamination. At Mupe Bay, particular care was taken in collecting small samples with minimal disruption to the locality (see Parfitt & Farrimond 1997 this volume) which had been previously damaged by extensive hammering (see Miles *et al.* 1994, note added in proof). At Osmington Mills, samples were taken as small core plugs using a hand-held drill with a metal 1-cm diameter coring bit, 3-cm long. A larger sample was also collected, without removal of the outer surface, to test for the effects of weathering.

Extraction and fractionation of oil

All of the samples were only lightly lithified and/or cemented by oil, such that disaggregation could be carried out with a pestle and mortar. Dried sediment powder was repetitively solvent-extracted in a centrifuge tube with methanol and dichloromethane (DCM). Samples were sonicated for 10 min before being centrifuged (e.g. 10 min at 3000 rpm). The extracts were combined and then rotary evaporated to dryness, and the extract yield recorded. A test of the stepwise extraction procedure on an Osmington Mills sample showed that a fourth stage using dichloromethane afforded less than 2% of total extract yield, therefore three steps were routinely employed.

Aliphatic hydrocarbon fractions were obtained by thin-layer chromatography (TLC) (glass plates, 20×20 cm; 0.5 mm of Keisgel silica; activated at 120°C for 24 h; petroleum ether developer).

Molecular analysis of aliphatic hydrocarbons

The aliphatic hydrocarbon fractions were analysed by gas chromatography (GC) using a Carlo Erba 5160 Mega Series instrument, fitted with an on-column injector, a flame ionization detector (FID) and a glass capillary column (OV-1; 25 m × 0.32 mm i.d.). Hydrogen was used as carrier gas, with a flow rate of approximately 1 ml min^{-1}. The temperature programme was from 50°C (held for 2 min) to 300°C at a rate of $4°C\,min^{-1}$. Data were acquired using a VG Multichrom data system.

Gas chromatography–mass spectroscopy (GC–MS) analysis of aliphatic fractions was performed on a Hewlett-Packard 5890 GC (using splitless injection of 45 s at 250°C) interfaced to a Hewlett-Packard 5970B quadrupole mass selective detector (electron voltage 70 eV; source temperature 200°C). The acquisition was controlled by a Hewlett-Packard series 900(216) computer in selective ion mode (SIM) for greater sensitivity. The sample was injected in DCM ($c.\,1\,\mu l$) and separation was performed on a fused-silica HP-1 capillary column (25 m × 0.2 mm i.d.; 0.11 μm film thickness). The GC was temperature programmed from an initial temperature of 40°C at a rate of $10°C\,min^{-1}$ up to 175°C, then at $6°C\,min^{-1}$ up to 225°C, and finally at $4°C\,min^{-1}$ up to 300°C, and this final temperature held for 20 min. Helium was the carrier gas (flow rate $1\,ml\,min^{-1}$: pressure 50 kPa). Peak identification was made using a combination of relative retention times and mass chromatographic responses, in comparison with a standard sample (Brent oil) and literature data.

Bulk oil composition (TLC–FID)

A separate extraction was performed for TLC–FID (Iatroscan). About 1 g of sediment was extracted with 7 ml of a 93:7 DCM:methanol mix, sonicated for 20 min and settled out for 4 h. Three microlitres of sample was placed (in a small 2 mm spot) by syringe on clean silica-coated rods (10 cm long in a rack of 10 rods). Two rods in each rack were spotted with a standard solution, and each sample extract was also analysed in duplicate.

The first developing tank used contained hexane and this solvent was allowed to rise to the top of the rods before removal from the tank. The rack was left to dry for 3 min and then placed in a second tank containing toluene which was allowed to elute to 50% of the way up the rods before their removal from the tank. They were then allowed to dry for 6 min before being placed in the third tank containing a DCM:methanol mix (95:5) and developed 30% of the way up the rods. The rack was then placed in an oven at 60°C for 1.5 min before analysis by Iatroscan (FID). The results were checked for accuracy (the results of duplicates had to be within 10% of one another for saturates, aromatics, resins and asphaltenes). Re-runs were performed where necessary, and the results corrected to the standard.

Principal components analysis (PCA)

PCA was performed on a data set comprising 36 biomarker peak heights (measured in the appropriate mass chromatograms; Table 2) for the 17 seep and three oil samples. The program was written by P. Yendle (University of Bristol), and has been tested for equivalence against commercial packages. The data were normalized (to eliminate analytical variance resulting from the absolute intensity differences between samples), autoscaled (i.e. variables divided by their standard deviation to give each compound comparable weight in the analysis) and centred, prior to analysis.

Table 2. *Biomarker compounds used in the principal components analysis. Also used for peak assignments in Fig. 4 and identification of variables in Fig. 6*

m/z 191 mass chromatogram
1. C_{28} Tricyclic terpanes (22S + R)
2. C_{29} Tricyclic terpanes (22S + R)
3. 22,29,30-Trisnorneohopane (Ts)
4. 22,29,30-Trisnorhopane (Tm)
5. $C_{29}\alpha\beta$ hopane
6. C_{29} Norneohopane (29Ts)
7. C_{30} Diahopane
8. $C_{29}\beta\alpha$ hopane
9. $C_{30}\alpha\beta$ hopane
10. $C_{30}\beta\alpha$ hopane
11. $C_{31}\alpha\beta$ hopane (22S)
12. $C_{31}\beta\alpha$ hopane
13. $C_{32}\alpha\beta$ hopane (22S)
14. $C_{32}\alpha\beta$ hopane (22R)
15. $C_{32}\beta\alpha$ hopane
16. $C_{33}\alpha\beta$ hopane (22S)
17. $C_{33}\alpha\beta$ hopane (22R)
18. $C_{34}\alpha\beta$ hopane (22S)
19. $C_{34}\alpha\beta$ hopane (22R)
20. $C_{35}\alpha\beta$ hopane (22S)
21. $C_{35}\alpha\beta$ hopane (22R)

m/z 217 mass chromatogram
22. $C_{27}\beta\alpha$ diasterane (20S)
23. $C_{27}\beta\alpha$ diasterane (20R)
24. $C_{27}\alpha\beta$ diasterane (20S)
25. $C_{27}\alpha\beta$ diasterane (20R)
26. $C_{28}\beta\alpha$ diasterane (20S, 24S)
27. $C_{28}\beta\alpha$ diasterane (20S, 24R)
28. $C_{28}\beta\alpha$ diasterane (20R, 24S)
29. $C_{28}\beta\alpha$ diasterane (20R, 24R)
30. $C_{29}\beta\alpha$ diasterane (20S)
31. $C_{27}\alpha\alpha\alpha$ cholestane (20R)
32. $C_{29}\beta\alpha$ diasterane (20R)
33. $C_{29}\alpha\alpha\alpha$ cholestane (20S)
34. $C_{29}\alpha\beta\beta$ cholestane (20R)
35. $C_{29}\alpha\beta\beta$ cholestane (20S)
36. $C_{29}\alpha\alpha\alpha$ cholestane (20R)

Results and Discussion

Levels of oil staining

The samples from the three different seeps display a broad range of oil contents (2.8–11.8% by weight; Table 1), although the outcrops are characterized by similar average levels of staining (Osmington Mills 6.3%; Stair Hole 8.8% (one sample only); Mupe Bay 6.3%). At Osmington Mills, there is only a very general relationship between surface colour and the oil content of the sand (Table 1), and the three samples taken from different depths within the outrop (up to 6 cm) show no significant difference in the level of staining. In the case of Mupe Bay, the dark-coloured clasts contain more oil than the matrix (see Cornford et al., 1988; Miles et al., 1993; Parfitt & Farrimond 1997 this volume); this despite the much finer grain size, and inferred lower porosity, of the clasts.

Bulk oil composition

A ternary plot showing the bulk chemical composition of reservoired and seep oils (Fig. 2) demonstrates that whilst the reservoired oils differ in composition from the seeps, there are also significant differences between seep oils from the different localities. The reservoired oils are richest in saturate hydrocarbons, especially the oil from the Kimmeridge field. The seep oils from Osmington Mills and that from Stair Hole contain lower amounts of saturates, and display significant variation in the relative content of aromatic hydrocarbons. Finally, the seep samples from Mupe Bay all contain low amounts of saturates, and the clast samples are relatively enriched in aromatics compared to the matrix samples and the oil from the underlying sandstone (Fig. 2). These bulk compositional differences between the seep oils is likely to be the result of biodegradation and water washing under surface (or near-surface) conditions. Studies of biodegradation under reservoir conditions, and laboratory simulations, indicate that aliphatic hydrocarbons are preferentially biodegraded, and that the polar constituents are, on average, most resistant to attack (e.g. Connan 1984). Lighter components, and particularly the aromatics, will have been lost through water washing; rainfall, surface water

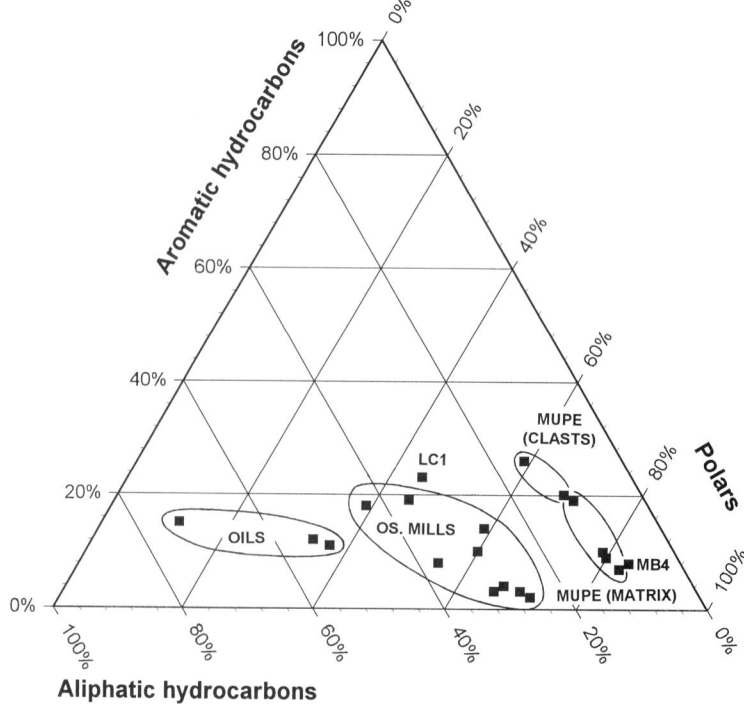

Fig. 2. Ternary plot of bulk oil composition determined by thin-layer chromatography–flame ionization detection (TLC–FID) (except three of the Mupe Bay samples which were determined by gravimetry of fractions afforded by conventional TLC).

run-off and groundwater movement causing solution and removal of the most water-soluble components. The differences in bulk composition noted, even in individual outcrops, are likely to be the result of the relative extents of biodegradation and water washing.

Molecular composition of the oils

The extent to which the seep oils have suffered biodegradation becomes more obvious when their molecular composition is compared to that of the reservoired oils. Gas chromatograms (GCs) of the aliphatic fractions of all the seep oils take the form of a broad unresolved complex mixture (UCM), with a series of biomarker peaks (predominantly steranes and hopanes) superimposed. This contrasts dramatically with aliphatic hydrocarbon GCs of the reservoired oils (Fig. 3), which are dominated by n-alkanes and acyclic isoprenoid alkanes. These components have been universally removed from the seep samples by biodegradation. The preferential removal of these compounds is typical of petroleum biodegradation under reservoir (Seifert & Moldowan 1979), laboratory (Connan 1984) and surface conditions (Atlas et al. 1981).

Although the GCs of all the seep oils look similar, differences in their detailed molecular composition can be observed when gas chromatography-mass spectrometry (GC–MS) is applied to study sterane and hopane distributions (Fig. 4). The seep oils from Osmington Mills and the clast samples (and lower

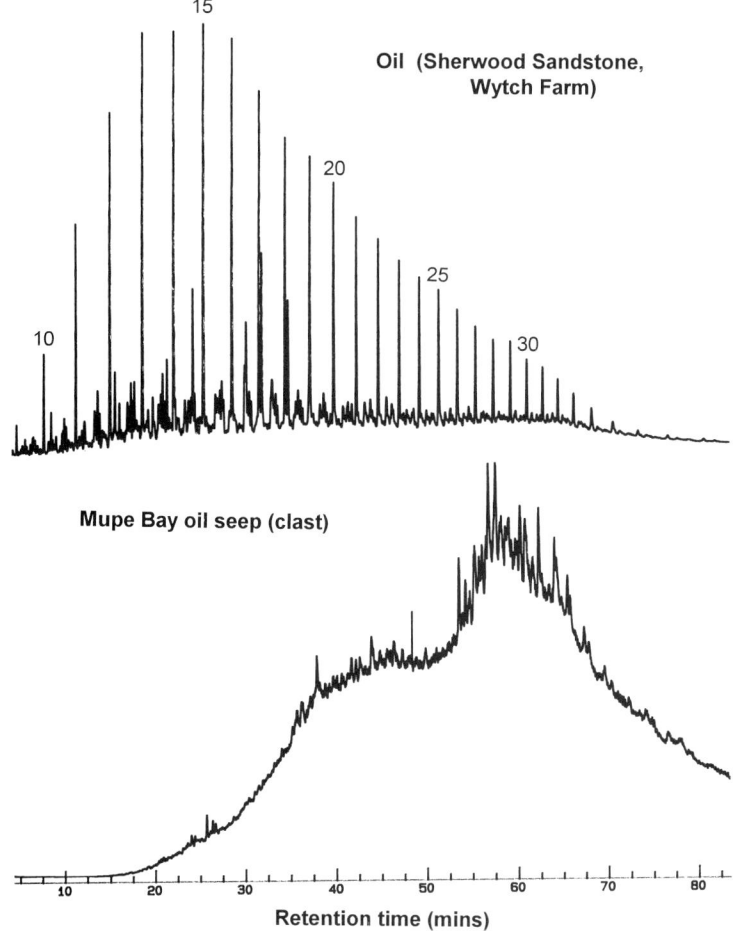

Fig. 3. Gas chromatograms comparing the aliphatic hydrocarbon distribution of a typical seep (a Mupe Bay clast) and an oil (Sherwood Sandstone). Selected n-alkanes are labelled with their carbon number.

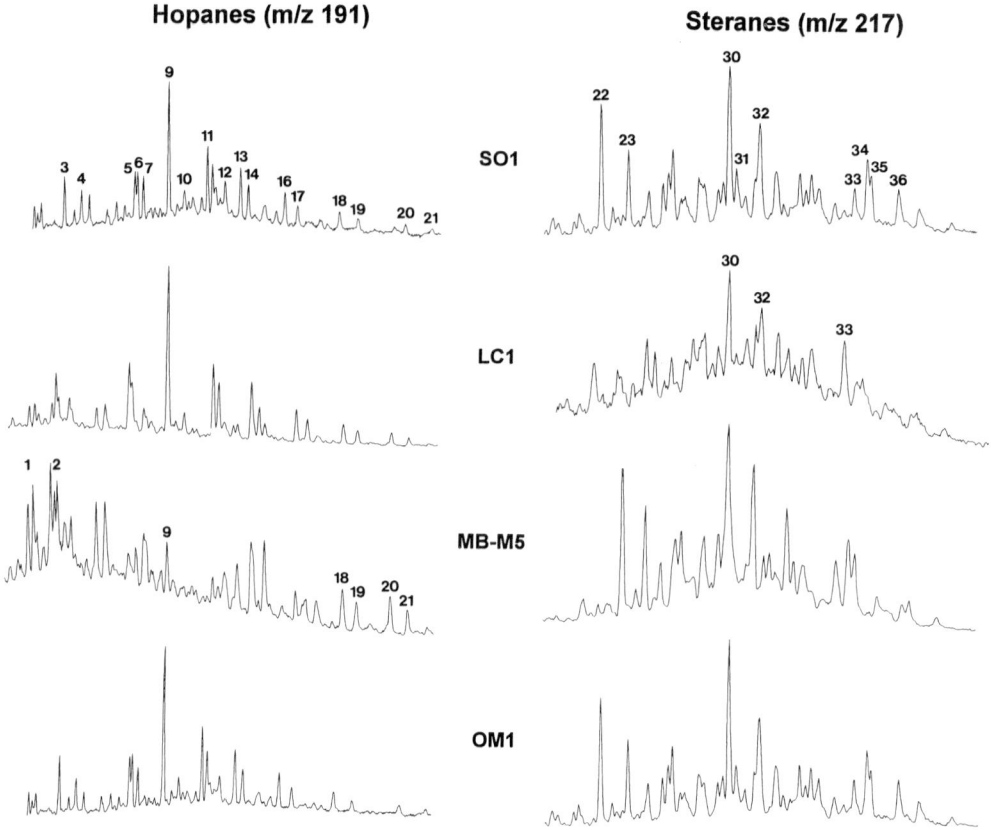

Fig. 4. Distributions of hopanes (m/z 191 mass chromatograms) and steranes (m/z 217 mass chromatograms) in the Sherwood Sandstone (Wytch Farm) oil and three seep samples. Peak assignments are given in Table 2.

sandstone) from Mupe Bay display apparently undegraded hopane and sterane distributions, directly comparable to those of the reservoired oils which have not suffered any biodegradation. However, the seep oil from Stair Hole and three of the four matrix samples from Mupe Bay display evidence of extensive biodegradation of steranes and/or hopanes.

In the case of the Stair Hole seep, the hopanes appear undegraded, with a distribution dominated by $C_{30}\alpha\beta$ hopane and a series of higher homologues (C_{31-35}) which gradually diminish in concentration. The steranes, however, have undergone extensive modification, with seemingly complete removal of the $C_{27-29}\alpha\alpha\alpha20R$ steranes, and substantial loss of the $\alpha\beta\beta$ steranes and diasteranes. The resulting sterane distribution is dominated by the $\alpha\alpha\alpha20S$ steranes, which in this case appear to be the most resistant to biodegradation.

The matrix samples from Mupe Bay (with the exception of a fresh sample from the slump) have undergone a very different pattern of biodegradation of the biomarker compounds. Here, the sterane distribution records only partial loss of the $\alpha\alpha\alpha20R$ isomers, and no apparent removal of any other steranes. This contrasts with the dramatic preferential loss of the $\alpha\beta$ hopanes in the C_{29}–C_{33} range, with less marked removal of the C_{34} and C_{35} compounds (Fig. 4; similar to the relative resistance to biodegradation of $C_{35}\alpha\beta$ hopanes in Monterey tar sands, observed by Requejo & Halpern 1989). Compounds such as the tricyclic terpanes and diahopane have not been notably affected.

25-Norhopanes were not detected in any of the samples, although these compounds are found in many biodegraded petroleums (see Peters & Moldowan 1993, and references therein). Their origin is uncertain. Although they appear to be formed by biodegradation (removal of the methyl group at C-10) of hopanes, it has also been suggested that they may be normal constituents of oils which are only observed after removal of the more abundant hopanes (e.g. Chosson *et al.* 1992). The presence or absence of these compounds may be associated with two different pathways for heavy biodegradation (Brooks *et al.* 1988): first, one where 25-norhopanes appear in association with loss of hopanes but prior to sterane degradation; and, second, where steranes are preferentially degraded prior to the hopanes, and 25-norhopanes are not observed (Peters & Moldowan 1993). In the present study the seep at Stair Hole fits the second pathway, but the Mupe Bay matrix oils agree with neither, having preferentially lost the hopanes but without the occurrence of 25-norhopanes. This latter observation is further evidence that 25-norhopanes are formed through demethylation of hopanes, but that this process may not occur under surface conditions.

Principal components analysis (PCA): the effects of biodegradation

A biomarker data set, comprising 36 variables (biomarker peak areas) for the 20 oil and seep samples, was analysed by PCA to establish the major sources of variance within the sterane and hopane distributions. PCA is a multivariate statistical approach which identifies relationships between samples and variables in complex data sets, and summarizes the main sources of variance as principal components (PCs), which are linear combinations of the original variables (see Davis 1986, for a discussion of PCA). The PCs are, by definition, orthogonal to each other, such that no variance explained by one PC will contribute to another. In this analysis, the first two PCs explained 63.6% and 24.4% of the variance within the scaled data set, respectively. This allows the relationships between the samples to be examined in a simple two-dimensional scores plot (PC1 v. PC2) which incorporates the effects of all the original variables, and displays the bulk (88%) of the scaled data variability (Fig. 5). This plot demonstrates that the bulk of the seep samples cluster in association with the reservoired oils indicating little or no biodegradation of their constituent biomarkers. However, the sample from Stair Hole lies well away from this main group (negative PC1 and PC2), and three of the matrix samples from Mupe Bay form another distinct group (positive PC1).

An explanation of what these two PCs represent is given by loadings plots (Fig. 6). These diagrams show the loadings of the different variables within the principal components; large loadings, either positive or negative, indicate a significant influence of the variable on that PC. Thus, PC1 essentially distinguishes samples which are relatively depleted in hopanes (positive PC1; especially the $C_{29-33}\alpha\beta$ compounds) from those relatively depleted in steranes (negative PC1). The scores plot (Fig. 5) graphically displays the different molecular effects of biodegradation of the Stair Hole (loss

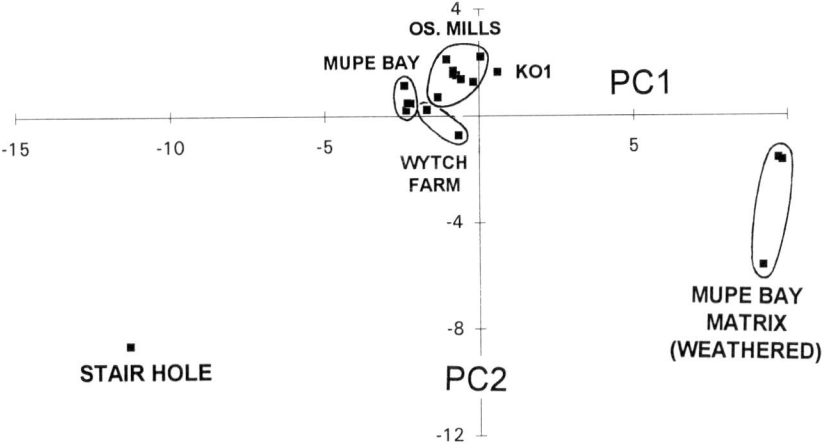

Fig. 5. Principal components analysis scores plot (PC1 v. PC2) showing the relationships between the samples in terms of the first and second principal components. PC1 explains 63.6% of the total variance in the scaled data set; PC2 explains a further 24.4%. PC1 essentially differentiates the samples in terms of their biodegradation (see Fig. 6); samples with high positive PC1 scores being relatively depleted in many of the hopanes, whilst the Stair Hole sample with large negative PC1 scores is relatively depleted in most of the steroids.

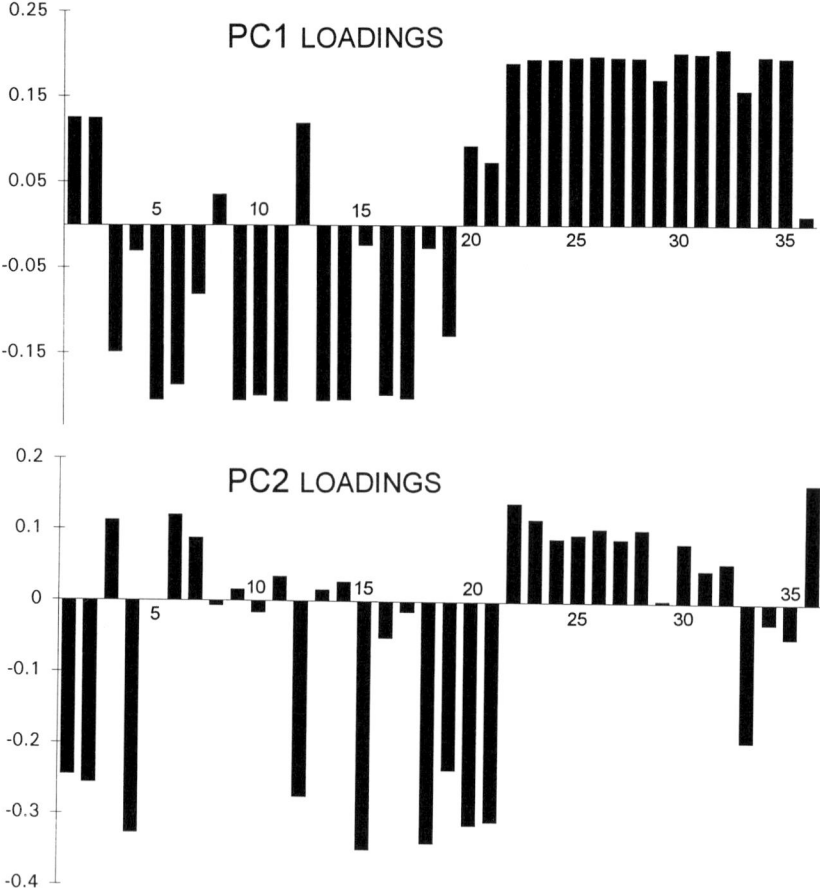

Fig. 6. Loadings plots showing the composition of the first two principal components, which record 63.6 and 24.4% of the total variance in the scaled data set, respectively. PC1 essentially comprises the relative abundance of most of the hopanes (negative PC1) v. most of the steranes (positive PC1).

of steranes) and Mupe Bay (matrix; loss of hopanes) seep oils. The remaining samples cluster together; differences in their molecular composition are unlikely to be the result of biodegradation, as there is as much variation between the reservoired oils (which have not suffered biodegradation) as between the various seeps. The Osmington Mills samples plot separately from those of Mupe Bay, indicating slight difference in the relative contents of hopanes and steranes in these seeps. It is notable that the three samples selected to examine the effects of surface weathering on the Osmington Mills outcrop are virtually superimposed on the scores plot (Fig. 5) indicating no compositional difference in terms of the hydrocarbon biomarkers. The biodegradation process in these outcrops clearly occurs below the outer few centimetres of the outcrop.

There are also no obvious differences in extents of biomarker biodegradation between the light-, medium- and dark-coloured samples, or between samples with different measured oil contents at Osmington Mills. The extent of biomarker biodegradation appears to be partly controlled by sediment grain size and probably porosity–permeability, although these latter were not measured. The factors controlling the observed difference in order of biomarker degradation are currently unclear, as the seep oils are believed to have the same source (Ebukanson & Kinghorn 1986), and those at Mupe Bay and Stair Hole are separated by only 2 miles and both occur in Wealden sands. Nevertheless, it is possible that different bacterial populations are degrading the petroleum at these sites. Literature reports indicate that steranes are usually degraded prior to the hopanes (e.g. Seifert et al. 1984; Chosson et al. 1992), such as at Stair Hole. The observation that $\alpha\beta$ hopanes are preferentially degraded in the Mupe Bay oil seep has implications for the application of biomarkers in monitoring biodegradation of oil spills, where the $C_{30}\alpha\beta$ hopane has been used as a conservative marker compound (e.g. Bragg et al. 1994).

Implications for oil correlation

Whilst all the seep samples have had their n-alkanes and acyclic isoprenoid alkanes removed (biodegradation ranking of 5; moderately degraded; Peters & Moldowan 1993), three matrix samples from Mupe Bay and the sample from Stair Hole have been more extensively affected (ranking 6 or 7; heavy biodegradation), with partial loss of the biomarker compounds. For these samples, preferential loss of the hopanes or steranes has resulted in drastic modification of various molecular parameters used in the correlation of oils (e.g. hopane:sterane ratio; tricyclic:hopane ratio; sterane carbon number distribution). However, the application of PCA is beneficial not only in recognizing these extensively altered samples, but also in identifying compositional differences between the seeps at Osmington Mills and Mupe Bay (clasts, fresh matrix sample and underlying sandstone) which are unlikely to be the result of biodegradation.

Selected molecular parameters often used in correlations (e.g. Mackenzie 1984; Peters & Moldowan 1993) between oils are shown in Table 3. The variability of these parameters within individual sites (excluding those samples

Table 3. *Molecular parameters often employed in oil correlation studies*

Sample	27/29 ($\alpha\alpha\alpha$ 20R)[1]	27/29 ($\beta\alpha$ dia)[2]	sterane/dia (C_{29})[3]	C_{30} hopanes/ C_{29} steranes[4]	C_{29} tris/ $C_{30}\alpha\beta$ hop[5]
Osmington Mills					
OM1	1.19	0.74	0.75	0.93	0.29
OM4	1.16	0.75	0.81	0.87	0.26
OM17	1.28	0.72	0.87	0.78	0.30
OM25	1.29	0.73	0.82	0.86	0.31
OM27	1.25	0.78	0.82	0.85	0.34
OM38	1.23	0.80	0.89	0.82	0.30
OM(S)	1.27	0.74	0.81	0.89	0.30
OM(M)	1.19	0.82	0.83	0.88	0.26
OM(D)	1.40	0.80	0.80	0.88	0.32
Average	**1.25**	**0.76**	**0.82**	**0.86**	**0.30**
Stair Hole (near Lulworth Cove)					
LC1	*0.66*	*0.52*	*3.07*	*0.73*	*0.23*
Mupe Bay					
MB-C1	0.78	0.72	1.09	0.66	0.23
MB-C7	0.83	0.72	1.05	0.68	0.22
MB-M2	0.81	0.74	1.11	0.64	0.23
MB-M5	*1.95*	*0.71*	*0.91*	*0.36*	*2.16*
MB3	*1.78*	*0.65*	*0.97*	*0.28*	*2.66*
MB6	*2.06*	*0.71*	*0.88*	*0.33*	*2.55*
MB4	0.83	0.69	1.14	0.66	0.22
Average*	**0.81**	**0.72**	**1.10**	**0.66**	**0.22**
Reservoired Oils					
SO1	1.30	0.81	0.79	0.84	0.42
BO1	1.08	0.87	1.04	0.69	0.22
KO1	0.92	0.81	0.80	0.85	0.33

* Average of those samples which have not suffered notable degradation of the biomarker compounds. Extensively degraded samples have data shown in italics.
[1] $C_{27}\alpha\alpha\alpha$ cholestane (20R)/$C_{29}\alpha\alpha\alpha$ cholestane (20R).
[2] $C_{27}\beta\alpha$ diasterane (20S + R)/$C_{29}\beta\alpha$ diasterane (20S + R).
[3] $C_{29}\alpha\alpha\alpha$ cholestane (20S + R) + $C_{29}\alpha\beta\beta$ cholestane (20S + R)/$C_{29}\beta\alpha$ diasterane (20S + R).
[4] $C_{30}\alpha\beta$ hopane + $C_{30}\beta\alpha$ hopane/$C_{29}\alpha\alpha\alpha$ cholestane (20S + R) + $C_{29}\alpha\beta\beta$ cholestane (20S + R).
[5] C_{29} tricyclic terpanes (22S + R)/$C_{30}\alpha\beta$ hopane.

known to be extensively biodegraded) is very small, but it is clear that the seep oils from Osmington Mills differ significantly in biomarker composition from those from Mupe Bay (Table 3). These observed differences are not consistent with biodegradation, and must relate to real differences in the biomarker composition of the oils at the two localities. Interestingly, the three reservoired oils also show significant variation in the selected correlation parameters. The seep oil at Osmington Mills appears closest in biomarker composition to the Wytch Farm oil reservoired in the Sherwood Sandstone (SO1), whilst the seep at Mupe Bay is closer in composition to the Wytch Farm oil reservoired in the Bridport Sands (BO1), although the sterane and diasterane carbon number ratios differ (Table 3). These results, having excluded the influence of biodegradation, indicate a certain degree of variation in detailed molecular composition of the oils in the Wessex Basin, which may reflect facies variation in the source rock.

Selected biomarker maturity parameters (Table 4) indicate that there is also some variability in the maturity of the different oils and seeps, although samples from individual seep localities show remarkably consistent values (excepting those known to be heavily modified by biodegradation). The samples from Osmington Mills display consistently higher maturity values than those from Mupe Bay (see Parfitt & Farrimond 1997 this volume, for a discussion of the maturity of Mupe Bay oil). The reservoired oils are also variable in their apparent maturity. For example, the oil from the Sherwood Sandstone is more mature than the oil from the Bridport Sand reservoir (Table 4). The Kimmeridge oil exhibits conflicting hopane and

Table 4. *Selected molecular maturity parameters*

Sample	$29Ts/2\alpha\beta^1$	$dia/30\alpha\beta^2$	$C_{29}\ \%\alpha\beta\beta^3$	$C_{29}\ \%20S^4$
Osmington Mills				
OM1	1.05	0.25	59	48
OM4	1.10	0.25	60	49
OM17	1.05	0.26	61	49
OM25	1.03	0.29	61	50
OM27	1.14	0.27	60	48
OM38	1.08	0.28	60	48
OM(S)	1.11	0.27	60	48
OM(M)	1.03	0.28	60	48
OM(D)	1.05	0.26	60	51
Average	**1.07**	**0.27**	**60**	**49**
Stair Hole (near Lulworth Cove)				
LC1	*0.70*	*0.15*	*40*	*70*
Mupe Bay				
MB-C1	0.74	0.16	54	43
MB-C7	0.71	0.14	55	44
MB-M2	0.73	0.16	56	42
MB-M5	*0.77*	*0.85*	*64*	*64*
MB3	*0.75*	*1.03*	*65*	*65*
MB6	*0.57*	*0.97*	*65*	*66*
MB4	0.83	0.16	54	45
Average*	**0.75**	**0.16**	**55**	**44**
Reservoired Oils				
SO1	0.98	0.34	59	48
BO1	0.71	0.21	56	46
KO1	0.96	0.34	54	42

*Average of those samples which have not suffered notable degradation of the biomarker compounds. Extensively degraded samples have data shown in italics.
[1] 22,29,30-trisnorneohopane/$C_{29}\alpha\beta$ hopane.
[2] C_{30} diahopane/$C_{30}\alpha\beta$ hopane.
[3] $C_{29}\alpha\beta\beta$ cholestane (20R+S)/$C_{29}\alpha\beta\beta$ cholestane (20R+S)+$C_{29}\alpha\alpha\alpha$ cholestane (20S+R) (%).
[4] $C_{29}\alpha\alpha\alpha$ cholestane (20S)/$C_{29}\alpha\alpha\alpha$ cholestane (20S+R) (%).

sterane maturity parameters, although the evidence from the hopane distributions indicates comparable or higher maturity than the Wytch Farm oils.

Conclusions

The seep oils on the Dorset coast have suffered moderate to heavy biodegradation, all samples having lost their *n*-alkanes and acyclic isoprenoid alkanes. Three samples from the conglomerate matrix at Mupe Bay, and the sample from Stair Hole, are the most extensively degraded, with partial loss of steranes and/or hopanes. However, these two localities are characterized by different pathways of biomarker biodegradation:

(1) at Stair Hole, the steranes have undergone extensive biodegradation, with complete removal of $\alpha\alpha\alpha$(20R) isomers, and partial loss of $\alpha\beta\beta$(20R and 20S) isomers and diasteranes. The hopanes appear undegraded;

(2) at Mupe Bay, conglomerate matrix samples have suffered extensive hopane biodegradation, with particularly notable loss of the $C_{29-33}\alpha\beta$ isomers, leaving tricyclic terpanes dominant, and relatively enhanced $C_{34-35}\alpha\beta$ hopanes. The sterane distributions are only slightly modified, with partial removal of the $\alpha\alpha\alpha$(20R) compounds.

Such divergent pathways have been noted before (Brooks *et al.* 1988, Western Canada Basin), although in the present study 25-norhopanes are absent from the oils of both pathways. The factors controlling which pathway is followed are uncertain, as the seep oils are believed to have the same Liassic source, and both occur in Wealden sands.

Biomarker parameters for the less extensively biodegraded samples indicate significant variability in source facies of the seep oils and locally reservoired oils. Maturity parameters also vary, suggesting the following order of increasing maturity: Mupe Bay ≈ Wytch Farm (Bridport) < Wytch Farm (Sherwood) < Osmington Mills.

The Natural Environment Research Council is gratefully acknowledged for a studentship (MAB). Technical assistance was provided by P. Donohoe and I. Harrison, and the artwork produced by C. Jeans and B. Brown. British Petroleum are thanked for the provision of oil samples.

References

ATLAS, R. M., BOEHM, P. D. & CALDER, J. A. 1981. Chemical and biological weathering of oil, from the Amoco Cadiz spillage, within the littoral zone. *Estuarine, Coastal and Shelf Science*, **12**, 589–608.

BIGGE, M. A. 1993. *Variability of Migrated Hydrocarbons in the Bencliff Grit Exposure, Osmington Mills, Dorset*. MSc dissertation, University of Newcastle upon Tyne.

BRAGG, J. R., PRINCE, R. C., HARNER, E. J. & ATLAS, R. M. 1994. Effectiveness of bioremediation for the *Exxon Valdez* oil spill. *Nature*, **368**, 413–418.

BROOKS, P. W., FOWLER, M. G. & MACQUEEN, R. W. 1988. Biological marker and conventional organic geochemistry of oil sands/heavy oils, Western Canada Basin. *Organic Geochemistry*, **12**, 519–538.

CHOSSON, P., CONNAN, J., DESSORT, D. & LANAU, C. 1992. In vitro biodegradation of steranes and terpanes: A clue to understanding geological situations. *In*: MOLDOWAN, J. M., ALBRECHT, P. & PHILP, R. P. (eds) *Biological Markers in Sediments and Petroleum*, Prentice-Hall, Englewood Cliffs, New Jersey, 320–349.

CONNAN, J. 1984. Biodegradation of crude oils in reservoirs. *In*: BROOKS, J. & WELTE, D. H. (eds) *Advances in Petroleum Geochemistry*, Vol. 1. Academic Press, London, 299–335.

CORBETT, P. W. M., STROMBERG, S. G., BRENCHLEY, P. J. & GEEHAN, G. 1994. Laminaset geometries in fine grained shallow marine sequences: core data from the Rannoch Formation (North Sea) and outcrop data from the Kennilworth Member (Utah, USA) and the Bencliff Grit (Dorset, UK). *Sedimentology*, **41**, 729–745.

CORNFORD, C., CHRISTIE, O., ENDERSON, U., JENSEN, P. & MYHR, M. 1988. Source rock and oil seep maturity in Dorset, southern England. *In: Advances in Organic Geochemistry 1987. Organic Geochemistry*, **13**, 399–409.

DAVIS, J. C. 1986. *Statistics and Data Analysis in Geology*, 2nd edition. Wiley, New York.

EBUKANSON, E. J. & KINGHORN, R. R. F. 1986. Oil and gas accumulations and their possible source rocks in southern England. *Journal of Petroleum Geology*, **9**, 413–428.

GOODWIN, N. S., PARK, P. J. D. & RAWLINSON, T. 1983. Crude oil biodegradation. *In*: BJORØY *et al.* (eds) *Advances in Organic Geochemistry 1981*. Wiley & Sons, New York, 650–658.

HESSELBO, S. P. 1997. Basal Wealden of Mupe Bay: a new model. *This volume*.

—— & ALLEN, P. A. 1981. Major erosion surfaces in the basal Wealden Beds, Lower Cretaceous, south Dorset. *Journal of the Geological Society, London*, **148**, 105–113.

LAFARGUE, E. & BARKER, C. 1988. Effect of water washing on crude oil compositions. *Bulletin of the American Association of Petroleum Geologists*, **72**, 263–276.

LEES, G. M. COX, P. T. 1937. The geological basis for the present search for oil in Great Britain by the D'Arcy Exploration Company Ltd. *Journal of the Geological Society, London*, **93**, 156–194.

MACKENZIE, A. S. 1984. Application of biological markers in petroleum geochemistry. *In*: BROOKS, J. & WELTE, D. H. (eds) *Advances in Petroleum Geochemistry*, Vol. 1. Academic Press, London, 115–214.

MILES, J. A., DOWNES, C. J. & COOK, S. E. 1993. The fossil oil seep in Mupe Bay, Dorset: a myth investigated. *Marine and Petroleum Geology*, **10**, 58–70.

——, —— & —— 1994. Reply to 'The Mupe Bay oil seep demythologized?' by Kinghorn *et al*. *Marine and Petroleum Geology*, **11**, 125–126.

PARFITT, M. A. 1993. *An Investigation of the Mupe Bay Oil Seep, Dorset*. MSc dissertation, University of Newcastle Upon Tyne.

—— & FARRIMOND, P. 1997. The Mupe Bay oil seep: a detailed organic geochemical study of a controversial outcrop. *This volume*.

PETERS, K. E. & MOLDOWAN, J. M. 1991. Effects of source, thermal maturity, and biodegradation on the distribution and isomerization of homohopanes in petroleum. *Organic Geochemistry*, **17**, 47–61.

—— & —— 1993. *The Biomarker Guide: Interpreting Molecular Fossils in Petroleum and Ancient Sediments*. Prentice-Hall, Englewood Cliffs, New Jersey.

REQUEJO, A. G. & HALPERN, H. I. 1989. An unusual hopane biodegradation sequence in tar sands from the Pt Arena (Monterey) Formation. *Nature*, **342**, 670–763.

SEIFERT, W. K. & MOLDOWAN, J. M. 1979. The effect of biodegradation on steranes and terpanes in crude oils. *Geochimica et Cosmochimica Acta*, **43**, 111–126.

——, —— & DEMAISON, G. J. 1984. Source correlation of biodegraded oils. *Organic Geochemistry*, **6**, 633–643.

SELLEY, R. C. 1992. Petroleum seepages and impregnations in Great Britain. *Marine and Petroleum Geology*, **9**, 226–244.

—— & STONELEY, R. 1987. Petroleum habitat in south Dorset. *In*: BROOKS, J. & GLENNIE, K. (eds) *Petroleum Geology of North West Europe*. Graham & Trotman, London, 139–148.

SEPHTON, M. 1992. *Migrated Hydrocarbons in the Wessex Basin*. MSc dissertation, University of Newcastle Upon Tyne.

STONELEY, R. & SELLEY, R. C. 1986. *A Field Guide to the Petroleum Geology of the Wessex Basin*. Imperial College, London.

TISSOT, B. P. & WELTE, D. H. 1984. *Petroleum Formation and Occurrence*. Springer, Berlin.

WIMBLEDON, W. A., ALLEN, P. & FLEET, A. J. 1996. Penecontemporaneous oil-seep in the Wealden (early Cretaceous) at Mupe Bay, Dorset, U.K. *Sedimentary Geology*, **102**, 213–220.

The Mupe Bay oil seep: a detailed organic geochemical study of a controversial outcrop

M. A. PARFITT[1,2] & P. FARRIMOND[1]

[1] *Fossil Fuels and Environmental Geochemistry (NRG), Drummond Building, University of Newcastle upon Tyne, Newcastle upon Tyne NE1 7RU, UK*
[2] *Present address: Department of Geology and Applied Geology, University of Glasgow, Hillhead, Glasgow G12 8BW, UK*

Abstract: The oil seep in an intraformational conglomerate at Mupe Bay has attracted much interest as providing possible evidence of oil generation in the Wessex Basin by the early Cretaceous (Wealden). Previous organic geochemical studies investigating whether oils of different maturity exist in the clasts and matrix of the conglomerate have been complicated by the effects of heavy biodegradation of the oil. In this work, we present detailed molecular organic geochemical data from a significantly larger suite of samples than has been previously studied. These have been screened using a multivariate statistical approach to identify those samples which have been least influenced by biodegradation. Conventional molecular maturity parameters applied to the subset of samples which have suffered no detectable modification to their biological marker (hopane and sterane) distributions indicate that the oil in the clasts is of exactly the same maturity as that in the matrix. However, this observation does not preclude two phases of staining, although it remains to be proven that the clasts contained oil at the time of their deposition.

The Mupe Bay oil seep on the South Dorset Coast (Fig. 1) is found in an unusual conglomerate (Wealden) containing heavily oil-stained and poorly consolidated sandstone clasts in a more lightly oil-stained, coarser sandstone matrix (Hesselbo & Allen 1991). The sandstone clasts are cemented by oil, and it has been proposed that they were eroded from a contemporary Cretaceous oil seep, with the staining in the coarser matrix coming from a later migration of oil from the same Liassic source rock (Selley & Stoneley 1987; Wimbledon *et al.* 1996; Hesselbo 1997 this volume). This model has been cited as critical evidence for oil generation and migration in the Wessex Basin by the early Cretaceous. Such a time constraint for the early generation of oil in the basin is significant, as it requires a very high geothermal gradient to be invoked in the Jurassic in order to bring the proposed source rocks to a level of sufficient maturity (Miles *et al.* 1993).

A preliminary geochemical study supported the early Cretaceous palaeo-seep theory by providing indications of a possible time gap between oil staining of the clast sands and the matrix sands (Cornford *et al.* 1988). Sterane and hopane biomarker maturity parameters suggested that the oil in the matrix is more mature than that in the clasts, implying that there were two phases of oil emplacement with a period of source rock burial (and hence maturation) in between. However, the seep oils at Mupe Bay, as with others cropping out along the Dorset coast, have suffered extensive alteration by microbial activity (biodegradation-weathering; Bigge & Farrimond 1997 this volume), which complicates the interpretation of biomarkers in terms of maturity and oil correlation. Within individual biological marker classes, the compounds which are most resistant to biodegradation tend in general to be those which are more thermally stable, making maturity differences particularly difficult to identify. In a recent study of the Mupe Bay outcrop, Miles *et al.* (1993) argued that the reported maturity differences (Cornford *et al.* 1988) are actually the result of differential biodegradation (see Bigge & Farrimond 1997 this volume); the oil in the more porous matrix being more extensively altered than that in the tighter clasts, and thus taking on certain molecular characteristics of a more mature petroleum charge.

The two previous organic geochemical studies of the Mupe Bay seep were performed on insufficiently large data sets (just two samples by Cornford *et al.* 1988; three samples by Miles *et al.* 1993) to divorce molecular differences due to the effects of biodegradation from those related to maturity. The aim of our work was to prove whether or not two phases of oil of different maturity exist in the outcrop by examining a larger sample set and rationalizing

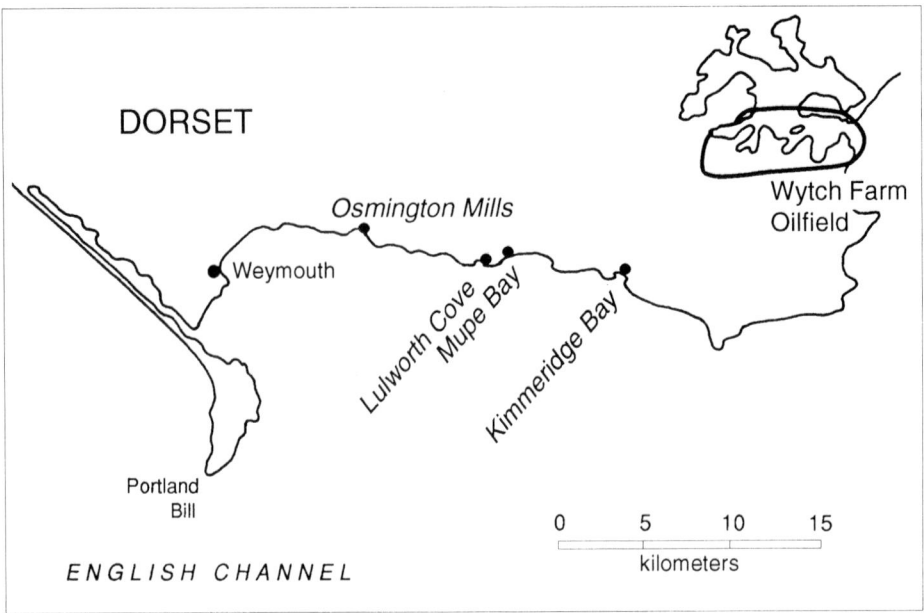

Fig. 1. Map showing the locality of the Mupe Bay outcrop.

the effects of biodegradation upon molecular maturity parameters using a multivariate statistical technique.

Field geology

The sandstones which host the Mupe Bay oil seep are found in a sequence of interbedded Lower Cretaceous, Neocomian (Wealden) sands and clays just north of the steps leading down onto the beach (NGR SY 843 798). Here, erosion and the steep northward dip of the beds towards the Isle of Wight–Purbeck Disturbance to the north (Cornford et al. 1988) have exposed an infilled fluvial channel (Kinghorn et al. 1994; Hesselbo 1997 this volume; cf. Hesselbo & Allen 1991) as an outcrop in the cliff and on the beach platform below. The oil-stained sandstones which infill the channel are floored by an intraformational sandstone conglomerate, and it is the detailed molecular composition of the seep oil(s) in the clasts and matrix of this bed which forms the basis for this paper.

The fluvial channel is cut into grey shales, with the conglomerate of the channel floor grading up into a cross-bedded sandstone of similar nature to the conglomerate matrix; this is then overlain by grey shales (Kinghorn et al. 1994). The conglomerate consists of poorly consolidated, heavily oil-stained sandstone clasts enclosed in a more lightly oil-stained and coarser pale brown sandstone matrix. A fine-grained sandstone, lying beneath the shales which underlie the conglomerate, is also lightly oil stained (Sephton 1992). Detailed accounts of the sedimentology of the outcrop will not be presented here but the reader is referred to Selley & Stoneley (1987), Cornford et al. (1988), Hesselbo & Allen (1991) and Miles et al. (1993, 1994).

Experimental

Sampling and preparation

Six clast and five matrix samples (Table 1) were collected from a few centimetres beneath the surface of a recent slump on the south side of the outcrop. This slump provided fresh material of unambiguous provenance (one large slump block recorded the clay–conglomerate erosive contact), and allowed comprehensive sampling with negligible effect on the outcrop. (Important, as this site has suffered under the hammer over recent years, and must be protected from further damage.) Along with the slump samples, one small (2 cm in diameter) clast was taken from the face of the outcrop for comparison with the slump samples, and a sample was taken from the oil-stained pale yellow sandstone underlying the claystone (Table 1). Before analysis, the outer 1–2 mm of each sample was removed to minimize contamination, and the samples then

Table 1. *Sample descriptions and extract yields (% rock by weight)*

Sample	Description	EOM (%)
C1	small *in situ* clast	8.9*
C2	slump clast	7.5
C3	slump clast	8.8
C4	slump clast	9.7
C5	large slump clast (edge peel)	5.7*
C6	large slump clast (mid-depth)	5.1
C7	large slump clast (core)	7.2
C7R	large slump clast (core) repeat	7.3*
M1	slump matrix	2.7
M2	slump matrix	9.4*
M3	slump matrix	2.6
M4	slump matrix	2.3*
M5	jarrocite spotted slump matrix	3.2
M5R	repeat of above	3.0
LS1	lower sandstone	4.8

* Average of two repeat extractions.

crushed lightly in a pestle and mortar. One clast (15 cm long) taken from the slump was sampled in four places: at the edge, midway to its core, and twice from its core, in order to assess any compositional variability which may have arisen from alteration of the rim during transport, or diffusion of oil from the matrix into the clasts.

Bulk compositional analysis

The seep oils were analysed using an Iatroscan TH-10 thin layer chromatography–flame ionization detector (TLC–FID) analyser interfaced to a Perkin–Elmer LC-100 Laboratory Computing Integrator, to determine their bulk chemical composition. Whole rock samples were crushed and placed in crimp-top vials together with distilled solvent (5 ml dichloromethane : methanol; 93 : 7 v : v). The vials were sealed and sonicated in an ultrasonic bath for 5 min, and the resulting solution analysed using a multiple solvent elution (petroleum ether, toluene, dichloromethane : methanol; 93 : 7) on Iatroscan rods. Detection was by a flame ionization detector with a H_2-air flame at $0.8 \, kg \, cm^{-2}$ H_2 pressure. The system was calibrated against a standard of squalane, 2,6-dimethylnaphthalene, anthracene and 1-undecanol.

Seep oil extraction and fractionation

The seep oils were extracted from their sand hosts by repetitive cold extraction with a 9 : 1 mixture of dichloromethane : methanol, and complete disaggregation in an ultrasonic bath. After each extraction, sonicated samples were centrifuged (3000 rpm for 10 min) to settle the particulate material, and the supernatant was collected. After combining all extracts of a single sample, excess solvent was removed by rotary evaporation, and the weight of extracted oil measured. The resulting oils were fractionated using thin-layer chromatography (Keiselgel 60G; 0.5 mm thickness; petroleum ether) into three fractions: 'aliphatic hydrocarbons' ($R_f = 0.6$–1.0), 'aromatic hydrocarbons' ($R_f = 0.08$–0.6) and 'polars' ($R_f = 0.0$–0.08).

Molecular analysis

Gas chromatography (GC) of the aliphatic fraction was performed on a Carlo Erba Mega Series 5160 instrument fitted with an on-column injector and a glass OV-1 capillary column (25 m × 0.32 mm i.d.; 0.15 μm film thickness). The oven was temperature programmed from 50 to 300°C at $4°C \, min^{-1}$, and hydrogen used as carrier gas. Aromatic fractions were analysed on an identical instrument fitted with a glass SE52 capillary column (15 m × 0.45 mm i.d.; 0.45 μm film thickness). The oven was again temperature programmed from 50 to 300°C at $4°C \, min^{-1}$, and hydrogen used as carrier gas. A VG Multichrom data system was used for data acquisition.

Gas chromatography–mass spectrometry (GC–MS) of the aliphatic fractions was performed using a Hewlett-Packard 5890 Series II GC linked to a Hewlett-Packard 5972 mass-selective detector. The GC was fitted with a split/splitless injector (280°C), a Hewlett-Packard 7673 Autosampler and a fused-silica HP-5 column (30 m × 0.25 mm i.d.; 0.25 μm film thickness). The oven was temperature programmed from 40 to 175°C at $10°C \, min^{-1}$, held for 1 min, from 175 to 225°C at $6°C \, min^{-1}$ and then from 225 to 300°C at $4°C \, min^{-1}$ where it was held for 27 min. Helium carrier gas was used at a flow rate of $1 \, ml \, min^{-1}$. The mass spectrometer was operated in selective ion monitoring mode to maximize signal response for the m/z 177, 191, 217, 218 and 259 ions, and controlled by a Vectra 486/33PC running Hewlett-Packard 'Windows' software. Compounds were identified using a combination of mass chromatographic responses, mass spectra and retention times, and by comparison with the literature.

Principal components analysis (PCA)

A data set comprising 49 variables (aliphatic biomarker peak height responses from appropriate mass chromatograms; Table 2) for 15

Table 2. *Biological marker compounds used in the principal components analysis (PCA). The numbers are also used to label peaks in Fig. 4*

Tricyclic terpanes and hopanes (m/z 191)
1. C_{20} Tricyclic terpane
2. C_{21} Tricyclic terpane
3. C_{23} Tricyclic terpane
4. C_{24} Tricyclic terpane
5. C_{25} Tricyclic terpane
6. C_{26} Tricyclic terpane + C_{24} Tetracyclic
7. C_{26} Tricyclic terpane
8. C_{28} Tricyclic terpane
9. C_{28} Tricyclic terpane
10. C_{29} Tricyclic terpane
11. C_{29} Tricyclic terpane
12. 22,29,30-Trisnorneohopane-II (Ts)
13. 22,29,30-Trisnorhopane (Tm) + C_{30} Tricyclic terpane
14. C_{30} Tricyclic terpane
15. C_{29} 17α(H), 21β(H)-30-norhopane
16. 18α(H)-30-norneohopane (C_{29} Ts)
17. C_{30} diahopane
18. C_{30} 17α(H), 21β(H) hopane
19. C_{30} 17β(H), 21α(H) hopane
20. C_{31} 17α(H), 21β(H) hopane (22S)
21. C_{31} 17α(H), 21β(H) hopane (22R)
22. C_{32} 17α(H), 21β(H) hopane (22S)
23. C_{32} 17α(H), 21β(H) hopane (22R)
24. C_{33} 17α(H), 21β(H) hopane (22S)
25. C_{33} 17α(H), 21β(H) hopane (22R)
26. C_{34} 17α(H), 21β(H) hopane (22S)
27. C_{34} 17α(H), 21β(H) hopane (22R)
28. C_{35} 17α(H), 21β(H) hopane (22S)
29. C_{35} 17α(H), 21β(H) hopane (22R)

Steranes ($m/z = 217$)
30. C_{29}13β(H), 17α(H)-diasterane (20S)
31. C_{29}13β(H), 17α(H) diasterane (20R)
32. C_{28}5α(H), 14α(H), 17α(H) sterane (20R)
33. C_{29}5α(H), 14α(H), 17α(H) sterane (20S)
34. C_{29}5α(H), 14β(H), 17β(H) sterane (20R)
35. C_{29}5α(H), 14β(H), 17β(H) sterane (20S)
36. C_{29}5α(H), 14α(H), 17α(H) sterane (20R)

$\alpha\beta\beta$ **Steranes** ($m/z = 218$ not illustrated)
37. C_{27} 5α(H), 14β(H), 17β(H) sterane (20R)
38. C_{27} 5α(H), 14β(H), 17β(H) sterane (20S)
39. C_{28} 5α(H), 14β(H), 17β(H) sterane (20R)
40. C_{28} 5α(H), 14β(H), 17β(H) sterane (20S)
41. C_{29} 5α(H), 14β(H), 17β(H) sterane (20R)
42. C_{29} 5α(H), 14β(H), 17β(H) sterane (20S)

Diasteranes ($m/z = 259$ not illustrated)
43. C_{27} 13β(H), 17α(H) diasterane (20S)
44. C_{27} 13β(H), 17α(H) diasterane (20R)
45. C_{28} 13β(H), 17α(H) diasterane (20S, 24S)
46. C_{28} 13β(H), 17α(H) diasterane (20S, 24R)
47. C_{28} 13β(H), 17α(H) diasterane (20R, 24S)
48. C_{29} 13β(H), 17α(H) diasterane (20S)
49. C_{29} 13β(H), 17α(H) diasterane (20R)

samples was analysed using a principal components analysis program written by P. Yendle (University of Bristol; tested for equivalence against commercial packages). Data were normalized to 100% for each sample (to eliminate concentration differences between samples), autoscaled (i.e. divided by the standard deviation of the variable, to allow comparable influence from each variable during analysis) and centred (by subtracting the average of each variable, corresponding to moving the co-ordinates to the centre of the data) before analysis. It should be noted that the normalization procedure causes closure (Davis 1986) which can introduce spurious correlations, as when large variables increase other variables must decrease. However, the results show no indication of closure being a problem in this study.

Results and discussion

Levels of oil staining

The striking colour difference between the black clasts and the pale brown matrix has often been cited as evidence of higher concentrations of oil in the clasts (e.g. Miles et al. 1994). Certainly, the average grain size of the matrix is much higher than that of the clast, and one might expect it to have a higher porosity, and thus oil content. In the present study, the clast samples contain a mean oil content of 7.5% (by weight; range from 5.1 to 9.7%), whilst the matrix samples contain only 2.7% (range 2.3–3.1%; Table 1), with the exception of an extremely heavily stained matrix sample (M2; 7.5–11.3% EOM) which is very coarse and heterogeneous. (Such a high value for the extract yield of the matrix must be considered anomalous; such high values have not been reported previously; Cornford et al. 1988; Sephton 1992; Miles et al. 1993.) The darker colour of the clasts may also be due in part to an opaque black substance on the grain surfaces (Sephton 1992; Miles et al. 1993) which is absent from the matrix. This material is unlikely to be a degradation product of the petroleum (Miles et al. 1993), as a concentrate (obtained through density separation) gave no response in Rock-Eval pyrolysis (Parfitt 1993).

Degradation of oil

The oil seep samples from Mupe Bay, like the seeps at other locations on the Dorset Coast (Bigge & Farrimond 1997 this volume), have suffered extensively from the effects of biodegradation and surface weathering (e.g. water washing). These processes have significantly altered the bulk chemical composition (i.e. relative proportion of aliphatic hydrocarbons, aromatic hydrocarbons and polar compounds) of the seeps compared to locally reservoired oils (Fig. 2). All the clast and matrix samples have lost a significant proportion of their aliphatic and aromatic compounds compared with the local unaltered oils. However, the matrix oils appear more degraded than the clast samples (with one exception), having had a greater proportion of aromatic hydrocarbons removed. Previous investigations have found that the aromatic

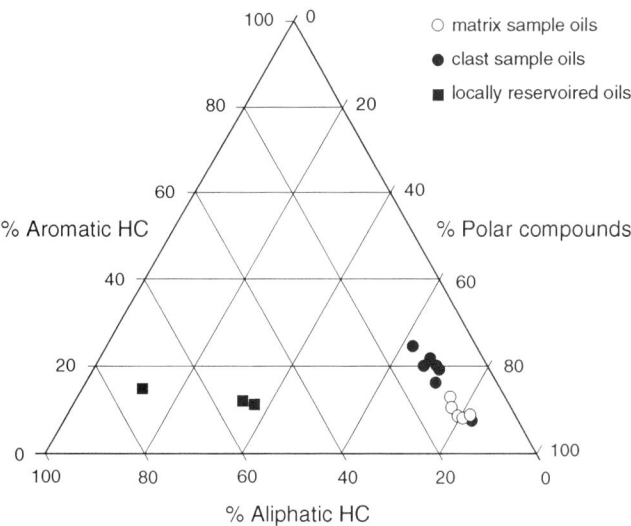

Fig. 2. Triangular plot comparing the bulk chemical compositions (% aliphatics, aromatics and polars) of the Mupe Bay samples and locally reservoired petroleums (two from Wytch Farm and one from the Kimmeridge field). A sample from the underlying sandstone at Mupe Bay coincides exactly with the clast sample which plots nearest the polar corner of the diagram.

hydrocarbons are preferentially degraded during subaerial weathering of organic matter in shales (Clayton & Swetland 1977), and the susceptibility to water washing of light aromatic hydrocarbons in oils is well known (e.g. Lafargue & Barker 1988).

The lower concentration of oil in the matrix, despite its coarser nature, may be at least partially reconciled by more extensive biodegradation. Indeed, this is consistent with the coarser grain size of the matrix which would facilitate biodegradation in comparison with the tighter clasts. This feature has been observed in previous organic geochemical studies (Cornford et al. 1988; Miles et al. 1993), and has led to complications in interpreting the relative maturities of the oil in clast and matrix samples.

Gas chromatography of aliphatic fractions shows that at a molecular level, the seep oil samples are characterized by unresolved complex mixtures (UCM) of components typical of biodegraded oils (Fig. 3). All of the n-alkanes and acyclic isoprenoids have been lost, leaving a clear series of biomarker peaks superimposed on the UCM. However, GC–MS analysis shows that in some samples even the hopanes and steranes have been partially degraded (Fig. 4). The most heavily biodegraded oil is seen in the matrix sample M5; this is a coarse-grained sand which is visibly the most weathered sample, its surface being coated with the yellow alteration mineral, jarrocite. In this sample, the hopanes have been extensively degraded, particularly the shorter chain $\alpha\beta$ hopanes. The chromatogram of this sample is dominated by the compounds which are most resistant to biodegradation, notably the tricyclic terpanes, Ts, diahopane and the C_{34} and $C_{35}\alpha\beta$ hopanes (Fig. 4). These changes make the use of molecular maturity parameters impossible. The remaining samples show less obvious evidence of hopane degradation and may, perhaps, be more successfully employed.

The steranes are less obviously biodegraded than the hopanes although, once again, the visibly weathered matrix sample (M5) is the most extensively modified. Here, there has been preferential removal of the $\alpha\alpha\alpha$ (20R) sterane isomers, most notably seen for the C_{29} homologue (Fig. 4). The preferential loss of these isomers has the effect of grossly modifying the most commonly employed sterane maturity parameters (the C-20 and nuclear isomerizations; Mackenzie 1984), a problem encountered previously in Mupe Bay studies by Cornford et al. (1988) and Miles et al. (1993).

The molecular characteristics of the most visibly weathered matrix sample (M5) equate to an oil which has been 'heavily biodegraded' (biodegradation ranking 6; Peters & Moldowan 1993), in which the steranes and hopanes are partially degraded, but the diasteranes remain intact. The remaining samples can be considered to be 'moderately to heavily biodegraded', with only minor degradation of the biomarkers apparent in the GC–MS data (Fig. 4). Whilst M5 has clearly suffered the most extensive surface degradation, with consequent modification of the biomarker distributions, it is important to identify any more subtle effects of biodegradation in the remaining samples which might compromise the use of conventional molecular maturity parameters. In order to do this, we employed the statistical technique of principal components analysis.

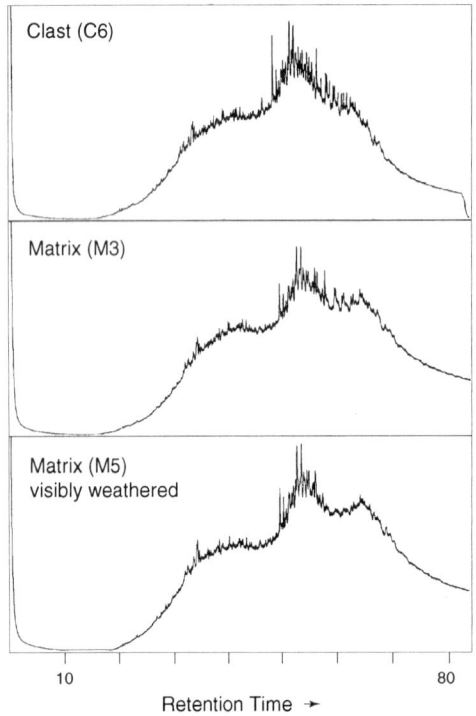

Fig. 3. Gas chromatograms of the aliphatic fractions of a typical clast sample, and two matrix samples (one of which is visibly more weathered).

Principal components analysis (PCA) to identify the least biodegraded oils

Principal components analysis (PCA) is a multivariate statistical technique which identifies the major sources of variability within complex data

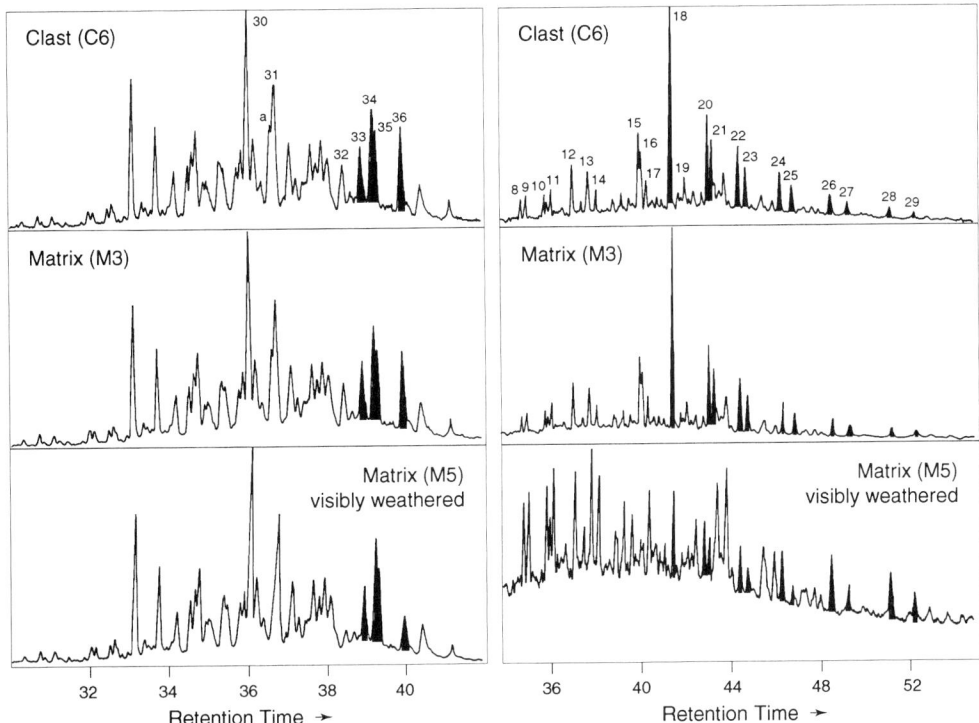

Fig. 4. Mass chromatograms showing the distributions of hopanes (m/z 191) and steranes (m/z 217) of a typical clast sample, and two matrix samples (one of which is visibly more weathered). Peak numbers refer to Table 2. ($a = C_{27}\alpha\alpha\alpha$ cholestane 20R; not used in principal components analysis). Shaded peaks correspond to the C_{30}–$C_{35}\alpha\beta$ hopanes (m/z 191) and the C_{29} steranes ($\alpha\alpha\alpha$20S, $\alpha\beta\beta$20R, $\alpha\beta\beta$20S and $\alpha\alpha\alpha$20R; m/z 217).

sets. These are the principal components (PCs) which are made up of linear combinations of the original variables. PCA can often describe the bulk of data variability in just a few principal components, allowing the data to be plotted graphically in two dimensions, whilst still retaining most of the information within the complex data set (Meglen 1992).

PCA was initially performed on a data set comprising 49 biomarker peak heights measured from the relevant GC–MS mass chromatograms from 15 analyses (seven clast samples, plus one analytical repeat; five matrix samples, plus one analytical repeat; and a sample from the underlying stained sand). In this analysis, the first principal component proved to hugely dominate, explaining 84% of the variance in the scaled data set. Its effect was to separate the extremely weathered matrix sample (M5) and its analytical repeat (M5R) from the remainder of the samples; hardly surprising, as we saw from simple visual comparison of the biomarker fingerprints (Fig. 4) that this sample had undergone more extensive biodegradation than the others. The loadings from this initial PCA analysis (not shown) indicated that this sample (and its repeat) differed from the remaining samples in containing greater relative proportions of tricyclic terpanes, Ts, diasteranes, $\alpha\beta\beta$ steranes, and reduced hopanes (especially $<C_{34}$) and $\alpha\alpha\alpha$ steranes. These effects are well documented as characteristics of biodegraded oils (Connan 1984).

Because the extremely weathered matrix sample dominated the PCA, we repeated the analysis without this sample (and its analytical repeat). This time the data set comprised the same 49 variables for seven clast samples (plus a replicate), four matrix samples and a single sample from the underlying stained sand. In this analysis, PCs 1 and 2 explained 47% and 20%, respectively, of the variance within the scaled data set. The relationships between the samples in terms of these two PCs are shown in a scores plot (Fig. 5). There is a general grouping of most samples towards the left-hand side of the plot (i.e. negative PC1 scores), with three obvious outliers – two clast samples (C4 and C5) and one

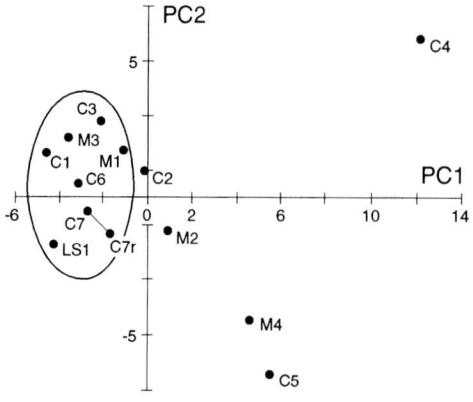

Fig. 5. A scores plot (PC1 v. PC2) showing the relationships between the samples (M5 excluded) in terms of the first and second principal components. PC1 explains 47% of the total variance in the scaled data set; PC2 explains a further 20%. PC1 is effectively a 'biodegradation parameter' (see loadings plot in Fig. 6), with the more extensively biodegraded samples plotting further to the right (positive PC1).

of the sample set (in terms of all the aliphatic biomarker data, and not just selected ratios) allows us to select those samples which have been least affected by biodegradation–weathering. These samples will give the most reliable maturity measurements using conventional molecular parameters.

Molecular maturity parameters

PCA (Fig. 5) indicates that the least extensively biodegraded clast samples are C1, C3, C6 and C7, whilst two matrix samples (M1 and M3) have suffered similar levels of biodegradation, with minimal alteration of the aliphatic biomarker distributions. Conventional molecular maturity parameters for these samples, and the lower oil-stained sand, are shown in Table 3. There are no apparent differences in the maturity levels of the clast and matrix samples (compare with the analytical reproducibility shown for C7). The oil in the underlying sand is also of the same maturity level.

The samples rejected from the PCA in recognition of their more extensive biodegradation generally show more elevated values for the molecular maturity parameters (Table 3), particularly those samples which lie furthest from the main grouping. These samples (with the exception of M5) did not exhibit gross differences in their biomarker distributions which could be identified easily by visual inspection of the chromatograms, showing the power of the PCA procedure in recognizing such anomalous samples. Inclusion of these samples could easily

matrix (M4). PC1 is again a 'biodegradation parameter' (Fig. 6), comprising the relative abundance of the more easily biodegradable biomarkers (hopanes (especially $<C_{34}$) and $\alpha\alpha\alpha$ steranes; negative PC1) v. the more resistant compounds (tricyclic terpanes, Ts, diasteranes and $\alpha\beta\beta$ steranes; positive PC1). Thus, the samples plotting to the left (negative PC1) side of the diagram are less extensively biodegraded than those to the right. This detailed comparison

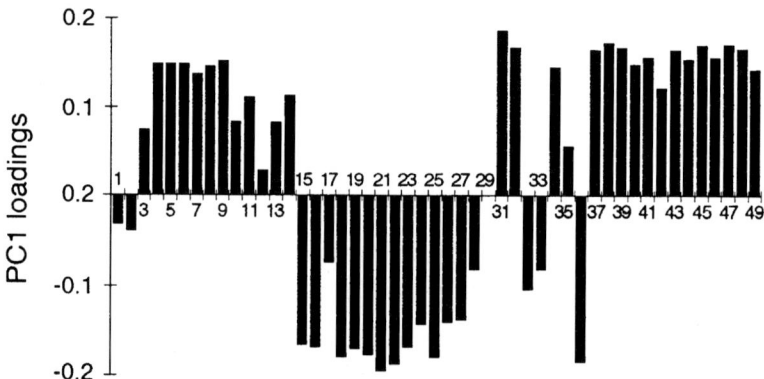

Fig. 6. Loadings plot showing the composition of the first principal component (PC1) in terms of the original variables. PC1 is effectively a 'biodegradation parameter'; relatively resistant compounds (tricyclic terpanes, diasteranes and most steranes) having positive loadings, and the more labile compounds ($\alpha\alpha\alpha$ steranes and most hopanes (diahopane and $C_{35}\alpha\beta$ being of less influence, as these are more resistant)) having negative loadings.

Table 3. *Molecular maturity parameters based upon steranes, triterpanes and aromatic steroid hydrocarbons, for oil extracted from clast and matrix samples from the Mupe Bay oil seep. Samples in bold have been identified by PCA as being the least biodegraded, with no significant modification of the hopane and sterane distributions, and thus give the most confident maturity values*

Sample	Hopane Maturity Parameters					Sterane Maturity Parameters			Aromatic maturity
	C_{30} dia/$C_{30}\alpha\beta$[1]	C_{23}tri/$C_{30}\alpha\beta$[2]	Hop/op+mor[3]	% 22S[4]	% 20S[5]	C_{29} diasteranes/ diasteranes+steranes[6]		C_{29} % $\alpha\beta\beta$[7]	parameter C-ring aromatization[8]

Sample	C_{30} dia/$C_{30}\alpha\beta$	C_{23}tri/$C_{30}\alpha\beta$	Hop/op+mor	% 22S	% 20S	C_{29} diasteranes/diasteranes+steranes	C_{29} % $\alpha\beta\beta$	parameter C-ring aromatization
Clast samples								
C1	**0.156**	**0.23**	**86.5**	**58.6**	**42.5**	**48.6**	**54.5**	—
C2	*0.137*	*0.25*	*87.2*	*58.3*	*44.3*	*49.9*	*56.3*	*0.57*
C3	**0.148**	**0.23**	**87.4**	**57.3**	**42.8**	**49.4**	**55.0**	**0.58**
C4	*0.179*	*0.27*	*86.9*	*58.6*	*46.8*	*51.6*	*61.5*	*0.59*
C5	*0.159*	*0.31*	*87.3*	*58.4*	*42.9*	*52.3*	*55.8*	*0.57*
C6	**0.150**	**0.23**	**86.6**	**55.0**	**42.5**	**49.2**	**54.8**	**0.58**
C7	**0.156**	**0.25**	**86.5**	**57.0**	**42.1**	**49.9**	**54.6**	**0.57**
C7R[9]	**0.153**	**0.26**	**86.7**	**58.3**	**44.7**	**49.7**	**54.8**	**0.57**
Matrix samples								
M1	**0.149**	**0.22**	**88.1**	**57.5**	**42.6**	**49.1**	**55.7**	**0.60**
M2	*0.143*	*0.27*	*87.9*	*57.0*	*43.7*	*51.3*	*56.2*	*0.57*
M3	**0.162**	**0.20**	**87.1**	**57.5**	**44.7**	**49.7**	**55.4**	**0.59**
M4	*0.141*	*0.30*	*86.5*	*57.5*	*44.1*	*51.6*	*56.3*	*0.61*
M5	*0.989*	*2.01*	*71.3*	*50.5*	*63.0*	*53.6*	*65.8*	*0.63*
M5R[10]	*0.956*	*2.12*	*72.6*	*52.0*	*67.3*	*53.5*	*66.9*	*0.62*
Lower Sandstone								
LS1	**0.151**	**0.23**	**86.2**	**57.0**	**42.0**	**48.6**	**55.6**	—

[1] C_{30} diahopane/hopane.
[2] C_{23} tricyclic (cheilanthane)/hopane.
[3] Hopane/hopane+moretane ($C_{30}\alpha\beta/(\alpha\beta+\beta\alpha)$%).
[4] Hopane isomerization at C-22 ($C_{31}\alpha\beta$ 22S/(22S+22R)%).
[5] Sterane isomerization at C-20 ($C_{29}\alpha\alpha\alpha$ 20S/(20S+20R)%).
[6] $C_{29}\beta\alpha$dia/($\beta\alpha$dia+$\alpha\alpha\alpha$+$\alpha\beta\beta$)%.
[7] Sterane nuclear isomerization ($C_{29}\alpha\beta\beta/(\alpha\beta\beta+\alpha\alpha\alpha)$%).
[8] Aromatization of C-ring aromatic steroid hydrocarbons; parameter from Mackenzie (1984); tri–tri+mono.
[9] Analytical repeat of clast sample.
[10] Analytical repeat of degraded matrix sample.

have led to a mis-interpretation of the average maturity of clast and matrix samples.

Aromatic hydrocarbons

Miles *et al.* (1993) turned to an aromatic hydrocarbon parameter to demonstrate the comparable maturity of clast and matrix samples, as the aromatic steroids are generally more resistant to biodegradation (Wardroper *et al.* 1984). The aromatic fractions show loss of all low molecular weight aromatic hydrocarbons (naphthalene, phenanthrene, alkylnaphthalenes and alkylphenanthrenes), probably through water washing (e.g. Lafargue & Barker 1988). Benzohopanes are present in all samples except the highly biodegraded matrix sample (M5).

Within the subset of least biodegraded samples, the parameter measuring the aromatization of C-ring monoaromatic steroid hydrocarbons (Mackenzie 1984; apparently a different parameter from that used by Miles *et al.* 1993) appears slightly higher for the matrix samples (M1 and M3) than the clast samples (C1, C3, C6 and C7). However, there is considerable variability in this value, particularly within the matrix samples (including the more heavily biodegraded ones). This might be ascribed to the aromatic hydrocarbons being more susceptible to the effects of biodegradation at outcrop than has previously been supposed, with slight preferential loss of monoaromatic steroids, particularly in the coarse-grained matrix. Despite this, M2 has a value as low as any of the clast samples, and we therefore agree with Miles *et al.* (1993), that there is no apparent maturity difference between the oils in the clast and the matrix.

Conclusions

A detailed molecular organic geochemical study of the Mupe Bay outcrop has revealed moderate to heavy biodegradation in both matrix and clast samples, which in many cases has modified biomarker (sterane and hopane) distributions, compromising their use in maturity assessment. However, through the application of principal components analysis, we have been able to identify a subset of the least heavily modified samples, within which the steranes and hopanes are apparently undegraded. Application of conventional aliphatic biomarker maturity parameters to these samples shows no difference in maturity between the clast and matrix samples. An aromatic maturity parameter is also comparable in matrix and clast samples.

Although the oil in the clasts and that in the matrix are of the same maturity, this does not preclude two phases of oil staining in the Mupe Bay outcrop, and thus the early Cretaceous generation of oil. The absolute maturity of the oil(s) is of little relevance, but it is important to identify whether or not the clasts contained oil when they were deposited. A final solution to the Mupe Bay controversy probably lies in a coherent combination of organic geochemistry (e.g. oil partitioning between clast and matrix) and sedimentology (e.g. oil v. mineral cements and clast structure–distribution).

The authors are grateful to the Natural Environment Research Council for a studentship (M. A. Parfitt), and to S. R. Larter and S. Haszeldine for their support in allowing the completion of this work. Technical assistance was provided by P. Donahoe, K. Noke and I. Harrison, and the work was guided by the results from a preliminary study by M. Sephton. C. Jeans and B. Brown are thanked for production of the artwork.

References

BIGGE, M. A. & FARRIMOND, P. 1997 Biodegradation of seep oils in the Wessex Basin – a complication for correlation. *This volume.*

CONNAN, J. 1984. Biodegradation of crude oils in reservoirs. *In*: BROOKS, J. & WELTE, D. H. (eds) *Advances in Petroleum Geochemistry*, Vol. 1. Academic Press, London, 299–335.

CORNFORD, C., CHRISTIE, O., ENDERSON, U., JENSEN, P. & MYHR, M. 1988. Source rock and oil seep maturity in Dorset, southern England. *In*: *Advances in Organic Geochemistry 1987. Organic Geochemistry*, **13**, 399–409.

CLAYTON, J. L. & SWETLAND, P. J. 1978. Subaerial weathering of sedimentary organic matter. *Geochimica et Cosmochimica Acta*, **42**, 305–312.

DAVIS, J. C. 1986. *Statistics and Data Analysis in Geology*, 2nd edition. Wiley, New York.

HESSELBO, S. P. 1997. Basal Wealden of Mupe Bay: a new model. *This volume.*

—— & ALLEN, P. A. 1991. Major erosion surfaces in the basal Wealden Beds, Lower Cretaceous, south Dorset. *Journal of the Geological Society, London*, **148**, 105–113.

KINGHORN, R. F., SELLEY, R. C. & STONELEY, R. 1994. The Mupe Bay oil seep demythologised? *Marine and Petroleum Geology*, **11**, 124.

LAFARGUE, E. & BARKER, C. 1988. Effect of water wash-ing on crude oil compositions. *Bulletin of the American Association of Petroleum Geologists*, **72**, 263–276.

MACKENZIE, A. S. 1984. Application of biological markers in petroleum geochemistry. *In*: BROOKS, J. & WELTE, D. H. (eds) *Advances in Petroleum Geochemistry*, Vol. 1. Academic Press, London, 115–214.

MEGLEN, R. R. 1992. Examining large databases, a chemometric approach using principal component analysis. *Marine Chemistry*, **39**, 217–237.

MILES, J. A., DOWNES, C. J. & COOK, S. E. 1993. The fossil oil seep in Mupe Bay, Dorset: a myth investigated. *Marine and Petroleum Geology*, **10**, 58–70.

——, —— & —— 1994. Reply to 'The Mupe Bay oil seep demythologized?' by Kinghorn et al. *Marine and Petroleum Geology*, **11**, 125–126.

PARFITT, M. A. 1993. *An Investigation of the Mupe Bay Oil Seep, Dorset*. MSc Dissertation, University of Newcastle Upon Tyne.

PETERS, K. E. & MOLDOWAN, J. M. 1993. *The Biomarker Guide: Interpreting Molecular Fossils in Petroleum and Ancient Sediments*. Prentice-Hall, Englewood Cliffs, New Jersey.

SELLEY, R. C. & STONELEY, R. 1987. Petroleum habitat in south Dorset. *In*: BROOKS, J. & GLENNIE, K. (eds) *Petroleum Geology of North West Europe*. Graham & Trotman, London, 139–148.

SEPHTON, M. 1992. *Migrated Hydrocarbons in the Wessex Basin*. MSc Dissertation, University of Newcastle Upon Tyne.

WARDROPER, A. M. K., HOFFMAN, C. F., MAXWELL, J. R., BARWISE, A. G. J, GOODWIN, N. S. & PARK, P. J. D. 1984. Crude oil biodegradation under simulated and natural conditions – II. Aromatic steroid hydrocarbons. *In: Advances in Organic Geochemistry 1983. Organic Geochemistry*, **6**, 605–617.

WIMBLEDON, W. A., ALLEN, P. & FLEET, A. J. 1996. Penecontemporaneous oil-seep in the Wealden (early Cretaceous) at Mupe Bay, Dorset, U.K. *Sedimentary Geology*, **102**, 213–220.

Reservoir architecture of the upper Sherwood Sandstone, Wytch Farm field, southern England

T. McKIE,[1] J. AGGETT[1] & A. J. C. HOGG[2]

[1] *Badley Ashton and Associates, Winceby House, Winceby, Horncastle, Linconshire LN9 6PB, UK*
[2] *BP Exploration Operating Company Ltd, Blackhill Road, Holton Heath, Poole, Dorset BH16 7LS, UK*

Abstract: The Sherwood Sandstone Group reservoir in the Wytch Farm field comprises a c. 150 m thick succession of arkosic sandstones deposited in a variety of fluvial, lacustrine and aeolian depositional systems. These systems show at least three orders of facies variability, which are interpreted to be the depositional response to climatic changes. These comprise a first-order evolutionary trend over the entire Sherwood Sandstone Group from perennial braidplain to ephemeral sheetflood systems to ephemeral lacustrine conditions. This trend culminated in deposition of the Mercia Mudstone Group, and reflects a long-term waning of sand supply and increasing 'flashiness' of the fluvial system. This trend is further subdivided into second-order cycles defined by five areally widespread floodplain and lacustrine deposits containing minimal development of fluvial sandstones. These represent widespread, episodic reductions in fluvial sediment supply and rising base level during more 'humid' climatic conditions. These horizons form the basis for the reservoir layering scheme. Each floodplain episode is increasingly more mud-rich upwards through the Sherwood section, and the sand-rich fluvial packages between become systematically more ephemeral in character. Third-order cycles are defined by thin (<2 m), but areally widespread floodplain and lacustrine horizons which are most readily identifiable in the upper half of the Sherwood section. The sandstones between these cycles are composed of aeolian and sheetflood deposits, but are incised by coarse-grained multistorey–multilateral channel deposits. The incisions are interpreted to be the result of fluvial erosion during dry climatic conditions when lake levels fell and the alluvial plain was devegetated. These incised fluvial deposits form the principal producing intervals in the upper part of the reservoir, particularly in the eastern part of the field. Higher frequency stratigraphic cycles are locally expressed by variations in ephemeral lake levels, palaeosol development and episodic development of wind-blown sand patches. At outcrop, the stratigraphically equivalent Otter Sandstone Formation (c. 100 km to the west) shows comparable evolutionary patterns, albeit with a subtly different facies make-up. The recognition of a hierarchy of climatically driven cycles within the reservoir permits high-resolution correlation and the recognition of subtle, but important, changes in sandbody geometry and connectivity within successive cycles.

The Wytch Farm Field was discovered beneath Poole Harbour, Dorset in 1974. It is the largest onshore oilfield in western Europe and contains an estimated 428×10^6 barrels (bbl) of oil-equivalent reserves with a proven 189×10^6 barrels remaining (Underhill & Stoneley this volume). The field is currently on plateau and delivers about 110 000 bbl/day, 17.6×10^6 scf of gas and 725 tonnes of liquefied petroleum gases (LPG) per day. Decline is forecast from mid-1997.

The geology of the Wytch Farm Field has been described by Colter & Harvard (1981), Dranfield *et al.* (1987) and Bowman *et al.* (1993). The field comprises a series of northerly dipping fault blocks formed during early Cretaceous extension. Oil generation and accumulation began when thermal subsidence in the late Cretaceous gave rise to mature source kitchens in the hanging walls of the extensional faults (Hawkes *et al.* 1997). The source rocks are believed to be marine mudrocks of the Lower Lias (Selley and Stoneley 1987). There are three reservoirs: the Middle Jurassic Frome Clay Limestone Member, the Upper Liassic shallow marine Bridport Sandstone; and the Triassic fluvio-lacustrine Sherwood Sandstone Group. The principal reservoir is the Sherwood Sandstone which contains some 397×10^6 bbl of recoverable oil (Underhill & Stoneley this volume) at about 1585 m true vertical depth sub-sea (Fig. 1).

Approximately one half of the field reserves are held in an offshore extension of the Sherwood Sandstone reservoir which is being developed by a series of record-breaking 'extended reach drilling' (ERD) wells being drilled at step-outs >8 km from onshore Poole Harbour (Fig. 1). The drilling and completion of the Wytch Farm ERD wells has been extensively

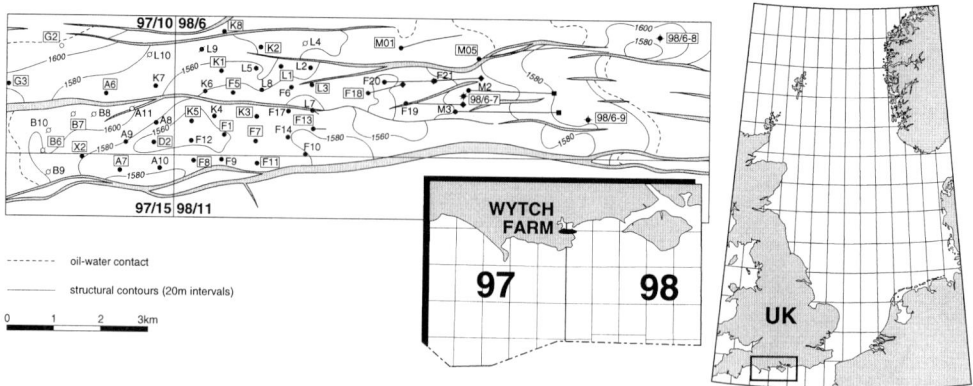

Fig. 1. Location of Wytch Farm field and well locations (with cored wells highlighted).

documented (papers presented at the 1995 SPE conference by Brodie *et al.* and Payne *et al.*; papers presented at the 1995 SPE conference by Harrison & Mitchell and Wood *et al.*; papers presented at the 1996 IADC/SPE conference by Bruce *et al.*, Jariwala *et al.* and Lenn *et al.*). The geotechnical planning of ERD wells is discussed by McClure *et al.* (1995) and Hogg *et al.* (1996). The high-resolution sequence stratigraphic description of the Sherwood Sandstone, and the subject of the present paper, has proved critical to the successful geo-steering and completion of these horizontal wells.

The database for the study consisted of core, wireline log and limited magnetostratigraphic data from the field itself (59 wells), and outcrop of the stratigraphically equivalent succession *c.* 100 km to the west in largely accessible sea cliffs. In addition, 123 thin sections were used to establish the principal controls on reservoir quality. Seismic resolution was insufficient to provide useful information for this study.

Sherwood stratigraphy

The Sherwood Sandstone comprises a broadly fining- and muddying-upward succession *c.* 150 m thick, which is overlain by >300 m of playa mudrocks of the Mercia Mudstone (Fig. 2). The Sherwood Sandstone can be divided into two intervals: a lower Sherwood section of high nett:gross fluvial deposits; and an upper Sherwood section with more common lacustrine and floodplain mudrocks. This study was largely focused on the upper Sherwood in order to understand the permeability architecture of the western part of the reservoir prior to miscible flooding, and to improve the stratigraphic understanding of the eastern part of the reservoir to aid the positioning of extended reach wells (Hogg *et al.* 1996).

Facies associations

The upper Sherwood section consists of arkosic sandstones and mudrocks which can be broadly subdivided into six facies associations with unique sedimentological characteristics. These comprise the following.

Floodplain deposits. These are characterized by up to 10 m of thinly interbedded (mm–cm scale) very fine- to fine-grained sandstones and mudrocks with local concentrations of mudclasts and

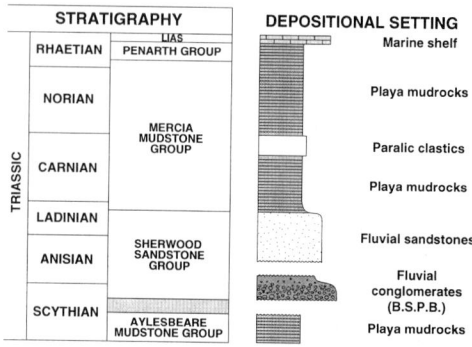

Fig. 2. Triassic stratigraphy in southwest England (modified after Hamblin *et al.* 1992). In the Wytch Farm area the fluvial conglomerates of the Budleigh Salterton Pebble Beds (B.S.P.B.) are absent and the Sherwood Sandstone rests unconformably on the Aylesbeare Mudstone Group.

extensive disruption by shallow, fibrous, branching root traces. Primary depositional structures (planar and current ripple laminae) are locally preserved, and wave ripples are rarely present. Palaeosol fabrics are limited to shallow root traces and rare vertisol structures in the mud-prone sections.

These deposits are interpreted as floodplain deposits which saw episodic sediment supply by sheetfloods or overbank flooding. Post-depositional processes involved desiccation and colonization by plants, possibly during the brief time when the post-flood water table was high. The vertisol structures suggest repeated wetting and drying of the sediment. This may have been in response to the ephemeral flooding and fluctuations in the water table (creating rare lacustrine conditions).

Sandflat and lacustrine deposits. These represent two end-member lithological groupings which are interbedded on a metre-scale. Sandflat deposits are characterized by fine-grained sandstones dominated by finely interbedded plane-bedded, ripple-laminated and adhesion-laminated fabrics which are disrupted by root traces. In contrast, the lacustrine deposits are heterolithic in lithology, with abundant wave ripple forms (cross-cut by root traces and desiccation cracks) and locally abundant *Taenidium* burrows (analogous to *Scoyenia* described from the Triassic of Greenland by Bromley & Asgaard 1979). The sediment is locally displaced by evaporite growth.

These interbedded associations are interpreted within a lacustrine context because of the ubiquitous presence of wave ripples and burrows in the heterolithics. The cross-cutting root traces and desiccation cracks imply that lake levels were highly variable. These lakes episodically retreated outside the area of the field to be replaced by the sandflat deposits. These record episodic sheet flooding (in the form of plane-bedded and ripple-laminated sandstones), but aeolian reworking into adhesion-laminated sandstones and the local presence of displacive evaporite fabrics suggest the formation of siliciclastic sabkhas. The overall setting of these lacustrine deposits may therefore be likened to the ephemeral, saline lakes described by Hardie *et al.* (1978), albeit with a larger areal extent than the modern examples.

Sheetflood deposits. These comprise thinly bedded (<1m), fine-grained sandstones which are typically interbedded with more mud-prone floodplain deposits. A wide range of depositional fabrics are present, including horizontal and low-angle planar laminae, current ripple laminae, dewatering structures, rare burrows and root trace disruption.

These sandstones are interpreted as the product of deposition under largely upper flow regime conditions by shallow, episodic sheet floods, and are similar in thickness and internal structure to modern ephemeral channel-fills described by Williams (1971) and Abdullatif (1989), and ancient examples described by Olsen (1989). Post-depositional processes involved sediment dewatering, brief colonization by burrowing animals and subsequent plant colonization.

Channel-fill deposits. These comprise fine- to very coarse-grained, dominantly cross-stratified, sandstones forming erosively based intervals *c.* 1–5 m thick. Coarse-grained lag deposits of quartz granules and intraclasts (reworked rhizocretion nodules and cobbles of floodplain mudrock) are common and are typically calcite cemented. Fining-upward trends dominate, but trendless and coarsening-upward motifs also occur. Single and multistorey examples are present.

These sandstones represent confined deposition within channels whose original planform geometry is unknown due to vertical and lateral amalgamation. However, their sand-rich nature would suggest that they were of low sinuosity. Their most distinguishing features in the upper Sherwood are the markedly coarser grain size (relative to the other associations) and the presence in basal lags of rip-up clasts of calcrete and mudrock, suggesting extensive floodplain erosion.

Aeolian deposits. These form intervals up to 8 m thick which can be mapped as kilometre-scale, lobate sandbodies. They are characteristically fine-grained and well sorted, lacking mudclasts or lag deposits, and have a distinctive 'bow'-shaped grain-size profile with gradational contacts with floodplain deposits above and below. Adhesion laminae are observed on the margins of these sandbodies, but in general they are characterized by the ubiquitous presence of rhizocretion structures defined by scattered and coalesced carbonate nodules (<1 cm in diameter). The two-dimensional slabbed core surfaces give a limited perspective on the three-dimensional scale of the rhizocretions, but they appear to be analogous in scale to the outcrop examples in the Otter Sandstone, which can exceed 2 m in length and be up to 0.15 m in diameter (cf. Purvis & Wright 1991).

These rhizocretion structures largely reflect the colonization of sandy groundwater aquifers by phreatophyte-like plants, irrespective of the original depositional setting of the sandbody. However, the gradational contacts with floodplain deposits and the sandbody geometry preclude a fluvial origin, whilst the consistent sorting and grain size, together with presence of adhesion structures, provide evidence to suggest an aeolian origin. The rhizocretion structures develop at specific stratigraphical levels within the sandbodies (Fig. 5), and are interpreted to represent episodes of vegetation cover on wind-blown sand patches.

Stratigraphic cycles

The Sherwood section in Wytch Farm can be rationalized into a number of facies 'cycles' which can be correlated across the field. These comprise:

- A 'first-order' evolutionary trend over the entire Sherwood section from perennial braidplain (cf. Dranfield et al. 1987) to sheetflood to ephemeral lacustrine conditions (Fig. 3). This trend culminated in deposition of the Mercia Mudstone Group and reflects a long-term reduction in slope, or rise in base level. The presence of paralic sediments within the Mercia Mudstone c. 100 km to the northwest (Warrington et al. 1980) may mark the culmination of this long-term base-level rise. This long-term trend is further subdivided into
- 'Second-order' cycles (Fig. 3) defined by five areally widespread, floodplain deposits (up to 10 m thick) which are heavily rooted with minimal development of fluvial sandstones. These can be correlated with lacustrine heterolithics to the northwest, and each becomes successively more mud-rich upwards through the Sherwood stratigraphy. These floodplain deposits bracket more sand-rich fluvial intervals, and successive intervals have finer and narrower grain-size ranges, show a greater proportion of lacustrine (and aeolian) facies and greater floodplain preservation (with fewer calcic paleosols). The second-order cycles form the basis for the reservoir zonation.
- 'Third-order' cycles are defined by thin (<2 m), but areally widespread floodplain–lacustrine horizons (Fig. 4). Within the fluvial deposits, between the third-order floodplain horizons, incised multistorey channel-fill deposits occur which in places vertically amalgamate and rework third-order (and locally second-order) floodplain deposits (Fig. 5). In the upper Sherwood the incised

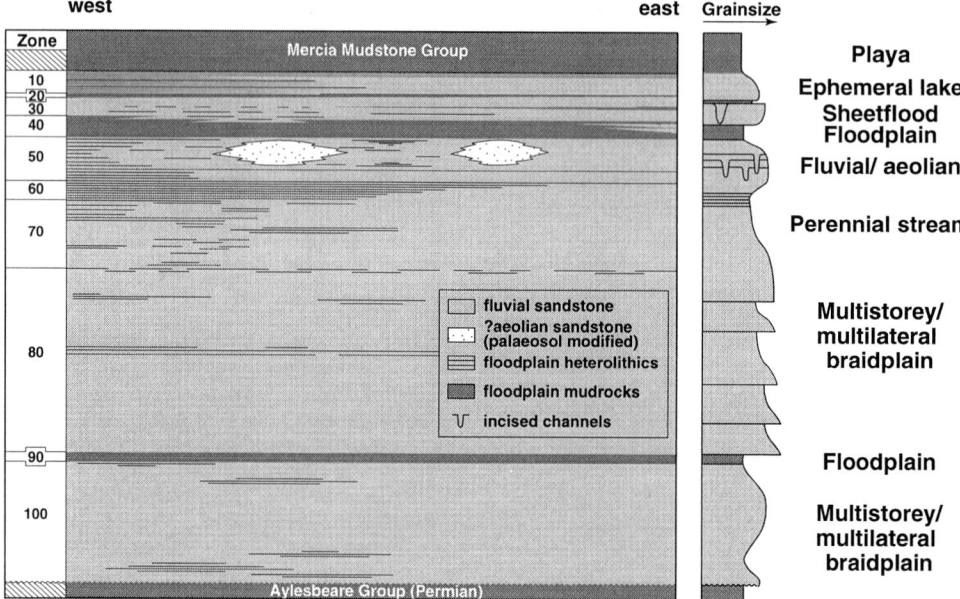

Fig. 3. Schematic reservoir layering scheme for the Sherwood Sandstone Group in the Wytch Farm reservoir. The panel depicts the broad lateral facies distribution across the field (cf. Fig. 1), and the height is approximately 150 m.

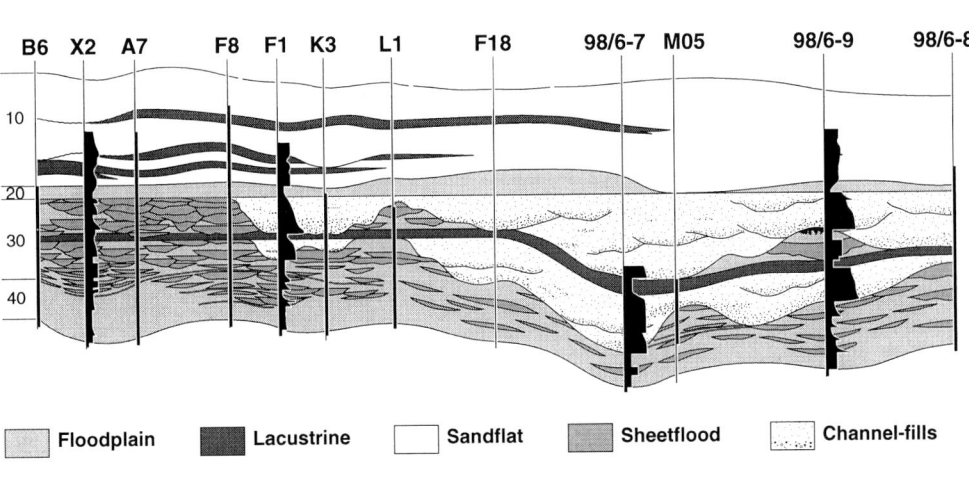

Fig. 4. Representative west-east transect across the field showing the facies architecture of zones 10–40, with grain-size profiles shown for selected wells. Features to note are: the lateral persistence of the floodplain deposits of zones 20 and 40; the successive eastward penetration of lacustrine horizons in Zone 10 (with local erosion), and the incised multistorey–multilateral channel complexes which erode and rework sheetflood and floodplain deposits in Zone 30. Zone 20 shows lacustrine facies to the north of the field. Floodplain and lacustrine facies are interpreted to record rapid base-level rise during 'wet' climatic phases, whilst fluvial incision is interpreted to be the result of base-level fall during 'dry' climatic phases (facilitated by devegetation).

systems represent the principal producing sandstones and are typically filled by coarse-grained sandstones (with abundant pebbles of hinterland detritus and rip-up clasts of mudstone and calcrete).

- The existence of a fourth-order 'cyclicity' may be inferred by the presence of desiccation cracks and rooted horizons within the lacustrine deposits, but these are impossible to correlate to demonstrate their areal extent, nor is their expression detectable outside the lacustrine facies.

The origin of these stratigraphic cycles is attributed to the interaction of tectonics on long-term accommodation and climate on base level and sediment yield. The first-order trend of Sherwood–Mercia deposition which ultimately resulted in marine incursion may record long-term tectonic subsidence, relative sea-level rise and reduced fluvial gradient. The cycle-bounding floodplain deposits all show evidence of lacustrine or high water table conditions and record higher frequency episodes of base-level rise, increased vegetation cover and reduced sand transport. Conversely, the more sand-prone intervals are interpreted to be the result of drier conditions (cf. Hall 1990) which reduced vegetation cover, lowered base level and released coarser-grained sediment into the fluvial drainage system and allowed aeolian reworking. In the third-order cycles, dry conditions resulted in fluvial incision and sediment by-passing. Within the upper Sherwood the climatic range of the second-order oscillations appears to have progressively narrowed, from lacustrine–fluvial–aeolian cycles in the lower section (Fig. 5) to purely intra-lacustrine cycles in the upper part (Fig. 4), possibly as a result of the first-order trend towards the Mercia playa mudrock system modulating the higher frequency cycles. The overall stratigraphic evolution of the Sherwood can therefore be viewed as the product of three 'nested' cycles of climate–base-level change with the depositional environment in each cycle modulated by its position in a lower frequency cycle. Climatic fluctuations have also been interpreted from within the Mercia Mudstone Group (Talbot *et al.* 1994), indicating that such variations probably occurred throughout the Triassic in this area.

At outcrop, the stratigraphically equivalent Otter Sandstone Formation (*c.* 100 km to the west) shows comparable evolutionary patterns,

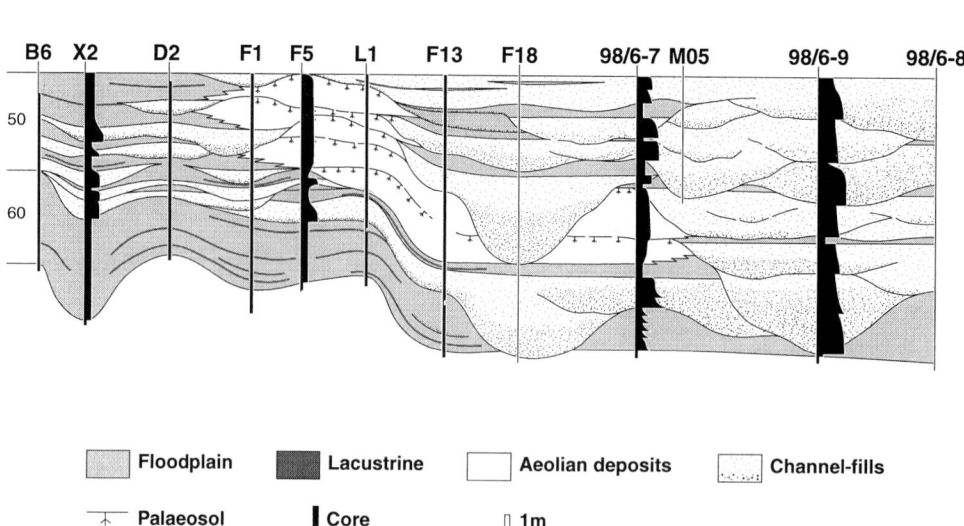

Fig. 5. Representative west–east transect across the field showing the facies architecture of zones 50 and 60, with grain size profiles for selected wells. Internal marker beds within these zones consist of lacustrine horizons within the floodplain deposits and calcic paleosols overprinting the aeolian deposits. The channel-fills contain lags of reworked calcrete and rip-up clasts (up to cobble grade) of floodplain mudrock. In contrast to Zone 30, aeolian deposits are interpreted to record the driest climatic conditions, whilst channel-filling must have occurred as the climate became wetter following calcrete stabilization of the aeolian deposits. Wet climatic episodes are recorded by lacustrine and floodplain deposits, but lacustrine facies are more weakly developed then in the overlying zones.

albeit with a subtly different facies stacking pattern. In this area the 'upper Sherwood' shows a c. 40 m thick transition from braidplain facies to Mercia playa facies via a succession of sheetflood deposits. These are in turn packaged into c. 4–7 m thick intervals punctuated by calcic paleosols in the lower part and floodplain mudrocks in the upper part. However, the sandbody types are sufficiently similar in their vertical facies character to form useful sandbody analogues for the subsurface. The second-order cycles in the Wytch Farm reservoir can be tentatively correlated to this exposure via the widespread floodplain horizons, but this requires further magnetostratigraphic data to substantiate.

Reservoir architecture

The reservoir zones erected for this study (Fig. 3) are based on the second-order floodplain deposits (zones 20, 40, 60 and 90) which define 1–10 m thick vertical transmissibility barriers enclosing reservoir sandstone intervals (zones 10, 30, 50, 70, 80 and 100). The sand-prone sections are characterized by the following.

- Zone 10 forms a 10 m thick, fining-upwards, perilacustrine section composed of sandflat and lacustrine heterolithic deposits. Lacustrine deposits encroach successively southeastwards, but are locally eroded by the sandflat deposits which rest sharply above (Fig. 4).
- Zone 30 comprises a 7 m thick interval of sheetflood sandstones interbedded with mudprone floodplain and lacustrine deposits. At least two incised fluvial systems occur in this interval (Fig. 4) separated by a minor lacustrine horizon. These form the principal producing intervals in this zone.
- Zone 50 is a 12 m thick section of channel-fills, sheetflood deposits and floodplain mudrocks, with locally developed, lobate sandbodies of aeolian origin. At least five incised fluvial systems dissect this interval, with well developed calcic paleosols on their interfluves (Fig. 5).
- Zone 70 is a 20 m thick interval of floodplain mudrocks and sand-prone channel-fills.

- Zone 80 (60 m) and 100 (30 m) are composed of high nett:gross, vertically and laterally amalgamated, multistorey channel-fill deposits capped by variably preserved floodplain deposits. Zone 100 rests unconformably on the Permian Aylesbeare Group mudstones (Fig. 3). These zones were not studied in detail.

Sandstone quality within these zones can be shown to be primarily a function of grain size and detrital clay content, with diagenetic overprinting by early calcite and 'burial' anhydrite, particularly in the coarser-grained sandstones. K-feldspar cementation is ubiquitous, but early K-feldspar (adularia) needles are associated with the aeolian sandstones in Zone 50. Heterogeneities with the sand-prone zones are provided by discontinuous floodplain–lacustrine deposits, channel abandonment plugs and lag deposits of reworked calcrete (which formed nucleation sites for later calcite cementation).

Discussion

The architecture of the Sherwood Sandstone in the subsurface and at outcrop demonstrates that extrinsic factors (principally climate) exerted a significant control on the alluvial architecture of what has formerly been viewed as a high nett:gross 'braidplain' system dominated by random channel avulsion. The general aspects of the sequence architecture which can be derived from this semi-arid, fluvio-lacustrine system are:

- episodes of base-level rise are recorded by lacustrine deposits whose up-dip correlative equivalents are mud-rich floodplain deposits;
- following maximum lacustrine flooding, subsequent fluvial deposits comprise fine-grained, poorly confined, sheetflood deposits and shallow, rooted, floodplain heterolithics;
- base-level lowering during dry phases, resulted in fluvial incision. However, the incisions are typically shallow (<10 m) and *wide* (tens of kilometres);
- interfluves can be marked by up to 4 m deep rhizocretionary calcrete formed by plants attempting to reach the lowered water table. However, this is not ubiquitous and locally a planar surface with no palaeosol development occurs (aeolian deflation surface?).

These features have also been observed in Triassic fluvio-lacustrine systems in Marnock Field (UK Continental Shelf–UKCS) and in the Devonian in Clair Field (cf. McKie & Garden 1996). This suggests that perilacustrine systems respond to lake-level fluctuations by fluvial aggradation and incision in a manner analogous to that of paralic systems. However, the fluvial incisions generally appear to have shallower relief, possibly reflecting the low relief of the playa basins into which these fluvial systems drained. This would have precluded significant nick-point erosion (Schumm 1993).

Conclusions

The recognition of a hierarchy of climatically driven cycles within the reservoir permits high-resolution correlation and the recognition of subtle, but important, changes in sandbody geometry and connectivity within successive cycles. This has led to the generation of a fine-scale reservoir model in which the deterministic components are maximized and the type of analogue data used to model the stochastic elements is tightly constrained. The principal producing intervals within the field are the products of sand deposition during relatively dry climatic conditions when sediment yield increased as a result of devegetation, but with sufficient runoff to maintain fluvial sediment transport. Aeolian transport during driest conditions probably produced local sediment recycling, but was unlikely to have contributed to significant sediment transport into the field area. In contrast, during relatively wet climatic conditions base level rose, resulting in lacustrine incursions and deposition of fine-grained, mud-rich, floodplain deposits. These form the principal flow barriers and baffles within the field (and provide the means for the correlation framework).

References

ABDULLATIF, O. M. 1989. Channel-fills and sheetflood facies sequences in the ephemeral terminal River Gash, Kassala, Sudan. *Sedimentary Geology*, **63**, 171–184.

BOWMAN, M. B. J., MCCLURE, N. M. & WILKINSON, D. W. 1993. Wytch Farm Oilfield: deterministic reservoir description of the Triassic Sherwood Sandstone. *In:* PARKER, J. R. (ed.) *Petroleum Geology of Northwest Europe: Proceedings of the 4th Conference*. Geological Society, London, 1513–1517.

BROMLEY, R. & ASGAARD, U. 1979. Triassic freshwater ichnocoenoses from Carlsberg Fjord, East Greenland. *Palaeogeography, Palaeoclimatology, Palaeoecology*, **28**, 39–80.

COLTER, V. S. & HAVARD, P. J. 1981. The Wytch Farm Oil Field, Dorset. *In:* ILLING, L. V. & HOBSON, G. D. (eds) *Petroleum Geology of the Continental Shelf of North-west Europe*. Heyden, London, 494–503.

DRANFIELD, P., BEGG, S. H. & CARTER, R. R. 1987. Wytch Farm Oilfield: reservoir characterisation of the Triassic Sherwood Sandstone for input to reser-voir simulation studies. *In:* BROOKS, J. & GLENNIE, K. W. (eds) *Petroleum Geology of North West Europe.* Graham & Trotman, London, 149–160.

HALL, S. A. 1990. Channel trenching and climatic change in the southern, U.S. Great Plains. *Geology,* **18**, 342–345.

HAMBLIN, R. J. P., CROSBY, A., BALSON, P. S., JONES, S. M., CHADWICK, R. A., PENN, I. E. & ARTHUR, M. J. 1992. *United Kingdom Offshore Regional Report: The Geology of the English Channel.* HMSO, London, for the British Geological Society.

HARDIE, L. A., SMOOT, J. P. & EUGSTER, H. P. 1978. Saline lakes and their deposits: a sedimentological approach *In:* MATTER, A. & TUCKER, M. E. (eds) *Modern and Ancient Lake Sediments.* International Association of Sedimentologists, Special Publications, **2**, 7–42.

HAWKES, P. W., FRASER, A. J. & EINCHCOMB, C. C. G. 1997. The tectono-stratigraphic development and exploration history of the Weald and Wessex Basins, southern England. *This volume.*

HOGG, A. J. C., MITCHELL, A. W. & YOUNG, S. 1996. Predicting well productivity from grain size analysis and logging while drilling. *Petroleum Geoscience,* **2**, 1–15.

MCCLURE, N. M. WILKINSON, D. W., FROST, D. P. & GEEHAN, G. W. 1995. Planning extended reach wells in Wytch Farm Field, UK. *Petroleum Geoscience,* **1**, 115–127.

MCKIE, T. & GARDEN, I. R. 1996. Hierarchical stratigraphic cycles in the Lower Clair Group, UKCS. *In:* HOWELL, J. A. & AITKEN, J. F. (eds) *High Resolution Sequence Stratigraphy: Innovations and Applications.* Geological Society, London, Special Publications, **104**, 139–157.

OLSEN, H. 1989. Sandstone-body structures and ephemeral stream processes in the Dinosaur Canyon Member, Moenave Formation (Lower Jurassic, Utah, USA). *Sedimentary Geology,* **61**, 207–221.

PURVIS, K. & WRIGHT, V. P. 1991. Calcretes related to phreatophytic vegetation from the Middle Triassic Otter Sandstone of South West England. *Sedimentology,* **38**, 539–551.

SELLEY, R. C. & STONELEY, R. 1987. Petroleum habitat in south Dorset. *In:* BROOKS, J. & GLENNIE, K. W. (eds) *Petroleum Geology of North West Europe.* Graham & Trotman, London, 139–148.

SCHUMM, S. A. 1993. River response to base level change: implications for sequence stratigraphy. *Journal of Geology,* **101**, 379–294.

TALBOT, M. R., HOLM, K. & WILLIAMS, M. A. J. 1994. Sedimentation in low-gradient desert margin systems: a comparison of the Late Triassic of north-west Somerset (England) and the late Quaternary of east-central Australia. *In:* ROSEN, M. R. (ed.) *Palaeoclimate and Basin Evolution of Playa Systems.* Geological Society of America, Special Paper, **289**, 97–117.

UNDERHILL, J. R. & PATERSON, S. 1998. Genesis of tectonic inversion structures: seismic evidence for the development of key structures along the Purbeck–Isle of Wight Disturbance. *Journal of the Geological Society, London,* in press.

—— & STONELEY, R. 1982. Introduction to the development, evolution and petroleum geology of the Wessex Basin. *This volume.*

WARRINGTON, G., AUDLEY-CHARLES, M. G., ELLIOTT, R. E., EVANS, W. B., IVIMEY-COOK, H. C., KENT, P. E., ROBINSON, P. L., SHOTTON, F. W. & TAYLOR, F. M. 1980. *A Correlation of Triassic Rocks in the British Isles.* Special Report of the Geological Society, London.

WILLIAMS, G. E. 1971. Flood deposits of the sand-bed ephemeral streams of Central Australia. *Sedimentology,* **17**, 1–40.

The Kimmeridge Bay oilfield: an enigma demystified

JONATHAN EVANS,[1] DAVID JENKINS[1] & JON GLUYAS[2]

[1] *BP Exploration, Sherwood House, Blackhill Road, Holton Heath Trading Estate, Poole, Dorset BH16 6LS, UK (e-mail: evansij@bp.com)*
[2] *Monument Oil and Gas Ltd, Kierren Cross, 11 Strand, London WC2N 5HR, UK*

Abstract: The Kimmeridge oilfield was the first commercial discovery in the Wessex Basin. It was discovered in 1959 and is still producing from a single well. To date over 3 million barrels of oil have been extracted. Production is likely to continue into the next century. Oil is trapped in the fractured Cornbrash Limestones in an anticline close to the main inversion axis. The Kimmeridge accumulation remains the only producing field in the hanging wall to the Purbeck Disturbance, making it an intriguing enigma within the basin. Many myths have been perpetuated regarding the size of the field and the results from drilling elsewhere within Kimmeridge Bay. By presenting data on the field here, we hope to dispel some of these myths.

The Kimmeridge oilfield lies in Kimmeridge Bay on the South Dorset coast. It is a shallow oil accumulation within a faulted inversion anticline which lies immediately to the south (downthrown) side of the main Purbeck Disturbance, which is the most important structural feature in the area (Stoneley 1982; Underhill & Paterson 1998; Underhill & Stoneley this volume). The field was discovered in 1959 and began producing in 1961. It is still producing from a single well at a rate of 100 bbl/day and, to date, has produced over 3 million barrels of oil. The 'nodding donkey' above Kimmeridge Bay is a local landmark (Fig. 1) and Kimmeridge is probably the most visited oilfield in the UK. Despite this notoriety, little information has been published about the oilfield, apart from short mentions in Brunstrum (1963) and Selley & Stoneley (1987). In this paper we present some historical and geological background to the field.

Exploration in Kimmeridge Bay

The first prospecting licences under the 1934 Petroleum Production Act were granted to the D'Arcy Exploration Company in December 1935. These licences included 2227 square miles in southern England. The main target for early exploration in southern England was the Jurassic (Lees & Cox 1937). The first structures to be tested were large anticlines in southern England such as Portsdown, Henfield and Kimmeridge.

Exploration in Kimmeridge Bay formed part of D'Arcy's exploration activity in the new licences. Drilling began in 1936 and the last exploration well was drilled in 1980. The Kimmeridge Field was discovered in 1959 and was developed under a Mining Licence (ML5) granted in 1964. The field is still producing under the terms of that licence which expires in 2014. To date, six wells have been drilled in Kimmeridge Bay (Fig. 2), these are as follows.

Broadbench-1

Well Broadbench-1 was drilled at the western end of Kimmeridge Bay in 1936–1937 to test the Corallian in the Kimmeridge Anticline. The well reached the Upper Corallian, finding an 'oblique joint in grey sandstone wet with light oil' at 252 m depth in the Sandsfoot Grit. This show was not adequately tested. The main objective, the Bencliff Grit, which is oil impregnated at outcrop near Osmington Mills 10 miles to the west (Miles *et al.* 1993), was not reached because of mechanical difficulties. Broadbench-1 was plugged and abandoned at 287 m in the Osmington Oolite.

Broadbench-2 (Kimmeridge-1)

The main objective of a second well on the Purbeck Anticline was to re-test the Corallian. In 1950 it was recommended that a well should be drilled to a depth of 760 m in order to test this target and also to investigate a high velocity layer (which is likely to have been the Great Oolite) recorded on a seismic refraction line along the northern flank of the anticline. However, this well was not sanctioned. A well to re-test the Corallian at Broadbench was proposed in 1953 and again in 1956 when a renewed phase of exploration in southern England began. This was sanctioned and Broadbench No. 2 was put into the programme for the Failing M.1 rig which was to drill a number

Fig. 1. Photograph showing the Kimmeridge-1 well and Kimmeridge Bay.

of shallow holes on the Weymouth Anticline. By now a tank firing range occupied the highest part of the anticline on land so a site was chosen just outside the firing range on the cliff top in Kimmeridge Bay at the same structural elevation as Broadbench No. 1.

Broadbench No. 2 (subsequently renamed as Kimmeridge No. 1) was drilled in February–March 1959. The well cored into the Cornbrash limestone at a depth of 512 m sub-sea (mSS). Some of the core oozed oil from partially leached calcite veins. Substantial mud losses to the formation were experienced at 520 mSS. An initial production test yielded 30 bbl/day. Two acid treatments were performed after which the well flowed briefly at 4300 bbl/day. After these encouraging tests the well was completed as a producing well in the Cornbrash. The well was originally intended to test the Oolites and the Lias but testing these horizons was postponed until Kimmeridge No. 3.

Kimmeridge No. 2

During 1960 Kimmeridge No. 2 was drilled 670 m to the east of Kimmeridge No. 1. It encountered the Top Cornbrash at 583 m and was terminated within the uppermost Forest Marble. Mud losses occurred within the Corallian, the lowermost Oxford Clay, Kellaways Beds, Cornbrash and Forest Marble. Tests conducted on sands within the Oxford Clay produced 7.5 bbl of oil in 40.5 h. After acid treatment, the Cornbrash produced

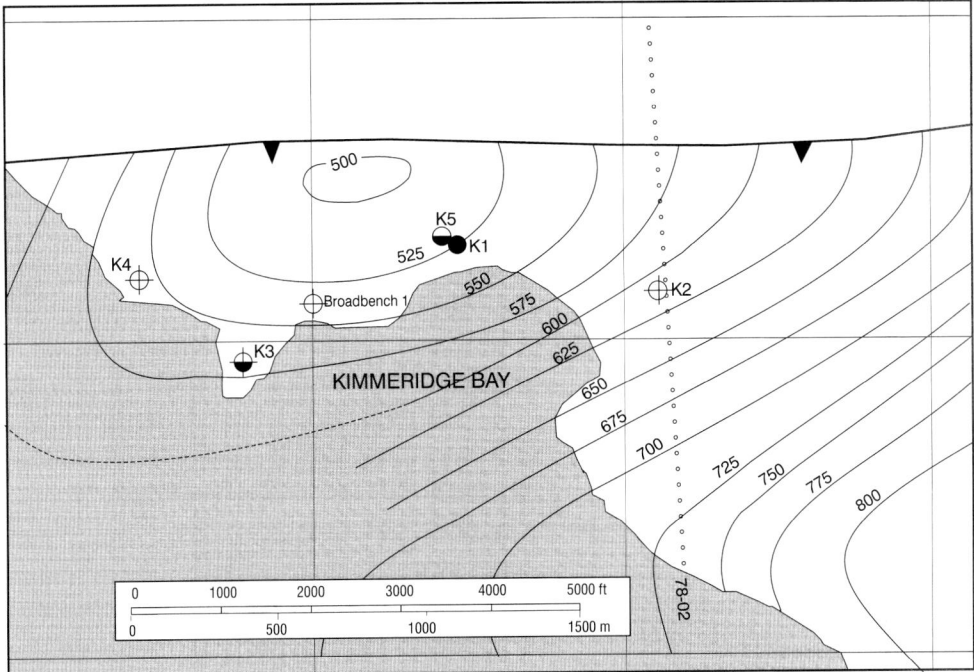

Fig. 2. Map of Top Cornbrash showing well locations. Contours are depth to Top Cornbrash in metres below sea level.

only 6 gallons of oil. The well was retained as an observation well. Pressure measurements indicate that this part of the Oxford Clay is in communication with the Cornbrash reservoir.

Kimmeridge No. 3

Kimmeridge No. 3 was drilled during 1959–1960, prior to Kimmeridge No. 2. It was drilled to complete the programme intended for Kimmeridge No. 1, by testing the Great Oolite, Inferior Oolite and Upper Lias, and to further develop the Cornbrash. The well lies 762 m southwest of Kimmeridge No. 1. During drilling, mud losses were encountered at the Top Kellaways Beds and within the Forest Marble. The well was cored through the Cornbrash with heavy mud losses. A test of the Forest Marble (559–575 mSS) produced 600 bbl/day of water. Tests on deeper formations were either tight or produced only water. There were some oil shows within the Bridport Sands around 942 mSS but tests could not be performed on this zone. The well was plugged back and perforated within the Cornbrash. After three acid stimulations, oil production from the Cornbrash was measured at 26.5 bbl/day with 4.5 bbl/day water. An oil–water contact was observed at 535.5 mSS. The well was used as an observation well until the end of 1992 when a further acid stimulation was performed and the well was then put on production for a short time.

Kimmeridge No. 4

Kimmeridge No. 4 was an appraisal well drilled in 1960 and located 412 m northwest of Kimmeridge No.1 to test the geological structure. The well was terminated at 262 mSS owing to a mechanical breakdown and was not tested. No mud losses were experienced. The well was not sufficiently deep to penetrate the Cornbrash reservoir.

Kimmeridge No. 5

Following the discovery and successful appraisal of the Wytch Farm oilfield (Colter & Havard 1981) which lies only a few miles to the northeast of Kimmeridge Bay, there was renewed interest in the prospectivity of the deeper reservoirs in the area. In 1980 Kimmeridge No. 5 was drilled as an exploration well to test the deeper potential of the Kimmeridge structure at Sherwood Sandstone level with the Bridport Sands as a secondary target. The well was drilled to the Aylesbeare Group (Permo-Triassic). Weak gas shows and minor fluorescence were recorded throughout the Jurassic. The Sherwood Sandstone was encountered 453 m deeper than prognosed at 2272.8 mSS. The sandstones had weak oil shows but reservoir quality was much poorer than in the Wytch Farm field.

Petroleum geology

Reservoir

The main reservoir in the Kimmeridge Field is provided by fractured Cornbrash Limestone. Some reserves may also be held within fractured Oxford Clay.

The Cornbrash Limestone is around 20 m thick and is a tight, occasionally fissured limestone. Cores from wells Kimmeridge-1 and Kimmeridge-3 cover almost the entire section. Core analysis data indicate an average porosity of around 1% and virtually zero permeability. Oil staining in the core is restricted to the open fractures. It is therefore assumed that production is from an extensive fracture system developed within the Cornbrash and also in the adjacent Kellaways Beds–Oxford Clay and the Forest Marble.

The Cornbrash reservoir is characterized by abnormally low reservoir pressures, well below hydrostatic. The initial pressure in Kimmeridge-1 was 400 PSI at 520 mSS. This pressure would be expected in a reservoir several hundred metres shallower. Deeper formations are normally pressured. Brunstrum (1963) suggested that oil may have been sealed in the fracture system prior to Miocene folding and that the fissures were physically enlarged during folding leading to reduced pressures. Some such explanation is necessary as there is no history of recent burial which might otherwise explain the pressures.

Source

In common with oil from the Wytch Farm and Wareham fields, the oil at Kimmeridge is believed to have been sourced from the Lower Lias (Selley & Stoneley 1987). Geochemical analysis of the oils supports this. The Kimmeridge Bay oil has a significantly lighter gravity (45° API) than Wytch Farm oil (35° API) which suggests either that it was expelled from more mature source rocks or that some fractionation occurred during migration.

Seal and trap

The Cornbrash reservoir is sealed by the overlying Oxford Clay.

The trap at Kimmeridge is believed to be a simple faulted anticline which lies just to the south of the main Purbeck Disturbance. The Purbeck Disturbance follows the line of a major fault along which most of the inversion movement has occurred in this area (Fig. 3). The structure is poorly imaged on seismic owing to the proximity to large faults and to the shoreline. The map shown in Fig. 2 is based on seismic data together with well data and outcrop mapping.

Surface facilities and processing

The Kimmeridge well site is normally unmanned and has minimal surface facilities (Fig. 1). The oil is now produced from the single well Kimmeridge-1 using a beam pump. Oil is pumped into two surface tanks which are capable of holding 850 barrels. At least twice a week a road tanker visits the site to empty the tanks. Oil is then taken by road to Furzebrook where it is held in storage tanks. Periodically the oil is transferred by pipeline to the main Wytch Farm Gathering Station near Corfe Castle where it is commingled with Wytch Farm crude oil and processed into stabilized crude, liquified petroleum gas (LPG) and dry gas which are exported.

Production history and oil reserves

The production history from start-up to present day is shown in Fig. 4. When full production commenced in 1960, it rose rapidly to 250 bbl/day and peaked at 500 bbl/day in 1972. Production then declined steadily to reach approximately 100 bbl/day in 1986 at which point it stabilized. Production continues at this rate today.

Production was initially by natural flow but artificial lift, in the form of a beam pump ('nodding donkey') was installed in 1964. Water production is minimal, however, gas production has increased gradually over time.

The initial rise in production rate from 200 bbl/day in 1961 to 400 bbl/day in 1972 was difficult to explain and led to many theories about the amount of oil at Kimmeridge and its

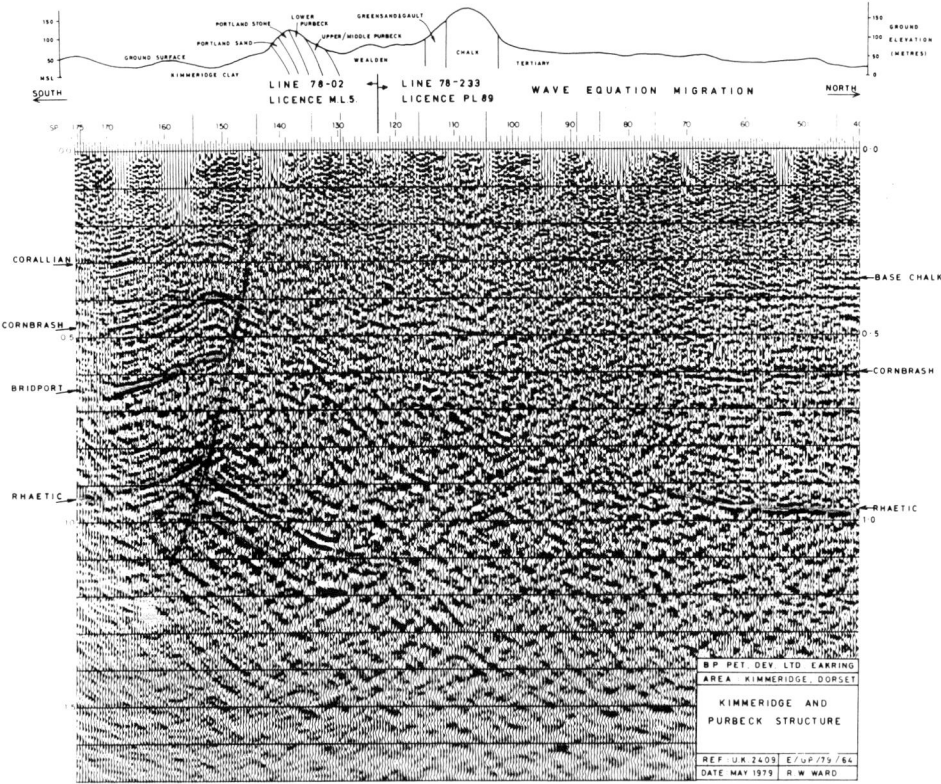

Fig. 3. Seismic line running north–south through the Kimmeridge field. Well Kimmeridge-2 lies at SP 158.

location. By 1975 cumulative production was 1.725 million bbl (Fig. 5) and no decline had been witnessed.

Mapping at that time indicated closure over an area of 120–160 acres. The oil column was known to be 47–57 feet thick. Depending on the porosity (including fractures) which they assumed for the Cornbrash, various geologists calculated the area of closure required to contain the cumulative oil produced up to that time. These estimates varied from more than 2500 acres (assuming 1% effective porosity) to 250 acres (assuming 10% porosity). Each estimate of reservoir led to a new theory to explain the disparity between the mapped closure and that calculated from the reservoir behaviour. The most notable theories included:

(1) additional reserves trapped in an unmapped offshore extension to the Kimmeridge Structure at Cornbrash level;
(2) the presence of a more porous facies within the Cornbrash, not penetrated by wells but in contact with the producing wells through the fracture system;
(3) supply of the Cornbrash reservoir from a second, deeper reservoir such as the Bridport Sands, possibly by continuing migration along faults;
(4) the fissure system continues outside the Cornbrash into the Oxford Clay (as shown by Kimmeridge-2) and is extensive enough to contain all of the observed oil.

Some of these were discounted at the time based on available data. For example communication between the Bridport and Cornbrash was considered unlikely given their different pressure characteristics (the Bridport being normally pressured). Subsequent production has shown a fairly normal decline behaviour (Fig. 5) suggesting that the reservoir contains a finite reserve. Based on the decline behaviour, the ultimate recoverable reserve is now estimated to be 3.5 million barrels.

The increase in gas–oil ratio (GOR) through time is consistent with solution gas drive – the formation of a natural gas cap through pressure depletion. However, no gas coning is observed in the Kimmeridge-1 well. It has been suggested

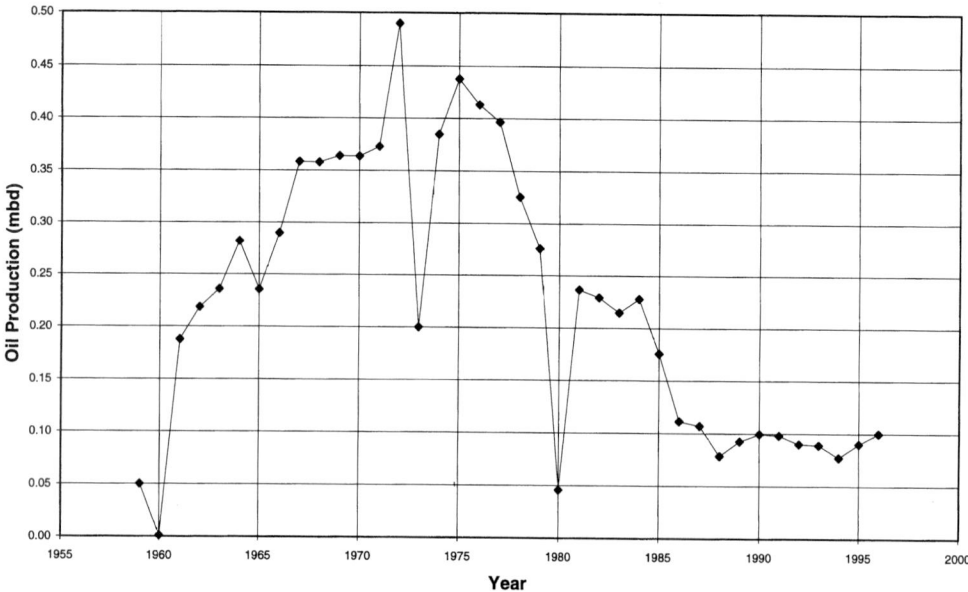

Fig. 4. Graph showing average daily production v. time.

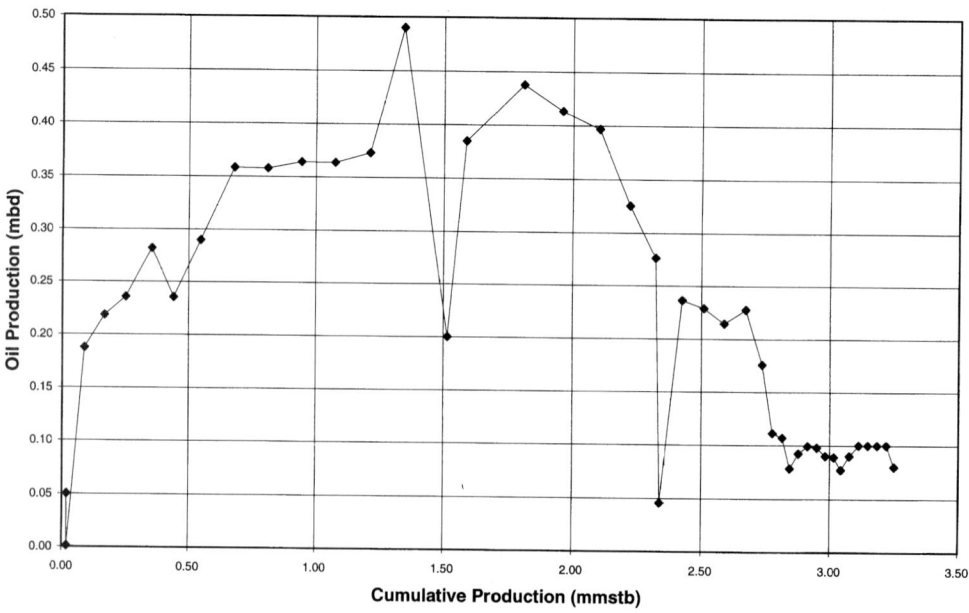

Fig. 5. Graph showing average daily production v. cumulative production.

that the increase in production rate during the early years of production may have been related to an increase in the GOR effectively lightening the produced fluids thereby making them easier to lift – a partial natural gas lift.

If a gas cap has formed during production, as seems likely, then this must be located outside the Cornbrash itself, probably in the overlying Oxford Clay. Based on cursory examination of the data now available, some combination of an extensive fracture system, extending out of the Cornbrash, possibly higher effective porosities and a larger structure than mapped onshore would seem to provide a plausible explanation of the production behaviour observed at Kimmeridge.

If the drive mechanism is assumed to be predominantly solution-gas then the recovery factor is likely to be between 15 and 30%. Back-calculation therefore suggests that the stock tank oil initially in place (STOIIP) is in the range 10–25 million barrels.

Conclusions

The Kimmeridge field provides the only commercial success from numerous exploration wells drilled to the south of the main inversion axis within the Wessex Basin. It remains the only successful example of the Tertiary Inversion anticline play within the Wessex Basin (Hawkes et al. 1997 this volume). It is a unique field in many ways being one of the first to be discovered in England and having been in almost continuous production from 1961 to the present day.

The size of the reservoir and nature of the trap are still not fully understood, however, production data suggest that the ultimate recoverable reserve will be around 3.5 million barrels which implies an original oil in place of at least 10 million barrels.

Despite being the most popular myths surrounding the field, the reservoir is neither fed from a deeper accumulation nor is migration into the field still occurring. Neither is oil being produced from a single isolated fractured zone which was only penetrated by the first well.

The facts do not support these stories. If migration was continuing then the production history would not be likely to show a conventional decline. The well results are consistent with a fractured reservoir contained within an anticlinal trap although the extent of the reservoir system and its effective porosity remain problematic. The outstanding enigma is why there is only one Kimmeridge-type success in the Wessex Basin. All of the other similar traps which have been tested seem to have either been breached or never to have been charged. Perhaps, like the Wytch Farm field to the north, it is the sole survivor of a much more significant play which was wiped out during inversion.

Thanks to BP Exploration Operating Company for permission to publish this paper, to J. R. Underhill for encouraging us to submit this paper for inclusion in this volume and for his patience in awaiting the manuscript! Finally thanks to the generations of BP geologists who have puzzled over the accumulation at Kimmeridge and whose combined thoughts we have attempted to summarize here.

References

BRUNSTRUM, R. G. W. 1963. Recently discovered oilfields in Britain. *In*: *6th World Petroleum Congress*, Section 1, 11–20.

COLTER, V. S. & HAVARD, D. J. 1981 The Wytch Farm oil field, Dorset. *In*: ILLING, L. V. & HOBSON, G. D. (eds) *Petroleum Geology of the Continental Shelf of North-west Europe*. Heyden, London, 494–503.

HAWKES, P. W., FRASER, A. J. & EINCHCOMB, C. C. G. 1997. The tectono-stratigraphic development and exploration history of the Weald and Wessex basins, southern England. *This volume*.

LEES, G. M. & COX, P. T. 1937 The geological basis for the present search for oil in Great Britain by the D'Arcy Exploration Company, Limited. *Quarterly Journal of the Geological Society, London*, **93**, 156–194.

MILES, J. A., DOWNES, C. J. & COOK, S. E. 1993 The fossil oil seep in Mupe Bay, Dorset: a myth investigated. *Marine and Petroleum Geology*, **10**, 58–70.

SELLEY, R. C. & STONELEY, R. C. 1987. Petroleum Habitat in south Dorset. *In*: BROOKS, J. & GLENNIE, K. (eds) *Petroleum Geology of North-west Europe*. Graham & Trotman, London, 139–148.

STONELEY, R. C. 1982. The structural development of the Wessex Basin. *Journal of the Geological Society, London*, **139**, 545–552.

UNDERHILL, J. R. & PATERSON, S. 1998. Genesis of tectonic inversion structures: seismic evidence for the development of key structures along the Purbeck–Isle of Wight Disturbance. *Journal of the Geological Society, London*, in press.

—— & STONELEY, R. 1982. Introduction to the development, evolution and petroleum geology of the Wessex Basin. *This volume*.

Index

Abbotsbury Cornbrash Formation 128, 130, 131, 151, 152
   biozonation 89, 94, 97
Abbotsbury Fault 34, 71, 76, 77, 79, 80, 81, 83
Abbotsbury–Ridgeway Fault 250, 251, 252, 253, 254, 255, 256, 257, 258, 262
acritarchs, spine lengths, conditions and 175, 178
aeolian deposits, Sherwood Sandstone Group 401–2
Albian unconformity 7, 246, 251, 252
alkanes, degradation 375–85, 390, 392, 393, 394, 395
allochthonous organic matter 169, 172, 174, 176, 177, 178, 179, 180, 182
Alpine Plate sequence 47, 61–3, 65
ammonite biostratigraphic schemes 167, 331, 337, 346
Ampthill Clay 153, 336
Ampthill Clay Equivalent 133, 136–7, 138, 139, 159
   biozonation 90, 94, 99
angular stratal relationships, Mupe Bay 349, 351–2
apatite fission track analysis (AFTA), palaeotemperature estimation by 194–6, 197, 200–12
Apoderoceras Bed 148
Aptian
   depositional sequences 332–46
   unconformities 7, 215, 217, 220, 223, 224, 232, 237, 280, 333
Apto-Albian unconformities 215, 220, 229, 230, 232–3, 234
Armorican Massif 224, 233
Ashdown Anticline 40–1
Ashdown Sands 59
Atlantic plate sequence 44, 45, 47–56, 64
autochthonous organic matter 169, 172, 174, 175, 178, 180
Aylesbeare Mudstone Group 5, 11, 69–71, 72, 83, 410

bank collapse 349, 351–3
Bargate Beds 336, 338, 339
basement-linked faults 243, 245–50, 262, 263, 265
Basin definition 1–3
Basin structure 285
Bay of Biscay 232–3, 234, 235, 237
Belemnite Marls 109, 112–13, 148, 168, 179
   biozonation 88, 93, 96
Belemnite Stone 148, 179, 182, 183, 184
Bencliff Grit 7, 11, 355–61, 368–70
   ichnology 358, 359, 361–5, 369
   oil seeps 373, 375
   palynofacies and micropalaeontology 364, 365–8, 369
Bencliff Grit Member 134, 135–6, 159, 355
Berriasian unconformity 220, 221, 222, 223, 224, 226, 227, 229, 230, 231, 232, 233, 234
   hydrocarbon prospectivity and 237, 238
Berry Member 89, 94, 97, 130, 152
bioremediation 374
biostratigraphic schemes 87–100, 106, 167, 331, 337, 346
   see also ichnology, palynofacies
Birchi Tabular Bed 148
Biscay plate sequence 45, 56–61, 65
bituminous shales 142

Black Ven Marls 88, 92, 96, 109, 112, 148
Blackstone 140
Blue Lias 87, 92, 95, 96, 110, 111, 148, 168
Blue Lias Formation 109–10
Boueti Bed 128, 129, 151
Bournemouth Bay 15, 16, 69, 76, 211
   basement-linked fault zone 243, 245–50, 263
Bovey Basin 326, 327
Brendon Hills 315, 316, 324, 327, 328
Bridport Sands 7, 11, 12, 14, 16, 64–5, 117, 118, 119, 206, 287
   biozonation 88, 89, 93, 97
   geological history 148, 150, 151
   reservoir 15, 16, 41, 42, 49, 63–4, 103, 148, 409, 411
   oil degradation 384
Bridport Sandstone Formation 10, 28, 51, 52
   oil reserves/reservoir 19, 22, 24, 25, 30, 34, 35, 399
   uplift/erosion 192, 194, 195, 196, 197
Bristol Channel 322–3, 326, 327, 328
Bristol Channel Basin 220, 314, 321, 328
Brittany Basin 228, 229
Budleigh Salterton Pebble Beds 5, 69, 71, 72, 75, 256
burial histories, reconstruction 205, 207

Calcareous Subgroup 136–7
Celtic Sea Basins 215, 217–19, 237
Central Bristol Channel 325, 326
Central Bristol Channel Fault 312, 313, 314, 322
Central Channel Basin 203, 283, 285, 288, 289, 290, 291
Central Channel Fault 271, 288, 291, 294, 295, 297
Central Channel High 3, 9, 130, 141, 143, 145, 146, 150
   Lias Group 111, 113, 115, 116, 120
   structure 283–4, 285–97
Central Somerset Basin 244, 322, 323, 324
Cerne-Winterbourne-Christchurch Trough 69, 76, 79, 81, 83
Chale Bay 340, 341, 342
Chalk 7, 14, 60, 61, 69, 217, 229, 270, 288, 321
Chalk Group 219, 224, 286, 287
Channel Basin 3, 9, 15, 118, 217, 219, 220, 284, 333, 336
   oil generation 14, 212, 235, 237
   salt tectonics 243, 244–5, 254, 260–5
   structure 243, 248, 254
   see also Central Channel Basin; English Channel
channel-fill deposits, Sherwood Sandstone Group 401
Chief Shell Beds Member 137, 138
Cimmerian unconformity 45, 222, 230
Cinder Bed 144, 158, 159
climatic changes, effects of 148, 157, 159, 403, 405
Clophill 336, 338, 341, 345
Coinstone Horizon 148, 175, 178, 179, 184
Compton Valence 83, 244
Corallian Group 11, 133–9, 153, 154, 157, 355
Corallian limestones 148, 334, 342
Corallian Sandstone 55, 65
Cornbrash 11, 25, 40, 41, 77, 288, 409
Cornbrash Limestone 63, 408, 410, 411, 413
Cornubia 104, 114
Cornubian Platform 216, 224, 231, 232, 233, 235, 237
   stratigraphic gap 220

Cornwall granites, thermal history 203
Cotham Member 107, 109
Cothelstone Fault 313
Cranborne–Fordingbridge High 3, 105, 136, 155, 157, 314
   Corallian Group 137, 139
   Kimmeridge Clay Formation 141
   Lias Group 114, 119
   Nothe Grit Formation 133
   Oolite groups 122, 124, 126, 129
   Portland Group 144
   Purbeck Group 145, 146
Crediton Trough 4, 5, 316, 321, 324, 326, 327, 328
Cretaceous
   biozonation 87–100
   lithostratigraphy 105–59
   megasequences 5, 7, 15
   sedimentation controls 79–81
   sequence boundaries 331–46
   thermal history 202–3, 211
   unconformities 215, 216, 217, 219–25, 231, 267, 269
   uplift *see* uplift

Day's Shell Bed 179, 182, 184
depositional sequences 332, 336, 340, 344–5
   tectonic influence 333–6, 343–4, 345, 346
detachment faulting 254–6
Digona Bed 128, 129
dinocyst zonal scheme, English Channel 95–100
Dorset 67, 71, 83, 212, 243, 261, 274
   geological history 147, 151, 153, 155, 156
   inversion tectonics 269, 274, 280
   stratigraphy 128, 220, 267–9
Dorset Basin 104, 105, 150, 151, 153, 155, 203
   stratigraphy 113, 114, 115, 116, 118, 122, 133, 135, 137, 138, 139, 159
Dorset Coast 112, 114, 120, 122, 142, 167, 300
   hydrocarbon seepages 40, 155, 166–8, 178, 211, 385, 387
   palynofacies analysis 175
   stratigraphy 112, 114, 120, 122, 133, 142
      Jurassic sequences 166–8, 173, 175–84
Dorset Halite 244, 245, 252, 272, 273, 280
Dorset High 124, 126, 129
Down Cliff Clay 115, 116, 117–19, 120, 121, 122, 179
   biozonation 88, 93, 97, 174
   sedimentation 148, 150
Down Cliff Sands 114, 117, 182, 287

Early Albian Carstone 157
en echelon structures 314, 326, 327
English Channel 103–7, 187, 188–97, 270
   biozonation 87–100
   lithostratigraphy 107–59
   see also Central Channel High; Channel Basin
English Channel Basin *see* Channel Basin
erosion, uplift and 190–7, 200, 205–10, 212
Exeter Volcanics 316, 321, 328
Eype Clays 88, 115, 117, 182
Eypemouth Fault 253, 260, 261, 262

Faringdon Sponge Gravels 334, 342
Faringdon–Baulking, stratigraphy 342, 343, 345
Fastnet Basin 220, 228, 230
fault block/horst plays 63
faults
   geometry 260–5, 290, 294–7, 299–309, 311
   history 311–28
   propagation 301, 302
   reverse faulting 83, 85, 252, 260, 262, 265, 278, 285
   types, WCHF 319, 320, 321, 325, 326
   see also inversion tectonics; salt tectonics; strike-slip faults
field-scale faults 302, 303, 307
firm-linked faulting 262, 263
Fleet Member 97, 130, 151, 152
flower structures 260, 273, 319, 325
folds, formation 277, 278, 280
Folkestone Formation 339, 340, 342
foraminiferid zonal scheme, English Channel 92
Forest Marble 53, 89, 94, 97, 128, 129, 130, 141, 151, 152
   oil exploration 408, 409, 410
   reservoirs 41, 49
Frome Clay 25, 53, 124, 126, 127, 128, 151, 159
   biozonation 89, 94, 97
Frome Clay Limestone Member 11, 15, 399
Fuller's Earth 11, 53, 64, 80, 129, 346
Fuller's Earth Clay 14, 89, 97, 121, 123–5, 126, 127, 151
Fuller's Earth Formation 29
   as top seal 22, 30, 34
   uplift/erosion 187, 188, 189, 190, 191, 193, 196
fuller's earth pits, stratigraphy 338
Fuller's Earth Rock 15, 89, 94, 97, 123, 125, 126, 151

gas 5, 55, 237, 412
   fields 40, 41, 199, 216
   seeps and shows 40–1, 56, 210, 211, 373
   see also hydrocarbons
Gault Clay 56, 60, 80, 81, 83, 287
geological history 67–86
Goban Spur Basin 220, 227, 228, 235
gravity anomalies, Channel Basin 269–71, 273, 275
Great Oolite 63, 64, 89, 94, 97, 148
Great Oolite carbonate platform 49, 54
Great Oolite Formation 127–8
Great Oolite Group 123–46, 151–2, 159
Great Oolite Limestone 65
Green Ammonite Beds 76, 109, 112, 113, 115, 182
   biozonation 88, 93, 96
Greensand Group 217, 219, 224
Grey Ledge 175, 178

halite 76, 80
   see also salt tectonics
Hampshire Basin 2, 3, 5, 40, 61, 62, 63, 203, 205
Hampshire–Dieppe High 3, 105, 150, 155, 157
   Corallian Group 139
   Kellaways Formation 131
   Kimmeridge Clay Formation 141
   Lias Group 110, 111, 113, 115, 118, 120
   Lower Corallian subgroup 135, 136
   Oolite Groups 122, 124, 125, 128, 129, 130
   Penarth Group 107, 109
   Portland Group 144
   Purbeck Group 145, 146
hanging-wall anticline, development 275

hard-linked faulting 263
hebridica lumachelle unit 129
Hesters Copse Member 125, 151
Humbly Grove oilfield 42, 103, 125, 127, 151, 152
Hummocky Limestone Bed 175, 179, 183
hydrocarbons 39–42, 61, 63–4, 119, 130, 142, 238
  accumulation 10, 14, 15, 25, 199, 210–11
  degradation 373–85, 387, 388, 391–6
  exploration 39–42, 64, 408, 409, 410
  generation 14, 199, 205, 210, 211, 212, 216
  kitchen areas 12, 14, 31, 32, 34, 399
    switched off 16, 35
  maturity 10, 12, 13, 14, 15, 16, 34–5, 116, 210–11, 235, 237
  migration 13, 15, 16, 34–5, 63, 211, 275
  prospectivity 63–4, 235–7, 238, 293
  source rocks 10, 15, 30, 31, 34, 65, 235–7, 238
  traps 10, 11–12, 30, 34, 35, 63, 64, 65, 85
  see also gas; oil; petroleum; reservoirs
Hythe Formation 336, 339, 340, 343

ichnology, Bencliff Grit 361–5, 369
Inferior Oolite 11, 51, 76, 79, 89, 93, 97, 272, 275, 276, 287, 288
  as reservoir seal 34
Inferior Oolite Bed 148
Inferior Oolite Group 121–3, 148, 150, 151, 273
intrabed-scale faults 302–3, 305, 306, 307
inversion tectonics 63, 64, 69, 81–5, 269–71, 285
  Channel Basin 263, 264, 265
  Lyme Bay 271–9, 280
  see also salt tectonics
Isle of Purbeck 143, 145
Isle of Wight 7, 51, 63–4, 67, 71, 76, 81, 83, 149, 243, 251, 328, 338
  stratigraphy 110, 113, 125, 333, 341–2, 343, 345
  seismic data 75, 80, 84, 85
Isle of Wight High 75, 76, 80, 81
Isle of Wight Monocline 85, 270, 341
Isle of Wight–Purbeck area 280

Junction Bed 111, 115–16, 117, 179, 182, 272
  biozonation 88, 93, 96, 97
  deposition 147, 148, 149, 150, 159
Jurassic
  palaeotemperatures 201–2, 211
  sedimentation controls 76–81
  stratigraphy 169–84
  uplift *see* uplift

Kellaways Beds, oil exploration 408, 409, 410
Kellaways Clay Member 130–1, 152, 159
Kellaways Formation 89, 94, 97, 130–1, 152
Kellaways Sand Member 131, 152, 153, 159
Kellaways Sandstone, reservoirs 41
Keuper Evaporites 321, 322
Kimmeridge 63
  oilfield 2, 11, 12, 15, 24, 25, 130, 373, 378, 407–13
    reservoir 41, 384–5
Kimmeridge Anticline 40, 41, 61
Kimmeridge Bay 299–309, 407–13
Kimmeridge Clay 11, 80, 217, 251, 287, 288
  hydrocarbon resources 30, 31, 103

Kimmeridge Clay Formation 15, 55, 61, 139, 140–2, 155, 156, 157, 300
  uplift/erosion 187, 188, 189, 190, 193, 196
Kimmeridge Oil Shale 11
kitchen areas 12, 14, 16, 31, 32, 34, 399
  switched off 16, 35

lacustrine deposits, Sherwood Sandstone Group 401
Langport Member 92, 107, 109
Lias Clay 61, 64
Lias Group 15, 49, 61, 109–21, 148
Lias mudstones, hydrocarbon resources 30, 31, 103
Lilstock Formation 95, 107
Litton Cheney Fault 76, 77, 79, 83, 251, 252, 253, 254, 258, 261
London Basin 62, 63
London Platform 81, 104, 105, 147, 152, 284, 338
  Great Oolite Group 125, 128
  Lias Group 114, 121, 149
  Penarth Group 107
Lopen Fault Belt 312, 313, 314, 320, 321, 322, 325, 326
Lower Calcareous Member 140
Lower Corallian Subgroup 134–6
Lower Fuller's Earth Clay 89, 93, 97, 123, 125, 151
Lower Greensand Group 7, 56, 60, 61, 80, 81, 287, 333, 336, 340, 341, 342
Lower Inferior Oolite 89, 93, 96, 97, 119, 121, 122, 151
Lower Kimmeridge Clay 90, 94, 95, 99, 140
Lower Lias 34, 109, 110, 111, 112, 148, 166, 308
Lower Purbeck Beds 91, 100, 157
Lulworth 24, 81, 203, 207, 373
Lulworth Banks Anticline 41
Lustleigh–Sticklepath Fault 284, 311
Lyme Bay 69, 114, 203, 211, 262, 263, 267–9
  tectonics 243, 244, 245, 256–60, 271–9, 280
Lyme–Portland Faults 271, 280
Lynton Anticline 315, 317, 323

Maidstone 333, 338, 340, 341, 343, 345
Mangerton Fault 68, 69, 260, 261, 262, 273
map-scale faults 302, 303, 307
margaritatus Clay 182
margaritatus Clay Bed 114
margaritatus Stone 114, 179, 182, 184
margaritatus zone 182
Marlstone Rock Bed 115, 116, 117, 119, 148, 149, 150
  biozonation 88, 93, 96
martinioides zone 332, 336, 339, 340, 341, 342, 343, 345
maximum flooding surfaces 166, 170, 174, 183, 184
Mercia Halite 79, 80
Mercia Mudstone 75, 76, 80, 85, 316, 325
Mercia Mudstone Group 5, 11, 72, 78, 269, 313, 314, 317, 318, 319, 320, 322, 327, 402, 403
  palaeogeography 27
  as regional seal 22, 30, 31, 287
  reservoirs 25, 30, 48, 404–5
  tectonics 35, 244, 245, 252, 279
microfossil biozonation, English Channel 87–100, 106
micropalaeontology, Bencliff Grit 364, 365–8, 369
Mid Dorset High 76, 80, 167
Mid Dorset Platform 69, 72, 76, 77, 79, 80, 81, 85
Middle Inferior Oolite 89, 97, 121, 122, 151
Middle Lias 110, 114, 148–9, 166
Middle Lias Clays 88, 93, 96, 114, 115

Middle Lias Sands 114, 115, 159
Middle Purbeck Beds, biozonation 91, 92
Mongrel Horizon 175
Monksilver Fault 274, 313, 315, 317, 322, 326, 327
Moray Firth 307, 308, 309
Mupe Bay 349–51, 388–96
  angular stratal relationships 349, 351–2
  oil seeps 12, 14, 373–388
    biodegradation 291–6, 387, 388, 394, 396
    palaeo 349, 350, 351, 352

Nash Point 299, 300, 308
Needles fault 246, 247, 248, 249
normal compaction relationship 191–4
Normandy, geological history 150, 151, 152, 153
North Celtic Sea Basin 221, 222, 226, 227–8, 237
North Curry Sandstone 317, 318, 324
Nothe Clay, palynofacies 365, 366–7, 369
Nothe Clay Member 134, 135
Nothe Clay–Bencliff Grit boundary 356, 368
Nothe Grit Formation 90, 94, 98, 133
Nutfield Beds 338, 339
nutfieldiensis zone 332, 336, 338, 339, 340, 341, 342, 343, 345, 346

oil 5, 41, 103
  accumulation 199, 399
  analysis 389–91
  degradation 373–85, 387, 388, 391–6
  failed wells 35
  generation 61, 235, 237, 351, 373, 387, 399
  maturity 12, 15, 65, 85, 235–9, 373, 396
  migration 10, 12, 14, 15, 210, 211, 237, 351, 352, 387, 413
  source rocks 10, 11, 15, 47, 49, 55, 65, 85, 399, 410
  see also hydrocarbons; petroleum; reservoirs
oil seeps 12, 56, 211, 373, 387
  biodegradation 373–85, 387, 388, 391–6
  Mupe Bay 349, 350, 351, 352, 387, 388
  Purbeck–Isle of Wight Fault Zone 210
oil shales 142, 156
oil spills, biodegradation 374, 375
oil staining 373, 378, 387, 391, 396, 410
Osmington Mills 11, 356, 357, 368, 376
  oil seeps 373, 375, 378, 379, 382, 384, 385
Osmington Mills Ironstone Member 138
Osmington Oolite Formation 90, 94, 98, 133, 136, 356, 368
ostracod zonal scheme, English Channel 87–92
Otter Sandstone 72, 75, 76, 85, 317, 318, 319, 320, 323, 324
Otter Sandstone Formation 403
Oxford Clay 49, 64, 203, 334, 410
  hydrocarbons 30, 31, 103
  oil 15, 49, 55, 61, 408, 409, 410, 412
Oxford Clay Formation 90, 94, 98, 131–3, 153
  uplift/erosion 187, 188, 189, 190, 193, 196

palaeo-oilseep, Mupe Bay 349, 350, 351, 352
palaeocurrent analysis, Bencliff Grit 360, 361, 369
palaeotemperatures
  estimation by AFTA and VR 200
  uplift and erosion related to 205–10

palaeothermal episodes 201–5, 211–12
palynofacies
  Bencliff Grit 361, 364, 365–8, 369
  and sequence stratigraphy 165–6, 168–84
Paper Shales 115–16
particulate organic matter (POM) 165, 169, 172–84
Pavior Bed 175, 178
Pavior horizon 184
Pays de Bray Fault 284, 285, 288, 327, 328
Penarth Group 5, 67, 76, 78, 111, 287, 317, 318, 319
  stratigraphy 107–9
Permian Breccias, thermal history 201–2
Permian–Lower Cretaceous megasequence 5–7, 15
Permian–Triassic rift model 327
Peterborough Member 90, 94, 98, 131, 132, 133, 153
Petrockstow Basin 326, 327
petroleum
  exploration 22, 24–5, 39–42
  geology, Kimmeridge oilfield 410–13
  see also hydrocarbons; oil
play types 63–4
Pliensbachian succession, Dorset Coast 167, 168, 173, 179–82, 183, 184
Porlock Basin 313, 326
Portland Beds 41, 157, 159, 287, 288, 292, 293
Portland Group 58, 140, 142–4, 155, 157, 158, 217
Portland Limestone Formation 11, 91, 100, 142, 144, 157
Portland Sand Formation 91, 95, 100, 142, 144, 157
Portland Sandstone 55, 58, 65, 342
Portland–Isle of Wight Fault Zone 210
Portland–Kimmeridge Fault System 56
Portland–South Wight Basin 56, 64
Portland–South Wight Trough 76, 77, 79, 80, 83
Portland–Wight, tectonics 203, 284, 291
Portland–Wight Basin 9, 19, 21, 30, 31, 33, 35, 103, 104, 105, 130, 203
  biozonation 91, 94
  bottom water conditions 111, 112
  Corallian Group 135, 137, 138, 139
  geological history 150, 153, 155, 157, 159
  hydrocarbons 14, 24
  Kellaways Formation 131
  Lias deposition 148, 149
  Lias Group 110, 112, 113, 115, 116, 118, 119, 120
  Nothe Grit Formation 133
  Oolite groups 122, 123, 126, 128, 129
  Portland Group 143
  Purbeck Group 145, 146
  stratigraphy 22, 23, 106, 107, 108
  thermal history 203
Portland–Wight Fault System 24, 29, 31, 32, 34, 285, 297
Portland–Wight Graben 245, 248
Portland–Wight Thrust 284, 297
Portsdown 3, 40, 81
Portsdown Fault Zone 141
power law 299
Poxwell Pericline 256, 265
Poyntington Fault 68, 69, 83
pre-Planorbis Beds 109, 110, 316
Preston Grit Member 134–5
prospect failure, reasons for 63–4
Purbeck Anhydrite, as reservoir seal 55

INDEX

Purbeck Beds 55, 144–7, 157, 158, 159, 287, 288
   reservoirs 41
Purbeck Disturbance 63, 77, 79, 81, 83, 85, 410
Purbeck Fault 150, 245, 246, 247, 248, 249, 250, 254, 256, 258
Purbeck Fault Zone 150
Purbeck Group 143, 144–7, 155, 217
Purbeck Limestones, oil seeps 373
Purbeck Monocline 250, 300
Purbeck–Isle of Wight Disturbance 3, 14, 85
Purbeck–Isle of Wight Fault System 7–8, 9, 12, 14, 15, 269
Purbeck–Isle of Wight Fault Zone 64, 153, 245, 246
   Corallian Group 137
   gas and oil seepages 210
   Kimmeridge Clay Formation 141
   Lias Group 110, 116
   Lower Corallian subgroup 136
   Oolite groups 122, 124, 126, 129, 130
   Portland Group 143
   Purbeck Group 145, 146
   uplift/erosion 196
Purbeck–Isle of Wight monocline 203

Quantock Anticline 315, 323
Quantock High 326
Quantock Hills 313, 315, 316, 318, 325, 327
Quantock Massif 316, 319, 322–3, 326
Quantock–Exmoor Upper Palaeozoic succession 313

Red Beds Member 137, 138
Redcliff Formation 133, 134, 355
Redcliff Oolite Formations, biozonation 90, 94, 98
Redhill 336–40, 342, 343, 344, 345
Redhill Sands 338, 339, 340, 341
regional uplift
   quantification 187, 188–97
   see also uplift
reservoirs 5, 7, 10–11, 15, 25, 30, 40, 41, 49, 55, 63–4, 119, 148, 412
   oils, biodegradation 374, 375, 376, 378, 379, 382, 384, 385, 391
   seals 10, 19, 22, 30, 31, 34, 49, 55, 208, 287, 410
   Sherwood Sandstone Group 15, 16, 19, 22, 24, 25, 42, 47, 63, 64, 103, 293
      architecture 399, 400–5
   Wytch Farm 64, 119, 287, 404
Ridgeway Fault 72, 83
Ringstead Formation 90, 94, 99, 133, 137, 138–9
Ringstead relay zone 254–6, 262
rotational bank collapse 349, 351–3

salt rollers 261
salt swells 252, 261
salt tectonics 30, 260–5, 274–9
   detachment 35, 250–60, 262, 275–7
   Triassic sequences 243–65
salt welds 253, 254, 256
sandflat deposits, Sherwood Sandstone Group 401
Sandgate Formation 336, 339, 340
Sandsfoot Clay Member 138
Sandsfoot Formation 90, 94, 99, 133, 137, 138, 139
Sandsfoot Grit Member 138, 159

sea-level changes
   effects of 147, 149, 151, 152, 153, 155, 156, 157, 159, 217, 232, 237
   on palynofacies 165, 169–72, 175, 178, 182, 183
   on sedimentation 166, 168, 403
seals 10, 19, 22, 30, 31, 34, 49, 55, 208, 287, 410
Second Mongrel Horizon 175
section restoration Central Channel High 288–97
sedimentary sequence analysis *see* depositional sequences
sedimentation controls 79–81, 148–50, 152, 153, 155, 156, 157, 168
sedimentological history 218
seep oils
   biodegradation 373–85, 388
   see also oil seeps
sequence boundaries, tectonic accentuation 331–46
sequence palynology 165–6, 169–84, 339–46
Shales with Beef 87, 92, 96, 109, 111, 148
sheetflood deposits, Sherwood Sandstone Group 401
Sherwood Sandstone Group 5, 10, 11, 12, 14, 16, 71–6, 85, 256, 269, 287, 316, 319, 323
   controls on 30–4
   hydrocarbons 19, 30, 275, 379, 380, 384, 385, 410
   palaeogeography 26, 27
   reservoirs 15, 16, 19, 22, 24, 25, 42, 47, 63, 64, 103, 293
      architecture 399, 404–5
   stratigraphy 47, 400–4
   tectonics 46, 195, 196, 197, 252
   thermal history 201–2
Sinemurian succession, Dorset Coast 167, 173, 175, 176, 177, 178, 179, 183, 184
soft-linked faulting 262, 263
Somerset, tectonic structure 318–22
Somerton Anticline 315, 317
sonic velocity quantification of uplift 188–97
South Celtic Sea Basin 220, 222, 227, 228, 234, 328
Stair Hole, oil seeps 373, 375, 376, 378, 380, 382, 383, 385
Starfish Bed 115, 182
Stewartby Member 90, 94, 98, 131–2, 133, 153
Sticklepath Fault 326, 327, 328
Stoborough 8, 25
stratigraphic framework 5–7, 268
stratigraphic gap 220, 221, 238
stratigraphic sequences *see* sequence palynology
strike-section geometries 9–10
strike-slip faults 83, 259–60, 262, 264, 311, 312, 313, 314, 315, 318, 321, 327, 328
   Lyme Bay 274, 276, 278, 279, 280
structure, Wessex Basin 3–5, 7–9, 67–9, 86, 218
subsidence analysis 225–33, 237

Table Ledge 148
tectonics 43–5
   controls on depositional sequences 76–86, 333–6, 343–4, 345, 346
   controls on sedimentation 148, 149, 150, 151, 152, 153, 155, 156, 157, 159
   see also inversion tectonics; salt tectonics; uplift
tectono-stratigraphic development 43–5, 47–63, 64–5
Terrestrial/Marine index (T/M index) 172, 173

Tertiary
  megasequence 7
  tectonics 63, 64, 67, 69, 81–5, 211, 212, 226, 237, 285
  thermal history 203–5, 211
thermal history 200, 201–12
thermal uplifting 235, 237
Thorncombe Sands 10, 88, 96, 114, 117, 174, 182
Timberscombe Fault 274, 313, 315, 317, 322, 327, 328
Tiverton Trough 312, 316, 321, 324, 326, 327, 328
Top Hythe Cherts 334
Top Hythe Pebble Beds 334, 339
traps 10, 11–12, 30, 34, 35, 63, 64, 65, 85
Triassic
  biozonation 87–100
  lithostratigraphy 105–59
  palaeotemperatures 201–2
  salt sequences 243–65
Trigonia clavellata Formation 94, 99, 133, 136, 137–8, 155

unconformities 71, 72, 80, 85, 227, 228, 229
  Albian 7, 246, 251, 252
  Aptian 7, 217, 220, 223, 224, 232, 237
  Apto-Albian 220, 229, 230, 232–3, 234
  Berriasian 220, 221, 222, 223, 224, 226, 227, 229, 230, 231, 232, 233, 234, 237, 238
  Cimmerian 45, 230, 232
  Cretaceous 215, 217, 219–25, 231, 267, 269
  Tertiary 226
uplift 187, 188–97, 224
  latest Jurassic–earliest Cretaceous 215–17, 224, 232–5, 237–8
    evolution of Wessex Basin during 236
    hydrocarbon prospectivity 235–7
    subsidence curves 227–32
  quantification 200, 205–10, 212
Upper Calcareous Member 140, 159
Upper Fuller's Earth Clay 89, 94, 97, 123–5, 126, 127, 151
Upper Greensand 80, 81, 287
Upper Inferior Oolite 89, 93, 121, 122, 151
Upper Kimmeridge Clay 91, 95, 99, 100, 140, 307
Upper Lias 115, 166
Upper Purbeck Beds 92, 146–7
Upton Member–Bencliff Grit boundary 356, 368
Upton Syncline 251
Upwey Syncline 251, 256

Vale of Pewsey Fault 284, 285, 297
Variscan
  basement structure, southern England 283–5
  inheritance 42–3
  sedimentation controls 68, 69
  tectonics 285, 294, 297, 315, 316, 321, 326
Variscan Front Thrust 284
vitrinite reflectance analysis (VR), palaeotemperature estimation by 194–6, 200, 201, 205–12

Wardour Fault System 3
Wardour–Portsdown Fault System 284, 285, 297
Wareham oilfield 2, 8, 12, 14, 15, 25, 42, 72, 373
washover sands 369
Watchet Fault 273, 313, 314
Watchet–Cothelstone Fault 68, 69, 83, 284
Watchet–Cothelstone–Hatch Fault System (WCHF) 3, 311, 312, 313, 314–18, 323–5
water washing, of oils 374, 375, 378, 391, 392, 396
Wattonensis Beds 89, 94, 124, 125, 126, 127, 151
Weald Basin 1, 2, 63, 64, 81, 104, 105, 233, 284, 333, 338, 351
  Corallian Group 135, 137, 139
  geological history 147, 150, 153, 155, 157, 159
  geological setting 42–7
  Great Oolite Group 124, 125, 127, 129
  Kellaways Formation 131
  Lias Group 110, 113, 114, 115, 116, 118, 120
  oil reserves 42
  plate sequences 47–63
  Portland Group 143, 144
  Purbeck Group 146
  sedimentary thicknesses 79
  stratigraphy 336
Wealden 83, 290, 295, 296
Wealden Beds 80, 81, 203, 288, 292, 293
Wealden deposits 217, 220, 223, 232
Wealden Group 7, 11, 217, 219, 220, 224, 232, 237, 286, 349
Wealden Sands, oil seeps 382
Westbury Formation 95, 107
Western Approaches Basin 224, 234, 274
Western Approaches Trough 215, 217, 225, 228, 229, 232, 237
Western Permian–Triassic Basin 316, 321, 324, 326
Weymouth 263, 356
Weymouth Anticline 77, 80, 203, 244, 251, 252–4, 265
  hydrocarbons 211, 408
Weymouth Bay 41, 79, 250, 260
Weymouth Member 90, 94, 98, 132, 133, 153
White Stone Bands 140, 141, 301, 302, 307
Wight–Bray Fault 122
Winterborne Trough 324, 328
Winterborne–Kingston Graben 244, 246
Winterborne–Kingston Trough 3, 126
Winterbourne–Kingston 51, 53, 212
Wiveliscombe Sandstones 313, 316, 317, 319
Wytch Farm 10, 11, 51, 69, 71, 72, 76, 81, 83, 149, 197, 402
  Kellaways Formation 131
  Kimmeridge Clay Formation 141
  Lias Group 111, 112, 119
  oil 42, 199, 210, 211, 410
    degradation 380, 385
  oilfield 2, 8, 11, 13, 14, 15, 19, 24, 25, 30, 48, 49, 103, 216, 237, 373, 399–400, 409
  reservoirs 64, 119, 287, 404
  trap type 34
  Oolite groups 122, 124, 126, 128
  thermal history 201, 202, 203, 209, 210, 211, 212
Wytch Farm area, uplift/erosion 193–4, 195, 196–7